Taxonomy of Angiosperms

Taxonomy of Angiosperms

S.N. Pandey Ph.D., F.B.S.

Principal, B.R.D. College
Formerly Sr. Reader, Botany Deptt., D.A.V. College
C.S.J.M. University, Kanpur

&

S.P. Misra

Formerly Principal, Khalsa G.D. College
Formerly Sr. Reader, Botany Deptt., Christ Church College
C.S.J.M. University, Kanpur

Ane Books Pvt. Ltd.

New Delhi ♦ Chennai

Taxonomy of Angiosperms

S.N. Pandey & S.P. Misra

Published by

Ane Books Pvt. Ltd.

4821, Parwana Bhawan, 1st Floor, 24 Ansari Road, Darya Ganj,
New Delhi - 110 002, Tel.: +91(011) 23276843-44, Fax: +91(011) 23276863
e-mail: kapoor@anebooks.com, Website: www.anebooks.com

Branch Office :

- Avantika Niwas, 1st Floor, 19 Doraiswamy Road, T. Nagar,
  Chennai - 600 017, Tel.: +91(044) 28141554, 28141209
  e-mail: anebookschennai@gmail.com, rathinam@anebooks.com

ISBN : 978-81-8052-176-8

Printed at : R.K. Offset Process, New Delhi.

Preface

Plant taxonomy is the oldest discipline of botany that got wedded early with the human civilisation and became inherent to it. The progress of civilisation marked a relative increase in man's knowledge about plants–their identification, naming and classification on the basis of his needs and greed. During the last three centuries, plant taxonomy has developed from a completely morphology–oriented static and descriptive discipline to a most dynamic area of study. Over the last few decades, the availability and usage of comparative and recombinant technology data from different branches of plant science and allied fields has made the subject multidisciplinary and synthetic that has led to the redefining the affinities between plant groups. It has substantially helped in the establishment of evolutionary trends and phylogenetic relationship and refinement of taxonomic concepts at various levels of taxonomic hierarchy and understanding of mechanisms that guide evolution and facilitate in devising a phenetic classification expressing natural relationships. Thus, all has led a modern systematist to become "chemocytohistomorphotaxometrician".

The present book is a concerted effort to collect, blend and pool the vast existing literature and present an up-to-date account of the subject in a lucid form. Every effort has been made to concise the matter comprehensively in simple language. The textual matter is spread over general account, modern trends, morphological features, diversity of plants in select dicotyledonous and monocotyledonous families and basic concepts of phytogeography with no sharp boundaries in its layout. The chapters chosen for inclusion in the text are as per need of the syllabi of various Indian Universities in general and UGC in particular. The text has been enriched by including some new topics such as Relevance of Taxonomy : Conservation and Sustainable Utilisation of Bioresources and Ecosystem Research; Hot Spots and Hottest Hot Spots; Plant Explorations; Indian Flora; and Invasions and Introduction. Chapter entitled Major Systems of Classification deals with systems of classification from Theophrastus to most recent APG (Angiosperm Phylogeny Group) System including some important systems, such as those proposed by Bentham and Hooker, Engler and Prantl, Rendle, Hutchinson, Cronquist, Dahlgren, Thorne etc. coupled with their merits and demerits. Chapter Morphological Features provides a perfect knowledge of the technical terms that come across the text Following broadly Cronquist's system of classification, an endeavour has been made for systematic treatment of selected dicotyledonous (51) and monocotyledonous (10) families dealing their distribution in the world and names of commonly distributed genera; range of vegetative and floral characters; economic importance; and systematics and phylogeny. Description of families is supported by illustrations depicting flowering twigs, floral structures, cross-section of ovary, any special structure and floral diagrams. Economic importance and ethnobotanical significance have also been incorporated. Comparative tables have been provided outlining distinctive characters of various families. The individual contribution of outstanding taxonomists has been given within the boxes.

Written in a student-friendly manner, the book will cater the need of undergraduate, honours and postgraduate students of all Indian Universities and those preparing for various competitive examinations persuing studies on plant systematics.

Authors acknowledge a deep sense of gratitude to various scientists, authors, publishers, books, journals, periodicals etc. for the help taken from them in the preparation of the book. Thanks are due to Mr. Sunil Saxena and Mr. JR Kapoor and staff of Ane Books Pvt. Ltd. for their sincere cooperation which enabled the book to see the light of the day.

Suggestions for the improvement of the book are welcome.

AUTHORS

Contents

Chapter 1

Historical Retrospect, Aims and Objectives and Scope of Plant Taxonomy

HISTORICAL RETROSPECT

History of plant taxonomy has witnessed various phases from early history to modern approaches.

Early History of Plant Taxonomy

Man, soon he set his foot on this planet, felt the need to have the knowledge about plants. His first priority was to know what plants he should make use as food, utilise for shelter, use as cure for ailments and avoid others. Initially, this information was accumulated and stored in his brain and passed on to generations orally, and remained restricted to small communities. With the man's knowledge of black and white, it could become possible to facilitate others to share and improve knowledge about this treasure of information and bring the mankind to reach a stage whereby this can easily be stored and utilised for far-reaching conclusions in plant taxonomy.

Plant hunt for basic necessities of life, such as food, shelter and clothing, is age old. The history of human civilisation is closely linked with the history of man's knowledge of plants. The progress of civilisation marked a relative increase in the knowledge of the plants. Despite the fact that early civilisations flourished in Babylonia, Egypt, China and India, written records of Indian botany in Sanskrit, appeared much before those of Greeks but they remained in obscurity. Vedic Period (2000 BC to 800 BC) finds mention of several crops such as wheat, barley, cotton etc. *Atharva Veda* (written around 2000 BC) describes medicinal uses of many plants.

The early history of development of botanical science is nothing but a history of development of plant taxonomy. **Theophrastus** (372-287 BC), *Father of Botany*, the Greek philosopher-naturalist, a pupil of **Plato** (426–347 BC) and successor of **Aristotle's** (384–322 BC) philosophy, made a pioneering attempt to place his knowledge of plants on a scientific footing. In his *Enquiry into plants,* he dealt with the plants at large and attempted to arrange them in several groups. He named and described some 500 plants in his *De Historia plantarum.* **Parasar** (250–120 BC), an Indian scholar, who compiled *Vrikshayurveda,* dealt several aspects of plant life from scientific standpoint and **Charaka** (1st century AD), another Indian harbalist in his *Charaka samhita,* recognised medicinal properties of a large number of Indian plants. **Pliny** the elder-**Caius Plinius Secundus** (23 AD to 79 AD), a Roman naturalist, compiled a 17 volumed monumental work *Historia naturalis* (77 AD), in nine volumes which were devoted to medicinal plants, incorporating all possible information of plants gathered upto that time and

Box 1.1 Theophrastus : Father of Botany

Theophrastus (372 BC to 287 BC), the successor of Aristotle (*Father of Biology*) in the peripheral school (those following the philosophy propagated by Aristotle), was a native of Erescus in Lesbos. His original name was Tyrtamus, but he later became known by nickname 'Theophrastus' given to him probably by Aristotle to indicate the grace of his conversation. After receiving his first introduction to philosophy in Lesbos from one Leucippus or Aclippus, he proceeded to Athens and became a member of the Platonic circle. Soon after the death of Plato, Theophrastus attached himself to Aristotle and after the death of Aristotle, he inherited his library and the garden. Ultimately he rose to become the head of the Lyceum at Athens. He is credited with having authored more than 200 works, most of which survived as fragments or as quotations in the work of other authors. However, two of his works *Enquiry into plants* and *The causes of plants,* have survived intact and are available in English translations. He described about 500 kinds of plants. He classified plants into four major groups : trees, shrubs, undershrubs and herbs and also made distinction between flowering and non-flowering plants, superior and inferior ovary, free and fused petals and fruit types. The fact that many cultivated plants do not breed freely, was well known to him. Many of the names, *e.g. Daucas, Crategus* and *Narcissus,* which are in use even today, were used by him in his *De Historia plantarum.* Alexander the great, patronaged him and send back materials recovered during his cronquest to Athens, enabling Theophrastus to write about exotic plants such as cotton, cinnamon and banana. Thus the knowledge of botany fluorished at the Lyceum in Athens during this truely **golden age of learning,** whose botanical advance Theophrastus was privilaged to stear.

Fig. 1.1 Theophrastus (372–287 BC).

added much to the same the plants collected by himself from his travel far and wide. **Discoroides-Pedanios Dioscoroides** (62–127 AD), now popularly called *Father of Medicine*, a Roman physician, compiled his work in his famous *Materia medica* and described nearly 600 medicinal plants mentioning their local names, providing sketches and giving their medicinal properties. The book proved as an asset for many literate people for the next 15 centuries and of so much of recognition that no drug was recognised genuine unless mentioned in *Materia medica*.

The period between 5th to 15th century AD (Middle Ages), the dark period of botanical investigation, witnessed little or no significant progress because Europe and Asia witnessed wars, famine and epidemics. Earlier manuscripts were copied, recopied or translated. **Albert Magnus** (1193–1280 AD), the best remembered naturalist of that period, wrote his *De vegetabilis* which dealt with medicinal plants and their description based on first hand information. He was first to recognise dicots (tunicate) and monocots (corticate) based on their stem structure and also separated vascular and non-vascular plants. The 15th century witnessed the onset of Renaissance in Europe. The technical innovations, mainly the invention of printing technology around 1440 and science of navigation, made the publishing of literature and its circulation cheap and readily available to people at distant places. The botanical-medical books, the herbals, written by herbalists, were first to become popular. A significant progress in the area was made when a few German herbalists carried their enquiries about plants still farther, making study of botany quite popular. Foremost among them was **Otto Brunfels** (1464–1534) who wrote a three-volumed *Herbarium vivae eicones* (1530–36) which was profusely illustrated with beautiful figures prepared from living plants. This marked the beginning of modern taxonomy. He was first to recognise the distinction between imperfecti and perfecti groups of plant kingdom on the basis of absence and presence of flowers, respectively. **Jerome Bock** (1498–1554), another German naturalist, published his *New kreterbuch* (1539) which contained accurate description of about 567 species of flowering plants, classifying them as herbs, shrubs and trees. His descriptions were based on first hand information mentioning the localities and habitats of plants and were better understood. With advancement in the knowledge, necessity to express plant characters through certain technical terms was felt. **Leonhard Fuchs** (1501–66), the pioneer to publish an illustrated glossary of technical terms, published *De Historia stirpium* (1542) that contained descriptions as well as illustrations of 487 species of medicinal plants. **Valerius Cordus** (1515–44) during his short span of life made extensive explorations in forests of Germany and Italy, which if continued might have made him to become the greatest of all herbalists. Unfortunately he died at the age of 29. His work *Historia plantarum*, published many years after his death, contained accurate description of 502 species of plant, of which 66 were apparently new. Perhaps he was the first to demonstrate how to describe plants in nature accurately.

The Italian botanist-physicist **Andrea Caesalpino** (1519–1603), while working as the Director of Botanical Garden at Bologna and later becoming Professor of Botany and Medicine at Bologna in 1563, prepared a herbarium of 768 well mounted plants which is still preserved as a relevant herbarium in Natural History Museum at Florence. Linnean Society still has a copy of his book *De Plantis libri* which appeared in 16 volumes in 1583 and contained descriptions of 1520 species of plants grouped as herbs and trees and further differentiated on the characters of phyllotaxy, ovary, fruits and seeds. The ***Dutch Big Three*–Rembert Dodoens, Carolus Clusius** and **Mathias de L'obel**, through their herbals, spread their knowledge to Holland and France. **William Turner's**, in his *Herbal* (1551–68), written in English, proved to be a valuable work that swept out many superstitions based on the belief that plants and plant parts cured that particular portion of the human body which they resembled, such as herbs with yellow sap would cure jaundice, walnut kernal would comfort the brain and maidenhair fern would prevent baldness. As a result, his *Herbal* won for Turner the title of *Father of British Botany*.

As a result of renewed interest in plants, especially after extensive explorations of Europe, Asia, Africa and New World, the list of plant species increased tremendously. In this field, then came Swiss botanists **Bauhin brothers**. The elder brother **Jean (Johna) Bauhin** (1541–1613) wrote a book entitled *Historia plantarum universalis*, that included description of about 5,000 plants with figures of more than 3,500 plants, which could be published after his death (1650–51). **Gaspard (Caspar) Bauhin** (1560–1624), the younger brother of **Jean Bauhin** and pupil of **Fuch's**, travelled extensively and prepared a herbarium of about 4000 specimens. He published three botanical treatises– *Phytopinax* (1596), *Prodromus theatri botanici* (1620) and *Pinax theatri botanici* (1623) containing a list of 6000 species of plants giving synonyms (other names given by earlier authors) and introducing **binomial nomenclature** for several species named by him. Bauhin brothers made use of the habit characters in classifying plants. Tragically the two brothers never collaborated but worked on identical lines independently. **Gaspard Bauhin** formulated the idea of the genus and sought to clarify in a single publication the confusion regarding multiplicity of names for all species known at that time. **Joaching Jung** (1587–1657), a German medical teacher, frequently known as *First Terminologist*, defined homologous and analogous characters with great clearity and was opposed to the fundamental grouping of trees, shrubs and herbs as followed by earlier naturalists. He suggested many new terms such as nodes and internodes, simple and compound leaves, leaf blade, stamens, styles and capitulum–composed of ray and disc florets. Although he could not publish any of his work but two of his pupils preserved records of his teaching.

John Ray (1627–1705), a British botanist who travelled extensively in Europe, laid foundation of modern plant taxonomy. He published most of his work incorporating his system of classification of which the most significant being *Methodus plantarum nova* (1682), the last edition of which was published in 1703 included 18,000 species and *Historia plantarum* (1686–1704) which was a three volume work containing an improved version of his earlier system of classification. Undoubtedly, he was a trend-setter for a natural system of classification of the flowering plants. He gave much thought in proposing a system of classification based on morphology-characters of flowers and seeds with emphasis on embryo characters and number of cotyledons and was first to recognise two major taxa namely, **Dicotyledons** and **Monocotyledons** of flowering plants. **John Ray** recognised two major subdivisions namely **Herbae** (Herbs) and **Arborae** (Trees) of plant kingdom. Herbae was divided into **Imperfectae** (Cryptogams) and **Perfectae** (Seed Plants). The arborae was divided into **Monocotyledons** and **Dicotyledons.** He also tried to group the plants into several families which he called classes. He placed 25 of his classes under **Dicotyledons** and **Monocotyledons.** John Ray's contemprary **Pierre Magnol** (1638–1735), based on morphological characters of root and stem, suggested a system of classification upto family level. Floral and seed morphological characters were used for the first time. The genus *Magnolia* is commemorated after his name. **JP de Tournefort** (1656–1708), a French botanist, is believed to be the founder of the concept of modern genera (although G Bauhin mentioned it earlier). He travelled extensively in Greece and Asia Minor and collected 1356 plants. Based on his collection of 698 genera spread over 10,145 species, he published *Elements de Botanique* (1694). *Flora of Environ of Paris* (1698) and *Institutions of herbarie* (1700). The illustration of plants in copper plates, were his other valuable and distinctive works. He tried to work out a system of classification of flowering plants on the line of John Ray's system. His work proved quite useful in identifying plants upto the level of species. **Rudolf Camerarius** (1665–1721), a German botanist and the Director of Botanical Garden at Tubingen in 1694, established the fact of sexuality in flowering plants. He concluded that stamens were male sex organs and seeds were produced only when pollen, developed in stamens, come in the vicinity of style and the ovary–the female sex organs. Without pollen, the ovaries in flowers remained sterile. Although his work is not evidenced by any publication but he communicated his findings to the reputed scientists and scholars at various academic institutions and provided a new conceptual approach of sexual system, a turning point in classification approach, which was amply exploited by Linnaeus for classifying flowering plants. He is popularly known as the pioneer in artificial crossing of plants and is rightly known as *First Artificial Plant Breeder*.

Later Progress in Plant Taxonomy

By the end of 16th century, the science of botany developed as an independent discipline which caused the need to study more and more plants and large number of characters in order to arrive at a satisfactory classification. The period from the 17th century onwards witnessed a great revolution in constructing suitable systems of classification of plants. Various systems of classification were suggested by different taxonomists during this period. Those who contributed significantly in the field of plant exploration, identification, naming and classifying plants and giving various systems include *Sexual* (*Artificial*) *System*–**Linnaeus** (1707–78); *Natural System*–**de Jussieu** family (1668–1789), de Candolle family (1778–1873), **Bentham** and **Hooker** (1800–1911); *Phylogenetic System*–**Eichler** (1839–87), **Engler** and **Prantl** (1844–1930), **Bessey** (1845–1915), **Hutchinson** (1884–1972), **Benson** (1957); *Contemporary Phylogenetic System*–**Takhtajan** (borne 1910), **Cronquist** (1919–92), **Thorne** (borne 1920), **Dahlgren** (1932–87) and **Angiosperm Phylogeny Group** (**APG**) : **Bremer, Chase** and **Stevens** (1998), **APG II** (2003), **AP web** (2003) etc.

Modern Approaches in Plant Taxonomy

Plant taxonomy has shown a considerable stride from herbarium records to databanks, recording information from all possible corners of plant world. But this advancement is at various levels of its development in different parts of the world because of extreme climatic diversity, floristic variability, degree of accessibility and economic disparity in the different regions. Tropical Asia and Tropical Africa have richest floristic diversity in the world but are among the poorest in their economic resources to pursue floristic study. European countries, although rich in economic resource and have ill defined floristic variability, have been explored tremendously. India, with great climatic, geographical and floristic variability and meager resources, has a vast scope of floristic study as evidenced in the past few decades. A small country like Bhutan, with an area of only 40,000 sq km, 72 per cent of its area under forest cover supporting 7,000 plant species, has only few botanists to study the flora. *Alfa taxonomy* (descriptive taxonomy) deals with the description and designation of species, typically on the basis of morphological characters. *Beta taxonomy* (*macrotaxonomy*) is concerned with higher arrangement of species into hierarchial systems of higher categories or taxa. *Gamma taxonomy* covers the aspect of taxonomy concerned with intraspecific populations and with phylogenetic trends.

According to Davis and Heywood (1963), plant taxonomic advancement can be spread over : *(i) exploratory* or *pioneer phase*–during which plants were collected and preserved as herbarium record, *(ii) consolidation* or *systematic phase*–during which, after careful observations, floras and monographs were prepared, *(iii) experimental* or *biosystematic phase*–during which, apart from herbarium records and variation studies, data from studies on transplant experiments, breeding behaviour and chromosomes, were also incorporated and *(iv) encyclopaedic* or *holotaxonomic phase*–during which, apart from attaining previous three phases, information available on all botanical fields, is assembled, analysed and synthesised by numerical taxonomy and applied in tracing phylogeny of various plant groups. Most tropical African and Asian countries are passing through exploratory phase and most parts of the Central Europe, North America and Japan are experiencing consolidation phase. Central Europe has reached experimental phase of plant taxonomy. Exploratory and consolidation phases of plant taxonomy are often kept under **alpha taxonomy** whereas encyclopaedic phase under **omega taxonomy**. Presently, work on encyclopaedic phase is being undertaken by a few workers that too is on isolated taxa. Thus omega taxonomy is within reach of a few groups only whereas for great majority of plants, mainly in tropics, even the alpha stage has not been crossed. Conclusively, total integration of all available information for the entire plant world seems to be a dream.

In recent years, the diversification of plant taxonomy has resulted in recognising distinct disciplines such as biosystematics, chemotaxonomy, phylosystematics (phylogenetics), numerical taxonomy (taxometrics), cladistics etc. Biosystematics deals with transplant experiments, breeding behaviour, chromosomes and variation in plants. Chemotaxonomy make use of the chemical evidence in solving taxonomic problems. Phylosystematics deals with application of phylogenetic criteria in classification. Numerical taxonomy utilises numerical or computer methods in analysing and synthesising information obtained from various related fields, attaining grouping based on a phenetic relationship. A prominent field of study-cladistics, which has emerged more recently, is based on phylogenetic systematics. It make use of a distinct methodology to analyse the phylogenetic data on lines parallel to taximetrics (numerical taxonomy) and constructs a cladogram and not the phenogram, as is constructed in taximetrics.

AIMS AND OBJECTIVES OF PLANT TAXONOMY

Plants play key role in sustenance of human life. Among huge variety of plants surrounding us, it is extremely difficult to pin point a specific plant of interest unless we note its similarities and differences with other plants. To have adequate knowledge about each and every plant, is almost a difficult task. But man has a wonderful capacity to discriminate rich variety of innumerable plants. In other words he identifies, name and classify them into manageable groups. The activities of a plant taxonomist are basic to all other botanical sciences. A plant taxonomist is aimed at identification, nomenclature and classification of plants in order to study them in an orderly and scientific manner. This facilitates a botanist to know what plant he is working on, before he passes on information about it to others. Among innumerable plants, showing differences and similarities, none of them are identical. A plant taxonomist should be aimed at providing a convenient method of identification and communication. For proper identification, well identified and arranged herbarium specimens, dichotomous keys, polyclaves and computer aided identification should be adopted as an aid.

Code–International Code of Botanical Nomenclature (ICBN), written and documented through the efforts of **International Association of Plant Taxonomy (IAPT)**, which decides the single correct name acceptable to the whole botanical community, greatly helps in naming a plant. One should also be aimed at providing a workable classification having the taxa arranged in the hierarchy. The plants showing similarities be placed in smaller groups as species, larger as genera, still large as families and so on, and provide an inventory of district, state, national, regional and world flora.

There are two main approaches to follow a plant taxonomic procedure. One way is **emperical approach** which is based on experience and involves observation of characters and the facts of the plants that ultimately lead to the construction of a basic classification. This is a natural system of classification, based on which most plants have been classified in the past. Micro-level observation and involvement of characters as much possible, will obviously facilitate a more acceptable system of classification. The other way is **interpretive approach** which involves construction of classification in such a way that it can interpret the evolution of a taxon as well. This is a phylogenetic system of classification which, although, difficult to achieve because of scanty knowledge of data from plant history, is more scientific. The modern plant taxonomy requires the synthesis of these two approaches and is aimed at providing a convenient method of identification and classification based on natural affinities of the plants as far as possible. And for all this, a taxonomist has to detect the evolution at work, *i.e.*, to reconstruct the evolutionary history of the plant kingdom and to determine the sequence of evolutionary change and character modification. The phylogenetic relationship between the groups is commonly depicted with the help of a **phylogram** in which the longest branches represent more advanced groups and those which are shorter and nearer the base, represent primitive ones. Additionally, the groups are represented by **baloons** of different sizes that are proportional to the number of species in the respective group. Such a phylogram is known as the **baloon diagram**. Some workers depict phylogenetic relationship in the form of a **phylogenetic tree** (with vertical axis representing the geological time scale) in which existing species reach the top. The **bubble diagram** may be the cross section of the top with primitive groups towards the centre and the advanced groups towards the periphery. For providing an integration of all available information, data can be gathered from all related fields, analysed statistically with the help of computers and synthesised so as to develop a system of classification based on overall similarity. And this gathering, analysing and synthesising all possible information to be used to construct a system of classification is an unending process because new information will continue to pour new challenges for the plant taxonomists.

Other objectives of plant taxonomy are to provide an information reference and methodology for information storage, retrieval exchange and utilisation, supply of valuable information about endangered species and genetic and ecological diversity. Attempts should also be made for providing new concepts, reinterpreting the old and developing new procedure for accurate taxonomic affinities in terms of phylogeny and phenetics and finally providing integrated databases including all species across the planet for which several organisations have joined their hands to develop and establish online searchable databases which provide information about taxon names, images, descriptions, synonyms, molecular information etc. The objective should also be to play an integrating and unifying role in the training of plant taxonomy learners. **Parataxonomists**, people with no previous higher education in botany or plant taxonomy but who receive basic training in specimen collection, herbarium preparation and identification, can also be employed to speed up the work of preparation of inventory of plants of various localities and finally of entire planet.

SCOPE OF PLANT TAXONOMY

The scope of plant taxonomy is diverse. It is benefitting not only to botanists but also to those who are working in other fields where plants are used as tool. Plant taxonomy, being multidisciplinary in nature, has close link with morphology, anatomy, embryology, palynology, palaeobotany, cytology, genetics, evolution, physiology, molecular biology, chemistry, geography,

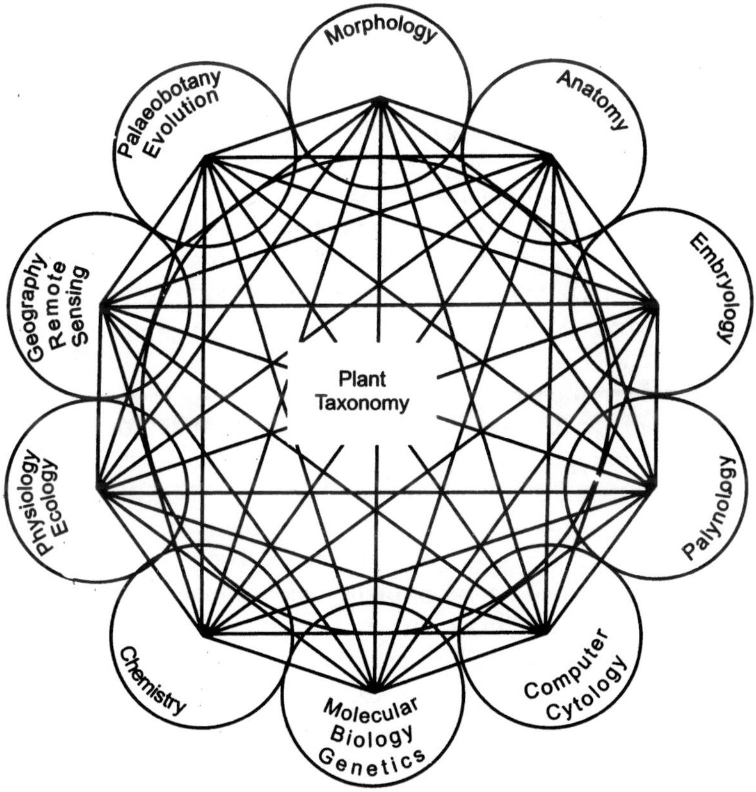

Fig. 1.2 Multidisciplinary nature of plant taxonomy.

ecology, remote sensing and computer science (fig. 1.2). It owes to many such disciplines which have played a significant role in its progress and is credited for great help to various branches related to plant sciences. Plant taxonomy has played a significant role in agriculture, horticulture and forestry in developing high yielding varieties. These hybrid varieties have greatly increased the quality and production of many crops, fruit plants and trees used for afforestation. This could become possible only with adequate knowledge of plant taxonomy because unless one identifies the plant of his interest, he can not perform a successful hybridization experiment. Pharmaceutical industries largely depend on many plant taxonomic information of their interest. Many other industries, that depend on plants or plant products to make use of them as raw material, require the basic knowledge of data derived from plant taxonomy. Ecologists, having basic knowledge of plant taxonomy, find it easy to choose suitable plants to be raised as soil binders to check desertification, fast growing trees for afforestation, mineral indicator plants for mineral exploration, energy plantation as renewable source of energy, raising dust filter plants to check air pollution, environmental assessment, effective conservation, management of resources and ecological research. The inventory of plants, books, monographs, floras and other published work on plant taxonomy greatly helps in selecting plants from their natural environment. Plant taxonomy also plays an important role in compiling fragmentary knowledge of scattered fossil record of various plants that existed in the past. The identification, naming and classification of this hidden treasure of fossil plants can only be done with the help of plant taxonomy. Conclusively, plant taxonomy appears to be the mother of all branches of botany.

Recently, the Ministry of Environment and Forests, Government of India has set up an All India Coordinated Project on Capacity Building in Taxonomy (AICPTAX) to explore, identify and describe large number of plant species and prepare an inventory. The scheme was reviewed by the Ministry in 2005 and it was decided to assist 28 botanical gardens of the country to make more achievements.

The scope of plant taxonomy has greatly expanded on account of our better understanding of the plant wealth with the help of **Geographic Information System (GIS)** and **Remote Sensing (RS)**. Remote sensing has proved useful for vegetation discrimination, plant species differentiation, assessment of vegetation types and their status, forest resource management, making priorities and strategies of plant resource management etc.

The electronic revolution, with the development of **Internet sites** (websites), in last few years, has brought the plant systematics to become dominant component of plant sciences. Internet has proved quite useful for locating plant taxonomists, gardens, herbaria, books, journals, families, genera, species, plant names and many more information related to plant taxonomy. These electronic databases, in the form of electronic lists, are constantly upgraded. Most of the prominent gardens, herbaria, botanical surveys, institutions, libraries, publishers, NGOs, research organisations and societies are already on internet with their websites that afford ready access to their literature and other informations. Many searchable databases such as *Index Nominum Genericorum* (ING)–for all published generic names of plants, *World Biodiversity Database* (WBD)–for published scientific names of plants (also animals) under *Species 2000 Project, International Plant Names Index* (IPNI)–for database of names and associated basic bibliographical details of all seed plants incorporating data from *Index Kewensis* and *Gray Cards Index*, are already operative. Missouri Botanical Garden maintains **TROPICOS** which is the world's largest botanical database of estimated 9,20,000 plant names and more than 18,00,000 specimen records. The version of the **Code**–International Code of Botanical Nomenclature–**ICBN** (St. Louis, 2000), with appropriate links to various articles for consulation, is also available on web. A **Draft Biocode** which is an attempt at a unified Code for all living organisms and **Phylo Code** which is aimed at recognising monophyletic clades, are also on web. These are available for discussions among biologists and for possible adoption in the first decade of current century. The three major contemporary systems of classification of angiosperms with their latest versions–Cronquist (1988), Takhtajan (1997) and Thorne (2003) are already on the web with links to various **News Groups** recognised by these workers. In view of **Angiosperm Phylogeny Group (APG)**, Thorne's system of classification has been undergoing considerable changes and its latest electronic version containing many nomenclatural addition (2003), is available on website of Rancho Santa Ana Botanic Garden, Claremont, USA. Such an approach is expected to go long way in maintaining communication of new ideas within short times.

❑❑❑

Chapter **2**

Fundamental Components (Principles) of Taxonomy and Concept of Taxonomic Characters

FUNDAMENTAL COMPONENTS (PRINCIPLES) OF TAXONOMY

For scientific study of an organism, it is necessary to identify it, have a name for its communication and assign it to a group to which it fits well–the three main functions of taxonomy. The literal meaning of **taxonomy** (Greek **taxus** = arrangement; **nomos** = law or rule) is the "*lawful arrangement*" or "*arrangement by rules*" of things. Plant taxonomy refers to classification of plants following certain rules or principles, after they have been identified and named. **AP de Candolle** (1813) was first to introduce the term plant taxonomy which has its synonymous names as systematic botany and plant systematics in general. Later the term became more inclusive and was used to refer to plant sciences dealing with identification, nomenclature and classification of plants (Lawrence, 1951). Keeping in view that identification and nomenclature are important prerequisite for any classification, taxonomy is often defined as '*science dealing with the study of classification including its bases, principles, rules and procedures*' (Davis and Heywood, 1963). Systematics, a term derived from the Latin word *systema* (organised whole), was recognised as a formal major field of study only during later half of the twentieth century. The term which formed the title of the famous work of **Linnaeus** *Systema naturae* (1735) appeared first in his *Genera plantarum* (1737). However, **Huxley** (1888) is often credited for the first use of the term systematics, as he used the term in his article *Nature* on the systematics of birds. **Simpson** (1961) has rightly defined systematics as '*scientific study of the kinds and diversity of organisms and of any and all relationships between them*'. In view of extended and enlarged scope of taxonomy in recent years, systematics has become synonym of taxonomy. **Stace** (1980) has given a broader definition of taxonomy to coincide with systematics as '*the study and description of variation in organisms, the investigation of causes and consequences of this variation and the manipulation of the data obtained to produce a system of classification*'. Since the term systematics is more inclusive, it is often preferred by most of the recent workers.

Fundamental Components of Plant Taxonomy

The three functions of plant taxonomy include identification, nomenclature and classification (fig. 2.1) but the main aim is to provide an ideal system of classification. Construction of an ideal system of classification necessiates a convenient method of identification, description, nomenclature and tracing the natural affinities of a taxon as far as possible. The term **taxon** (pl. **taxa**), which was first used by **Adolf Meyer** (1926) for animal group, was later proposed for plant system by **Herman J Lam** (1948). **Taxon** is a taxonomic group of any rank, *e.g.* a species, a genus, a family and so on.

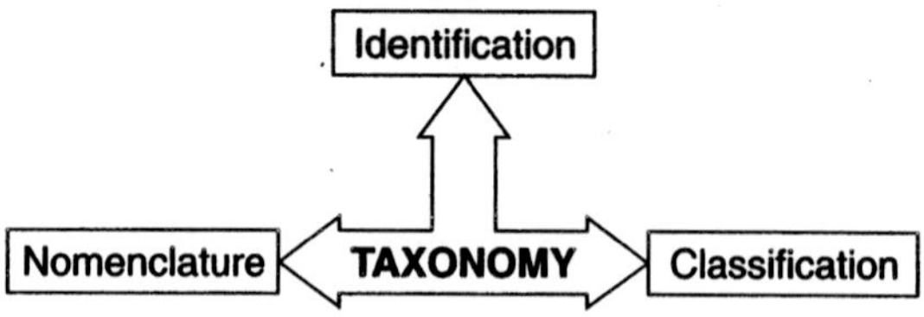

Fig. 2.1 Fundamental components of plant taxonomy.

Identification : Identification or determination is to recognise an unknown specimen and is the first and foremost aspect of the study. The identification of an unknown taxon is based on overall similarities and differences with an already known taxon and is generally made by comparing the unidentified taxon directly with that of another taxon which has been authoritatively predetermined or indirectly with the help of the literature on keys, descriptions, illustrations etc. Generally the characters derived from gross morphology are sufficient for provisional determination but other characters are also considered. In case specimen in question does not agree or compare with existing predetermined specimen, it is taken to be a new to the science and this is the most exciting situation for a taxonomist to discover a "new" taxon. Floras, monographs, manuals and identifications are important sources of literature. After an unknown specimen has been provisionally identified with the help of key, its identification can be further confirmed with the detailed description of the taxon .provided in the literature source employed. The literature regarding any particular specimen or taxa can also be obtained from the **Black and Atwood's** *Guide to the flora of the world.* The various methods of identification include :

1. Expert Determination. In terms of reliability or accuracy, it is the best method of identification. Experts have prepared treatments such as monographs, revisions, synopses etc. of taxa in question. It is probable that expert's more concepts may be incorporated in recent floras or mannuals. Experts are engaged in botanical gardens, herbaria, museums, universities, research organisations, colleges etc. Despite being reliable, the method presents problem of requiring valuable time by experts and creating delays for identification.

2. Recognition. In reliability, recognition approaches expert determination as it is based on extensive past experience of the identifier with plant or plant group in question.

3. Comparison. This method involves comparison of a plant in question with identified and named specimens, illustrations, photographs or descriptions. Despite being reliable, the method is time consuming or sometimes becoming virtually impossible due to lack of suitable material for comparison. The reliability of the method depends on accuracy and authenticity of descriptions, photographs or illustrations of named specimens used for comparison.

4. Use of Keys, Synopses, Outlines etc. By far the most widely used, this method does not require long time and experience in comparison and recognition.

In recent years, the method of taking photographs of plants and plant parts, uploading this picture on the website and making an enquiry from concerned electronic lists and newsgroups to look at the photograph and send their comments to enquirer, is gaining popularity.

Nomenclature. Nomenclature deals with the determination of a correct name for a taxon following a precise and universal system of rules used and accepted by botanists globally. Once a taxon has been identified, it becomes necessary to assign it a correct name–the botanical name. Nomenclature, an important aspect of plant taxonomy, is an effective tool of communication of information about the taxon, say for example a plant. Nomenclature of plants is governed by **International Code of Botanical Nomenclature (ICBN)** through its rules and recommendations framed by International Botanical Congress held at regular intervals. The Code is updated every six years or so and helps in picking up a single correct name out of various scientific names available for a taxon, with particular circumscription, position and rank. An already existing list of botanical names is provided in the Code which avoids inconvenient name changes for certain taxa. Naming of cultivated plants is governed by **International Code of Nomenclature for Cultivated Plants (ICNCP)**, slightly modified from ICBN. The naming of bacteria is governed by **International Code of Nomenclature of Bacteria (ICNB)**, now called **Bacteriological Code (BC)**. A separate Code, **International Code of Virus Classification and Nomenclature (ICVCN)**, whose draft is being currently developed from the most recent rules of virus classification and nomenclature framed by the International Committee for the Taxonomy of Viruses (ICTV), exists for viruses.

The electronic revolution caused the need of a common database for the living organisms for global communication. As a result, the first **Draft-Biocode** was prepared in 1995. After successive reviews by the International Committee of Bionomenclature **Draft Biocode (1997)** was prepared on the web. **Phylo Code** is based on the concepts of phylogenetic systematics and omits all ranks except species and clades based on the concept of recognition of monophyletic groups. Phylo Code's latest version (July 2002) is also available on the web.

Classification. Keeping in view an enormous number of species of plants on this planet, it is impossible to study them individually. And thus, it becomes necessary to study them in an organised manner and place them into small or large groups based on their similarities. These groups, the taxa, are then arranged in a sequential order into categories according to their levels. Each category is given a name following the rules of nomenclature. The similar plants are grouped under species, similar species under a genus, similar genera under a family and so on in sequence of increasing inclusiveness. Thus, classification is an arrangment of organisms into groups on the basis of similarities by rank or position and placing them within the taxonomic hierarchy following certain criteria. It also includes the determination of position or rank for new as well as old taxa that have been recommended, splitted, united, transferred or altered in rank. Once established, a classification provides an important mechanism of source of storage, retrieval and usage of information of organisms.

Basically there are three kinds of classifications : artificial, natural and phylogenetic. They are not mutually exclusive but often overlap in practice.

1. **Artificial Classification.** Artificial classification of plants is based on arbitrary and easily observable one or few characters such as their habit, colour, number, form etc., often irrespective of their natural affinity. Early systems of classification such as suggested by **Theophrastus** that utilised habit of plants and **Linnaeus** that utilised the number of stamens and some more characters, were mostly artificial.

2. **Natural Classification.** Natural classification of plants is based on overall similarity in grouping taxa, considering most gross morphological characters. Initiated by **M Adanson**, natural classification of plants culminated in the most extensively used **Bentham** and **Hooker's** system. Usage of morphological characters in delimiting the overall similarity, the concept used in eighteenth and nineteenth century, has undergone considerable refinement in recent years and now overall similarity is judged on the basis of the features derived from all available fields of taxonomic information, *i.e.*, phenetic relationship.

Classification based on phenetic relationship utilises data from morphology, anatomy, embryology, ultrastructure, phytochemistry and many other fields of study. Such classification, strongly advocated by **Sneath** and **Sokal** (1973), could not gain much favour with major systems of classification of higher plants. However, the phenetic relationship, to decide realignments within the systems of classification, has been prominantly used in modern phylogenetic systems. It is presumed that the larger the number of characters shared by different taxa, more closely they are related to each other.

3. **Phylogenetic Classification.** Phylogenetic classification is based on the known or inferred evolutionary descent of a group of plants. The relationship is depicted either through a **phylogram, phylogenetic tree** or

cladogram exhibiting presumed evolution of groups. While constructing a phylogenetic classification, care is taken that all descendents of a common ancestor be kept in the same group, *i.e.*, the group should be monophyletic. If some, descendent have been left out, rendering the group paraphyletic, they are brought back into the group so that its monophyletic nature is maintained, *e.g.* merger of Asclepiadeaceae with Apocynaceae and merger of Capparaceae with Brassicaceae in recent classifications. Also, if the group is polyphyletic–with members from more than one phyletic lines, it is splitted to create monophyletic taxa, *e.g.* genus *Arenaria* splitted into *Arenaria* and *Minuarites*, the **Cladistics** approach practiced by **Cladists.** More often phylogenetic classifications are termed as vertical classifications.

The term evolutionary taxonomic classification is sometimes preferred over phylogenetic classification but former differs from latter in that in evolutionary taxonomic classifications the gaps in variation patterns of phylogenetically adjacent groups are regarded as more important in recognising groups. It accepts the gaps due to leaving out of certain descendents of a common ancester if they are not significant, *i.e.*, recognising paraphyletic groups and thus failing to provide a true picture of genealogical history. Basis of choosing the significant characters in the evolution and consequently in the classification, the **eclecticism** practiced by **eclecticist**, depends on experts, authorities and institutions of systematics, *e.g.* classifications advocated by Simpson (1961), Ashlock (1979), Sauessy (1990) and Mayer and Ashlock (1991). The contemporary phylogenetic systems of classification as suggested by Takhtajan, Cronquist, Thorne and Dahlgren are largely based on phenetic relationships.

The form of classification is determined by the purpose for which it is constructed, *e.g.* natural system of classification serves greatest number of purposes such as compilation of floras and arranging plants in botanical gardens and herbaria and is, therefore, called **general classification** because it is most generally used. A particular classification can be called better than other only in relation to its purpose.

Description of plants and phylogeny are among other components of plant taxonomy. Description of a plant involves listing its features in short–consisting of only those taxonomic characters which distinguish the plant from other related ones. This process is known as **diagnosis** and characters as **diagnostic characters**. The diagnostic characters of a plant determine its circumscription. The description is recorded in semitechnical language following a set pattern such as habit, root, stem, leaves, flowers and its parts, fruits, seeds etc. Each character has an appropriate **character-state** such as for flower colour (character) may thus be red, yellow, purple etc. (state). Specific terms are used for each character-state, *e.g.*, petiole (character) may thus be petiolate or sessile (state) to enable a proper documentation of data. It is easy to describe a fresh plant specimen. The dry specimens, particularly for dissection, need to be softened in hot water or in a wetting agent to make them suitable to be described. After a plant has been classified, it needs to be assigned proper place according to its phylogeny. **Phylogeny** refers to the study of genealogy and evolutionary history of a plant or a taxonomic group. **Genealogy** is the study of ancestral relationships and lineages.

CONCEPT OF TAXONOMIC CHARACTERS

Surrounded by innumerable incomprehensible objects, almost impossible to know about each of them, man has succeeded in discriminating them using certain attributes or characters common to large number of objects, *i.e.*, he classified them into manageable groups. Man has always been in practice to discriminate between the objects around him based on his liking or disliking. He proved able enough to pin point a thing (a group of alike things) of his interest by noticing the similarities and differences with other things for precision and better communication. Since the advent of the civilisation, this process in man's life begins from the cradle and ends to the grave.

Plants, even of the same taxon, are never exactly alike. They exhibit variation in their form and structure. The criteria to delimit them from each other is based on characters. Since each individual exhibits a large number of characters, selection of characters poses a serious challenge to their classification as upto what extent they may be applied to decide the status of a taxon. The decision for their adoption, *i.e.*, selection of characters, is made only after careful and exhaustive study. Only those characters, which could easily be recognised for working, are selected.

In living organisms, a character is mainly concerned with variation observed. The deviation in selection of characters may lead to discrepencies in different classifications or even in one and the same classification thus, causing the need of some kind of agreement in selection of characters. The terms resemblences and differences; similarity and dissimilarity etc., are chiefly applied with respect to characters. Attempts have always been made to search and select such characters that may provide sound evidence for construction of natural and more acceptable classification of living organisms in general and plants in particular, where number of characters observed and employed, may determine the nature of classification of a taxon. Despite vast amount of data resulted, it has been possible for some taxonomists to use only a few characters while others could employ hundreds of them. Pre-Darwinian classifications have derived characters mainly from exomorphology while Post-Darwinian classifications, that have attracted much attention and popularity, are largely based on traditional exomorphological characters fortified with characters from anatomy, embryology, cytology, physiology etc.

Characters : What Are They ?

A character may be defined as '*anything that can be considered as a variable independent of any other thing considered at the same time*' (Cain and Harrison, 1958). According to Heywood (1969) a unit character is '*any attribute (or descriptive phase) referring to form, structure or behaviour which the taxonomist separates from the whole organism for particular purpose such as comparison or interpretation*'. Sneath and Sokal (1962) defined unit character as '*taxonomic character of two or more states, which within study at hand can not be further divided logically* ...'. In practice it is rather difficult to spread over characters into units due to lack of skill and perception. The comparison of characters may be objective but their selection is largely subjective. And for the purpose of numerical taxonomy Sneath and Sokal (1962) have recommended the usage of unit's character controlled by a single gene. For expression in taxonomy, unit character may be defined as taxonomic character of two or more states which can not be logically divided further into components (except for sub-divisions which become necessary due to alterations in the method of coding the states). Theoretically, these definitions of characters are logical and have simple explanation but often there are certain practical problems in ascertaining what should be regarded as unit character, especially in numerical taxonomy. It is convenient to deal with unit characters which can be treated quantitatively but their choice is largely a matter of experience. Over all similarities of an organism can be judged by a combination of characters. Natural selection operates on the basis of combination of characters.

As a matter of fact, characters may be considered as those features which can be expressed, measured, counted and determined, *e.g.* ratio of cotyledons; size and shape of floral parts; number of sepals, petals (or perianth) and stamens; ovules and seeds–their structure and weight; number of teeth on leaf margin; number of transverse septa in anthers etc. A character may be regarded as simple when it has single expression. When any character is expressed by way of assemblage of various aspects, it may be called compound (multiple) character, *e.g.* leaf apex is obovate having some measurement and this expression of character comprises components of length–maximum and minimum and width-maximum and minimum, width at various intervals, length and width ratio etc. Obviously a single character–obovate can be spread over smaller units.

Kinds and Nature of Character

Kind and nature of character selection depends mainly on type of plant and status of a taxon undertaken for study. In taxonomic procedure, selection of character may be employed for two main purposes, (i) identification, characterisation and delimitation of species and (ii) categorisation of these species into higher taxa. Accordingly, characters may be **analytical** or **synthetic**. Analytical characters are mostly diagnostic ones. They are useful in identification, characterisation and delimitation of lower taxa such as species. On the other hand synthetic characters are useful in grouping these taxa in higher ranks as they gradually decrease in number as one proceeds from lower to higher level. Obviously, only a few characters become increasingly constant at higher levels in taxonomic hierarchy, *e.g.* characters like sympetalous (gamopetalous) vs. polypetalous corolla, superior vs. inferior gynoecium etc., are quite constant. They do not vary within smaller groups or within populations of the same species. Such characters are termed as **constitutive** or **organisational** characters. **Diagnostic** or **key characters,** used for identification and circumscription, are often restricted in their occurrence and thus alone, they are enough to reach a correct diagnosis. They must be included in floras and mannuals as they are most useful and easy for the identification.

It has also been noted that there is no uniformity in distribution of characters amongst various parts of different plants as demonstrated by counting and measuring plant parts from different angles, observations of colour, texture, sculpturing of seeds etc. This variation may be **qualitative** (including presence or absence) for characters relating to form and structure or **quantitative**–for features assessed by size, length, breadth etc. and both of these are used in taxonomic descriptions. In the qualitative description characters such as presence or absence of corolla, position of ovary, union of stamens etc. are considered. In the quantitative description, on the other hand, the number of parts and their range such as seeds 12 to 18; stamens 8 to 10 etc.; size expressed as length, breadth and its range such as leaves 5 to 6 cm long and 3 to 4 cm broad; shape, colour, sculpturing and certain other such characters are considered. Qualitative characters find more use in distinguishing taxa of specific or higher ranks whereas quantitative characters are generally useful to separate lower taxonomic categories at infra-specific levels. Theoretically, all characters are useful in classification, but their use, as often experienced, appears to be limited to a small number as certain characters are considered to be **good**, *i.e.*, more reliable than others. The characters considered as good are not subject to wide variation within the study samples, do not have high intrinsic genetic variability and are not easily susceptible to environmental modification, *i.e.*, exhibit consistency or agree with the correlations of characters existing in a natural system of classification. Conclusively, '*a taxonomic character is only as good as its constancy*'.

Character Preference (Character Weighing)

Although each plant specimen bears a large number of characters but, out of them, only a few are always chosen in accordance with the specific type of study. During selection of characters for classification, taxonomists use their knowledge, experience and personal skill, taking into consideration the status of taxon and its interrelationship with other taxa. Heywood (1963) used the term **weight** for this selection of character from virtually unlimited number of characters from the point of view of its relevant importance over other characters. Thus, character weighing refers to search of those characters which are more constant and can be given more weight. Character weighing is done for the reason because constant characters are most easily observed; show highest correlation with others in a natural group; are believed to have some known or inferred phylogenetic importance or are shown to be more conservative and were left outs while constructing a natural group. The plant characters, given weight, may be **a priori**, *i.e.*, selection of characters that are conservative and genetically stable, indicating phylogenetic correlations between taxa. They are used to construct phylogenetic or vertical classification. Such features showing high degree of importance, realised to be of importance after construction of taxa, are given **a posteriori** or ***correlation weighing***, the only type of weighing which should be used for classification. They are employed for a natural or horizontal classification. Characters abstracted by **a posteriori** weighing will be those that diagnose a group and these markers will probably be most likely the characters to be used for keys. The characters showing highest average correlation with other similar characters are called **indicator characters** or simply **indicators**. The indicator characters are maximally correlated with other characters and, therefore, have the highest taxonomic value. For example, monocarpellary gynoecium with marginal placentation in family Leguminosae is a marker or indicator character. A set of such indicator characters constitutes the so-called **taxonomic coefficient**. **Rejection** (residual weighing), useful for natural classification, refers to rejection of the characters that are found more variable than the others and are often not useful for diagnosis or for construction of keys, despites in some cases they may be specialised characters for a particular mode of life.

Biological Characters

Biological characters are those that are related to some vital functions or advantages. These include mostly floral characters such as a sympetaly, zygomorphy, epigyny etc. Such characters are considered to be involved in progressive evolution of angiosperm marked by functional relationships, such as economy in production of reproductive parts and adaptation to insect pollination. **Fortuitous characters** show no functional relationship.

Characters in Phylogenetic Methods (Cladistics)

The key factor in determination of the phylogenetic position of a particular group is the **pleisiomorphic** (primitive) or **apomorphic** (advanced) characters that the group contains. During the second decade of the ninteenth century it was realised that certain morphological characters are statistically correlated and that these can be used in the study of evolution. It is believed that correlation between characters arises because rates of their evolution have been correlated. Sporne (1974) has, however, argued that correlation can be shown to occur even though rates of evolution of characters vary and within any taxonomic group, primitive characters may be expected to show positive correlation merely because of their distribution is not random. As such, primitive members of that group have retained a relatively high proportion of ancestral (primitive) characters whereas advanced members have dispensed with a relatively high proportion of these characters, either by loss or replacement with different advanced characters. And thus, the distribution of primitive characters is displaced towards primitive members with higher proportion of primitive characters than the average for the groups as a whole. Departure from the random distribution can be statistically calculated so as to establish correlation among characters. These calculations led Sporne (1974) to prepare a list of 24 characters in dicotyledons and 14 in monocotyledons that exhibit positive correlation. On the basis of their distribution, these characters have been categorised as **magnoloid** (primitive) and **amarylloid** (advanced), respectively and on this basis Sporne calculated an **advancement index** for each family and gave a circular diagram for different families of angiosperms in which he placed the most primitive families near the centre and most advanced along the periphery. That the earliest members of angiosperms are extinct, is evident from the fact that none of the present day families has the zero advancement index but all living families have advanced in some respect, *e.g. Tetracentron* has primitive vesseless wood but the pollen grains are advanced, being tricolpate.

To keep the organisms with two superficial resemblances in separate groups, the understanding of two terms **homology** and **analogy**, first used and defined by Owen (1848), is quite helpful in evolutionary biology. Homology is defined as the occurrence of the same organ in different organisms under every variety of forms and functions whereas analogy as the occurrence of a part or an organ in one organism which has the same function as another part or an organ in a different organism. Darwin (1859) was first to apply these terms. Simpson (1961) has defined homology as the resemblance due to inheritance from a common ancestry and analogy as the functional similarity and not due to inheritance from a common ancestry. The homology between the two organisms can result only from their having evolved from a common ancestor, the ancestor must also contain the same feature or features for which the two organisms are homologous.

If a character shared by two organisms is not traced to a common ancestor, the similarity may be the result of **homoplasy** which may be either **parallelism** or **convergence.** Parallelism results due to the organisms having a common acestor but the character was not present in their common acestor. Whereas convergence results due to two characters in different ancestors evolving into identical characters and both situations, because similar character is derived and not traced to a common ancestor, represent **false synapomorphy**. Example of parellelism is the development of vessels in Gnetales and dicotyledons. In phylogenetic methods, convergent genera are separated to make groups natural and monophyletic. Convergence commonly results due to adaptation to similar habits, *e.g.* gross similarity between certain succulent species of Euphorbiaceae and Cactaceae, similar modes of pollination, *e.g.* pollinia in Asclepiadaceae and Orchidaceae; similar modes of dispersal, *e.g.* seed dispersal in Malvaceae, Asteraceae and Asclepiadaceae etc.

At the end of ninteenth century, a gradual shifting from natural to phylogenetic systems of classification resulted changes in principles of taxonomy. But it is not easy to ignore Cronquist's (1968) certain basic principles of taxonomic characters :

1. ***Taxa are properly established on the basis of multiple correlation of characters.*** It is probably the most important principle of taxonomy emphasising significance of correlation of characters. It states that no character of a taxon should be considered singly for the establishment of its relationship with other taxa instead it should always be considered along with other characters.
2. ***Taxonomic importance of a character lies as to how well it correlates with other characters. This means the taxonomic importance of a character is determined by a posteriori rather than a priori.*** And thus, a character chosen should exhibit maximum correlation with other characters. The importance of a character depends upon the degree of correlation, *e.g.* a character with no obvious correlation, is often taken to be anomalous one and is usually not of taxonomic importance. That is why zygomorphic flower in Ranunculaceae, only five stamens in Tiliaceae, alternate leaves in Oleaceae, actinomorphic corolla in Scrophulariaceae, simple leaves in Bignoniaceae and numerous such taxonomically insignificant characters are taken to be anomalous. They have no correlation with other general characters of the taxa, although, they might have been considered important from their diagnostic point of view but the relationship must be considered on overall resemblances.
3. ***An important feature of taxonomy is its predictive value.*** Anomalous characters often find their use in predicting probable relationship of the taxa with other groups where these characters are considered as normal characters.

Chapter **3**

Plant Nomenclature

EARLY ATTEMPTS IN NAMING OF PLANTS

The art of naming a taxonomic unit of any rank is, in fact, a science of application of names–**nomenclature** (in Latin *nomen* = name; *clare* = call), *i.e.*, the science of providing distinct and proper names to plants and taxonomic groups to enable their recognition and to differentiate them from other plants and taxonomic groups. The purpose of giving names to plants and taxonomic groups is not to indicate their characters or history, but to supply a means of referring to them and to indicate their taxonomic ranks. Plants are given names for two main reasons, (*i*) as an aid to communication and (*ii*) to indicate their relationship. In practice, nomenclature is often combined with identification.

Since pre-historic times, man has always been interested in plants because of their usefulness in some or the other way. In all nations, tribes and languages, names have been given to the plants in various ways. Often one and the same plant has been given different names in different languages in different countries, even in different parts of one and the same country or one and the same common name is applied for several different plants. And thus, for use of such names of plants, no uniform accord could be attained.

Common (Vernacular) Names

Many plants have common (vernacular) or local names which have a limited usage. Widely distributed plants and the plants with high economic value have a larger number of common names. For example **pansy** (*Viola tricolor* L.), grown commonly as an ornamental plant in most of the European and American countries, has more than 150 names, including about 50 common names in English. Likewise *Verbascum* has over 140 and *Plantago* has more than 50 common names. The common names of the plants are quite indefinite and are restricted to the people of one language or even one region of the same country. In multilingual country like India, *Ipomoea batatas* (English name **sweet potato**), in Hindi and Punjabi is known by its common name **Shakarkand**; in Bengali, **Lal Aalu**; in Orriya, **Kanda**; in Telegu, **Chelaganda** and in Marathi, **Ratalu**. In ancient Indian literature on medicinal plants–the **Ayurveda**, *Mangifera indica* L. (English name **mango**) is known by over 50 different names, all in Sanskrit language. As such, because of limited usage, no plant can be recognised for wider use on the basis of its common name throughout the country or in different countries of the world. Since common names are neither methodical nor universal, often misleading and inadequate, they have thus proven unsatisfactory. As such it became necessary to name a plant for wider use in such a way that it becomes easy to identify it correctly world over.

Scientific Naming of Plants

In order to overcome many problems, such as common names *(i)* not being available for all species known to man, *(ii)* being restricted in the usage and applicable in a single or few languages in restricted locality, *(iii)* not providing proper information or indicating generic and family relationship, *(iv)* two or more unrelated species given the same name and *(v)* several names given to one and the same species in same language in the same or different locality, professional botanists have given scientific names to all known plants for international communication in a methodical way.

Despite the fact that much more botanical literature, especially in the area of medicinal plants, was written in Sanskrit during ancient Indian period, there was no obvious proposal of naming the plants in Sanskrit. Botanists of the world agreed upon that the scientific names, regardless of their origin, should be in Latin. There are various reasons as to why plant names be given in Latin.

Some important reasons are :

1. During middle ages, Latin was a widely used language (lingua fransa–spoken and written) throughout the European countries where a lot of literature was produced in this language. Since Europe, particularly Greece, has dominated the world in the field of science in general and botany in particular during 1600 BC to 185 AD, Latin became an acceptable language of botany and other allied sciences.
2. Latin is a dead language and as such meaning and interpretation are not subject to changes unlike other languages such as English. It is not being used by any country or nation at present and thus its use causes no national bias.
3. It is from Latin that most of the European languages have evolved.
4. Latin is specific and exact in meaning and in the grammatical sense of the word as is commonly obvious, *e.g.* white translated as *album*–neuter, *alba*–feminine or *albus*–masculine.
5. Latin has been found to be suitable for descriptive phases of natural science because it employs Roman alphabet which fits well in the text of most languages.

Polynomial Nomenclature

For several centuries, the names of plants appeared as polynomials–long descriptive phrases often difficult to be used and remember easily. In polynomials, a single name, thus, had to be supplemented with many descriptive terms or adjectives to facilitate more or less neat description of the plant. For example, presently named *Sida acuta* was named by Pulkonat as *Chrysophyllum foliis ovalis, supeme glabris, parallele striatis, subtus tomentosonitidis* and a species of willow, by Clusius in his herbal, was named as *Salix pumila angustifolia altera*. Polynomials have been used by many early taxonomists including Linnaeus. Linnaeus, in his *Species plantarum*, intended to use polynomials but it was only for the sake of convenience that he used additional **trivial names** (specific epithets) for these polynomials.

Binomial Nomenclature

In efforts to shorten the names, early attempts were made by Gaspard (Caspar) Bauhin in his *Pinax theatri botanici* (1623) to introduce the concept of **binomial nomenclature**, which after more than hundred years, was adopted by Linnaeus who started using trivial names (specific epithets) to replace descriptive terms or adjectives used in polynomials. Linnaeus suggested a brilliant system of binomial nomenclature which he employed firmly in the first edition of his *Species plantarum* (1753)–a two volume catalogue of plant kingdom, and later throughout his work. The book marked the first consistent use of binomial nomenclature and diagnosed some 6000 species in 1000 genera and is considered to be the most important work in the botanical world. The date 1st May, 1753 when the *Species plantarum* was published, is marked as the beginning date of the present day binomial nomenclature.

Thus, Linnaeus (Carl von Linn, 1753) replaced the polynomials by binomials in which the name of a plant species consisted of two parts, the first–the name of the genus and the second-specific epithet (trivial name). The first name, the generic name, is substantive (noun) and starts with a capital letter whereas the second, following the generic name, the specific epithet, is an adjective (or noun) and starts with a small letter and indicates which of the several members of the genus has been considered. One name given to the plants of one type is known as generic name and all other plants of similar type with slight variations are denoted by small explanatory words, the specific epithets. For example, the botanical name of **mango** is *Mangifera indica* and in this, first term–*Mangifera* designates the genus of the **mango** plant whereas the second term–*indica* designates a particular species of the genus *Mangifera*. Generic name *Mangifera* alone, indicates that it does have one or more species but specific epithet, *indica* alone has no bearing. There is only one botanical name, known as its legitimate name, for a plant. Botanical names are always printed in *italics* and handwritten or typed names are always underlined.

Linnaeus set forth the elementary rules of nomenclature in his *Critica botanica* (1737) which he further amplified in his *Philosophica botanica* (1751). These rules of botanical nomenclature served as guide to botanists. In his *Theorie elementaire de la botanique* (1813), AP de Candolle gave explicit explanations and procedures of binomial nomenclature which were mainly taken from Linnaeus's guiding rules. Steudel, in his first edition of *Nomenclator botanicus* (1821), gave Latin names to all flowering plants, along with their synonyms, known to him by that time.

Box 3.1 Carolus Linnaeus : Father of Taxonomy

Carolus Linnaeus (1707–78), also known as Carl Von Linne, Carl Linnaeus and Carle Linne was borne in Rashult, Sweden on May 23, 1707. Botany got attached to him at birth because Linnaeus is Latin to Linn or Linden tree (*Tilia* sp.). His father wanted his son to become priest but Linnaeus preferred to choose University of Lund in 1927 to learn medicine and later shifted to the University of Uppsala in 1729 where in recognition of his enthusiasm for plants he was introduced to botanist Professor Rudbeck. Linnaeus published his first paper on sexuality in plants in 1729 which got wide publicity. As a result he was appointed as Demonstrator and subsequently promoted as Docent. In 1730 he published *Hortus upplandicus*, enumerating plants in Uppsala Botanic Garden according to the system given by Tournefort. On finding it difficult to fit increasing number of plants according to Tournefort's system, he published a revised system of *Hortus upplandicus*. He classified plants according to sexual system in *Systema naturae* (1735) and later in *Classes plantarum* (1738), a work in which he described some 7,300 species of plants. In 1735, he went to The Netherlands and obtained MD degree and later became personal physician of George Clifford, a prosperous man, who gave Linnaeus an opportunity to study numerous tropical and temperate plants grown by him in his garden. Clifford financed Linnaeus in publication of his *Hortus cliffortianus* and *Genera plantarum* (1737). Following death of Professor Rudbeck, Linnaeus was appointed as Professor of Medicine and Botany at University of Uppsala where he held this position till his death in 1778. During this course, he published his *Species plantarum* on 1 May, 1753, the date marked as the beginning date of present day binomial nomenclature and the publication as the most important work in the botanical world.

Fig. 3.1 Carolus Linnaeus (1707–78).

In recognition of his contributions he was given the title of **Knight of the Polar Star** in 1753 and in 1961 **Patent of Nobility**, and from this date he came to be known as Carl Von Linne.

Index Kewensis

Publication of *Index Kewensis–Plantarum phanerogarum*, a comprehensive list of scientific names of flowering plants, was started (in 1893–95) from Royal Botanic Gardens, Kew, England (the largest Botanic Garden in the world). Steudel's second edition of *Nomenclator botanicus* (1840), which contained the Latin names of all flowering plants, together with their synonyms, formed the basis of *Index Kewensis.* The entire work (from 1846 to 1927), initiated by Sir JD Hooker, the then Director of the Kew Botanic Garden, in collaboration with Benjamin, Daydan and Jackson, needs special mention. These publications have been compiled since 1893 onwards. Hooker received financial assistance for these publications from the donations of Charles Darwin, the famous phylogenesist and close friend of Hooker. In its first publication, that included 375,000 scientific names–both generic and specific, known since the time of Linnaeus, were arranged alphabetically with mention of authority. Attempts were made to provide a brief description of the plant's native country, place of publication and name of the author of the published account. The compilation is of immense importance as it greatly helps in tackling taxonomic problems.

Uninomial Nomenclature

There was a general feeling among the botanists that while strictly following binomial system of nomenclature as suggested by Linnaeus, certain practical difficulties arise. They found that the rules were slightly complicated, and following the rules, it was difficult to shift one individual species from one genus to another genus, carrying the specific epithet of previous name and sometimes often duplication of names in the same taxon. The rules suggested that due to change in position of the individual in taxonomic hierarchy, both generic name and specific epithet are to be changed. To simplify the system and overcome these problems, Heywood (1974) suggested that the two names–generic and specific, may be united together into a single unit or hyphenated (-) to constitute a single unit, *e.g.* when *Ranunculus ficaria* is shifted to genus *Ficaria*, it need not to be *ficaria* but as *Ranunculus ficaria*, regardless whether it belongs to *Ficaria, Ranunculus* or any other genus. Thus in case of shifting of one species to another taxon, no necessity is caused to introduce a change in generic name or specific epithet. It was also suggested that for each taxonomic group, two numbers may be given which would remain unchanged throughout, irrespective of any change in its position and relationship. This number, termed as **reference number**, incorporates summary of individual's characters. The coding of reference numbers with description of characters was also suggested. The reference numbers are under preparation at international level. Since knowledge about plants is greatly expanding and making it impossible to remember all the names and characters regarding plants, the reference numbering method would prove more convenient to taxonomists in future. This system of giving one name to the individual, with certain advantages for different purposes, is termed as uninomial system of nomenclature.

INTERNATIONAL CODE OF BOTANICAL NOMENCLATURE

Discovery of new plants recorded from extensive explorations caused concern over procedures for their naming. Moreover, in different countries, nomenclature followed different patterns and caused personal and national jealousies. And thus, a need for international accord was felt. Arguments and criticism and raising objections against every new thing, as is human nature, raised difference of opinions despite the rules for nomenclature were welcomed by botanists worldwide. The first organised effort towards the development of uniform botanical nomenclature was made by Alphonse de Candolle, son of AP de Candolle, who circulated a copy of his draft of laws of nomenclature *Lois de la nomenclature botanique* which was put in the first International Botanical Congress held at Paris in August 1867. His draft of laws of nomenclature, also known as **de Candolle Rules**, was adopted in this Congress and was published as Paris Code. Linnaeus (1753) was made starting point for plant nomenclature and rule of priority (Article No. 15) was made fundamental. Following this principle of priority O Kuntze in his *Reverisio generum plantarum* (1891) changed 1000 generic names and more than 30,000 species names. The subsequent meetings of International Botanical Congress were held in 1892 (Rochester Code), 1905 (Vienna Code), 1907 (American Code) and 1910 (Brussels Code), but a general agreement regarding rules of plant nomenclature, which could be accepted internationally, was reached in 1930 (Cambridge Code) at 5th International Botanical Congress at Cambridge during which for the first time in the botanical history, a code of nomenclature– **International Code of Botanical Nomenclature (ICBN)**, which was accepted internationally, came into being. The modifications or ammendments, as suggested by International Botanical Congresses at subsequent years, have regularly been incorporated in the Code (ICBN). Various codes, which resulted during different International Botanical Congresses are mentioned below :

1. Paris Code (1867). The first International Botanical Congress was held at Paris in August 1867. It was aimed at standardisation and legislation of proper nomenclatural practices. About 150 botanists from American and European countries were invited to attend the congress. Laws of botanical nomenclature (*Lois de la nomenclature botanique*) were discussed, published (1868) and adopted with slight revision. And it was resolved that laws, as adopted by assembly, shall be recommended as the best guideline for nomenclature of vegetable kingdom. Since laws were adopted at French capital, they were called Paris Code or **de Candolle's Rules** (as they were prepared by Alphose de Candolle). According to the Paris Code, the starting point for all nomenclature was fixed with Linnaeus. However, it did not specify any date or any work. The rule of priority was considered as basic with no provisions of exceptions. Emphasis was given that publication be valid and attention was given to author citation and terms applied to categories of plants. Despite the fact that Paris code provided guidelines for taxonomic activities in most of the countries to a considerable extent, their application showed many inherent defects. Gradually, American and British botanists deviated from the rules and started practising unwritten law, named *Kew Rule* with the provision that if a species was transferred to another genus, the specific epithet need not be transferred to the new genus instead the author was given freedom to use a new epithet in the new combination.

2. Rochester Code (1892). A, batch of botanists headed by NL Britton, met at Rochester, New York, (USA) in 1892 and framed a set of rules based on modifications of the Paris Code. These modified rules, called Rochester Code, included some new recommendations listed below :

(1) Establishment of the type concept to ascertain the correct application of names;
(2) Strict adherence to the principle of priority;
(3) Acceptance of alternate binomials resulting from employment of the principles of priority, even if the specific epithet repeats the generic name (referred to as **Tautonym**);
(4) Interpretation of priority to the precedence of a name in a publication in addition to the date of publication; and
(5) Each new species published to be based upon a designated herbarium specimen to be known as the **type specimen** for the new binomial.

3. Vienna Code (1905). The rules adopted at the third International Botanical Congress, held at Vienna in June 1905, are substantially the same as adopted in Paris in 1868 but with several clear improvements. The new changes include :

(1) Establishment of Linnaeus's *Species plantarum* (1753) as the starting-point for naming vascular plants.
(2) *Nomina generica conservanda*, by which the generic names having a wide use would be conserved over earlier but less well-known.
(3) Banning of tautonyms.
(4) Requirement that names of new taxa be accompanied by Latin diagnosis.

The prescriptions were divided into **principles**, **rules** and the **recommendations**. The principles are the foundation of the rules and recommendations. The rules govern the names or forms of nomenclature, and anything contrary to a rule can not be maintained. The recommendations were on the secondary points for greater uniformity and clearness in nomenclature and may or may not be accepted.

4. American Code (1907). The advocates of the Rochester Code attempted to have their principles included in the proposed revision at Vienna but failed to do so. The dissatisfied American adherents of Rochester Code refused to accept the Vienna Code and published the new American Code, a slight modification of Rochester Code in 1907 and thus two opposing schools of thought of Rochester Code, one under the leadership of NL Britton and his students who followed the provisions of the American Code and the other under the direction of Asa Gray's pupil, who followed the International Code, were created.

The first provision of American Code was that advocates would not subscribe to the principles of *nomina generica conservanda* or of the requirement of Latin diagnosis but they accepted the type concept. Another provision of the Code was that "a binomial may not be used again for a plant in any way if it has been employed previously for another plant, even though the previous use may have been illegitimate".

5. Brussels Code (1912). At the fourth International Botanical Congress held at Brussels in 1910, the most significant decision was the establishment of different starting points for priority of names of non-vascular plants, the recognition of value of type concept and classification of phrascology of the Vienna Code.

6. Cambridge Code (1935). The fifth International Botanical Congress held at Cambridge, England in 1930 was successful in reconciling the basic differences between the Vienna Code and American Code and bringing about harmony among major botanical fractions. The new rules regulated at this Congress constituted the Code of nomenclature truely international in name and function. Following were the approved provisions.

(1) The type concept should be pursued;
(2) There should be a list of *nomina generica conservanda*;
(3) Tautonyms are not admissible; and
(4) Latin diagnosis is required after January 1, 1932.

7. Amsterdam Code (1947). At the sixth International Botanical Congress, Amsterdam (1935), a few major changes in rules were made, and it was resolved that "from January 1, 1935, names of new groups of recent plants (except the bacteria), are considered as validity published only when they are accompained by a Latin diagnosis." An attempt to select a list of *nomina specifica conservanda* was thwarted by an overwhelming vote.

8. Stockholm Code (1954). The seventh International Botanical Congress, helt at Stockholm in 1950, introduced certain number of definitions on types. The word **taxon** (pl. **taxa**) was introduced for the first time to designate any taxonomic group or entity.

9. Paris Code (1956). The eighth International Botanical Congress held at Paris in 1954, emphasised on types (Appendix IV), but the rule of Latin came under fire. It was resolved that the Code should be published in English, French and German languages. The separation of **Preamble** and **Principles** from the Rules and Recommendations was major feature of the Code. Appendix II (*nomina generica conservanda at rejicianda*) was amended and supplemented.

10. Montreal Code (1961). The ninth International Botanical Congress held at Montreal, Canada (1959), appointed a special committee to study the question of conservation of family names (*nomina familiarum conservanda* for angiospermae as Appendix II), was introduced. In the Code it was specified that naming of fossil plants should follow the same line as that of recent plants.

11. Edinburgh Code (1966). In the tenth International Botanical Congress held at Edinburgh in August 1964, the report of the special committee appointed during the ninth Congress, some points of which are given below, were submitted.

(1) AL de Jussieu's *Genera plantarum* (1789) is the starting point for family names. Although, many of the names included in the list are not a matter of dispute, but their inclusion in the list forms a ready reference that provides the correct names of the families together with a type genus for each family.

(2) In the list of *nomina familiarum conservanda*, a few names have spelling somewhat different from long-established names, *e.g.* Cannabaceae for Cannabinaceae, Capparaceae for Capparidaceae, Haloragaceae for Haloragidaceae, Melastomataceae for Melastomaceae and so on.

(3) The rule of Latin for new diagnoses was finally settled and no amendments were proposed.

A committee was set up to prepare a glossary of the main technical terms that come across the Code. And, this resulted in *An annonated glossary of botanical nomenclature.*

12. Seattle Code (1972). The eleventh International Botanical Congress held at Seattle in August 1969 proposed Seattle Code (edited by SA Stafleu) which was published in 1972. Most of the proposals submitted to this Congress were concerned with refinements and increased precision. The chief issues in this Code included tautonymous (where specific epithet exactly repeats the generic name with or without transcribed symbol) designations of taxa between genus and ·pecies and below species. The perennial questions of superfluous names and reorganisation of the rules for hybrids were other issues. The word **a onym** (automatically established names) was also introduced in this code.

13. Leningrad Code (1978). The recommendations of the twelfth International Botanical Congress held at Leningrad, Russia in July 1975, came out in 1978 indicating only few differences from the Seattle Code. The changes are discussed below :

(1) The concept of organ genera is eliminated for fossil plants.

(2) The Code does not apply to names of organisms related (as bacteria) and does apply to all other organisms treated as plants.

(3) The principle of autonymic typification is extended to those names of taxa above the family rank that are ultimately based on the generic names and the application of the priority principle is recommended while selecting the names thus typified.

(4) A name or combination published before 1953, without indicating the rank, is considered validity published but imperative in question of priority, except homonym and certain names to be accepted at the varietal rank.

(5) Article of the previous Code is modified on the basis of type method and Articles 70–71 dealing with discardant elements and montrosities were deleted, but the Article numbers are retained to facilitate the use of the Code.

(6) The section on orthography is thoroughly rewritten.

(7) Individual paragraphs on all Articles and Recommendations are numbered in a decimal-like system, some being rearranged.

14. Sydney Code (1983). The thirteenth International Botanical Congress held at Sydney, Australia in 1981, gave a Code aimed at the provision of a stable method of naming taxonomic groups, avoiding and rejecting the use of names which may cause error or ambiguity or throw science into confusion. Next in importance is the avoidance of useless creation of names. Other considerations, such as absolute grammatical correctness, regularity or euphony of names, more or less prevailing custom, regard for persons etc. notwithstanding their undeniable importance, are relatively accessary.

15. Berlin Code (1988). This Code includes the proposals made at fourteenth International Botanical Congress held at Berlin in 1987. *Nomina specifica conservanda* was for the first time introduced at this Congress. Two species names *Lycopersicum esculentum* P Miller and *Triticum aestivum* Linn have been conserved against the rule of priority, as these names are widely used and any change may create confusion. Articles 66 and 67 were removed.

16. Tokyo Code (1994). The proceeding of fifteenth International Botanical Congress held at Yokohama, Tokyo in 1993, have been translated into Chinese, French, German, Japanese, Russian and Slovak. In this Code extensive renumbering had taken place and, therefore, its preface included a tabulation, comparing the placement of its provisions with those of Berlin Code.

17. St. Louis Code (2000). This Code includes the proceedings of the sixteenth International Botanical Congress held at St Louis, Missouri in July-August, 1999, superseding Tokyo Congress. The Code was published in 2000 (Greuter *et al.* 2000). Like Tokyo Code, it was anticipated that this Code too, will become available in several languages in due course. The detailed provision of the Code is spread over Rules, set out in Articles and Recommendations. The object of the Rules is to put the nomenclature of the past into order and to provide for that of the future. Names, contrary to a rule, can not be maintained. The Recommendations deal with subsidiary points. Their object is to bring about great uniformity and clearness, especially in future nomenclature. Names, contrary to Recommendation can not, on that account, be rejected, but they are not examples to be followed. The Rules and Recommendations apply to all organisms whether fossil or non-fossil, including fungi but excluding bacteria. Bacterial nomenclature is governed by a separate Code called International Code of Nomenclature of Bacteria (ICNB).

Presently, International Code of Botanical Nomenclature (ICBN) published by the International Association of Plant Taxonomy (IAPT), governs the botanical nomenclature. Whereas naming of cultivated plants is governed by International Code of Nomenclature for Cultivated Plants (ICNCP), which is largely based on ICBN. During the last decade, for convenient handling of combined database for all plants, attempts have been made to develop a unified code for the vegetable kingdom. Although developing **Phylo Code** is an concerted effort in this direction but it is expected that it will take a long time before it is accepted.

PUBLICATION OF INTERNATIONAL CODE OF BOTANICAL NOMENCLATURE (ICBN)

The publication of Code is based on the realisation that botany requires a precise and simple system of nomenclature used by botanists in all countries and aims at providing a stable method of naming plants and taxonomic groups, avoiding and rejecting the use of names which may cause error or ambiguity or throw science into confusion. **Preamble** highlights the philosophy of Botanical Code. The Code is spread over three divisions **Principles, Rules** and **Regulations**, and **Provisions for the Governance of the Code**.

The Code also inçludes appendices such as (I) names of hybrids, (IIA) *nomina familiarum algarum, fungorum, pteridophytorum at fossilium conservanda at rejicienda,* (IIB) *nomina familiarum bryophytorum at spermatophytorum conservanda,* (IIIA) *nomina generica conservanda at rejicienda,* (IIIB) *nomina specifica conservanda rejicienda,* (IV) *nomina utique rejicienda* (A–Algae, B–Fungi, C–Bryophyta, D–Pteridophyta, E–Spermatophyta) and (V) *Opera utique opressa.* The last three (IIIB, IV and V) are useful appendices and are included in the Tokyo Code. Of these the appendix IIIB includes the names of conserved and rejected specific names; the IV list of names and all combinations based on these names which are ruled as rejected under Article 56 and none is to be used and V abounds the list of the publication (and category of taxa therein) which are not validity published according to the Code.

Preamble

International Code of Botanical Nomenclature's preamble, with set of eleven points, highlights the philosophy of the Botanical Code. These include :

(1) Botany requires a precise and simple system of nomenclature used by botanists in all countries, dealing on the one hand with the terms which denote the ranks of taxonomic groups or units and on the other hand with the scientific names which are applied to the individual taxonomic groups of plants. The purpose of giving a name to a taxonomic groups is not to indicate its characters or history, but to supply a means of referring to it and to indicate its taxonomic rank. This **Code** aims at the provision of a stable method of naming taxonomic groups, avoiding and rejecting the use of names which may cause error or ambiguity or throw science into confusion. Next in importance is the avoidance of the useless creation of names. Other considerations, such as absolute grammatical correctness, regularity or euphony of names, more or less prevailing custom, regard for persons, etc., not withstanding their undeniable importance, are relatively accessory.

(2) The Principles form the basis of the system of botanical nomenclature.
(3) The detailed Provisions are divided into Rules, set out in the Articles and Recommendations. Examples (Ex.) are added to the rules and recommendations to illustrate them.
(4) The object of the Rules is to put the nomenclature of the past into order and to provide for that of the future; names contrary to a rule cannot be maintained.
(5) The Recommendations deal with subsidiary points, their object being to bring about greater uniformity and clarity, especially in future nomenclature; names contrary to a recommendation cannot, on that account, be rejected, but they are not examples to be followed.
(6) The Provisions regulating the governance of this Code form its last division.
(7) The Rules and Recommendations apply to all organisms traditionally treated as plants, whether fossil or non-fossil, *e.g.* blue-green algae (Cyanobacteria), fungi–including chytrids, oomycetes and slime moulds, photosynthetic protists and taxonomically related non-photosynthetic groups.
(8) The International Code of Nomenclature for Cultivated Plants is prepared under the authority of the International Commission for the Nomenclature of Cultivated Plants and deals with the use and formation of names for special plant categories in agricultural, forestry and horticultural nomenclature.
(9) The only proper reasons for changing a name are either a more profound knowledge of the facts resulting from adequate taxonomic study or the necessity of giving up a nomenclature that is contrary to the rules.
(10) In the absence of a relevant Rules or where the consequences of Rules are doubtful, established custom is followed.
(11) This edition of the Code supersedes all previous editions.

(I) Principles of ICBN

The International Code of Botanical Nomenclature is based on the following set of six Principles (Division I), which are the philosophical basis of the Code and provide guidelines for the taxonomists who propose amendments or deliberate on the suggestions for modification of the Code :

(1) Botanical Nomenclature is independent of Zoological Nomenclature. The Code applies equally to the names of taxonomic groups treated as plants whether or not these groups were originally so treated.
(2) The application of names of taxonomic groups is determined by means of nomenclatural types.
(3) Nomenclature of a taxonomic group is based upon priority of publication.

(4) Each taxonomic group with a particular circumscription, position and rank can bear only one correct name, the earliest that is in accordance with the rules.
(5) Scientific names of taxonomic groups are treated as Latin, regardless of derivation.
(6) The Rules of nomenclature are retroactive, unless expressly limited.

(II) Rules and Recommendations of ICBN

The Rules and Recommendations of International Code of Botanical Nomenclature are included in the Division II (Chapter I–VII and Articles 1–62). Some important ones are discussed below :

1. The Rank of Taxa

According to current Code, taxa and their ranks are covered under Chapter 1 (Articles 1–5). The term **taxon** (pl. **taxa**), a taxonomic group of any rank, has been introduced for the first time in 1956 edition of the Code (Article 1). The system of nomenclature provides a hierarchial arrangement of ranks and every plant is treated as belonging to a number of successively higher taxa, each assigned a particular rank with species as a basic unit. Mango thus belongs to *Mangifera indica* (species rank), *Mangifera* (genus rank), Anacardiaceae (family rank) and so on. The seven principal-obligatory ranks in descending sequence are : kingdom (regnum), division or phylum (divisio, phylum), class (classis), order (ordo), family (familia), genus (genus) and species (species). The ending of the name indicates its rank, *e.g.* ending-bionta denotes a kingdom, –phyta a division, –phytina a subdivision, –opsida a class, –opsidae or –ideae a subclass, –ales an order, –ineae a suborder, and –aceae a family. The rank of species is basic, one or more species make up a genus, one or more genera make up a family and so on. The Article 4 of the Code permits that if a greater number of taxa is required, the terms for these are made either by adding a prefix (sub–) to the terms denoting the ranks or by introducing supplementary terms.

(1) Family. The name of a family ends in –aceae (Article 18). There are, however, eight families of angiosperms whose orginal names, not in accordance with the rules, have been given alternate names in accordance with ICBN rules. These include Cruciferae–alternate name Brassicaceae, Guttiferae–Clusiaceae, Leguminosae–Fabaceae, Umbelliferae–Apiacae, Compositae–Asteraceae, Labiatae–Lamiaceae, Palmae–Arecaceae and Gramineae–Poaceae. Use of their old name is sanctioned because of old traditional usage but use of their new names is encouraged. Under unique exception to Article 18 of the Code, the name of Leguminosae is sanctioned as alternate of Fabaceae only as long as it includes all the three subfamilies : Faboideae, (Papilionoideae), Caesalpinoideae and Mimosoideae, but if these are upgraded as families then the name of Papilionaceae is conserved against Leguminosae for the first of getting the name Fabaceae and the two alternate names allowed now are Papilionaceae and Fabaceae.

(2) Genus. The generic name (Article 20) is a uninomial singular word treated as substantive noun and is capitalised, *i.e.*, its first letter is capital. The generic name may have any of the three masculine, faminine or neuter genders. The gender is usually recognisable by the name ending, *e.g.*, masculine by the ending–us or –pogon, faminine by –a or –is and neuter by –um. It is a general rule that for one kind of plants only one generic name is given. Names may be in Latin or may have Latin termination. Long names are abbreviated and adjectives (used as nouns) are avoided. The name may indicate, if possible, with the help of formation or ending of the name, the affinities of the genus. Usually names do not consist of two words until joined by a hyphen (–), *e.g.* originally published *Sebastiano–Schaueria.* During early attempts, plants were named only in Latin but gradual increase in knowledge of a large number of plants swelled so much that it became necessary to invent some new generic names and give them Latin appearance by slight modifications in their spelling.

The generic names may owe to any source but the common sources are discussed below :

(i) *Commemoration of a Person.* Such names are given directly to commemorate a distinguished person, a botanist or a patron of botany, *e.g. Bauhinia* for Bauhin, *Victoria* for Queen Victoria of England, *Washingtonia* for King George Washington, *Zinobia* for Queen Zinobia of Palmyra, *Linnaea* for Linnaeus, *Grayia* for Asa Gray, *Jeffersonia* for Jeffron, *Theophrasta* for Theophrastus, *Dioscorea* for Dioscorides, *Magnolia* for Magnol, *Hookera* for Hooker and so on. Such names always take a feminine form and are constructed by adding –ia if name of a person ends in a consonant (*Fuchsia* after Fuchs), –a, if it ends in a bowel (*Ottoa* after Otto), but if it ends in –a, –ea is added (*Collaea* after Colla). If name ends in –er, both –a or –ia are permitted (*Kernera* for Kerner : *Sesleria* for Sesler). For Latinised personal names ending with –us, this termination is dropped before appropriate ending (*Linnea* after Linnaeus).

(ii) *Based on a Place.* Such names are based on the place, *e.g. Araucaria* after Arauco–a province of Chile, *Salvadora* for EL Salvadore, *Arabis* for Arabia, *Sibiraea* for Siberia and so on.

(iii) *Based on an Important Character.* Such names are based on certain characteristics, *e.g. Cercocarpus* (coiled fruit), *Xanthoxylum* (yellow wood), *Hepatica* (liver-like), *Trifolium* (trifoliate leaves), *Gymnocladus* (naked tree), *Phytelephas* (plant elephant) and so on.

(iv) *Aborginal Names.* Such names are taken from a language other than Latin, without alteration of ending, *e.g. Narcissus* (a Greek name for daffodils) named after the famous Greek God Narcissus, *Ginkgo* a Chinese, *Vanda* a Sanskrit, *Sasa* of Japanese and so on or formed by Latinisation of common names, *e.g. Nelumbo* from Ceylonese, *Gantelbua* from South Indian, *Ravenala* from Medagascar, *Tsuga* from Japanese, *Ailanthus* from Moluccan, *Catalpa* from Red Indian, *Pandanus* from Malay and so on.

(v) *Combination of Words.* Some names have resulted by combination of two or more Greek or Latin words, usually expressive of some diagnostic feature of the plant, *e.g. Hygrophila* (marsh loving), *Polygala* (more milk), *Asterocantha* (spreading spines).

Some generic names owe fanciful, mythological, ethical or poetic origin, *e.g. Putranjiva* (putra-ratna-jiva) and *Theobroma* (god's food); *Nymphaea* (lovely water-nymphs) etc.

(3) Species. The name of a species (Article 23) is a binomial, consisting of two words–a generic name followed by a specific epithet, beginning with lower case initial letter (small letter) and is often derived from any source or composed arbitarily. But if specific epithet is derived from a person's name, former generic name or a common name, depending on the discretion of the user, the initial letter of the specific epithet can be an upper case initial letter (capital letter), although Code discourages such usage for specific epithets. Specific epithet is often an adjective (or sometimes noun). Sometimes the specific epithet is a noun and is derived from a common Greek or Latin word, *e.g. Capsella Bursa-pastoris* (Bursa-pastoris means Shephard's purse). Such commemorative names may be spelled with capital initial under special circumstances. The specific epithets when they are adjective, should agree grammatically with generic name. Among specific epithets, masculine is indicated by *albus, japonicus, niger, repens, chinensis, pubescence, bromides, vulgaris* etc., feminine by *alba, japonica, nigra, chinense* etc. and neuter by *album, japonicum, nigrum, vulgare* etc. If the specific epithet is in honour of a worker, it may end with –ae, *e.g. Mariae* or else it is after the name of the author, it is to end with –i or ii, *e.g. Baileyi.*

Specific epithets owe their origin commonly from the following sources :

(i) *Name of a Person.* Specific epithets named after a person may be of genetive (possesive) or an adjectival nature. In the former the specific epithet takes its form depending on the ending of the person's name. For example, the names ending in a bowel or –er, the letter –i is added for male person –*roylei* (*Impatiens roylei*) after Royle, *hookeri* (*Iris hookeri*) after Hooker, –ae for female person–*toddiae* (*Rosa toddiae*) after Todd and –orum for more than one persons with the same surname–*hookerorum* after Hooker and Hooker. And if the name ends in –a, then –e is added –*senguptae* (*Euphorbia senguptae*) after Sengupta. If the name ends in a consonant, –ii is added for male person (*wallichii* after Wallich), –iae for a female person (*wilsoniae* after Wilson) and –iorum for more than one persons with same surname of which at least one is male (*verlotiorum* after Verlot brothers) and –iarum if both are female (*brauniarum* for Braun sisters). The names of the persons already in Latin, *e.g.* Linnaeus, the Latin ending, *e.g.* –us in this case, is dropped before adding appropriate genetive ending. The genetive typed specific epithets are not related to the gender of the genus. In latter, *i.e.*, the specific epithets of adjectival nature, the epithet takes its ending from the gender of the genus after adding –ian if name of a person ends in consonant, adding –an if the name ends in a bowel except when it ends in –a, wherein –n is added.

(ii) *Based on Place.* The specific epithets may also be formed by using the place name as an adjective. In this case, genus determines the ending after the edition of -ian or -ie and then the relevant gender ending is determined by the genus. A specific epithet may also be formed by adding -ensis (for masculine and faminine gender, such as *Hedera nepalensis, Rubus canadensis*) or -ens (for neuter genera such as *Lingustrum nepalense*) to the place name. Some other specific epithets based on place, are *Mangifera indica* (India), *Rosa indica* (India), *Iris kashmiriana* (Kashmir), *Delphinium kashmirianum* (Kashmir), *Argemome mexicana* (Mexico) and so on.

(iii) *Based on an Important Character.* Specific epithets based on some distinguishing character are always in an adjectival form and derive their gender from the genus, *e.g.* based on white plant part-*alba* (*Rosa alba*), *album* (*Chenopodium album*) or *albus* (*Melilotus albus*), for cultivated plants-*sativa* (*Oryza sativa*), *sativum* (*Allium sativum*) or *sativus* (*Lythrus sativus*) depending on the gender of the genus to which specific epithet is assigned or other characters, *e.g. bicolor* (two-coloured), *repens* (creeping), *aquatica* (in water), *cordifolia* (cordate-leaved), *grandiflora* (large-flowered), *hirta* (hairy) and so on.

(iv) *Noun in Opposition.* Such epithet may carry its own gender, usually in the nominative case. In binomial, *Pyrus malus* is based on the Greek name malus for common apple. Similarly in *Allium cepa*, cepa is the Latin name for onion.

Some specific epithets, that are descriptive adjectives and are based on various criteria, are : relative abundance of plant, *e.g. rara* (rare), *vulgaris* (common); utility of the plant, *e.g. edulis* or *esculenta* (edible); season and duration, *e.g. hiberna* (of winter), *autmnalis* (of autmn), *binnes* (*binneal*) and so on. There are some specific epithets that carry mythological, ethenic or superstitious denotions, *e.g. Ficus religiosa, Rauvolfia serpentina* etc.

Both, generic names and specific epithets, when printed, they are in *italics* or boldface and when written or typed, are underlined. A generic name, after it has been spelled out in a specific name, if used for the other species in continuation, it may be abbreviated using initial capital, *e.g. Solanum nigrum, S. melongena, S. tuberosum* etc.

(4) Names of Infraspecific Taxa. The name of a taxon below the rank of a species (infraspecific taxon) when containing the nomenclatural type of next higher taxon, can not be indicated by epithets such as *typicus, originalis* etc. In such cases specific epithet should be repeated unaltered in infraspecific taxon. For example the combination of *Lobelia spicata* var. *originalis* Mc Vaugh which includes the type of *Lobelia spicata* should be named as *Lobelia spicata* Lam. var. *spicata*.

2. The Type Method

According to current Code, the names of the different taxonomic groups are based on **type method** by which certain representative of the group is the sources of name for the group. This representative is called the **nomenclatural type** or simply the **type** and methodology as **typification**. Chapter II-Section 2 (Articles 7-10) of the Code have greatly emphasised on typification of taxa so as to bring about stabilisation. The application of naming of taxonomic groups is determined by means of nomenclatural types, where a type is that element to which name of taxon is permanently attached either as a correct name or as a synonym but not necessarily the most typical or the representative element of a taxon. Thus, the tea family name (Theaceae) is derived from synonym *Thea*, although, the correct name for the genus is *Camellia*. Similarly *Mimosa* is the type for family *Mimosaceae*, but unlike most representatives of the family which have pentamerous flowers, the genus *Mimosa* has tetramerous flowers.

The single element of a taxon to which its name is permanently attached and on which the descriptive matter, that satisfies the conditions of availability or valid publication, is based, is called the type. The type method (type concept) is a method of nomenclature that attaches a name permanently to a type specimen, *i.e.*, a name which is retained by the type throughout all subsequent taxonomic acts. This means when a species is described new, the author must indicate which is the type of specimen on which new species is based.

The author must designate a certain specimen or an acceptable substitute for a specimen on the type of that specimen. In case of names of species or infra-specific names, the type is an individual specimen, which is the type specimen on which the new species is based. Article 7 of the Code states that the application of names of the taxa of the rank of family or below is determined by means of nomenclatural type. Article 8 of the Code deals with various nomenclatural type, Article 9 of the Code states that the nomenclatural type of a species or taxon below the rank of a species is a single specimen. Article 10 of the Code states that the nomenclatural type of any taxon between genus and species is species, that of a family or of any taxon between family and genus is the genus. Some of the important nomenclatural types are discussed below :

(i) *Holotype* or ***The Type***. This is the one specimen or other element used by the author or designated by him as the nomenclatural type (Article 7.3). The holotype is chosen by the original author from a single gathering made by a plant collector at one time and expressed definitly at the time of original publication. It automatically fixes the application of the same.

(ii) *Isotype*. An isotype is any duplicate (part of a single gathering by a plant collector at a time) of the type or holotype, such as if several samples of a tree are collected at the same time, one specimen may be chosen as the type or holotype, the rest are isotypes. For example, if a new species is based on a single gathering that consists of five specimens of the same plant with the same field number placed on five separate herbarium sheets, the original author will designate one of them as holotype and the remaining four as isotypes. According to Article 10, if all these (five) specimens are mounted on the same herbarium sheet, all of them may be the part of the holotype.

(iii) *Paratype*. A paratype is a specimen from a collection (cited with the original description) other than the type, holotype or isotype, but on which the description of the new taxon has been based. For example, if a new taxon is collected without flowers and fruits in one season and the same taxon is collected with flowers and fruits in a different season, then the two collections will be assigned different field numbers, and out of these two collections, the author will select one collection as holotype and isotype while the collection of the second lot of season will be called paratype.

(iv) *Syntype.* A syntype, according to Article 7.7, is one of the two or more specimens cited by the author when no holotype was designated or it is any one of the two or more specimens originally designated as types. Duplicate of a syntype is designated as **isosyntype**.

(v) *Lectotype.* A lectotype, according to Article 7.5, is a specimen or other element selected from the original material to serve as a nomenclatural type when no holotype was designated at the time of publication or as long as it is missing. In case, when a holotype is lost or destroyed, the lectotype is selected from the isotypes or when two or more specimens have been designated by author as types of a species, the lectotype must be chosen from among these types.

(vi) *Neotype.* A neotype, according to Article 7.8, is a specimen or illustration to serve as a nomenclatural type as long as all of the material on which the name of the taxon was based, is missing.

(vii) *Epitype.* An epitype is a specimen or illustration selected to serve as an interpretive type when the holotype, lectotype or previously designated neotype or all the original material associated with a validly published name, is demonstrably ambiguous and can not be critically identified for the purposes of the precise application of the name of a taxon. Upon designation of an epitype, the holotype, lectotype, or neotype that the epitype supports, must be explicity cited. In most cases in which no holotype was designated, there will also be no paratypes because all cited specimens will be syntypes. But, when an author has designated two or more specimens as types, any remaining cited specimens are paratyes and not syntypes.

(viii) *Tototype.* A tototype refers to the name given to the specimen collected from the same locality from where the holotype was originally collected and is the contingent from the same population.

In cases where the type of a name is a culture permanently preserved in a metabolically inactive state, any living isolate from that case should be referred to as **ex-type (ex typo)**, **ex-holotype (ex holotypo)** or **ex-isotype (ex isotypo)** etc., so as to make it clear that they are derived from the type but are not themselves the nomenclatural type.

When, within a species, an infraspecific variant is recognised for the first time, it automatically establishes two infraspecific taxa : the one which includes the type specimen of the species must have the same epithet as that of the species, *e.g. Acacia nilotica* ssp *nilotica* and such a name is called **autonym** and the specimen an **autotype**. The variant taxon with its own holotype is differentiated by an epithet different from the specific epithet, *e.g. Acacia nilotica* ssp *indica*.

The application of type method, *i.e.*, **typification** is a methodology different from **typology**, the concept which was very much in vogue before Darwin put forward his idea about variations. Typology is a concept based on the idea that does not recognise variation within the taxa and believes that an idealised specimen or pattern can represent a natural taxon.

3. Typification

According to Code it is strictly recommended that a holotype should be designated and deposited in a permanent responsible herbarium or institution and that it is scrupulously conserved. Publication on or after January 1, 1958 of the name of a new taxon of the rank of the genus or below is valid only when the type of the name is indicated or same published on or after January 1, 1990, an indication of the type must include one of the words **typus** or **holotypus** or its abbreviation or even its equivalent in a modern language. In case of name of a new species or infraspecific taxon on or after January 1, 1990, whose type is a specimen or unpublished illustration, the herbarium or the institution in which the type is conserved must be specified. On or after January 1, 2000, lectotypification or neotypification of a name of a species or infraspecific taxon is not affected unless indicated by use of the term **lectotypus** or **neotypus**, its abbreviation or its equivalent in a modern language. The specimen selected as a type should belong to a single gathering. That is why the specimen of *Echinocereus sapendronsis* (Raudonat & Rischer, 1995), which according to label was taken from the same cultivated individual at different times and preserved in alcohol in a single jar, belonging to more than one gathering, as compared to that of a holotype consisting of a complete plant with roots, a detached branch, an entire flower, a flower cut in halves and two fruits, was not accepted as a type and Raudonat & Rischer's name is thus not validly published. The principle of typification does not apply to the names of taxa above the rank of family.

4. Author Citation

For a name to be complete, accurate and easily verifiable, it should be accompained by the name of the author or authors who first published the name validly. This facilitates in verifying the dates of publication and imparting precision in botanical nomenclature. Commonly, names of the authors are abbreviated, *e.g.* Linn. or L. for Carolus Linnaeus, Benth. for G Betham, Hook. for William Hooker, Hook. f. for Sir J.D. Hooker (f stands for filius, the son : Sir JD Hooker was son of William Hooker), R.Br. for Robert Brown, Lamk. for JP Lamark, DC. for AP de Candolle, Wall. for Wallich, A.DC. for Alphonse de Candolle, Scop. for GA Scopoli, Pers. for HC Persoon and so on. The names of the authors are never underlined or typed in italics, *e.g. Vitex* Linn., *V. trifolia* Linn., *V. trifolia* var. *simplicifolia* Cham. The name of a single author proposing a new name follows the name of a species (or any other taxon). These citations can indicate bibliographic references that greatly help in recognition of homonyms. For example, *Utricularia caerulea* Linn. and *Utricularia caerulea* Clarice are two names referring to two different taxa and for them it would have been impossible to recognise this if the citation of author's names appended to the respective plant names were not given.

In cases, where names of two or more authors are associated with a name for variety of reasons, the different situations are exhibited by citing the names of the authors differently.

1. When two or more authors publish a new species or propose a new name jointly, their names are linked by *et* or an ampers and (&) *e.g. Delphinium viscosum* Hook. f. *et* Thompson.
2. When more than three authors are involved in proposing a new name jointly, citation is normally restricted to the first author and is followed by *et al.*
3. The names of two authors are linked by *ex.* when the first author had proposed a name but the name was validly published by the second author, the first author who suggested the name but could not publish it validly or failed to satisfy all or some of the norms of the Code, then the name of the first author is connected to the name of the second author, *e.g. Acalypha racemosa* Wall. *ex.* Baill.
4. Whenever the name of a taxon is changed by the transfer from one genus to another or by upgradation or downgradation of the level of the taxon, the original epithet should be retained. The name of the taxon providing epithet is termed as **basionym** and the name of the original author or authors, whose epithet is being used in the changed name, is placed within parantheses and the name of author or authors who got the name changed outside the parantheses, *i.e.*, the doubled citation, *e.g. Cyanodon dactylon* (Linn.) Pers., based on basisnym *Panicum dactylon* Linn., the original name for the species.
5. The names of the authors are linked using *in* when the first author published a new species or a name in publication of another author, *e.g. Carex kashmirensis* Clarke *in* Hook. f. In this case Clarke published this new species in the *Flora of British India* whose author was Sir JD Hooker.
6. The names of two authors are linked using *emend* (**emandavit**–person making correction) when the second author makes some change in the diagnosis or in circumscription of a taxon without altering the type, *e.g. Phyllanthus* Linn. *emend* Mull.
7. To indicate pre-starting point of author, square brackets are used, *e.g.* the generic name of *Lupinus* was defectively published by Tournefort in 1719, but since it happens to be earlier than 1753, the starting date of botanical nomenclature based on *Species plantarum* of Linnaeus, the appropriate citation for the genus is *Lupinus* [Tourne.] Linn.
8. If a taxon is of garden origin, then while citing the name, it should be ascribed to *hort.* (*hortulanorum*) and connected to the name of the author who published it by an *ex* such as *Geanaria dwaklarii hort. ex* Hook.

While naming an infraspecific taxon, the authority is cited for both specific epithet and infraspecific epithet, *e.g. Acacia nilotica* (Linn.) Del. ssp *indica* (Benth.) Brenan. But in case of an autonym, the infraspecific epithet does not bears the author's name as it is based on the same type as the species, *e.g. Acacia nilotica* (Linn.) Del. ssp *nilotica.*

5. Latin Diagnosis

Names of all new species (or other taxa new to science) published on January 1, 1935 onwards should have a Latin diagnosis, *i.e.*, Latin translation of diagnostic features. Full description of the species (in any language) can accompany the Latin diagnosis. A description (in any language) not accompanied by a Latin diagnosis is allowed for publications before January 1, 1935. For publications before January 1, 1908, an illustration with analysis without any accompanying description is valid. Thus description in any language is essential from January 1, 1908 onwards and this accompanied by a Latin diagnosis from January 1, 1935. In case of name changes and new names of already known species, a full reference to the original publication should be made.

6. Publication of Names

The name of the first published taxon should be such that it should fulfill certain requirements as to become legitimate name for consideration when deciding on a correct name. A validly published name is that which has been properly formulated and its nature is indicated by a proper abbreviation after the name of the author or authors and its formulation be such as sp. nov. for *species nova*-a species new to science, *Tragopogon kashmirianus* G Singh sp. nov. (published in 1976) for *combinatio nova*-a name change involving the epithet of the basionym with name of the original author within the paranthesis *Vallisnaria natans* (Lour.) Hara comb. nov. (published in 1974), comb. etc. stat. nov. for *combinatio et status nova*-when a new combination also involves the change of status, and in such case the combination of epithet of basionym will accordingly be used in the combination intended, nom. nov. for *nomen novum*-when the original name is replaced and its epithet can not be used in the new name *Caragona opulens* Kom. va. *licentiana* (Hand-Mazz) Yakovl. comb. et stat. nov. (published in 1988) etc. In future references, these first published abbreviations, are replaced by the name of the publication, page number and the year of the publication for full citation or atleast the year of publication.

7. Effective (Valid) Publication

In general, the publication becomes effective by distribution in printed form, through marketing, exchange or gift to concerned people or atleast the places of botanical importance such as libraries of botanical gardens, herbaria, research institutes, universities, colleges etc. The new name is not considered valid if communicated at a public meeting, by placing of names in collections or gardens open to the public, by the issue of microfilm made from manuscripts, typescripts or other unpublished material, by publication on-line or by dissemination of distruable electronic media such as cassets, CD's etc. Publications has to fulfill certain conditions for validity such as those that have been published on or after 1.1.1935 must be accompained by a Latin description or diagnosis or by a reference to previously and effectively published Latin description or diagnosis of the taxon. The publication in newspapers (1.1.1953 onwards) and seed exchange lists (1.1.1977 onwards) is not valid. Publication of handwritten material, reproduced by some mechanical or graphic process such as lithography, offset or metallic etching before 1.1.1953 is valid. A name of a new taxon of fossil plants published on or before 1.1.1996, accompanied with Latin or English description or diagnosis or by a valid previous description or diagnosis is treated valid. According to St Louis Code (2000), new names of plants and fungi would have to be registered after 1.1.2000.

8. Rejection of Names

The process of selection of correct name for a taxon involves the identification of **illegitimate names** not satisfying the rules of botanical nomenclature. A legitimate name must not be rejected because it or its epithet is inappropriate or disagreeable or because other is preferable or better known or because it has lost its original meaning, *e.g.* the name *Scilla peruviana* L. (1753) is not to be rejected merely because the species does not grow in Peru. The following types of names fall under the category of illegitimate names :

(i) *Tautonym.* This is the name where specific epithet exactly repeats the generic name with or without transcribed symbol, *e.g. Malus malus.*

(ii) *Autonyms.* It is an automatically created tautonym for infrageneric or infraspecific taxa.

(iii) *Synonym.* It is a different name used for the same taxonomic group or taxon.

(iv) *Typnonym.* This is the name which has older valid name based on the same types.

(v) *Metanym.* This is the name which has valid name based on another member of the same group.

(vi) *Homonym.* This is a preoccupied name, *i.e.*, identical names to two different taxa.

(vii) *Hypnonym.* This is a name belonging to natural group that has not been determined.

(viii) *Basionym.* It is the original name of a taxon subsequently replaced by another as a result of change in rank citation of a taxon.

(ix) *Isonym.* When same name, based on the same type, has been published independently at different times by different authors, then only earliest of these, the isonym has nomenclatural status.

Any one or more of the following situation leads to the rejection of a name.

1. Name not properly formulated, effectively published, lacking typification or without a Latin diagnosis.
2. *Nomen nudum* (Abbreviated *nom. nud*). A name that does not fulfil the criteria set by the Code as legally prescribed scientific name (with no proper and complete accompanying description).
3. **Tautonyms**, *i.e.*, binomials with identical generic and specific names such as *Malus malus*. In *Cajanus cajan* the words are not exactly identical and hence it is not a tautonym and is thus legitimate. However, repitition of a specific epithet in an infraspecific epithet, *e.g. Acacia nilotica* ssp *nilotica* does not make it a tautonym.
4. **Later Homonyms**, *i.e.*, use of the same name for two different species given at a later date even if the earlier homonym–one published at an earlier date, *e.g. Zizyphus jujuba* had long been used as the correct name for the cultivated fruit jujub Lamk., 1789, but this name was ascertained to be a later homonym of a related species *Z. jujuba* Mill. 1768 and thus binomial *Z. jujuba* Lamk., 1789 is thus rejected and *jujub* correctly named as *Z. mauritiana* Lamk., 1789.
5. *Later Isonym*, Among isonyms–same name based on the same type published independently at different times by different authors, the earliest, cited from its original place of valid publication has nomenclatural status whereas later isonyms are disregarded, *e.g.* Baker (1892) and Christensen (1905) independently published the name *Alsophila kalbreyeri* as a substitute of *A. podophylla* Baker (1891), *A. kalbreyeri*, being a later isonym, is discarded.
6. *Name superfluum* (Abbreviated as *nom. superfl.*). A name published appearing nomenclaturally superfluous, *i.e.*, if the taxon to which it was applied–as circumscribed by its author–included the type of a name or epithet which ought to have been adopted under the rules, *e.g. Physkium natans* Lour. (1790), thus when transferred to genus *Vallisnaria*, the epithet *natans* should have been retained but de Jussie used the name *Vallisnaria physkium* Juss. (1826), a name becoming superfluous.
7. *Nomen ambiguum* (Abbreviated as *nom. ambig.*) If a name used in different sense by different authors becomes a source of permanent error, is rejected, *e.g.* name *Rosa villosa* Linn. is rejected because it has been applied to several different species.
8. *Nomen confusum* (Abbreviated as *nom. confus.*). A name based on a type consisting of two or more entirely discarded elements causing differently in selecting a satisfactory holotype, is discarded, *e.g.* the characters of genus *Actinotinus* were derived from two genera *Viburnum* and *Aesculus*, due to the insertion of the inflorescence of *Viburnum* in the terminal bud of an *Aesculus* by a collector, and thus the name *Actinotinus* was abandoned.

9. *Nomen dubium* (abbreviated as *nom. dub.*). A dubious name, *i.e.*, a name of uncertain application because of making it impossible to establish the taxon to which it should be referred, *e.g.* Linnaeus (1753) attributed the name *Rhinanthus crista-galli* to a group of several varieties which he later described under separate names, rejecting the name *Rhinanthus crista-galli.* Although, the use of the name continued for a considerable period but finally Sehwarj (1939) listed it under *nomen. dubium.*
10. **Names Based on Monstrosity.** A name based on monstrosity, *e.g.* generic name *Uropedium* Lindl. (1846), based on a monstrosity of the species, now referred to as *Phragmidium caudatum* (Lindl.) Royle, 1896, is rejected.

9. Principle of Priority

Principle of priority (Chapter II, Section 3, Articles 11 and 12 of the Code) is concerned with the selection of a single correct name for a taxonomic group. It is the most important principle that has served to provide a stable method of naming taxonomic groups. The method involves identification of illegitimate and legitimate names and selecting a correct name from among legitimate ones. In case more than one legitimate name exists for a taxon, the earliest legitimate name in the same rank forms the correct name except for nine families as named earlier, for which alternate names have been given. For the taxa at the level of species or below, the correct name is either the earliest legitimate name or a combination based on the earliest legitimate basionym, unless the combination becomes a tautonym or later homonym, rendering it legitimate. Although, the Code recommends that while selecting from among typified names above the rank of family, the principle of priority should be followed. It is not mandatory and will not affect the legitimacy of names selected otherwise.

Strict adherence to the principle of priority was recommended in the Rochester Code (1892) in view of huge collection of plant specimens from practically all over the world that required identification and nomenclature. This was done by many botanists but quite independently, as a result, the plant species occurring widely in various countries, received different names at the hands of these botanists. The same species was named differently in different parts of the world by different botanists. *e.g.* Indian Coral Tree (*Erythrina variegata* L.) has as many as 200 scientific names (Kurkoff, 1972). Many economically important grasses, ornamental plants and plants of medicinal value have numerous names, most of them are confusing. An example of nomenclatural history of a genus is given below :

The common apple was first described by Linnaeus (1753) under the name *Pyrus malus.* The species was subsequently transferred to the genus *Malus*, but the combination *Malus malus* (Linn.) Brit. (1888) could not be taken as a correct name because it becomes a tautonym. The two other binomials under *Malus* available for apple include *M. pumila* Mill. (1768) and

M. domestica Borkh. (1803), and thus *M. pumila* Mill. (1768), being earlier of the two is selected as the correct name and citation written as :

Malus pumila Mill. (1768)
Pyrus malus Linn. (1753)
M. domestica Borkh. (1803)
M. malus (Linn.) Britt. (1888)–Tautonym

Application of principle of priority finds many problems. The principle starts with the *Species plantarum* of Linnaeus published on 1.5.1753, the starting date for **limitation** of principle of priority for seed plants (spermatophyta), pteridophytes, sphagnaceae, hepaticae, most algae, slime moulds and lichens, for mosses (excluding sphagnaceae) 1.1.1801, for fungi 31.12.1801, fossils 31.12.1820, algae (Nostocaceae) 1.1.1886 and algae (Oedogoniaceae) 1.1.1900. The publications before these dates, for respective groups, are ignored while deciding the priority.

Similarly, by adhering strictly to the principle of priority, several changes were brought about to provide disadvantageous changes in the nomenclature of genera, families and intermediate taxa, for which Code provides lists of names that are conserved (*nomina conservanda*), earlier legitimate names that are rejected (*nomina rejicienda*) and provision of conservation of names of species (*Triticum aestivum* Linn. is the first species name conserved by Berlin Code–1988).

10. Retention of Names of Taxa which are Divided

When a genus is divided into two or more genera, the original generic name must be retained for the genus including the type species.

Similarly when a species is divided into two or more species, the original specific name must be retained for the species containing the type. The same rule also applies to infraspecific taxa.

11. Changes in Name of Taxa

Changes in the name of taxa may be required *(i)* by transference of the taxon or *(ii)* by its union with another taxon of the same rank or *(iii)* by change of its rank. In the first case, when a subdivision of a genus is transferred to another genus without change of rank, its original name (if legitimate) must be retained. When a species is transferred to another genus without change of rank its original name (if legitimate) is retained. The same rule applies to infraspecific taxa. In the second case, when two or more taxa of the same rank are united into one, the oldest legitimate name is retained for the combined taxon. If the names of the taxa are of the same date, the author who first unites them has the right to choose any one of the names and his choice must be followed. In the third case when the rank of a genus or a taxon below the rank of genus is changed, the correct name is the earliest legitimate one available in the new rank.

(III) Provisions for the Governance of the Code

Provisions for the governance of the Code are looked after by Nomenclatural Committees. Permanent nomenclatural committees are established under the auspices of the International Association for Plant Taxonomy. Members of these committees are elected by an International Botanical Congress. The committees have power to co-opt and establish sub-committees. Office bearers of the committees, as may be desired, are elected. This entire machinery undertakes proper implementation of Provisions of the Governance of the Code.

NAMES OF HYBRIDS

Hybridity is indicated by use of the multiplication sign (×) or by the addition of the prefix *notho*–to the term denoting the rank of the taxon, the principal ranks being **nothogenus** and **nothospecies**. A hybrid between the named taxa may be indicated by multiplication sign (×) between the names of taxa, the whole expression is then called a **hybrid formula**, *e.g.* a hybrid between two species of *Salix* : *S. aurita* and *S. caprea* is named as *S. aurita* × *S. caprea*. When parents are not definitely known, hybrids may be given altogether a new name, *e.g. Musa* × *paradisiaca* L. indicates that this species has a hybrid origin. Bigeneric hybrids are designated by a **condensed formula** such as × *Asplenophyllites* because it is a hybrid between species of two genera *Asplenium* and *Phyllites*. Variants of interspecific hybrids are named **nothosubspecies** and **nothovarieties**, *e.g. Salix rubens* nothovar. *basfordiana*. The grafts between two species are indicated by a plus sign (+) between two grafted species, *e.g. Rosa webbiana* + *R. floribunda*.

NAMES OF CULTIVATED PLANTS

Names of cultivated plants are governed by International Code of Nomenclature for Cultivated Plants (ICNCP). In ICNCP, most of the rules are taken from ICBN with additional recognition of the rank **cultivar** (abbreviated cv. for cultivated varieties). The names of cultivar is not Latin and is not written in italic, but rather a common name starting with capital letter either proceeded by cv., *e.g. Rosa floribunda* cv. or Blessings or within single quotation marks, *e.g. Rosa floribunda* Blessings. Cultivars may also be written directly under a common name, *e.g.* Hybrid Tea Rose 'Red Lion' or directly under a genus, *e.g. Hosta* 'Decorata' or under a hybrid, *e.g. Rosa* × *paulii* 'Rosea'. The correct nothogeneric name for plants such as derived from the *Triticum* × *Secale* crosses is × *Triticosecale* Wittmack ex. A Camus. At species level, no correct name is available for the commonest crop Triticales and it is recommended that the crop Triticales be named by appending the cultivar name to the nothogeneric name such as × *Triticosecale* Newton. Since 1.1.1959, new cultivar name should have a description published in any language and these names must not be the same as the botanical or common name of a genus or a species. The cultivar names such as Rose, Onion etc. are not permitted as cultivar names. Recommendations are for registering the cultivar names with proper **registering authority** for various plant groups, *e.g.* rose, orchids or genera to prevent duplication or misuse.

DRAFT BIOCODE

Draft BioCode, an attempt at a unified Code for all living organisms, infact, a combination of all biological Codes, has been appreciated for some time and an explanatory meeting on the subject was held at Egham, UK in March 1994. In view of the crucial importance of scientific names of organisms in global communication, the decisions taken included not only agreement to take steps towards harmonisation of existing terminology and procedure, but also the desirability of working towards a unified system of biological nomenclature, the Draft BioCode, as the first public expression of these objectives. The first draft, which was prepared in 1995, was reviewed successively and the fourth draft, named Draft BioCode (1997) prepared by the International Committee for Bionomenclature (ICB) and published by Greuter, *et al.* (1998) which is now available on the web : http://www.rom.on.ca/biodiversity/biocode/biocode 97 html from the Royle Ontario Museum.

In Draft BioCode, largely based on the pattern of ICBN, no examples are listed, notes omitted at the present stage, a considerable number of articles and paragraphs have been dropped, has only 41 Articles (in St. Louis Code of ICBN there are 62 Articles) etc.

PHYLOCODE

The concept of PhyloCode was first introduced by de Queiroz and Gautheir (1992). The development of PhyloCode, phylogenetic nomenclature on the philosophy of phylogenetic taxonomy replacing the multirank Linnaean system with rankless system recognising only species and '**Clades**', resulted from 1990 onwards and three symposia, the first in 1995, second in 1996 and the third in St Louis, Missouri, USA (1999) entitled "*Overview and Practical Implications of Phylogenetic Nomenclature*". The starting date of PhyloCode has not yet been determined but in the Draft Code it is cited as 1.1.2000.

PhyloCode grew out of recognition that the current Linnaean system of nomenclature is not well suited to govern the naming of clades and species that compose tree of life and are the most significant entities above the organism level rank. Their assignment is subjective and biologically meaningless whereas PhyloCode will provide the rules to express the purpose of naming the parts of the tree of life (both clades and species) by explicit reference to phylogeny. As such PhyloCode extends tree thinking to nomenclature and is designed so that it can be used concurrently with the pre-existing codes or after rules governing species names are added, as the sole code governing the names of taxa, if the scientific community decides that it should be so. The PhyloCode is intended to cover all biological entities, living as well as fossils. It is based on the principle that the primary purpose of a taxon name is to provide a means of referring unambiguously to taxon, not to indicate its relationship.

Rules. PhyloCode provides rules for naming clades. It is expected to provide rules for naming species soon. Till then the naming of the species will be governed by pre-existing codes. In this system, categories 'clades' and 'species', are not the ranks but are considered as different kinds of biological entities. A species is a segment of population lineage, whereas a clade is monophyletic group of species and both of them, which are discovered by systematics, rather than being created, are products of evolution and even if not named have an objective existence. According to the PhyloCode, supraspecific name is given a phylogenetic definition (in pre-existing codes supraspecific names have types) and is applied to the clade which fits that definition, irrespective of its hypothesised composition. Species, specimens and synapomorphies cited within these definitions are called **specifiers** as they specify the clade to which name applies and function somewhat like types in providing reference points that determine the application of a name. Under the PhyloCode, synonyms are the names whose phylogenetic definitions specify the same clade, regardless of prior associations with particular ranks. But under the pre-existing codes, synonyms are names of the same rank based on the types within the group of concern, regardless of prior associations with particular clades. The PhyloCode prescribes registration of established names and thus reduces the frequency of accidental homonyms.

Inspite of hierarchial relation between the taxa, PhyloCode is rankless, as assignment of rank is not part of the naming process and, thus, has no bearing on the spelling or application of taxon name. The PhyloCode allows systematists to restrict the application of names with respect to clade composition and if one wishes to ensure that a name refers to a clade that either excludes or includes a particular subtaxa, the system has the provision of including a qualifying clause in the definition specifying conditions under which the name can not be used.

Advantages. Phylogenetic nomenclature was presumed to have many advantages over the traditional system. Based on pre-existing codes, it is often difficult to name a newly discovered clade, which either requires the use of an unconventional intermediate rank such as superfamily or shifting of less or more inclusive clades to lower or higher ranks, causing instability and making delay untill an entire classification is developed. PhyloCode facilitates in naming of new clades loaded with latest information of phylogeny obtained from molecular biology and technology as they are discovered and not waiting till a full classification is developed. Presently this Code has rules only for clades but when extended to species, it will improve nomenclatural stability as well, by removing the linkage to a genus name. But when species will be named, there will be absence of hierarchial information in species name, making one unable to infer phylogenetic relationship from the species name in the way one can infer genus assignment names governed by the pre-existing codes. This is because PhyloCode is aimed at the name of a taxon to provide a means of referring unambiguously to a taxon and not to indicate its relationship. But under PhyloCode, the phylogenetic position can easily be indicated by associating the species name with the names of one or more clades to which it belongs. The abandonment of ranks in PhyloCode is one more advantage that will eliminate the error caused by many researchers who treat taxa at the same rank as if they are comparable in some biological meaningful way and due to this they commit the error of counting clades or species that possess properties relevant to the question of interest or tracing the evolution of these properties on a phylogenetic tree, leading to faulty reasoning. Thus abandonment of ranks eliminates most subjective aspect of taxonomy as arbitrary nature of ranking is not widely appreciated by non-taxonomists.

The latest version of PhyloCode is available at web.(http://www.ohiou.edu./phylocode.July 2002.) Despite strong opposition by Nixon *et al.*, (2003) that "The PhyloCode is fatally flawed and the Linnaean system can be easily fixed", the PhyloCode is being maintained by the Society for Phylogenetic Nomenclature (SPN) through its own committees for governance.

❑❑❑

Chapter **4**

Hierarchy of Characters in Plant Taxonomy

TAXONOMIC HIERARCHY

The present day vegetation is the outcome of the evolutionary changes in the past, during which a variety of forms evolved and flourished. Many common features still exist in diverse groups and these may be attributed to the continuity of descent in evolutionary sequence from common ancestors, giving way to a problem for their placement in organised groups based on the similarities and differences of characters. It would be a total chaos to study and document information about such a large number of species of plants if there were no proper mechanisms for grouping them in an organised manner. And this caused the need to form certain categories and ranks at different levels based on artificial characters, overall morphology, phylogeny and phenetic characters. This scientifically oriented framework of selection basis of delimitation, based on a set of principles, is called **taxonomic hierarchy**.

The plants are first recognised and assembled into groups in various categories on the basis of certain resemblances. The ratio of similarities and differences is the deciding factor for framing various taxonomic ranks. The smaller groups are assembled into larger and more inclusive group. The process is repeated until finally all organisms are assembled into a single largest and most inclusive group. The taxonomic groups are termed as **taxa** (singular **taxon**). These taxa of various levels (species, genus, family and so on), are arranged in order of their successive inclusiveness (fig. 4.1).

The groups, thus formed and arranged, are next assigned to various **categories**, having a fixed sequence of arrangement, the taxonomic hierarchy. The most inclusive group is assigned the highest category (generally a **division**) and the least inclusive the lowest category (usually a **species**). Species is a taxon with largest proportion of similarities of characters and also breeding truely amongst its individuals. The species together, are placed

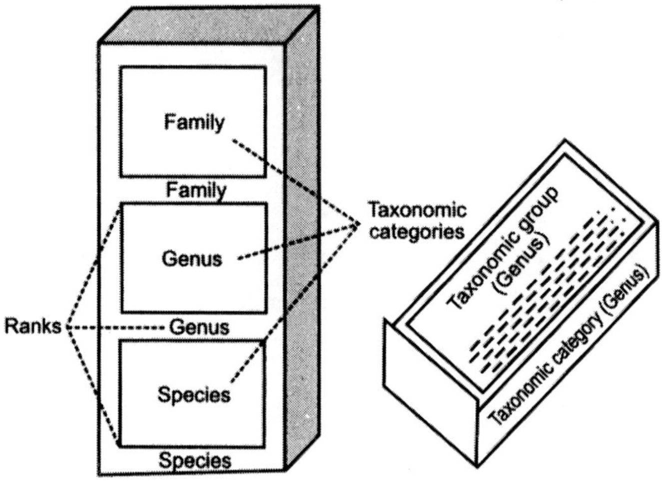

Fig. 4.1 Various taxonomic groups.

in larger and more inclusive category **genus** (plural **genera**), genera together in still larger and somewhat more inclusive category **family** and likewise families together in **order** and so on.

The names are assigned to the taxonomic groups in such a way that the name gives an understanding of the category in which it is assigned, *e.g.* Primulales (order) Primulaceae (family), *Primula* (genus), *P. vulgaris* (species) (fig. 4.2). Once all groups have been assigned, categorised and named, the

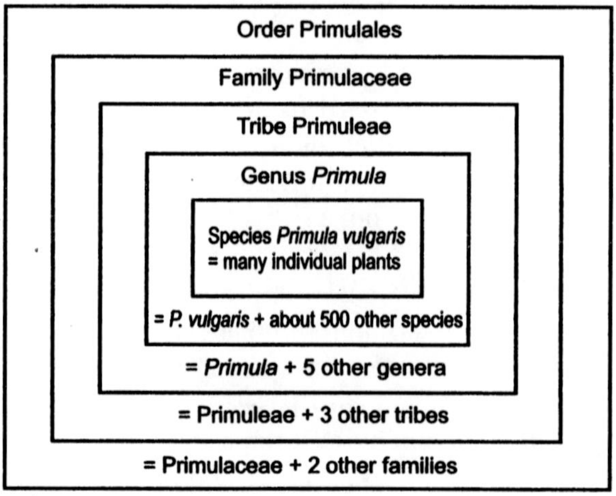

Fig. 4.2 Box-in-box representation of taxonomic categories.

process of classification is complete. In other words the **taxonomic structure** of the whole largest and most inclusive group has been achieved. Since the arrangement of the categories to which the groups are assigned is hierarchial, the system of classification achieved is known are **hierarchial classification**.

In sixteenth century, Bauhin brothers had already pointed out the distinction between species and genus, which was not recognised by earlier workers. Conclusively, Bauhin brothers made a pioneering attempt in recognising distinction between smaller (species) and larger (genus) groups in plant classification. Linnaeus (1737) furnished description to the genera and designated species with **trivial names** (specific epithet), included under each genus. Bernard de Jussieu and AL de Jussieu recognised still larger groups, the orders, which in present day classification, are equivalent to families.

The exploration of more and more areas during subsequent two centuries resulted enormous number of species that caused the construct still larger groups to accomodate them. As such taxonomists recognised several groups at various levels. With recent micro-level studies on the living population of plants, several smaller groups below the rank of species (**subspecies, variety, race** etc.), have come into being. Massion (1950), was first to recognise different hierarchial categories which function as levels or ranks in a hierarchy to which they are assigned. Since categories are defined by taxonomists, they are purely artificial, subjective and unrealistic. They are defined only by their position relative to other categories whereas groups placed in categories represent the discrete sets of plants in nature and are thus objective or non-arbitrary.

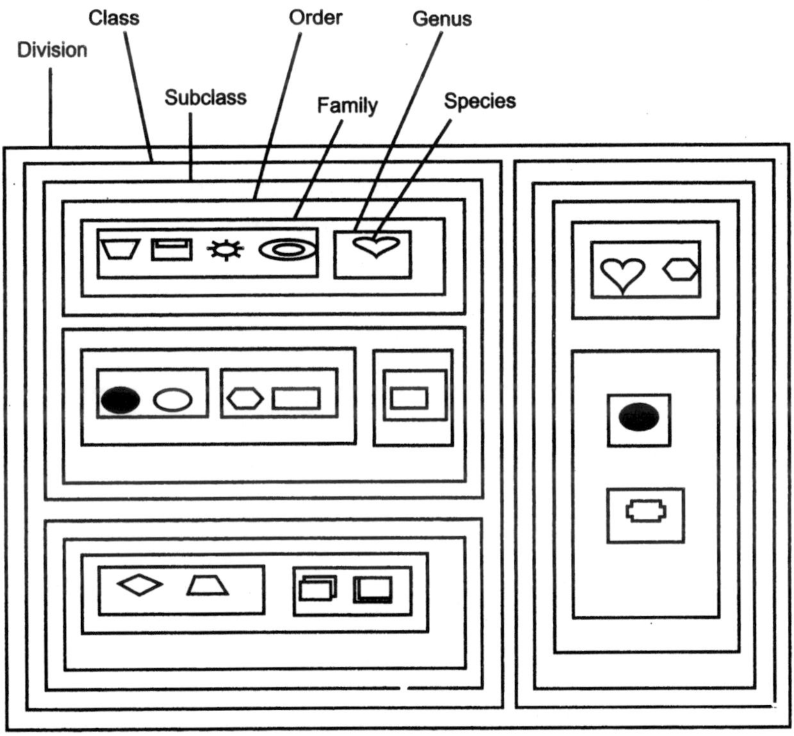

Fig. 4.3 Process of assembling taxonomic groups according to the hierarchical system as depicted by box-in-box method. In the example, there are 18 species grouped into 10 Genera, 6 Families, 4 Orders, 3 Subclasses, 2 Classes and 1 Division.

The level of a category is decided by levels of groups included under it, *e.g.* the groups that are more inclusive (such as order) are considered to be at higher levels than those which are less inclusive (such as species). The range of similarities converges towards larger group, *i.e.*, from species towards order. Lower groups are included in higher groups and must possess all characters on which assignment of the higher group is based, *e.g.* a species, a genus and a family, all must possess some common characters of the order in which they are included. The concept of categories, groups and taxonomic structure has been illustrated in the figures 4.2, 4.3 and 4.4.

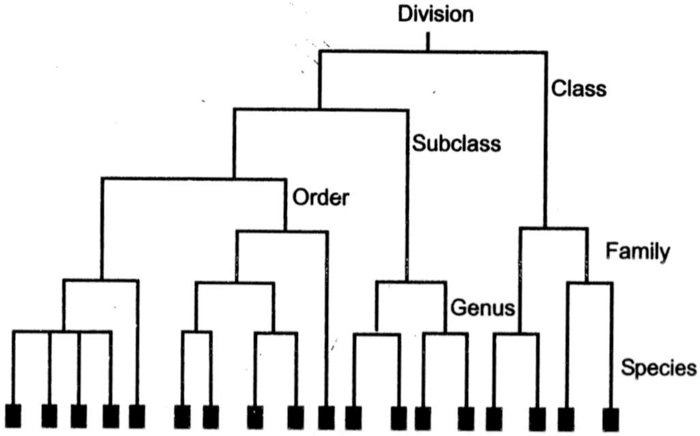

Fig. 4.4 Dendrogram method for depicting hierarchical system based on same hypothetical diagram as in fig. 4.3.

In box-in-box system, the size of boxes varies according to nature of the groups. The boxes constitute the categories whereas the plants included in them are termed as the units. Taking an example of plants of similar kind included in a species *Primula vulgaris*, a unit (rank of the species) in hierarchy, the category is a species (fig. 4.2). The categories are like selves of an almirah at different heights (fig. 4.1), having no significance when empty and importance and meaning after something has been placed in them.

TAXONOMIC GROUPS, CATEGORIES AND RANKS

Names for the groups, assigned to these categories, have been given following International Code of Botanical Nomenclature (ICBN). Taxonomic groups, categories and ranks are inseparable, once a hierarchial classification has been achieved, *e.g. Primula vulgaris* is thus nothing else but a species and *Primula* is nothing other than a genus. But the differences do exist in concept and application. Conclusively, the **taxonomic group** is any assemblage of individuals or small discrete groups sharing some common characters. **Taxonomic category** is a level or **rank** assigned to that group. **Taxonomic unit** may be defined as any group of individuals sharing maximum number of characters among them and also the greatest continuity of characters within them, but at the same time it is discontinuous from any other group in respect of those characters. Thus, species is taken to be the basic unit of classification while **biotype** is considered to be the basic unit in biosystematic studies.

Taxonomic Categories

Each taxonomic category represents a group of plants but none of these is defined in a strict manner. Their delimitation varies. Each is subjective in character. The categories of higher level such as **Division, Class, Order** etc. fall under **major categories** while those at lower level such as **Species, Variety** etc. come under **minor categories**. The major category may bear any name which is not a part of the name of the plant included under it whreas minor category is considered to be one whose name is also a part of the name of a particular plant.

1. Major Taxonomic Categories. The term **Division** has been prescribed, by rules, to represent the highest category within the plant kingdom. It is more or less equivalent to the term **Phylum** prescribed for animals. The number of divisions within the plant kingdom has remained variable. Eichler (1883) included all plants known to him into five divisions. Engler and Prantl (1887–1915), in their system, recognised as many as 14 major divisions. In general all seed plants are grouped under one divisions, the **Spermatophyta**, that is characterised by the dominant sporophytic generation and presence of ovule that develops into seed. The ending for the name of a division has been recommended as '**phyta**', *e.g.* Embryophyta, Pteridophyta, Bryophyta, Thallophyta etc. A division may be subdivided into two or more **subdivisions**, *e.g.* division Spermatophyta is often subdivided into subdivisions Gymnospermae and Angiospermae. The recommended ending for the name of subdivision is '**phytina**', *e.g.* Pterophytina.

The term **Class** is recommended for the category next to division in descending order. The names applied to class, as usual practice, are Latin and ordinarily have ending **–opsida**, *e.g.* Magnoliopsida or **–ae**, *e.g.* Magnoliatae, Angiospermae etc. A class may be further subdivided into subclasses, *e.g.* subdivision Angiospermae is composed of two classes-Monocotyledons and Dicotyledons and class Dicotyledons is subdivided into Polypetalae, Gamopetalae and Monochlamydeae. Despite there are no strict rules for selecting names of subclasses but recommended ending is '**opsideae**'. Addition of categories such as **superclass, infraclass** etc. has been practiced by some authors (Takhtajan, 1969; Thorne, 1976 etc.).

The term **Order** is in line next to class. Certain earlier authors, *viz.*, Bentham and Hooker (1862–83), used this term in lieu of the term family which is taken to be illigitimate by the rules of nomenclature. The names of the orders have the ending–'**ales**', *e.g.* Ranales and are generally based on one of the constituent families, *e.g.* Malvales (constituent family Malvaceae). But certain order names, *e.g.* Centrospermae, Caryophyllineae, Tubiflorae etc., adopted before the existence of currently accepted rules, being old and in long use, have been allowed. An order is considered to be more homogeneous, *i.e.*, having a degree of phyletic unity than the categories at higher ranks. An order may also be subdivided in **suborders** with name ending–'**ineae**'.

The term **Family**, smallest of major categories, is the most commonly encountered in the taxonomic studies. Family names are generally based on any of the constituent genera with suffix–'**aceae**', *e.g.* Ranunculaceae (constituent genus *Ranunculus*) and Malvaceae (constituent genus *Malva*). Certain family names, with irregular terminations, such as Cruciferae, Leguminosae, Compositae, Umbelliferae, Palmae, Labiatae etc., have been allowed by rules, as exception because these names were in use long before the present rules of nomenclature were framed and accepted. But those who wish to follow the rules, even for these families which do not have suffix–**ceae**, have been provided the alternate names for them. The alternate name for family Cruciferae is Brassicaceae, for Umbelliferae → Apiaceae, Guttiferae → Clusiaceae, Palmae → Arecaceae, Gramineae → Poaceae, Compositae → Asteraceae, Labiatae → Lamiaceae and Leguminosae → Fabaceae (if three subfamilies of Leguminosae are raised to the family level then the alternate name given is used for Papilionaceae and others are named as Caesalpiniaceae and Mimosaceae). If the family is larger enough, it is subdivided into **subfamilies**,.whose names are based on constituent genera and end with–'**oideae**'. When family Leguminosae is considered as comprising three subfamilies, their names are Faboideae, Caesalpinioideae and Mimosoideae. Large subfamilies are divided into **tribes**, which have names based on any one constituent genus with suffix–'**eae**', *e.g.* Caesalpineae, Andropogoneae etc.

2. **Minor Taxonomic Categories.** The name of any minor category which may be a genus, a species or any other category or subordinate rank, forms a part of the name of the plant, *e.g. Citrus* (genus) *medica* (species) *acida* (variety). Among minor categories, **Genus** is the category of the highest magnitude but is subordinate to the lowest rank of major category rank, family. A family may include one (monotypic) or more genera. The various genera within the family are closely related and are descendents of a common stock. They are precisely delimited by a set of well defined characters. The generic name of a plant is the first of the two words comprising a binomial, *e.g. Hibiscus* (genus) *rosa-sinensis* (species) and its first letter is always capital. As a group of plants of common affinity, genera were recognised as early in eighteenth century, much before the present concept of genus was established. JP de Tournefort (1716) recognised the genus as the fundamental category of classification. Plants having two or three common characters used to be regarded as belonging to the same genus. Carolus Linnaeus (1737) was the first to furnish generic description and form limits of the groups. The genera described by him and by Hooker and other taxonomists were delimited on the basis of their morphological characters alone and it was held that their components, *i.e.*, species, have more characters in common with each other than those of the components of the other genera of the same

family. This concept is largely followed by several taxonomists even today. Therefore, very often, it is experienced that the characters that are taxonomically important in separating components of one genus may not be equally so important for another genus of the same family. Despite having this shortcoming, the concept has served the practical need of identification and nomenclature of the plants.

Some botanists are of the opinion that the genus is more than a mere taxonomic category and the components of the genus indicate phyletic relationship among themselves as well as with other genera of the same family. And thus, the genus is considered as a biological category, and the taxonomic studies of the genus should consider not merely the morphological similarities but also the origin; migration; genetical, cytological, physiologic and ecologic behaviours; and geologic history of its components and need re-evaluation of existing genera using these tenets. A large genus may be divided into **subgenera**, the subgenus into **sections**, section into **subsections** and subsection into **series**.

Species, the next subordinate category, has been considered to be the basic unit of all taxonomic work, mainly due to the reason that this category was treated as special creation and, therefore, quite immutable entity.

According to International Code of Botanical Nomenclature (ICBN) '*plant species are convenient classificatory units defined by trained botanists using all information available*'.

Species as a Basic Unit of Taxonomy. That the species constitutes the basic unit of classification or for that matter, taxonomy (systematics), can be substantiated by the following facts :

1. In majority of cases, there is no infraspecific names, especially in families such as Apiaceae (Umbelliferae) and Liliaceae. It is conclusive of the fact that the species is the basic unit of classification.
2. Unlike other taxa, species can be described and recognised without relating to taxa of other ranks. This enables selecting herbarium sheets into different species without difficulty, without knowing or bothering to know how many genera are covered by these sheets. One can not recognise genera or describe them without reference to the included species. Thus it becomes clear that species is the only category that deals directly with plants.
3. The species is unique in being non-arbitrary to both inclusion and exclusion in terms of morphological discontinuity or restriction of gene exchange. If all members of a group are continuous by an appropriate criterion or the group is discontinuous from any other group by same criterion, the group is non-arbitrary to inclusion, and if group shows internal discontinuity or the group does not show discontinuity with other groups, it is arbitrary to inclusion. All higher taxa, although, non-arbitrary to inclusion are arbitrary for inclusion, *i.e.*, they exhibit internal discontinuity as species exhibit external discontinuity form part of these taxa.

Ideal Species. An ideal species, *i.e.*, a species that can easily be distinguished and poses no problem in its identity, is very rare. An ideal species is supposed to fulfil the following characteristics :

1. it poses no taxonomic problems and is easily recognised as a distinct identity on the basis of morphological features,
2. it exhibits no discontinuity of variation within, *i.e.*, it contains no subspecies, varieties or formas,
3. it is genetically isolated from other species,
4. it is sexually reproducing, and
5. it is atleast partially outbreeding.

Infraspecific Categories

Infraspecific category is the category below the rank of the species. These categories were limited in earlier taxonomic work and if taken into account, were considered as abnormalities. They were described as appendages to the species. With the development of biosystematic studies, the species is considered to be represented by the initial binomial but all subsequently described variants are ascribed to that binomial, *e.g.* when one infraspecific epithet is present, a trinomial (*Herniaria hirsuta* var. *diandra*) and if more polynomial (*Amaranthus hybridus* subsp. *cruentus* var. *paniculatus*).

For infraspecific categories, the rules of nomenclature provide as **subspecies (subsp.)**, **variety (var.)**, **subvariety (subvar.)**, **forma (f.) forma biologica, forma specialis** and **forma individuum**, of these, the three most commonly used include subspecies (subsp.), variety (var.) and forma (f.). Subspecies has been considered next to the species in lower rank and has been used by various workers with different circumscriptions on account of various reasons. A species of small magnitude distinguished by less significant morphological features and a major morphological variation of a species that has distinct geographical distribution other than occupied by other subspecies of the same species or counter parts of ecotype, cause to recognise a subspecies. The variety, next subordinate rank to subspecies, has been used rather indiscriminately either as a morphological varient of the species without regard of distribution, a morphological variant having its own geographical distribution, a morphological variant sharing areas common to one or more varieties of the same species or a variant representing only a colour or habit phase. The forma is the smallest category used in the taxonomic literature and is usually applied to trivial variants occurring among the individuals of any population. Categories smaller than forma rarely appear in taxonomic works and are not being considered here.

SPECIES CONCEPT

The species are so important, not just because they exist in huge numbers, but each species turns out to be different from every other in its morphological and other characters. Mayer (1942) in his *Systematics and*

Origin of Species stated "*they represent an important level of integration in living nature. This recognition is fundamental to pure biology. An inventory of the species of animals and plants of the world is the base line of further research in biology*" and further he states "*whether he realizes or not, every biologist–even he who works on the molecular level-works with the species or parts of species*".

History of biology reveals that two names–Linnaeus and Darwin, are closely associated with the concept of species, the former believed in the constancy and shape delimitation of species while the latter believed in the variation and overlapping (over-changing) pattern of species. With the publication of Darwin's *Origin of Species* (1859), the biologists were clearly divided in two camps, namely followers of Darwin and the followers of Linnaeus. The discovery of Mendel's laws resulted in an even more unrealistic concept of species.

For quite a long time, the concept of species remained a matter of dispute and great discussions among biologists. Most of the early biologists regarded species as aggregate of individuals, unconnected except by descent. Mayer (1942) regarded them under practical, morphological, genetic, sterility and biological and subsequently in 1957 considered three basic concepts, *viz*., typological, non-dimensional and multi-dimensional species concept.

Various species concepts are as follows :

1. Nominalistic Species Concept. According to this concept, species can be defined by the language of formal relations and not by property of their organisms. The concept suggests that only individuals exist and species are man-made abstractions. According to Bessey (1908) "*Nature produces individual and nothing more* ....". For the purpose of nomenclature, all organisms must be referable to species as species is a category in taxonomic hierarchy that may correspond to a specific name in the binomial system of nomenclature. Appearing to be logically sound, the concept is scientifically irrelevant because ultimate aim is to place a particular group of individuals in a species. The concept remains only of academic interest now.

2. Typological Species Concept. Aristotle (384–322 BC) suggested that "*for every natural group, there is an inner essence that makes them what they are*". This typological species concept was first proposed by John Ray (1686) and further elaborated by Linnaeus in his *Critica botanica* (1737). Linnaeus refuted the idea of **transmutation** of species, an ancient Greek idea, which persisted as late as seventeenth century. It was believed by many botanists that transmutation of wheat may result in barley, crocus in gladiolus and so on and suggested that although there is some variation within a species, the species by themselves are fixed. This **fixity of species**, created by Almightly Creator, denotes that "*species is a group of plants which breed true within their limits of variation*". In later course, Linnaeus moved away from the idea of fixity of species and got convinced that species can arise by hybridization, which he imagined in his later publication *Fundamenta fructifications* (1762) that at the time of creation, there arose as many genera as were the individuals and these in the course of time were fertilised by others and thus arose species until so many were produced as now exist. These species were sometimes fertilised by other species of the same genus giving rise to varieties. Typological concept of species has nothing to do with typification which is a methodology of nomenclature, providing names to taxonomic groups.

This species concept, now only of historical importance, has been rejected universally due to the facts that in strict sense typological-morphological concept does not meet requirements to cope with infraspecific variations in many stages of the plants, in being incapable of distinguishing absence of visible morphological differences between natural populations which are nevertheless distinct and reproductively isolated and in treating species as random aggregates of individuals which have the "*essential properties of the type of the species*" and "*agrees with diagnosis*".

3. Non-dimensional Species Concept. The essence of non-dimensional species concept, introduced by John Ray (1686) and further elaborated by Linnaeus (1737), is the relationship of two co-existing natural populations at a single locality at the same time, *i.e.*, **sympatric** and **synchronus**. According to this concept the relationship of two different species can be defined as reproductive isolation. Since long, it has been realised that species is more than an aggregate of individuals held together by a biological bond. Interbreeding within the species is more conspicuous as compared to reproductive isolation against other species.

4. Multi-dimensional Species Concept. Multi-dimensional species concept considers species as groups of populations that inter-breed with each other, actually or potentially and as such they are not clearly demacrated from other similar groupings. The so-called genetic species concept shows that at this level, all the three basic concepts can be expressed in genetic terms.

Thus it becomes clear that both morphological and genetic species concepts are attemps to deal directly with the discontinuities in nature.

5. Taxonomic Species Concept. Darwin (1859), who recognised continuous and discontinuous variations, developed his taxonomic species concept based on morphology, as such it is also called **morphological species concept.** According to this concept the "*species is regarded as assemblage of individuals with morphological features in common and separable from other such assemblage by correlated morphological discontinuity in a number of features*". Supporters of this view believe in the concept of continuous and discontinuous variations. According to this concept, the individual species shows continuous variation, shares certain characters and exhibits distinct continuity with the individuals belonging to another species with respect to all or some characters. The concept was modified by Du Rietz (1930) who developed morpho-geographical species concept by incorporating geographical distribution of populations. He defined species as–"*the smallest population that is permanently separated from other populations by distinct discontinuity in a species of biotypes*". The populations that are recognised as distinct species and occur in separate geographical areas, are generally quite stable and remain so even when they are grown together. But few species pairs, *e.g.*, *Platanus orientalis* of the Mediterranian region, *P. occidentalis* of North East United States and *Catalpa ovata* of Japan and China and *C. bignonioides* of America, which are morphologically quite distinct and well adapted to their respective places, when grow together, readily inter-breed and form intermediate fertile hybrids bridging the discontinuity gap between the species. This phenomenon is called **vicariance** or **vicarism** and the species pairs as **vicarious pairs** or **vicariants.**

Both morphological and morpho-geographical taxonomic species concepts have been widely accepted by taxonomists, even by those who incorporated data from cytology, genetics and ecology, but they firmly believe that "*species recognised must be delimited by morphological characters*".

Taxonomic species concept is advantageous in that it is useful for general taxonomic purposes, especially for the field and herbarium identification of plants. The concept is most widely applied and majority of species have been recognised using this concept. The morphological and geographical features used in application in this concept can easily be seen in the populations. The concept is being applied in cryptic form even by those experimental taxonomists who do not recognise it. The majority of species recognised through this concept, correspond to those species which have been established after experimental confirmation.

Taxonomic species concept also has inherent drawbacks such as the concept is highly subjective and different sets of characters are used for different groups of plants. It needs much practical experience because only after considerable observation and experience can a taxonomist decide the characters that are reliable in a particular taxonomic group. The concept does not take into account the genetic relationship between the plants.

6. Biological Species Concept. Biological species concept, also known as **Newer Species Concept**, was first formulated by Jordan (1905) in all of its consequences. It combines elements of the typological and nominalistic species concepts and formulates that species have independent reality and are typified by the statistics of populations of individuals. The concept differs from both, by stressing the populational aspect and genetic cohesion of the species. The concept pointed out that species receives its reality from the historically evolved and shared information of its gene pool. Proponents of this concept advocate their case on the contention that there are objective discontinuities in nature which delimit the units, the species. These discontinuities are formed by restriction of gene flow between inbreeding populations which are actually potent.

According to this widely held biological species concept, given by Mayer (1969), "*a species is a group of interbreeding natural populations that are reproductively isolated from other such groups*". Species consists of individuals that share common gene pool, *i.e.*, total set of genes and alleles present in a population, different and separated from other distinct individuals belonging to other species and gene pool. The biological species are maintained by reproductive isolation mechanism which are not necessarily absolute.

Among various other definitions of species of higher plants, Grant (1971) defined biological species as the "*reproductively isolated from other such groups*". His view was based on the applicability of the biological species definition to plant groups and suggested that the groups of non-interbreeding populations be considered biological species. Lewin (1981) considers three, more or less distinct, ways of considering the species concept.

(i) The species as perceived by the organisms themselves whether or not we recognise, their specific distinction fall under the category of Biological Species.

(ii) The species, as we perceive it, with our limited but extending field of experience and techniques, are included under Taxonomic Species.

(iii) The species designed a specific name, which does not necessarily correspond to taxonomic species; are kept under Nominal Species.

Biological species concept is advantageous in that it is objective and the same criterion is used for ali other groups of plants. It has a scientific basis as the populations exhibiting reproductive isolation do not mix among themselves and the morphological differences are maintained even if the species grow in the same locality. The concept is based on the analysis of features and does not require experience to put it into practice.

The concept finds some difficulties when applied to plants as a good majority of plants exhibit only vegetative reproduction and the concept is difficult and time consuming to carry out fertility-sterility tests experimentally, the main criteria of the concept.

7. Evolutionary Species Concept. The concept was developed by Meglitsc (1954), Simpson (1961) and Wiley (1978). According to Wiley (1978) "*an evolutionary species is a single lineage of ancestor-descendent populations which maintains its identity from other such lineages and which has its own evolutionary tendencies and historical fate*". Grant (1971) has defined it as a spacio-temporal lineage of populations that evolved separately from other lineage and has its own ecological niche. According to him the concept appreciates the dynamic nature of population but is not subject to the 'interbreeding pitfalls' of the biological species concept and as such can be applied to both sexual and asexual groups. Based on specific criterion, Grant (1981) recognised several terms to distinguish species such as : **semispecies** or **multispecies**.

Semispecies or **Multispecies** are sympatric species, which are capable of gene exchange. Multispecies, while genetically discrete, are not genetically isolated species groups.

Compilospecies. Compilospecies are the problematic cases of species that have originated from intermediates, formed by gene exchanges between allopatric as well as sympatric species or have multiple origin from different intermediates.

Microspecies is a term which is used in place of amphiapomicts to designate various apomictic forms.

Sibling Species are morphologically similar populations that are reproductively isolated.

8. Ecological Species Concept. Van Valen Leigh (1976), defines species "*as a lineage which occupies an adaptive zone minimally different from that of any other lineages outside its range*".

9. Selection Species Concept. This concept defines species "*as a system of genetically similar individuals and populations maintained as a cohesive unit by a set of selection pressures that balance the disruptive forces imposed by the environmental factors, mutations or genetic recombinations*".

Despite many concepts, no single concept of a natural species has yet satisfied the needs of every practicing taxonomist. Irrespective of the definitions or the concept of species, the unit species has to be fit in into binomial system of nomenclature. To fit in, it must have a Latin name composed of two words, the generic name with its first letter in capital and the specific epithet starting with small letter, *e.g. Mangifera* (generic name) *indica* (specific epithet).

❑❑❑

Chapter 5

Taxonomic Literature and Indian Flora

TAXONOMIC LITERATURE

A vast literature, concerned with the plant taxonomy existing globally, would require thousands of volumes to review. Still, plants are being studied in various parts of the world. Some of the most precious ancient works remained unpublished. Well known unique works have been published from time to time but much remains to be explored. Printed literature in the form of books was available towards the close of 15th century. With more and more investigations and overall advancement, based on the modern approach to the subject, the taxonomic literature is swelling day-by-day. The taxonomic literature exists in the form of illustrations, articles in periodicals, research papers in journals, monographs, floras and voluminous books. Even sometimes letters exchanged between experts in the area or other fellow workers assume significance as the literature on taxonomy. The growing literature of worldwide interest is published mainly in popular international languages such as English, Russian, German, French, Spanish etc., but the ancient literature is in Latin. It would not be possible to enumerate the vast taxonomic literature but an attempt is being made to furnish information of some important aspects mainly being used for determination of various plant taxa and their classification.

Unfortunately, India lags far behind as even after 60 years of independence, the botanists have no other authentic literature on the plant wealth of this country except the monumental work *Flora of British India* (7 volumes) compiled by Sir JD Hooker (1872–97). It is sad to note that Indian taxonomists have to depend on the plant systematists of western countries, especially Kew Garden, England for confirmation of their plant identification. The scenario is changing gradually but slowly. Efforts of Botanical Survey of India (established in 1890), after 1954, in the area of documentation of plant diversity in the form of national, regional, state and district floras and other publications, are noteworthy. During the years 2005–06 alone, BSI collected 4,500 specimens of plants from different parts of country of which 3,500 specimens belonging to 385 species were identified. A manual of CITES (Convention on International Trade in Endangered Species of Wild Fauna and

Flora) plants in India, *i.e.*, the **Negative List of Export**, is on way of preparation.

Libraries form important centres of various kinds of literature incorporating descriptions, illustrations, identification keys etc. for proper identification of unknown plants and gather knowledge about various aspects of the plants. Thus libraries are important in taxonomic work since they provide vide variety of taxonomic literature vital to the practicing taxonomists. The literature on taxonomy is one of the oldest literatures in science. While dealing chapter taxonomic literature, the text has been spread over relevant headings, but since some of the publications fall under more than one category, proper demarcation could not be made. The major forms of literature with names of important ones are accounted here.

Floras, Manuals etc.

A flora is an inventory of plants of a defined geographical region. The coverage of a flora may range from a small patch of forest, a valley, a mountain range, a city, district, state, nation or even to a continent. Since it is difficult to make an inventory of all plants growing all over the world, limited areas are selected. A flora may be fairly illustrative or simply synoptic but ideally it should provide systematic enumeration, comprehensive description, key for identification and if possible illustrations of various plant species included in it. Depending on their scope and the area covered, the floras may be categorised as :

Local Flora. It covers a limited geographical area, usually a city, district, state, country, a valley or a small mountain range, *e.g. Flora of Delhi* by JK Maheshwari (1963), *Flora Simlensis* by H Collet (1921), *Flora of Tamil Nadu* by KM Mathew (1983), *Flora of Texas* by RG Reeves (1972) etc.

Regional Flora. It includes a larger geographical area, usually a large country or a botanical region, *e.g. Flora of British India* by Sir JD Hooker (1872–97), *Flora of Malaysia* by CG Steenis (1948), *Flora Iranica* by KH Rochinger (1963) etc. A flora covering a country is more appropriately called **National Flora**. Botanical Survey of India (BSI) has recently taken up the publication of *Flora of India.*

Continental Flora. It covers the entire continent, *e.g. Flora Europea* by TG Tutin, *et al.* (1964–80), *Flora Australiensis* by G Bentham (1863–78) etc.

Comprehensive Treatments have much broader scope and are the joint efforts of comprehending several important taxonomic works which took place worldwide, *e.g. Genera plantarum* of G Bentham and JD Hooker (1862–83), *Die naturarlichen pflanzenfamilian* of A Engler and KA Prantil (1887–1915), *Das pflanzenreich* of A Engler etc.

A **manual** differs slightly from flora in that it emphasises mainly on providing suitable keys for identification, descriptions and glossary. They are more exhaustive than floras and generally cover specialised groups of plants, *e.g. Manual of Cultivated Plants* by LH Bailey (1949), *Manual of Cultivated trees* *and Shrubs Hardy in North America* by A Rehder (1940), *Manual of Aquatic Plants* by NC Fossel (1957) etc. A manual differs from a monograph in that the latter has a detailed taxonomic treatment of a taxonomic group.

Both floras and manuals are of immense importance in conceiving knowledge about wild and cultivated or both types of plants growing in a particular area. Most of the floras provide keys either for families or genera. Some flora, *e.g. Flora of Upper Gangetic Plain* by JF Duthie (1903), provide the keys upto species level. These keys help in initial identification of plants upto the level indicated therein. Further confirmation has to be done by reference to other relevant literature.

Botanical Survey of India is actively engaged in documentation of plant diversity in the form of district, state, regional and national flora (*Flora of India*). By March 2006, *Flora of Medak District, Andhra Pradesh* was more or less ready for publication; *Flora of Uttar Pradesh*, Vol II : Taxonomic description of 199 species was completed; *Flora of Cold Deserts of Western Himalayas*, Vol II : five species completed and *Flora of India : Lamiaceae*, was underway of revision.

SF Blake (Part I, 1941; Part II, 1961) has elucidated the list of floras more comprehensively in his *Geographical Guide to Floras of the World*. Frodin (1984) has also given a comprehensive list of world floras in his *Guide to the Standard Floras of the World*. One can also know more about them in Lawrence (1951), Core (1955), Heywood (1963), Reed (1969), Volumes of *Taxon* etc.

Monographs, Revisions, Conspectus, Synopsis etc.

A **Monograph** is a comprehensive treatment of a taxonomic group, generally a genus or a family, dealing compilation of worldwide available all significant information concerning the taxon. The geographical scope is worldwide because it is impossible to discuss a taxon without including all its members and often all its species, subspecies, varieties and formas. It also incorporates an exhaustive review of literature, as also a report on author's research work. An ideal monograph contains all information related to nomenclature, designated types, keys, exhaustive description, full synonymy and citation of specimens examined. Since preparation of monographs involves vast disciplines (biosystematics) to be undertaken, it takes several years of patience for the study to be completed. This is the reason that till today there are relatively few monographs on record, *e.g. The Genus Pinus* by TN Mirov (1967), *The Genus Datura* by AF Blakslee *et al.*, (1959), *The Genus Iris* by WR Dykes (1913), *Monograph of the Genus Avena* by BR Baum (1977).

A **revision** is generally less comprehensive than a monograph and is often based only on herbarium studies. It incorporates less introductory material and includes a synoptic literature review. Although it includes complete synonymy but the descriptions are shorter and often confined to diagnostic characters. The numbers of revisions is more than thousand worldover.

A **conspectus**, an effective outline of the revision, enlists all taxa with all or major synonyms with or without short diagnosis and with a brief mention of the geographical range. An ideal example of a conspectum is *Species plantarum* of C Linnaeus (1753).

A **synopsis** is a list of taxa with much abbreviated diagnostic distinguishing statements that are often in the form of keys.

Illustrations (Icones)

Illustrations (Icones) are usually published along with the text in Floras and Monographs but may sometimes be compiled exclusively. They are often detailed analysis that serves as useful tools for identification. As such many species of plants based on published illustrations only, without any accompanying description or diagnosis before 1.1.1980, have been accepted as validly published. The two principal compilations of icones are *Hooker's Icones* and *Wights Icones.* Others worth mentioning are : Illustrations of plants from Europe (Hegi, 1906–1931), North America (Gleason, 1963), Pacific coast trees (Mc Minn and Maino, 1946), Germany (Garcke, 1972), Korea (Lee, 1979) etc.

Regulations

Since 1867 onwards, International Botanical Congresses are being held periodically at usually four years of interval. A legislation has been agreed upon to recommend universally adaptable principles, rules and recommendations for naming the plants and publish the Code, subject to revision at every congress.

Periodicals (Journals)

Floras, Manuals and Monographs are published after a lot of taxonomic input. They require several decades for incorporating information on the results of ongoing research in the field with the help of journals or periodicals which play a significant role for continuance of taxonomic study. They provide a continuous update on additional taxa described or reported from a region and also information about nomenclatural changes and other taxonomic information. Reference to a publication in them includes volume number (which remains the same in all issues published within a year); issue number (numbered within a volume–a monthly journal will have 12 issues, quarterly 4, half-yearly 2 and annual 1); and page numbers on which the particular article appears.

Various herbaria, botanical gardens, botanical societies and many other agencies worldwide, publish the findings of their studies in the form of research papers, short articles, notes, abstracts etc. in these journals published at regular intervals. The numbers of journals may exceed several thousands. Lawrence and Bensen has compiled an excellent list of periodicals known to them by that time about published taxonomic literature. These periodicals are the authentic and learned treatises on concerned plant taxa and are quite useful in taxonomic study and research. Some of the common journals devoted largely to taxonomic research are : *Taxon* (International Association of Plant Taxonomy, Berlin), *Kew Bulletin* (Royal Botanic Gardens, Kew, England), *Plant Systematic and Evolution* (Denmark), *Botanical Journal of Linnaeun Society* (London), *Journal of the Arnold Arboretum* (Harvard), *Bulletin Botanical Survey of India* (Kolkata), *Botanical Magzine* (Tokyo), *Systematic Botany* (New York), *American Journal of Botany* (USA), *Botanical Gazzette* (USA), *Botanical Society of Bengal Bulletin* (India) etc.

Index

An index is an alphabetic catalogue of taxa with reference to their publication. It is prepared to depict a synoptical background of the work done in respect of names of flowering plants published throughout the world. They are volumes of compiled work that continuously compile abstracts of papers published. These include *Index Kewensis* (Royal Botanic Gardens, Kew, England, 1893), *Index Londiensis* (London, 1935), *Flowering Plant Index of Illustrations and Information* compiled by RT Issacson (1979), *Index Nominum Genaricum* (1986), *Gray Herbarium Card Index* (1893–1962)–*Supplement* in 1978, *Index Holminesis* (1969) etc.

Box 5.1 Index Kewensis

Index Kewensis–Plantarum phanerogarum (2 volumes, 15 supplements), Oxford (1893–supplements to date), the most comprehensive list of the scientific names of flowering (seed) plants, started its publication in 1893 from Royal Botanic Gardens, Kew, England–the largest botanical garden in the world. The entire work (from 1846 to 1927) was initiated by Sir JD Hooker who remained the Director of this prestigious Garden in collaboration with Benjamin, Daydon and Jackson. *Index Kewensis* enlists names of new and changed names of flowering plants with their bibliographic references to the place of first publication. In its first publication it included 3,75,000 scientific names known since the time of Linnaeus. Around ninteen eighties, the data was transferred to a computer database which continues to expand at the rate of about 6000 records per year. To make this data generally available, the whole *Index Kewensis* has been published in CD–ROM in 1993 that contains almost 9,68,000 records. *Index Kewensis* is an indispensable reference, most useful to any study of taxonomy of flowering plants.

Dictionaries and Glossaries

Dictionaries and glossaries are the alphabetical lists of difficult botanical terms which are in usage in the literature. They also provide valuable information concerning genera and families providing name of author, distribution, family of the genus and number of species it includes, *e.g.*

Dictionary of Flowering Plants and Ferns by JC Willis–6th edition (1973), *Botanical Dictionary*, 1960 (Moscow), *Dictionary of Gardening* by Chittenden (1951), *The Wealth of India–A Dictionary of Indian Raw Materials and Industrial Products*, 1948 (CSIR, New Delhi), *Glossary of Botanical Terms Commonly used in Range Research* by Dayton (Washington–1931), *Glossary of Botanic Terms with their Derivation and Ascent* by Jackson (1928), *Dictionary of Economic Plants in India* by Singh *et al.* (1983) etc. They are valuable in taxonomic studies.

Pictorial Encyclopaedias, Keys etc.

Pictorial encyclopaedias provide all available information on cultivated plants throughout the world. More famous among them are *Standard Cyclopaedia of Horticulture* by Baily (1914–17), *Indian Trees* by Brittan (1908), *North American Trees* by Covens (1950) etc.

Keys, provided in most of the manuals and floras in addition to the description, are not considered here. Keys prepared for general purpose of identification of plant families are widely used, particularly in determining plants from rich and diverse botanical provinces, *e.g. Key to the Families of Flowering Plants of the World* by Hutchinson (1967), *The Identification of Flowering Plant Families* by Davis and Cullens (1965), *The Identification of Flowering Plant Families* by Hansen and Rahn (1969), *Determination of Angiosperm Families by Punch-Card System* by Simpson and Janos (1974), *Punch Card Key to the Dicot Families of South India* by Naik (1977) etc.

Specific Literature on Cultivated Plants; Wild Plants; Trees, Shrubs and Grasses

Literature on **cultivated plants** has always been predominant since ancient period. This was due to man's interest of plants providing food, medicines and many more desirable plant products for his needs and greed. Some of the important ones are listed as : *A Manual of Cultivated Plants* by Bailey (1924), *Origin of Cultivated Plants* by de Candolle (1885), *The Mango : Its Cultivated Varieites* by Woodrow (1904), *Roses* by Pemberton (1920), *Coconut : Cultivation* by John (1955), *South Indian Bananas* by Jacob (1934), *The Indian Cottons* by Gammie (1907), *Field and Garden Crops of the North-West Provinces and Oudh* by Duthie and Fuller (1882–93), *Some Bananas at Assam* by Dutta (1952) etc.

Literature on **wild plants** appeared because of man's anxiety on plants growing around him. As a result of many ongoing explorations, plant taxonomists reported a large number of plant species, many of them were wild. Among literature on wild plants, worthmentioning to include are : *Wild Flowers* by Kosch (1964), *Wild Flowers of the West* by Clements (1927), *Some Useful Wild Plants of Delhi Province* by Singh (1945), *New or Noteworthy Plants from the Upper Gangetic plain* by Raizada (1950), *Our Early Wild Flowers* by Keeler (1916) etc.

Literature on **trees** and **shrubs** includes many publications on forest trees and shrubs and these growing elsewhere. Trees produce vast number of valuable commercial products and are grown as ornamentals whereas shrubs are valued for their economic products, especially drugs and used as fuel wood. Some important publications on trees and shrubs are : *Indian Trees* by Brandis (1906), *The Silviculture of Indian Trees* by Troup (1921), *Field Manual of Trees* by Schaffner (1926), *Flowering Trees of India* by Randhawa (1957), *Beautiful Trees of India* Randhawa (1958), *Shade and Ornamental Trees of California* by Pratt (1921), *Trees and Shrubs of the Rocky Mountain Region* by Longyear (1927), *Some Beautiful Climbers and Shrubs (Bombay)* by Bor and Raizada (1954), *Trees and Shrubs from Mexico* by Standly (1920–26), *A List of Trees and Shrubs of Indore State* by Biscoe (1910).

Literature on **grasses** comprises publications on various grasses. Although grasses include important undergrowth of forests and grow elsewhere, yet little is known about them. Some important publications on grasses include : *Some Uttar Pradesh Grasses and Common Grasses of United Provinces* by Bor (1941), *The World Grasses* by Bews (1940), *Grasses of North America* by Beal (1896), *The Gramineae* by Arhar (1934), *First Book of Grasses* by Chase (1922), *The Bambuseae of British India* by Gamble (1896), *A Key to the Genera of the Common Grasses of Lucknow and its Environment* by Pal (1966), *Grazing Lands in Himachal Pradesh* by Raina (1966).

As stated by Lawrence (1951) that "*Taxonomy demands an almost encylopaedic of world's pertinent literature, taxonomists must identify the various sources of taxonomic literature. The closer we reach the taxonomic literature, more we know about plants and better shall be our understanding of the floras*". India needs close cooperation among taxonomists in various botanical institutes such as Botanical Survey of India (Kolkata), National Botanical Research Institute (Lucknow), Forest Research Institute (Dehradun), Central Arid Zone Research Institute (Jodhpur), Indian Fodder and Grassland Research Institute (Jhansi), various herbaria and many universities and colleges carrying out research in plant taxonomy.

INDIAN FLORA

India, an important subcontinent of world's largest continent–Asia, with richest plant wealth, remained unexplored till 1565 when Garcia d' Orta published the first account of Indian plants in the Portuguese language which was later translated in Latin by Clusius. The work entitle *Coloquios dos Simples e Drogas da India* attracted European botanists to India. Subsequently *Hortus Indicus Malabaricus* (Hendrick von Rheede–1670 to 1703), *Flora Indica* (Burmann, 1768), *Plants of Coast of Coromendel* (Roxburgh–1820 to 24), *Plants Asiaticae Rariores* (Wallich–1830 to 32), *Icones Plantarum Indiae* (Robert Wight–1838 to 53) and a few others appeared.

Sir JD Hooker, in 1897, on compilation of the seventh and the last volume of *Flora of British India* (1872–97), expressed a wish that his work would facilitate the preparation and compilation of local floras and monographs of the genera of India which he considered "*the richest and most varied botanical area on the surface of the globe, where ample work is still to be done*". Inspired with his pioneering work, the main theme of botany, in ninteenth century, became the comprehensive study of floras of various parts of the country. Simultaneously, the establishment of Botanical Survey of India in 1890 and the organisation of extensive botanical explorations in then British India, caused to emerge many local and regional floras namely *Flora of Bombay Presidency* by Cooke (1901–08) and Talbot (1909–11), *Flora of the Presidency of Madras* by Gamble and Fischer (1915–36), *Flora of Bengal* by Prain (1903), *Flora of Upper Gangetic Plain* by Duthie (1903–22), *Flora of Nilgiris and Pulney Hill-tops* by Fyson (1915), *Flora of Bihar and Orissa* by Haines (1921–25), *Flora of Rajasthan* by Blatter and Hallberg (1918–21). Apart from these floras, several other shorter floristic accounts on limited regions and monographic studies on plant groups, such as orchids by King and Pantling (1898), Asiatic palms by Beccari (1908–31) etc. and genera like *Dalbergia* by Prain (1904), *Impatiens* by Hooker (1904–11), *Dioscorea* by Prain and Burkill (1936–38) etc., were made.

Most of such scientific pursuits of extensive botanical explorations and floristic studies were kept in abeyance during World War I (1914–18). Undoubtedly, all such contributions have enriched our knowledge of Indian flora considerably but the range of floristic work covered on such a big country is so meagre that the main task of preparation of modern flora of India is considered to be far beyond fulfilment of the goal. Influenced with biological species concept and natural variation in species, Turril *et al.* (1964), Davis and Heywood (1965) and many others gave importance to the data from other branches of botany in collaboration with the basic structure of herbarium taxonomy that greatly helped in proper understanding of many genera and families and their affinities and systematic position. This also helped in assignment of many taxa to proper taxonomic hierarchy which could not be picked up by Hooker and others. Even Hutchinson, who followed pure external morphological characters incorporating understanding of their phytogeography, comparative analysis of characters of regional floras and critical study of variation among species, could not do so.

In later half of the ninteenth century, extensive botanical explorations of various parts of the country, were made that resulted some valuable floristic accounts of various regions namely Khandala by Santapau (1963), Lahul by Rau (1960), Rajasthan by Deb (1961), North Garhwal by Rau (1961), Jammu and Kashmir by Rao (1961), Saurashtra by Santapau (1962), Andaman and Nicobar by Thothathri (1962), Sikkim by Rolla (1963), Delhi by Maheshwari (1963), Jodhpur division by Rolla *et al.* (1965) and Tirap district–NEFA by Panigrahi *et al.* (1966). Besides, revision work on several families and plant groups were done by various botanists. These contributions, apart from reprinting of old floras by Botanical Survey of India, created a considerable enthusiasm among various emerging workers in the area to undertake floristic work in different parts of the country.

Botanical Survey of India (established in 13 February, 1890), with its headquarters at Kolkata, which remained quiescent till 1954 and is now under Ministry of Environment and Forests, Govt. of India, has now made a concerted effort towards its basic objective to explore the plant resource of the country and to identify the plant species with economic virtues. The main objective of BSI is exploration and documentation of phytodiversity of the country in general and protected areas, hot spots, fragile ecosystems and sacred grooves in particular and publication of National, State and District Floras. During 2005–06 about 3,500 specimens belonging to 385 species were identified by different circles/units of BSI. Twelve herbarium consultation tours/study tours were also undertaken in connection with revisionary and floristic studies of *Flora of India; Orchidaceae* and *Lamiaceae; Monocot Flora of Dibang Valley* (Arunachal Pradesh); *Flora of Uttar Pradesh Vol. II; Flora of Tawang District* (Arunachal Pradesh); *Coastal and Marine Flora of Malvan Marine Sanctuary* (Goa); *Floristic Diversity of Upper Subansin District* (Arunachal Pradesh); *Flora of Sikkim* (Rubiaceae); *Flora of Nongkhyllem Wildlife Sanctuary* (Meghalaya) etc., during which many new species/varieites were discovered and some plants were collected after 30 years or more which were listed as rare, endangered or threatened in the Red Data Book of IUCN (International Council for Conservation of Nature and Natural Resources).

As signatory to CBD (Convention on Biological Diversity), India is committed to exploration and preparation of inventory of plants (also animals) and as such an All India Coordinated Project on Capacity Building in Taxonomy (AICOPTAX) has been set up by Ministry of Environment and Forests. The scheme was reviewed on 20th May, 2005 at Coimbatore in which it was decided that financial assistance should be given to 28 botanical gardens of the country to function more effectively in this area.

❑❑❑

Chapter **6**

Major Systems of Classification

Plant classification is the arrangement of plants into groups and categories for effective organisation, a clear understanding and proper study. According to Radford (1986) '*classification is the arrangement of groups of plants with particular circumscriptions by rank and position according to artificial criteria, phenetic similarities or phylogenetic relationships*'. The review of the historical background of plant classification throws light on that the classifications until seventeenth century (see Chapter 1) were based on features studied without microscopic aid. Linnaeus and earlier botanists tried to classify plant kingdom using a single or few characters chosen arbitrarily as they thought only about the convenience of following a system of classification solely to identify a particular plant. Such systems are, therefore, called **artificial systems** of classification. Later on, plant taxonomists, based on the idea that the plants belonged to some natural groups, tried to designate and distinguish such groups and attempted to classify plant kingdom accordingly. Such systems are known as **natural systems** of classification. Under these systems, such natural grouping gives the idea that the individuals under one particular group are closely related to one another, although, they believed in the fixity of species and simultaneous creation of all groups by God. After the publication of Darwin's *Origin of Species* dealing with the theory of organic evolution (Darwin and Wallace, 1858), the plant taxonomists began to think about the origin of each of the natural group from a more primitive group or from an individual of a primitive group. This gave birth to the idea that some of the natural groups are more primitive and some are recent or advanced and recent groups have been derived from some comparatively primitive groups. The systems that incorporated this idea, attempted to find out relation between the different groups, *i.e.*, to trace the phylogeny of the groups and are called **phylogenetic systems.** The systems of plant classification continued to be modified by incorporating data obtained from related branches such as palaeobotany, biochemistry and ultrastructure using electron microscopy and gave rise to

contemporary phylogenetic systems of classification. In recent years, workers of the **Angiosperm Phylogeny Group (APG)** suggested that angiosperms should not be divided into monocotyledons and dicotyledons, instead be divided into number of monophyletic groups directly and resulted in the development of **APG Classification** in 1998, grouping 462 families into 40 putative monophyletic orders and a small number of informal monophyletic higher groups. Under these informal groups, there were also listed a number of families without assignment to order and a large number of unassigned families, listed both at the beginning and at the end of APG classification. Some outstanding systems of classification with their salient features, brief outline of classification, merits and demerits and comparisons are discussed here.

REVIEW OF HISTORY OF CLASSIFICATION

A review of history of classification reveals that no written records of the activities of our pre-literate ancestors are available as they were practical taxonomists having acquired the knowledge as to which plants could be used as food or for other purposes. Such classifications of plants developed by isolated communities through the need of the society, without the influence of science, are termed as **folk taxonomics**, often parallel to modern taxonomy as illustrated by equivalence of common English names grasses and sedges with Poaceae and Cyperace.

Early civilisations, that flourished in Babylonia, Egypt, China and India, practiced a short of plant taxonomy as people knew fairly good number of plants particularly those of food and medicinal value. One of the earliest Indian works dealing with plants in a scientific manner is *Vrikshayurveda* (science of plant life) compiled by Parasara (250 BC to 120 BC), a manuscript, discovered few decades ago. The book, apart from other information about the plants, also provided a system of classification wherein he organised the plants into several families (**ganas**) on the basis of morphological features, not known to the European classification untill the eighteenth century. For example, the **Svastika-ganiyam** (evidently the Cruciferae, new name Brassicaceae) were differentiated and named so because of calyx resembling to **Swastika** (卐), ovary superior, 4 free sepals and petals, 6 stames of which 2 are shorter and 2 carpels forming bilocular fruit; **Shamiganyam** (evidently the Leguminosae, new name Fabaceae) were distinguished and named so because of hypogynous flowers, 5 petals of different sizes, gamosepalous calyx and the fruit being legume; **Tripusaganyam** (evidently the Cucurbitaceae) were circumscribed so because of epigynous and sometimes bisexual flowers, 5 sepals, 5 united petals, 3 stamens and a trilocular ovary with 3 rows of seeds and so on. Among other Indian works *Charaka Samhita* written by Charaka (1st Century AD) deals with organising trees without flowers, trees with flowers, herbs which wither after fructification and other herbs with spreading stems, as separate groups. This huge treatise on Indian medicines is largely based on much earlier treatise published by Angnivesh. In 1897, it was translated in English by AC Kaviratna.

Early taxonomists centred round largely on plants of medicinal value. It was Andrea Cesalpino (1519–1603), the **first plant taxonomist** who described 1520 species of plants, grouped them as herbs and trees and further differentiated them on the basis of fruit and seed characters in his work *De Plantis libri* (1583) in 16 volumes. The contributions of Joaching Jung (1587–1657), the **first terminologist**, Gaspard (Caspar) Bauhin (1560–1624), **Legislateur en botanique**, John Ray (1627–1705), JP de Tournefort (1656–1708), **father of genus concept** and Camerarius (1694), who established the fact of sexuality in plants, are also noteworthy.

ARTIFICIAL SYSTEMS

The earliest systems, from 300 BC upto a considerable long period, were artificial systems. At that time, very little was known about the plants. These systems were based on one or a few characters, such as those propounded by herbalists were based on habit–trees, shrubs, herbs etc. In this sequence of the artificial systems of classification of plants, name of Theophrastus is worth mentioning.

Theophrastus (372 BC–287 BC)

Theophrastus–often called **Father of Botany**, a pupil of the great philosopher Aristotle, classified plants into four groups : **herbs, subshrubs, shrubs** and **trees** on the basis of their habit. He gave names and descriptions to some 500 plants in his *Historia plantarum*, the oldest botanical work in existence. The names given by him received general recognition and were used widely. Some of them, such as Daucan (*Daucus*) and Asparagos (*Asparagus*) still appear as generic names in modern classifications. He also pointed out differences between dicotyledons and monocotyledons and recognised different types of insertion of floral appendages, thus laying the foundation for the study of floral morphology. He is credited with having authored more than 200 works, most of which survive as fragments or quotations. But two of his great botanical works, *Enquiry into plants* and *Causes of plants*, translated in English in 1916 and 1927, respectively, are available.

Carolus Linnaeus (1707–1778)

Carolus Linnaeus, **Father of Taxonomy**, a great Swedish naturalist, also known as Carl Linnaeus, Carl Linne or Carl Von Linne, became interested in the study of natural history since his childhood. He renewed the thought regarding sexuality in flowering plants which was established by Camerarius in 1694. Linnaeus published his first paper on the sexuality in plants in 1729 and later *Hortus Uplandicus* (1730) wherein he enumerated the plants of the Uppsala Botanical Garden. He widened his knowledge at Haderwijk, The Netherlands while pursuing M.D. degree. During this course, he met Hermann Boerhaav and John Frederick, the latter financing his *Systema* *naturae* (1735). In 1737 he published his famous *Hortus cliffortianus* based on collection of tropical and temperate plants in the garden of George Clifford at Hartecamp. It was at Clifford's expense that he published several more manuscripts such as *Genera plantarum* (1737) and *Classes plantarum* (1738). Linnaeus's *Philosophia botanica* appeared in 1751 which contained a revised version of his system published previously in his *Systema naturae* (1737) and *Glasses plantarum* (1738), a work in which some 7,300 species were described and arranged according to his sexual system of classification. He published his best known *Species plantarum* in 1753. In recognition of his contribution, Linnaeus was made **Knight of the Polar Star** in 1753 and **Patent of Nobility** in 1761, and from this date (1761) he came to be known as Carl Von Linne.

Linnaeus's *Genera plantarum* (1737) is of great importance in plant taxonomy as a source of descriptions of 935 genera. It was published in five editions plus two supplements called *Mantissae* and in all a total of 1336 genera were diagnosed. These were grouped into 24 classes which were based almost entirely on the number and morphological arrangement of their stamens as : 1. Monandria (Stamen one)–*Canna, Salicornia*; 2. Diandria (Stamens two)–*Olea, Veronica*; 3. Triandria (Stamens three)–many grasses; 4. Tetrandia (Stamens four)–*Protea, Galium*; 5. Pentandria (Stamens five)–*Ipomea, Campanula*; 6. Hexandria (Stamens six)–*Narcissus, Lilium*; 7. Heptandria (Stamens seven)–*Trientalis, Aesculus*; 8. Octandria (Stamens eight)–*Vaccinium, Diera*; 9. Enneandria (Stamen nine)–*Laurus, Butomus*; 10. Decandria (Stamens ten)–*Rhododendron, Oxalis*; 11. Dodecandria (Stamens 11–19)–*Asarum, Euphorbia*; 12. Icosandria (Stamens 20 or more on the calyx)–*Opuntia*; 13. Polyandria (Stamens 20 or more on the receptacle)–*Ranunculus*; 14. Didynamia (Stamens didynamous)–many forms of Labiatae (= Lamiaceae); 15. Tetradynamia (Stamens tetradynamous)–*Brassica*; 16. Monadelphia (Stamens monaldelphous)–Malvaceae, 17. Diadelphia (Stamens diadelphous)–Legumes; 18. Polyadelphia (Stamen polyadelphous)–*Theobroma, Hypericum*; 19. Syngenesia (Stamens syngenesious)–many Composits (= Asteraceae); 20. Gynandria (Stamens united to gynoecium)–Orchids; 21. Monoecia (Plants monoecious)–*Morus, Carex*; 22. Dioecia (Plants dioecious)– *Salix*; 23. Polygamia (Plant polygamous)–*Acer, Nyssa*; and 24. Cryptogamia (Flowerless plants)–*Pteris, Riccia, Nostoc*, lichens.

These classes were further subdivided, on the basis of carpel characteristics, into orders such as Monogynia, Digynia etc. Such a system of classification based on stamens and carpels, resulted in the artificial grouping of unrelated taxa and separation of relatives. He called his sexual system as *Nuptiae plantarum* (marriages of plants) which he wrote in metamorphical terms in his *Systema naturae* as "*husband and wife have same bed*" or "*five husbands in the same marriage*" etc., referring to the arrangement of stamens and carpels in the flower.

The system presented by Linnaeus, was very simple and remained popular for over 75 years. Linnaeus knew that his system was more convenient than natural, but it was the need of the day as there was a tremendous increase in the number of plants known to man, causing the necessity of quick identification and placement. This was really the merit of sexual system of Linnaeus. Since the system proved to be of great help in identifying a plant, many flora writers, *e.g.* William Roxburgh in his *Flora indica*, arranged the plant following Linnaeus's classification. The only disadvantage in the system lies that it gives no indication as to the natural or phylogenetic relationship among different plant taxa and plants with no affinity at all, are placed together in an artificially created group.

NATURAL SYSTEMS

Despite the fact that Linnaeus had provided a readily referable cataloguing scheme for a large number of plants known by that time, it soon became evident that unrelated plants came together in such groupings. The French never really adopted Linnaeus's system and, based on their intellectual ferment and own experience in the botanical gardens of Paris (*Jardin de botanique de Paris*), soon developed their own, the first natural system of classification. The leading botanists of that time which took the lead in this area are :

Michel Adanson (1727–1806)

A French botanist, not impressed with artificial choice of characters, Michel Adanson, while exploring plant wealth in Africa, developed ways to logically deal with large number of characters in his classification. He recognised 58 **natural orders**, equivalent to modern orders and families, based on their natural affinity in his two-volume work *Families des plantes* (1763). He is credited for giving birth to the present day **Numerical Taxonomy** which has now developed into **Neo-Adansonian principles**, *i.e.*, use of as many characters as possible for the most useful classification (phenetics).

Jean BP Lamarck (1744–1829)

A French naturalist, better known for early **theory of evolution by inheritance of acquired characters**, authored *Flore Francaise* (1778), which, in addition to key for identification of plants, contained principles concerning the natural grouping of species, orders and families.

de Jussieu Family

Of the four well known botanists of this prominent French family (the three brothers Antoine (1686–1758), Bernard (1699–1776) and the youngest Joseph (1704–79)), Berdard de Jussieu, contemporary of Linnaeus and Professor of Botany at the Royal Gardens, Versailles, arranged the plants according to a system, more or less his own, with modifications in the later arrangements of Linnaeus. He suggested a classification, based on the number of cotyledons and presence or absence of petals and their fusion wherein he divided the flowering plants into monocots and dicots but did not publish it. It was left to his nephew Antonie Laurent de Jussieu (1748–1836) to publish this classification along with his own changes in *Genera plantarum* (1789), the book often designated as the **beginning of natural systems of classification.** In this classification, the plants were spread over three groups : Acotyledones, Monocotyledones and Dicotyledones, the latter, on the basis of corolla characteristics and ovary position, was further divided in Apetalae, Monopetalae, Polypetalae and Diclines irregularis. Acotyledones, apart from cryptogams, included some hydrophytes whose reproduction was not known then. Diclines irregularis contained, Amentiferae, Nettles, Euphorbias as also the Gymnosperms. He divided plants into 15 classes which were further subdivided into 100 natural orders corresponding to many of present day families. Botanists of that period readily accepted this system over that of Linnaeus as the plants which looked alike were grouped together.

de Candolle Family

The Swiss de Candolle family of botanists was headed by Augustin Pyramus de Candolle (1778–1841), who in an effort to do a world flora, published several books–the most important one being *Theorie elementaire de la botanique* (1813) wherein he proposed a new classification scheme, outlined important principles and **introduced the term taxonomy**. From 1816 untill his death AP de Candolle worked in Geneva and undertook monumental work towards describing all known species of vascular plants under the title *Prodromus systematis naturalis regni vegetabilis*, the first volume of which appeared in 1824 and seven volumes of were published during his life time.

Fig. 6.1 AP de Candolle (1778–1841).

The plan of AP de Candolle's system was published in 1819 wherein he delimited 161 natural orders (equivalent to families) as against 100 of AL de Jussieu, including 58975 plant species grouped primarily on the basis of presence or absence of vascular structures. In this work he set forth the principles of a natural system of classification and further developed morphological approach of classification. The system suggested by him was similar to that of AL de Jussieu in many respects but it was certainly an improvement over that of the latter, particularly in the treatment of the dicotyledons. He spread over plant kingdom into Vasculares (Vascular bundles present) and Cellulares (no vascular bundles) to which he included two classes : Foliaceae (Mosses, Liverworts) and Aphyllae (Algae, Fungi, Lichens). Also Vasculares were spread over in two classes : Exogenae (Dicotyledons) and Endogenae to which he included Phanerogamae (Monocotyledons) and Cryptogamae. He divided Exogenae, giving special treatment, into Diplochlamydeae and Monochlamydeae (also including gymnosperms). The Diplochlamydeae was further subdivided into Thalamiflorae, Calyciflorae and Corolliflorae. It depicted catkin bearing dicots, such as Betulaceae, Fagaceae etc. as being related to gymnosperms. His system of Dicotyledons began with Ranunculaceae.

The chief merit of AP de Candolle's system was in the classification of Dicotyledons but one of the serious drawbacks was the inclusion of vascular cryptogams among Monocots, *i.e.*, to provide ferns a place co-ordinate with Monocots and in contrary to AL de Jussieu, Gymnosperms were given a place, although among Dicotyledons. The importance of anatomical features was highlighted and was successfully employed in his system. The system dominated till 1860, despite the fact that about two dozen systems were proposed during this period (1825–45).

His son Alphonse de Candolle and grandson Cashmir de Candolle continued his work. Alphonse de Candolle (1806–93) published the rest of the ten volumes of *Prodrumus* ......, the last in 1873 with the help of specialists and the last edition of *Theorie elementaire* ...... (1844) in which he included 213 natural orders.

Stephen Endlicher (1804–49)

Stephen Endlicher (fig. 6.2) published his *Genera plantarum secundum ordinis disposita* (1836–40) describing 6,835 genera and arranging them in a system of his own. It was widely used in continental countries. He divided plant kingdom in Thallophyta and Cormophyta. The Thallophyta included plants with simple structure, without stem, root, vessels and clearly defined sex organs, *e.g.* Algae, Fungi and Lichens. The Cormophyta included plants with stem, roots, vessels and clearly distinguishable sex organs. In Cormophyta, he included Bryophytes, Pteridophytes and Seed plants. He could not recognise the Gymnospermae as a single distinct taxon but placed the Zamiae (cycads) with the Pteriodophyta and Coniferae with Dicotyledons.

Fig. 6.2 Stephen Endlicher (1805–49).

Robert Brown (1773–1858)

Robert Brown, an English botanist, did not propose a classification of his own but demonstrated that Gymnosperms were a group discrete from Dicotyledons and had naked ovules. He also made clarification about floral morphology and pollination of Orchidaceae and Asclepiadaceae, morphology of grass flowers, structure of cyathium in Euphorbiaceae and established several families. His name appeared fairly large in chronicles of systematic botany because he had often been referred to as **botanicorum facile princeps**. Brown was the first keeper of the Botanical Department of the British Museum. He published his *Prodromus florae noval Hollandiae* containing description of Australian plants collected by Banks and Solander. In 1827, he published a paper entitled *The female flower in Cycadaceae and Coniferae* and in this paper he announced the important discovery of the distinction between Angiosperms and Gymnosperms.

Bentham (1800–84) and Hooker (1817–1911)

George Bentham (1800–84) and Joseph Dalton Hooker (1817–1911), two British botanists, associated with Royal Botanic Gardens, Kew, England (**largest botanical garden in the world**), made extensive survey of the seed plants throughout the world and jointly produced the most elaborate natural system of classification of seed plants in their three volume work *Genera plantarum* (1862–83). This monumental work required about 21 years to complete. The first volume of the book appeared in July, 1862 and the last in April, 1883. It comprised descriptions of all genera of seed plants known at that time. They recognised Dicotyledons, Gymnosperms and Monocotyledons as distinct groups. In all, 202 natural orders (now treated as families) were recognised. These were grouped into cohorts (now treated as orders) and series. The descriptions were based on original studies of the specimens and dissections of the plants made by themselves and did not represent a mere compilation of the existing literature. This was the greatest taxonomic work ever produced. The *Genera plantarum*, listing 97,205 species, proved to be a very useful reference book. The system of classification given in this book became very popular in Great Britain and other countries of the world including America. The book included the first rate descriptions of the natural orders (families), together with names of all species under each genus, the synonyms, localities, references to literature and keys for identification of natural orders (families) and genera.

Bentham and Hooker have taken into account Lindley, de Jussieu and others, but their system was patterned directly on the system developed by de Candolle, a close friend of Bentham and was a refinement over it. The system, however, differed from de Candolle's classification in the delimitation of families and genera on the basis of natural affinities and was pre-Darwinian in concept. Every genus known to them was described in details with help of the material collected from and herbaria of most part of the world. Thus they provided a firsthand and accurate information in respect to all genera. Bentham and Hooker described 97,205 species belonging to 7,569 genera of 202 natural orders (families) of phanerogams or seed plants. The term order was named as **Cohor** or **Cohort** and modern systematics's family as **order (natural order)** and the larger genera were further divided into subgenera and sections.

The refinement over de Candolle's system included creation of a new series **Disciflorae** placing it in between **Thalamiflorae** and **Calyciflorae** in their **Polypetalae**, revision of apetalous taxa etc. Gymnosperms were treated as a third taxon, collateral with and placed between Dicotyledons and Monocotyledons. Various genera were grouped under order (natural order or present day family) and different orders under cohors or cohorts on the basis of the overall similarities and differences in the data available to them. Certain orders, which could not be satisfactorily placed in any cohort, were

Box 6.1 Bentham and Hooker

George Bentham (1800–84) was a self-trained botanist. He wrote many significant monographs on families like Labiatae (= Lamiaceae), Ericaceae, Polygonaceae and Scrophulariaceae. He is also credited for writing and publishing his *Handbook of British Flora* (1858) and *Flora Australiensis* in 7 volumes (1863–78). This all could become possible because Bentham was a very critical botanist who gave full time attention to his study. He was on intimate terms with many taxonomists like Robert Brown, de Candolle etc. He joined the staff of Kew in 1854.

Fig. 6.3 George Bentham (1800–84).

Sir **Joseph Dalton Hooker** (1817–1911), son of Sir William Jackson Hooker (1785–1865), the then Director of Kew Gardens, England. JD Hooker got actively engaged in study of plants since his younger days during which with his father, he also got frequent opportunities of making long-distance botanical excursions in different countries to explore plants. During these excursions he came in contact with famous taxonomist Asa Gray with whome he exchanged the ideas. Hooker's magnificient collections from Himalayas, Lebanon, Antarctic and Atlas mountains, added valuable herbarium specimens to Kew. His brilliancy as plant explorer and phytogeographer soon made him popular among European taxonomists. He later on took over the Director of Kew in 1865 succeeding his father and continued till 1885. After retirement he was given the responsibility of preparing *Index Kewensis* volumes (1893) and listing names of all known species and their synonyms. He is also credited for the publication of *Flora of British India*, 1872–97 (in 7 volumes), *Student's Flora of British Islands* (1870) and also revised later editions of *Handbook of British Flora*, which remained a major British Flora until 1952.

Fig. 6.4 J.D. Hooker (1817–1911).

treated as **ordines anomali** (anomalous orders). Their system of classification has been retained in British and Indian herbaria and in these countries many of the botanists have followed Bentham and Hooker's system of classification in the Floras. In many botanical gardens plants have been arranged following their system.

Broad outline of Bentham and Hooker's system with number of genera and species in major taxa as given in *Genera plantarum* (1862–83) is :

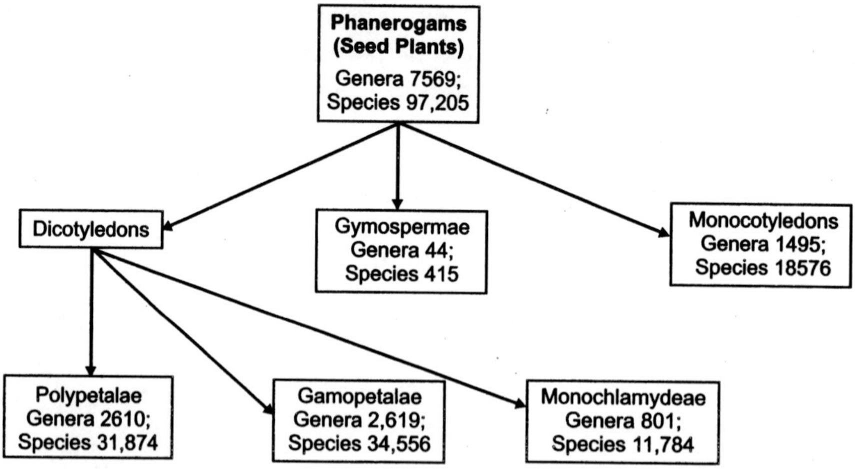

Following is the summary of Bentham and Hooker's system with distinguishing characters and number and names of cohorts or series in various taxa.

Table 6.1 Outline of the system of classification presented by Bentham and Hooker in *Genera plantarum* (1862–83).

PHANEROGAMS OR SEED PLANTS

GROUP I. DICOTYLEDONS : Seed with 2 cotyledons, flowers pentamerous or tetramerous, leaves net-veined, 14 series, 25 cohorts and 165 natural orders (families).

A. Polypetalae : Sepals and petals distinct, petals free.

Series	(1) Thalamiflorae	Flowers hypogynous, stamens many, disc absent *6 Cohorts* : Ranales, Parietales, Polygalineae, Caryophyllineae, Guttiferales and Malvales.
	(2) Disciflorae	Flowers hypogynous, disc present–below the ovary. *4 Cohorts* : Geraniales, Olacales, Celastrales and Sapindales
	(3) Calyciflorae	Flowers perigynous or epigynous *5 Cohorts* : Rosales, Myrtales, Passiflorales, Ficoidales and Umbellales

B. Gamopetalae : Sepals and petals distinct, petals united.

Series	(1) Inferae	Ovary inferior *3 Cohorts* : Rubiales, Asterales and Campanales

(2) Heteromerae	Ovary superior, stamens in one or two whorls, carpels more than 2 *3 Cohorts* : Ericales, Primulales and Ebenales
(3) Bicarpellatae	Ovary superior, stamens in one whorl, carpels 2 *4 Cohorts* : Gentianales, Polemoniales, Personales and Lamiales

C. Monochlamydeae : Flowers apetalous; perianth lacking or if present–not differentiated into sepals and petals.

Series	(1) Curvembryeae	Embryo coiled, ovule usually 1
	(2) Multiovulatae aquaticae	Aquatic plants, ovules many
	(3) Multiovulatae terrestres	Terrestrial plants, ovules many
	(4) Microembryeae	Embryo minute
	(5) Daphnales	Carpel 1, ovule 1
	(6) Achlamydosporae	Ovary inferior, unilocular, ovules 1–3
	(7) Unisexuales	Flowers unisexual
	(8) Ordines anomali	Relationship uncertain

GROUP II. GYMNOSPERMAE : (Ovules naked); 3 Natural orders (families)

GROUP III. MONOCOTYLEDONS : Flowers trimerous, venation parallel; 7 series, 34 natural orders (families)

Series	(1) Microspermae	Ovary inferior, seeds minute
	(2) Epigynae	Ovary inferior, seeds large
	(3) Coronarieae	Ovary superior, carpels united, perianth coloured
	(4) Calycinae	Ovary superior, carpels united, perianth green
	(5) Nudiflorae	Ovary superior, perianth absent
	(6) Apocarpae	Ovary superior, carpels more than 1, free
	(7) Glumaceae	Ovary superior, perianth reduced, flowers enclosed in glumes

Summary of Bentham and Hooker's System of Classification

GROUP : DICOTYLEDONS

(Leaves with reticulate venation, tap root system, flowers pentamerous, seeds with two cotyledons; 14 series, 25 cohorts and 165 natural orders –(families)).

A. POLYPETALAE

(Flowers with perianth in two whorls, corolla consisting of separate petals)

Series I. Thalamiflorae : (Sepals distinct, stamens many and hypogynous, disc absent)

Order 1. Ranales : (Carpels free, floral parts arranged in spiral, cyclic or spirocyclic manner)

Families :	1. Ranunculaceae	2. Dilleniaceae	3. Calycanthaceae
	4. Magnoliaceae	5. Annonaceae	6. Menispermaceae
	7. Berberideae	8. Nymphaeaceae	

Order 2. Parietales : (Carpels united to form an unilocular ovary, parietal placentation)

Families :	1. Sarraceniaceae	2. Papaveraceae	3. Cruciferae
	4. Capparideae	5. Resedaceae	6. Cistineae
	7. Violaceae	8. Canellaceae	9. Bixineae

Order 3. Polygalineae : (Syncarpous, ovary 2-locular)
Families : 1. Pittosporeae 2. Tremandreae 3. Polygaleae
4. Vochysiaceae

Order 4. Caryophyllineae : (Free central placentation)
Families : 1. Frankeniaceae 2. Caryophylleae 3. Portulaceae
4. Tamariscineae

Order 5. Guttiferales : (Sepals imbricate, stamens many)
Families : 1. Elatineae 2. Hypericineae 3. Guttiferae
4. Ternstroemiaceae 5. Dipterocarpeae 6. Chlaenaceae

Order 6. Malvales : (Sepals valvate, stamens usually numerous, ovary with 3-8 carpels, axile placentation)
Families : 1. Malvaceae 2. Sterculiaceae 3. Tiliaceae

Series II. Disciflorae : (Calyx consisting of united or distinct sepals which may be free or adnate to ovary, stamens usually definite in number, hypogynous, disc present).

Order 7. Geraniales : (Pentamerous flowers, disc annular or broken into glands, androecium obdiplostemonous, multicarpellary ovary, syncarpous, axile placentation, ovules pendulous, rephe usually ventral).
Families : 1. Lineae 2. Humiriaceae 3. Malpighiaceae
4. Zygophylleae 5. Geraniaceae 6. Rutaceae
7. Simarubeae 8. Ochnaceae 9. Burseraceae
10. Meliaceae 11. Chailletiaceae

Order 8. Olacales : (Androecium diplostemonous, ovules pendulous, raphe usually dorsal).
Families : 1. Olacineae 2. Ilicineae 3. Cyrilleae

Order 9. Celastrales : (Androecium haplostemonous, ovules erect, raphe ventral).
Families : 1. Celastrineae 2. Stackhousieae 3. Rhamneae
4. Ampelideae

Order 10. Sapindales : (Usually compound leaves–exstipulate, androecium diplostemonous, ovules ascending, often 1-2 in each locule, raphe ventral or inverted, flowers usually in panicles).
Families : 1. Sapindaceae 2. Sabiaceae 3. Anacardiaceae
Ordines anomali : 1. Coriarieae 2. Moringeae

Series III. Calyciflorae : (Sepals united, rarely free, petals in one ore two series, perigynous or epigynous, ovary generally inferior).

Order 11. Rosales : (Indefinite stamens–often twice or more the number of petals, styles distinct).
Families : 1. Connaraceae 2. Leguminosae 3. Rosaceae
4. Saxifrageae 5. Crassulaceae 6. Droseraceae
7. Hamamelideae 8. Bruniaceae 9. Halorageae

Order 12. Myrtales : (Stamens, usually definite, flowers perigynous or epigynous, axile placentation).
Families : 1. Rhizophoraceae 2. Combretaceae 3. Myrtaceae
4. Melastomaceae 5. Lythrarieae 6. Onagrarieae

Order 13. Passiflorales : (Leaves often much lobed and divided, tendril climbers, stamens 5, parietal placentation)

Families : 1. Samydaceae 2. Loaseae 3. Turneraceae 4. Passifloreae 5. Cucurbitaceae 6. Begoniaceae 7. Datisceae

Order 14. Ficoidales : (Perianth undifferentiated, Ovules in basal or axile placentation).

Families : 1. Cacteae 2. Ficoideae

Order 15. Umbellales : (Umbel inflorescence)

Families : 1. Umbelliferae 2. Araliaceae 3. Cornaceae

B. GAMOPETALAE

(Flowers with perianth in two whorls, corolla partially or completely fused).

Series I. Inferae : (ovary inferior, stamens as many as the petals and alternating with them).

Order 1. Rubiales : (Stamens epipetalous, anthers distinct, ovary with 2-many locules–each with 1-many ovules).

Families : 1. Caprifoliaceae 2. Rubiaceae

Order 2. Asterales : (Stamens epipetalous, anthers distinct or united, ovary with 1 locule with 1 ovule).

Families : 1. Valerianeae 2. Dipsaceae 3. Calycereae 4. Compositae

Order 3. Campanales : (Stamens not adnate to corolla, ovary with 2-6 locules–each with many ovules).

Families : 1. Stylidieae 2. Goodenovieae 3. Campanulaceae

Series II. Heteromerae : (Stamens in two or more series corresponding to petal number, epipetalous or free from corolla, ovary superior, carpels more than two).

Order 4. Ericales : (Stamens twice the number of corolla lobes or isomerous and alternating with them).

Families : 1. Ericaceae 2. Vaccinieae 3. Monotropeae 4. Epacrideae 5. Diapensiaceae 6. Lennoaceae

Order 5. Primulales : (Stamens as many as corolla lobes and opposite them, ovary with 1 locule).

Families : 1. Plumbagineae 2. Primulaceae 3. Myrsineae

Order 6. Ebenales : (Stamens as many as corolla lobes or many and opposite them, ovary with 2-many locules).

Families : 1. Sapotaceae 2. Ebenaceae 3. Styraceae

Series III. Bicarpellatae : (Stamens as many as corolla lobes or fewer and alternating with them, carpels two, ovary superior).

Order 7. Gentianales : (Leaves opposite, corolla actinomorphic).

Families : 1. Oleaceae 2. Salvadoraceae 3. Apocynaceae 4. Asclepiadaceae 5. Loganiaceae 6. Gentianaceae

Order 8. Polemoniales : (Leaves alternate, corolla actinomorphic)

Families : 1. Polemoniaceae 2. Hydrophyllaceae 3. Boragineae 4. Convolvulaceae 5. Solanaceae

Order 9. Personales : (Corolla zygomorphic, posterior stamen often reduced to staminode, ovules usualy more than 4).

Families : 1. Scrophularineae 2. Orobanchaceae 3. Lentibularieae 4. Columelliaceae 5. Gesneraceae 6. Bignoniaceae 7. Pedalineae 8. Acanthaceae

Order 10. Lamiales : (Corolla zygomorphic, posterior stamen reduced, ovules 4)

Families : 1. Myoporineae 2. Selagineae 3. Verbenaceae 4. Labiatae

Ordines anomali : 1. Plantagineae

C. MONOCHLAMYDEAE OR INCOMPLETAE

(Presence of one or incomplete whorl of perianth or perianth altogether absent, perianth sepaloid, petals absent).

It has no sub-division into cohorts (orders) but includes eight series.

Series 1. Curvembryae : (Ovule generally one, embryo coiled, seeds mostly with mealy endosperm).

Families : 1. Nyctagineae 2. Illecebraceae 3. Amarantaceae 4. Chenopodiaceae 5. Phytolaccaceae 6. Batideae 7. Polygonaceae

Series 2. Multiovulatae Aquaticae : (Ovary syncarpous, ovules many, plants aquatic).

Family : 1. Podostemaceae

Series 3. Multiovulatae Terrestres : (Ovary syncarpous, ovules many, plants terrestrial).

Families : 1. Nepenthaceae 2. Cytinaceae 3. Aristolochieae

Series 4. Microembryae : (Embryo very minute, ovary apocarpous or syncarpous with single ovule).

Families : 1. Piperaceae 2. Chloranthaceae 3. Myristiceae 4. Monimiaceae

Series 5. Daphnales : (Plants, woody or herbaceous, perianth sepaloid, stamens perigynous, ovary with one carpel and 2-4 ovuled).

Families : 1. Laurineae 2. Proteaceae 3. Thymelaeaceae, 4. Penaeaceae 5. Elaeagnaceae

Series 6. Achlamydosporeae : (Ovary usually inferior locules one, with 1-3 ovules, seeds endospermic, without testa).

Families : 1. Loranthaceae 2. Santalaceae 3. Balanophoreae

Series 7. Unisexuales : (Flowers unisexual, perianth usually absent).

Families : 1. Euphorbiaceae 2. Balanopseae 3. Urticaceae 4. Platanaceae 5. Leitnerieae 6. Juglandeae 7. Myricaceae 8. Casuarineae 9. Cupuliferae

Ordines anomali (families of uncertain relationship)

Families : 1. Salicaceae 2. Lacistemaceae 3. Empetraceae 4. Ceratophyllaceae

GROUP : GYMNOSPERMAE

(Plants naked seeded; 3 natural orders–(families)).

Families : 1. Gnetaceae 2. Coniferae 3. Cycadaceae

GROUP : MONOCOTYLEDONS

(Leaves with parallel venation, flowers trimerous, seeds with one cotyledon; 7 series and 34 natural orders–(families)).

Series 1. Microspermae : (Inner perianth lobes petaloid, ovary inferior, seeds minute–many, exalbuminous).

Families : 1. Hydrocharideae 2. Burmanniaceae 3. Orchideae

Series 2. Epigynae : (Perianth biseriate, inner whorl petaloid, ovary inferior, seeds large, few to many).

Families : 1. Scitamineae 2. Bromeliaceae 3. Haemodoraceae 4. Irideae 5. Amaryllideae 6. Taccaceae 7. Dioscoreaceae

Series 3. Coronarieae : (Perianth biseriate, inner whorl petaloid, ovary superior, seeds albuminous).

Families : 1. Roxburghiaceae 2. Liliaceae 3. Pontederiaceae 4. Philydraceae 5. Xyridaceae 6. Mayaceae 7. Commelinaceae 8. Rapateaceae

Series 4. Calycinae : (Perianth biseriate, sepaloid, ovary superior, seeds albuminous).

Families : 1. Flagellariaceae 2. Juncaceae 3. Palmae

Series 5. Nudiflorae : (Perianth usually absent or reduced, ovary superior, seeds albuminous).

Families : 1. Pandaneae 2. Cyclanthaceae 3. Typhaceae 4. Aroideae 5. Lemnaceae

Series 6. Apocarpae : (Perianth in 1-2 series or absent, ovary superior, carpels free one or more, apocarpous, needs exalbuminous).

Families : 1. Triurideae 2. Alismaceae 3. Naiadaceae

Series 7. Glumaceae : (Flowers, minute, perianth reduced, glumaceous, bracts large, scaly, ovary or locules with 1 ovule).

Families : 1. Eriocauleae 2. Centrolepideae 3. Restiaceae 4. Cyperaceae 5. Gramineae

---

Merits of the System

Despite the fact that the system does not incorporate phylogeny and is more than 125 years old, it still enjoys reputation of being a first rate natural system of repute due to following merits :

1. The system is of great practical value for identification of plants as it is very easy to follow for routine identification.
2. Many countries of the world including Great Britain and India widely follow it for the arrangement of plant specimens in the herbaria.
3. The system is based on careful comparative examination of actual specimens of all living genera of phanerogams (seed plants) and not a mere compilation of known facts from literature.
4. Unlike de Candolle, the Gymnosperms are not placed among Dicotyledons but they are treated as an independent group.

5. Dicotyledons are placed before the Monocotyledons–a position approved by all present-day taxonomists.
6. According to this system, Monocotyledons are derived from Dicotyledons and in Monocotyledons more importance has been given to the relative position of the ovary and perianth characteristics.
7. Although the system is not phylogenetic one, Ranales (in the broader sense including families now separated and included under order Magnoliales) is placed at the beginning of Dicotyledons, as also regarded primitive by most of the leading taxonomists.
8. Description of natural orders (families) and genera is precise. Larger genera have been spread over sub-genera so as to facilitate identification.
9. Keys to the identification are very useful.
10. The arrangement of taxa is based on overall natural affinities decided on the basis of morphological characters which can be easily studied with the naked eye or with a hand lens.
11. The system is easily workable especially in field applications, such as floristic study and plant explorations in larger areas.
12. The grouping of taxa is based on combination of characters, rather than a single character, *e.g. Delphinium* with fused petals has been kept in Ranunculaceae, along with the related genera and placed in Polypetalae and some Gamopetalous genera of Cucurbitaceae are retained along with the polypetalous ones and placed in Polypetalae.
13. Heteromerae has been rightly placed before Bicarpellatae.
14. Greater emphasis has been given to divide Dicotyledons, a taxa with largest number of species. Using perianth characters and contrast between free and fused (united) petals, Dicotyledons have been subdivided in Polypetalae (petals free), Gamopetalae (petals united) and Monochlamydeae (perianth undifferentiated into calyx and corolla).
15. Provision of dividing certain taxa on the basis of aquatic characteristics, *e.g.* Multiovulate Aquaticae and terrestrial habit, *e.g.* Multiovulatae Terrestres in Monochlamydeae.

Demerits of the System

The system being pre-Darwinian in its approach, has the following demerits :

1. Despite the fact that the classification was published after Darwin published his evolutionary theory, the system does not incorporate phylogeny.
2. Gymnosperms are placed for the sake of convenience between Dicotyledons and Monocotyledons whereas their proper position is before Dicotyledons as they constitute a taxa independent from Angiosperms.

3. Monochlamydeae is an unnatural assemblage of taxa belonging to elsewhere, as is indicated by separation of closely related families, such as Caryophylleae, Chenopodiaceae, Illecebraceae. They are related to such an extent that in major contemporary systems they have been placed in one order.
4. In Monocotyledons, Liliaceae (under Coronarieae) and Amaryllideae (under Epigynae), are placed under different series. These are generally regarded as closely related and are often included in the same order. Cronquist merged Amaryllideae with Liliaceae.
5. The system is ignorant about the affinities of the families placed under Ordines Anomali, as a result these families were tentatively grouped together. However, Takhtajan (1987) and Cronquist (1988) placed Ceratophyllaceae (included in Ordines Anomali) under subclass Magnoliidae and other three families of Ordines Anomali under the Dilleniidae.
6. Unisexuales is treated as a loose assemblage of diverse families sharing only one major character, *i.e.*, unisexual flowers. Cronquist separates these families under two distinct subclasses Hamamelidae and Rosidae and Takhtajan (1987) under Hamamelididae and Dileniidae.
7. Many large families such as Euphorbiaceae, Urticaceae, Liliaceae and Saxifragaceae, are unnatural assemblages and represent polyphyletic groups. And these have been splitted by subsequent taxonomists into smaller, natural and monophyletic families.
8. Orchidaceae is an advanced family with inferior ovary and zygomorphic flowers, but in the system, the family has been placed towards the beginning of Monocotyledons.
9. In Gamopetalae, Inferae, with an inferior ovary, has been placed before the other two series having a superior ovary. Now inferior ovary is considered to have been derived from a superior ovary.
10. According to the system, Angiosperms are divided into Dicotyledons and Monocotyledons, whereas modern phylogenetic systems place paleoherb families before Monocotyledons and Magnoliids before Eudicots.
11. Giving stress on relative position of the ovary and perianth, classifying Monocotyledons appears unnatural and unphylogenetic.
12. The system does not throw light on the origin of angiosperms.
13. In use of terminology of different ranks of taxa, there is no uniformity.

TRANSITION FROM NATURAL SYSTEMS TO PHYLOGENETIC SYSTEMS

The transitional period from natural systems to phylogenetic systems exhibited the systems of classifications that were not phylogenetic in true sense, rather they were attempts to arrange earlier natural systems in the light of prevalent phylogenetic theories. The systems in which the plant groups were arranged either haphazardly, in a linear sequence or in a stepwise progressive manner, were gradually replaced and those, that were influenced by evolutionary theory of Darwin and Wallace, were welcomed. Post-Darwinian systems, based on the supposed nature of a primitive angiospermic flower, can be arranged into two main groups which were later called : the **Englerian School** and the **Ranalian School**. The first of these concepts was proposed by a German taxonomist Alexander Braun in his *Flora der prowing Braderburg*. This influenced his successor A.W. Eichler in Berlin.

Eichler (1839–87)

AW Eichler (1839–87), a German botanist, modified to some extent the system given by Alexander Braun (1859) and proposed his classification in a book entitled *Blutendiagramme* (1875–78). The system was later elaborated in *Syllabus der vorlesungen* (1883), in which he, for the first time, separated Gymnospermae from Angiospermae and covered the entire plant kingdom. The plant kingdom was divided into two subgroups Cryptogamae and Phanerogamae and the latter was further subdivided into Gymnospermae and Angiospermae. Angiospermae was divided into two classes : Monocotyledons and Dicotyledons and in Dicotyledons only two groups Choripetalae and Sympetalae were recognised. Gymnopermae was given separate identity and placed before Angiospermae. Monochlamydeae of Bentham Hooker was abolished and dispersed among Choripetalae and Sympetalae. Monocotyledonae was placed before Dicotyledonae. Since he studied the morphology in terms of the doctrine of evolution, the system was an effort towards constructing a classification based on phylogenetic principles.

Fig. 6.5 A.W. Eichler (1839–87).

Engler (1844–1930) and Prantl (1849–93)

Adolph Engler (1844–1930) and Karl AE Prantl (1849–93), the two German botanists, jointly published the classification of the plant kingdom in a twenty three volume monumental work *Die Naürlichen pflanzenfamilien* (1887–1915). Engler was Professor of Botany at the University of Berlin and later Director, Berlin Botanic Garden. The system is commonly known under Engler's name, who first published the classification upto the family level under the title *Syllabus der pflangenfamilien* in 1892 which was constantly revised by him and after his death, by his followers till latest (12th edition) appearing in 1954 (ed. H Melchior and E Werdarmann) and 1964 (ed.

M Melchior). In its last edition Dicotyledons were placed before Monocotyledons.

Engler and Prantl's system was based on Eichler's (1839–87) system but differing in details. Their system, often considered as the **beginning in phylogenetic schemes**, was not strictly phylogenetic in the modern sense as it was an arrangement in linear sequence, starting with simplest groups in order of progressing complexity. Their work included the keys and the descriptions of all known genera of plants. The seed plants–Embryophyta Siphonogama were divided into Gymnosperms and Angiosperms and the latter into Archichlamydeae (petals absent or free) and Metachlamydeae (petals fused). They placed Monocotyledons before Dicotyledons and considered orchids to be more highly evolved than the grasses, apetalous Dicotyledons (Amentiferae) as primitive from those whose flowers possessed petals and those with simple unisexual flowers–the views, however, are not acceptable to most of the recent phylogenetists. The amalgamation of Polypetalae and Monochlamydeae and placing Gymnosperms before Angiosperms was certainly an improvement over Bentham and Hooker's system.

Fig. 6.6 Adolf Engler (1844–1930).

Fig. 6.7 Karl Prantil (1849–93).

In their scheme of classification, the plant kingdom was spread over 13 divisions (11th edition of *Syllabus der pflanzenfamilien*) of which 13th division dealt with Embryophyta Siphonogama. Divisions 1 to eleven dealt with Thallophyta and the 12h with Embryophyta Asiphonogama. The ouline of classification of 13th division Embryophyta Siphonogama is as follows :

Table 6.2 An outline of the system of classification proposed by Engler and Prantl in 11th edition of Syllabus der Pflanzenfamilien.

DIVISION 13. EMBRYOPHYTA SIPHONOGAMA

SUBDIVISION 1. GYMNOSPERMAE

SUBDIVISION 2. ANGIOSPERMAE

Class 1. Monocotyledoneae

11 Orders, 45 Families

(Orders : 1. Pandanales–**Pandanaceae first family**, 2. Helobiae, 3. Triuridales, 4. Glumiflorae, 5. Principes, 6. Synanthae, 7. Spathiflorae, 8. Farinosae, 9. Liliflorae, 10. Scitamineae, 11. Microspermae–**Orchidaceae–last family**)

Class 2. Dicotyledonae
44 Orders, 258 Families
Subclass 1. Archichlamydeae (Petals absent)
33 Orders, 201 Families

(Orders : 1. Verticellatae–**Only family casuarinaceae; first family,** 2. Piperales, 3. Salicales, 4. Hydrostachyales, 5. Garryales, 6. Myricales, 7. Balanopsidales, 8. Leitneriales, 9. Juglandales, 10. Julianiales, 11. Batidales, 12. Fagales, 13. Urticales, 14. Podostemonales, 15. Proteales, 16. Santalales, 17. Aristolochiales, 18. Balanophorales, 19. Polygonales, 20. Centrospermae, 21. Ranales, 22. Rhoedales, 23. Sarraceniales, 24. Rosales, 25. Pandanales, 26. Geraniales, 27. Sapindales, 28. Rhamnales, 29. Malvales, 30. Parietales, 31. Opuntiales, 32. Myrtiflorae, 33. Umbelliflorae–**Cornaceae last family**.)

Subclass 2. Metachlamydeae (Petals united)
11 Orders, 57 Families

(Orders : 1. Diapensiales–**Only family Diapensiaceae; first family,** 2. Ericales, 3. Primulales, 4. Plumbaginales, 5. Ebenales, 6. Contortae 7. Tubiflorae, 8. Plantaginales, 9. Rubiales, 10. Cucurbitales, 11. Campanulatae–**Compositae–last family**).

Melchior, in 12th edition (1964) of Engler's *Syllabus der pflanzenfamilien* made profound changes and rearranged the entire group Embryophyta Siphonogama wherein Dicotyledonae were treated as more primitive than Monocotyledoneae and placed before them, various orders were re-splitted and families were rearranged within the orders. In the 12th edition (last revision), the number of orders was increased from 55 to 62 (48 of Dicotyledonae and 14 of Monocotyledoneae) and of families from 303 to 344 (291 of the Dicotyledonae and 53 of Monocotyledoneae).

Englerian system, being improved than that of Bentham and Hooker's system, soon replaced Bentham and Hooker's classification in serveral European, American and continental network. Although the system is now obsolete in the light of current phylogenetic view, the system continued to be most dominant in most of the floras and herbaria for many years.

Engler was assisted by Karl Prantl and others who provided keys and descriptions of all families of plant kingdom, *i.e.*, from Algae to Compositae. Prantl and others provided information from other branches of botany and related disciplines.

Merits of the System

Engler and Prantl's system of classification has the following improvements over Bentham and Hooker's system and merits of its own :

1. The system was first to incorporate the idea of organic evolution.
2. It was a first major step towards phylogenetic system of classification.
3. The system covered the entire plant kingdom.

4. If provides descriptions and identification keys down to the level of family in *Syllabus der pflanzenfamilien*, genus in *Die Naturlichen pflanzenfamilien* and even species for many of the families in *Das pflanzenreich*, providing information on anatomy and geography.
5. Abolition of Monochlamydeae has resulted in bringing together many closely related families such as Illecebraceae is merged with Caryophyllaceae. Families Chenopodiaceae and Caryophyllaceae are placed in the same order Centrospermae.
6. Gymnosperms are separated and placed rightly before Angiosperms.
7. Many large unnatural families of Bentham and Hooker have been splitted into smaller natural families such as family Urticaceae is split into Urticaceae and Moraceae.
8. Compositae, at the end of dicotyledons and orchidaceae at the end of monocotyledons, have rightly been placed.
9. Their keeping Monocotyledons before Dicotyledons (eudicots) is still in practice in several recent systems.
10. Consideration of gamopetalous condition as advanced over polypetalous condition, fits well with the current phyletic views.
11. The system, in being throughly prepared, has been widely used in text books, floras and herbaria throughout the world.
12. The terms cohor (cohort) and natural orders have been replaced by the appropriate terms orders and families, respectively.
13. The closely related families Amaryllidaceae and Liliaceae have been grouped under the same order Liliflorae.
14. The families with inferior ovary in Archichlamydeae and Metachlamydeae are placed in the last and an advance is marked from hypogyny to epigyny.

Demerits of the System

Many drawbacks of Engler and Prantl's system have come into light with better understanding of the phylogenetic concepts. These drawbacks resulted primarily from the fact that they applied the concept of **simplicity representing primitiveness**, even to the Angiosperms where evolutionary reduction is the major phenomenon, not commonly seen in the lower groups of the plants. Following are some of the demerits of the system :

1. The classification is not truely phylogenetic.
2. Many of the ideas of Engler and Prantl are outdated in the modern sense.
3. Monocots are placed before Dicots whereas in recent systems Paleoherbs and sometimes Magnoliids are placed before monocots.
4. They considered Amentiferae, having families Betulaceae, Juglandaceae and Fagaceae, with reduced unisexual flowers, possessing few floral members and flowers borne in catkins, primitive. Based on studies on wood anatomy, floral anatomy and

palynology, it has now been established that Amentiferae is an advanced group as simplicity in flowers is due to evolutionary reduction and not primitiveness. According to Loconte (1996), the phylogenetic tree based on this hypothesis, may exceed six times more than the shortest tree.

5. Consideration of angiosperms as a polyphyletic group does not tally with present view of monophyletic origin.
6. Araceae are included in the order Spathiflorae which is placed before Liliflorae that also includes Liliaceae, but now Araceae in monocots are believed to have evolved from Liliaceae.
7. Helobiae, placed after a relatively advanced group Pandales is now considered as a primitive group of families like Butomaceae, Alismaceae and Potamogetonaceae.
8. Ranales, placed much lower down (after Amentiferae), are now considered as a primitive group with bisexual flowers, spirally arranged floral parts and numerous floral members.
9. Incorporation of the concept of derivation of free central placentation from the pariental placentation and the latter from the axile does not seem to be logical because free central placentation is now believed to have been evolved from the axile placentation through the disappearance of septa.
10. Consideration of derivation of bisexual flowers from unisexual flowers and entomophily from anemophily does not match with present phylogenetic views.
11. Liliaceae was considered as a large unnatural assemblage which now has been splitted into several smaller families like Liliaceae, Alliaceae, Asparagaceae, Asphodelaceae in recent classification of Judd, *et al.* (2002), APG II (2003) and Thorne (2003).

INTENTIONAL PHYLOGENETIC SYSTEMS

The natural systems rearranged in light of phylogenetic consideration, soon opened the door to systems that reflect evolutionary development. The second concept of the primitive flower, well known as **Ranalian Concept** was originated by Charles E Bessey (1915) which was later modified by Hutchinson (1926). The concept was based on the fact that the angiospermic flowers arose from gymnosperms with a bisexual strobilus bearing megasporophylls above and microsporophylls below. The lower sporophylls, by progressive sterilisation, developed into sepals and petals (perianth) whereas the upper ones developed into stamens and carpels. Strobilus axis shortened to form the receptacle. The primitive flower thus had numerous free perianth parts with many spirally arranged free stamens and carpels. The nearest approach to this type is found *in Magnolia* (Magnoliaceae–Ranales *sensulato*) living today. The theoretical basis for this view has been provided by Arber and Parkin (1907). According to these authors the ancestor of angiosperm was an undiscovered gymnosperm, distantly related to that rare fossil *Cycadeoidea dacotensis* (Bennettitales), a fossil plant that has borne **anthostrobili** in the axil of its cycad-like leaves.

Charles Bessey (1845–1915)

CA Bessey (1845–1915), an American botanist and a student of Asa Gray, while working at the University of Nebraska, USA, as Professor of Botany, **laid the foundation of modern phylogenetic classification.** He was the first American to make a major contribution to plant classification. Also, he is credited as **first botanist to develop intentional phylogenetic classification.** In many respects, his system was a modification of Bentham and Hooker's system, realigned according to the evolutionary principles, as amplified in 1897 and finally in 1915. Bessey considered Angiosperms to have evolved monophyletically from Cycadophyta belonging to implied Bennettitalean ancestory. He was first to consider that the large-sized bisexual flowers of Magnoliaceae with spirally arranged floral parts represent the most primitive condition in Angiosperms –a theory followed by many subsequent authors.

Fig. 6.8 Charles Bessey (1845–1915).

He divided Angiosperms (Anthophyta) into two major classes : Alternifoliae (Monocotyledoneae) and Oppositifoliae (Dicotyledoneae) and each of these two classes was subdivided into Strobiloideae and Cotyloidae. While arranging various taxa in these groups he used the data available from histological, ontogenetical (developmental) and morphological studies of homologies according to his 28 Dicta (guiding principles) published in *Annals of the Missouri Botanical Garden* under the little *The phylogenetic taxonomy of flowering plants* (1915).

Table 6.3 Outline of the classification of Anthophyta (Angiosperms) proposed by Charles A. Bessey (1915).

CLASS 1. ALTERNIFOLIAE (MONOCOTYLEDONEAE)

Subclass 1. Strobiloideae

5 Orders, 31 Families

(Orders : 1. Alismatales, 2. Liliales, 3. Arales, 4. Palmales, 5. Graminales)

Subclass 2. Cotyloideae

3 Orders; 14 Families

(Orders : 1. Hydrales, 2. Iridales, 3. Orchidales)

CLASS 2. OPPOSITIFOLIAE (DICOTYLEDONEAE)

Subclass 1. Strobiloideae

4 Superorders, 14 Ordes, 146 Families

Superorder 1. Apopetalae-Polycarpellatae

7 Orders; 104 Families

(Orders : 1. Ranales, 2. Malvales, 3. Sarraceniales, 4. Geraniales, 5. Guttiferales, 6. Rhoeadales, 7. Caryophyllales)

Superorder 2. Sympetalae–Polycarpellatae
3 Orders; 16 Families

(Orders : 1. Ebenales, 2. Ericales, 3. Primulales)

Superorder 3. Sympetalae–Dicarpellatae
4 Orders; 26 Families

(Orders : 1. Gentianales, 2. Polemoniales, 3. Scrophulariales, 4. Lamiales)

Subclass 2. Cotyloideae

2 Superorers, 10 Orders, 109 Families

Superorder 1. Apopetalae
7 Orders; 86 Families

(Orders : 1. Rosales, 2. Myrtales, 3. Loasales, 4. Cactales, 5. Celastrales, 6. Sapindales, 7. Umbellales)

Superorder 2. Sympetalae
3 Orders; 23 Families

(Orders : 1. Rubiales, 2. Campanulales, 3. Asterales)

In Bessey's *dicta*, most primitive modern plants were the Ranales that gave rise to monocots but unfortunately he placed monocots before dicots. He belived that both monocots and dicots arose from within Ranales, along two major lines, the **Ranalian concept of evolution**. On being shown in the diagramatic form, his system took the form of a cactus plant, which came to be known as **Bessey's Cactus** or as **Opuntia Besseyi**. Bessey believed in **strobiloid theory** of the origin of the flower. According to him the flowers have originated from a vegetative shoot with spiral phyllomes, of which some got modified to form sterile perianth, fertile stamens and carpels. The two evolutionary lines from such a flower formed (1) the **strobiloideae** (Ranalian line) with connation of like parts and (2) the **cotyloideae** (Rosalian line) with connation of unlike parts. Ranales in dicots and Alismatales in monocots were considered to be the most primitive orders–a fact recognised by most subsequent authors.

Bessey's Dicta (1915)

A. General Dicta :

1. Evolution is not always upward, but often it involves degradation and degeneration.
2. In general, homogeneous structures (with many and similar parts) are lower and heterogeneous structures (with fewer and dissimilar parts) are higher.
3. Evolution does not necessarily involve all organs of the plant equally in any particular period and one organ may be advancing while another is retrograding.
4. Upward development is sometimes through an increase in complexity and sometimes by a simplification of an organ or a set of organs.

5. Evolution has been generally consistent and when a particular progression or retrogression has set in, it is persisted into the end of the phylum.
6. In a phylum, the holophytic (chlorophyll-green) plants precede the colourless (hysterophytic) plants and their latter are derived from the former.
7. Plant relationships are up and down the genetic lines and these must constitute the framework of phylogenetic taxonomy.

B. Special Dicta (With Reference to Flowering Plants) :

8. The stem structure with collateral vascular bundles arranged in a cylinder is more primitive than that with scattered bundles and the latter are to be regarded as derived from the former.
9. Woody stems (as of trees) are more primitive than herbaceous stem and herbs have been held to have been derived from trees.
10. The simple, unbranched stem is an earlier type, from which branching stems have been derived.
11. Historically, the arrangement of leaves in pairs on the stem is held to have preceded the spiral arrangement in which the leaves are solitary at the nodes.
12. Historically, simple leaves preceded branched (compound) leaves.
13. Historically, leaves were first persistent (evergreen) and later deciduous.
14. The reticulate venation of leaves is the normal structure and the parallel venation of some leaves is a special modification derived from it.
15. The polymerous flower structure precedes and the oligomerous structure follows from it and this is accompanied by a progressive sterilisation of sporophylls.
16. Petaly is the normal perianth structure and apetaly is the result of perianth reduction (aphanisis).
17. The apochlamydeous perianth is earlier and the gamochlamydeous perianth is derived from it by a symphysis of the members of the perianth whorl.
18. Actinomorphy is an earlier structure than zygomorphy and the latter resulted from a change from a similar to a dissimilar growth of the member of the perianth whorl.
19. Hypogyny is a more primitive structure from which epigyny was derived later.
20. Apocarpy is the primitive structure and from it syncarpy was derived later.
21. Polycarphy is the earlier condition and oligocarpy was derived from it later.
22. The endospermous seed is primitive and lower while the seed without endosperm is derived and higher.

23. Consequently, the seed with a small embryo (in endosperm) is more primitive than the seed with a large embryo (in scanty or no endosperm).
24. In earlier (primitive) flowers, there are many stamens (polystemonous) while in later flowers, there are fewer stamens (oligostemonous).
25. The stamens of primitive flowers are separated (apostemonous) while those of derived flowers are often united (synstemonous).
26. The condition of powdery pollen is more primitive than that with coherent or massed pollen.
27. Flowers with both stamens and carpels (monoclinous) precede those in which these occur on separate flowers (diclinous).
28. In diclinous plants, the monoecious condition is the earlier and the dioecious later.

Merits of the System

Bessey's system has been followed by a large number of supporters in the recent years on account of the following merits :

1. The system gains support due to a more acceptable choice of the Ranales as the primitive order and the concept developed there upon of the primitive flower.
2. According to him the families with superior ovary precede with families with inferior ovary.
3. He abolished Monochlamydeae in his system and distributed the families of Monochlamydeae in Oppositifoliae of Dicotyledons.
4. Although initially he followed Bentham and Hooker's classification, he, however, did not place the Gymnosperms in between Dicotyledons and Monocotyledons.

Demerits of the System

In light of availability of more information of phylogenetic nature, now, Bessey's classification has the following demerits :

1. A serious objection to this system has been due to two evolutionary lines considered by him.
2. Monocots have been placed before Dicots. This is a major demerit of this system.
3. The system has overemphasised hypogyny, perigyny and epigyny.

Hans Hallier (1868–1932)

Hans Hallier (1868–1932), a German botanist, developed a phylogenetic classification resembling Bessey's system starting with Ranales, although he worked independently. He took into consideration greater ontogenetical observations and palaeobotanical and anatomical evidences, most of which Bessey could not consider. He considered seed plants consisting of two phyla, each being monophyletic and derived angiosperms from an extinct tribe of cycads (Lardizabalaceae) situated near or descended from Bennettitales, admitting affinities with Marattiales. Hallier considered dicots primitive than monocots and derived all dicots through the order of his Proterogenes (which included Ranales but excluded Magnoliaceae). Thus, he separated Mangoliales from Ranales and placed it in a separate order.

Wettstein (1862–1931)

Wettstein (1862–1931), an Austrian taxonomist, published his classification in *Handbuch der systematischen botanik* (1930). The system resembled that of Engler in considering unisexual flower as primitive. He treated dicots primitive than monocots and considered Helobiae to be primitive and Pandanales advanced. Many of his phylogenetic conclusions have been adopted in subsequent classifications.

Alfred Barton Rendle (1865–1938)

Alfred Barton Rendle (1865–1938), an English botanist associated with British Museum of Natural History, published his two volume *Classification of Flowering Plants* (1904, 1925) wherein he classified plants in the manner as Engler did but with some differences. He considered monocots as more primitive than dicots and Amentiferae a primitive group under dicots. He did not claim his system as strictly phylogenetic (like Engler) but for the sake of convenience, as the orders were grouped according to difference in characters of floral structure, he divided dicots into Monochlamydeae, Dialypetalae and Sympetalae. Monochlamydeae included orders and families with simple flowers, some of which being reduced forms having affinities with higher grades or representing lines of development from earlier extinct groups. Dialypetalae included plants with flowers having free petals with orders arranged in ascending sequence beginning with the Ranales having many spirally arranged floral parts and ending in Umbelliferae with cyclic and reduced floral parts. Sympetalae included plants with flowers having fused petals with orders representing still higher grades of floral structures to have been evolved from different dialypetalous groups. In monocots, Palmae were separated as a distinct order and Lemnaceae considered to be advanced over Araceae. Rendle considered polyphyletic origin for the components of each three groups and emphasised anemophily and entomophily as the phyletic criteria. Also herbaceous habit was considered to be more advanced than woody habit in most of the casses. He placed grasses subordinate to lilies.

John Hutchinson (1884–1972)

John Hutchinson, a British botanist associated with Royal Botanic Gardens, Kew, England, proposed a phylogenetic system of classification of angiosperms that first appeared in his book *The Families of Flowering Plants*–the first volume on Dicotyledones appearing in 1926 and the second on Monocotyledones in 1934. The classification was revised periodically and the 3rd edition (the final treatment) of the book appeared in 1973 (Oxford University Press, London). The system suggested by him was more or less parallel to that of Bessey and Bentham and Hooker but included much of his own which he gained after a vast study of herbaria at Kew and with a special knowledge of African plants.

Hutchinson's system of classification dealt only with the flowering plants included under phylum Angiospermae, as distinct from phylum Gymnospermae. Angiosperms were considered monophyletic in origin from some hypothetical Bennettitalean ancestral group which he named **proangiosperms**. In this initial subdivisions, the angiosperms were regarded as evolved along two separate evolutionary lines : **Herbaceae**–including most of the predominantly herbaceous families derived from Ranales *sensu stricto* and **Lignosae**–including fundamental woody representatives derived from Magnoliales. The Monocotyledones were derived from the stocks ancestral to Ranales (fig. 6.9). He recognised a number of smaller groups bound together

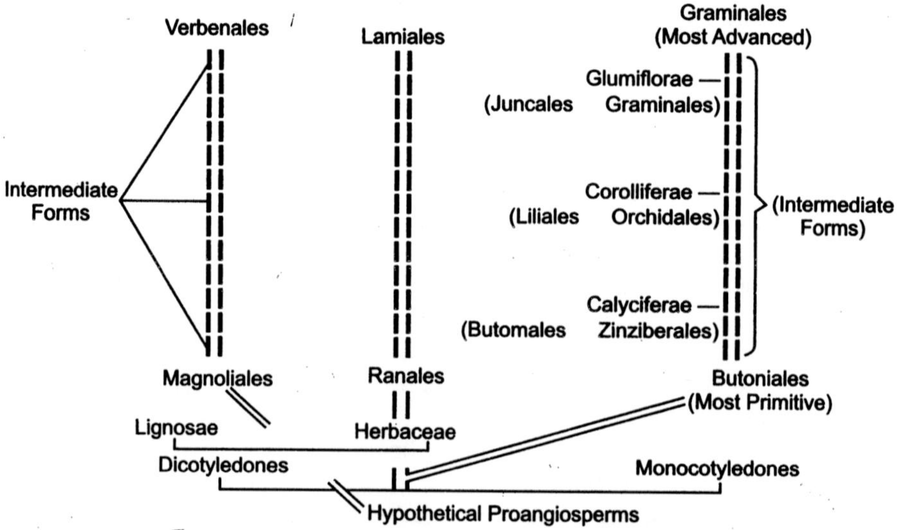

Fig. 6.9 Outline of Hutchinson's system of classification showing probable origin of dicots and monocots.

by a set of characters. He established Magnoliales as an order distinct from Ranales and considered these to have evolved in parallel lines. He considered Magnoliaceae as the most primitive family of the living angiosperms and dicotyledones to be more primitive than monocotyledones. Phylum Angiospermae was divided into subphyla Dicotyledones and Monocotyledones.

Hutchinson's classification was based on **24 principles** incorporated under : general principles for the classification of plants, principles relating to the general habit of the plants and principles relating to the general structure of flowering plants and principles relating to the flower and fruits of the plants. These principles are oulined below :

General principles for the classification of flowering plants : Hutchinson's revised classification (1959) is based on twenty four principles that are comparable to Bessey's dicta. These are as follows :

Other things being equal, it may be stated that :

1. Evolution is both upwards and downwards, the former tending towards preservation and the latter to their reduction and

suppression (or characters). Examples of former are towards the sympetalous condition and epigyny and of latter towards the apetalous state of many flowers and unisexuality in flowering plants.

2. Evolution does not necessarily involve all organs of the plant at the same time. One organ or set of organs may be advancing while another set may be stationary or retrograting.
3. Evolution has generally been consistent and when a particular progression or retrogression has set in, it is persisted into the end of the phylum, *e.g.* the strong tendency to zygomorphy of the corolla coupled with the reduction in the number of stamens in Engler's hypogynous Metachlamydeae and the vast tendency to perigyny and epigyny in Archichlamydeae and Metachlamydeae as seen in the families Umbelliferae and Rubiaceae, respectively.

Principles relating to the general habit of plants :

4. In certain groups, trees and shrubs are probably more primitive than herbs, *e.g.* Mimosaceae and Caesalpiniaceae (trees and shrubs) as compared with the derived family Papilionaceae (Fabaceae) (often herbaceous).
5. Trees and shrubs are older than climbers in any one family or genus, the latter habit being acquired through particular environment.

Box 6.2 John Hutchinson

John Hutchinson (1884–1972) was a British botanist associated with the Royal Botanic Gardens, Kew, England. He also served as keeper of Kew herbarium for many years. Hutchinson developed a system of classification of angiosperms which first appeared in this book entitled *The Families of Flowering Plants*–the first volume on Dicotyledones appearing in 1926 and the second on Monocotyledones in 1934. The classification was revised periodically–second edition in 1959 and the 3rd by his son in 1973, after his demise. Hutchinson also published his valuable works such as *Flora of West Tropical Africa* (1927–29), *Common Wild Flowers* (1945), *A Botanist in South Africa* (1946), *Evolution and Classification of Rhododendrons* (1946), *British Flowering Plants* (1948), *More Common Wild Flowers* (1950), *British Wild Flowers* (1955), *The Genera of Flowering Plants* (1964–67), *Key to the Families of Flowering Plants of the World* (1968) and *Evolution and Phylogeny of Flowering Plants* (1969). His desire of an ambitious plan of revising *Genera plantarum* of Bentham and Hooker, under the title of *The Genera of Flowering Plants,* could not be fulfilled and only two volumes of this work were published first in 1964 and the second 1967.

Fig. 6.10 John Hutchinson (1884–1972).

6. Perennials are older than biennials and from them annuals are derived, *e.g.* there are only a few annuals in the primitive family Ranunculaceae while there are many more in the more advanced and natural family Cruciferae.
7. Aquatic flowering plants are derived from terrestrial ancestors, epiphytes, saprophytes and parasites are more recent than plants of normal habit.

Principles relating to general structure of flowering plants :

8. Dicotyledones with collateral vascular bundles arranged in a cylinder are more primitive in origin than Monocotyledones with scattered vascular bundles, although it does not necessarily follow that Monocots are directly derived from Dicots.
9. The spiral arrangement of leaves on the stem and of the floral leaves is more primitive than the opposite and whorled arrangement.
10. Simple leaves are usually more primitive than compound leaves.

Principles relating to the flowers and fruits of plants :

11. Unisexual flowers are more advanced than bisexual and dioecious plants are more recent than monoecious.
12. Solitary flowers are more primitive than those in inflorescence, the highest forms of the latter being the number and capitulum (Umbelliferae and Compositae).
13. Spirally imbricate floral parts are more primitive than whorled and valvate. Example–*Magnolia* and *Clematis.*
14. Many parted flowers (polymerous) precede and the type with few parts (oligomerous) follows from it, being accompanied by progressive sterilisation of reproductive parts. Example—*Magnolia* and *Cheiranthus.*
15. Apetalous flowers are derived from petaliferous ones, the former condition being as a result of reduction.
16. Polypetaly (free petals) is more primitive than gamopetaly (fused petals).
17. Actinomorphy is more primitive than zygomorphy.
18. Hypogyny is the most primitive structure and from it perigyny and epigyny have been derived.
19. Apocarpy (free carpels) is more primitive than syncarpy (connate carpels). Sometimes, however, when the carpels have remained loosely united during evolution, they may again become quite free, *e.g.* Asclepiadaceae.
20. Polycarpy (many carpels) precede oligocarpy (few carpels), *e.g. Ranunculus* and *Nigella.*
21. Endospermic seeds with small embryo are older than non-endospermic seeds having a large embryo, *e.g.* Ranunculaceae and Rosaceae.
22. Indefinite stamens indicate greater primitiveness than an androecium with a few stamens only (exception Malvaceae), *Ranunculus* and *Cheiranthus.* This condition may however, be

reversed within the confines of a single family, *e.g.* Papaveraceae, where bees feed on the pollen.

23. Separate stamens precede connate ones, *e.g.* Tiliaceae and Malvaceae, Campanulaceae and Lobeliaceae.
24. Aggregate fruits are more highly evolved than single fruits. Generally, capsule precedes berry or drupe.

Table 6.4 General Trends of Characters (after Hutchinson 1959)

	Primitive	Advanced
1.	Tropical habitat	Temperate habitat
2.	Woody plants	Herbaceous or climbing plants
3.	Wood without vessel (Homoxyly)	Wood with vessels (Heteroxyly)
4.	Perennial habit	Biennial or annual habit
5.	Terrestrial habit	Aquatic, epiphytic, saprophytic or parasitic habit
6.	Cylindric arrangement of vascular bundles	Scattered vascular bundles
7.	Chlorophyll present	Chlorophyll absent
8.	Evergreen plants	Deciduous plants
9.	Stipules present	Stipules absent
10.	Node unilacunar with two traces	Node various
11.	Spiral phyllotaxy	Opposite or whorled phyllotaxy
12.	Leaves simple	Leaves compound
13.	Flowers hermaphrodite	Flowers unisexual
14.	Flowers solitary	Flowers in inflorescences
15.	Flowers entomophilous	Flowers anemophilous
16.	Floral parts spirally imbricate	Floral parts whorled or valvate
17.	Flowers polymerous	Flowers oligomerous
18.	Perianth undifferentiated	Perianth differentiated into calyx and corolla or further reduced
19.	Flowers with petals	Flowers apetalous
20.	Petals distinct	Petals united
21.	Actinomorphy	Zygomorphy
22.	Hypogyny	Perigyny and epigyny
23.	Stamens many	Stamens few
24.	Stamens separate	Stamens united
25.	Stamens 3-veined microsporophylls with linear embedded sporangia	Stamens variously modified
26.	Pollen monocolpate	Pollen tricolpate
27.	Carpels many	Carpels few
28.	Carpels distinct	Carpels united
29.	Carpels conduplicate, unsealed, styleless	Carpels variously modified
30.	Placentation laminar	Placentation various

31.	Fruits single	Fruits aggregate
32.	Fruit a follicle	Fruit various
33.	Seeds large, embryo small, endosperm abundant	Seeds small, embryo large, endosperm scanty or none
34.	Cotyledons two	Cotyledons one or three (or more)
35.	Ovule anatropous	Ovule various
36.	Integuments two	Integuments one or more
37.	Endosperm nuclear	Endosperm cellular or helobial
38.	Ovule crassinucellar	Ovule tenuinucellar
39.	Sieve cells slender and long with scattered sieve areas	Shorter sieve-tube elements with specialised sieve plates
40.	Chromosome number low; $n = 7$	High number

Hutchinson, in Dicotyledones, abolished groups Polypetalae, Gamopetalae and Monochlamydeae of Bentham and Hooker, instead he recognised two evolutionary lines : division Lignosae (fundamentally woody group) and division Herbaceae (fundamentally herbaceous group). The Lignosae, within dicots, started with family Magnoliaceae and ended with Verbenaceae. The Herbaceae, within dicots, started with Paeniaceae and ended with Labiatae (= Lamiaceae). Within Monocotyledones, he recognised three evolutionary lines : division Calyciferae (calyx-bearers), division Corolliferae (corolla-bearers) and division Glumiflorae (glume-bearers).

Hutchinson recognised a total of 411 families (342 in Dicotyledones and 69 in Monocotyledones). Lignosae includes 54 orders, Herbaceae 29 among Dicotyledones whereas among Monocotyledones, Calyciferae includes 12, Corolliferae 14 and Glumiflorae 3 orders. Although in his first publication (1926), the Dicotyledones included two subclasses, the Archichlamydeae (petals free or absent) having 59 orders and 213 families and the Metachlamydeae (petals united) 23 orders and 51 families but in the last revision (1973), minor alteration was made in the Dicotyledones wherein Lythrales was transferred from the Herbaceae to Lignosae under the order Myrtales. Similarily minor alteration was done in the Monocotyledones. Thus, there are 82 orders and 343 families of Dicotyledones and 29 orders and 69 families of the Monocotyledones. An outline of his classification and linear sequence of orders according to 1973 revision is depicted in fig. 6.11.

Whereas Hutchinson considered the woody habit to be primitive in dicots, in monocots he considered the herbaceous habit to primitive and woody forms derived from herbaceous ones. He also considered Monocotyledones to be a monophyletic group derived from Ranales, Butomales having a link with Helleboraceae and Alismatales with Ranunculaceae. The presence of endosperm in seeds of Ranunculaceae and its absence in Alismataceae and Butomaceae, otherwise considered closer, is explained by him to be the result of aquatic habit in Alismataceae and Butomaceae. **Phylograms** showing probable phylogeny of various orders in Dicotyledones and Monocotyledons are depicted in fig. (6.12).

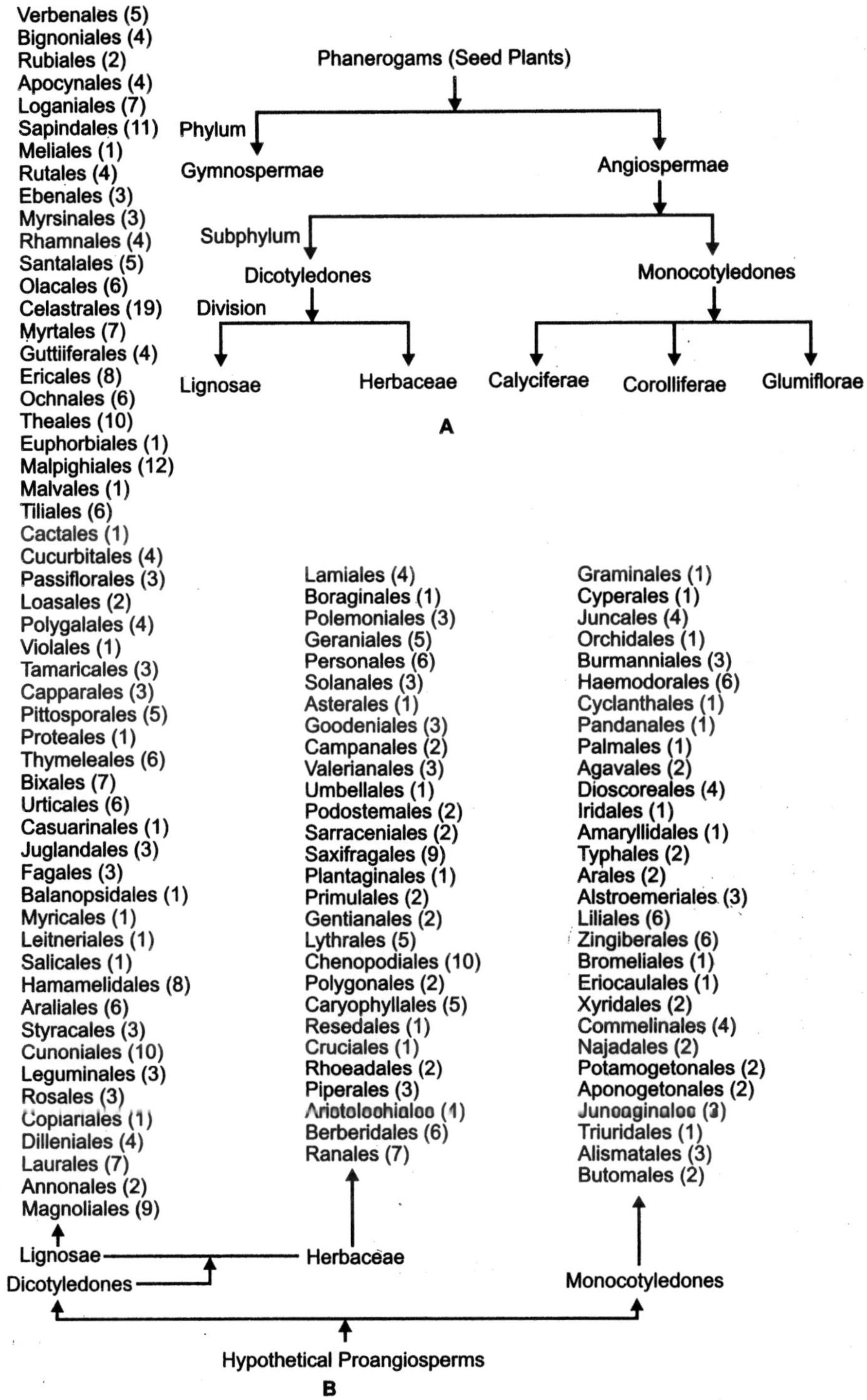

Fig. 6.11 Hutchinson's system of classification (1973) A. Outline of the system; B. Linear sequence of the orders. (The number of families indicated in parantheses)

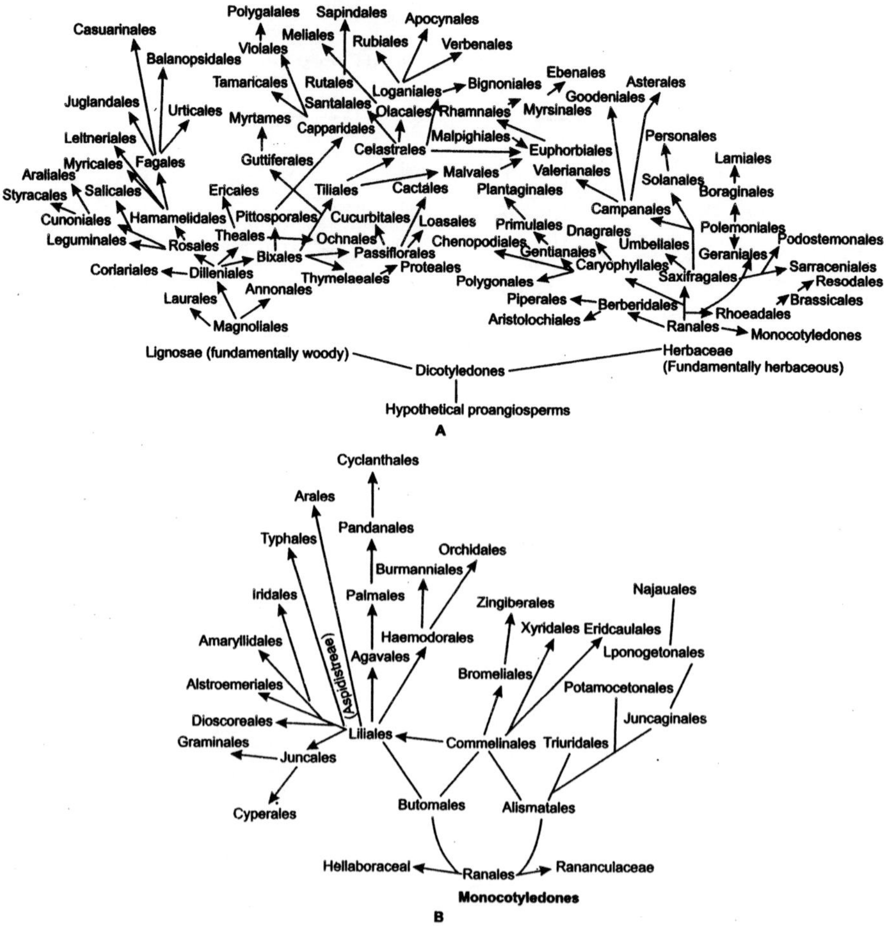

Fig. 6.12 Hutchinson's diagrams (phylograms) showing phylogeny and relationships of orders of A. Dicotyledones and B. Monocotyledones in his 1973 classification.

Table 6.5 An outline of Hutchinson's system of classification of phylum Angiospermae.

SUB PHYLUM—DICOTYLEDONES

DIVISION 1. LIGNOSAE

Order 1. **Magnoliales**

Families :	1. Magnoliaceae	2. Illiciaceae
	3. Winteraceae	4. Canellaceae
	5. Schisandraceae	6. Himantandraceae
	7. Lactoridaceae	8. Trochodendraceae
	9. Cercidiphyllaceae	

Order 2. **Annonales**

Families :	10. Annonaceae	11. Eupomatiaceae

Order 3. **Laurales**
Families : 12. Monimiaceae, 13. Austrobaileyaceae, 14. Trimeniaceae, 15. Lauraceae, 16. Gomortegaceae, 17. Hernandiaceae, 18. Myristicaceae

Order 4. **Dilleniales**
Families : 19. Dilleniaceae, 20. Connaraceae, 21. Crossosomataceae, 22. Brunelliaceae

Order 5. **Coriariales**
Family : 23. Coriariaceae

Order 6. **Rosales**
Families : 24. Rosaceae, 25. Dichapetalaceae, 26. Calycanthaceae

Order 7. **Leguminales**
Families : 27. Caesalpiniaceae, 28. Mimosaceae, 29. Papilionaceae

Order 8. **Cunoniales**
Families : 30. Pterostemonaceae, 31. Cunoniaceae, 32. Philadelphaceae, 33. Hydrangeaceae, 34. Grossulariaceae, 35. Oliniaceae, 36. Greyiaceae, 37. Escalloniaceae, 38. Baueraceae, 39. Crypteroniaceae

Order 9. **Styracales**
Families : 40. Lissocarpaceae, 41. Styracaceae, 42. Symplocaceae

Order 10. **Araliales**
Families : 43. Cornaceae, 44. Alangiaceae, 45. Garryaceae, 46. Nyssaceae, 47. Araliaceae, 48. Caprifoliaceae

Order 11. **Hamamelidales**
Families : 49. Tetracentranaceae, 50. Hamamelidaceae, 51. Myrothamnaceae, 52. Platanaceae, 53. Stachyuraceae, 54. Buxaceae, 55. Daphniphyllaceae, 56. Bruniaceae

Order 12. **Salicales**
Family : 57. Salicaceae

Order 13. **Leitneriales**
Family : 58. Leitneriaceae

Order 14. **Myricales**
Family : 59. Myricaceae

Order 15. **Balanopsidales**
Family : 60. Balanopsidaceae

Order 16. **Fagales**
Families : 61. Betulaceae, 62. Fagaceae, 63. Corylaceae

Order 17. **Juglandales**
Families : 64. Rhoipteleaceae 65. Juglandaceae
66. Picrodendraceae

Order 18. **Casuarinales**
Family : 67. Casuarinaceae

Order 19. **Urticales**
Families : 68. Ulmaceae 69. Cannabiaceae
70. Moraceae 71. Urticaceae
72. Barbeyaceae 73. Eucommiaceae

Order 20. **Bixales**
Families : 74. Bixaceae 75. Cistaceae
76. Flacourtiaceae 77. Cochlospermaceae
78. Hoplestigmataceae 79. Achatocarpaceae
80. Lacistemaceae

Order 21. **Thymelaeales**
Families : 81. Gonystylaceae 82. Aquilariaceae
83. Geissolomataceae 84. Penaeaceae
85. Thymelaeaceae 86. Nyctaginaceae

Order 22. **Proteales**
Family : 87. Proteaceae

Order 23. **Pittosporales**
Families : 88. Pittosporaceae 89. Byblidaceae
90. Stegnospermaceae 91. Vivianiaceae
92. Tremandraceae

Order 24. **Capparidales**
Families : 93. Capparidaceae 94. Moringaceae
95. Tovariaceae

Order 25. **Tamaricales**
Families : 96. Frankeniaceae 97. Tamaricaceae
98. Fouquieriaceae

Order 26. **Violales**
Family : 99. Violaceae

Order 27. **Polygalales**
Families : 100. Polygalaceae 101. Krameriaceae
102. Trigoniaceae 103. Vochysiaceae

Order 28. **Loasales**
Families : 104. Turneraceae 105. Loasaceae

Order 29. **Passiflorales**
Families : 106. Malesherbiaceae 107. Passifloraceae
108. Achariaceae

Order 30. **Cucurbitales**
Families : 109. Cucurbitaceae 110. Begoniaceae
111. Datiscaceae 112. Caricaceae

Order 31. **Cactales**
Family : 113. Cactaceae

Order 32. **Tiliales**

Family : 114. Dirachmaceae 115. Scytopetalaceae
116. Tiliaceae 117. Sterculiaceae
118. Peridiscaceae 119. Bombacaceae

Order 33. **Malvales**

Family : 120. Malvaceae

Order 34. **Malpighiales**

Families : 121. Ixonanthaceae 122. Malpighiaceae
123. Humiriaceae 124. Linaceae
125. Irvingiaceae 126. Huaceae
127. Ledocarpaceae 128. Erythroxylaceae
129. Ctenolophonaceae 130. Lepidobotryaceae
131. Balanitaceae 132. Zygophyllaceae

Order 35. **Euphorbiales**

Family : 133. Euphorbiaceae

Order 36. **Theales**

Families : 134. Bonnetiaceae 135. Theaceae
136. Sauraniaceae 137. Actinidiaceae
138. Pellicieraceae 139. Pentaphylacaceae
140. Tetrameristaceae 141. Marcgraviaceae
142. Caryocaraceae 143. Medusagynaceae

Order 37. **Ochnales**

Families : 144. Strasburgeriaceae 145. Ochnaceae
146. Rhodolaenaceae 147. Sphaerosepalaceae
148. Dipterocarpaceae 149. Ancistrocladaceae

Order 38. **Ericales**

Families : 150. Clethraceae 151. Pyrolaceae
152. Ericaceae 153. Epacridaceae
154. Diapensiaceae 155. Monotropaceae
156. Lennoaceae 157. Vacciniaceae

Order 39. **Guttiferales**

Families : 158. Hypericaceae 159. Clusiaceae
160. Eucryphiaceae 161. Quiinaceae

Order 40. **Myrtales**

Families : 162. Myrtaceae 163. Lecythidaceae
164. Rhizophoraceae 165. Sonneratiaceae
166. Punicaceae 167. Combretaceae
168. Melastomataceae

Order 41. **Celastrales**

Families : 169. Pandaceae 170. Aquifoliaceae
171. Salvadoraceae 172. Koeberliniaceae
173. Cneoraceae 174. Cardiopteridaceae
175. Cyrillaceae 176. Icacinaceae
177. Empetraceae 178. Aextoxicaceae
179. Pentadiplandraceae 180. Celastraceae
181. Corynocarpaceae 182. Stackhousiaceae

183. Goupiaceae
184. Hippocrateaceae
185. Erythropalaceae
186. Cupusiaceae
187. Scyphostegiaceae

Order 42. **Olacales**

Families : 188. Olacaceae
189. Opiliaceae
190. Octoknemaceae
191. Aptandraceae
192. Dipentodontaceae
193. Medusandraceae

Order 43. **Santalales**

Families : 194. Loranthaceae
195. Grubbiaceae
196. Santalaceae
197. Myzodendraceae
198. Balanophoraceae

Order 44. **Rhamnales**

Families : 199. Heteropyxidaceae
200. Elaeagnaceae
201. Rhamnaceae
202. Vitaceae

Order 45. **Myrsinales**

Families : 203. Myrsinaceae
204. Theophrastaceae
205. Aegicerataceae

Order 46. **Ebenales**

Families : 206. Ebenaceae
207. Sapotaceae
208. Sarcospermataceae

Order 47. **Rutales**

Families : 209. Rutaceae
210. Simaroubaceae
211. Burseraceae
212. Averrhoaceae

Order 48. **Meliales**

Family : 213. Meliaceae

Order 49. **Sapindales**

Families : 214. Melianthaceae
215. Sapindaceae
216. Podoaceae
217. Sabiaceae
218. Anacardiaceae
219. Aceraceae
220. Hippocastanaceae
221. Staphyleaceae
222. Akaniaceae
223. Julianiaceae
224. Didiereaceae

Order 50. **Loganiales**

Families : 225. Potaliaceae
226. Loganiaceae
227. Buddleiaceae
228. Antoniaceae
229. Spigeliaceae
230. Strychnaceae
231. Oleaceae

Order 51. **Apocynales**

Families : 232. Plocospermaceae
233. Apocynaceae
234. Periplocaceae
235. Asclepiadaceae

Order 52. **Rubiales**

Families : 236. Dialypetalanthaceae
237. Rubiaceae

Order 53. **Bignoniales**

Families : 238. Cobaeaceae
239. Bignoniaceae
240. Pedaliaceae
241. Martyniaceae

Order 54. **Verbenales**

Families : 242. Ehretiaceae 243. Verbenaceae
244. Stilbeaceae 245. Chloranthaceae
246. Phrymaceae

DIVISION 2. HERBACEAE

Order 55. **Ranales**

Families : 247. Paeoniaceae 248. Helleboraceae
249. Ranunculaceae 250. Nymphaeaceae
251. Podophyllaceae 252. Ceratophyllaceae
253. Cabombaceae

Order 56. **Berberidales**

Families : 254. Sargentodoxaceae 255. Lardizabalaceae
256. Menispermaceae 257. Nandinaceae
258. Circaesteraceae 259. Berberidaceae

Order 57. **Aristolochiales**

Families : 260. Aristolochiaceae 261. Hydnoraceae
262. Cytinaceae 263. Nepenthaceae

Order 58. **Piperales**

Families : 264. Piperaceae 265. Saururaceae
266. Choranthaceae

Order 59. **Rhoeadales**

Families : 267. Papaveraceae 268. Fumariaceae

Order 60. **Cruciales**

Family : 269. Cruciferae

Order 61. **Resedales**

Family : 270. Resedaceae

Order 62. **Caryophyllales**

Families : 271. Elatinaceae 272. Molluginaceae
273. Caryophyllacaceae 274. Ficoidaceae
275. Portulacaceae

Order 63. **Polygonales**

Families : 276. Polygonaceae 277. Illecebraceae

Order 64. **Chenopodiales**

Families : 278. Barbeuiaceae 279. Phytolaccaceae
280. Gyrostemonaceae 281. Agdestidaceae
282. Petiveriaceae 283. Chenopodiaceae
284. Amaranthaceae 285. Cyanocrambaceae
286. Batidaceae 287. Basellaceae

Order 65. **Lythrales**

Families : 288. Lythraceae 289. Onagraceae
290. Trapaceae 291. Haloragidaceae
292. Callitrichaceae

Order 66. **Gentianales**

Families : 293. Gentianaceae 294. Menyanthaceae

Order 67. **Primulales**
Families : 295. Primulaceae 296. Plumbaginaceae
Order 68. **Plantaginales**
Family : 297. Plantaginaceae
Order 69. **Saxifragales**
Families : 298. Crassulaceae 299. Cephalotaceae
300. Saxifragaceae 301. Eremosynaceae
302. Vahliaceae 303. Francoaceae
304. Donatiaceae 305. Parnassiaceae
306. Adoxaceae
Order 70. **Sarraceniales**
Families : 307. Droseraceae 308. Sarraceniaceae
Order 71. **Podostemales**
Families : 309. Podostemaceae 310. Hydrostachyaceae
Order 72. **Umbellales**
Family : 311. Umbelliferae
Order 73. **Valerianales**
Families : 312. Valerianaceae 313. Dipsacaceae
314. Calyceraceae
Order 74. **Campanales**
Families : 315. Campanulaceae 316. Lobeliaceae
Order 75. **Goodeniales**
Families : 317. Goodeniaceae 318. Brunoniaceae
319. Stylidiaceae
Order 76. **Asterales**
Family : 320. Compositae
Order 77. **Solanales**
Families : 321. Solanaceae 322. Convolvulaceae
323. Nolanaceae
Order 78. **Personales**
Families : 324. Scrophulariaceae 325. Acanthaceae
326. Gesneriaceae 327. Orobanchaceae
328. Lentibulariaceae 329. Columelliaceae
Order 79. **Geraniales**
Families : 330. Geraniaceae 331. Limnanthaceae
332. Oxalidaceae 333. Tropaeolaceae
334. Balsaminaceae
Order 80. **Polemoniales**
Families : 335. Polemoniaceae 336. Hydrophyllaceae
337. Cuscutaceae
Order 81. **Boraginales**
Family : 338. Boraginaceae
Order 82. **Lamiales**
Families : 339. Myoporaceae 340. Selaginaceae
341. Globulariaceae 342. Labiatae

SUB PHYLUM—MONOCOTYLEDONES

DIVISION 1. CALYCIFERAE

Order 83. **Butomales**
Families : 343. Butomaceae 344. Hydrocharitaceae
Order 84. **Alismatales**
Families : 345. Alismataceae 346. Scheuchzeriaceae
347. Petrosaviaceae
Order 85. **Triuridales**
Families : 348. Triuridaceae
Order 86. **Juncaginales**
Families : 349. Juncaginaceae 350. Lilaeaceae
351. Posidoniaceae
Order 87. **Aponogetonales**
Families : 352. Aponogetonaceae 353. Zosteraceae
Order 88. **Potamogetonales**
Families : 354. Potamogetonaceae 355. Ruppiaceae
Order 89. **Najadales**
Families : 356. Zannichelliaceae 357. Najadaceae
Order 90. **Commelinales**
Families : 358. Commelinaceae 359. Cartonemataceae,
360. Flagellariaceae 361. Mayacaceae
Order 91. **Xyridales**
Families : 362. Xyridaceae 363. Rapateaceae
Order 92. **Eriocaulales**
Family : 364. Ericaulaceae
Order 93. **Bromeliales**
Family : 365. Bromeliaceae
Order 94. **Zingiberales**
Families : 366. Musaceae 367. Strelitziaceae
368. Lowiaceae 369. Zingiberaceae
370. Cannaceae 371. Marantaceae

DIVISION 2. COROLLIFERAE

Order 95. **Liliales**
Families : 372. Liliaceae 373. Tecophilaeaceae
374. Trilliaceae 375. Pontenderiaceae
376. Smilacaceae 377. Ruscaceae
Order 96. **Alstroemeriales**
Families : 378. Alstroemeriaceae 379. Petermanniaceae
380. Philesiaceae
Order 97. **Arales**
Families : 381. Araceae 382. Lemnaceae
Order 98. **Typhales**
Families : 383. Sparganiaceae 384. Typhaceae

Order 99. **Amaryllidales**
Family : 385. Amaryllidaceae
Order 100. **Iridales**
Family : 386. Iridaceae
Order 101. **Dioscoreales**
Families : 387. Stenomeridaceae 388. Trichopodaceae
389. Roxburghiaceae 390. Dioscoreaceae
Order 102. **Agavales**
Families : 391. Xanthorrhoeaceae 392. Agavaceae
Order 103. **Palmales**
Family : 393. Palmae
Order 104. **Pandanales**
Family : 394. Pandanaceae
Order 105. **Cyclanthales**
Family : 395. Cyclanthaceae
Order 106. **Haemodorales**
Families : 396. Haemodoraceae 397. Hypoxidaceae
398. Velloziaceae 399. Apostasiaceae
400. Taccaceae 401. Philydraceae
Order 107. **Burmanniales**
Families : 402. Burmanniaceae 403. Thismiaceae
404. Corsiaceae
Order 108. **Orchidales**
Family : 405. Orchidaceae

DIVISION 3. GLUMIFLORAE

Order 109. **Juncales**
Families : 406. Juncaceae 407. Thurniaceae
408. Centrolepidaceae 409. Retionaceae
Order 110. **Cyperales**
Family : 410. Cyperaceae
Order 111. **Graminales**
Family : 411. Gramineae

Merits of the System

Hutchinson's system of classification is based on a number of sound phylogenetic principles and studies on a large number of plants at his disposal at Kew Garden and Herbaria and his extensive survey and critical study of plants, particularly in South Africa. His system has the following improvement over systems suggested earlier :

1. The classification is an intentional phylogenetic one, based on sound basis of phylogeny. It is an improvement over Engler and Prantl's system.
2. The system is based on phylogenetic principles and in conformity with the modern views which is generally recognised by all workers.

3. In this system, the orders and families are of small size and comprise only very closely related genera.
4. For widespread separation, Hutchinson has increased the number of the families to 411. For example Euphorbiaceae of Bentham and Hooker are split into Euphorbiaceae, Ricinaceae and Buxaceae and family Urticaceae into Urticaceae, Moraceae, Ulmaceae and Cannabinaceae.
5. The placement of Magnoliales in Lignosae and Ranales in Herbaceae of Dicotyledones in the beginning in the evolutionary series is in agreement with prevalent views.
6. The abolition of Polypetalae, Gamopetalae and Monochlamydeae of Bentham and Hooker and Archichlamydeae and Metachlamydeae and rearrangement of taxa on set of characters and not on one or few characters as in earlier system is more logical.
7. Description standard is very high and useful keys are provided for the identification of families.
8. The phylograms for dicots and monocots are much more superior than Bessey's cactus.
9. The classification of Monocotyledones, for which even keys are provided, is sounder and generally appreciated world over.
10. His derivation of Monocotyledones from Dicotyledones is widely agreed upon by many of the taxonomists.
11. Placement of Alismatales in the beginning of Monocotyledones finds general acceptace.
12. He provided a detailed classification, upto the generic level, given the keys and descriptions and accommodated all genera in two volumes of his *The Genera of Flowering Plants.*

Demerits of the System

Hutchinson system of classification has not been liked by many of the taxonomists because it did not proceed well beyond family level. Also he gave much importance to habit. Some of the demerits of the system are given below :

1. Description of families is very brief.
2. The system is not much useful from the point of view of plant identification, in floras and herbaria as it does not proceed well beyond family level in most of the taxa.
3. Division of Dicotyledones into Lignosae and Herbaceae, which is most artificial, seems to be superficial and illogical.
4. Many closely related natural groups (families) have been vivisected unnecessarily, *e.g.* Araliaceae in Lignosae and Apiaceae in Herbaceae. Similarily, Lamiaceae and Verbenace, which are very closely related, and are often placed in the same order in contemporary systems of classification, have been separated on the basis of habit. They have been separated not only in distinct orders but in separate divisions–Herbaceae and Lignosae, respectively.

5. Hutchinson could not provide satisfactory explanation for majority of his evolutionary concepts.
6. He divided Dicotyledones into Lignosae and Herbaceae on the basis of woody and herbaceous evolutionary lines, respectively and considered herbaceous habit derived from woody habit but later Bailey, Sinnot and others, on the basis of overall resemblances and critical anatomical studies, have shown that reverse, *i.e.*, woody habit derived from herbaceous habit can also occur occasionally. The two evolutionary lines, therefore, can not be considered distinct.
7. Monocotyledones are placed after Dicotyledones but in recent classifications they are placed between primitive angiosperms and Eudicots.
8. The system recognises Magnoliaceae as the most primitive family of living Dicotyledones but most contemporary phylogenetists consider vessel-less Winteraceae or Paleoherbs as most primitive forms.
9. Placing family Calycanthaceae in Rosales, which is related to Laurales, appears non-convincing.
10. In practice of splitting unnatural families into natural ones, Hutchinson has also splitted Ranunculaceae into Ranunculaceae and Helleboraceae on the basis of achene and follicle fruit, respectively, but studies on floral anatomy have shown that evolutionary stages in the reduction of ovule number can be seen in many genera of Helleboraceae and Ranunculaceae show traces which would have gone to now aborted ovules. Thus, Ranunculaceae (of Bentham and Hooker) represents a monophyletic group and need not to be split.
11. Hutchinson's family Liliaceae, a large unnatural assemblage, has been split into several smaller monophyletic families such as Liliaceae, Alliaceae, Asparagaceae and Asphodelaceae in recent classification of Judd, *et al.* (2002), APG II (2003) and Thorne (2003).

CONTEMPORARY PHYLOGENETIC SYSTEMS

A number of contemporary workers are involved in improving systems of classification based on new information available from various sources. Recent data from palaeobotany, phytochemistry and ultrastructure and their numerical analysis by improved techniques, have proved to be quite useful in developing systems of classification that have several features in common, though differing in some concepts. The contemporary phylogenetists have now generally agreed upon that angiosperms are a monophyletic group with dicots being more primitive than monocots and vessel-less Winteraceae and Paleoherbs as among the basal living angiosperms. There is, however, a positive trend of frequent updating of systems of classification in electronic versions. Since the last decade, the Angiosperm Phylogeny Group (APG) has been working actively towards realisation of monophyletic groups.

Armen Takhtajan (borne 1910)

Armen Takhtajan (fig. 6.13), a leading Russian plant taxonomist and Chief of the Department of Higher Plants in VL Komarov Botanic Institute, USSR Academy of Sciences, Leningrad (now named St. Petersburg) is an international authority on phytogeography and origin and phylogeny of flowering plants. He was the President of the 12th International Botanical Congress held at Leningrad in 1975. He developed a preliminary phylogenetic diagram of the orders of higher plants presented in 1942 in a paper on structural types of gynoecium and placentation. The actual system of classification, based on phylogenetic principles (1953), was first published in 1954 in the Russian language but came to be known outside the Soviet Union only after its English translation *Origin of Angiospermous Plants* published in 1958. Later he traced the evolution of angiosperms and elaborated his system in *Die Evolution der Angiospermen* (1959) and proposed a new system in *Systema et Phylogenia Magnoliophytorum* (1966) in Russian. The English translation of which as *Flowering Plants–Origin and Dispersal* (1969), by C Jeffrey, made his classification quite popular. Subsequently major revisions of his classification were made in 1980 and 1987 and the last version was published in 1997. The number of various taxa recognised by Takhtajan in 1980, 1987 and 1997 are given in Table 6.6.

Fig. 6.13 Armen Takhtajan (borne 1910).

Table 6.6 Number of Subclasses, Superorders, Orders and Families in Magnoliopsida (dicots) and Liliopsida (monocots) according to Takhtajan's Classification in 1980, 1987 and 1997.

Year	Class	Number of Taxa			
		Subclass	Superorder	Order	Family
1980	Magnoliopsida	7	20	71	333
	Liliopsida	3	8	21	77
1987	Magnoliopsida	8	37	128	429
	Liliopsida	4	16	38	104

(Contd....)

1997	Magnoliopsida	11	52	175	458
	Liliopsida	6	16	57	131

Takhtajan classified angiosperms upto the family level. He belongs to Besseyan School and is strongly influenced by Hutchinson, Hallier and other more progressive German botanists. He believes in the **monophyletic origin** of angiosperms–the group having evolved from the seed ferns Lyginopteridophyta. According to him, the angiosperms are of **neotenous origin** (retention of juvenile characters in the adult plant also–**paedomorphosis**). He divided division Magnoliophyta (in which he placed angiosperms) into two classes : **Magnoliopsida** (dicots) and **Liliopsida** (monocots). Takhtajan considered Magnoliopsida as primitive and Liliopsida to have been derived from Magnoliales under Magnoliopsida. The two classes have again been divided in turn into subclasses, superorders, orders and families. He accounted 13,000 genera (10,000 in Magnoliopsida and 3,000 in Liliopsida) and 2,50,000 species (1,90,000 in Magnoliopsida and 60,000 in Liliopsida). The outline of the classification of angiosperms proposed by Takhtajan in 1997 is given in Table 6.7.

Table 6.7 Outline of the system of classification of Angiosperms proposed by Takhtajan (1997) upto the level of super-orders (S. Orders).

DIVISION. MAGNOLIOPHYTA (ANGIOSPERMS)

Class 1. MAGNOLIOPSIDA (DICOTYLEDONS)

Subclass 1. Magnoliidae

S. Order : 1. Magnolianae 2. Lauranae 3. Rafflesianae 4. Balanophoranae

Subclass 2. Nymphaeidae

S. Order : 1. Nymphaeanae 2. Ceratophyllianae

Subclass 3. Nelumbonidae

S. Order : 1. Nelumbonaceae

Subclass 4. Ranunculidae

S. Order : 1. Ranunculanae

Subclass 5. Caryophyllidae

S. Order : 1. Caryophyllanae 2. Gyrostemonanae 3. Polygonanae 4. Plumbaginanae

Subclass 6. Hamamelididae

S. Order : 1. Trochodendranae 2. Myrothamnanae 3. Hamamelidanae 4. Barbeyanae 5. Buxanae 6. Faganae 7. Casuarinanae 8. Juglandanae

Subclass 7. Dilleniidae

S. Order : 1. Dillenianae 2. Theanae 3. Sarracinanae 4. Nepenthanae 5. Ericanae 6. Primulanae 7. Violanae 8. Malvanae 9. Urticanae 10. Euphorbianae

Subclass 8. Rosidae

S. Order : 1. Saxifraganae 2. Rosanae 3. Rhizophoranae 4. Myrtanae 5. Fabanae 6. Rutanae 7. Geranianae 8. Corynspermanae 9. Celastranae 10. Santalanae 11. Rhamnanae 12. Vitanae

Subclass 9. Cornidae

S. Order : 1. Cornanae 2. Aralianae 3. Dipsacanae

Subclass 10. Asteridae

S. Order : 1. Campanulanae 2. Asternae

Subclass 11. Lamiidae

S. Order : 1. Gentiananae 2. Solananeae 3. Oleanae 4. Lamianae

Class 2. LILIOPSIDA (MONOCOTYLEDONS)

Subclass 1. Liliidae

S. Order : 1. Lilianae 2. Dioscoreanae

Subclass 2. Commelinidae

S. Order : 1. Bromelianae 2. Pontederinae 3. Zingiberanae 4. Commelinanae 5. Hydatellanae 6. Juncanae 7. Roanae

Subclass 3. Arecidae

S. Order : 1. Arecanae

Subclass 4. Alismatidae

S. Order : 1. Alismatanae

Subclass 5. Triurididae

S. Order : 1. Triuridanae

Subclass 6. Aridae

S. Order : 1. Aranae 2. Cyclanthanae 3. Pandananae 4. Typhanae

---

Takhtajan considers that angiosperms arose under environmental stress, probably as a result of adaptation to moderate seasonal drought on rocky mountain slopes, in an area with monsoon climate. For many years he considered Winteraceae along with Degeneriaceae to represent the most primitive angiosperms but finally he believed Degeneriaceae as the most primitive family, placed under order Magnoliales. Takhtajan shifted Winteraceae to a separate order Winterales after Magnoliales. In his earlier classification (1987), he had placed Winteraceae towards the beginning in the phyletic line after orders Eupomatiales and Annonales. His keeping of Winteraceae coincides with the view of many contemporary authors who

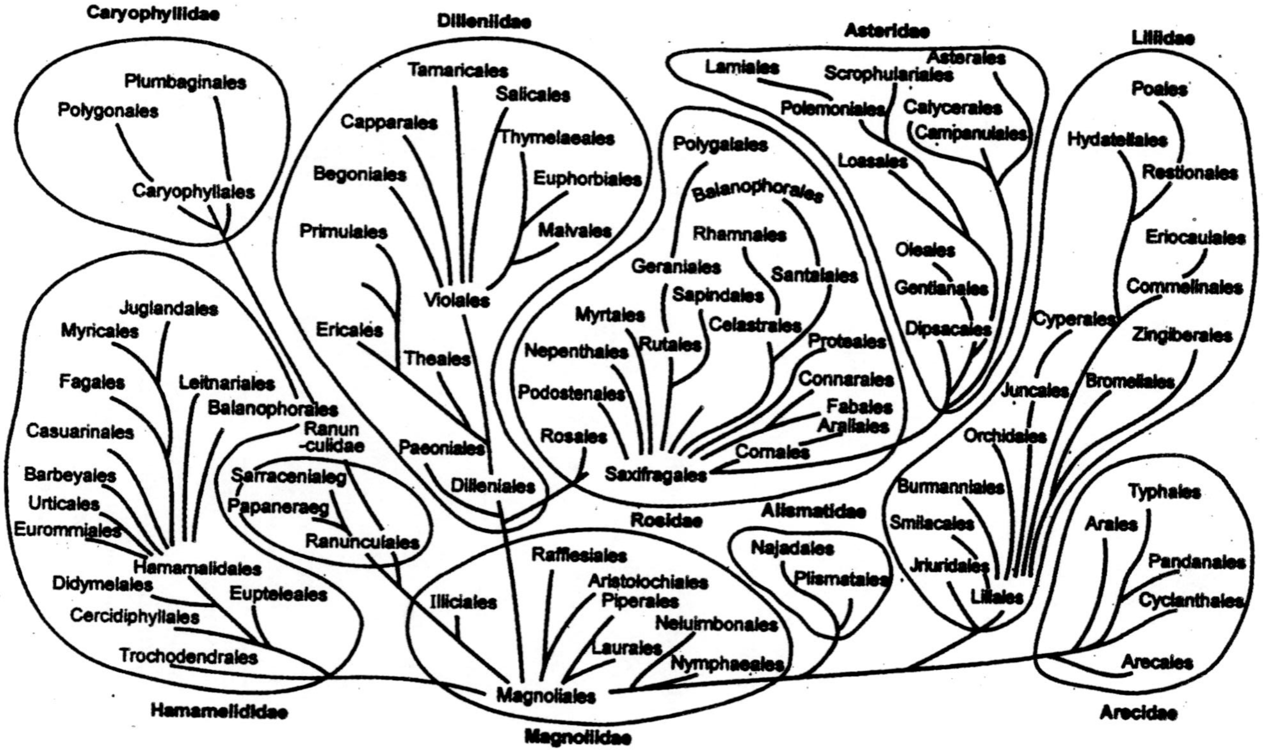

Fig. 6.14 Bubble diagram of Takhtajan showing the probable relationship between different subclasses and orders of dicotyledons and monocotyledons (based on Takhtajan, 1980, 1987, 1997).. Classification not included in the bubble diagram.

regard Winteraceae or paleoherbs (mainly Amborellaceae) among the most primitive groups of the living angiosperms.

Among subclasses, Takhtajan considers Magnoliidae to be the most primitive, forming the basal group from which all other subclasses have been derived (fig. 6.14). Among Magnoliopsida (dicots), he considers Asteridae as the most advanced. The three subclasses : Liliidae, Alismatidae and Arecidae are considered to be primitive than the other subclasses of Liliopsida (monocots) and to have been derived from Magnoliidae. Among Magnoliophyta (angiosperms), Winteraceae are considered to be most primitive and Poaceae the most advanced.

Takhtajan's Principles

Takhtajan system is based on 67 phyletic principles for evaluating the relative degree of advancement of flowering plants. The summary of these principles is given as :

1. **Growth habit** : The evolution of angiosperms began with small, relatively weakly branched woody forms. Large trees have been regarded to be originated from primitive small woody angiosperms. Thus woody plants are primitive than herbaceous and deciduous woody plants are considered evolved from evergreen plants.
2. **Leaves** : Palmate venation is considered to be evolved from pinnate venation. Parallel venation is advanced than reticulate and alternate leaf arrangement is the most primitive.
3. **Stomatal apparatus** : The general trend in evolution is from stomata with subsidiary cells to those lacking subsidiary cells.
4. **Nodal structure** : The unilacunar nodal structure has been derived from tri-pentalacunar type.
5. **Wood anatomy** : The evolution of vessels is from vessels with scalariform perforations to vessels with simple perforations.
6. **Inflorescence** : Cymose inflorescence is primitive; by repeated branching, the simple cyme gives rise to compound cyme and from the latter evolved the raceme.
7. **Floral structure** : During the course of evolution, due to progressive shortening of floral axis, there was gradual transition from spiral to cyclic arrangement and to the fixation of the number of floral parts.
8. **Androecium** : The stamens of the earliest angiosperms were leaf-like pinnate microsporophylls with marginally situated microsporangia.
9. **Pollen grains** : The main trend of evolution of dicot pollen is from monocolpate to tricolpate and from tricolpate to triporate.
10. **Gynoecium** : Apocarpous condition is most primitive; primitive carpels are conduplicate and with large number of ovules.
11. **Ovules** : Unitegmic ovules arose from the bitegmic; tenuinucellate ovules evolved from the primitive crassinucellate onces. Thus bitegmic ovules are primitive than unitegmic. Mesogamic and chalazogamic conditions have evolved from porogamic condition.

12. **Pollination** : Anemophily arose from entomophily.
13. **Gametophyte** : The tetrasporic type embryo sac evolved from the primitive monosporic. 8-nucleate *Polygonum*—type female gametophyte.
14. **Seeds** : Primitive seeds were 5–10 mm long, with abundant endosperm, minute undifferentiated embryo; seed coat with multiplicate integuments and well developed sarcotesta, is the basic primitive type.
15. **Fruits** : Many seeded follicle, developing from a multicarpellate, apocarpous gynoecia, is the most primitive and basic fruit type.

Merits of the System

In light of recent information on phylogeny and phenetics, the later classification of Takhtajan given in 1997, shows several improvements. Not only this, in his latest version, he has also retained many merits achieved in his earlier versions of 1980 and 1987. Important merits of his system are :

1. If reflects evolutionary kinship, better than the earlier systems.
2. It is based on 67 clear phyletic principles designed as against 24 of Hutchinson.
3. It considers angiosperms to be monophyletic and monocots to have originated from primitive dicots.
4. Angiosperms are treated as one division Magnoliophyta divided into Magnoliopsida (dicots) and Liliopsida (monocots).
5. The subclass Magnoliidae is considered as the most basic group from which other subclasses have originated.
6. Families are small homogeneous units, comprising closely related genera.
7. Subclasses are not based on a single character but on consideration of all known information.

Demerits of the System

Despite Takhtajan tried to remove deficiencies of his system in 1997 version, some of the defect are still left. These include :

1. The main drawback of his system lies in his derivation of monocots from the stock ancestral to Nymphaeales.
2. The extremely narrowly defined taxa have resulted in the unwarranted splitting of related groups.
3. The system prevails a controversy for Takhtajan's idea of Cradle of Angiosperms.
4. The system, although very sound and highly phylogenetic, is not helpful in identification and for adoption of herbaria, as classification is upto family level.
5. Bubble diagram for his latest version of 1997 is not available.
6. Monocots are placed after dicots, whereas the recent systems of classification place them between primitive angiosperms and Eudicots.

7. The families Winteraceae and Canellaceae are in two separate orders whereas multigene analysis (Solti, *et al.* 1999; Zanis, *et al.* 2002, 03) have provided 99–100 per cent **bootstrap** or **jackknife** support in their relationship. In APG II and AP web. Also these are placed in the same order and under the same suborder by Thorne (2003).

Arthur Cronquist (1919–92)

Arthur Cronquist (1919–92), an internationally renowned American taxonomist, first proposed his system of classification in 1957. Later in 1966, in collaboration with Takhtajan and Zimmerman, he produced a broad classification of Embryobionta. In 1968, he proposed a detailed classification of angiosperms in his book entitled *The Evolution and Classification of Flowering Plants.* Cronquist further elaborated his system in 1981 in his book entitled *An Integrated System of Classification of Flowering Plants* and the final revision of the system was published in the second edition (1988) of his book *The Evolution and Classification of Flowering Plants.*

Cronquist's system of classification is quite elaborative and is first of its kind after Bessey. It parallels with Takhtajan to a considerable extent but differs in details. Like Takhtajan, his classification is based on evidences derived from all sources, emphasising phylogenetic features of plant habit, stem, leaves, stomatal apparatus, node, flower, pollination, pollen grain, placentation, ovule and embryo. His system is based on a conceptual synthesis which best reflects the totality of similarities and differences among angiosperms. In his system, the phylogenetic interpretations of seggregation of taxa are mostly original.

Cronquist recognised a total of 83 orders, 383 families and about 2,15,000 species under Magnoliophyta (Angiosperms) of which Magnoliopsida (Dicots) includes 64 orders, 318 families and about 1,65,000 species and Liliopsida (Monocots) 19 orders, 65 families and about 50,000 species. In Magnoliopsida, the Ranunculidae of Takhtajan are merged with Magnoliidae and Lamiidae are not given a separate rank at subclass level but are retained in Asteridae. In Liliopsida, Zingiberidae are treated separate from Liliidae and Triuridales are kept under Alismatidae. Cronquist kept Winteraceae (and not Degeneriaceae as done by Takhtajan) at the beginning of Magnoliposida as was done by Thorne (1992) and included along with Degeneriaceae, Magnoliaceae, Annonaceae etc. in the same order Magnoliales. Paeoniaceae, unlike other authors, are not separated into a distinct order Paeoniales but instead are shifted to order Delleniales under Dilleniidae.

In contrast to Takhtajan's system, Nelumbonaceae are placed in Nyphaeales (and not as separate order Nelumbodales), Typhales in Commelinidae (and not in Arecidae) and sympetalous families of Magnoliopsida in a large subclass Asteridae (and not three subclasses Asteri-

Box 6.3 **Arthur Cronquist**

Arthur Cronquist was born on 19 March, 1919 in San Jose, California and grew up in Portland, Oregon and Poatello, Idaho. Since his childhood be became interested in the natural history of mountains. For his doctoral degree he went to the University of Minnesota, doing a revision of the North American species of *Erigeron*. In 1943, he joined as Assistant Curator at New York Botanical Garden. In between, during the periods of 1946–1951, he left the garden to teach botany at the University of Georgid (1946–48) and Washington State University (1948–51). Finally he returned back to the garden in 1952 and got greatly impressed by Henry A Gleason. He became interested in floristics and published *New Britton and Brown Illustrated Flora* in 1952 in collaboration with Henry A Gleason. To this *Flora*, Cronquist contributed the section on Asteraceae and *Manual of Vascular Plants of Northeastern United States and Adjacent Canada* (1963)–an encyclopaedic reference and field book whose intellectual genealogy can be traced to National Lord Britton and Adison Brown's original *Illustrated Flora* of 1996–1998 and is known to botany students as the **Green Bible**. He also jointly published the book The *Natural Geography of Plants* (1964) with Henry A Gleason. Cronquist made extensive surveys throughout western intermountaneous region of North America and published *Intermountain Flora* Vol. I in 1972 in collaboration with other taxonomists.

Fig. 6.15 Arthur Cronquist (1919–92).

In later 1950's, he got associated with Arman Takhtajan of Russia that proved to be critically a stimulus in the development of his system of classification. As an educator, Cronquist wrote two finest botany textbooks in 1960's. His *Introductory Botany* and *Basic Botany* gained wide popularity and got translated into several languages. By this time he had moved into the category of senior statesman for botany. He delivered several lectures in different parts of world, was sought as a consultant and wrote numerous articles on evolution and phylogeny and botany in general. He contributed significantly to national and international organisations and in 1973 was nominated as President, Botanical Society of America, honorary vice-president of XII International Botanical Congress, Leningrad (1975) and received the Medal for Botany in 1986 from Linnean Society of London. He died on March 22, 1992 while he was studying plant specimens in the herbarium of Brigham Young University, Utah.

dae, Cornidae and Lamiidae). Urticales are included along wind-pollinated families under Hamamelididae (and not with its related order Malvales and Euphorbiales) and Malvales and Euphorbiales are kept in separate subclasses Dilleniidae and Rosidae, respectively (and not in the same subclass Dilleniidae). In Liliopsida, Amaryllidaceae has been merged with Liliaceae. The two families Amaryllidaceae and Liliaceae have been placed in two separate orders Amaryllidales and Liliales, respectively. Cronquist believed in aquatic origin of Liliopsidan members from primitive vessel-less ancestors resembling present day Nymphaeales. He has provided a synoptic arrangement of taxa facilitating identification upto family level.

The system, besides providing probable evolutionary trends among angiosperms, has blending of philosophical concepts behind them. The problems, such as the phenomenon of evolutionary parallelism and relationship between phylogeny and taxonomy, are well explained in his system. He has discussed a wide range of characters important to phylogenetic system and reviewed various taxa with reference to display to such characters. Cronquist has provided synoptical keys designed to bring the taxa into an alignment suited to his system and gave charts showing relationship of orders within various subclasses. He has considered seed ferns (Pteridosperms) as probable ancestors of angiosperms which later followed parallel development and has rightly pointed out that "*many of the evolutionary trends bear little apparent relation to survival value and that there are some reversals.*"

Following Takhtajan, the angiosperms are given the name Magnoliophyta which is divided into two classes : Magnoliopsida (dicots) and Liliopsida (monocots) of unequal size. Cronquist recognises six subclasses in Magnoliopsida (dicots) and five in Liliopsida (monocots). Superorders, however, have not been recognised.

Class : Magnoliopsida (Dicots)	Class : Liliopsida (Monocots)
1. Magnoliidae	1. Alismatidae
2. Hamamelidae	2. Arecidae
3. Caryophyllidae	3. Commelinidae
4. Dilleniidae	4. Zingiberidae
5. Rosidae	5. Liliidae
6. Asteridae	

Outline of the Classification (1988) Upto the Level of Families

All taxonomic circumscriptions in the system are based on common or 'typical' key characters. The key characters, used to divide the two classes : Magnoliopsida and Liliopsida, are given below :

Key Characters	Magnoliopsida	Liliopsida
Number of cotyledons	Usually 2	Usually 1
Root System	Primary and adventitious	Adventitious only
Leaves	Usually veined reticulately	Usually veined parallely
Vascular Cambium	Often present	Absent
Primary Vascular Bundles	Arranged in a ring	Often scattered
Pollen	Various types	Monosulcate
Floral Parts	Usually in sets of 4s or 5s	Usually in sets of 3

Class : MAGNOLIOPSIDA

The key characters used to circumscribe six subclasses : Magnoliidae, Hamamelidae, Caryophyllidae, Dilleniidae, Rosidae and Asteridae, are as follows :

Subclass	Characteristic features	Number of Orders	Number of Families	Number of Species
Magnoliidae	Archaic, flowers many-parted	8	39	11,000
Hamamelidae	Ancient, floral parts reduced, mostly wind pollinated	11	24	3,400
Caryophyllidae	Herbs with betalains and 'centered' placentation types	3	14	11,000
Dilleniidae	Some sympetalous, few apocarpous	13	78	25,000
Rosidae	Polypetalous, often numerous stamens	18	114	58,000
Asteridae	Mostly sympetalous	11	49	56,000

The subclass Magnoliidae consists mainly of the families of the traditional Ranales and is regarded as the basal complex from which other angiosperms are considered to have been derived. The Hamamelidae is regarded to consist of a group of wind-pollinated families with reduced, mainly apetalous flowers, that are often borne in catkins (the traditional Amentiferae). The families included in Caryophyllidae have a strong tendency towards free central placentation and have betalains (nitrogenous anthocyanins and anthoxanthins) in many of them. The Dilleniidae and Rosidae, that are morphologically very similar, are considered to be parallel groups. The Asteridae includes the higher sympetalous families, that are mostly derived from the Rosidae (fig. 6.16).

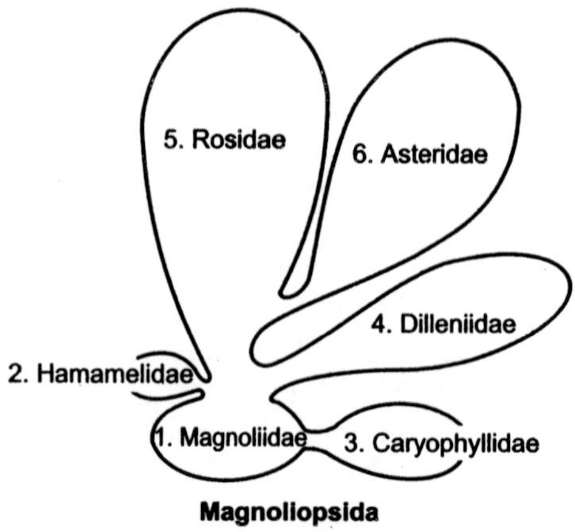

Fig. 6.16 Depiction of phyletic relationships among six subclasses of class Magnoliopsida (Redrawn from Cronquist, 1988).

The Magnoliidae includes extant dicot elements that carry the greatest number of archaic features (similar to 'original' flowering plants) and the Asteridae (sunflowers and relatives) as the most derived or specialised element of the subclass Magnoliidae. The size of **balloons** is roughly proportional to the number of species included in respective subclass (fig. 6.16).

Subclass 1. Magnoliidae

Magnoliidae includes 8 orders, 39 families and about 11,000 species. These families show a tendency towards retention of primitive features and regardless of classification system, most are placed at the base of the dicot lineage. Primitive features are :

1. A woody dicot-like plant : may be arborescent, but more likely a fruticose, shrubby plant of seasonally dry, sub-tropical habit.
2. Leaves persistent (evergreen), simple, alternate, entire, pinnately veined, glabrous and probably stipulate.
3. Flowers solitary, terminal, perfect, actinomorphic, with no connation or adnation. The numerous floral parts spirally arranged on an elongated floral axis.
4. Perianth many, parted with poor differentiation between appendages of the corolla and calyx (tepals).
5. Androecium included many stamens that were broad and laminar with no distinct filament.
6. Gynoecium apocarpous and produced follicle.
7. The flowers probably adapted for insect pollination (entomophilous), most likely by beetles, as opposed to wind pollination (anemophilous).

Nearly similar characters are observed in the first largest order Magnoliales which includes ten families. The terminal group–the order Papaverales, on the other hand, is made up of herbaceous plants. The orders and families included in this subclass are :

Subclass 1. MAGNOLIIDAE

Order 1. **Magnoliales**

Families : 1. Winteraceae 2. Degeneriaceae 3. Himantandraceae
4. Eupomatiaceae 5. Austrobaileyaceae
6. Magnoliaceae 7. Lactoridaceae 8. Annonaceae
9. Myristicaceae 10. Canellaceae

Order 2. **Laurales**

Families : 1. Amborellaceae 2. Trimeniaceae 3. Monimiaceae
4. Gomortegaceae 5. Calycanthaceae 6. Idiospermaceae
7. Lauraceae 8. Hernandiaceae

Order 3. **Piperales**

Families : 1. Chloranthaceae 2. Saururaceae 3. Piperaceae

Order 4. **Aristolochiales**

Family : 1. Aristolochiaceae

Order 5. **Illiciales**

Families : 1. Illiciaceae 2. Schisandraceae

Order 6. **Nymphaeales**

Families : 1. Nelumbonaceae 2. Nymphaeaceae 3. Barclayaceae
4. Cabombaceae 5. Ceratophyllaceae

Order 7. **Ranunculales**

Families : 1. Ranunculaceae 2. Circaeasteraceae
3. Berberidaceae 4. Sargentodoxaceae
5. Lardizabalaceae 6. Menispermaceae
7. Coriariaceae 8. Sabiaceae

Order 8. **Papaverales**

Families : 1. Papaveraceae 2. Fumariaceae

Subclass 2. Hamamelidae

Hamamelidae, smallest subclass of Magnoliopsida, includes 11 orders, 24 families and about 3,400 species with the orders Urticales and Fagales alone including about 90 per cent of species. The fossil record of this group, which is weighed in taxa at higher ranks (orders with few families and families with few species), extends to the Lower Cretaceous (100 million years ago) and it is believed that the group is either marked by high levels of extinction, thereby producing an array of well separated, more or less relictual taxa or perhaps polyphyletic (an artificial assemblage with elements drawn from several ancestral types). The families are typically woody (Urticaceae being an exception) and characterised by a suite of structural adaptations for wind pollination (anemophily). These features are :

1. Unisexual (imperfect) flowers with either dioecious or monoecious plants.
2. Perianth reduced, weakly developed or absent (apetalous).
3. Inflorescence with numerous, reduced, often apetalous flowers with the most specialised type—the catkin or ament (older name of the group was thus **Amentiferae**). The orders and the families included in this subclass are :

Subclass 2. HAMAMELIDAE

Order 1. **Trochodendrales**

Families : 1. Tetracentraceae 2. Trochodendraceae

Order 2. **Hamamelidales**

Families : 1. Cercidiphyllaceae 2. Eupteleaceae 3. Platanaceae
4. Hamamelidaceae 5. Myrothamnaceae

Order 3. **Daphniphyllales**

Family : 1. Daphniphyllaceae

Order 4. **Didymelales**

Family : 1. Didymelaceae

Order 5. **Eucommiales**

Family : 1. Eucommiaceae

Order 6. **Urticales**

Families : 1. Barbeyaceae 2. Ulmaceae 3. Cannabaceae
4. Moraceae 5. Cecropiaceae 6. Urticaceae
7. Physenaceae

Order 7. **Leitneriales**

Family : 1. Leitneriaceae

Order 8. **Juglandales**

Families : 1. Rhoipteleaceae 2. Juglandaceae

Order 9. **Myricales**

Family : 1. Myricaceae

Order 10. **Fagales**

Families : 1. Balanopaceae 2. Ticodendraceae 3. Fagaceae
4. Nothofagaceae 5. Betulaceae

Order 11. **Casuarinales**

Family : 1. Casuarinaceae

Under this subclass, the order Trochodendrales is placed at the base and the order Casuarinales–family Casuarinaceae with only genus *Casuarina* (Australian Pine) takes the most derived or specialised position. The other two monotypic families Trochodendraceae and Tetracentraceae, placed in Trochodendrales, lack vessel elements and tend toward apocarpy whereas Casuarinaceae, also being monotypic, shows multiple (fruiting pistillate inflorescence) fruiting structure.

Subclass 3. Caryophyllidae

Caryophyllidae includes only 3 orders, 14 families and about 11,000 species. The large and diverse order Caryophyllales, which includes about 90 per cent of the families, also carries relatively distinct set of features such as :

1. Majority of families of Caryophyllidae lack the anthocyanin pigments, typical of most angiosperms. These are replaced by betalains, a class of pigment that is found only in this order (and in some fungi).
2. The perianth tends to be uniseriate via apetaly but often showy via petaloid sepals.
3. The placentation pattern is often basal or free central, thereby tending to place the ovules in the centre of the ovary–the archaic term **Centrospermae** denotes this tendency.
4. The order is marked by a suite of unusual embryological features, such as presence of **perisperm**, a nutritive tissue found in the seed. It is derived from sporophytic (diploid) tissue and not the endosperm (triploid). Also the embryo tends to occupy a peripheral position in the seed and often produces a characteristic 'beaked' (protrusion of radical) asymmetry.

The order and families included in this subclass are :

Subclass 3. CARYOPHYLLIDAE

Order 1. **Caryophyllales**

Families : 1. Phytolaccaceae 2. Achatocarpaceae
3. Nyctaginaceae 4. Aizoaceae 5. Didiereaceae
6. Cactaceae 7. Chenopodiaceae
8. Amaranthaceae 9. Portulacaceae 10. Basellaceae
11. Molluginaceae 12. Caryophyllaceae

Order 2. **Polygonales**

Family : 1. Polygonaceae

Order 3. **Plumbaginales**

Family : 1. Plumbaginaceae

Subclass 4. Dilleniidae

Dilleniidae includes 13 orders, 78 families and about 25,000 species. Five orders : the Violales (5,000 species), Capparales (4,000 species), Ericales (4,000 species), Theales (3,500 species) and Malvales (3,000 to 3,500 species) alone include about 75 per cent of the species under this subclass.

The subclass includes a diverse set of orders that lack distinct set of key characters. The general characteristic features of Dilleniidae are :

1. All members are syncarpous except Dilleniales, which stands as a structural link to the subclass Magnoliidae.
2. More than one third of species show parietal placentation.
3. About one third of the species are sympetalous.

The orders and families included in this subclass are :

Subclass 4. DILLENIIDAE

Order 1. **Dilleniales**

Families : 1. Dilleniaceae 2. Paeoniaceae

Order 2. **Theales**

Families : 1. Ochnaceae 2. Sphaerosepalaceae 3. Sarcolaenaceae 4. Dipterocarpaceae 5. Caryocaraceae 6. Theaceae 7. Actinidiaceae 8. Scytopetalaceae 9. Pentaphylacaceae 10. Tetrameristaceae 11. Pellicieraceae 12. Oncothecaceae 13. Marcgraviaceae 14. Quiinaceae 15. Elatinaceae 16. Paracryphiaceae 17. Medussagynaceae 18. Clusiaceae (Guttiferae)

Order 3. **Malvales**

Families : 1. Elaeocarpaceae 2. Tiliaceae 3. Sterculiaceae 4. Bombacaceae 5. Malvaceae

Order 4. **Lecythidales**

Family : 1. Lecythidaceae

Order 5. **Nepenthales**

Families : 1. Sarraceniaceae 2. Nepenthaceae 3. Droseraceae

Order 6. **Violales**

Families : 1. Flacourtiaceae 2. Peridiscaceae 3. Bixaceae 4. Cistaceae 5. Huaceae 6. Lacistemataceae 7. Scyphostegiaceae 8. Stachyuraceae 9. Violaceae 10. Tamaricaceae 11. Frankeniaceae 12. Dioneophyllaceae 13. Ancistrocladaceae 14. Turneraceae 15. Malesherbiaceae 16. Passifloraceae

17. Achariaceae 18. Caricaceae 19. Fouquieriaceae
20. Hoplestigmataceae 21. Cucurbitaceae
22. Datiscaceae 23. Begoniaceae
24. Loasaceae

Order 7. **Salicales**
Family : 1. Salicaceae

Order 8. **Capparales**
Families : 1. Tovariaceae 2. Capparaceae
3. Brassicaceae (Cruciferae) 4. Moringaceae
5. Resedaceae

Order 9. **Batales**
Families : 1. Gyrostemonaceae 2. Bataceae

Order 10. **Ericales**
Families : 1. Cyrillaceae 2. Clethraceae 3. Grubbiaceae
4. Empetraceae 5. Epacridaceae 6. Ericaceae
7. Pyrolaceae 8. Monotropaceae

Order 11. **Diapensiales**
Family : 1. Diapensiaceae

Order 12. **Ebenales**
Families : 1. Sapotaceae 2. Ebenaceae 3. Styracaceae
4. Lissocarpaceae 5. Symplocaceae

Order 13. **Primulales**
Families : 1. Theophrastaceae 2. Myrsinaceae 3. Primulaceae

On account of lack of connation in the gynoecium, as expressed by peony (*Paeonia*–Paeoniaceae), the order Dilleniales, with two of its families (Dilleniaceae and Paeoniaceae), is placed at the base of this subclass and forms structural link to the basal flowering plant subclass Magnoliidae.

Subclass 5. Rosidae

Rosidae includes 18 orders, 114 families and over 58,000 species. This is regarded as the largest subclass due to number of families. It is more or less tied with subclass Asteridae with regards to species diversity. The five orders : Fabales, Myrtales, Euphorbiales, Rosales and Sapindales of this subclass alone include about 75 per cent of the species.

Like Dilleniidae, this subclass also includes a diverse set of orders that lack a distinct set of key characters. The general characteristic features of Rosidae are :

1. Leaves simple or pinnately compound.
2. Petals usually distinct or sometimes reduced, often connate at the base, rarely sympetalous.
3. Mostly possessing various types of nectary discs, many of which are staminoidal in origin and positioned at the base of the ovary.
4. Usually syncarpous–exception Rosales, where ovary is apocarpous and Fabales and Proteales where it is monocarpous.

5. Placentation of various types–most often axile, with generally fewer ovules per locule than found in Dilleniidae.

The orders and families included in this subclass are :

Subclass 5. ROSIDAE

Order 1. **Rosales**

Families : 1. Brunelliaceae 2. Connaraceae 3. Eucryphiaceae 4. Cunoniaceae 5. Davidsoniaceae 6. Dialypetalanthaceae 7. Pittosporaceae 8. Byblidaceae 9. Hydrangeaceae 10. Columelliaceae 11. Grossulariaceae 12. Greyiaceae 13. Bruniaceae 14. Anisophyllaceae 15. Alseuosmiaceae 16. Crassulaceae 17. Cephalotaceae 18. Saxifragaceae 19. Rosaceae 20. Neuradaceae 21. Crossosomataceae 22. Chrysobalanaceae 23. Surianaceae 24. Rhabdodendraceae

Order 2. **Fabales**

Families : 1. Mimosaceae 2. Caesalpiniaceae 3. Fabaceae (Papilionaceae)

Order 3. **Proteales**

Families : 1. Elaeagnaceae 2. Proteaceae

Order 4. **Podestemales**

Family : 1. Podostemaceae

Order 5. **Haloragales**

Families : 1. Haloragaceae 2. Gunneraceae

Order 6. **Myrtales**

Families : 1. Sonneratiaceae 2. Lythraceae 3. Rhynchocalycaceae 4. Alzateaceae 5. Penaeaceae 6. Crypteroniaceae 7. Thymelaeaceae 8. Trapaceae 9. Myrtaceae 10. Punicaceae 11. Onagraceae 12. Oliniaceae 13. Melastomataceae 14. Combretaceae

Order 7. **Rhizophorales**

Family : 1. Rhizophoraceae

Order 8. **Cornales**

Families : 1. Alangiaceae 2. Cornaceae 3. Garryaceae

Order 9. **Santalales**

Families : 1. Medusandraceae 2. Dipentodontaceae 3. Olacaceae 4. Opiliaceae 5. Santalaceae 6. Missodendraceae 7. Loranthaceae, 8. Viscaceae 9. Eremolepidaceae 10. Balanophoraceae

Order 10. **Rafflesiales**

Families : 1. Hydnoraceae 2. Mitrastemonaceae 3. Rafflesiaceae

Order 11. **Celastrales**

Families : 1. Geissolomataceae 2. Celastraceae 3. Hippocrateaceae 4. Stackhousiaceae 5. Salvadoraceae 6. Tepuianthaceae 7. Aquifoliaceae 8. Icacinaceae 9. Aextoxicaceae 10. Cardiopteridaceae 11. Corynocarpaceae 12. Dichapetalaceae

Order 12. **Euphorbiales**

Families : 1. Buxaceae 2. Simmondsiaceae 3. Pandaceae 4. Euphorbiaceae

Order 13. **Rhamnales**

Families : 1. Rhamnaceae 2. Leeaceae 3. Vitaceae

Order 14. **Linales**

Families : 1. Erythroxylaceae 2. Humiriaceae 3. Ixonanthaceae 4. Hugoniaceae 5. Linaceae

Order 15. **Polygalales**

Families : 1. Malpighiaceae 2. Vochysiaceae 3. Trigoniaceae 4. Tremandraceae 5. Polygalaceae 6. Xanthophyllaceae 7. Krameriaceae

Order 16. **Sapindales**

Families : 1. Staphyleaceae 2. Melianthaceae 3. Bretschneideraceae 4. Akaniaceae 5. Sapindaceae 6. Hippocastanaceae 7. Aceraceae 8. Burseraceae 9. Anacardiaceae 10. Julianiaceae 11. Simaroubaceae 12. Cneroraceae 13. Meliaceae 14. Rutaceae 15. Zygophyllaceae

Order 17. **Geraniales**

Families : 1. Oxalidaceae 2. Geraniaceae 3. Limnanthaceae 4. Tropaeolaceae 5. Balsaminaceae

Order 18. **Apiales**

Families : 1. Araliaceae (Thorne) 2. Apiaceae (Umbelliferae)

Subclasses Dilleniidae and Rosidae have been kept apart because each seems to constitute a natural group separately derived from the ancestral Magnoliidae, rather than because of any definitive distinguishing characters and that is why, conceptually, the two groups hold the position of subclasses separately rather than they are combined into one or any attempt has been made towards abandoning any of the two during organisation of Magnoliopsida into subclasses.

Subclass 6. Asteridae

Asteridae includes 11 orders, 49 families and nearly 56,000 species. Species diversity is comparable to subclass Rosidae, but unlike subclasses Dilleniidae and Rosidae, the family level is fewer in this subclass. The family Asteraceae (Compositae) alone includes about one-third of the species of this subclass. In terms of class-level phylogenetic position, level of unique adaptive modification and species diversity, family Asteraceae (Compositae) is comparable to monocot family Orchidaceae. The general characteristic features of Asteridae are :

1. Members are well marked by the presence of sympetaly.
2. The subclass can be distinguished from sympetalous Dilleniidae by an androecium with stamens either equal to or less than in number of corolla lobes and positioned between the corolla lobes.
3. Gynoecium bicarpellate and stamens epipetalous.

Cronquist regards the subclass Asteridae as the most advanced in the Magnoliopsida (Dicots) and probably the most recently evolved (only subclass Caryophyllidae may be more recent). Unlike any other subclass, Asteridae exploit specialised pollinators and specialised means of presenting pollen. It seems likely that the rise of Asteridae is closely correlated with the evolution of the insects capable of recognising the complex floral structure. Thus, this class has been roughly organised into two lineages or clusters of orders, one of which is marked by a dominance of hypogyny and the other, terminated by the Asterales, by epigyny.

The orders and families included in this subclass are :

Subclass 6. ASTERIDAE

Order 1. **Gentianales**

Families : 1. Loganiaceae 2. Gentianaceae 3. Saccifoliaceae 4. Apocynaceae 5. Asclepiadaceae

Order 2. **Solanales**

Families : 1. Duckeodendraceae 2. Nolanaceae 3. Solanaceae 4. Convolvulaceae 5. Cuscutaceae 6. Retziaceae 7. Menyanthaceae 8. Polemoniaceae 9. Hydrophyllaceae

Order 3. **Lamiales**

Families : 1. Lennoaceae 2. Boraginaceae 3. Verbenaceae 4. Lamiaceae (Labiatae)

Order 4. **Callitrichales**

Families : 1. Hippuridaceae 2. Callitrichaceae 3. Hydrostachyaceae

Order 5. **Plantaginales**

Family : 1. Plantaginaceae

Order 6. **Scrophulariales**

Families : 1. Buddlejaceae 2. Oleaceae 3. Scrophulariaceae 4. Globulariaceae 5. Myoporaceae 6. Orobanchaceae 7. Gesneriaceae 8. Acanthaceae 9. Pedaliaceae 10. Bignoniaceae 11. Mendonciaceae 12. Lentibulariaceae

Order 7. **Campanulales**

Families : 1. Pentaphragmataceae 2. Sphenocleaceae 3. Campanulaceae 4. Stylidiaceae 5. Donatiaceae 6. Brunoniaceae 7. Goodeniaceae

Order 8. **Rubiales**

Families : 1. Rubiaceae 2. Theligonaceae

Order 9. **Dipsacales**

Families : 1. Caprifoliaceae 2. Adoxaceae 3. Valerianaceae 4. Dipsacaceae

Order 10. **Calycerales**

Family : 1. Calyceraceae

Order 11. **Asterales**

Family : 1. Asteraceae (Compositae)

Class : LILIOPSIDA

The two classes Magnoliopsida and Liliopsida under Magnoliophyta, have diverged early in the evolutionary history and hence they are fairly distinct, and in terms of flowering plant identification they represent the first level of recognition. In class Liliopsida, Alismatidae chiefly includes the aquatic members whereas the other four subclasses include those families, the forms of which are basically land plants with typically large, often petiolate leaves, usually arborescent or have flowers crowded into a spadix. The families under subclass Commelinidae are characterised by floral reduction and are mostly wind-pollinated (anemophilous). The members of the subclass Liliidae are insect pollinated (entomophilous) fig. (6.17).

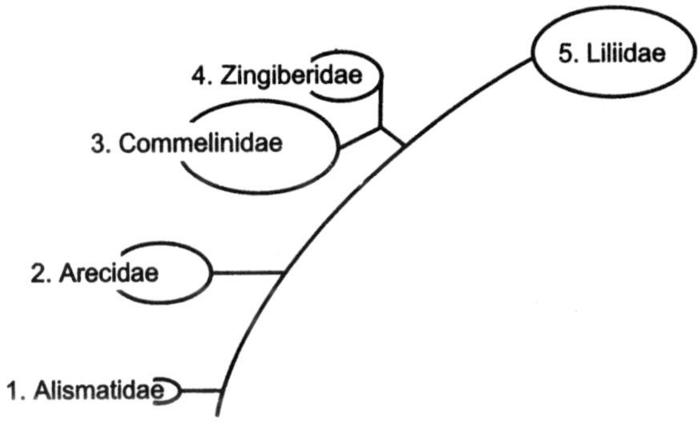

Fig. 6.17 Phyletic relationship among various subclasses within class Liliopsida (Redrawn from Cronquist, 1988).

The alignment features the Alismatidae as the extanct Monocot element, that carries the greatest number of archaic features (similar to the 'original' monocots) and the Liliidae (Orchids and relatives), are considered as the most derived or specialised elements of the class Liliopsida. The size of **balloons** is roughly proportional to the species included per subclass.

The key characters to circumscribe five subclasses : Alismatidae, Arecidae, Commelinidae, Zingiberidae and Liliidae, are :

Subclass	Characteristic Features	Number of Orders	Number of Families	Number of Species
Alismatidae	Aquatics with archaic, many-parted flowers	4	16	500
Arecidae	Inflorescences specialised	4	5	5,600
Commelinidae	Mostly herbs with reduced flowers	7	16	16,200
Zingiberidae	Tropical herbs, epigynous	2	9	3,800
Liliidae	Flowers showy with petaloid sepals	2	19	25,000

Liliopsida, a group close to modern Nymphaeales, was considered to have arisen from aquatic ancestors. All five subclasses of this class were considered to have some ecological cohesence. Arecidae includes often woody plants. The Commelinidae and Liliidae, which probably arose from a common ancestor, show partly divergent and partly parallel series of advances.

Subclass 1. Alismatidae

Alismatidae includes 4 orders, 16 families and about 500 species. It is the smallest subclass of the flowering plants (both in Magnoliopsida and Liliopsida) in terms of species diversity. Cronquist has considered the characteristic features of the gynoecium and apocarpy as a central unifying feature for Alismatidae. Aquatic adaptations of all orders except Triuridales, also provide a tentative link to the archaic dicot element (Magnoliidae–Nymphaeales) placed by Cronquist, as a possible extant link to the monocots.

The orders and families in this subclass are :

Subclass 1. ALISMATIDAE

Order 1. **Alismatales**

Families : 1. Butomaceae 2. Limnocharitaceae 3. Alismataceae

Order 2. **Hydrocharitales**

Family : 1. Hydrocharitaceae

Order 3. **Najadales**

Families : 1. Aponogetonaceae 2. Scheuchzeriaceae
3. Juncaginaceae 4. Potamogetonaceae
5. Ruppiaceae 6. Najadaceae
7. Zannichelliaceae 8. Posidoniaceae
9. Cymodoceae 10. Zosteraceae

Order 4. **Triuridales**

Families : 1. Petrosaviaceaceaceae 2. Triuridaceae

Subclass 2. Arecidae

Arecidae includes 4 orders, 5 families and about 5,600 species. More than half of the species of this subclass belong to the monotypic order Arecales. In terms of both structural and ecological adaptations, this is perhaps, the most diverse subclass of flowering plants as it includes **smallest flowering plant** (*Wolffia*–Lemnaceae–Arales), the **largest monocots** (*Areca, Cocos*–Arecaceae–Arecales), the **largest flowering plant fruits-seeds** (*Lodoicea maldivica*–Arecaceae–Arecales) and the **largest leaves** (*Raphia regalis*–Arecaceae–Arecales).

The orders and families included in this subclass are :

Subclass 2. ARECIDAE

Order 1. **Arecales**
Family : 1. Arecaceae (Palmae)
Order 2. **Cyclanthales**
Family : 1. Cyclanthaceae
Order 3. **Pandanales**
Family : 1. Pandanaceae
Order 4. **Arales**
Families : 1. Acoraceae 2. Araceae 3. Lemnaceae

Subclass 3. Commelinidae

Commelinidae includes 7 orders, 16 families and about 16,200 species. Large percentage of the species (about 15,000) of this subclass is included within a single family, the grass family or Poaceae (Gramineae), which is the most important family of flowering plants from the point of view of both economic and ecological importance, atleast from the perspectiveness of mammals and more specifically of humans. The order Commelinales, which forms the basal or archaic group, shows floral adaptations for entomophilous mode of pollination with showy flowers and well developed corolla. The more derived orders, on the other hand, are anemophilous. Thus, according to Cronquist, "*The trend of phyletic pattern within Commelinidae is towards floral reduction and other adaptive changes for wind pollination*".

The orders and families included in this subclass are :

Subclass 3. COMMELINIDAE

Order 1. **Commelinidales**
Families : 1. Rapateaceae 2. Xyridaceae
3. Mayacaceae 4. Commelinaceae
Order 2. **Eriocaulales**
Family : 1. Eriocaulaceae
Order 3. **Restionales**
Families : 1. Flagellariaceae 2. Joinvilleaceae
3. Restionaceae 4. Centrolepidaceae
Order 4. **Juncales**
Families : 1. Juncaceae 2. Thurniaceae
Order 5. **Cyperales**
Families : 1. Cyperaceae 2. Poaceae (Gramineae)
Order 6. **Hydatellales**
Family : 1. Hydatellaceae

Order 7. **Typhales**

Families : 1. Sparganiaceae 2. Typhaceae

Subclass 4. Zingiberidae

Zingiberidae includes 2 orders, 9 families and about 3,800 species. Its most of the members are often associated with the orders of subclasses Commelinidae and Liliidae. Cronquist opined that these plants are "**discardant**" when so treated, *i.e.*, Bromeliales does not "**fit**" well within the Commelinidae and Zingiberales which are a bit anomalous with the context of the Liliidae. If both the orders are excluded from these subclasses, it serves to "**sharpen**" the definition or circumscription of both. It, however, does not create a unit that is not recognised by other current system of flowering plant classifications.

The orders and families included in this subclass are :

Subclass 4. ZINGIBERIDAE

Order 1. **Bromeliales**

Family : 1. Bromeliaceae

Order 2. **Zingiberales**

Families : 1. Strelitziaceae 2. Heliconiaceae 3. Musaceae
4. Lowiaceae 5. Zingiberaceae 6. Costaceae
7. Cannaceae 8. Marantaceae

Within the subclass Zingiberidae, the order Bromeliales (monotypic) includes the epihytes or xerophytes which are characterised by narrow, parallel-veined leaves that are often leathery and without petiole and the flowers usually actinomorphic with 6 stamens. The other order Zingiberales includes mesophytes of the forest floor or emergent aquatics with broadly expanded, petiolate leaves that carry a distinctive pattern of venation and usually zygomorphic flowers with 4 or 5 (rarely 6) stamens.

Subclass 5. Liliidae

Liliidae includes 2 orders, 19 families and about 25,000 species. The members of this subclass are characterised by large, showy flowers that tend to show petaloid sepals, *i.e.*, the perianth is usually made up of 6 non-green petal like tepals. As is the case in members of the class Liliopsida, in general, the Cronquist's system differs from other current systems of flowering plant classification with regard to alignment of these taxa.

The orders and the families included in this subclass are :

Subclass 5. LILIIDAE

Order 1. **Liliales**

Families : 1. Philydraceae 2. Pontederiaceae
3. Haemodoraceae 4. Cyanastraceae 5. Liliaceae
6. Iridaceae 7. Velloziaceae 8. Aloeaceae
9. Agavaceae 10. Xanthorhoeaceae
11. Habguanaceae 12. Taccaceae 13. Stemonaceae
14. Smilacaceae 15. Dioscoreaceae

Order 2. **Orchidales**

Families : 1. Geosiridaceae 2. Burmanniaceae 3. Corsiaceae 4. Orchidaceae

The order Liliales is characterised by plants which are usually mycotrophic (associated with fungi), with mostly actinomorphic flowers having inferior or superior ovary and seeds of ordinary size and structure, usually with endosperm and well developed embryo. The other order Orchidales is characterised by plants which also are usually mycotrophic, sometimes without chlorophyll, flowers strongly zygomorphic, ovary inferior and seeds very numerous and minute with undifferentiated embryo lacking endosperm.

Merits of the System

Following are the merits of Cronquist's system of classification :

1. Cronquist's system is by far the latest, exhaustive and broad based. The system was revised completely in 1981 and 1988 and was presented in a comprehensive form.
2. The system is completely a phylogenetic one and shows general agreement with major contemporary systems.
3. Apart from phylogenetic consideration, the system also incorporates the knowledge obtained from other branches of plant science and hence is largely accepted.
4. The system is based on morphological, anatomical, ecological, embryological, palynological, cytological, serological, chemical as well as ultrastructural evidences.
5. Synoptical keys for each group, upto the level of family, are distinct and very useful.
6. The system is the outcome of his long experience in plant systematics, teaching and his personal observations as Curator of New York Botanic Garden (USA).
7. The system finds great use for a student or a professional botanist who has enough aquaintance with modern broad scale taxonomy.
8. The text, being in English, has been readily adopted in books and floristic projects in USA.
9. The placing of Winteraceae at the beginning of dicots, is generally favoured by most authors.
10. Nomenclature adopted in classification is in accordance with International Code of Botanical Nomenclature (ICBN).
11. Placement of Magnoliidae as most primitive group of angiosperms, dicots before monocots, Magnoliales at the beginning of Magnoliidae and Butomaceae at the beginning of Liliopsida, finds general agreement with others.
12. Compositae have been placed towards the end of dicots and Orchidaceae at the end of monocots, as is generally agreed by many workers.
13. Relationship among various groups has been depicted by diagrams, that provide valuable information on relative advancement, cladistic relationships and size of various subclasses.

14. The separation of Brassicaceae and Capparaceae has gained support from chloroplast sequence data (Hall, Sytsma and Iltis, 2002), consistent with morphological data.

Demerits of the System

Following are the demerits of Cronquist's system of classification :

1. Data obtained from studies on biochemistry, cytology, palynology etc., used for the first time in his system of classification, still await adoption in major universities and herbaria of the world.
2. Cronquist's system of classification, at some levels, has shown reliance on single character, such as centrifugal stamens, free-central placentation etc.
3. The arrangement of certain families in Liliales has been strongly criticised, particularly the submergence of Amayrillidaceae into Liliaceae.
4. The system is not very useful for identification and adoption by the herbaria because identification keys for the genera, their distribution and descriptions, are not provided.
5. His considering angiosperm as a division, is not at par with Dahlgren (1983, 1989) and Thorne (1981, 2003) who consider it a class.
6. In the system, Asteridae represents a loose assemblage of several diverse sympetalous families.
7. Cronquist placed Typhales in Commelinidae, which according to Clifford's numerical studies in 1977, finds a better place in Arecidae.
8. Superorder, as a rank above order, is not recognised, thus showing a significant departure from other contemporary systems.
9. Placement of Hamamelidae as an ancient side-branch of Magnoliidae, has not been recognised by Enrendorfer (1983) who considered it remnants of a transition from Magnoliidae to Dilleniidae–Rosidae–Asteridae.
10. Behnke (1977) and Behnke and Barthlott (1983) are of the opinion that Polygonales and Plumbaginales, with 5-type plastids should be removed from Caryophyllidae and placed in Rosidae and only Caryophyllales, with PIII-type plastids, be retained to Caryophyllidae.
11. Placement of Urticales in Hamamelidae with wind-pollinated families, has not been advocated by Dahlgren (1983, 1989) who considers their place close to Malvales and Euphorbiales.
12. Cronquist (1988), while recognising Physenaceae, as a family under order Urticales, was not sure about its exact placement.
13. Monocots are placed after dicots whereas in recent classifications they are placed between primitive angiosperms and Eudicots.
14. Cronquist placed Dioneophyllaceae in order Violales before family Ancistrocladaceae whereas Metcalfe and Chalk (1983), on the basis of a unique combination of anatomical characters, suggested that it should be given an isolated taxonomic position.

15. Cronquist believed monocots to have been arose from vessel-less ancestors resembling present day Nymphaeales whereas recent workers do not believe in the aquatic ancestry of monocots.
16. Cronquist placed Winteraceae towards the beginning of Magnoliales and Canellaceae towards the end. The multigene analysis (Soltis *et al.*, 1999; Zanis *et al.*, 2002, 03) have provided 99–100 per cent **bootstrap** or **jackknife support** in the relationship of these families and the two are accordingly placed in a separate order in APG II and AP web and under the same suborder in Thorne (2003). Morphological studies by Doyle and Endress (2000) have also supported the affinities between these two families.

Table 6.8 Comparison of Bentham and Hooker's; Engler and Prantl's; Hutchinson's and Cronquist's Systems of Classification

	Bentham and Hooker (1862)	Engler and Prantl (1931)	Hutchinson (1959)	Cronquist (1968)
1.	Important publication—*Genera plantarum* (1862–83)	Important publication—*Die naturlichen pflanzenfamilien* (1887–1915)	Important publication—*Families of Flowering Plants* (1926, 1959)	Important publication—*Evolution and Classification of Flowering Plants* (1968)
2.	Natural system	Phylogenetic system	Phylogenetic system	Contemporary phylogenetic system
3.	Based on de Candolle's system (1818); An elaboration and slight modification of Jussieu's arrangement (1779)	Based on Eichler's system (1875–77)	Based on Bessey's system (1883)	Based on and parallels to Takhtajans system (1954)
4.	Criteria of classification—Based on single and mostly artificial characters	Criteria of classification—Based on their own experiences and assumptions that in flowering plants the absence of perianth is a primitive character	Criteria of classification—Based on 24 principles and combination of characters	Criteria of classification—Based largely on literature, personal observations and discussions with other botanists to incorporate morphological, phylogenetic and ecological characters coupled with the knowledge obtained from other branches of science

(Contd....)

5.	Published as two volumes in Latin; Number of families : 202	Appeared in 20 volumes in German; Number of families : 280	Appeared in 2 volumes in English; Number of families : 411	Second edition appeared in 1988; Number of families : 382
6.	Deals with the classification of seed plants	Deals with classification of entire plant kingdom : from algae to angiospermae	Deals with the classification of flowering plants	Deals with the classification of flowering plants
7.	System is imbued with the idea of fixity of species. Differences are more emphasised	They considered Darwinian theory of descent as their guide	System is based on considerable knowledge of phylogeny. Resemblances more emphasised	Phenomenon of evolutionary parallelism and relationship between phylogeny and taxonomy are well explained.
8.	Seed plants are divided into Dicots, Gymnosperms and Monocots	Seed Plants, the 14th Division : Embryophyta Siphonogama, is divided into Monocots, Dicots and Gymnosperms	Seed Plants, the Angiosperms are divided into two Phyla : Dicots and Monocots	Seed plants, the Magnoliophyta, are divided into two classes : Magnoliopsida (Dicots) and Liliopsida (Monocots)
9.	Dicots are further divided into Polypetalae, Gamopetalae and Monochlamydeae	Dicots are further divided into Archichlamydeae and Sympetalae. Sympetalae includes both Polypetalae and Monochlamydeae	Dicots are further divided into two divisions : Lignosae and Herbaceae	Dicots (Magnoliopsida) are further divided into six subclasses : Magnoliidae, Hamamelidae, Caryophyllidae, Dilleniidae, Rosidae and Asteridae that include 64 orders and 318 families
10.	Dicots begin with order Ranales (Ranunculaceae).	In dicots, Amentiferae (Casuarinaceae, Piperaceae, Salicaceae etc.) are regarded as primitive and are placed before Ranunculaceae, Magnoliaceae etc.	Lignosae : woody habit starts with Magnoliaceae and ends with Verbenaceae whereas Herbaceae starts with Ranunculaceae and ends with Lamiaceae (Labiatae).	The first suborder of dicots, Magnoliidae starts with Winteraceae and the last subclass, Asteridae ends with Asteraceae (Compositae)

(Contd....)

11.	Monocots are placed after dicots. Gymnosperms are kept between dicots and monocots	Monocots are placed before dicots, the idea which gets little support from fossil record. Gymnosperms are considered as a distinct group from the angiosperms.	Monocots are placed after dicots from which they have been considered to have been derived at an early stage. Gymnosperms considered as a distinct group from angiosperms.	Monocots (Liliopsida) are placed after dicots (Magnoliopsida). Gymnosperms are considered as a distict group from angiosperms.
12.	Monocots are divided into seven series, beginning with Microspermae and ending in Glumaceae	Monocots are divided into 11 orders, beginning with Pandanales and ending in Microspermae	Monocots are divided into three distinct groups, *viz.*, Calyciflorae, Corolliflorae and Glumiflorae	Monocots (Liliopsida) are divided into five subclasses *viz.*, Alismatidae, Arecidae, Commelinidae, Zingiberidae and Lilidae
13.	In monocots, the first family is Hydrocharitaceae and the last one is Gramineae (Poaceae)	In monocots, the first family is Typhaceae and the last one is Orchidaceae	In monocots, the first family is Butomaceae and the last one is Gramineae (Poaceae)	In monocots (Liliopsida), the first family is Butomaceae and the last one is Orchidaceae
14.	In dicots, families with free petals are included in Polypetalae and those with fused petals in Gamopetalae. Monochlamydeae included the families without proper distinction of calyx and corolla, *i.e.*, with perianth.	Among dicots Polypetalae and Monochlamydeae, as considered by Bentham and Hooker, are amalgamated under the single group Archichlamydeae whereas Sympetalae includes those families in which petals are fused	Among dicots there are no groups such as Polypetalae, Gamopetalae, Monochlamydeae, Archichlamydeae or Sympetalae, but based on resemblances, the families with free or fused petals or with perianth, are distributed among Lignosae (woody) and Herbaceae (Herbaceous) families	Among dicots (Magnolipsida), the families showing retention of primitive characters, *i.e.*, reduced and apetalous flowers, are kept before families with free or fused petals.

(Contd....)

15.	Among monocots, family Gramineae (Poaceae), is considered as the most advanced and is placed at the end in series Glumaceae	Among monocots, family Orchidaceae of the order Microspermae is considered more highly evolved than the grass family Gramineae (Poaceae)	Among monocots, family Gramineae of order Graminales has been given the highest place as the last family under Glumiflorae	Among monocots, family Orchidaceae of order Orchidales, has been given the place as the last family
16.	The system is useful for practical purposes, such as for arranging plants in botanical gardens and specimens in the herbaria	The system is a model of original description of the individuals and inclusion of large number of plants with keys as an aid to identification	The system is useful for theoretical purposes such as for documentation.	The system is quite useful because it has clearly defined evolutionary principles

Rolf MT Dahlgren (1932–87)

Rolf MT Dahlgren (1932–87), Professor at the Botanical Museum of the University of Copenhagen, Denmark, first proposed his system and new method of illustrating phylogenetic relationships in a text book *A Textbook of Angiosperm Taxonomy* in Danish in 1974 (Dahlgren *et al.*, 1974). The revised system in English, and subsequent revisions were published in 1975, 1980,

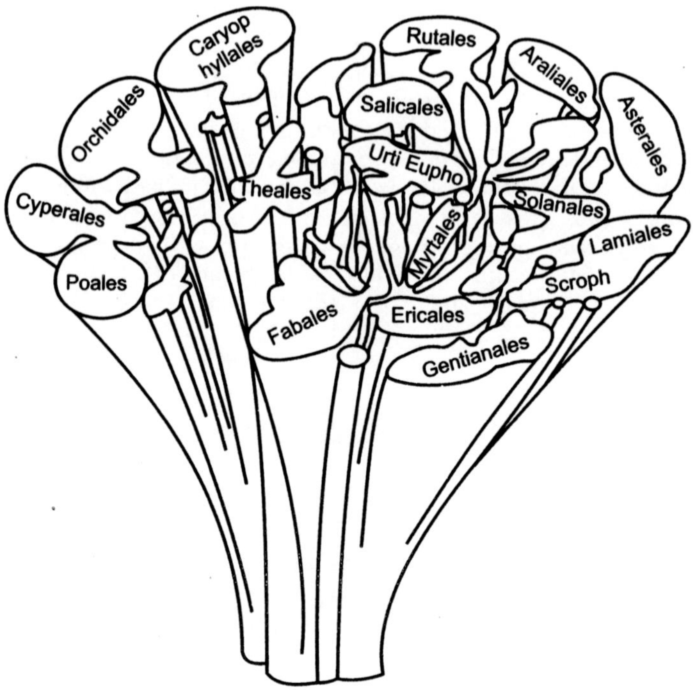

Fig. 6.18 Dahlgen's phylogenic shrub-in cross-section (after Kubitzki, 1977).

1981 and 1983. A useful detailed treatment of Monocotyledons was presented in a book *The Families of Monocotyledons* (Dahlgren, *et al.*) in 1985. In the beginning, the system was proposed as an imiginatory phylogenetic shrub in cross-section (fig. 6.18).

His diagram, a cross-section through the top of an imaginary phylogenetic shrub, became very popular for mapping the distribution of character-states in various orders of angiosperms and is popularly known as **Dahlgrenogram** (fig. 6.19).

The Dahlgrenogram is, in fact, a bubble diagram in which various orders are represented as bubbles. The size of the bubble is relative to the number of species in the order. The relative position of the bubbles reflect their phylogenetic affinities. The orders are combined into superorders so as to form bubble complexes. Gertrud Dahlgren (wife of Rolf MT Dahlgren), while presenting a revision (in 1989) of Dahlgren's system, made significant changes in the superorders Theanae, Malvanae, Rutanae and Cornanae in dicotyledons and minor alterations in monocotyledons. In monocotyledons, the minor changes included recognising Acoraceae as a family in Arales, merging Sparganiaceae in Typhaceae, Thismiaceae in Iridac, coupled with shifting of a few families. Gertrud Dahlgren also included the position of these families in the bubble diagram.

In construction of classification, Dahlgren believed that angiosperms are of monophyletic origin and have evolved from one particular line of gymnosperms and that none of the extant group of flowering plant is ancestral to any present-day group. Among dicotyledons, Magnoliannae was considered most primitive and among monocotyledons, Alismatanae as the most primitive. Among two subclasses, Liliidae stand out as a possibly monophyletic group by virtue of their single cotyledon and characteristic triangular protein bodies in the sieve tube plastids (Behnke, 1969), in addition to the less constant differences from Magnoliidae.

After the tragic death of Dahlgren in a car accident in 1997, his wife Gertrud Dahlgren continued his work and finally published the last Dahlgrenogram for dicotyledons followed by a classification of monocotyledons, both in 1989 incorporating the latest ideas of Dahlgren and bringing up an updated classification of angiosperms. She also changed the ending of superorders from–**florae** to **anae** because of the term–**florae** restricted its application to only flowering plants and the change to–**anae** was for uniformity according to nomenclatural rules. Although the practice of using–**anae** was initially started by Takhtajan but it has now been incorporated by Thorne (since 1992) who earlier like Dahlgren preferred to use–**florae**. Gertrud Dahlgren followed this upto 1991 with the mapping of various embryological character–states.

Dahlgren's classification is quite similar to the earlier versions of Thorne in using the name Magnoliopsida for angiosperms, Magnoliidae for dicots and Liliidae for monocots. The realignments are based on a large number of phyletic characteristics, mainly phytochemistry, ultrastructure and embryology of various taxa. The angiosperm (Magnoliopsida) represents a

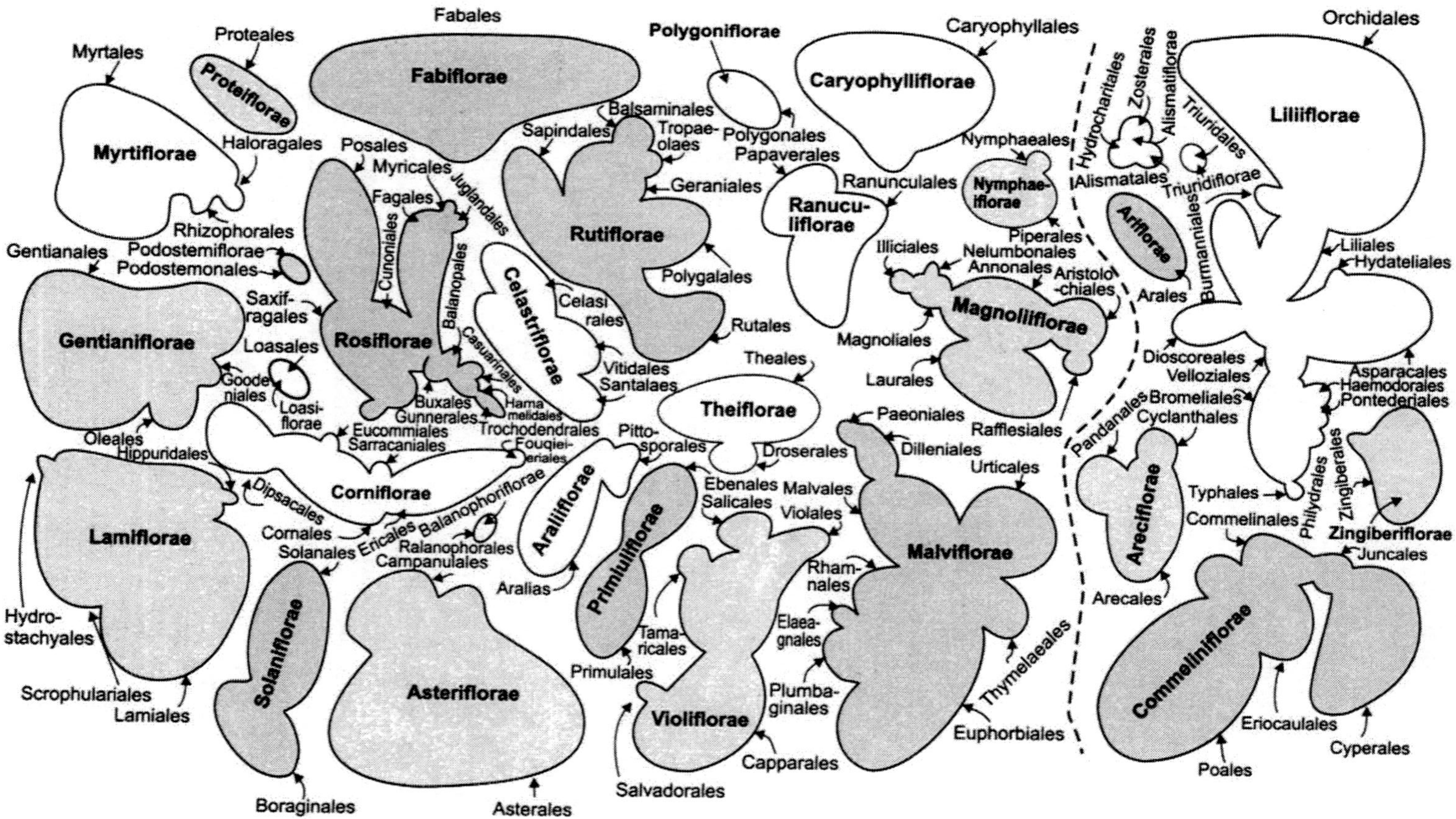

Fig. 6.19 Superorders of Angiosperms–illustrated as clusters, as seen by Dahlgren (1980).

class which has been divided into two subclasses : Magnoliidae (dicots) which has further been divided into 25 superorders, 87 orders and 343 families and Liliidae (monocots) which has further been divided into 10 superorders, 24 orders and 104 families (Table 6.9). Several hundred such embryological character-state maps have been developed by Dahlgren and his associates. According to Dahlgren, recognition of dicotyledons and monocotyledons would not be allowed if one follows rigid **cladistic** approaches but he nevertheless considered monocotyledons as a unique group worthy of subclass rank.

Table 6.9 Outline of the updated Dahlgren's system of classification of angiosperms as presented by his wife Gertrud Dahlgren (1889). (Dicots : 25 superorders, 87 orders and 343 families; Monocots : 10 superorders, 24 orders and 104 families; The number of orders given in parantheses).

CLASS. MAGNOLIOPSIDA (ANGIOSPERMS)

Subclass 1. Magnoliidae (Dicotyledons)

Superorder :		
	1. Magnolianae (10)	2. Nymphaeanae (2)
	3. Ranunculanae (2)	4. Caryophyllanae (1)
	5. Polygonanae (1)	6. Plumbaginanae (1)
	7. Malvanae (7)	8. Violanae (7)
	9. Theanae (4)	10. Primulanae (2)
	11. Rosanae (15)	12. Proteanae (2)
	13. Myrtanae (2)	14. Rutanae (9)
	15. Vitanae (1)	16. Santalanae (1)
	17. Balanophoranae (1)	18. Aralianae (2)
	19. Asteranae (2)	20. Solananae (2)
	21. Ericanae (1)	22. Cornanae (3)
	23. Loasanae (1)	24. Gentiananae (3)
	25. Lamianae (3)	

Subclass 2. Liliidae (Monocotyledons)

Superorder :		
	1. Alismatanae (2)	2. Triuridanae (1)
	3. Aranae (1)	4. Lilianae (6)
	5. Bromelianae (6)	6. Zingiberanae (1)
	7. Commelinanae (3)	8. Arecanae (2)
	9. Cyclanthanae (1)	10. Pandananae (1)

Merits of the System

Dahlgren's system of classification has several advancements over the previous systems. The following are the merits of the classification.

1. The system is highly phylogenetic because it has taken into account the additional data based on phytochemistry, embryology and anatomy.

2. Like Thorne and other systems, the angiosperms are given a more agreeable rank of a class.
3. No family of angiosperms is left unplaced.
4. Dahlgren has attempted to clarify most of the doubts and objections raised by many previous phylogenetists.
5. His presenting Dahlenogram in the form of a bubble diagram is quite useful in giving the idea about relationship of the superorders, orders and even families.
6. Dahlgrenogram represented as a bubble diagram gives an idea of the relative number of species in various groups in angiosperms.
7. Dahlgrenogram finds great use in plotting and comparing the character-states in angiosperms.
8. The use of ending–**anae** is in line with Takhatajan and Thorne.
9. The use of superorder rank, similar to Thorne and Takhtajan, has given more realistic arrangement of families and orders.
10. The separation of Brassicaceae and Capparaceae, is supported by findings on chloroplast DNA sequence data (Hall, Systma and Iiltis, 2002), consistent with morphological data.
11. The separation of genus *Acorus* from Araceae into a distinct family Acoraceae, has recently been followed by Takhtajan (1997), Thorne (2000, 03) and APG II, who have also separated family and placed it under a distinct order Acorales. The genus is distinct from Araceae in having ensiform leaves, glandular tapetum and the type of endothecial cells.
12. The placement of Cornales, closer to Ericales, is justified by Judd, *et al.*, (2002). It is also followed up in Thorne (2003) and APG II (2003).
13. The merger of Scrophulariaceae with Lamiales, has been followed in the recent system of APG II.
14. Dicots (Magnoliopsida) are discussed before monocots (Liliopsida).
15. The angiosperm system is illustrated as a phylogenetic shrub depicting the relationship between the orders.

Demerits of the System

Despite several improvements incorporated in Dahlgren's system over previous ones, it suffers from the following drawbacks :

1. The system does not proceed below family level. Thus, it is not useful for arranging specimens in herbaria or use in floras.
2. Dahlgren has divided angiosperms into dicots (Magnoliopsida) and monocots (Liliopsida) whereas in recent APG II (2003) classification and Thorne (2000, 03), the primitive angiosperms are placed separately.
3. He has placed Asteranae, Cornanae and Aralianae before Lamianae whereas data from molecular studies justifies placement of the group (with circumscription somewhat similar to Euasterales II) after Lamiidae (comparable to Euasterids I of APG II).

4. Dahlgren has placed family Ceratophyllaceae under order Nymphaeales but studies of Zanis *et al.* (2002) and Whitlock *et al.* (2002) have shown that this family is a sister group of monocots (as indicated by microsporogenesis and structure of leaf margin). Accordingly in APG II, it is placed just before monocots.
5. Monocots (Liliopsida) are placed after dicots (Magnoliopsida) whereas in recent classifications they are placed between primitive angiosperms and Eudicots.
6. Dahlgren had placed the family Acoraceae under order Arales but molecular studies (Chase *et al.* 2000 and Fuse and Tamura, 2000) suggest that it should be placed before the rest of the monocots.
7. He placed family Winteraceae in a separate order much after Canellaceae but the multigene analyses (Soltis *et al.* 1999; Zanis *et al.* 2002, 03) have provided 99–100 per cent **bootstrap** or **jackknife support** in their relationship and accordingly the two are placed in the same order under APG II and APG web and under the same suborder in Thorne (2000, 03). Morphological studies of Doyle and Endress (2000) also support the affinities between these two families.
8. He has regarded Budlejaceae and Myoporaceae as distinct from Scrophulariaceae but morphological studies of Bremer *et al.* (2001) and molecular data (three gene analysis) by Olmstead *et al.* (2001) support the merger of these two families, which is followed by APG II (2003) and Thorne (2003).

Robert F Thorne (borne 1920)

Robert F Thorne, an American taxonomist, associated with the Rancho Santa Ana Botanic Garden, has developed and periodically revised a system of a classification that closely approaches the system of Dahlgren in giving angiosperms a rank of the class and dicots and monocots as subclasses. He further divided them in superorders, orders, suborders and families. There was a considerable parallel development in general approach of arrangement of orders and families with other three contemporaries Cronquist, Takhtajan and Dahlgren but the system differs in that angiosperms are ranked as a class (in contrary to Cronquist and Takhtajan). He first put forward his classification in 1968 and proposed eight revisions upto 2003. Since 1992 he, like Gertrud Takhtajan, preferred ending–**anae** over–**florae**. In realignment of taxa and recognising subfamilies more frequently, he incorporated the role of phytochemistry. Also, he made use of the data obtained from comparative morphology, cytology, embryology, palaeobotany, phytogeography, pollen and seed morphology, ultrastructure and host-parasite relationships.

Previously, he gave the name Annonopsida to angiosperms and Annonidae to dicots, replacing Magnoliflorae for Annoniflorae and Magnoliales by Annonales. Since 1992, he adopted the generally accepted names Magnoliopsida (angiosperms), Magnoliidae (dicots) and Magnoliales.

Since 1999, he preferred the name Angiospermae (Magnoliopsida to angiosperms) and instead of dividing it into dicots and monocots (unlike other contemporaries), divided Angiospermae directly into 10 subclasses which were further divided into 33 superorders, 90 orders, 489 families (having estimated 13,309 genera and 2,54,980 species). This last revision of 2003 is available on website of Rancho Santa Ana Botanic Garden at *www.rsab.org/publications/angiosp.htm.* Upto 1999 he recognised only 31 superorders, 74 orders and 475 families in angiosperms (available at *http://www.inform.umd.edu/PBI0/fam/thorneangiosp99.html*) maintained by Dr. JL Reveal of the University of Maryland. The latest version (2003) has, although, maintained the number of subclasses but subclass Dilleniidae (the largest subclass with 160 families) has been abolished and its members are distributed within subclasses mainly Rosidae and restructed Hamamelididae mainly from order Hamamelidales of Rosidae of 1999 version. Since 1992, Thorne has been in practice of putting up electronic revisions of his system of classification of angiosperms. Recognised upto 1992, he has abandoned the traditional division of angiosperms in dicots and monocots in conformation with current cladistic approach and since 1999 placed monocot taxa in between primitive angiosperms and advanced ones (Eudicots of APG).

Table 6.10 Outline of the system of classification proposed by Thorne (2003) (available at www.rsab.org/publications/angiosp.htm)

CLASS : ANGIOSPERMAE

(10 subclasses, 33 superorders)

Subclass 1. Magnoliidae (3 superorders, 7, orders, 32 families, 307 genera, 9095 species)

Superorder : 1. Magnolianae 2. Nymphaeanae 3. Rafflesianae

Subclass 2. Alismatidae (3 superorders, 6 orders, 19 families, 177 genera, 3530 species)

Superorder : 1. Acoranae 2. Aranae 3. Alismatanae

Subclass 3. Liliidae (3 superorders, 9 orders, 54 families, 1257 genera, 28540 species)

Superorder : 1. Pandananae 2. Taccanae 3. Lilianae

Subclass 4. Commelinidae (2 superorders, 9 orders, 35 families, 1293 genera, 25730 species)

Superorder : 1. Arecanae 2. Commelinanae

Subclass 5. Ranunculidae (2 superorders, 9 orders, 18 families, 299 genera, 6330 species)

Superorder : 1. Proteanae 2. Ranunculanae

Subclass 6. Hamamelididae (1 superorder, 6 orders, 26 families, 16 genera, 4695 species)

Superorder : 1. Hamamelidanae

Subclass 7. Caryophyllidae (1 superorder, 5 orders, 34 families, 688 genera, 11630 species)

Superorder : 1. Caryophyllanae

Subclass 8. Rosidae (12 superorders, 20 orders, 146 families, 3758 genera, 72360 species)

Superorder : 1. Santalanae 2. Balanophoranae 3. Dillenianae 4. Rosanae 5. Celastranae 6. Oxalidanae 7. Brassicanae 8. Rutanae 9. Myrtanae 10. Violanae 11. Hypericanae 12. Geranianae

Subclass 9. Asteridae (4 superorders, 15 orders, 81 families, 2733 genera, 44680 species)

Superorder : 1. Cornanae 2. Ericanae 3. Aralianae 4. Asteranae

Subclass 10. Lamiidae (2 superorders, 4 orders, 44 families, 2635 genera, 48340 species)

Superorder : 1. Solananae 2. Laminanae

Thorne's diagram, indicating the relationship between the different groups, is a phylogenetic shrub-viewed from above with centre of the diagram left empty to indicate extinct early angiosperms. Those, that are nearer the centre, represent primitive groups and those nearer the periphery, the advanced groups. The relative number of species is indicated by baloons of different sizes. Such a diagram has not kept pace with latest electronic revisions at websites. The diagrammatic representation, showing evolutionary relationships among different taxa of angiosperms, is given in fig. (6.20).

Thorne's system of classification of angiosperms (now known by class name Angiospermae) published in 2000 in *Botanical Review* recognised 10 subclasses, somewhat approaching the informal group of APG. Out of these 10 subclasses, 7 belong to traditional dicots and 3, placed after Magnoliidae and before Ranunculidae, to traditional monocots. This more realistic phylogenetic arrangement of different taxa brings the system more closer to APG. The system recognised 31 superorders, 74 orders, 471 families in angiosperms. There were also 7 uncertain genera out of which 4 were assigned tentatively to monotypic families. In those orders where it was felt necessary, the classification is elaborate and suborders and subfamilies have also been recognised.

The classification had undergone a major revision in 2003 after APG II has been published. This system has some significant changes from 1999 (and 2000) version. Some significant parallels in Thorne (2003) and APG II are :

1. Placing Amborellaceae, Chloranthaceae and Austrobaileyaceae towards the beginning of angiosperms.
2. Recognising Canellales, Piperales, Laurales as independent orders within Magnoliidae.
3. Shifting Triuridaceae and Stemonaceae closer to Pandanaceae.

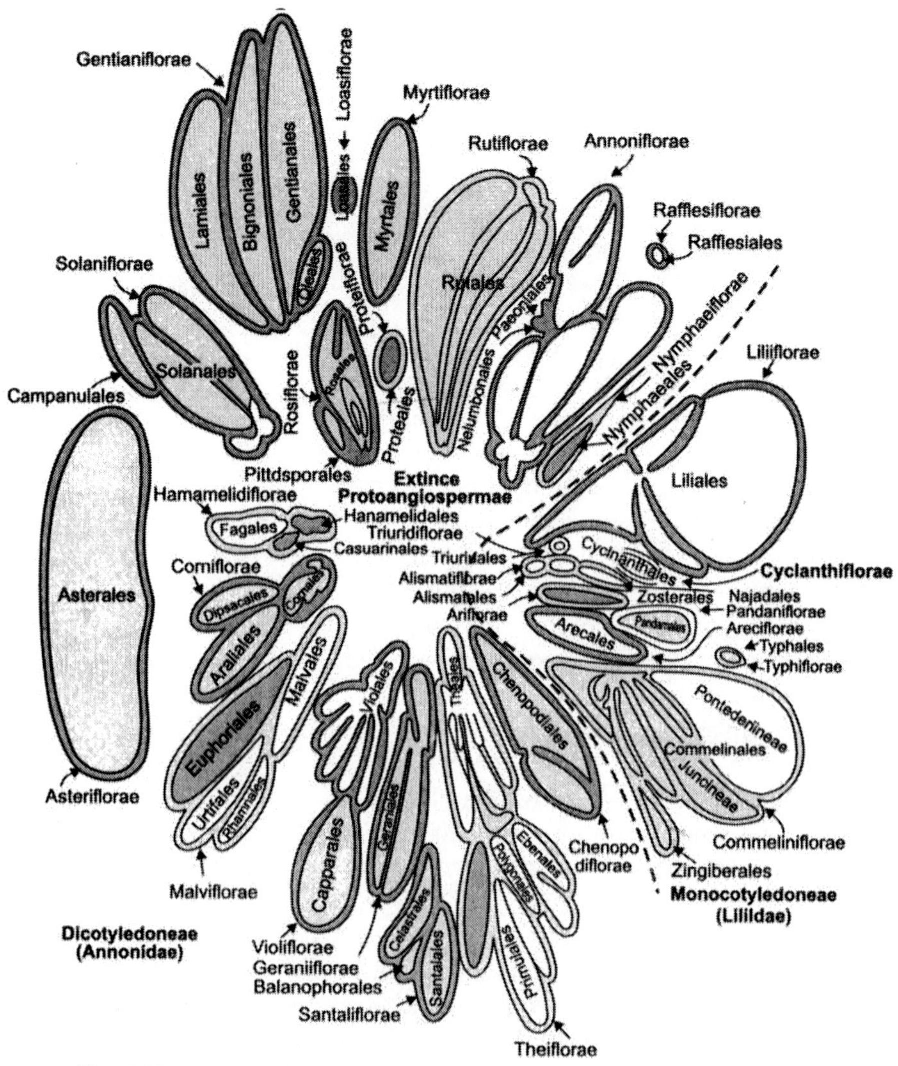

Fig. 6.20 A diagrammatic representation showing interrelationships among different taxa of angiosperms (Thorne, 1983).

4. Bringing families : Proteaceae, Platanaceae, Buxaceae and Didymelaceae closer.
5. Changing the position of Hamamelididaceae, Saxifragaceae, Vitaceae away from Rosidae to Hamamelididae.
6. Placing Ericaceae and related families under Asteridae and away from Rosidae (and abolishing Dilleniidae).
7. Recognising Garryales as distinct order, including Garryaceae, Aucubaceae and Eucommiaceae.

In 2003 version, Thorne also introduced the concept of assigning the degree of confidence in hierarchial level, *i.e.*, circumscription and alignment of taxa, *e.g.* A denotes for those representing limited confidence, B for probably correct and C for considerable confidence in assignment.

Merits of the System

Thorne's classification has kept pace with recent developments and is being regularly updated. It has the following merits over the previous systems of classification :

1. Thorne's system is highly phylogenetic as it incorporates the recent information and evidences from molecular systematics and chemotaxonomy and balancing with evidence from other sources.
2. Like Dahlgren and other recent systems. Thorne's system recognises an agreeable rank of class to angiosperms.
3. The classification is more exhaustive than other contemporary systems.
4. Orders and families, where necessary, have been divided into suborders and subfamilies, respectively.
5. Thorne's system, unlike APG II, has found a place for all unplaced families of APG.
6. Abolition of traditional groups monocots and dicots and dividing angiosperms directly into various subclasses is in line with the recent phylogenetic concept.
7. The circumscription of subclasses paralleles with informal groups of APG.
8. The system is superior over APG II in that formal group names are given to all supraordinal ranks.
9. Placement of Amborellaceae, Chloranthaceae and Austrobaileyaceae at the beginning of angiosperms is generally favoured in recent cladistic schemes of APG.
10. The subclass Magnoliidae, at the beginning of angiosperms, includes both traditional Magnoliids as well as Paleoherbs, both of which are now generally regarded as among the most primitive angiosperms.
11. Recognition of superorders with name ending–**anae** has resulted in more realistic arrangement of orders within subclasses.
12. In agreement with APG, the monocot families have been placed between primitive dicots and more advanced dicots and not towards the end of angiosperms as followed by Takhtajan, Cronquist and Dahlgren.
13. Order Urticales, along with Euphorbiaceae, has been shifted under Rosidae, where it seems to be better placed.
14. Families Winteraceae and Canellaceae have been brought together under the same order, and rather same suborder on the basis of their affinity strongly supported by morphological studies and multigene analyses.
15. The separation of Brassicaceae and Capparaceae has been supported by data from chloroplast DNA sequence (Hall, Systma and Iltis, 2002), consistent with data from morphological studies.

16. Family Liliaceae (of Hutchinson and earlier workers) has been split into a number of monophyletic families such as Liliaceae, Asparagaceae etc., in line with the arrangement in APG classification.
17. The Cornales and Ericales are placed together under Asterales. This is in agreement parallel with APG.
18. The concept of assigning the degree of confidence (A, B or C) in the hierarchial level and circumscription and alignment, has proved quite useful for better understanding of phylogenetic affinities.
19. Merger of Budlejaceae in Scrophulariaceae is supported by morphological studies (Bremer *et al.*, 2001) and also molecular studies (three gene analysis) by Olmstead *et al.* (2001).
20. Shifting Triuridaceae and Stemonaceae closer to Pandanaceae is in agreement with the arrangement in recent APG schemes. Their placement under Pandanaceae is justified on the basis of evidence from 18 S rDNA sequencing (Chase *et al.*, 2000). Triurididae, as an independent subclass, is not justified as evidenced by recent studies.
21. Merger of Asclepiadaceae with Apocynaceae has been supported by molecular analyses by Judd *et al.* (1994) and later by Sennblad and Bremer (1998). According to Judd *et al.* (2002), recognition of distict Asclepiadaceae would render Apocynaceae paraphyletic.

Demerits of the System

Despite several improvements over previous classifications, system has some of the following demerits :

1. Inspite, the system being highly phylogenetic, it is not very useful for identification and adoption in herbaria and floras because identification keys for genera, their distribution and descriptions are not given.
2. The system places Asteridae before Lamiidae but data from molecular studies justify the placement of Asteridae (with circumscription somewhat similar to Euasterids II) after Lamiidae (comparable to Euasterids I of APG II).
3. The system places Ceratophyllaceae after monocots under Ranunculidae whereas in recent versions of APG Ceratophyllaceae are placed before monocots. Studies of Zanis *et al.* (2002) and Whitlock *et al.* (2002) have shown that the family is a sister group of monocots as also shown by studies on microsporogenesis and structure of leaf margin.
4. Grewiaceae have been recognised as an independent family but recent APG classification (Judd *et al.*, APG II and APweb) places all members of Tiliaceae, Bombacaceae and Sterculiaceae under Malvaceae.

5. The system places Nymphaeaceae after the usual families of Magnoloid complex despite molecular studies (Qui *et al.*, 2000; Soltis *et al.*, 2000) strongly support the placement of Nymphaeaceae in basal angiosperm clades.
6. The system separates *Cabomba* and *Brassenia* under Cabombaceae on the basis of trimerous flowers, with distinct sepals and petals, in having 2–3 free carpels and fruit a follicle but the cladistic analyses support their placement under Nymphaeaceae as also by APG II, APweb and Judd *et al.* (2002). The separation of Cabombaceae would render Nymphaeaceae as paraphyletic.
7. The system is not very clear about the affinities of five genera (*Emblingia, Guametela, Haptanthus, Heteranthia* and *Pteleocarpa*) of angiosperms.

Angiosperm Phylogeny Group (APG)

The first serious attempt towards developing a **Cladistic classification** was made by Bremer and Wanntrop (1978, 81). They suggested that Angiosperms should be treated as subclass Magnoliidae of class Pinatae (seed plants). The Magnoliidae should not be divided into monocotyledons and dicotyledons because it will make the group paraphyletic, instead it be divided directly into number of monophyletic groups (superorders). The proposal was not taken seriously as all major systems of classification recognised monocots and dicots as separate groups.

With further utilisation of molecular data and development of powerful tools of data handling, there has been a considerable revival of cladistic concepts. This led to the development of **Angiosperm Phylogeny Group (APG) Classification.** And all this could become possible by the dedicated efforts of nearly two decades by a group, **Angiosperm Phylogeny Group** of dedicated workers (K Bremer, MW Chase, B Bremer, A Backlund, B Briggs, MHG Gustafsson, SB Judd, FA Kellogg, PF Stevens, M Thulin and several others) who published the APG classification of angiosperms in 1998. They arranged 462 families into 40 putative monophyletic orders and a small number of informal monophyletic higher groups : monocots, commelinoids, eudicots, core eudicots, rosids, eurosids I, eurosids II, asterids, euasterids I and euasterids II. Under these informal groups, there were also listed a number of families without assignment to orders, and a large number of unassigned families were listed both at the beginning and at the end of the system of classification. In the beginning of the system, there were 4 orders with no supraordinal grouping (rank above the order) into informal groups and at the end, there were 25 families of uncertain position for which no conclusive data existed regarding placement anywhere within the system.

Recent cladistic analyses are revealing the phylogeny of flowering plants in increasing detail. Also there is support for the monophyly of many major groups above the family level. With many elements of the branching sequence of phylogeny established, a revised suprafamilal system of classification of flowering plants becomes both feasible and desirable. That a simplest division of angiosperms into monocots and dicots does not reflect phylogenetic history, was realised because of cladistic information.

Judd, *et al.* (1999) presented some modifications in APG classification by recognising a total of 51 orders and shifting some families to these orders from informal groups where they were placed directly earlier (in 1998) classification. Their book, however, lists only major families and as such about 200 families have been left out but the revision presented in the 2nd edition (2002) has further improvements in line of their strategy of APG and except for minor differences, is largely similar to APG II classification. The families that are not listed in the book, presumably their placement is the same as in APG II.

Another revision of APG (APG II, 2003) (table 6.11) with continuous upgradation by PF Stevens (2003) on Angiosperm Phylogeny web–*http://www.mobot.org/MOBOT/research/AP web,* has resulted in considerable refinement in the APG scheme, with more and more families (and some orders) coming out of list of unplaced taxa. Angiosperm Phylogeny Group II classification (2003) recognises in all 47 families placed under 45 orders of angiosperms, of which 44, considered more or less monophyletic, are placed in 11 informal groups and one order is unassigned at the beginning of angiosperms.

Table 6.11 Broad outline of APG II (2003) classification of Angiosperm Phylogeny Group
MAGNOLIOPHYTA

S. No.	Group	Order	S. No.	Group	Order
		Austrobaileyales			
1.	**Magnoliids**	1. Magnoliales 2. Laurales 3. Canellales 4. Piperales 5. Ceratophyllales	6.	**Rosids**	1. Crossosomatales 2. Geraniales 3. Myrtales
2.	**Monocots**	1. Acorales 2. Alismatales 3. Asparagales 4. Dioscoreales 5. Liliales 6. Pandanales	7.	**Eurosids I**	1. Celastrales 2. Malpighiales 3. Oxalidales 4. Fabales 5. Rosales 6. Cucurbitales 7. Fagales
3.	**Commelinids**	1. Arecales 2. Poales 3. Commelinales 4. Zingiberales	8.	**Eurosids II**	1. Brassicales 2. Malvales 3. Sapindales
4.	**Eudicots**	1. Ranunculales 2. Proteales	9.	**Asterids**	1. Cornales 2. Ericales

(Contd....)

5.	**Core Eudicots***	1. Gunnerales 2. Caryophyllales 3. Santalales 4. Saxifragales	10.	**Euasterids I**	1. Garryales 2. Gentianales 3. Lamiales 4. Solanales
			11.	**Euasterids II**	1. Aquifoliales 2. Apiales 3. Asterales 4. Dipsacales

Such proposition includes :

1. **Unplaced families** : Amborellaceae, Cabombaceae, Chloranthaceae, Nymphaeaceae.
2. **Taxa of uncertain position** : *Aneulophus*, Balanophoraceae, *Centroplacus*, Cynomoriaceae, Cytinaceae, Haplostigmataceae, *Leptaulus*, Medusandraceae, Metteniusaceae, Mitrastemonaceae, Pottingeriaceae, Rafflesiaceae, *Soauxia* and *Trichostephanus*. The unplaced families in groups include 2 in Monocots, 1 in Commelinids, 5 in Eudicots, 3 in Core Eudicots, 2 in Rosids, 3 in Eurosids I, 2 in Eurosids II, 3 in Euasterids I and 10 in Euasterids II.

APG II classification is largely based on molecular data from studies on chloroplast DNA using restriction site polymorphism (cp RFLP), analysis of chloroplast gene for subunit F of NADP dehydrogenase (ndh F, in the small copy region), for 'a' and 'b' subunits of RNA polymerase II (rpoA and rpoC2 in a large single copy region) for 'b' subunit of ATP synthetase I. Data obtained from studies on *rbc*L sequences have supported the monophyletic nature of Caryophyllidae, the union of family pairs : Asclepiadaceae–Apocynaceae, Araliaceae–Apiaceae and Brassicaceae–Capparaceae and polyphyletic nature of Saxifragaceae and Capprifoliaceae. The phylogeny of seed plants is being revealed in increasing detail by recent cladistic analysis of many major groups above the family level (suprafamilial groups). Data obtained from molecular studies strongly support suprafamilial classification of seed plants. The APG II classification organises seed plants into a selected number of monophyletic suprafamilial groups.

The history of APG classification reveals several interesting facts. A few trends that are emerging fast include convinient placing of monocots under two groups : the commelinids and the rest of the monocots, and giving them place after primitive dicots (and possibly Magnoliids). In 1998 edition of APG classification, there were 81 unplaced families of which 11 were placed towards the beginning, 25 towards the end and 45 in the informal eleven groups. In addition, 18 families, classified in four orders and placed in the beginning, did not have any taxon as supraordinal rank (rank above the level of the order). In APG II (2003), the number of these families has gone down to only 40 and all four orders that were placed in the beginning were adjusted under magnoliids (a combination of non-monocot paleoherbs and magnoloid complex of Judd *et al.*, 2002), recognising only one unplaced order in the beginning. The list of unplaced families in the beginning has been reduced to only 4 and uncertain families towards the end to just 9. Stevens, in the

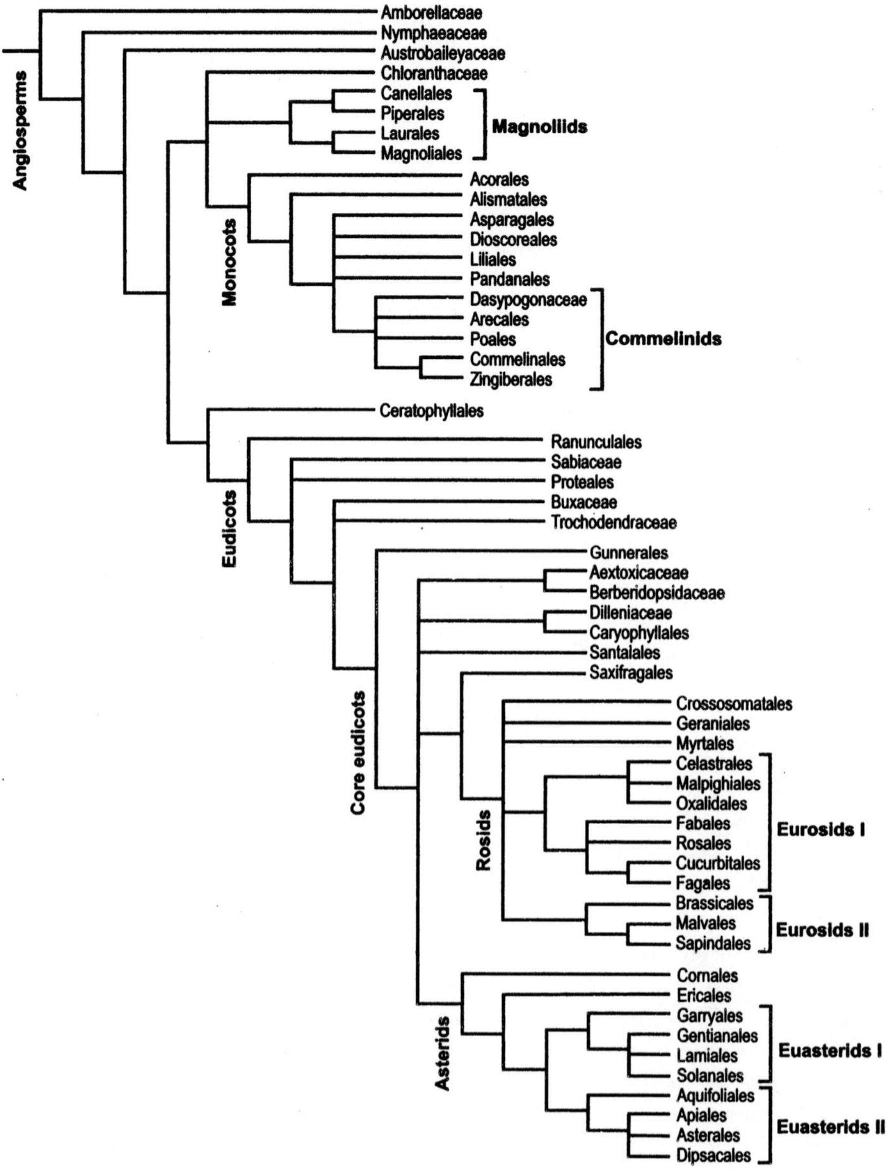

Fig. 6.21 Inter-relationship of orders and some families presented in APG II classification (2003), having **bootstrap support** of more than 50 per cent.

periodically updated APweb (version 4 May, 2003), has reduced the number of unplaced families to 32, none in the beginning and 10 towards the end. He has also recognised 9 additional orders of which 2 accommodate unplaced families of APG II (2003) in the beginning of angiosperms, 1 in Commelinids, 2 in Eudicots, 3 in core Eudicots and 1 in Eurosids I. The publication of Judd *et al.* (2002) covers only major families of angiosperms and as such many unplaced families are unattached (not listed), but presumably the arrangement of APG II is accepted for these.

Despite Thorne, not belonging to Angiosperm Phylogeny Group, has kept pace with the recent molecular developments. He is trying to balance hierarchial classification with concept of monophyly, so sacred with APG. The question arises-should primitive angiosperms include only paleoherbs (Nymphaeaceae, Cabombaceae, Piperaceae, Amphorelaceae etc.) and true magnoliids (Magnoliales, Laurales etc.) placed after commelinids as in the classification of Judd *et al.* (1999) and Stevens (2003) or should the primitive angiosperms include paleoherbs as well as true magnoliids and be placed before monocots as in Judd *et al.* (2002), Thorne (2003) and APG II (2003), remains unanswered. Significantly, Thorne (1999, 2000, 03) has come up with a major revision of his classification, bringing it on lines of APG and maintaining the hierarchial structure and finding a place for almost all families with only five genera remaining unplaced. Notably his 10 subclasses are more or less complementary to the 11 informal groups of APG, the relationship somewhat reversed in Asterids.

The true phylogenetic picture of angiosperms is still to come out. There seems a considerable unanimity in removing some of the primitive herbaceous families of Magnoloid complex (such as Amborellaceae, Cabombaceae, Paeoniaceae, Austrobaileyaceae, Trimeniaceae, Illiciaceae and Schizandraceae) and place them towards the beginning of angiosperms, rendering both monocots and dicots as truely monophyletic groups. At present, the position of Magnoliids (after removal of herbaceous families) is, however, uncertain whereas APG II places them before monocots, APweb (authored by Stevens, who also contributed to APG II), placed this group after monocots. Judd *et al.*, 2002 (who also contributed to APG II), in first edition (1999) of their book, had placed Magnoliids, like APweb, after monocots, but in their second edition (2002), like APG II, Magnoliid complex is placed before monocots.

The system of classification of Judd *et al.*, differs from APG II in recognising 3 orders, Amborellales, Nymphaeales and Austrobaileyales for **basal families**, shifting Ceratophyllales (although with uncertain position) to the end of Magnoliid complex, recognising Nartheciales as a distinct order (APG II includes Nartheciaceae under order Dioscoreales) under monocots, including Saxifragales under Rosid clade (APG II includes under core Eudicots), recognising order Vitales under Rosid clade (Vitaceae unplaced in Rosids in APG II), shifting Myrtales from Rosid clade to either Eurosids I or II (position uncertain), recognising Zygophyllales as a distinct order under Eurosids I (Zygophyllaceae unplaced in Eurosid I in APG II) and shifting the sequence of some orders.

Although the jinx of dicot-monocot grouping of angiosperms has been broken but at present it is not sure about the position of Magnoliid complex, that includes some of the most primitive representatives of angiosperms. The position of Piperales seems to be more or less stabilised towards the end of Magnoliids but Ceratophyllaceae is still to find a stable position. Judd *et al.* (2002), place family under order Ceratophyllales after Piperales within Magnoliid complex with uncertain position but Stevens in APweb and APG II places it before monocots, towards the beginning of angiosperms.

Merits of the System

Angiosperm Phylogeny Group (APG) classification is the result of concerted efforts of dedicated workers of APG. It is a newly emerging and fast evolving system, that over last five years, has undergone a dramatic modification of the previous systems of classification. There are several merits of APG (and APweb) systems :

1. The system is based on sound phylogenetic principle of constructing taxa only where monophyly is established.
2. APG system is based on a synthesis of information based mainly on morphology, anatomy, embryology, phytochemistry and more strongly molecular studies.
3. The formal groups, mostly upto the level of order, are given only where monophyly is firmly established.
4. The traditional division of angiosperm into monocots and dicots, has been abandoned.
5. Various monocot taxa have been placed in between primitive angiosperms and eudicots, thus overcoming problem of polyphyly in the earlier recognised monocots and dicots.
6. There is a constant effort to construct supraordinal monophyletic clades.
7. In order to show the general affinities between various groups of angiosperms, based on molecular as also on information from other fields, a number of cladograms are being presented.
8. The families with several primitive features are placed towards beginning of angiosperms, *e.g.* Amborellaceae, a unique family in angiosperms in having granular and non-tectate ectexine, is placed at the start.
9. The four unplaced families in APG towards the beginning of angiosperms, have been accommodated under orders in APweb.
10. The number of unplaced families in various informal groups and uncertain families towards the end have been sufficiently reduced in APG and APweb systems, forming ordinal places for many unplaced families of APG (1998).
11. Winteraceae and Canellaceae, based on multigene analyses and morphological studies, are brought together under the same order.
12. The Liliaceae of Hutchinson and earlier workers, have been spread over several monophyletic families such as Liliaceae, Alliaceae, Asparagaceae, Asphodelaceae etc.
13. The merger of Asclepiadaceae with Apocynaceae has been supported by molecular evidences (Judd, *et al.*, 1994) and Sennblad and Bremer (1998). According to Judd *et al.* (2002), recognition of Asclepiadaceae, as a distinct family, would render Apocynaceae as polyphyletic.
14. The merger of Budlejaceae and Myoporaceae with Scrophulariaceae has been supported by morphological studies (Bremer *et al.*, 2001) and molecular analysis—three gene analysis by Olmstead *et al.* (2001).

15. Circumscription of Agavaceae has further strengthened to include other genera (*Hosta, Camassia* and *Chlorogalum*) which also have bimodal karyotypes and such placement has also been adopted by Judd *et al.* (2002) and Thorne and APweb.
16. The circumscription of Malvaceae has been broadened to also include Tiliaceae, Sterculiaceae and Bombacaceae, based on morphological studies and molecular evidence, thus making family Malvaceae as monophyletic.

Demerits of the System

Despite the fact that APG classification is still evolving and continuously improving, it is expected that the system will take considerable time before it is stabilised and is tested by various parameters. A few obvious shortcoming of the system are :

1. The system of classification has not proceeded below the family level. As such it is not useful in practice and for adoption in herbaria and floras.
2. Despite best efforts of assembling families into more or less monophyletic orders, a large number of unplaced families and few unplaced genera still remain to be placed suitably.
3. Although most of the orders have been assembled into informal groups but no formal names, in accordance with Botanical Code, have been given to these groups.
4. APG II (2003) places all presumed primitive families of angiosperms before monocots but APweb has transferred magnoliids to a position after monocots and Commelinids.
5. Family Chloranthaceae is generally placed in Magnoliidae in other contemporary phylogenetic systems. The family has primitive floral morphology and is placed towards the beginning of angiosperms but in APG II, it has been placed after commelinids and before magnoliids in APweb without justification of the shifting.
6. The system provides angiosperms the rank of a division, but there are no formal taxa between the rank of an order and division. This is rather an unusual phenomenon of a classification.
7. APG classification merges family Capparaceae with Brassicaceae, but chloroplast DNA data points to the separation of these two families as also family Cleomaceae. Thorne (2003) also recognises Capparaceae, Brassicaceae and Cleomaceae as distinct families.
8. In several cases, results from chloroplast phylogeny and nuclear phylogeny do not agree, which warn not to over rely on these attributes while ascertaining their suprafamilial status.

Chapter 7

Herbaria

Herbarium is a place where plant material is preserved using various techniques and arranged in the sequence of an accepted system of classification to facilitate identification and nomenclature and preserve the record of the plants. The specimens, after collection, are properly dried and pressed, taking care that they retain all essential parts necessary for identification. Certain plants, which are either succulents or otherwise unsuitable for pressing and drying technique, may be fixed in suitable liquid preservatives such as formaldehyde (2–5 per cent) or formaline aceto alcohol –(FAA–5 : 5 : 90).

Thus, herbarium is a depository, usually of dried plant specimens, to preserve for all times a series of specimens and notes that will provide the maximum information about the plants. It was Luca Ghini, also credited for developing Pisa Botanical Garden in Italy, in 1544, who initiated the art of herbarium making by pressing and sewing specimens on sheets of paper. This art was disseminated throughout Europe by his students who mounted the sheets and bound them into book volumes. Linnaeus preferred to mount the specimens on single sheets, storing them horizontally–a practice followed even today.

Following equipments and operations are needed to successfully establish a herbarium.

Field Equipments

While proceeding for field collections–a significant part of the taxonomic study, following essential articles (fig. 7.1) are needed for plant collection and exploration of vegetation of an area.

1. Digger or **Trowel and Pruning Knife** or **Shear.** For obtaining a herbaceous whole plant, its deep rooted parts or rhizomes, corms, bulbs, tubers etc. are required to be taken out with the help of a digger or a trowel. These underground parts should be made as a part of the specimen because they furnish useful diagnostic features and provide perfectness to the specimen. Suitable shape and size of digger or trowel may be choosen according to soil types and deepness of sub-aerial parts.

Sharp prunning and cutting, knives with heavy sheet, dandalian knife, hammer with chisel blade, secateur, prunning shears, are required for cutting branches and twigs of shrubs or trees. The tools must be very sharp as to avoid disappointment during the act. A sharp bent sickle, with a long handle, will prove useful to obtain twigs from tall trees. The length of the handle can be extended by adding additional handle rods.

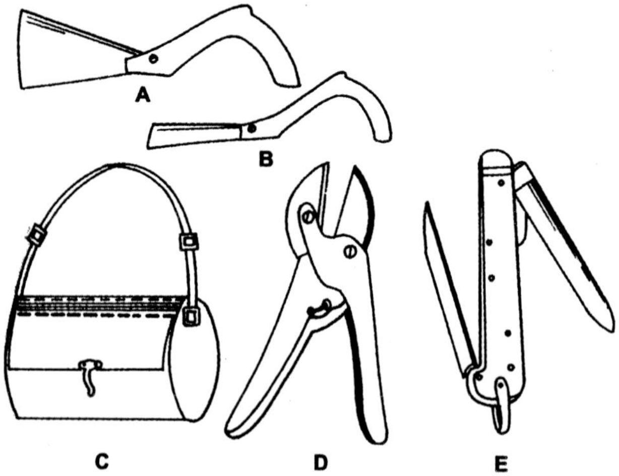

Fig. 7.1 Field equipments. A, B-trowel, C-vasculum, D-prunning shear and E-knife.

2. Vasculum. Vasculum (fig. 7.1C) is a container made of tin or aluminium sheet, with a hinged tight lid, useful for handling and keeping the specimens fresh before pressing them. A standard vasculum is somewhat oval on the ends and is usually 45 to 60 cm long. It is painted white to reflect sun rays in order to protect specimens from heat and locate it quickly in greenery. Vasculum is provided with a strap for getting it hanged on the shoulders. It helps retain freshness of specimens untill they are preserved. Recyclable polybags are also useful for collection of specimens and to allow them remain separate and isolated and fresh for a long time.

3. Field Notebook. Recording of data at the time of collection, is important while on a field trip. Special type of bound pocket sized notebooks, having serially numbered detachible sheets with tags (fig. 7.2), are used for

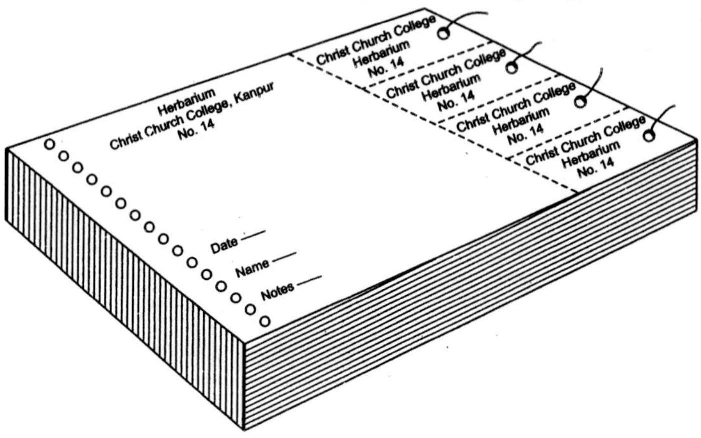

Fig. 7.2 Field notebook.

recording the data, such as field number of the specimen, date of collection, locality of specimen, the habitat, the botanical name of the plant (if known), the flower colour, abundance, variations, associated plants, vernacular name, ecological condition, topography, branching system, root system, plant size etc. These sheets at the time of collection, are tied by tags to the respective specimens or the bags in which they have been kept.

For study of smaller plants and flowers, a pocket magnifying lens will prove as an aid. For taking coloured picture of each plant in its natural environment, a camera or a video camera may prove helpful.

4. Plant Press and Folders. Plant press or folder, used for pressing, drying and preserving flat specimens, usually comprises a pair of metal, wood or strong plywood frame of 30 × 45 cm (fig. 7.3A). Between the frames are kept blotting papers or old newspapers which absorb moisture quickly and uniformly from plants. In between the blotters or old newspapers, are placed corrugated sheets as ventilators when plants are dried by artificial heat. These provide passage for dry hot air through the press. The press is then tied up tightly by means of a distantly placed strap or cord (fig. 7.3B). Depending on moisture content, the blotters can be replaced by new ones. The beauty of the herbarium depends on proper placing the specimen and care in the method of their pressing.

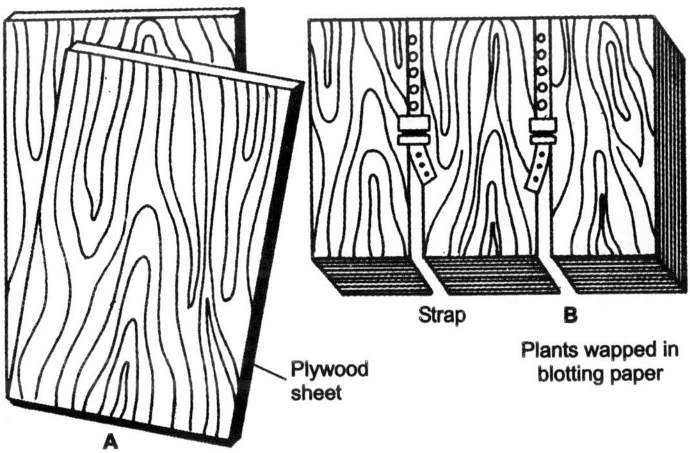

Fig. 7.3 Herbarium press. A-plywood sheets and B-plant press with specimens.

Herbarium Techniques

Herbarium techniques involve a series of operations, such as collection, pressing, drying and poisoning, mounting and stitching, labelling, filling and deposition. These comprise following steps.

1. Collection. While collecting the plant specimens for making herbarium, following points should be kept in mind :

(i) The plant specimen should bear flowers and fruits, if present.

(ii) Herbaceous small plant specimen should be collected with roots or other underground parts, if possible.

(iii) For woody plants, a twig of about 25 cm with leaves and flowers, will form an ideal material.

(iv) To avoid any chance of loss or damage, atleast four specimens of the plant should be collected.

(v) Note sheet's number and data recorded in the field notebook, should immediately be tagged with the specimen.

(vi) Soon after the specimens are collected, they should be pressed in the field itself or just after, in the rest camp.

2. Pressing. The collected specimens should carefully be displayed by placing them in the centre on the pressing sheets (blotting papers of rough quality or old newspapers). If specimens are large enough, they should be bent giving them shape of V, N or W. Extra branches and leaves can be removed to avoid overlapping of organs. The specimens should be placed for pressing in the manner that the bundles are uniform in thickness in the middle and on the sides, thus exerting a uniform pressure. The pressing sheets loaded with the specimens should be kept one above the other horizontally and then tightly bound in a plant press (fig. 7.3B).

3. Drying and Poisoning. For effective drying, the press is opened the following day and the drying papers are replaced by fresh ones. The changing of drying papers is repeated everyday for about fortnight, or until the plant specimens appear perfectly dried. In the humid climate, the changing of papers is done twice a day to have good results. If the wet papers are not replaced by fresh drying papers, the specimens get mouldy and spoiled. Drying in intense sunlight causes plant specimens to lose their original colour and become black. Artificial heat may be given if the weather is too humid.

The dried specimen is then poisoned, usually with a saturated solution of mercuric chloride ($HgCl_2$) in alcohol or sprit. Lauryl pentachloride is a safer substitute of mercuric chloride. The specimen is either dipped in the solution by holding it with tong or applying the solution by means of a brush. Care should be taken that the solution does not come in contact with the skin. After sometime, when the alcohol evaporates and the specimen becomes dry, it is mounted on herbarium sheet and labelled (fig. 7.4).

4. Mounting and Stitching. After drying and poisoning, the plant specimen is mounted on a herbarium sheet (fig. 7.4) of standard size, *i.e.*, 29 × 42 cm. The herbarium sheets are usually made up of thick hand-made durable card sheets. The dried specimens are glued on herbarium sheets and the stem or branches and the petioles of the leaves are stitched. Narrow strip of adhesive liners, strips of brown gummed paper or adhesive coated cellophane tapes may be used instead of stitching by strong threads. It is advisable to mount one specimen on each herbarium sheet. Dissected and loose parts, such as flowers, fruits and seeds, are kept in paper packets and pasted to the mounted sheet.

5. Labelling. After mounting the specimen, a label is glued on the lower right hand corner of the sheet, to provide information gathered from the field notebook. The format and size of the label, which is sometimes printed and

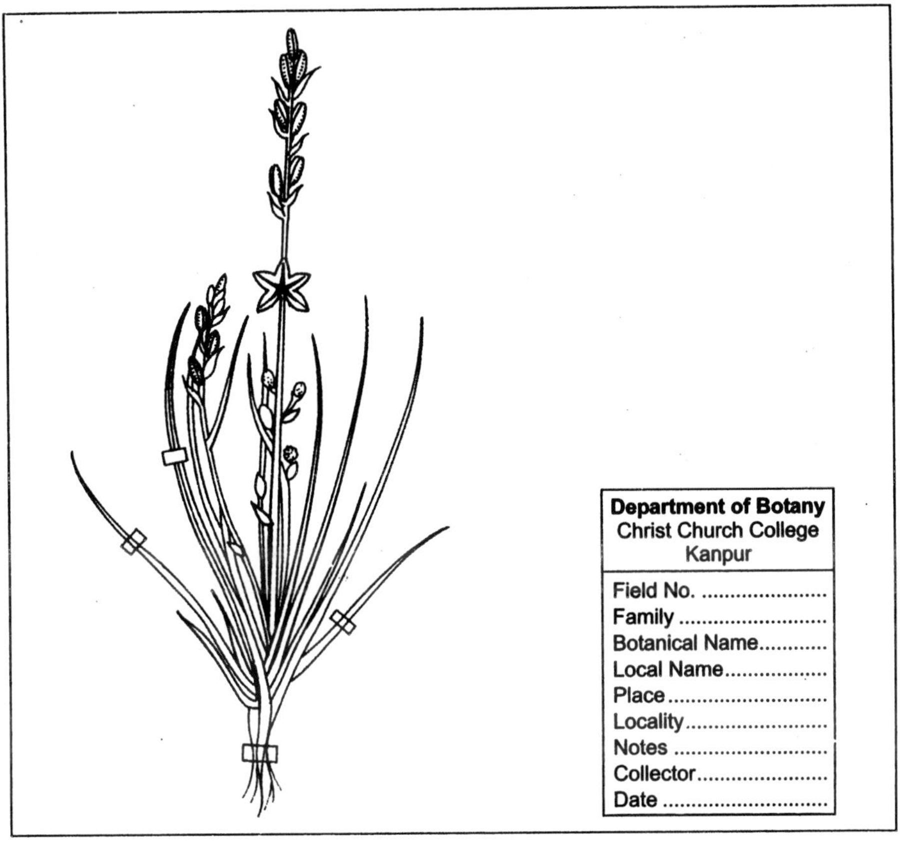

Fig. 7.4 A herbarium sheet.

awaits for information to be filled, is about 4.5 × 7.5 cm. The label should furnish following minimum information about the specimen plant.

(i) Name of the organisation with which specimen plant originated
(ii) Name of the family to which specimen plant belongs
(iii) Botanical name of the plant with authority
(iv) Locality of collection including place
(v) The date of collection
(vi) The habitat of plant specimen
(vii) Field notes and collection number
(viii) Name of collector
(ix) Vernacular name and local use of plant.

6. Filling and Storing. Plant specimens, which have been properly mounted and identified, are filled systematically in special wooden or steel cabinets with pigeon holes alloted for different families. The herbarium sheets loaded with specimens are filed inside folders which are of various colour schemes indicating species, genus, family, geographical area etc. In most of Indian herbaria, as also in Kew, the specimen plants are arranged and stored following Bentham and Hooker's system of classification, although Engler and Prantl's system is also recommended.

The permanent collection of herbarium specimens require its proper protection and care against damages by fungi and insects. A periodical fumigation with chemicals and repoisoning them by brushing with solution of mercuric chloride or lauryl pentachlorophenate and using insect repellents such as Dichloro-diethyl trichlorothane (DDT), would save the herbarium from damage and check the loss of valuable plant specimens.

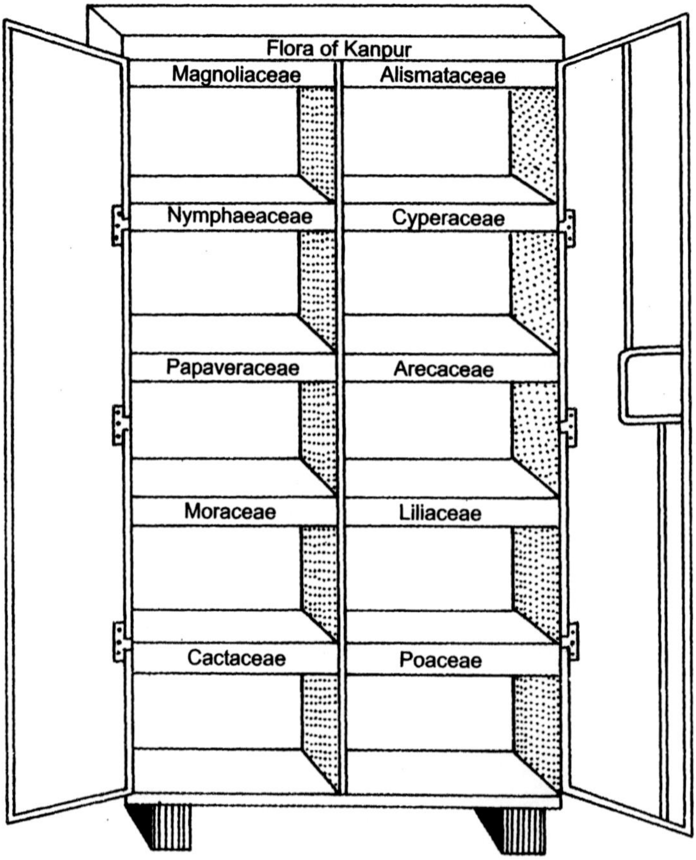

Fig. 7.5 A herbarium almirah.

Role of a Herbarium

Herbaria function as a safe place for storing pressed plant specimens, especially Types and have become major centres of taxonomic study. From isolated personal collections, they have grown into large institutions of national and international stature with millions of specimens from different parts of the world. They also form an important link for research in other fields of study. The classification of world flora is mainly based on herbarium material and associated literature. Recently, herbaria have gained importance as sources of information on endangered species and are of primary interest for conservation strategies. The role played by herbaria includes :

1. Primary role of a herbarium is to act as a repository of dried plant specimens, safeguard them against loss and destruction by fungi, insects etc. and make them available for study.
2. Several herbaria of repute, keep Type Specimens–the principle proof of the existence of a species, in safe custody, often in rooms with restricted access.
3. As original documents upon which knowledge of taxonomic characters rests, herbarium specimens greatly help in developing floras, manuals and monographs.
4. The majority of world's herbaria, have a wide-ranging collection of specimens and extend facility of identification on-site or sending specimens and getting them identified by experts. Some advanced herbaria have on-line identification facilities of specimens. Those engaged in taxonomic studies, can personally identify their collection by comparison with already identified herbarium specimens.
5. Voucher specimens (specimens for research study), preserved in various herbaria, provide an index of specimens on which studies on chromosomes, phytochemistry, ultrastructure micro-morphology etc., have been undertaken. On finding a contradictory or doubtful report, the voucher specimen can be critically examined in order to arrive at a more satisfactory conclusion.
6. Most herbaria have specimens collected from different parts of the world and, thus, their scrutiny can provide information on the geographical distribution of taxa.
7. Many of the herbaria extend facilities for training in herbarium techniques, organising field trips and even expeditions to remote areas.

Major Herbaria of World

The plant specimens collected over the years throughout the world with great care, are available for the reference and the study. The systematic development of herbaria started during eighteenth century when Linnaeus in Sweden and Roxburgh in India made efforts in this direction. Since then a large number of herbaria, 850 institutions, including 50 major herbaria (Lanjouw and Staffeu, 1952, 64), have been established throughout the world. A detailed index of world herbarium–*Index Herbariorum*, edited by Patricia Holmgren (Holmgren, Holmgren and Barnett, 1990), lists word's important herbaria. Some of the more important herbaria, their standard abbreviations and approximate number of specimens they contain, are given in tabie 7.1.

Table 7.1 Major Herbaria of the World, their Standard Abbreviations and Approximate Number of Specimens they Contain.

Herbarium	Standard Abbreviation	No. of Specimens Over
1. Royal Botanic Gardens, Kew, UK	K	7,000,000
2. British Museum of Natural History, London, UK.	BM	5,200,000
3. Museum of Natural History, Paris, France.	P	9,377,000
4. New York Botanical Garden, New York, USA.	NY	6,500,000
5. Conservatory and Botanical Garden, Geneva, Switzerland.	G	6,000,000
6. Missouri Botanical Garden, Saint Louis, Missouri, USA.	MO	5,250,000
7. VL Komarov Botanical Institute of Azerbaijan (Formerly Leningrad), Russia.	LE	5,700,000
8. Naturhistorika Riksmuseet, Stockholm, Sweden.	S	4,400,000
9. US National Herbarium (Smithsonian), Washington, USA.	US	4,340,000
10. Royal Botanic Garden, Edinburgh, UK.	E	1,500,000

Major Herbaria of India

In India, there is a wide range of herbaria : local, regional, institutional and national. Local herbaria serve the purpose of a small area, regional of the region and institutional, established by the Universities and Colleges, for teaching and research. The Central National Herbaria (CAL) of Indian Botanic Garden, Botanical Survey of India, Kolkata, in India, has over 2 million specimens. A few important herbaria, their standard abbreviations and approximate number of specimens they contain are given in table 7.2.

Table 7.2 Major Herbaria of India, their Standard Abbreviations and Approximate Number of Specimens they Contain.

Herbarium	Standard Abbreviation	No. of Specimens Over
1. Central National Herbarium, Botanical Survey of India (BSI), Kolkata.	CAL	2,000,000
2. Herbarium of the Forest Research Institute, Dehradun.	DD	300,000
3. Madras Herbarium, Coimbatore.	MH	150,000

(Contd....)

4. Herbarium of National Botanical Research Institute, Lucknow.	LWG	1,15,000
5. Herbarium of BSI, Central Circle, Allahabad.	BSA	50,000
6. Herbarium of BSI, Eastern Circle, Shillong.	ASSAM	1,15,000
7. Herbarium of BSI, Dehradun.	BSD	65,000
8. Herbarium of BSI, Pune.	BSI	1,25,000
9. Industrial Section of BSI, Indian Museum, Kolkata.	BSIS	40,000

Some other places that maintain their own herbaria are : Indian Agricultural Research Institute, New Delhi; Agricultural College and Research Institute, Coimbatore; Osmania University, Hyderabad; St. Xaviers College, Fort, Mumbai etc.

Chapter 8

Select Botanic Gardens

A garden is a place for growing flowering plants, fruits, vegetables etc. as a means of recreation and as a centre of aesthetic beauty owned by an individual or an organisation. But a botanic garden is a place where plants of differnt kinds and varieties, collected from various parts of the country and the world, are grown in a scientific and systematic manner. Botanic garden functions as an institution for scientific activities, both pure and applied, and as a place of aesthetic value to bring awareness in general public about plants.

History of Development of Botanic Gardens

The history of development of gardens is closely related to the history of human civilisation. Ancient Indian history reveals mention about gardens such as *'Panchvati'* and *'Ashok Vatika'* in our great epic *Ramayana.* Many rulers in India, developed gardens around their palaces for aesthetic beauty. Famous garden Shikander Bagh (now National Botanical Research Institute), in Lucknow, was established and later developed by Nawabs of Awadh.

The development of botanical gardens in the western countries, started as the centres of cultural activities whereas in the eastern countries as the centres for the training of medical students. In the ancient Mediterranean civilisation, gardens were the prominent features in the vicinity of places of worship and palaces. The 'Hanging Gardens' of Babylon are counted among the wonders of the ancient world.

The relevance of botanical gardens to the development of taxonomy dates back to the establishment of world's first botanic garden by an Italian Professor of Botany, Luca Ghini at Pisa, Italy in 1543. Luca Ghini, the first Director of this garden, was later succeeded by Andrea Caesalpino (1554-58). Many of the botanic gardens have been instrumental in motivating several well known authors to develop their own systems of classification while arranging plants grown in the gardens, *e.g.* Linnaeus while working at Uppasala, Bernard de Jussieu at Versailles, Hooker at Kew Gardens, England etc. The post-Linnean period witnessed accelerated exploration of plant wealth, giving rise to many botanic gardens in different parts of the world.

Functions of Botanic Gardens

Botanical gardens have been instrumental in many plant taxonomic activities. They have played a key role in introduction and acclimatisation of exotic plants. Many of them have helped in motivating several well known authors to develop their own systems of classification while trying to arrange the plants grown in the garden, *e.g.* Linnaeus while working at Uppasala, Bernad de Jussieu at Versailles and Bentham and Hooker at Kew etc. An ideal botanic garden is expected to perform the following functions :

1. Raising, introducing and acclimatising wide variety of global trees, shrubs, herbs and climbers of cultivated plants of ornamental, medicinal and other economic values. Tropical gardens need indoor growing space, such as screen houses or glass houses for specific plants.
2. Developing certain geographic formations, such as desert plants, alpine plants, marshy plants, aquatic plants etc. Water plants need a lake, impounded water reservoir or large water tank.
3. Conducting explorations, floristic studies and research in plant taxonomy and other related disciplines.
4. Supply of authentic plant material to organisations engaged in research and development.
5. Provide aesthetic appeal and attract people to observe general plant diversity, as also to curious plants, such as 'Big Banyan Tree' (*Ficus benghalensis*) in Indian Botanical Garden, Sibpur, Howrah, as an identity of the garden.
6. Integration in research projects to incorporate information obtained from related branches, such as anatomy, embryology, ecology, cytology, phytochemistry, physiology etc., in plant taxonomy with the help of research laboratories.
7. *Ex situ* conservation of genetic diversity, as also in conserving rare and endangered species of plants and germplasm collection of rare plant species.
8. Maintaining herbarium as a record of plants and access to identification for those who need it.
9. Offering literature about plant taxonomy and other related fields by its library.
10. Extending public services by providing information to general people for identification of native and exotic species of plants and methods of their propagation.
11. Facilitate domestication and conservation of wild plant species of economic virtues.
12. Exploration of possibilities of economic values in plant heitherto not explored.

13. Provision of sale of plants, seeds or plant propagules to garden lovers.
14. Offering seed exchange with those interested.
15. Provision of technical advice/consultation service for improvement of other gardens.
16. Networking of database of plant resources and to document and publish it.
17. Disseminating general information of the achievements of the garden among common people through popular literature.
18. Organising conferences, seminars, popular lectures, flower shows etc. to create mass awareness of plants.

MAJOR BOTANIC GARDENS OF THE WORLD

The history of botanic gardens of the world presents a fascinating insight. By the 12th century about 500 gardens were developed in Europe alone. During the course and later, many pleasure gardens were converted in botanical gardens. By now, a large number of botanic gardens have been developed worldover. Of these, about 800 important ones are documented in the International Directory of Botanical Gardens published by Henderson (1983). For exhaustive texts regarding botanic gardens of the world till 1969, one can go through the *Great Botanic Gardens of the World* (Hyams and Mac Quitty, 1969). In the light of conservation strategy, a botanic garden is an area set aside and maintained by an organisation for growing various groups of plants for study, aesthetic, conservation, economic, educational, recreational and scientific purposes. Some of the major botanic gardens are discussed/listed as follow :

Royal Botanic Gardens, Kew, England

More popularly known as 'Kew Gardens', this is the finest and largest botanic garden in the world. It also serves as the most important botanical research and resource centre in the world. The garden was developed around 1600 by the Kew House owned by Richard Bennet. The widow of the Prince of Wales commissioned the Garden in 1759 and William Atton took over as its Superintendent. Sir Joseph Banks introduced large collections from various parts of the world to this Garden. After the death of Princess of Augusta in 1772, the Garden was united with the gardens of Palace of Richmond and the extensive area and the management of the two gardens was transferred from the crown to the government in 1838 when it became named as Royal Botanic Gardens, which was originally established in 1760 but officially opened in 1841 when Sir William Jackson Hooker became its first official Director. He was responsible mainly for the advancement of the Garden. Later Sir JD Hooker took over as Director, succeeding his father. He authored several important publications. John Hutchinson also worked in this garden and developed his famous system of classification.

Under the able guidance of these authorities and many others, the Garden has grown into a premier research and educational institute with excellent herbarium (largest in the world) and library. Originally the Garden covered an area of 120 ha. The outstation of the Garden, Wakehurst in West Sussex, a rural estate of 202 ha with Elezabethan mansion, was acquired in 1965. The Royal Botanic Gardens, Kew, has directed and financed the development of Wakehurst to the extent that this place now makes a vital contribution in maintaining international reputation of Living Collections Department (LCD). The rich and diverse plant collections coupled with *in situ* conservation policies, which are maintained, add greatly to the LCD's activities. The High Weald area of Sussex offers varied topography, higher rainfall and more diverse and moisture retentive soil, and thus facilitates plantation of a great diversity of plants, many of which fail to thrive at Kew.

The living collection of Royal Botanic Gardens at Kew and Wakehurst is probably the largest and most diverse in the world and presents a multilevel encyclopaedic reference collection reflecting global phytodiversity. It is proving a source of reference in all aspects of botanical and horticultural science within Kew, Great Britain and throughout the world. The living collections of Kew, are most diverse with 351 families, 5465 genera and more than 28,000 species, growing successfully in the Garden. The arboretum (where only trees or woody species are grown), covers the area with 7,000 species of mature temperate trees. About 13,000 species of tropical plants are maintained following indoor glasshouse methods including Aroid House, Palm House, Filmy Fern House etc. Several curious plants such as *Victoria amazonica* from South America and *Welwitschia mirabilis* from Angola, are also growing here. A *Rhododendron* garden, introduced by Sir JD Hooker and Alpine House, Rose Garden, Bamboo Garden and a Lily Pond by T Dyer, add beauty to the Garden. The Jodrell Laboratory at Kew, provides research facilities to botanists for various fields. Some other remarkable features of the Garden include Chelsea Physic Garden Demonstration, Garden and world's best Rock Garden famous for its Himalayan alpine plants.

Kew Herbarium has about 275,000 Type Specimens. It is world's most famous herbarium maintaining over 7 million specimens of vascular plants and fungi from every country of the world. Kew Library is very extensive and rich and contains more than 750,000 books and journals. *Kew Bulletin* and *Index Kewensis*, are its two continuing premier publications. The Kew Herbarium runs an International Diploma Course in Herbarium Techniques. The General Catalogue contains over 1,22,000 records available throughout RBG, Kew on network.

Kew maintains database on plant names, taxonomic literature, economic botany, plants of arid lands and on plant groups of special economic and conservation values and makes about 10,000 identifications a year through its Herbarium Service. The Garden is also involved in major phytodiversity programmes in many parts of the world.

Orto Botanico of Padua, Paris

It is regarded as the oldest botanical garden in Europe, founded in 1545 by University of Padua. The Garden was aimed primarily to meet the needs of illustrative lectures on medicinal plants. It occupies an area of about 20,000 m sq. with beautifully designed squares, rectangulars and circles decorated with various grasses, alliums, irids, paeonias, a number of succulents, hydrophytes etc. The bolustrade is decorated with busts of great botanists. The Garden has played a significant role in the introduction of several plants in Europe in the past. The speciality of this Garden is the elegance and Halian taste, which has been wedded to the service of science. The elegance and beauty of the Garden is no less than Kew Gardens.

Pisa Botanic Garden, Italy

Almost as old (or probably older) as Padua, Pisa Botanic Garden was developed by the first Director Luca Ghini in 1544. Luca Ghini was succeeded by Andrea Caesalpino (1554-1558). It is credited as the first modern botanic garden. John Ray and Tournefort were closely associated with this Garden. The Garden was known for several important trees such as *Magnolia grandiflora, Aesculus hippocastanum, Liriodendron tulipifera* etc., which grew well here. The Garden does not exist today but the records of its design demonstrate geometric outlay of planting that are characteristic of several continental gardens even today.

Uppasala Botanic Garden, Scandinavia

The University of Uppasala has carefully preserved its Botanic Garden as a monument to Karl Linne known as Linnaeus who got perfected here and published his classic *Introduction to the Floral Nuptiales.* The Garden's layout is spread over Rock Garden, Experimental Plants Garden, Nursery, Plant Collection Division etc. The plants have been labelled with painted slate plates. The systematic gardening enables Garden a living work of botanical reference.

Missouri Botanic Garden, USA

It is one of the top three botanical gardens in the world. Marked as a National Historical Landmark, it is a centre for botanical research, education and horticultural display. The Garden was founded by an Englishman–Henry Shaw and opened for public in 1859 with the active help of Asa Gray and Sir William Hooker and Englemann. The Garden covers an area of 79 acres and operates world's most active Tropical Botanical Research Programme. The Garden is famous for strategies of conservation and sustainable living. The special features of the Garden are Climatron (for providing suitable climate to plants introduced), a Japanese Garden and Chinese, English, German and Victorian Gardens. The Herbarium, 6th in the world, of the garden has over 5·3 million specimens of collections dating back to 1700, providing exchange facility. The Garden shares information via website TROPICOS, the world's largest database containing more than 9,20,000 scientific plant names and 18,00,000 specimen records. The reference library of the Garden has over 2,20,000 volumes, including many rare books. *Flora of North America* and *Flora of China* are its progressing and ongoing contributions.

Berlin Botanic Garden and Museum, Berlin-Dahlem, Germany

The Garden, set up in 1679, developed largely due to the untiring efforts of Adolph Engler and L Diels, the Directors. Much of the Garden was destroyed during World War II but it was rebuilt with the efforts of Rober Pilger, its then Director. Today, Garden comprises 126 acres with about 20,000 different species of plants. There are about 1800 species of trees and shrubs in its arboretum and taxonomy section. About 1000 species of herbaceous plants are arranged according to the classification by Adolph Engler. The section of plant geography, is one of the biggest of its kind in the world, depicting whole of the Northern Hemisphere. The main tropical green house is one of the largest in the world, featuring tall trees with epiphytes, rich ground vegetation and woody climbers which give an idea of the vast variety of tropical vegetation.

Montreal Botanic Garden, Canada

The Garden stands as a monument to greatest Canadian Botanist Marie Victorian, EC. The Garden supports 20,000 species of different kinds of plants. Salient features of the Garden are the Physic Garden–collection of medicinal plants. Poisonous native plants, causing allergic diseases, are also planted. Water Garden has 10 pools to support various charming aquatics of Canada, Asia and Europe.

Moscow Botanic Garden, Yalta, USSR

The Garden, in suburb of Moscow, is spread over 900 acres. It has a vast planted woodland area and a greenhouse on a crescent plan. The Garden is aimed mainly in exploring economic values of the plants grown and raise ornamental plants. There are about 1800 species of trees, 3000 herbs, 6000 decorative cultivars and 2000 crop cultivars.

There is one more garden-Nipitisky in Yalta, established in 1812, covering an area of 272 acres which plays great role in conservation of indigenous plants, particularly of economic virtues. It has a subtropic garden in which plants have been well arranged and labelled illustratively. Its Halian garden is one of finest with pergolas of water reservoirs.

Kebun Raya Botanic Garden, Bogor, Java

The Garden is rated as one among six best botanic gardens of the world due to its beauty and extent of scientific and economic services. It supports about 6000 species of trees and shrubs and more than 5000 species of other flowering plants. Among specialities of the Garden are torch ginger plant, gigantic lianas, climbing aroids etc.

MAJOR BOTANIC GARDENS OF INDIA

India has a large number of beautiful botanic gardens. Some major botanic gardens are :

Indian Botanic Garden, Sibpur, Howrah

Indian Botanic Garden is one of the greatest botanic gardens in the world and one of the first to be established in the trópics. It is the largest and oldest garden in India located at Sibpur near Kolkata on the west bank of river Hooghly. The Garden was founded by Lt. Col. Robert Kyd in July 1787, with an aim to establish a stock of plants which may be disseminated and prove beneficial to the inhabitants and not for collecting rare plants as things of mere curiosity for furnishing articles for the gratification of luxury. This garden, then known as Royal Botanic Garden, has long been known for its distinct role in the introduction and acclimatisation of plants and development as well as cultivation of spices, tea, jute, cinchona, teak and mahogany in India. Lt. Kyd was succeeded by William Roxburgh, considered as **Father of Indian Botany**. He held the post of the second Director till 1813, during which he founded the famous Herbarium of this garden. Dr. K Biswas was the first Indian appointed as Superintendent of this Garden in 1937. In 1963, the Garden came under the control of Botanical Survey of India (BSI) as its section.

The Garden covers an area of 110 ha of land and contains representative collection of world with about 12,000 living plants, mostly of tropics. The main attraction of the Garden is 'Big Banyan Tree', considered to be more than 200 years old and has over 1600 aerial prop roots actually rooted to the ground, with no trace of main trunk, giving the tree area look like a miniature forest with a circumference of about 404 m and maximum height of about 30 m. The palm houses, nurseries, various curious plants, including Amazon Lily (*Victoria amazonica*) etc. add to the importance and beauty of the Garden. It also has well-maintained Bambusatum, Orchid House, Pinetum, Cactus House, Fernery etc. The garden has typical creeper (*Porana paniculata*) and the branching palm (*Hyphaenethebaca*). One of the latest inclusions to this Palm House is the Double Coconut (*Lodoicea maldivica*), palm like tree '*Coco de mer*' of Seychelles Island, producing two-lobed edible nut, largest and heaviest known fruit in the plant world.

The Garden with its herbarium (Central National Herbarium) is second to none in the East. It is the outcome of untiring efforts of many eminent botanist including W Roxburgh, N Wallich, W Griffith, T Anderson, David Prain, D Chatterjee etc., who had been associated with this Garden in various capacities and played a significant role in plant explorations and floristic studies of the country since its inception. The Herbarium continues to serve as centre of multiple plant taxonomic studies. The collection of plant specimens during Wallich's time were enormous and these, after identification, were extended by him in catalogue : the Wallichian Catalogue that serves as a useful record for authenticated specimens from India and adjacent countries. The Herbarium has a rich collection of plants over 2 million including more than 10,000 Type Specimens of Indian species of flowering plants.

National Botanic Garden (National Botanical Research Institute), Lucknow

The Garden was established in 1789 by the emperor Nawab Saadat Ali Khan and was named as Shikander Bagh by Nawab Wajid Ali Khan in rememberance of his beloved wife Beghum Shikander Mahal. It was later converted into a Botanical Garden in 1948 with Prof. K.N. Kaul as its first Director. Now (since 1978), it is known as National Botanical Research Institute, Lucknow, one of the premier National Laboratories of repute taken over in 1953 by Council of Scientific and Industrial Research (CSIR), New Delhi.

The Garden is situated on the banks of river Gomti and occupies an area of 25 ha in the heart of the city and additionally 120 ha at its sub-centre at Banthra, 20 km from Lucknow where economic plants are grown on large scale. The garden area is nicely decorated and is spread over with the decoration of flowering plants, Palm House, Fern House, Cactus House, Rosarium, Orchid House, Hydroponics, Bonsai Division, medicinal plants and orchards of mango, guava, *Citrus* etc. Attached with the Garden is a herbarium with collection of over 1,15,000 specimens of plants from India and adjacent countries, a rich library and laboratories carrying out extensive research in various fields of Botany. It has National Facility of an ideal botanic garden, Economic Botany Information Service, Herbarium and Banthra Research Station. It publishes NBRI News Letter (Quarterly), Applied Botany Abstracts (Quarterly), its annual report, extension bulletins, folders, manuals, survey reports, reviews and proceedings.

The Botanic Garden of NBRI is well known all over the world. It is the third largest and one of the oldest botanic gardens in India, besides the Indian Botanic Garden, Sibpur (Kolkata) and Lalbaugh Garden, Banglore, occupying the first and the second positions, respectively. The Garden serves as a National Facility to conserve the indigenous and exotic flora and fulfils the basic function of making available, for study, research and use, at one place, diverse kinds of plant species. The Garden is highly reputed for its well identified and aesthetically displayed plant wealth in a well designed landscape to capture a living nucleus of phytodiversity for the posterity. A repository of germplasm collection of various tropical and subtropical plant species comprises more than 7,000 taxa, representing over 210 families. The Garden has rich genetic treasure with the collection of trees, shrubs, climbers and herbs of ornamental, economic, medicinal and aromatic virtues, hailing from the indigenous and exotic sources. The Gardens performs all important functions which are expected from a model botanic garden.

Lloyd Botanic Garden, Darjeeling

This beautiful Garden, situated at an attitude of 2100 m in the heart of Queen of Hills, Darjeeling (WB), was established as a branch of the then Royal Botanic Garden, Kolkata in 1878 on 40 acres of land donated generously by William Lloyd, after whom it was named. Initially it served as a nursery for cultivation of many exotic ornamental plants. It now has a rich collection of many indigenous and exotic plants from the temperate regions. The Garden is spread over divisions such as an Orchid House–having more than 47 species of orchids, a large and a small conservatory, a Cactus House–with 40 species of cactii, a Herbarium-with over 3,230 species of plants belonging to about 173 families, a Library, a living fossil and a Rock Garden–botanist's paradise conserving some rare species of plants, most of them of ethnobotanical importance. Most significant spots on this track are the areas housing two living fossils : Maiden hair tree–*Ginkgo biloba* and Dwan red wood-*Metasequoia glyptostroboides* (Phantom tree). The conservatories host some of the most endangered plant species from different parts of the world. About 118 year's old massive serpentine woody climber *Wisteria chinensis*, looking like Anaconda Snake, is the most curious plant in this Garden. There are about 117 species of trees, 195 species of shrubs, 60 species of herbs and innumerable species of Bryophytes and Pteridophytes and most of the species of Gymnosperms found in India.

Lalbaugh or **Mysore State Botanic Garden, Banglore**

Lalbaugh Garden, Banglore is the second largest Botanic Garden, after Indian Botanic Garden, Sibpur (Kolkata), in India. It is also one of the oldest gardens, which was named as 'Lalbaugh' by Hyder Ali in 1760 because of its beautiful rose garden and red flowers. Its first Director was Major Waugh (1799-1819), who introduced a number of exotic plants brought into it from various parts of the world. This Garden was converted into a proper botanic garden in 1856. Rao Bahadur HC Jayaraja was its first Indian Director. This Garden is famous for its beautiful layout and a big centre of horticultural activities. It has a grape orchard, economic plant's garden, a herbal garden, a tropical nursery and well-equipped laboratories for soil and seed testing. The Garden occupies an approximate area of 240 acres. It houses the State Vegetable Breeding Station, the tree nursery, the school garden and numerous beautiful trees, shrubs, herbs and climbers of botanical and public interest. The Garden is recognised as one of the oldest and best for its layout, grandeur, maintenance and scenic beauty. Marshal Tito, the former President of Russia, during his visit to India in 1955, commented "India is a garden and the Lalbaugh is the heart of it."

Botanic Garden of Forest Research Institute, Dehradun

The Garden was established at Dehradun in 1934 by CE Parkinson. It covers an area of about 20 acres. The Garden is the major centre of research in plant introduction. The Garden has about 700 different species of plants brought into it from different countries of the world, a green house, a cactus house, a plant introduction nursery and a big herbarium attached to it.

Botanic Garden at Bhaba Atomic Research Centre (BARC)

A large garden attached to BARC at Trombay, performs many of the functions of a botanic garden.

Botanical Garden, Saharanpur

The Garden was established in 1779 when the muslim ruler Zabita Khan decided to spend the revenue of seven villages on the maintenance of a garden. The Garden covers an area of 40 acres with section for Linnean garden and agriculture and horticulture nurseries. In the beginning, the Garden naturalised several plants such as potato, tobacco, pineapple, guava, chillies, papaya, sapota etc. In 1876, Duthie took over as the Superintendent of the Garden and trained its staff for plant collection which resulted in the compilation of the *Flora of Upper Gangetic Plains.* The collection of mosses represented the rich flora of North West Himalayas. The Garden is now a horticultural research station of UP Government.

Garden of Indian Agricultural Research Institute (IARI), New Delhi

The Garden occupies an area of about 60 acres. It is maintained chiefly as an experimental garden of IARI while carrying on other functions of a botanic garden.

❑❑❑

Chapter 9

Botanical Survey of India

Botanical Survey of India (BSI) is a government organisation engaged in the survey, study, utilisation and conservation of the plant wealth of the country. The basic objectives of BSI are to explore the plant resources of the country and to identify plant species with economic virtues. It remained quiescent till 1954, when Government of India as a part of scientific development of the country reorganised this department. During successive plan periods, the functional base of BSI was further expanded to include various new areas such as to make an inventory of endemic, rare and threatened plant species; to evolve conservation strategies; studies on fragile ecosystems and protected areas like Sanctuaries, National Parks and Biosphere Reserves; monitoring of changes in floristic components; conservation, multiplication and maintenance of plant genetic resources, endemic and threatened species, wild ornamentals etc., in Botanic Gardens and Orchidaria; ethnobotanical and geobotanical studies and development of National Database on Herbarium (including Type Specimens) and live collections; plant genetic resources; plant distribution and nomenclature. Headquarter of BSI is located at Kolkata. Apart from various other existing circles, BSI has opened its Deccan Circle in Hyderabad on 15th December, 2005.

Historical Account

The Botanical Survey of India was established on 13th February, 1890 for exploring the plant wealth of the country with Sir George King, then Superintendent of the Royal Botanic Garden, Sibpur, Howrah as ex-officio Director. The country was divided into four botanical regions, viz., East, West, North and South, each under the charge of Regional Director. During its early stage, the department conducted extensive surveys in different parts of the country. Later, several of its functions were transferred to other departments and survey work slowed down. With the retirement, in 1939, of Dr CC Calder, the last incumbent to the post of Director, the activities of the department were reduced to the minimum. It was reorganised in 1952, and on the recommendation of the Officer on Special Duty, the Government of India set up a Headquarter Organisation in Calcutta (now Kolkata) and four regional circles at Dehradun, Coimbatore, Poona and Shillong. A Central Botanical Laboratory (now National Botanical Research Institute) was started at the same time in Lucknow. In 1957, the famous Sibpur Herbarium at the Indian Botanic Garden, Howrah, was transferred to the BSI for its further development into a National Herbarium. The plans were drawn to develop the Industrial Section of the Indian Museum into a Modern Botanical Museum. A Central Circle was opened in 1962 at Allahabad. In 1963, the Indian Botanic Garden at Sibpur, Howrah, was transferred to BSI. Two more Circles namely, the Andaman and Nicobar Circle and the Arid Zone Circle, were started at Port Blair and Jodhpur, respectively in 1972. Arunachal Field Station was started at New Itanagar in 1977 while Sikkim-Himalaya Circle in 1979 and Deccan Circle, Hyderabad in 2005.

Activities

India has a rich and varied flora represented in different types of vegetation, such as tropical, sub-tropical, temperate, alpine and desert vegetation. It is also considered as one of the seven centres of origin of cultivated plants. Out of 25,000 species of vascular plants recorded, India has about 15,000 species, estimated so far. The Botanical Survey of India, regularly sends exploration parties to various parts of the country including remote forest areas for recording the occurrence and uses of plant wealth. Since, reorganisation, BSI has made plant collections through numerous botanical explorations/expeditions and collected about 3,00,000 field numbers of plants. Its nine Regional Herbaria serve as repositories of plant specimens and hold among them over 20,00,000 specimens. Nearly 250 new plants (genera and species) have so far been discovered and described while more than 500 new records of plants have been found for our country. The main activities of BSI are :

1. Exploration, inventorisation and documentation of plant diversity in general and protected areas, hot spots, fragile ecosystems and sacred grooves in particular.
2. Publication of National, State and District Floras.
3. Publishing of Flora of India in the form of Fasciles for higher plants.
4. Survey, inventorisation and documentation of lower plants like lichens, mosses and ferns, which are also important constituent of our natural flora.
5. Monitoring plant diversity to evaluate qualitative changes in species rich and sensitive areas.
6. *Ex situ* conservation of critically threatened taxa in botanical gardens.
7. Analysis and evaluation of the impact of the developmental activities on the environment by monitoring of vegetation and plant populations, as plants act as indicator of healthy or degraded environment. On such studies, BSI is involved in Sutlej-Beas Link Canal, Tehri Dam, Silent Valley, Lalpur Dam, Idukki Dam Project, etc.

8. Development of the Central National Herbarium and the various Regional Herbaria as repositories of the Types and other authentic specimens.
9. Development of Indian Botanic Garden, the Regional Experimental Gardens and the National Orchidaria for the study, introduction and conservation of flora.
10. Development of the Central Botanical Laboratory and Botanical Section of Indian Museum for experimental studies of plants and popularising their role in our lives.
11. Assessment and care of threatened plant species and studies on little known or new uses of plants particularly among rural and tribal societies.
12. Identification of species with traditional economic uses and preparation of protocols for their conservation and sustainable utilisation.
13. Qualitative and quantitative study of the diversity of selected economically useful plant species.
14. Preparation of Pollen Atlas of Indian plants.
15. To complete a National Database of plant species, herbarium specimens, live specimens, illustrations, relatives of cultivated species and economically important species.
16. Assisting in teaching and research activities in the country.
17. Study of marine plant resources of the coastal waters, both at macro and micro-levels.
18. Evaluation of plants in relation to environmental pollution.
19. Multiplication of threatened plant species through tissue culture and seed banks.
20. Conservation of areas of biological interest.
21. Educating people to promote consciousness about the paramount role of plants in environment.

Organisation

BSI Directorate is headed by a Director. The Directorate has scientific, technical and administrative units that look after respective works. Besides the Directorate, there are Central Botanical Laboratory, Central National Herbarium and Indian Botanic Garden at Howrah (WB). Industrial Section and Indian Museum is located at Kolkata. There are nine regional Circles/Field Stations (till 1984) : Southern circle—Coimbatore; Northern Circle-Dehradun; Eastern Circle-Shillong; Western Circle-Pune; Central Circle-Allahabad; Arid Zone Circle-Jodhpur; Andaman and Nicobar Circle-Port Blair; Sikkim-Himalays Circle-Gangtok and Arunachal Field Station-New Itanagar. On 15th December, 2005, BSI has opened its Deccan Circle in Hyderabad. Most of these Circles/Field Stations have well developed herbaria, library, drawing and photographic sections. A Regional Botanist, at the Royal Botanic Gardens, Kew, England, serves as Liaison Officer.

The Headquarter's Organisation is the central coordinating office of BSI. In addition to administrative wing, there are several divisions, such as Ecology, Cryptogams, Plant Chemistry, Flora, Pharmacognosy, Publication, Documentation and Technical.

The Central Botanical Laboratory, Howrah, came into existence in 1954 to conduct research on various disciplines of Botany and thereby providing information on living plants of India for utilisation and exploration of the plant wealth in horticulture and forestry, trade and industry and medicine. Cytology, Economic Botany and Plant Physiology, are the three major sections of the Laboratory that undertake chromosome survey of Indian flora; study and evaluation of less known economic/medicinal plants and plants in use by tribals; and studies on dormancy, viability and germination behaviour of seeds of economically important plants.

Central National Herbarium, Howrah, started by W Roxburgh in 1793, is now India's largest Herbarium with more than 20,00,000 specimens collected from all parts of the world. Herbarium has over 10,000 Type Specimens and almost the same number of Wallichian collections (collected by Wallich) that are housed in air-conditioned chamber. Microfiches of the Linnean and several other important herbaria are also available. Further, there is a large collection of icones, of which 2583 hand-coloured illustrations forming the Roxburgh icones, takes the pride of the place. There is a continuous addition of graphic material in the form of photonegatives of Types and authentic materials obtained from Kew Herbarium. With its rich collection and library, this Herbarium is the hub of taxonomic activities in the country. (Indian Botanic Garden, Howrah, has already been discussed in chapter 8).

Industrial Section, Indian Museum, Kolkata has been a part of BSI since 1911. Its gallaries contain 15,000 exhibits arranged in various bays and sections, classified according to usage : crude drugs, food, fodder, dyes, tannins, gums, timbers, narcotics, beverages etc. This section is also engaged in visual education through exhibitions, mobide demonstrations and lectures etc. on the importance of plants and the essential need for their conservation.

Till 1984, there were nine Regional Circles/Field Stations of BSI :

1. **Northern Circle, Dehradun** includes the states of Jammu and Kashmir, Punjab, Haryana, Himachal Pradesh, Delhi and 24 districts of Uttar Pradesh and Uttarakhand–covering the hill areas. There are more than 65,000 specimens representing about 5,000 species, mostly from hitherto poorly explored NW Himalayan areas and 14 Type Specimens in the Herbarium. High Altitude Flowering Plants and Illustrations, are the two major contributions published from the Circle. Studies of the Valley of Flowers, Flora of Corbett National Park and State Flora of Himachal Pradesh, are the important scientific programmes of the Circle. In Experimental Garden, developed at Pauri and Khirsu, more than 150 species have been introduced.

2. **Eastern Circle, Shillong** covers the states of Assam, Manipur, Meghalaya, Nagaland and Tripura. The Circle has the National Orchidarium at Shillong and Experimental Garden at Barapani. There are more than

1,15,000 specimens in this Regional Herbarium. Detailed studies on the Monocot Flora of Assam, Flora of Jowai, Nongpoh, Cyperaceae of North East India, Orchids of Khasi and Jaintia Hills and Assam plains, have been completed and mostly published. Programmes on cultivation of orchids in collaboration with North Eastern Council and studies on the vegetation and Flora of Namdapha in Arunachal Pradesh, under Man and Biosphere (MAB) programme, are in progress.

3. Southern Circle, Coimbatore covers the states of Andhra Pradesh, Kerala and Tamil Nadu. There are more than 1,90,000 specimens including 1608 Types. Several new taxa including 5 genera have been published. Flora of many areas/districts, especially of Nilgiris, have been studied and published. The National Orchidarium and Experimental Garden at Yercaud provide scope for studies on orchid systematics, hybridisation and culture through introduction and acclimatisation. The State Flora of Tamil Nadu has been worked out. The Circle has undertaken Idukki Dam Project. The preliminary assessment of importance and significance of the flora and vegetation of the Silent Valley, Kerala, was made by the Circle.

4. Western Circle, Pune covers the States of Maharashtra, Karnataka, Union territories of Goa, Lakshdweep and Mimicoy islands. The Regional Herbarium has more than 1,25,000 specimens. One new genus and 30 new species have been discovered and published. Flora of Rajasthan and Goa has been worked out. The Circle has also carried out studies on the improvement of colchicine content in species of *Iphigenia*. Experimental Garden at Mundhwa with more than 800 species including 25 rare and threatened species, has been maintained successfully. The Circle is engaged in the preparation of the Flora of Maharashtra and selected areas of Karnataka.

5. Central Circle, Allahabad covers the states of Madhya Pradesh and the plain districts of Uttar Pradesh. The Herbarium has a collection of more than 50,000 specimens. The Circle has carried out extensive and intensive botanical exploration of Madhya Pradesh, particularly in tribal areas. The Experimental Garden within the campus of the Circle has several plants of local medicinal and horticultural interest. Studies on District Floras of Durg, Raipur, Rajnandgaon, Damoh and Bilaspur, have been completed and work on other districts of Madhya Pradesh is in progress.

6. Arid Zone Circle, Jodhpur, set up in 1972 to study flora and vegetation of the regions which are by and large arid, includes areas like districts of Bhatinda and Firozpur of Punjab; districts of Hissar, Mahendragarh and Gurgaon of Haryana; Delhi Ridge; districts of Mathura, Agra and Jhansi of Uttar Pradesh; districts of Gwalior, Datia, Shivpuri, Guna, Mandsor, Ratlam and Jhabua of Madhya Pradesh; districts of Kutch, Banaskantha, Mehsana, Sabarkantha, Ahmedabad; all districts of Saurashtra from Gujarat and entire state of Rajasthan. The District Floras of Tonk and Banswara have been completed. Study on State Flora of Rajasthan are in progress.

7. Andaman and Nicobar Circle, Port Blair, started in 1972, has a Herbarium with more than 20,000 specimens. Explorations were conducted in different islands of Andaman and Car Nicobar, Kamorta, Nancowry and

Great Nicobar islands of the Nicobars. Two genera and a number of new species/records have been described. An Experimental Garden-cum-Arboretum at Dhanikari, about 17 km from Port Blair, has several endemic plant species. Enumeration of plants of the Islands is the main scientific work of the circle.

8. Sikkim-Himalayas Circle, Gangtok, started in 1979, undertakes exploration and study of the rich vegetation and flora of Sikkim and adjacent areas. The Herbarium has more than 5,000 specimens.

9. Arunachal Field Station, New Itanagar, started in 1977, undertakes study of the unexplored areas of North Eastern India. The herbarium has more than 6,000 specimens collected principally from Subansiri and Kameng districts.

Achievements

BSI has contributed significantly in exploration of plant resources of the country. It has succeeded in identifying a large number of plant species with economic virtues. Some other achievements made by BSI are :

Work on Funded Projects : BSI also works on various funded projects such as MAB (Man and Biosphere) Programme under which it operates : (i) All India Coordinated Research Project on Ethnobiology with main Centre at Howrah and Centres at Shillong, Port Blair and Coimbatore, (ii) Environmental and Ecological aspects of Idukki Multipurpose River Valley Project, Kerala, (iii) Environmental and Ecological Studies on Namdapha Reserve Forests, Arunachal Pradesh and (iv) Ethnobotany of Ambujh Marh, Madhya Pradesh. BSI also operates NES (North East Council) Joint Project on multiplication and propagation of orchids at Shillong. A Project POSSCEF on study survey and conservation of endangered species of Indian flora, is run at main centre, Howrah and Regional Centres at Coimbatore, Shillong and Dehradun.

Preservation of Phytodiversity : BSI contributes significantly in monitoring phytodiversity to evaluate qualitative changes in species rich and sensitive areas in Silent Valley (Kerala), Namdapha (Arunachal Pradesh), Similipal (Orissa), Nanda Devi (Uttarakhand), Dachigam (Kashmir), Valley of Flowers (Uttarakhand) etc. Sustainable utilisation of our plant resources is the watch word for BSI as such multipronged efforts are made to conserve plants. Practical steps for regulating trade in endangered species, are regularly suggested to the Agriculture Department. Preserving the diversity of flora, essential for maintaining the natural equilibrium in various ecosystems, is a major area of interest of the BSI. Sanctuaries have been set up in forest areas for selected plants like the curious carnivorous pitcher plant (*Nepenthes khasiana*) in Meghalaya and beautiful Rhododendrons and Orchids in Sikkim and Darjeeling, Himalayas.

Academic Coordination : BSI maintains close collaboration with Universities, Research Organisations such as CSIR (Council of Scientific and Industrial Research), ICAR (Indian Council of Agricultural Research), FRI (Forest Research Institutes), Forest Departments of States and several Herbaria and Botanic Gardens of the world. One Regional Botanist of BSI works in the Herbarium of Royal Botanic Gardens, Kew, England to undertake critical taxonomic works and helps Indian workers on plant taxonomy and systematic botany. It also deputes scientists abroad for higher training in the area under various exchange programmes. BSI also assists in teaching and research activities in the country by identifying selected botanical specimens and also by supplying samples of authentic materials. In collaboration with UGC (University Grants Commission), BSI has a programme to set up a chain of herbaria in educational institutions in order to make the students in rural areas easily accessible to well-established herbaria.

Publications : Phytodiversity is documented and published in various forms.

Series-1 deals with National Flora in the form of Fascicles (Various families, tribes etc.)

Series-2 deals with State Flora Analyses etc. (Flora of Tamil Nadu; Flora of Himachal Pradesh; Flora of Karnataka etc.)

Series-3 deals with District Flora (Flora of Jowai, Vol I-1981; Flora of Jowai Vol II-1983; Flora of Tonk District, Rajasthan–1983; Flora of Banswara District, Rajasthan–1983 etc.)

Records of the BSI deals with Flora of Khandla of Western Ghats of India (3rd Rev. Ed.). Vol. XVI (i)–1967; Materials for the Flora of Bhutan, Vol. (2)–1973; Flora of the Punjab Plains, Vol XXI (i)–1978; Orchids of Nilgiris, Vol. XXII–1982; Hypnobryales suborder Leskeineae of the Himalayas, Vol. XXIII–1983 etc.

Bulletin of BSI–a quarterly journal, deals with plant sciences mainly relating to the study of Floristics, Vegetation, Taxonomy, Ecology, Cytology and Economic Botany.

Special Publications : Some common special publications are :

1. High Altitude Flowering Plants of West Himalayas, 1975.
2. Threatened Plants of India–A State of Art Report, 1980.
3. Conservation of Tropical Plant Resources, 1983.
4. An Assessment of Threatened Plants of India, 1983.
5. Key Works to the Taxonomy of Flowering Plants of India, Vol I-1984.
6. An Annonated Bibliography of Taxonomic Botany of Peninsular India, 1981.

❑❑❑

Chapter 10

Origin and Evolution of Angiosperms

Angiosperms, more commonly known as flowering plants, form the most dominant group of the present day plants with atleast 2,54,990 species (Thorne, 2003)–a number exceeding all other groups of the plants combined together. Not only in number of species, angiosperms also constitute the most diverse group of the plant kingdom. Also, they exhibit a far greater range of habitats than any other group of land plants. They represent a group of plants in which there was an early differentiation along several lines. The great diversity exhibited by the plants of this group is the result of their adaptive specialisation in response to great climatic changes. Some forms became highly elaborated in their vegetative as well as floral features while others became specialised by reduction, retrogression or suppression. The former forms are designated as progressively evolved and the latter, by some, to be primitive. Not all families under the angiosperms are equivalent in diversity. Certain families, such as Asteraceae, Orchidaceae, Poaceae etc. have much larger number of species than others (Stebbins, 1981).

Definition and Characters of Angiosperms

The term *angiosperm*, in Greek, literally means *vessel seed, i.e.,* plants bearing seeds which are borne within the ovary. Thus, angiosperms may be defined as *a group of flowering plants whose seeds are enclosed within the carpel or carpels and stem bears vessels* (Eames, 1961). They share a unique combination of characters. Such important characters of the group include :

1. Presence of ovules enclosed within carpel or carpels.
2. Presence of vessels in the xylem; vessel elements with scalariform perforations; sieve tubes with companion cells. Vessels are absent in Winteraceae and some others.
3. Occurrence of double fertilisation resulting in triploid endosperm.
4. Presence of a complex reproductive structure, the flower.
5. Presence of parallel or reticulate venation with intercostal veins.
6. Presence of highly reduced male and female gametophytes.

7. Germination of pollen grains on the stigma; in having extexine and non-lamellate endexine; ectexine differentiated into a foot layer, collumellar layer and tectum (tectum absent in Amborellaceae and some others).

The combination of these charactes (with certain exceptions) is unique to angiosperms and not found in any other group of seed plants.

ORIGIN OF ANGIOSPERMS

The mystery of the origin of angiosperms remained a puzzle for the botanists for a long time. More than hundred twenty five years ago, Darwin called the origin of angiosperms an "abominable mystery", which is largely because of poor record of the earliest angiosperms that lived in habitats not best suited for fossilisation. In the fossil record, they appeared rather suddenly, with no obvious ancestors for a period of about 80-90 million years, prior to their appearance. Not even fossil leaves or pollens were known from this earlier period. There has been considerable difference of opinion about many facts of the origin of angiosperms. Some main aspects of the problem are : (i) What is their age ? (ii) Which is the place of their origin ? (iii) Are they monophyletic or polyphyletic and (iv) What are their possible ancestors ?

(1) Age of Angiosperms

The period of origin of angiosperms, as to how old they are? is a matter of considerable debate as there are several views regarding their age. For determination of first trace of angiosperms, one has to depend almost entirely upon fossil evidence. For a long time, the earliest well-documented angiosperm fossil was considered to be the form-genus *Clavitopollenites* described (Couper, 1958) from Barremian and Aptian strata of Early Cretaceous (132 to 112 mya-million years ago) of Southern England, a monosulcate pollen with distinctly sculptured exine-resembling the pollen of extant genus *Ascarina*. A similar but unaperturate pollen, recorded by Brenner and Bickoff (1992), from the Valanginian (CA 141 mya) of Helez Formation of Israel is now considered to be the oldest record of angiosperm fossil (Taylor and Hickey, 1996). The number and diversity of angiosperm fossils increased suddenly. By the end of Early (Lower) Cretaceous (CA 100 mya) period major groups of angiosperms, including herbaceous Magnoliidae, Magnoliales, Laurales, Winteroids and Liliopsida, were well represented but in Late (Upper) Cretaceous, at least 50 per cent of the species in the fossil flora were angiosperms. By the end of the Cretaceous, many extant angiosperm families had appeared and subsequently increased exponentially constituting the most dominant land flora that continues till today (table 10.1).

Table 10.1 Mesozoic Era-as a Part of Geological Time Scale.

Time	Era	Period	Epoch	Stage
m years (mya)				
__74__				Maestrichtian
__83__				Campanian
__87__				Santonian
__89__			Upper	Coniacian
__90__				Turonian
__97__				Cenomanian
__112__		Cretaceous		Albian
__125__				Aptian
__132__	Mesozoic			Barremian
__135__			Lower	Hauterivian
__141__				Valanginian
__146__				Berriasian
			Upper	
		Jurassic	Middle	
__208__			Lower	
			Upper	
		Triassic	Middle	
235			Lower	

There appears an incomplete and ambiguous picture in the trail in reverse direction as many claims of angiosperm records, made before the Cretaceous, were largely rejected. A pre-Cretaceous *Eucommiidites*, described by Erdtman (1948) as a tricolpate dicotyledonous pollen grain from Jurassic, had been found to have bilateral symmetry instead of radial symmetry of angiosperms (Hughes, 1961) and granular exine with gymnospermous laminated endexine (Doyle *et al.* 1975). Moreover, this pollen grain was also discovered in the micropyle of ovule of the female cone of uncertain but clearly of gymnospermous affinities (Brenner, 1963). Several other pollens from the Jurassic, attributed to Nyphaeaceae, ultimately turned out to be gymnospermous. More or less similar stories are there about *Fercula*—a leaf described by Harris (1932) from upper Triassic of Greenland, *Pannaulika*—a dicot-like leaf form described by Cornet (1993) from Late Triassic from the Virginia-North Corolina border, etc. in which cases their angiosperm affinities could not be established. However, Sun *et al.* (1998, 2002) have described fossils of *Archaefructus* from upper Jurassic (nearly 124 mya) of China, with clearly defined spirally arranged conduplicate carpel enclosing ovules (and a follicle fruit), a feature not reported in earlier angiosperms, is considered to be the oldest record of angiosperm flower. The inconclusive pre-Cretaceous record of angiosperms indicates that they arose in the Late Jurassic or very

Early Createceous (Taylor, 1981) nearly 130-135 mya (Jones and Luchsinger, 1986).

Melville (1983), who strongly advocated his gonophyll theory, believed that angiosperms arose around 240 mya in the Permian and took about 140 my before they spread widely in the Cretaceous. The Glossopteridae, that gave rise to angiosperms, met with a disaster in the Triassic and disappeared. This disaster slowed down the progress of the angiosperms until the Cretaceous when they tended towards exponential increase. This idea, however, could gain little favour. In recent past (Troitsky *et al.* 1991; Doyle and Donoghue, 1993; and Crane *et al.* 1995) there has been an increasing realisation to distinguish two dates—**One** (i) in Triassic when the 'stem angiosperms' ('angiophytes' *sensu* Doyle and Donoghue, 1993 or 'proangiosperms' *sensu* Troitsky *et al.* 1991) separated from sister groups, *i.e.*, Gnetales, Bennettitales and Pentoxylales and the **second** (ii) in the Late Jurassic when the 'crown group of angiosperms' (Crown angiophytes) split into extant subgroups (fig. 10.1).

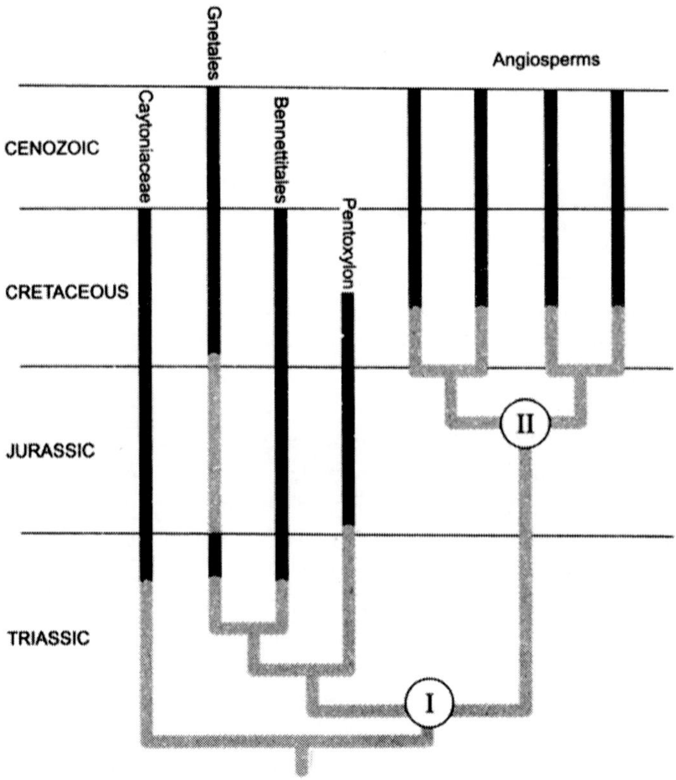

Fig. 10.1 Phylogenetic tree of anthophytes (angiosperm lineage and sister groups). Point (I) marks when angiosperm lineage separated from sister groups in the Triassic, and (II) marks the splitting of crown angiosperms into extant subgroups in the Late Jurassic. Light coloured lines represent conclusions for which fossil record is not available (after Doyle and Do.ıoghue, 1993).

Molecular Dating. A number of attempts, so far made, to estimate the time of divergence of angiosperms by applying a molecular clock to nucleotide sequence data, gave the contradictory results which were mostly pointing to much earlier origin of angiosperms. The first attempt made in this direction was by Martin *et al.* (1989) who used nine angiosperm sequences from *gap*C, the nuclear gene encoding GADPH (cytosolic glyceraldehyde-3-phosphate dehydrogenase). The observed number of non-synonymous substitution between each pair of species was compared to estimated rates of substitutions per site per year inferred from known divergence times (*e.g.* plant-animals, plants-yeast, mammal-chicken, human-rat). The results that implied separation of monocots and dicots at 276 + 33 mya, a dicot radiation at 276 + 33 mya and cereal grass divergence at 103 + 22 mya, were questioned by several authors because in the study only single gene was used. Later Wolfe *et al.* (1989) made an attempt to date the monocot-dicot split, using a large number of genes in chloroplast genome with involving a three-tiered approach and suggested Late Triassic (200 mya) as the likely estimate of monocot-dicot split. In their later approach, Martin *et al.* (1993) provided new data to support Carboniferous origin (300 mya) of angiosperms by using both *rbc*L and *gap*C sequences for this study. Systma and Baum (1996) concluded that the results strongly caution use of the molecular clock for dating unless extensive sampling of taxa and genes with quite different molecular evolution is completed. Thus, for the resolution of angiosperm phylogeny, one has to wait for a more complete molecular data and its proper appraisal.

(2) Place of the Origin of Angiosperms

Earlier, it was believed that angiosperms arose in Arctic region (Seward, 1931), with subsequent southwards migration. Axelrod (1970) believed in **upland theory**, suggesting that flowering plants evolved in mild uplands at low altitudes. Smith (1970) located the geographic region of South-East Asia, adjacent to Malaysia as the site where angiosperms evolved when Gondwana and Laurasia were undergoing initial fragmentation. According to Stebbins (1974), their origin occurred in exposed habitats in areas of marginal ecotonal regions of tropical rain forests with equable climate, **museums of angiosperms**. Takhtajan (1969), rejecting the Arctic (or Antarctic) origin of angiosperms, predicted their centre of origin to be in South-Eastern Asia and suggested that the **Cradle of the angiosperms** occurs somewhere between Assam (India) and Fiji (including Myanmar, Thailand, Indo-China and Malaysia), due to high abundance of many primitive families found in that region of the Pacific basin. Takhtajan (1980) believed in neotenous origin of angiosperms and suggested that angiosperms arose under environmental stress, probably as a result of adaptation to moderate seasonal drought on rocky mountain slopes in areas with monsoon climate. Retallack and Dilcher (1981) are of the opinion that the earliest angiosperms were probably woody and small-leaved plants that occurred in the Rift Valley system adjoining

Africa and South America and of these some adapted to the coastal environments and became widespread, following changing sea levels during the Early Cretaceous. Many authors (Hickey and Doyle, 1977; Upchurch and Wolfe, 1987; Hickey and Taylor, 1992 etc.) have suggested that early angiosperms lived along stream and lake margins, the **lowland theory**, later they appeared in more stable backswamp and channel sites and lastly on river terraces. Taylor and Hickey (1996) suggested that ancestral angiosperms were perennial rhizomatous herbs. These evolved along rivers and streams on sites of relatively high disturbance with moderate amounts of alluviation. These sites would have been characterised by high nutrient levels coupled with frequent loss of plant cover due to periodic disturbances.

(3) Are Angiosperms Monophyletic or Polyphyletic?

The two terms, monophyletic and polyphyletic, frequently used in the classification of angiosperms, need a clear understanding. Davis and Heywood (1963) regard the use of these terms to be purely relative, depending upon how far back in the ancestry one is prepared to go, *i.e.*, if life arose only once on earth, all organisms are ultimately **monophyletic** (derived from the common or same ancestor) or **polyphyletic** (not derived from a common or same ancestor). Simpson (1961) defines monophyly as 'the derivation of a taxon through one or more lineages from one immediately ancestral taxon of the same or lower rank, *e.g.* a family derived from a family, genus or species, Heslop-Harrison (1958), following Wernham (1912) regard, 'a monophyletic group as one derived from an ancestral form which would be regarded as belonging to the taxon in question (family from family, genus from genus and so on)'. According to Simpson (1961), "*the level of polyphyly is specified by the highest ranking taxa, two or more of which are immediately ancestral to the taxon concerned*".

The question of phyly of angiosperms still remains unsolved largely because of inadequate fossil records.

(i) Polyphyletic Origin. Angiosperms have undergone intensive evolution in many directions under varied environmental conditions, and as a result early forms have attained an extraordinary degree of morphological and ecological diversity, as is evident from their fossil record. Several phylogenisists, including Engler (1892), Games (1961), Eames (1961), Meeuse (1963), Hughes (1994), considered angiosperms to be polyphyletic, *i.e.*, monocotyledons and dicotyledons having evolved separately from different primitive stocks at different times due to considerable diversity in the Early Cretaceous and the extant angiosperms and attained their present status through parallel or convergent evolution. This theory of polyphyletic origin of angiosperms is supported by the fossil records of variety of perianth and the nature of carpel in both dicotyledons and monocotyledons. Polyphylesis is further supported by the fact that primitive orders of both, dicotyledons and monocotyledons do not exhibit any close relationship in their characters.

(ii) Monophyletic Origin. Morphological and taxonomic evidence, on contrary, leads to the conclusion that the angiosperms are one natural monophyletic branch of development. Many of the recent authors, including Hutchinson (1959, 73), Cronquist (1981, 88), Thorne (1983, 92, 2000), Dahlgren (1980, 89), Takhtajan (1987, 97), Judd *et al.* (2002), Bremer *et al.* (APG II, 2003) and Stevens (AP web, 2003), believe in monophyletic origin of angiosperms, *i.e.*, monocots having evolved from primitive dicots. Monophyletic origin of angiosperms is also supported by unique combination of characters (see characters of angiosperms discussed earlier), exhibited by both dicots and monocots. This view is also supported by the statistical analytic studies by Sporne (1974), as it is higly improbable that such a unique combination of characters could have arisen more than once independently from gymnosperm ancestors.

Interstingly Melville (1983) considered angiosperms to be monophyletic but the explanation offered by him clubs him with the proponents of polyphyletic origin. According to him angiosperms arose from several different genera of Glossopteridaceae. He is of the opinion that the species is not always to be considered as the ancestor for determining a monophyletic nature; a species from other species is monophyletic as is a genus from a genus and a family from a family. The principle, he believes, is that to be monophyletic–a taxon of any rank must be derived solely from another taxon of the same rank. Glossopteridae and Angiospermidae belong to the same rank–subclass and both taxa consist of minor lineages that may be like a rope with many strands, a condition called **pachyphyletic**, the explanation conforming to the concept of minimum monophyly—now an accepted criteria for monophyly.

(4) Probable Ancestors of Angiosperms

To trace the nature of probable ancestors of angiosperms is perhaps one of the most difficult tasks. The subject is most controversial and highly debatable. In absence of direct fossil evidence, various plant groups have been considered as possible ancestrals to angiosperms by one authority or the other. But none of these hypotheses appear to be tenable or based on sound footing. Some authors consider even *Isoetes* origin of monocots and algae (*Chara*) as probable ancestor of angiosperms.

***Isoetes* Origin of Monocotyledon.** Proposed by Campbell (1928), this view is essentially based on marked similarities between *Isoetes* and monocots especially *Najus flexilis*. He advocated that monocots have been derived from *Isoetes* via Ophioglossaceae. Both leptosporangiate and eusporangiate ferns and their allies such as Lycopodiales, Isoetales etc., have been shown to possess many common features with those of monocots. *Isoetes*, an aquatic or amphibious genus, is predominantly herbaceous, and geophilous, and like most of the Filicineae, it is found in humid tropics, a habitat which is also characteristic to a large number of monocots. The marked similarity in habit and in the embryo and the anatomy of older sporophyte of *Isoetes* and some other lower aquatic monocots, have also been pointed out by Campbell. He suggested that the herbaceous character in angiosperm is primitive, inherited from filicinean ancestors.

VARIOUS THEORIES OF ORIGIN OF ANGIOSPERMS

Various theories about origin of angiosperms have revolved around 'Euanthial theory' (Anthostrobilus theory) and 'Pseudanthial theory'. Some other theories about herbaceous ancestry for the angiosperms have also recently received attion. They have made the question of origin of angiosperms rather more ambiguous. Some important theories have been accounted here.

(I) Euanthial Theory (Anthostrobilus Theory)

Arber and Parkins (1907) proposed that an angiosperm flower is derived from an unbranched bisexual strobilus bearing spirally arranged pollen and ovulate organs, similar to the hermaphrodite reproductive structures of some extinct Bennettitalean gymnosperms. The carpel is thus regarded as a modified megasporophyll–**phyllosporous** origin of carpel. The hermaphrodite flower of Magnoliales has been considered to have evolved from such a structure. In agreement to this general principle, various other authors have also made an attempt to identify different gymnosperm groups as possible angiosperm ancestors :

(i) Cycadeoidales (Bennettitales). Also known as Bennettitalean theory, it was first proposed by Saporta and Marison (1885) and later by Arber and Parkin (1907). According to this theory, Bennettitales have been proposed as possible ancestors of angiosperms, on the basis of resemblance in the structure between the strobili of the Mesozoic genus *Cycadeoidea* (fig. 10.2) and the flower of *Magnolia.* Both these structures are bisexual and contain an elongated axis having protective bracts, microsporophylls and megasporophylls, arranged succesively from below upwards.

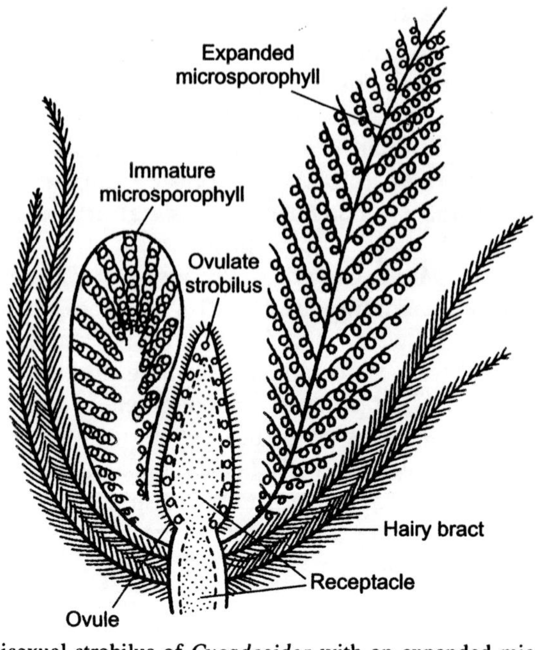

Fig. 10.2 Bisexual strobilus of *Cycadeoidea* with an expanded microsporophyll.

The group Bennettitales, now better known as Cycadeoidales, appeared in Triassic and disappeared in the Cretaceous. Their potential as ancestors of angiosperms, was largely built upon the studies of Wieland (1906, 16). Lemesle (1946) considered Cycadeoidales to be ancestral to angiosperms, primarily because of *Cycadeoidea* in having hermaphrodite nature with an elongated receptacle with perianth-like bracts, a whorl of pollen bearing microsporophylls surrounding ovuliferous region having numerous ovules and interseminal scales packed together (fig. 10.2). There were, however, signs of abscission at the base of the male structures which would have shed, exposing the ovuliferous region. The plant, believed to look-like Cycads, bore a short compact trunk and crown of pinnate compound leaves. Earlier, it was suggested that the microsporophylls opened at maturity but subsequent studies of Crepet (1974) revealed that microsporophylls were pinnate, and distal tips of pinnae were fused, the opening of the region was not structurally possible and they later disintegrated internally. The ovules were terminal, in contrast to their position in carpels of angiosperms.

(ii) Caytoniaceae. Caytonialean theory was proposed by H Hamshaw Thomas (1925, 36) on the basis of comparative studies of modern flowering plants with those of Caytoniales (Pteridosperms or seed ferns) which include certain fossil plants of the Middle Jurassic, discovered from Cayton Bay in Yorkshire and subsequently from Greenland, England and Canada. Opinion has strongly inclined in the recent years towards the probability that angiosperms arose from Caytoniaceae, which appeared in the Late Triassic and disappeared towards the end of the Cretaceous. The leaves (*Sagenopteris*) were borne on the twigs and not the trunk (fig. 10.3A). The leaves had two pairs of leaflets (rarely 3 to 6 leaflets) and were not veined. Male structure (*Caytonanthus*) had rachis with branching pinnae, each pinna with synangium of four microsporangia (fig. 10.3C). The seed-bearing structure (*Caytonia*) had rachis with two rows of stalked cupules, each cupule had several ovules borne in such a way that it is recurved, with a lip-like projection (often called stigmatic surface) near the point of attachment (fig. 10.3D).

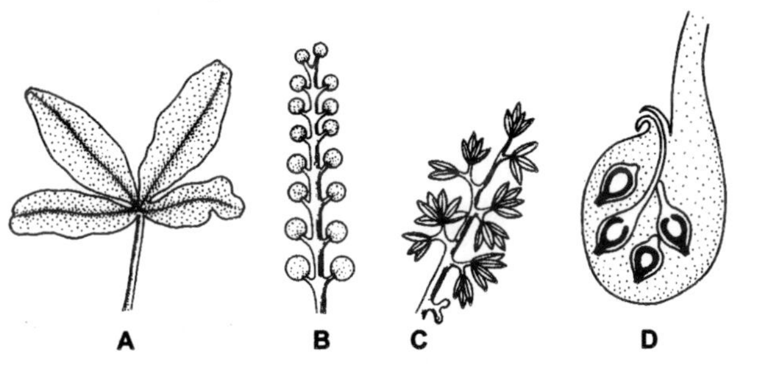

Fig. 10.3 A–*Sagenopteris phillipsi*, B–*Gristhorpia*-pinnately arranged cupules C–*Caytonanthus*-branched stamens, D–L.S. cupule of *Caytonia*.

The discovery of pollen grains within the ovules, gave an idea to suggest their true gymnosperm position, however, rather than being angiosperm ancestors. Krassilov (1977) and Doyle (1978) regarded the cupule as homologus to the carpel. Gaussen (1946) and Stebbins (1974) considered it the outer integument of the ovule. Cladistic studies of Doyle and Donoghue

(1987) support the Caytoniales–Angiosperm lineage. Thorne (1996) arrived on a conclusion that angiosperms evolved probably during the Late Jurassic from some group of seed ferns.

(iii) Cycadales. Palm-like habit of Cycadales, the ovules borne on leaf-life microsporophylls and trends in reduction of sporophyll blade, as seen in various species of *Cycas*, prompted Sporne (1971) to suggest possible links between Cycadales and angiosperms. Although it may be difficult to assume Cycadales as ancestral to angiosperms yet the fact that Cycadales have been derived from 'seed ferns' (Pteridosperms), and resemble angiosperms, further supports the origin of angiosperms from Pteridosperms.

(II) Pseudanthial Theory

Commonly associated with the Englerian School, Pseudanthial theory was first suggested by Richard Von Wettestein (1907). According to him angiosperms were derived from the Gnetopsida, represented by *Ephedra, Gnetum* and *Welwitschia*, which were formerly placed in the order Gnetales. Gnetopsida exhibits more angiosperm characteristics than any other group of gymnosperms. These characteristics include presence of vessels, reticulate dicot-like venation in leaves (in *Gnetum*), male flower with perianth and bracts, strong reduction in gametophytic phase and fusion of second male gametophyte with the ventral canal nucleus. *Ephedra* (fig. 10.4), in its general appearance, resembles *Casuarina*. According to Wettstein, the compound strobili of Gnetales are homologous to the infloresence and wind-pollinated Amentiferae. He regarded the showy insect–pollinated bisexual flowers of *Magnolia* as **pseudanthia**, derived by aggregation of unsexual units, the carpels, thus representing a modified branch (**Stachyosporous** origin of carpels).

Despite number of features, such as difference in origin of vessels (Bailey, 1944) in angiosperms (from tracheids with scalariform pitting) and

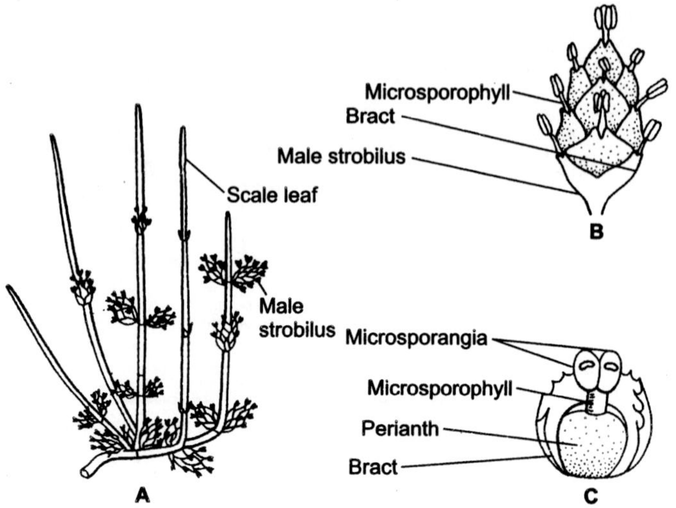

Fig. 10.4 *Ephedra* : A–staminate shoot, B–male strobilus, C–male flower.

Gnetopsida (from tracheids with circular pitting), several vesselless living angiosperms (Winteraceae) and Amentiferae are now regarded as advanced (due to floral reduction) and tricolpate pollen grains (considered advanced) and Gnetopsida a very young group. The theory has been strongly supported by Young (1981). He challenged the view that first angiosperms were vesselless and assumed that the vessels were lost in several early lines. Muhammad and Sattler (1982) reported scalariform perforations in vessel-element of *Gnetum*, suggesting that angiosperms may be derived from Gnetales after all. Carlquist (1996) concluded that this claim from *Gnetum* does not hold when large samples are examined.

According to this theory, **basal group** included amentiferous-hamamelid orders such as Casuarinales, Fagales, Myricales and Juglandales. It is notable that Wettstein (1907) also included in this basal group, Piperaceae and Chloranthaceae, the families which have been inviting considerable attention in recent years.

The significance of Gnetopsids in angiosperm phylogeny has further got support by the discovery of *Welwitschia*–like fossil *Archaestrobilus cupulanthus* described by Cornet (1996) from the Late Triassic of Texas. The plant had similarly constructed male and famale spikes, each bearing hundreds of spirally arranged macrocupules. The male spikes were borne in clusters of three whereas female spikes occurred singly, each female macrocupule with an axillary curled (tubular) bract-like organ having a narrow shaft and an expanded funnel-shaped apex. The macrocupule contained an ovule (or seed) surrounded by sterile scales and three to four minute bracts attached near the base and surrounding it. Each male macrocupule contained filament-like appendages instead of sterile scales within and on outside it was crowded with numerous bivalved microsporophylls, each with four pollen sacs attached to an inflated stalk. Gland-like structures, present on the outside of the female macrocupule, that resembled the stalks bearing pollen sacs on the male macrocupule, suggest an origin from a bisexual flower. The plant *Archaestrobilus cupulanthus* is considered as a gnetophyte more primitive than extant Gnetales.

In general, *Ephedra* is regarded to be the most primitive of the three living genera of Gnetopsida. Cornet (1996) is of the opinion that *Archaestrobilus* possessed characters, such as radial symmetry of floral parts (spirally arranged and not opposite), that may be plesiomorphic even for *Ephedra*. He considers that Bennettitales, Gnetales, Pentoxylales and angiosperms had a common ancestry sometime before Late Triassic. Gnetales are close relatives of angiosperms and Bennettitales that underwent drastic floral reduction and aggregation in response to wind pollination. Taylor and Hickey (1996) have given a hypothesis for the derivation of the flower of Chloranthaceae from the inflorescence unit (anthion) of Gnetopsids, with considerable reduction in reproductive parts.

(III) Anthocorm Theory

It is a modified version of Pseudanthial theory and was proposed by Neumayer (1924) and was later strongly advocated by Meeuse (1963, 72).

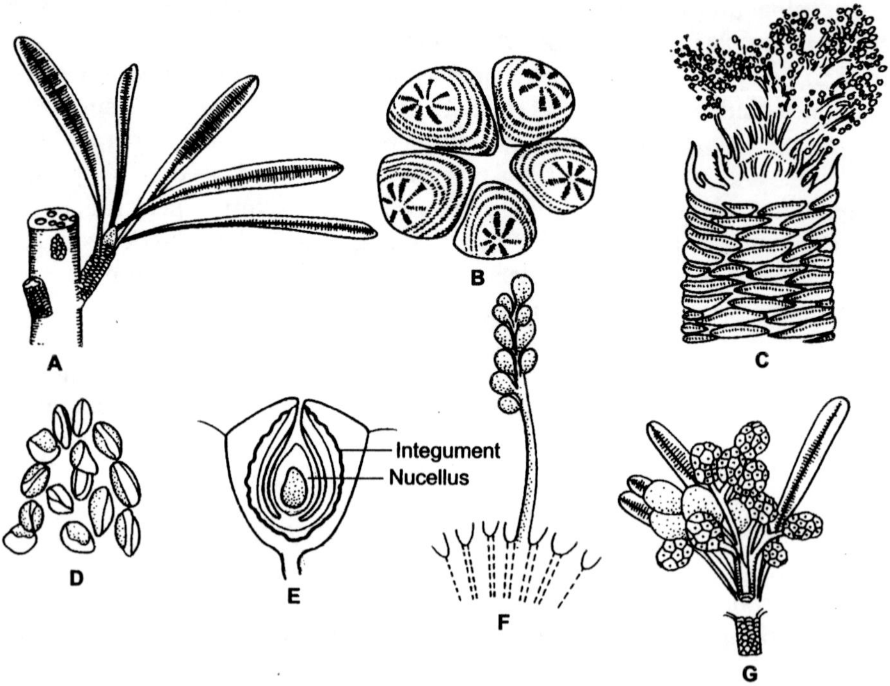

Fig. 10.5 *Pentoxylon*. A–stem and leafy shoot, B–T.S. of stem showing endocentric steles (*Pentoxylon sahnii*), C–male flower (*Sahnia nipanensis*), D–microspores, E–L.S. of ovule, F–a part of male flower (enlarged), G–female shoot (*Carnoconites*).

According to this theory, the angiosperm flower, *i.e.*, 'functional reproductive unit', has several separate origins (*i.e.*, angiosperms are polyphyletic). In many Magnoliidae and their dicotyledon derivatives, there are modified pluriaxial systems (**holanthocorms**) that arose from Gnetopsids via Piperales, whereas modification of an originally unisexual system (**gonoclad** or **anthoid**) gave rise to flowers of Chloranthaceae. Meeuse (1963) postulated the separate origin of monocots from fossil order Pentoxylales through the order Pandanales. Pentoxylales was described from Jurasssic of Rajmahal Hills, India and New Zealand. The stem (*Pentoxylon*) had five conducting strands; the pollen bearing organ (*Sahnia*) was pinnate-free above and fused into a cup below; and the seed bearing structure (*Carnoconites*) was similar to mulberry with about 20 short-stalked seeds. Each seed had an outer fleshy sarcotesta and the inner hard sclerotesta, the sarcotesta was considered homologous to cupule of seed ferns. The carpel of angiosperms was regarded as a unit composite structure being an ovule-bearing branch fused with supporting bract. Interestingly, Taylor and Hickey (1996) did not include *Pentoxylon* as a member of **Anthophytes**, which include angiosperm lineage and its sister groups Bennettitales and Gnetopsida. They considered *Pentoxylon* lacking key anthophyte characteristics such as distal, medial and proximal positioning of female, male and sterile organs on the reproductive axis and also the enclosure of ovules by bract-derived organs (fig. 10.5).

(IV) Gonophyll Theory

The theory was developed by Melville (1962, 83), largely on the basis of a study of the venaton pattern. He derived angiosperms from Glossopteridae that formed important elements in the Gondwana Flora of Gondawana land. He was of the view that angiosperm flower was derived from **gonophyll**–a fertile branching axis adnate to a leaf. As evident in simple glossopterids, such as *Scutum* and *Ottokaria*, the fertile branch consisted of a bivalved scale (having two wings) called **scutella** with terminal ovules on dichotomous groups of branches. The folding of scutella, along the cluster of its ovules, forms the angiosperm condition. An indication of this closure is found in the Permian fossil *Breytenia*. In *Lidgettonia*, the fertile branch consisted of four to eight disc-like structures bearing several seeds (fig. 10.6).

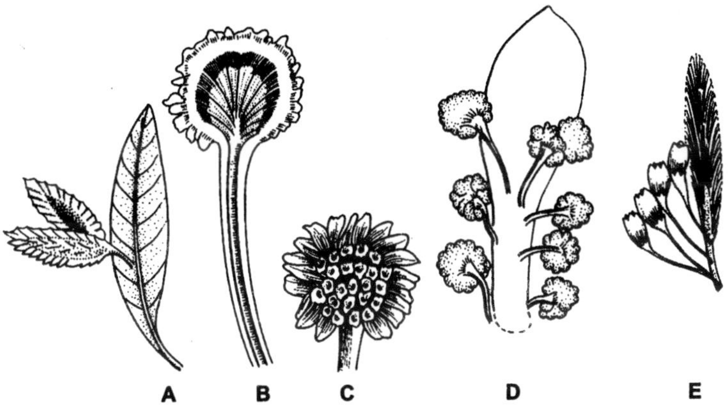

Fig. 10.6 A–cupule of *Scutum*, B and C–Cupules of *Ottokaria* sp., D–fertile branch of *Lidgettonia*, E–fertile branch of *Denkania*.

In *Denkania*, described from Raniganj (India), about six seed bearing **cupules** are attached to long stalks arising from the midrib of fertile scale. The leaves of *Glossopteris* are lanceolate with distinct venation, the fertile region is cone-like with a transition from leaves to fertile scales, spirally arranged and conforming to *Anthostrobilus*. In *Mudgea*, there is a suggestion of **anthofasciculi**, *i.e.*, leafy structures with two fertile branches–one male and the other female, forming the angiosperm flowers as found in *Ranunculus*.

Melville was of the opinion that angiosperms arose 240 my ago in the Permian and took around 140 my before they spread widely in the Cretaceous. But his justification brought him to stand in the row of proponents of polyphyletic origin of angiosperms.

(V) Herbaceous Origin Hypothesis

The hypothesis resembles pseudanthial theory but the ancestral plant is considered to be a perennial rhizomatous herb instead of a tree. The term **paleoherbs** was first used by Donoghue and Doyle (1989) for a group of derivative (not ancestral) forms of Magnoliidae having the characteristics such as anomocytic stomata, two whorls of perianth and trimerous flowers including families like Lactoridaceae, Aristolochiaeceae, Cabombaceae, Nymphaeaceae, Piperaceae etc. According to this hypothesis, ancestral forms of angiosperms were small herbaceous plants. They had rhizomatous to scrambling perennial habit and simple reticulately veined leaves, with a primary venation pattern that would have been indifferently pinnate to palmate and dichotomously branched secondary veins. The anatomy of vegetative parts included sieve-tube elements and elongate tracheary elements with both circular-bordered and scalariform pitting and oblique end walls. The flowers occurred in cymose to racemose infloresence; the carpels were free, ascidiate (ovules attached proximally to the closure) with one or two orthotropous, bitegmic, crassinucellate ovule and dicotyledonous embryo; and small monosulcate pollen with perforate to reticulate aperturing. Donoghue and Doyle cite extreme rarity of fossil angiosperm wood and abundance of leaf impressions in early fossils.

Phylogenetic studies have given a way for emerging consensus (Taylor and Hickey, 1996) that gnetopsids represent the closest living relatives of angiosperms, whereas the closest fossil group is Bennettitales. And thus, the angiosperm lineage, together with Gnetopsids and Bennettitales, constitute **Anthophytes**. The group is believed to have splitted in the Late Triassic, the angiosperm lineage continuing as **Angiophytes** upto the Late Jurassic when it further splitted into **Stem Angiophytes**, that included the early extinct angiosperms, and **Crown Angiophytes**, that constituted the extant groups of angiosperms.

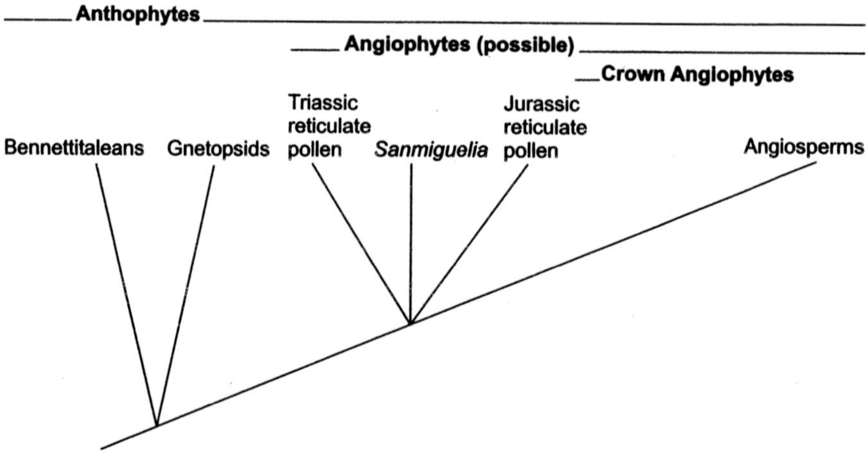

Fig. 10.7 Consensus phylogeny of Anthophytes (after Taylor and Hickey, 1996). Notably *Pentoxylon* has been excluded from sister groups (now only Bennettitaleans and Gnetopsids) of angiophytes.

Krassilov (1977), who believed in polyphyletic origin of angiosperms, identified three Jurassic groups : Caytoniales, Zcekanowskiales and Dirhopalostachyaceae as proangiosperms. Pollen germinating on tip, according to him, would be rather disappointing because these plants then be classified as angiosperms and be excluded from discussion of their ancestors.

He evolved Laurales-Rosales series from Caytoniales. Zcekanowskiales with bivalved capsules provided with stigmatic bands, showed links with monocots. Dirhopalostachyaceae, with paired ovules exposed on shield-like lateral appendages, probably evolved in Hamamelidales.

A more recent theory **Transitional-Combinational Theory** (Stuessy, 2004) for origin of angiosperms, suggests that angiosperms evolved slowly from the seed ferns in the Jurassic by first developing the carpel, later accomplishing the double fertilisation and then developing the flower. The three fundamental transitions might have taken more than 100 million years to complete. As presented by fossil record, living angiosperms did not appear until Early Cretaceous when the final combination of these three important angiosperm features occurred. The theory is an effort to explain discrepancy between fossil and molecular phylogenetic data, latter suggesting pre-Cretaceous origin of angiosperms when DNA (and protein) sequences showed first changes accompanying carpel evolution, much before than final combination of all three angiosperm features. According to this theory, except for seed ferns, no other gymnosperm had direct phylogenetic connections to modern angiosperms.

OLDEST ANGIOSPERM

Using the oldest and most complete fossil angiosperm on record, David Dilcher–a paleobotanist associated with the Florida Museum of Natural History at the University of Florida, recently declared about the discovery of a new **basal angiosperm** family of aquatic plant, Archaefructaceae, published in the journal *Science* with coauthors Ge Sun of Jilin University, Qiang Ji of the Geological Institute of the Chinese Academy of Geosciences and others (Sun *et al.* 2002). The family is based on the only genus *Archaefructus* with two species : *Archaefructus lianogensis* and *A. sinensis* (fig. 10.9). Both of the

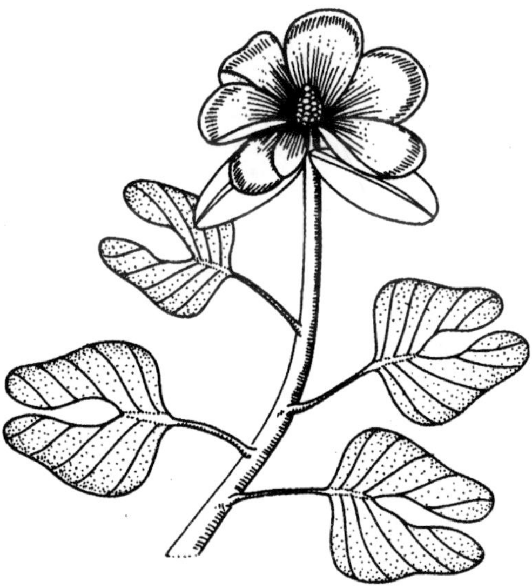

Fig. 10.8 Reconstruction of a leafy branch with flower of *Archaeanthus linnenbergeri* from middle Cretaceous, (after Dilcher and Crane, 1984).

them were probably aquatic herbs and living atleast 124 mya. *Archaefructus* has perfect flowers rather unlike those of living angiosperms–there being no perianth, with very elongated receptacle and paired stamens. The fruits are small follicles formed from conduplicate carpels helically arranged with adaxial elongate stigmatic crests conspicuous on each carpel. Earlier, in 1982, Dilcher and Crane, had described *Archaeanthus linnenbergeri* (fig. 10.8) from uppermost Albian/mid-Cenomanian (about 110 mya) of Middle Cretaceous as a primitive flowering plant with simple bilobed leaves and terminal flower with numerous free carpels producing follicle fruit.

Archaefructus plant was about 50 cm high, rooted in lake bottom and partially supported by water. Its thin stems reached surface of the water. The leaves of the plant were possibly submerged. Its pollen and seed organs were extended above the water and seeds probably dispersed on the water and floated towards the shore where they germinated in shallower areas. This is considered to be the oldest record of an angiosperm plant which is placed in a distinct family Archaefructaceae, probably sister family to all living families of angiosperms but at the same time it leaves a problem as to how to relate it to living angiosperms.

Fig. 10.9 A reconstruction of *Archaefructus sinensis* believed to be the oldest angiosperm fossil, nearly 124 mya old.

BASAL LIVING ANGIOSPERMS

It is increasingly believed that angiosperms evolved in very Late Jurassic or very Early Cretaceous. For nearly a century, there was general agreement that the early angiosperms were woody shrubs or small trees (herbaceous habit being derived). They had simple evergreen entire and pinnately veined leaves with stipules. With reference to most primitive living angiosperms, there have been two opposing schools of thought : **Englerian school**—considering Amentiferae (particularly Casuarinaceae) to be most primitive dicots and

Besseyan school—considering bisexual flowers of Magnoliales to be the most primitive. In view of recent studies, the course of evolution within angiosperms is being thoroughly examined with newer tools of research. During the last few years *paleoherbs* are emerging as the strong contenders. The *candidate basal groups* considered as primitive include Casuarinaceae, Magnoliaceae, Winteraceae, Degeneriaceae and Calycanthaceae. APG II (2003) accepted Magnoliidae and Nyphaeidae as the true basal group/complex and instead of Degeneriaceae, placed Amborellaceae at the beginning.

PALEOHERBS

Last decade of the past century has witnessed the strong development of herbaceous origin hypothesis for angiosperms (Taylor and Hickey, 1996) and originally developed *Paleoherb hypothesis.* Now most primitive angiosperms are considered to be rhizomatous or scrambling perennial herbs with simple net-veined leaves, flowers arranged in racemose or cymose inflorescence and free carpels containing one or two ovules. Paleoherbs fall under a number of angiosperm families. Thorne, earlier (2000), had placed all of them under Magnoliales along with Magnoliaceae and Winteraceae but in his latest revision (2003), he has placed Amborellaceae and Chloranthaceae (together with Trimeniaceae and Austrobaileyaceae) under Chloranthales, the first order of Magnoliidae (and accordingly angiosperms), the families arranged in that order. The family Ceratophyllaceae has been placed after monocot families, towards the begining of Ranunculidae. In the classification schemes of Judd *et al.* (2002), APG II (2003) and APweb (Stevens, 2003), Amborellaceae have been placed at the beginning of angiosperms. The position of other two families : Chloranthaceae and Ceratophyllaceae, is, however, not settled as Judd *et al.* and AP web consider both— Chloranthaceae towards the end of basal families before Magnoliid complex and Ceratophyllaceae towards the end of Magnoliid complex as having uncertain position. Like Throne, APG II places Amborellaceae and Chloranthaceae at the beginning of angiosperms (but as unplaced families), whereas Ceratophyllaceae is placed before Magnoliids.

Apart from Chloranthaceae, Ceratophyllaceae and Amborellaceae, other families that are considered belonging to paleoherbs and share plesiomorphic features include Saurutaceae, Piperaceae, Aristolochiaceae, Barclyaceae, Cabombaceae and Nymphaeaceae. All these families belonging to paleoherbs share characters of herbaceous habit, tectate collumellate monosulcate pollen, apocarpous gynoecia and simple floral units.

ORIGIN OF MONCOTYLEDONS

Despite it is now largely believed that monocots evolved from dicots monophyletically, for a long time it was thought that monocots arose before dicots (Engler, 1892) polyphyletically (Meeuse, 1963). Bailey (1944) and

Cheadle (1953) believed that vessels in monocots and dicots originated and specialised independently, and thus monocots arose from vessel-less dicots. Cronquist did not agree with the concept of the independent origin of vessels in two groups, instead he considered monocots to have an aquatic origin from ancestors resembling modern Nymphaeales. This was strongly refuted by Kosakai, Moosley and Cheadle (1970) on the basis of their studies made on vessels. They were of the opinion that it is difficult to believe that putatively primitive Alismataceae evolved advanced vessles in an aquatic environment yet gave rise to terrestrial monocots with more primitive vessel elements in the metaxylem of roots and they thus favoured the origin of Alismataceae from terrestrial forms.

Hutchinson (1973), believed that monocots arose from *Ranalian stock* along two lines : one (Ranunculoideae) giving rise to Alismatales and the other (Helleboroideae) giving rise to Butomales. Takhtajan (1980, 87) suggested a common origin for Nymphaeales and Alismatales from a hypothetical terrestrial herbaceous group of Magnoliidae. Dahlgren *et al.*, (1985) suggested that monocots appeared in the Early Cretaceous some 110 mya ago when the ancestors of Magnoliiflorae must have already acquired some of the present attributes of that group but were less differentiated : some other dicot groups had already branched off from the ancestral stock. Thorne (1996) is of the view that monocots appear to be very early offshoot of the most primitive dicots. Chase *et al.* (1993), in their *rbc*L sequence studies found the monocots to be monophyletic and derived from monosulcate Magnoliidae. *Acorus*, Melanthiaceae and *Butomus* are regarded to be the least specialised monocots.

EVOLUTION OF ANGIOSPERMS
(General Evolutionary Trends in Angiosperms)

Earliest evidence of angiosperms comes from fossil pollen (and fragmentary flowers and fruits) from Early Cretaceous, about 130 mya. These resemble pollen grains of living Chloranthaceae. DNA evidence suggests that a shrub *Amborella* (Amborellaceae), and water lilies diverged first in angiosperm evolution. The pollen and the form of flower of the earliest fossil are similar to those of living members of the Chloranthaceae, a small family from tropical America, east Asia and New Zealand. The flowers of these plants are simple and tiny, upto 3 mm long, some genera bearing hemaphrodite flowers, others unisexual-borne in inflorescence. Each flower contains one or three stamens and/or one carpel and some have a bract (small modified leaf). They have no sepals or petals. *Hermaphrodite* species are pollinated by unspecialised insects and unisexual species by wind. A range of large and small flowers appeared in later course. Putting all molecular evidences, obtained from analysis of DNA from genes in the nucleus, plastids and mitochondria together, it appears that a shrub, *Amborella* (Amborellaceae) from New Caledonia, that bears unisexual flowers, and water lilies which are herbaceous aquatics bearing large hermaphrodite flowers, diverged from other angiosperms earlier than any other group.

By now it has become clear that angiosperms are a monophyletic group. To trace the evolution of angiosperms, it is important to distinguish clearly between the origin of the line leading to modern group (*i.e.*, when this lineage splits from its sister lineage) and the origin of the least inclusive *stem lineage* which has been referred to as the *angiophytes* (Doyle and Donoghue, 1993) to help to distinguish from the *crown—clade* angiosperms. As has already been discussed that extraordinarily complete Early Cretaceous fossil *Archaefructus* of entirely extinct family Archaefructaceae, was described from China (Ge Q Sun Ji, DL Dilcher, S Zheng, KC Nicon and X Wang, 2002) is now reckoned to be a new basal angiosperm family.

Basic Evolutionary Trends

Different groups of angiosperms, reveal different lines of evolution. Many of the trends are recognised from comparative studies of living and fossil plants during attaining diversity. General trends in evolution of angiosperms include :

Evolution of Vegetative Structures

Flowering plants are enormously variable in their vegetative structures but the earliest ones and their living representatives are mainly shrubs or trees and all living gymnosperms are woody. It is likely that herbaceous plants are derived from woody forms originally, but some woody plants, including monocot trees, may be secondarily evolved from herbaceous plants. There is difference of opinion about evolution of leaf structure, phyllotaxy, stomatal apparatus etc. of angiosperms.

1. Xylem Evolution. Based on the studies of flowering plants, it is almost clear that all fossil and living gymnosperms possess tracheids with circular-bordered pittings. This leads to the conclusion that the tracheid is the most primitive type of tracheary element in the angiosperms. They have given rise to vessel elements with the transformation of tracheids with scalariform pitting followed by becoming smaller and broader and perforation plates more horizontal. Further reduction in the number of scalariform bars resulted shortest-broadest vessel elements with simple perforation plate and transverse end wall in most advanced forms.

Based on survey of wood anatomy, Carlquist (1996), has identified a number of distinct evolutionary trends in angiosperms and has concluded that vessels have originated numerous times in dicotyledons. It is presumed that vessel elements arose independently in Gnetales, dicotyledons, monocotyledons, certain ferns, *Selaginella* and *Equisetum* through parallel evolution. Vessel elements evolved early on, but a few angiosperms, including *Amborella* and the Winteraceae, have none and in water-lilies the xylem cells are somewhat intermediate between tracheids and vessels.

2. Nodal Anatomy. Depending upon number of leaf gaps, the nodes of angiosperms are either multilacunar, bilacunar or unilacunar. In ferns and majority of gymnosperms they are unilacunar but in Cycadales and *Gnetum* they are multilacunar. It is generally believed that in gymnosperms, multilacunar type has been derived from the unilacunar type in the course of evolution but in angiosperms nodal structure might have evolved from the gymnosperms with either multilacunar or unilacunar nodes. In addition to these two types, a third type, trilacunar, has been described for some angiosperms. Different authors, based on their studies, have considered all three types as primitive.

Evolution of Floral Structures

The earliest flowering plants have flowers which were simple and tiny upto 3 mm long, hermaphrodite or unisexual with no sepals and petals (some have a bract, one or three stamens and/or one carpel). During the course of early evolution, active adaptive radiation, *i.e.*, their spread and diversification through the Cretaceous period, parallel to other seed plant groups, resulted in appearance of some other living families. Outer whorls of flower became differentiated early on, the outermost whorl, referred to as calyx even if there is only one whorl, probably evolving from bract and when there are two different whorls, the inner whorl is referred to as the corolla, but some probably evolved from a second whorl of sepals, others from stamens losing their anthers and becoming organs purely for attracting insects. Some water lilies and nympheoids (*Nymphaea*) have intermediate organs, stamens with flattened coloured filaments between the fertile stamens and larger entirely sterile petals. Cultivated *double flowers* such as of roses have some stamens replaced by petals. Transitional stages of converting fertile stamens to fully developed petals can be seen in many flowers, *e.g.*, *Rosa* and *Nymphaea*. One main early line of evolution was towards a fix number in each whorl, often five in dicots (rarely four) and three in the monocots, making the flowers radially symmetrical, *i.e.*, actinomorphic.

There was a specialisation for pollination, by beetles along one line of evolution including magnolias, and by wind along other line leading to the catkin-bearing trees. The insect-pollinated group had hermaphrodite flowers of, at first, increasing, size and complexity, through later insect-pollinated flowers were smaller. By the end of Cretaceous period, many families with actinomorphic insect-pollinated flowers had appeared, such as buttercups, pinks and heathers. The wind-pollinated group retained small unisexual flowers, with few parts borne in inflorescences, usually with male and female on the same plant (monoecious), although such breeding system includes less than five per cent of living species, such as oaks, beeches, birches, hazels etc.

At the end of Cretaceous and beginning of the Tertiary period (between 75 and 50 mya), there was a second and much larger adaptive radiation which witnessed the increasing dominance of angiosperms and the appearance of the majority of modern plant families. This was closely associated with the adaptive radiation of specialist insect pollinators and leaf feeders–the butterflies and moths, long tongued flies and bees. Innovations included the fusion of the lower parts of the petals and/or the sepals into a tube and fusion of the carpels. Zygomorphic flowers appeared, usually facing sideways rather than upwards and with lower petal or petals forming a lip as a landing plateform for insects. Some plants, such as members of thyme family, have fused petals and zygomorphic flowers. These were adaptations for specialised insect pollination and atleast in part serve to exclude other visitors. Adaptations for pollination by birds or bats appeared later, mainly derived from specialist-pollinated flowers.

The general processes involved in attaining adaptive radiation, *i.e.*, the spread and diversification of angiosperms, included fusion of different floral parts, reduction in floral parts, change in symmetry (actinomorphy to zygomorphy), elaboration (in number of flowers) and remoration (evolutionary retrogression). The main trends are outlined in table 10.2.

Table 10.2 Evolutionary Trends in Flowers from the End of Cretaceous to the Early Tertiary (Many trends occurred in parallel in different families; some also happened in reverse).

	From		To
1.	Radial symmetry	→	Bilateral symmetry.
2.	Separate petals	→	Fused petals.
3.	Large number of floral parts	→	Fixed small number of floral parts.
4.	Large flowers	→	Small flowers.
5.	Carpels inserted above the petals	→	Carpels inserted below the petals.
6.	Carpels free	→	Carpels fused.
7.	Flowers with pollen as food reward	→	Flowers with nectar as food reward.
8.	Pollination by unspecialised insects	→	Pollination by specialised bees; butter-flies/moths; long tongued flies; birds or bats.
9.	Insects-pollinated species, hermaphrodite flowers	→	Unisexual flowers.
10.	Insect-pollinated species	→	Wind-pollinated species.

Aggregation of flowers in inflorescence is an adaptative trend in many angiosperm families but it has taken an extreme specialisation in daisy family (Asteraceae) in which there are often two different forms of flower in each inflorescence, the whole aggregation, sometimes of more than 100 flowers, resembling a single flower (capitulum infloresence).

A few flowers exhibit extreme specialisation in their pollination. For example, in many orchids, each flower needs a single visit for the successful formation of hundreds of seeds, so they become highly selective in the species that they use. Some tropical species have a long spur growing out of the back of the flower with nectar at its base, which only moths with long thin tongues can reach. The European bee orchid group and a few other groups of orchids in other parts of the world, have flowers that resemble the female bee, both in appearance and in scent, each orchid species (*e.g. Ophrys*) resembling a different bee; they are visited and successfully pollinated by the males of that bee species. The figs, a large group of tropical trees and stranglers, have flowers in an inverted infloresence with flowers opening into a central cavity, surrounded by a solid mass–hypanthodium. The inflorescence can only be penetrated by small wasps through a whole at the tip. The wasps lay their eggs, grow and mate in the inflorescence before collecting pollens and dispersing to another fig. Plant and wasps are totally independent but a kind of interdependence is seen in *Yucca* and moth. Such cases substantiate the parallel evolution of insects and flowers.

More generalised flowers, much more common than the specialists, particularly among the dominant vegetation, are usually pollinated either by wind or by a range of insects and sometimes birds as well. Pollinating insects (and animals) can vary manifold in number from year to year and place to place, so specialisation to one or a few species can be risky.

Specialisation for wind pollination, has nearly always involved a reduction in size and number of floral parts. Some families are nearly all wind-pollinated, *e.g.* grasses, but wind-pollinated species occur in numerous, mainly insect-pollinated families and have evolved many times. Evolution in the reverse direction, from wind to insect-pollination, has rarely occurred.

Evolution of Fruits and Seeds

There occurs some controversy over which fruit type is primitive. Fossil evidence gives few clues, but it is likely that the early fruits were fleshy, as many of the representatives of primitive groups have fleshy fruits, sometimes with outgrowths from the seed (an aril). Trends in the evolution of fruits have gone in many directions–towards larger or smaller fruits or seeds and towards dry dehiscent fruits. But main trend appears to have been for a reduction in size and in number of seeds per fruit. There has been parallel evolution of fruit types in many plant families, *i.e.*, evolution along similar paths in unrelated groups.

Primitive seeds are likely to have been fairly large and rich in endosperm features retained in some specialised large-seeded plants such as coconuts. The small seeds, mainly filled by embryo, seen in many families, are likely to be derived. Some of the most successful families such as grass family (Poaceae) and daisy family (Asteraceae) have one-seeded fruits, but another large family, the Orchidaceae have many hundred or thousands of tiny seeds per pod, although with very poor germination percentage.

Chapter **11**

Variation and Speciation in Plants

VARIATION

No two individuals are alike. Variation is apparent among species of the plants and between the individuals of the same species. It is now universally agreed upon the fact that different species are not fixed. The concept of variation, first proposed by Lamarck (1744–1829), was further developed by Darwin, culminating in his famous book *Origin of Species* (1859). Plant systematics, a unique natural science concerned with individual, population and taxon relationships for the purpose of classification, is based on the premise that in tremendous variation in the plant world, there exist conceptual discrete units, that are usually named as species. These units can be categorised, described, named and classified on the further premise that logical relationship, developed through evolution, exists among them. Variation provides raw material for evolution, as without variation natural selection would not have been possible.

VARIATION : ADAPTATION FOR SURVIVAL

The current patterns of variation, as discrete units, are the product of basic evolutionary mechanisms such as mutation, recombination and selection acting upon different environments. Variation and the resultant change in organisms, through times, depended largely on evolution. The population viability results due to developmental variation, environmentally induced variation or genetic variation, caused in them. Variations may be **continuous**–without a clear separation of a series into discrete units or **discontinuous**–with few or no intermediates in the series and, therefore, separable as discrete units. Discontinuous variations result in increase in discrete units that can be recognised, classified, circumscribed and named. The discontinuity in population results mainly from **isolation.**

The evolutionary success and potentiality for further diversification of a taxon or a species depends upon survival of the seedling and adult plant under a variety of environmental conditions, cross pollination, development and dispersal of seeds and establishement of seedlings (Stebbin, 1974). The adaptations for survival may be accomplished by environmental modifications of phenotypes, *i.e.*, ecotypic differentiation. Local population, comprising of individuals that have similar environmental tolerance and are specialised to meet environmental conditions, are termed as ecological races or ecotypes. The term ecotype was coined by Gote Turesson (1912) to describe '*genetic variation within a species*'. Existence of ecotypes has been demonstrated in many woody and herbaceous species in relation to geographical variation of several ecological factors, such as light, heat, moisture, soil properties and so on. It is through such ecotypes that species become competent to occupy wide ranges. It is not clear as to why some plants have this ability for wide dispersal and others lack it. It seems that the environmental modification on being incorporated into the genetic make-up, becomes permanent and brings about further differentiation within the population and their ecotypes. However, some species appear to be genetically conservative and can not adapt new environmental set up because for most adaptive changes, alteration of genotypes is required. Adaptation to cross pollination involves adaptive syndromes of characters favouring pollination by the most effective pollinator. Although these adaptations originate as mere modifications due to environment but are quickly incorporated into the genotypes. Adaptation for development and dispersal of seeds and establishment of seedlings involves much complexity and diversity. These three–development, dispersal of seed and establishment of seedlings, separately demand somewhat conflicting requirements, *e.g.* selection for increased seed number may bring about profound changes of gynoecium whereas adaptation for protection of developing seeds may involve changes in the floral structure and their surrounding bracts. Major categories of flowering plants have evolved due to these and similar adaptations which got retained in them. Development of adaptations for protection against insects and other predators, by producing repellent chemical substances and also symbiotic relations with animals, have only recently been recognised.

Environmental Variations

All variations in plants, are the outcome of plastic response of the individual to environmental factors involving climatic, edaphic and biotic that have two-fold effects–direct effects upon the phenotype and indirect effects on the genotype of individual. And thus, the individuals are the outcome of their genetic constitution and of environmental influences during their development. Traditionally, the characters that are greatly modified by environmental conditions, are considered as bad taxonomic characters. However, it is now apparent that any distinction between genetic and environmental impacts upon phenotype is artificial and rather meaningless (Dobzhansky, 1970). According to Dobzhansky, the species are maintained by reproductive isolating mechanism, *i.e.*, any property that prevents gene flow between species such that species living in close proximity remain distinct.

Physiological studies of environmental modification indicate that it is not random or biologically insignificant but is frequently adaptive and of considerable evolutionary and ecological importance. Taking an example of two species of *Lysimachia* (Primulaceae), Turesson (1922) found that these respond differently to light intensity. One of them shows phenotypic response while the other exhibits genotypic differentiation, *i.e.*, both altered genotypes and phenotypes may occur mixed in a single population (Nelson, 1965). The balance of the two mechanisms, however, may vary between populations and also between species (Jain, 1969). Keeping in view these considerations, Wells (1969) has criticised as *misleading distinction between environmentally induced phenotypic modification and genetically fixed ecological races.*

Recent studies in ecology, physiology and genetics, reveal that phenotypic variation is of considerable ecological and evolutionary significance, and this according to Snaydon (1973) may be an opportunate time to reconsider the taxonomic status of phenotypic variation. As a matter of fact, environmentally variable characters have so far been rejected in taxonomic consideration and stable floral characters have been given more importance. Snaydon (1973) emphasises adequate study and incorporation of vegetative characters, although these being much variable in response to environmental impacts in construction of keys. So far, taxonomists used only environmentally stable characters because phenotype closely reflects genotype. Within population, genetic variation is apparently determined by complex interactions between genetic and ecological factors. Here environmental factors also partly determine the genetic factors, such as breeding system, rate of mutation and dominance. Impact of ecological factors, both environmental and biotic, upon genetic variation within populations, is probably more than what has previously been assumed. The genetic variation within population, which has usually been studied at the morphological level, needs to be studied at the physiological and biochemical levels and its incorporation in taxonomy. Conclusively, the ecological information about various plant taxa is not large and taxonomic delimitation between plant genera, families etc., does not generally reflect differences in ecological behaviour. But in certain plant groups, there appears to be good correlation between taxonomic limits and ecological behaviour. The plants are producers, fungi are consumers and bacteria are decomposers in ecosystems and within each of these groups there are also broadly taxonomic and ecological groupings, *e.g.* taxonomic grouping in algae, bryophytes, pteridophytes, gymnosperms and angiosperms which correspond to ecological differences. Many of the angiosperm families, such as Salicaceae, Chenopodiaceae, Orobanchaceae, Ericaceae, Plumbaginaceae, Lentibulariaceae and Potamogetonaceae, are ecologically quite homogeneous.

Developmental Variation

Often there occurs a distinct change in attributes during different stages of development, *e.g.* juvenile leaves of *Eucalyptus, Salix* and *Populus* are often different (commonly in colour) from the mature leaves, sometimes causing confusion, but may prove equally useful when both types of leaves are available from a plant. The leaves of *Phaseolus*, that arise first, are opposite and simple, the later arising are alternate and pinnately compound. This is because in plant's life, seedling stage is most critical and the characters present at this stage surely have survival value. Takhtajan proposed a **neotenous origin** for angiosperms on the assumption of juvenile simple leaves of seed ferns having persisted in the adult forms, which he considered direct progenitors of angiosperms.

Genetic Variation

Apart from environmental variations, populations normally have variation by random gene mutation, that is shuttled from generation to generation.

Gene Mutation. Mutation, as defined by Mayr (1942) "*a discontinuous chromosomal change with a genetic effect*", may be at the gene level–gene mutation, at the level of chromosome–chromosomal mutation or even at the level of genome–genomic mutation. Gene mutation primarily brings about change in developmental processes and thus has specific effects on individual phenotype. Chromosomal and genomic mutations alter the whole series of processes or affect the relationship between them. Stebbins (1959) opines that origin of biological species is mainly because of them. The alterations in the organism, brought about by gene mutation, are of great concern to the taxonomists. The changes brought about due to gene mutation may be phenotypic, such as zygomorphic corolla of *Antirrhinum* (Scrophulariaceae) to actinomorphic one, gamosepalous calyx in *Silene* (Caryophyllaceae) to polysepalous one, or the spurred corolla of *Aquilegia* (Ranunculaceae) to spurless one. Among other phenotypic changes caused by single gene effects include alteration of habit from annual to biennial etc. Natural selection favours all these changes and contributes to evolutionary change. Most of the mutations, however, are disadvantageous to the plants.

Polymorphism, *i.e.*, the presence of a number of different alleles resulting in more than one phenotype in a population, is most common genetic variation that brings about change in flower colour forms (*Oxalis, Oldenlandia, Biophytum*), chemical forms (*Eucalyptus*) and so on. Such features may be adaptive or not, uniform or cryptic and discontinuous or more rarely continuous. The genetic basis of these changes may be genic, chromosomal–segmental (inversions or translocations) or chromosomal. Pleiotropic gene action, *i.e.*, many mutations, possibly most of them affecting several characteristics of the adult phenotype, may produce different but related effects on different organs, *e.g.* in *Nicotiana*, the gene that produces long petioles, also produces more acuminate leaf-tips, longer calyces with narrower sepal lobes, corollas with more acuminate lobes, longer anthers and more attenuated capsules (Stebbins, 1959).

Recombination. In production of adaptive variation in higher plants, gene recombination plays the role which is as important as mutation. It is largely determined by the breeding system. The whole series of mechanism, promoting the cross-fertilisation and the genetic system of self-incompatibility, has selective advantage in promoting gene recombination, self-fertilisation and apomixis, being secondary and is supposed to have been derived from flexible cross-fertilisation systems. Thus it becomes necessary for a taxonomist to have adequate knowledge of the breeding behaviour of populations. In certain plant groups sexual reproduction is replaced by a non-sexual method, *i.e.*, apomixis that may be by vegetative reproduction or agamospermy.

BREEDING SYSTEMS AND TAXONOMY

Breeding systems have been treated as inbreeding systems, outbreeding systems and apomixis.

Inbreeders. In course of time, the habitual breeders seggregate into several pure lines or genetically heterozygous populations, and if the genetic differences are also reflected in their phenotypic characters, each line will be easily distinguishable. The pure lines are less capable for evolution in response to environmental changes. On hybridisation, they give rise to hybrids, that on seggregation, produce new lines. As such, this aspect is commonly used in selection of crop plants for uniformity in certain desired morphological characters. Since many of these inbreedings lack consistancy, on account of their occasional outbreeding and large number, all such inbreedings are not recognised as species. Moreover, there are no hard and fast rules in dealing with the recognition of species in inbreeding groups, as is the case in apomicts, for their taxonomic recognition of species. The recognition of inbreeding, particularly in populations in which they appear to be constant and occupy considerable area, helps taxonomists to understand pattern of variation in these groups (Davis and Heywood, 1963). Naik (1976) has described cryptic speciation between morphologically similar inbreeding populations of *Urginea indica* (Liliaceae), which lack genetic variability. However, due to genetic differences, such as polyploidy, they may be isolated from each other, giving rise to **biological species**.

Outbreeders. The taxonomic separation of outbreeders, that show continuous pattern of morphological differences, pose difficulties and cause a taxonomist to seek for a discontinuity and circumscribe a taxon accordingly and if he fails to do so, he prefers to keep them under one name (Baker, 1959).

In nature, inbreeding (self-compatibility) and outbreeding (self-incompatibility) are not reported simultaneously in the same species. And if it occurs, it is better to recognise the two races as distinct species, *e.g.* in *Primula* (Primulaceae) and *Armeria* (Asteraceae).

Apomicts. Many of the taxonomists, including Love and Love (1956), Valentine (1960), Kawano (1963) etc., have associated themselves in problems related with the classification of apomicts, but the subject still remains controversial. Many grasses show transition between complete sexuality and complete apomixis, *i.e.*, facultative apomicts, but there are other plants in which no sexual phase is known, *i.e.*, obligate apomicts. Vegetatively reproducing apomicts, such as bulbils, tubers, rhizomes etc., taking over the whole reproduction, are common in nature. Since all individuals produced vegetatively, have the same genotype they belong to the same biotype, there being no new genotype and variation. Apomicts produced by agamospermy produce embryos and seeds but no sexual process is involved because meiosis and fertilisation are discarded and the resultant embryo is genotypically identical with the maternal plant. Thus, apomicts neither form populations nor do they exchange genes between their biotypes, instead each biotype builds its own population by self-duplication. Thus, the population may contain many biotypes, different biotypes maturing at different times. To recognise differences between the biotypes is difficult, except to a specialist (Valentine, 1960), but since each biotype is true breeding, a finer discrimination may be possible in them than in amphimictic groups. This enables to describe a large number of species, but the actual number will depend on the nature of apomictic mechanism involved and the extent of interest taken by the taxonomists to this group. The degree of variation shown by facultative apomicts is such that no useful classification of their variation is possible. Some common examples of such groups in active state of evolution can be traced in *Rubus fruticosus* and *Poa alpina*. On the other hand, obligate apomicts, such as found in *Ranunculus auricomus* and *Alchemilla* species, allow recognition of taxa of greater stability.

Variance Analysis

"*No two individuals in a population are similar*", needs some objective analysis for useful comparison. **Mean** is the simplest tool that can be calculated by adding a series of values and dividing the total by the number of values. Mean can be calculated as :

$$\overline{X} = \frac{\Sigma x_i}{n}$$

where X represents the mean, Σ summation of all the values of x, x_i represents the individual values of an attribute under study and n represents the number of values. Thus four plants of a species with a height of 16 cm, 15 cm, 13 cm and 12 cm would have a mean of 14 $(16 + 15 + 13 + 12/4)$. The extent of variation within a population of a species is best represented by **variance** which would have been the minimum had the various individuals not been far from this mean and if many individuals were far removed from mean, the variance would have been the higher. The variance may either be calculated for a population or a sample from the population.

ISOLATING MECHANISMS

Isolation, a distinct gap between two populations, is the key factor preventing intermixing of distinct species through prevention of hybridization. Discontinuous variation primarily results from isolation, which plays a significant role in establishing and widening the gap between the populations, allows evolution to take its destined course without any disturbance. Based on the mechanisms, isolation operates before or after sexual fusion, *i.e.*, it can be spread over : prezygotic mechanisms or postzygotic mechanisms.

Prezygotic Mechanisms

These isolating mechanisms, operating before sexual fusion, are pre-pollination mechanisms and post-pollination mechanisms.

1. Pre-pollination Mechanisms

Pre-pollination mechanisms of isolation are mainly of the following types :

(1) Geographical Isolation. It refers to geographical separation of two species by a gap larger than their pollen and seed dispersal, *e.g. Platanus orientalis* (Mediterranean region) and *P. occidentalis* (North America) are well established species but readily interbreed when brought into the same area (**vicarious** species).

(2) Ecological Isolation. It refers to two species occurring in the same region but flowering at different seasons, *e.g. Sambucus racemosa* and *S. nigra* flower nearly seven weeks apart.

(3) Temporal Isolation. It refers to two species flowering during the same period but during different times of the day, *e.g. Agrostis tenuis* flowers in the afternoon whereas *A. stolonifera* flowers in the morning.

(4) Ethological Isolation. It refers to two species being interfertile but have different pollinators, *e.g.* humming birds are attracted to red flowers and hawk-moths to white flowers.

(5) Mechanical Isolation. It refers to pollination between two related species prevented by structural differences between flowers, *e.g.* between flowers of *Ophrys insectifera* and *O. apifera.*

2. Post-pollination Mechanisms

Post-pollination mechanisms of isolation are mainly of the following types :

(1) Gametophytic Isolation. It is the commonest isolating mechanism wherein cross-pollination takes place but the pollen tube fails to germinate or if it germinates, it can not reach and penetrate embryo sac.

(2) Gametic Isolation. Such cases are reported in several crop plants wherein the pollen tube releases the male gametes into the embryo sac, but gametic and/or endospermic fusion does not occur.

Post-zygotic Mechanisms

These isolating mechanisms, operating after sexual fusion, are of the following types :

(1) Hybrid Inviability. In this case, mature seed is formed and manages to germinate but the F_1 hybrid dies before the flowering stage is reached, *e.g.* hybrid inviability resulting in crosses between *Papaver dubium* and *P. rhoeas.*

(2) F_1 Hybrid Sterility. In some cases, F_1 hybrids are fully viable and reach the flowering stage but flowers generally abort or abortion takes place as late as F_2 embryo formation, with the result that the F_1 hybrid fails to produce viable seeds.

(3) F_2 Hybrid Sterility (Inviability of F_2 Hybrid). In some cases F_2 hybrid dies much before it reaches the flowering stage or it fails to produce seeds.

(4) Seed Incompatibility. In this case the zygote or even immature embryo is formed but fails to develop. As such a mature embryo is formed, *e.g.* seed incompatibility resulted during cross between *Primula elatior* and *P. veris.*

SPECIATION

It is generally accepted that the Late Cretaceous and major part of Tertiary period were a time of relatively constant climates. In the Cretaceous, the angiosperms were confined to the areas lying between the latitudes 45°N and 45°S. It was Jurassic period when the angiosperms first appeared on the globe. They are pictured as originating and slowly diversifying for millions of years by replacing the non-flowering vegetation in the northern part of the globe. It was an age of natural evolution by the inherent process of change with time that caused distribution of flowering plants far more generalised than it is today. Angiosperms established themselves only in the Late Cretaceous period. Temperature and other climatic gradients everywhere were more gradual, mountains were lower with less pronounced climatic effect and the land surfaces of the globe were less scattered as the continents are believed to have drifted from one another in later period. The continental drift caused separation or isolation of plants and animal populations and this geographical isolation has reached its maximum today. This all, caused local speciation by the effects of seggregated and isolated evolution, probably the first kind of specialisation superimposed on the earlier generalised distribution of flowering plants (Good, 1974).

The picture changed in almost every respect towards the end of Tertiary, as during that time there were periodical catastrophies which consisted of a drastic and sudden change in temperature. This resulted in all kinds of minor and secondary variations which culminated in the **Ice Age**. It seems that equatorial temperature values were but little affected and thus gentler gradient from the equator to poles was replaced by steep gradient that culminated at altitudes in the arctic and antarctic conditions. Climatic change was anticipated by a period of intense mountain building, particularly in Miocene period, dating back to 30 million years ago. The effect of mountain building also affected the climate. In all these changes, such as the appreciable part of the earth's crust raised into cooler layers of the atmosphere, the newly elevated mountain ranges intercepting the moisture laden winds from the oceans and condemning many interior parts to aridity, brought all sort of alterations in climatic aspects and every kind of external condition which could modify plant life. The distribution factors took new dimensions and the potentiality of dispersal became less significant, whereas the importance of range of tolerance increased. Climate changes not only enforced migration but also determined and governed the direction of evolution. That is why, world today, does not have uniform flora but there are several floristic regions, characterised in their own way by peculiar association of plant types.

The world's floristic regions are determined on the basis of geographical distribution of plant genera. These are mapped individually, and from comparison of distribution, number of natural units emerge which form distinctive characteristic of peculiar region. Collinson (1977) has suggested three main sources of evidences of biogeographical theories namely autecology, fossil record and climatic change. Autecological evidences may throw light on conditions which might limit spread of a particular species, the fossil record furnishes evidence of past distribution and the climatic changes that might have directed the evolution.

Production of New Species

New species may develop through the mechanism of **abrupt speciation** (an example of sympatric speciation) commonly met in genera such as *Tragopogon* and *Senecis* or **gradual speciation** (a phenomenon more common in nature). In abrupt speciation species are often well isolated as any chance of hybridisation fails to culminate into successful hybrids because of genomic differences. In few cases, however, hybridisation may be accompanied by chromosome duplication so as to form allopolyploids. These allopolyploids show normal pairing at meiosis and thus represent well-isolated phenotypically as well as genotypically distinct species.

Gradual Speciation

Phenomenon of gradual speciation is more common in nature and involves **phyletic evolution** wherein one species might evolve into something different from its ancestor over a period of time, *i.e.*, **phyletic speciation**. But sometimes a population belonging to a single species might differentiate into two evolutionary lines through **divergent evolution**, *i.e.* **addictive speciation**.

I. Phylectic Speciation. Phyletic speciation is the sequential production of species within a single evolutionary lineage, *e.g.* species A, over a period of time, changes into species B and C into D, without ever splitting. The new species thus produced are variously called **succesional species,**

palaeospecies and **allochronic species**. In the process, the species that have become extinct are termed as **taxonomic extinct species** and the process as **taxonomic extinction**.

The concept of phyletic speciation remained controversial. Wiley (1981), although agreed with the phyletic concept transformation but rejected the phyletic concept on the grounds that it has never been satisfactorily demonstrated. Arbitrary species result in arbitrary speciation mechanisms and its recognition is an arbitrary practice. Mayr (1942) argued that delimitation of species, not belonging to the same time-scale, is difficult.

II. Addictive Speciation. Addictive speciation is the commonest mode of speciation adding to the diversity of living organism. In this case two previously independent species fuse to form a third-new species, themselves becoming extinct. Mayr (1963) called it **reductive speciation**. Likewise hybridisation produces new species but this always leads to an addition in the number of species. It is rather impossible that two evolutionary species can fuse to produce a third-new species and themselves become extinct. Although, this may occur in a particular region but not over the entire range of the species. Various kinds of addictive speciations are :

(1) Allopatric Speciation. Allopatric speciation occurs when the new species evolves in geographical isolation from the parent species. As indicated in fig. 11.1A, the species range becomes subdivided by a barrier such as a new mountain range or the change in the course of a river. The gene flow between the two subpopulations becomes impossible allowing evolution to proceed independently in each of them. Natural selection may favour different genotypes on either side of the barrier and random genetic drift and mutations could contribute to divergence. Over time, the divergence may proceed to the point that, where the two populations to meet again and interbreed and thus speciation would complete (fig. 11.1A). This form of speciation may also take place from the development of new species along the boundaries of a large central population (marginal populations or races), during environmental differentiation.

(2) Allo-parapatric Speciation. Allo-parapatric speciation occurs when two populations of an ancestral species separated, differentiate to a degree that is not sufficient for lineage independence and then develop lineage independence during a period of parapatry, *i.e.*, when two populations of an ancestral species differentiate despite the fact that no complete disjunction has occurred. Allo-parapatric speciation differs from allopatric speciation in that it is completed after a period of sympatry, *i.e.*, production of new species, with no geographical separation of populations and the process of attaining lineage independence, is potentially reversible because it is possible that two partly differentiated populations could form a single evolutionary lineage exhibiting clonal variation after they meet rather than the period of sympatry reinforcing differences between them most readily in small populations at the extreme edge of a species range. It is notable that the isolated population is smaller than the parent population. The peripheral population could become isolated, *e.g.* during contraction of the main species range in response to changes caused in the climate. The isolated population would be subject to the founder effect, *i.e.*, establishment of a new population by one or a few

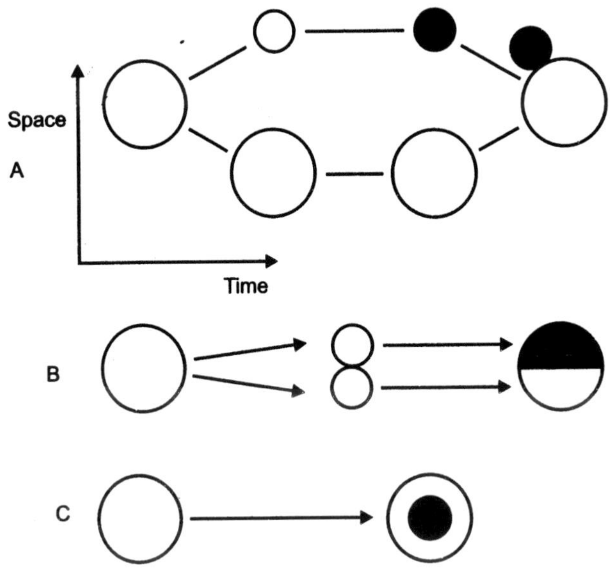

Fig. 11.1 The degree and type of isolation required by different models of speciation. (A) Allopatric speciation : the new species forms geographically separated from its parent species; (B) Parapatric speciation : the new species forms in an adjacent population; (C) Sympatric speciation : a new species emerges from within the range of the parent species.

individual/individuals, resulting in low levels of genetic variation and often a disappropriate number of alleles that are rare in the parent population. Thus, the isolated population could be genetically different from the parent population. The synergetic effect of a small typical population and extreme environmental conditions can cause rapid and extensive genetic reorganisation through random genetic drift coupled with natural selection, called **genetic revolution**. Islands have witnessed distinct peripheral isolates.

(3) Parapatric Speciation. This form of speciation (fig. 11.1B) occurs where speciating populations are continguous and hence only partially geographically isolated and thus are able to meet across a common boundary during the process of speciation. Where a species occupies a large geographical range, it may become adapted to different parts of that range. The intermediate or hybrid types will be found but the large distances involved prevent the two types from merging completely (Fig. 11.1B).

(4) Sympatric Speciation. Sympatric speciation refers to a situation where there is geographical separation between the speciating populations. Theoretically, all individuals are able to meet each other during the process of speciation. This model usually needs a change in preference of host, food or hibitat in order to prevent new species being swamped by gene flow. Whether this kind of speciation occurs at all, is a contentious issue. Theoretically, it can take place where there is polymorphism, *i.e.*, a population has alleles in the population conferring adaptation to two different habitats or niches. Reproductive isolation could then arise if the two morphs had a preference for their habitat (Fig. 11.1C).

Polyploidy, *i.e.*, spontaneous duplication of entire genome induced sympatric speciation is reported in plants resulting in an individual with a multiple of the original chromosome. The resultant individual is often large and vigorous. The polyploid plant is no longer sexually compatible with the parent population. However, it is able to establish a distinct population that may occupy a different habitat. For example, sand dune grass *Spartina towensendii*, a polyploid derived from the original *S. anglica*, is more vigorous than the parent and has colonised large areas of sand dune in Britain.

(5) Statispatric Speciation. Statispatric speciation is similar to parapatric speciation except that it results from spontaneous chromosomal modifications. The resultant chromosome arrangement must by fully viable in the homozygous state but of reduced viabillity in the heterozygous state.

The problem faced by the palaeobotanists is that, the full details of how species arise over time, is not well understood from the fossil data. According to Heningial philosophy and modern cladistic approach, all speciation events involve dichotomous (or polychotomous) splits. This type of speciation which results from lineage splitting into two (or more) separate species is referred to as **cladogenesis** (fig. 11.2). In contrast to this, many fossil records suggest that speciation can occur over time without any necessity for split to

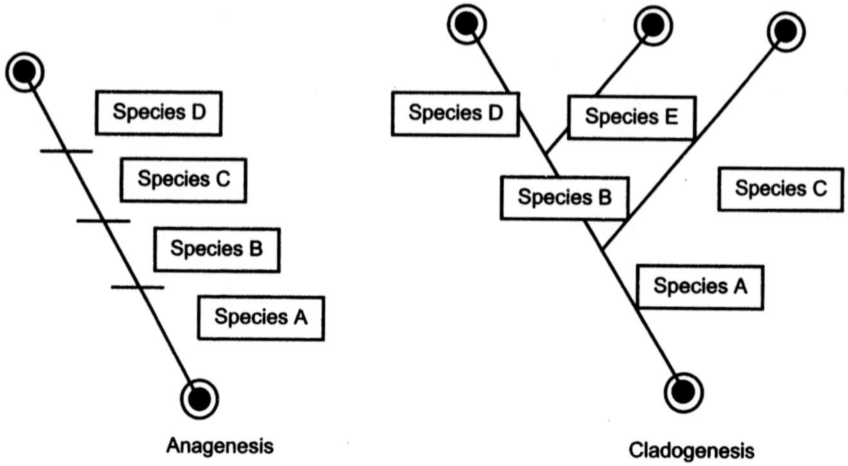

Fig. 11.2 Anagenesis and cladogenesis.

occur. This type of speciation, that results from accumulation of change over time in a single population is termed as **anagenesis** (fig. 11.2). According to Darwinian concept, evolutionary change is likely to be a gradual process, with new species often forming over geologically active periods of time, which is referred to as **phyletic gradualism.** If it is so, any near complete fossil sequence would be expected to show a gradual evolving sequence of forms, in which case, it would be quite difficult for any taxonomist to decide where to draw the line between species. Unfortunately very few fossil series are complete enough to permit assessment of phyletic gradualism and this incompleteness of fossil record often poses a drawback in including fossil data in cladistic analyses. Many data sets are incomplete and most extant species are still unknown to biologists.

Chapter **12**

Relevance of Taxonomy

Plant taxonomy has close link with many of the aspects of human welfare. Its scope is diverse and natue of study is multidisciplinary. It has benefitted not only the botanists but also those who are working on other fields where knowledge about plants is necessary. Plant taxonomy has played a significant role in agriculture, horticulture, forestry, conservation of bioresources, sustainable use of biodiversity, ecosystem research etc. It has helped greatly in exploration of plant resources of economic virtues. Many of the industries, such as pharmaceuticals, largely depend on the knowledge derived from plant taxonomy. Plant taxonomy plays a vital role in conservation of phytodiversity both *in situ* and *ex situ*. In lack of taxonomic information, sustainable utilisation of bioresources would be like a soldier without adequate knowledge about weapons. An ecologist, having basic knowledge of plant taxonomy, would find it easier to choose suitable plants to be raised as soil binders to check soil erosion and desertification; fast growing trees for afforestation; mineral indicator plants for mineral exploration; good varieties and suitable species of plants for agriculture and horticulture and for energy plantation; conservation of bioresources; ecosystem research etc. Inventory on plants, books, monographs, floras and other published work on taxonomy facilitates selecting the plants from their natural environment. India started making use of taxonomic knowledge in exploring plants of medicinal value and their conservation quite early, as evidenced by *Vrikshayurveda* compiled by Parasara, perhaps before the beginning of Christian era. *Atharva Veda* (written around 2000 BC) describes conservation of medicinal plants. Indian herbalist Charaka (1 century AD), in his *Charaka Samhita* had emphasised the sustainable use of several plants of medicinal value. Efforts made by Botanical Survey of India in conservation of phytodiversity, are appreciable. To make use of the knowledge of plant taxonomy in conservation, ecosystem management, sustainable use of bioresources of economic virtues and for many other purposes, Ministry of Environment and Forests, Government of India has recently set up an All India Coordinated Project on Capacity Building in Taxonomy (AICOPTAX) to explore, identify and prepare an inventory. By 2005–06, 47,000 species of plants have been identified and described, of which many are of economic virtues. A sound taxonomic base is a pre-requisite for effective conservation of bioresources, sustainable utilisation of biodiversity, ecological research and proper management of ecosystems.

Taxonomy and Conservation of Bioresources

Conservation of bioresources in India, is age old. Pre-historic Aryans were worshippers of nature and preferred sylvan surroundings. The epic *Vedas* and *Upanishads* were composed by our sages singing "May the God, the Water, the Plants and Forest Trees accept our prayers ...... May the blessing of our simul (tree) protect us for ever". The *Puranas* have said "One tree is equal to 10 sons ......" This all indicates that our ancestors had some or the other taxonomic background.

Taxonomic issues that need to be addressed in 21st century, are mostly those that require interfacing of systematics and other disciplines such as biorespecting, conservation biology, ecosystem management, sustainable use of bioresources, bioremediation, assessment of conservation status of species and role of species in communities and ecosystems. Conservation of bioresources refers to protection, upliftment and scientific management of genes, species and ecosystems/habitats. And unless we have adequate knowledge about taxonomic features of a species, how can we make the conservation effective. Both *in situ* (on site) and *ex situ* (off site) conservation strategies require adequate knowledge of taxonomy of species to be conserved. All measures of *in situ* conservation of biodiversity, such as Protected Areas, National Parks, Wildlife Sanctuaries, Biosphere Reserves, Sacred Grooves etc., are based on sound taxonomic studies. Taxonomic studies facilitate in achieving the goal of conservation of wild flora and fauna in their natural habitats and foster economic development which is ecologically sustainable. The Ministry of Environment and Forests is extending financial assistance to the respective State Governments for conservation and management of 14 Biosphere Reserves designated so far. In addition, research and development projects are also supported in these Reserves. Similarily, *ex situ* conservation and management strategies for biodiversity such as Botanical Gardens, Zoos and National Bureau of Plant Genetic Resources (NBPGR) and National Bureau of Animal Genetic Resources (NBAGR), also need solid background of taxonomy. Under the scheme of AICOPTAX, the Ministry of Environment and Forests, in 1992, initiated assistance to Botanical Gardens for *ex situ* conservation of rare and endemic species of plants.

Keeping in view that Botanical Gardens play a key role in conservation of plants, the Botanical Survey of India has maintained and developed Indian Botanic Garden at Sibpur, Howrah, the largest in the country, and a number of smaller gardens spread over the country. All these gardens give special emphasis to the conservation of rare plants involving multiplication of threatened plants through tissue culture and seed banks. BSI has published two special bulletins-*Threatened Plants of India* (1980) and *An Assessment of Threatened Plants of India* (1983).

Taxonomy and Sustainable Utilisation of Bioresources

The concept of sustainable development, *i.e.*, the development that meets the needs of the present without compromising the ability of future generations to fulfil their own needs, in context of bioresource utilisation, is inherent in Indian cultural ethos. In their hymns to earth, the sages of *Atharva Veda* chanted "*What of thee, I dig out let that quickly grow over. Let me not hit thy vitals or thy heart.*" It is evident that ancient Indians were quite aware with sustainable utilisation of bioresources and had recognition of bioresources of their choice such as trees of ***banyan, pipal, bel*** etc. India has a long history of sustainable use of bioresources. A greater emphasis in this direction was given after the Brundetland Report, **Our Common Future** (World Commission on Environment and Development), placed on the world's political agenda, creating need for taxonomic base in context of sustainable use of biodiversity.

Sustainable utilisation of our plant resources is the watch word of Botanical Survey of India. Multipronged efforts are being made to conserve the phytodiversity of the country. Practical steps for regulating trade in endangered species are regularly suggested to the Agriculture Department. BSI is also engaged in documentation of indigenous plant resources, taxonomic study of aromatic and cosmetic yielding plants, pharmacognostic studies on flora etc.

Taxonomy and Ecosystem Research

In view of rapid loss of biodiversity, it is urgently needed that the areas rich in biological diversity are studied and steps are taken to conserve them. In India, this practice of maintaining ecosystems in proper health, is age old. Mention is there of dense forests in Naimsaranya, Chitrakoot, Dandkaranya and Panchvati in great epic *Ramayana*. But by the time *Mahabharata* was compiled, onslaughts had been made on the forests. We read the burning of Khandav Vana. But conservation minded ancestors having taxonomic background, arrested such vandalism and saved several useful species of plants even though some in course of time, such as 'Kadam', disappeared from localities like Vrindavan.

Botanical Survey of India has taken keen interest in protecting and preserving biorich and biodiverse ecosystems, such as Silent Valley (Kerala), Namdapha (Arunachal Pradesh), Similipal (Orissa), Nanda Devi (Uttarakhand), Dachigam (Jammu and Kashmir), Keoladeo (Rajasthan), Valley of Flower (Uttarakhand) etc. Silent Valley and Eastern Himalayas have been declared hot spots of biodiversity. Some reserved parts of these hot spots have been declared as hottest hot spots. Chilika Lake (Orissa) has been declared as one of the most important Ramsar sites. Ecosystem research in these areas, involving study of vegetation (community structure), population dynamics of species and recording of flora, has greatly helped the government in planning future programmes for their conservation.

Ecosystem research programme, initiated by Ministry of Environment and Forests, is an inter-disciplinary programme incorporating taxonomic studies as a thirst area. During explorations and floristic studies, several new species (and few new genera) were recorded. Some plant species were collected after 30 years or so, *e.g. Taxocarpus cancanensis* (Apocynaceae)–a rare, endangered and threatened species from Maharashtra, was recorded after 50 years.

Apart from conservation, sustainable utilisation of bioresources and ecosystem research, taxonomy is also helpful in study of endemic species and species under threat in various categories of Red Data Book of International Union of Conservation of Nature and Natural Resources (IUCN) and planning strategies of their conservation.

❑❑❑

Chapter 13

Taxonomic Evidences

During the last three centuries, plant taxonomy has developed from a completely morphology-oriented static and descriptive discipline to a most dynamic area of study. Over the last few decades, the affinities between plant groups have been redefined as more and more information is accumulated from different sources. The applicaticn of comparative data from morphology, anatomy, embryology, cytology, molecular biology, palynology, phytochemistry, ecology, phytogeography, physiology, genetics and computer applications has led to the establishment of evolutionary trends, phylogenetic relationships and refinement of taxonomic concepts at all levels of taxonomic hierarchy to the understanding of mechanisms that guide evolution and help in devising a phenetic classification which expresses natural relationships. This all has led a modern systematist to become **Chemocytohistomorpho-taxonometrician.**

COMPARATIVE MORPHOLOGY
(Morphology in Relation to Taxonomy)

Traditionally, plant taxonomy has depended mainly upon comparative morphological features because these are easily observable in the living plants as well as in the herbarium specimens, requiring only keen observation, a hand lens or, at the most, a dissecting microscope as equipment. These external features show a wide degree of variation which help in delimitation and identification of taxa. Moreover, having been in use for several centuries, there is a well knit terminology to describe the variations among plants. Morphology continues to be mainstay for plant classification.

In recent years, certain new trends have emerged in morphotaxonomic research. One of them is to exploit unexplored characters or the characters which have been neglected in the past. Although floral morphology has largely been used for classifications, other morphological features have also contributed a lot in scientific grouping of the plants. Taking into consideration the habit of plants, Hutchinson (1926, 73) recognised **Herbaceae** (herbaceous) and **Lignosae** (woody) divisions within dicots. In contrary to previous belief that trees or shrubs represented the most primitive condition within angiosperm, it is now considered that **Paleoherbs** are the most primitive angiosperms. Characteristics of underground parts have played an important role in identification of various species of *Iris, Allium, Ranunculus* etc. Leaves form an important criterion for identification of palms, *Salix, Populus* etc. Separation of *Azadirachta* from *Melia* is largely based on leaf characters. Stipules are an important source for delimitation of *Viola* from *Salix*. Floral characters, such as inflorescence, flowers and their components and associated structures form the traditional basis for delimitation of taxa. Characters of fruits have also been explored for making diagnostic keys. Coode (1967) used only fruit characters in delimitation of species of *Valerianella* whereas Singh *et al.* (1972) considered fruit morphology as a criterion for identification of Indian genera of Asteraceae (Liguliflorae). In Asteraceae, the shape of *cypsella* (usually called achene), presence or absence of pappus (hairy, scaly or with bristles), presence or absence of beak and its length, number of ribs are some important features of identification. Seggregation of genera (*Silene, Melandrium* and *Cerastium*) in Caryophyllaceae, is largely based on number of capsule valves. Hairy outgrowths on the seeds, their length and colour also provide useful data in distinction of genera and species in Malvaceae, Convolvulaceae, Acanthaceae and Asclepiadaceae.

MICROSCOPIC MORPHOLOGY

Micromorphological features have played an increasingly important role in elucidating phylogenetic relationships in many taxa. The finer details (ultrastructures) are brought out using electron microscope. Epidermal features, such as panicot type of epidermal cells, cells containing silica and bicellular hairs are the features to recognise grasses. Stomatal features, such as their types, characteristicts to many families, are : *anomocytic*–ordinary epidermal cells surrounding the stomata (Ranunculaceae), *anisocytic*–three subsidiary cells of unequal size surrounding the stomata (Brassicaceae), *diacytic*–two subsidiary cells at right angles to the guard cells (Caryophyllaceae), *paracytic*–two subsidiary cells more parallel to guard cells (Rubiaceae), *actinocytic*–stomata surrounded by a ring of radiating cells, *cyclocytic*–with more than one concentric ring of subsidiary cells, *tetracentric*–with four subsidiary cells, and *graminaceous* type (Poaceae). Stace (1989) recognised 35 types of stomata in vascular plants.

The type of trichomes causing pubescence can be a useful morphotaxonomic character. Keeping in view variation in trichomes in different genera, Inamdar (1967), based on ontogeny and structure of trichomes, confirmed the systematic position of *Nyctanthes* in the family Oleaceae. Trichomes have also been found to be valuable tool in classification of certain members of Capparaceae (Capparidaceae), Caesalpiniaceae and the Indian species of *Malvastrum*. Lersten (1974), described secretory trichomes (colleters) in several genera of Rubiaceae. Trichomes have been of considerable help in Cruciferae (Brassicaceae) especially in genera of *Arabis* and *Arabidopsis* as also in a large genus *Astragalus* (with more than 200 species). The Himalayan species *Hedera napalensis* (having scaly trichomes) is distinguished from European relative *H. helix* (having stellate trichomes) based on this fact. In Combretaceae, trichomes play an important role in classification of genera, species and even varieties (Stace, 1973).

The use of scanning electron microscopy (SEM) in morphological studies, has opened up new vistas for taxonomists giving way to new dimensions of plant surface details. Based on studies on lodicules of Poaceae, Stebbins (1956) recognised four subfamilies : Festucoidae, Chloridoidae, Panicoidae and Bambusoidae.

In general, other morphological feature that are helpful in solving taxonomic problems include growth habit, growth pattern, seedling characters, seed surface patterns, fine relief of cell wall (secondary sculpturing) etc.

ANATOMY IN RELATION TO TAXONOMY

Anatomical characters of vegetative and floral parts, have successfully been employed in solving many taxonomic problems and for elucidation of phylogenetic relationships of various taxa. Value of anatomical features in solving taxonomic problems in dicots was realised by Metcalf and Chalk (1950). Anatomical work of taxonomic significance was largely undertaken by Bailey and his students. In context of primitive angiosperms, Carlquist (1996) has discussed the trends of xylem evolution. Anatomical features are of value in systematics because :

1. Anatomical features can be well exploited taxonomically with the help of only fragmentary material, such as a piece of wood.
2. When morphological characters prove to be of no help in the preliminary diagnosis, the anatomical study of herbarium material may prove helpful.
3. The range of anatomical data is very extensive.
4. The anatomical data prove of great help in ascertaining evolutionary trends in higher taxonomic categories and determining relationship between different taxa.

ANATOMY OF VEGETATIVE STRUCTURES

Among anatomy of vegetative parts, leaf anatomy, nodal anatomy and wood anatomy are chiefly included.

Leaf Anatomy

Leaf anatomy with special reference to nature and thickness of epidermis; structure and types of mesophyll, storage parenchyma, mid vein structure, bundle sheath; pattern of sclerenchyma; chloroplast structure and arrangement etc., provide various characters of taxonomic significance. The classification of several genera of grass family has been revised on the basis of *kranz anatomy* (dense thick-walled chlorenchymatous bundle sheath, mesophyll simple) in C-4 plants. Melville (1962, 83) developed his **gonophyll theory** on the basis of venation patterns of leaf and floral parts. Hickey and Doyle (1977) rejected *Sanmiguelia* and *Fercula* of Triassic as being angiosperms largely on the basis of venation pattern of leaves but Cornet (1986, 89) points the discovery of *Sanmiguelia* from Upper Triassic of Texas to presumed angiosperm incorporating character of both dicots and monocots. Various species of *Cyperus* (Cyperaceae) have largely been identified on the basis of leaf anatomy. Petiole anatomy of 64 species of *Baphia* of Leguminosae has been studied by Soladoye (1982). He observed that certain features provide evidences useful for intra-generic classification.

Nodal Anatomy

Nodal anatomy has gained much significance in taxonomy and phylogeny of angiosperms in recent years. The number of vascular traces entering the leaf bases and associated gaps (lacunae) left in the vascular cylinder of stem at each node may be : *unilacunar*—node may have a single gap from single leaf trace or three leaf traces or *trilacunar*—node may have three gaps associated with three leaf traces. Separation of *Illicium* from Winteraceae is largely because of unilacunar node in the genus. Unilacunar nodes occur in Laurales, Caryophyllales, Ericales, Ebenales, Primulales, Myrtales, some families of Tubiflorae and majority of families of Asteridae. Trilacunar nodes are common in majority of dicots. *Multilacunar nodes* occur in Magnoliales, Piperales, Trochodendrales, Asterales and Umbellales.

A fourth type of node *unilacunar two trace*, discovered by Marsden and Bailey (1955), is now considered as the basic type found in angiosperms. Sinnott (1914) previously thought it to be trilacunar. Based on nodal structure, the subfamily Icacinoideae of the family Icacinaceae has been spread over into two distinct groups–one with trilacunar nodes while other with unilacunar node (Bailey and Howard, 1941.)

Wood Anatomy

Wood anatomy has been used at all taxonomic levels because of the conservative nature of anatomical features of the secondary wood. Wood in angiosperm trees and shrubs has secondary xylem, consisting mainly of tracheids and vessels. Vessels are absent in gymnosperms but present in angiosperms. Along with other lines of evidences, wood anatomy has been successfully used in deciding systematic position of Amborellaceae, Tetracentraceae, Trochodendraceae and Winteraceae, all included under Magnoliales of angiosperms, leading to conclude that angiosperm ancestors were vesselless. It is commonly believed that angiosperms have undergone marked evolution from tracheids to long narrow vessel elements. Studies on wood anatomy have contributed largely in deriving conclusion that Amentiferae (Englerian group), including angiosperm families like Salicaceae, Betulaceae, Fagaceae, Juglandaceae etc., constitute a relatively advanced group, and that Gnetales (Gymnosperms) are not ancestral to angiosperms. Bailey (1944) concluded that, in both, angiosperms and gymnosperms, vessels arose independantly. The separation of *Paeonia* into a distinct family Paeoniaceae and *Austrobaileya* into a separate family Austrobaileyaceae, has largely been done on the basis of wood anatomy.

Apart from structure and abundance and distribution of vessels, sculpturing on vessel walls is of great taxonomic value. Thus, solitary vessels are considered to be primitive to aggregate groupings, such as pore clusters, pore multiples and pore chains and also diffuse-porous woods primitive to ring-porus woods. Various other wood characters also throw light on evolutionary sequence among many groups. Absence of wood parenchyma, is considered primitive atleast in some cases (Winteraceae). Vascular (medullary) rays, with cells radially elongated (homogenous rays), are considered as advanced to heterogenous rays, *i.e.*, the rays with both vertically and radially elongated cells. Woods with non-stratified cells, are considered primitive to storied structures. Presence or absence of latex vessels, resins, gums, crystals etc. in the wood, are also the features of taxonomic significance. Sclereids have been used as diagnostic tools in several taxa, such as Nymphaeaceae, Coronaraceae, Oleaceae, Theaceae, Unoniaceae and few genera of Araceae, Acanthaceae, Ericaceae and Melastomaceae. In dicots, sclereids are more common in woody forms than in herbaceous ones. They are extremely rare in monocots, except in certain genera of Araceae, Arecaceae, Agavaceae and a few other families. Many cellular contents, *e.g.* albuminoids (*Laportea*), starch grains (*Solanum tuberosum*), protein bodies (Cactaceae), large silica bodies in epidermal cells (Arecaceae, Musaceae etc.), calcium oxalate crystals (*Allium, Eichhornia*), calcium carbonate crystals–cystoliths (Cannabinaceae, Moraceae, Urticaceae), tanniniferous cells (Raptaceae, Xyridaceae) etc., also have some bearing in systematics.

At generic and intra-generic level, wood anatomy has contributed to provide evidence for placing taxa of uncertain affinities. Anatomical evidences have been able to delimit different subgenera and sections of *Quercus* (Fagaceae). Placing genus *Myristica* and related genera, close to Lauraceae, could be possible from the evidence of wood structure (Eames, 1961). While on the same basis Myristicaceae are considered unrelated to Annonaceae and Eupomatiaceae, on the other hand, the placement of Calycanthaceae in or near Rosales and Myrtales, is not supported by wood structure. The relationship and possible derivation of *Casuarina* from Hamamelidaceae is supported by data from comparative wood anatomy.

ANATOMY OF FLORAL STRUCTURES

The reproductive organs, the flowers, being highly conservative, throw much light on understanding of the phylogeny of angiosperms. Floral anatomy has been one of the thoroughly explored area that has provided a handful information for systematics. It has been studied in details by Eames (1935), Puri (1951–62), Murty and Puri (1980). Puri (1951) has played a significant role in solving many morphological problems on the basis of floral anatomy. He proposed that floral evolution has undergone a considerable modification through : (i) reduction, (ii) amplification, (iii) cohesion, and (iv) adnation of vascular bundles. Studies on floral anatomy have been able to resolve systematic positions of many taxa, such as (*i*) confirmation of the origin of the families Annonaceae, Calycanthaceae and Menispermaceae from Ranunculaceae; (*ii*) transfer of *Hydrocotyle asiatica* to genus *Centella* (*C. asiatica*); (*iii*) separation of *Paeonia* from Ranunculaceae and its inclusion under a separate family Paeoniaceae; (*iv*) derivation of Polemoniaceae and Caryophyllaceae from a caryophyllaceous stock; (*v*) inclusion of Solanaceae and Scrophulariaceae together under one single order Scrophulariales; (*vi*) separation of Cyperales and Poales from a single common order etc.

EMBRYOLOGY IN RELATION TO TAXONOMY

As early as the second and third quarter of the 19th century, Hofmeister and Strasburger indicated the possibility of using embryological characters in plant taxonomy but the data could be used only after 90 years, *i.e.*, by the middle of 20th century when Maheshwari (1950) took the lead in this area. Later Cave (1962), Kapil (1962), Johri (1963, 67), Davis (1966), Bhojwani and Bhatnagar (1979) and many others published their detailed treatments. It took many years of laborious and painstaking research due to very few useful outputs. Embryological features of taxonomic importance include microsporogenesis, structure and development of ovule, embryo sac development, development of endosperm and embryo etc. One of the main reasons for relatively less application of embryology in taxonomy is that the data are scarce and not well known (Davis and Heywood, 1963).

Application of Embryological Data in Taxonomy

Embryological data are of so much value that delimitation of angiosperms themselves is based on embryological characters–enclosed ovules; double fertilisation and triple fusion and post fertilised triploid endosperm, supporting their probable monophyletic origin. Further delimitation of monocots and dicots is largely based on the characteristics of embryo with two cotyledons (dicot) and with one (monocot), almost constant for both the groups. But there are certain groups which have well marked embryological features that can be used as tool in taxonomic studies to solve the problems above the family level, at the level of family and some specific cases.

Above the Family Level

There are certain orders of angiosperms whose members are morphologically well defined and exhibit a set of uniform embryological features.

1. **Caryophyllales** is more widely known as Centrospermae (Cronquist, 1968). The order, embryologically very distinct, is characterised by trinucleate pollens; bitegmic crassinucellate ovules (that are campylotropous or amphitropous) commonly with inner integument longer than the outer and very often with a space between the integuments towards the chalazal end; seed with a curved and peripheral embryo, more or less surrounding the food storage tissue which consists mainly of perisperm with little or no endosperm.
2. **Polygonales and Plumbaginales** are related to each other and also to Caryophyllales in having similar embryological features to a great extent.

3. **Ericales**, a homogenous group of sympetalous families, includes members, most of them sharing many common embryological features including presence of pollen tetrads.
4. **Helobiae** is a monocot order (Engler) characterised by the **helobial** type of endosperm. Despite the fact, the order is raised to the rank of subclass by some authors, the relationship of the constituent order is scarcely disputed.
5. **Orchidales** is characterised by undifferentiated embryo and very little or no endosperm.
6. **Gentianales** is characterised by lack of integumentary tapetum and nuclear endosperm which helps it to be distinguished from other sympetalous orders. Some of its families like Buddlejaceae and Oleaceae (formerly piaced under Gentianales) have now been removed from it, since they possess integumentary tapetum and cellular endosperm.
7. **Lecythidales** was formerly considered to be related to Myrtales. It has been found to differ from Myrtales in series of embryological features in its various members.

At the Family Level

The evidences derived from embryological studies have strongly supported proper placement of certain disputed families, once regarded to have a doubtful systematic position. A number of angiosperm families are characterised by unique embryological features found almost in their all members. These include :

1. **Podostemaceae** include perennial aquatic herbs which have unique embryological features–formation of *pseudoembryosac* due to disintegration of the nucellar tissue. The members of this family also possess paired pollen grains, bitegmic tenuinucellate ovules, bisporic embryo sac, solanad type of embryogeny, prominent suspensor haustoria, absence of triple fusion and consequently no endosperm.
2. **Cyperaceae** are characterised by the formation of only one microspore per microspore mother cell. Although microspore mother cell undergoes meiosis and results four microspore nuclei but only one of them gives rise to pollen grain. Such degeneration of three microspore nuclei is also seen in few members of Epacridaceae but Cyperaceae are distinct from these members in pollen shedding at the 3-celled stage, as against the 2-celled stage in some members of Epacridaceae. The embryological evidences have been able to conclude that this family is not close to either Poaceae or Juncaceae.
3. **Onagraceae** are characterised by *Oenothera* type of embryo sac, not found in any other family except as an abnormality. *Oenolhera* type of embryo sac is four nucleate and is derived from micropylar megaspore of the tetrad formed.

4. **Lemnaceae** are considered presumably to have been derived from Araceae or Helobiales but embryologic studies of SC Maheshwari (1954, 56, 58) suggest that it has evolved from Araceae (and not related to Helobiales) on account of the facts that both Lemnaceae and Araceae possess bitegmic ovules with micropylar cap of nucellus, cellular endosperm with a chalazal haustorium, irregular sequence in the development of embryo and short and stocky suspensor and integumentary operculum in the seed.
5. **Loranthaceae** are traditionally spread over two subfamilies : Loranthoidae and Viscoidae, largely on the basis of presence of a **calyculus** below the perianth in Loranthoidae and absence in Viscoidae. Maheshwari (1964) reported that in Loranthoidae, there occured triradiate pollen grains; *Polygonum* type of embryo sac; biseriate early embryogeny; presence of embryo and suspensor and viscid layer outside the vascular supply in the fruit whereas in Viscoidae there were spherical pollen grains; *Allium* type of embryo sac; many tiered early embryogeny; absence of embryo and suspensor and viscid layer inside the vascular supply of fruit. And thus, he advocated separation of the two as distinct families : Loranthaceae and Viscaceae, which is also accepted by Takhtajan (1980, 87, 97), Dahlgren (1980), Cronquist (1980, 88) and Thorne (1981, 92). Embryological data have also helped in solving phylogenetic positions of some other families as well, *e.g.* Empetraceae, Cactaceae, Garryaceae, Liliaceae etc.

At the Generic Level (Specific Examples)

Some specific examples of the embryological data having been quite useful in interpretation of taxonomic affinities include :

1. **Trapa** was earlier (Bentham and Hooker, 1883) placed under Onagraceae but was later removed from Onagraceae and placed under Trapaceae (Engler and Diels, 1936; Hutchinson, 1959, 73) on the basis of distinct aquatic habit, swollen petiole, two types of leaves, semiepigynous disc and spiny fruit. The embryological features that caused removal of *Trapa* from Onagraceae and its inclusion in Trapaceae, are better explained in Table 13.1.

Table 13.1 Comparison of Embryological Characters of Genus *Trapa* and Family Onagraceae

Character	*Trapa*	Onagraceae
Pollen grain	Pyramidal with three much folded meridional crests.	Blunty triangular and basin-shaped.
Ovary	Semi-inferior, bilocular with a single pendulous anatropous ovule in each chamber.	Inferior, mostly trilocular with many ovules per chamber on an axile placenta.
Embryo sac	*Polygonum* type	*Oenothera* type
Endosperm	Absent	Present (Nuclear)
Embryo	Solanad type	Onagrad type

(Contd....)

Suspenor	Well developed suspensor haustorium	Short and inconspicuous
Cotyledons	One cotyledon (extremely reduced)	Cotyledons equal
Fruit	Large, one-seeded drupe with prominent spines	Loculicidal capsule

2. **Paeonia** was earlier (Bentham and Hooker, 1883) included in Ranunculaceae but Worsdell (1908) suggested its removal from the family and placing to a distinct family Paeoniaceae. This was supported due to some distinct characters, such as certrifugal stamens (Corner, 1946), floral anatomy (Eames, 1961) and chromosonal data (Gregory, 1941). In all modern systems of classification, *Paeonia* has been placed in a distinct monogeneric family Paeoniaceae. The embryological features, that support removal of *Paeonia* from Ranunculaceae, are better explained in Table 13.2.

Table 13.2 Comparison of Embryological Characters of Genus *Paeonia* and Family Ranunculaceae

Character	*Paeonia*	Ranunculaceae
Stamen	Spirally arranged centrifugal.	Spirally arranged centripetal.
Anther	Multilayered endothecium, mostly two layered tapetum	One layered endothecium and one layered tapetum
Pollen	Reticulately pitted exine, large and elongate generative cell	Granular, papillate or smooth exine, small lenticular generative cell
Female archesporium	Multicelled, many megaspore mother cells function	Uni- or multicelled, one cell functions
Antipodal cells	Persistent, non-polyploid.	Persistent (ephemeral in *Adonis*), nuclei one or more than one polyploid
Embryogeny	Unique	Onagrad or, rarely, Solanad type
Seed	Arillate	Non-arillate
Fruit	Foilicle	Achene

3. **Exocarpos** is traditionally placed under the family Santalaceae but Gagnepain and Bourean (1947) suggested its removal to a distinct family Exocarpaceae near Taxaceae under Gymnosperms on the basis of presence of articulate pedicel, 'naked ovule' and presence of pollen chamber. Mohan Ram (1959), based on embryological studies, concluded that the flower of *Exocarpos* shows the usual angiospermous features–the anther with distinct endothecium and glandular tapetum, pollen grains shed at the two-celled stage,

Polygonum type of embryo sac, cellular endosperm and the transverse division of the zygote, confirming that the genus is undoubtedly an angiosperm (member of Santalaceae), there being no justification for its removal as a distinct family. As such *Exocarpos* is placed in Santalaceae in all major systems of classification.

4. **Butomus**, based on embryological data, finds its retention in the family Butomaceae and removal of *Parnassion* from Saxifragaceae as a separate family Parnassiaceae.

PALYNOLOGY IN RELATION TO TAXONOMY

Pollen morphological characters were fully appreciated as early as 1830-40 but as regards role of palynology in solving taxonomic problems, the first account appeared as *Pollen Morphology and Taxonomy* (Erdtman, 1952). Palynological characters have been able to reposition several disputed genera and interpret problems related to the origin and evolution of many taxa (Nair, 1980) and provide classification of angiosperms (Cronquist 1981).

Various characters of pollen used in taxonomy include : variation in shape, size and symmetry; pollen association; pollen nuclear membrane; number and position of furrows; type; number, position and structure of apertures; and architecture of pollen wall-exine stratification, sculpture and structure.

. The taxonomic and evolutionary significance of pollen morphology may be at specific, generic or higher levels, especially in identification, differenttiation and classification of various taxa. Pollen wall has been a subject of considerable importance in establishing evolutionary history of angiosperms. Based on pollen type, there can be two categories :

1. **Stenopalynous (Unipalynous).** In many taxa, the type of pollen is characteristic and constant, *i.e.*, they have a single morphological pollen type, as in Asclepiadaceae, Cruciferae (= Brassicaceae), Poaceae, Labiatae (Lamiaceae) etc. The stenopalynous taxa are generally considered to be more natural.

2. **Eurypalynous (Multipalynous).** There are many taxa in which the types of pollen may vary considerably in size, aperture, stratification of exine etc. These include Asteraceae, Acanthaceae, Rubiaceae, Verbenaceae etc. They are heterogenous atleast in certain instances. Such eurypalynous (multipalynous) taxa are of considerable significance in systematic palynology.

Phylogenetic Consideration

Palynological data suggest two distinct phylogenetic stocks in dicots : **monocolpate** (have a single furrow on one side of the pollen, remote from the point of contact in the tetrad) represented by the Magnoliaceae and **tricolpate** (with three meridionally placed furrows) represented by Ranunculaceae. Majority of monocots have monocolpate pollen grains. Due to this reason, monocots are considered to be more closely related to Magnolian stock and palynologically more ancient than the Ranalian stock where monocolpate elements are completely absent or replaced by new apertural forms. Kuprinova (1948) believed that the Helobiae are not related to other monocots. Thus it is clear that monocolpate condition is widely spread in primitive dicots and majority of monocots.

Fossil studies, over the last three decades, have revealed that monocolpate pollen of *Clavitopollenites* described (Couper, 1958) from Barremian and Aptian strata of the Early Cretaceous of Southern England (132-112 mya) to be the oldest recorded angiosperm fossil with distinct sculptured exine resembling the living genus *Ascarina*. A similar but unaperturate pollen, recorded by Brenner and Bickoff (1992) from the Valanginian (Ca 135 mya) from the Helez formation of Israel, is now considered being the oldest record of angiosperm fossils (Taylor and Hickey, 1996) which led to the belief that the earliest angiosperm pollen were without an opening, the monocolpate developing later.

Claims of angiosperm records from strata, earlier than Cretaceous, were largely rejected. According to Brenner's (1996) new model for the evolutionary sequence of angiosperm pollen types, the earliest angiosperm pollen (from Valenginian or earlier) was small, circular, tectate collumellate and with no aperture. In the Hauterivian, there was possible occurrence of thickening of the intine coupled with thickened endexine and sulcus evolution. A considerable diversification of these monocolpate pollens occurred in the Barremian. Tricolpate pollen evolved in the northern Gondwana in the Lower Aptian. Multicolpate and multiporate pollen arose at a later stage.

Pollen and Pollination

The pollen grains of anemophilous (wind pollinated) plants are usually small, rounded, smooth, thin-walled and dry whereas those of entomophilous (insect pollinated) and ornithophilous (bird pollinated) plants are large, sculptured and often coated with adhesive waxy or oily substances. Anemophilous pollen grains are found in *Populus*, Poaceae, Cyperaceae, Betulaceae and several other families whereas entomophilous pollen grains are found in many families with ornamental plants. The pollen grains of Asteraceae, though are generally highly elaborate, but have undergone simplification towards loss of sculpuring in several wind pollinated genera. The vestigial scattered patches of adhesive layer on pollen of wind pollinated genera have been considered as evidence of the derivation of anemophily from entomophily.

Pollen and Delimitation of Taxa

An example of role of pollen grains in delimitation of taxa is *Nelumbo* whose separation from Nymphaeaceae into distinct family Nelumbonaceae, is largely suported by the tricolpate pollens of *Nelumbo* as against the monocolpate pollens in Nymphaeaceae. The exine ornamentation patterns have been of great help in delimitation, identification and classification of various taxa, especially at lower levels. The separation from Malvaceae of *Bombax, Eriodendron* and *Adansonia* as a separate family Bombacaceae, is justified by differences in exine, among other things. In Malvaceae, the exine is characteristically spiny but in Bombacaceae, the exine is smooth and non-spiny. *Trapa*, once grouped in Onagraceae, because of presence of characteristic meridonial crest in its pollen, has been separated as a distinct family Trapaceae. Apertural characters have also been used in delimitation of lower level taxa (genera and species), *e.g.* in the genus *Sesamum* in which there are 11 colpi in *S. indicum*, 9 in *S. prostratum* and 8 in *S. leciniatum*. Normally pollen grains separate from tetrad, there are, however, taxa like Winteraceae, Onagraceae, Juncaceae etc. where they remain associated as tetrads or exhibit extreme cohesion into pollinia as seen in Asclepiadaceae and Orchidaceae.

CYTOLOGY IN RELATION TO TAXONOMY

Cytology has made an outstanding contribution in solving many taxonomic problems. The study of the cell, more appropriately of chromosomes, has played a singificant role in elucidating many problems in plant systematics in last few decades. Characters of cytotaxonomic importance commonly explored include : number, size and morphology of chromosomes coupled with their banding patterns and behaviour at meiosis.

Chromosome Number

Extensive records of chromosome number are available in the works dating back to Darlington and Janaki Amal (1945) to present International Association of Plant Taxonomy (IAPT)—*Index to Plant Chromosomes* in its series *Regnum vegetabile*. The chromosome number (represented as 2n or 2x) in angiosperms exhibits considerable variation—the lowest number (n=2) in *Haplopappus gracilis* (Asteraceae) to the highest (n=132) in *Poa littorosa* (Poaceae). Such a range of variation (n=2 to n=132), within a quarter million species of angiosperms, is not so important from cyto-taxonomic point of view. However, isolated role of studies on chromosomes, have been used in taxonomic delimitations.

Certain groups of plants possess a constant number of chromosomes in all their constituent members, such as genus *Quercus* which has n=12 chromosomes in all known species, on the basis of which they can be delimited from other related genera. Raven (1975) and many others believe the original basic number for angiosperms is n=7, which is characteristic of all major groups of both dicots and monocots except for Caryophyllidae with n=9. A reduced number (n=6) is characteristic of Ericales and n=5 for four unrelated families : Paeoniaceae, Frankeniaceae, Limnenthiaceae and Callitrichaceae. The family, Ranunculaceae is dominated by genera with large chromosome number (and n=8). The two genera *Thalictrum* and *Aquilegia*, which were orginally placed into separate subfamilies or tribes (and even two separate families Ranunculaceae and Helleboraceae by Hutchinson, (1959, 73) along with other achene bearing and follicle bearing genera, respectively), are distinct in having small chromosomes number (and n=7) and as such have been seggregated into a distinct tribe. *Paeonia*, with very large chromosomes number (and n=5) has been separated into a distinct family Paeoniaceae–a placement also supported by morphological, anatomical and embryological data. Some significant records in other families include : Rosaceae with n=7 in subfamily Pomoideae, whereas other subfamilies possess n=7, n=8 or n=9. Similarily in Poaceae, subfamily Bambusoideae has n=12, whereas Poaideae has n=7.

Studies at generic level reveal that *Spartina* was for long placed in the tribe Chlorideae with n=10, although its chromosomes (n=7) were at variance. For this, Merchant (1968) explained the genus to have in fact n=10, and thus secured its placement in Chlorideae.

The duplication of chromosome number leading to polyploidy, has also been able to prove of taxonomic value in grass genus *Vulpia*, Compositae genus *Senecio* etc.

Chromosome Morphology

Delaunay (1926) formulated the concept of karyotype which refers to the number, size and form of chromosomes of a plant or species. Karyotype studies have helped realignments in the family Liliaceae and Amaryllidaceae which are distinguished chiefly on the basis of ovary position. Earlier the genus *Yucca*, which has superior ovary, was placed under Liliaceae, while *Agave*, which has inferior ovary was placed under Amaryllidaceae. Cytological investigations revealed that each of these two genera has 5 long and 25 very short chromosomes (bimodal karyotype). Their karyotype similarity justifies their inclusion in a single family Agavaceae, as has also been done by Hutchinson. Similarily, the two genera of Brassicaceae : *Physaria* and *Lesquevella* were recognised by many as a single genus but cytological investigations supported that these two genera should remain separated. Rudall *et al.* (1997) recommended transfer of *Hosta* (placed in Hostaceae : Hesperocallidaceae by Thorne, 1999), *Camassia* and *Chlorogalum* (both placed under Liliaceae by Hutchinson, 1973 : Hyacinthaceae by Thorne, 1999) to family Agavaceae on the basis of their possessing bimodal karyotype, a suggestion incorporated by Judd *et al.* (2002) and Thorne (2003). Studies on the genus *Leontodon* by Rousi (1973), showed that data regarding the basic number, chromosome length, centromeric position and the occurrence of satellites, furnish evidence for the relegation of the former genus *Thrincia* (n=4) as a section of subgenus *Apargia* along with section *Asterothrix* (n=4, n=7). The subgenus *Leontodon* is distinct with n=6 or n=7, and a different chromosome morphology. Earlier, Cyperaceae and Juncaceae were placed far apart due to distinct floral organisation but both, based on possessing small chromosomes and lacking distinctive centromeres (holocentric chromosome), are now considered to be closely related. An interesting investigation of top onion, variously recognised as *Allium cepa* var. *viviparum* or *A. fistulosum* var.

proliferum (as also those of *Allium cepa* and *A. fistulosum*), made by Schubert, Ohle and Hanelt (1983), was subjected to Giemsa banding pattern and silver-staining technique and it was concluded that some chromosomes of top onion resemble *A. cepa* and others resemble *A. fistulosum*. Of the two satellites, one resembled either species. Since the top onion is pseudodiploid with no homologous pair, the study confirmed that it is a hybrid between *A. cepa* and *A. fistulosum* and thus it would be better known as *A.* × *proliferum* (Moench) Schrad. (based on *Cepa proliferum* Moench), and not as a variety of either species. It is interesting to note that the top onion owes its existence to the bulbils, which are produced in place of an inflorescence and ensure the multiplication of the hybrid, which is otherwise sterile.

Chromosome Behaviour

The pairing behaviour of chromosomes during meiosis can be an useful criterion for taxonomic delimitation at species level. Based on genomic analysis, it has been confirmed that the hexaploid *Senecio cambrensis* is allohexaploid between tetraploid *S. vulgaris* and diploid *S. squalidus*. Likewise the tetraploid *Tragopogon mirus* is the resultant of hybridisation between the two diploid species : *T. dubius* and *T. porrifolius*. The most interesting case, however, is the common bread wheat (*Triticum aestivum*), a hexaploid with AABBDD genome (A derived from diploid *T. monococcum*, B from *Aegilops speltoides*) both genomes being represented in the tetraploid *A. dicoccum*, the genome D is derived from the diploid *Aegilops squarrosa*.

ECOLOGY AND PHYTOGEOGRAPHY IN RELATION TO TAXONOMY

Very often taxonomic treatment at specific and sub-specific levels, is not very satisfactory. A number of such examples from Indian flora, where ecologically and taxonomically doubtful and often overlapping characters have been used in classification, show that taxonomy at this level should be based on more critical observations. Besides, morphological characters should be related to and taken in conjunction with the ecology and distribution of the individual.

The effects of environmental factors on phenotypic plasticity and ecotype variations have been considered using suitable examples. Morphological and physiological differences between ecotype populations are often closely related in differential environmental factors and are adaptive in nature. It is also necessary to know the range of plasticity of characters before ascribing taxonomic values. Taxonomic decisions at this level should be based upon a sufficiently large population size and taking into consideration, as far as possible, as many characters.

Ecological evidences of taxonomic value can be explored from habitat, abiotic relations, biotic relations, spatial relations and temporal relations. Ecological evidentary characters have proved helpful in understanding the distribution of taxa, speciation within the taxa and plant adaptations. Despite many families are ecologically quite homogenous, they have provided explanation about the origin of angiosperms and also about development of classification of various taxa, mainly below the level of the genus–particularly taxonomic status of species.

The holistic view of floristic studies is used by Alpha Taxonomy, Biosystematics, Ecofloristics and Phytogeography so as to know and explain a relatively stable assemblage of angiosperms within a region. It seeks to understand how the competing elements, controlled by a particular environment, balance each other. Alpha Taxonomy is the starting point of floristics and Encyclopaedic Taxonomy is the result of true floristics.

In order to make correlation between ecology and taxonomic studies, the following points are quite necessary :

1. Taxonomist should emphasise the range of plasticity of characters along with any correlation that the plasticity may have with the environment.
2. The concept of sub-species is almost universally accepted for geographical and/or ecological races which have morphological differences but that are insufficiently distinct because they may have transitional forms. Problems regarding species and sub-species arise where the taxa are highly disjunct or they are connected by hybridisation or where differentiation is very complex with ecotypes within ecotypes.
3. Taxonomic descriptions are highly conservative and are usually based on typical plants. Since the geographic and ecological variations are very complex in nature and are dependent upon the environment, a method of description and classification has to be developed which is more meaningful.
4. Functional characters should be more widely considered. These will be of the ecological value for taxonomic classification. The recent advances in biochemical and numerical taxonomy, using multivariate techniques, should be able to bring about more meaningful description and classification at this level.
5. Ecology is concerned with individual, population, community and ecosystem. The concept of species is, therefore, much more dynamic for an ecologist than it is for a taxonomist, for, very often a species is a heterogenous assemblage. Ecotypes rather than species, are often the indicators of ecological conditions. It follows from this that inter-communication between ecologists and taxonomists have necessarily to be more at intraspecific level than at higher taxonomic hierarchies.

Conclusively, linkage between ecology and taxonomy is more appropriate and is likely to be more rewarding in settling specific taxonomic problems.

PHYTOCHEMISTRY IN RELATION TO TAXONOMY (Chemotaxonomy of Plants)

Phytochemistry (chemistry of plants) is an expanding area of study that seeks to utilise chemical information to improve upon classification of plants. Phytochemical characters often correlate well enough with other types of characters and are helpful in providing clearer insights into taxonomic relationship. Recognition of chemical evidences started in fact since man first began to classify plants as edible and non-edible, obviously based on their chemical differences. Later, around 15th century, the chemical information about medicinal plants, gathered by Herbalists, was published in Herbals. It was mainly concerned with application of physiologically active secondary metabolites such as alkaloids of plants. Chemical knowledge about the plants greatly increased during 18th and 19th centuries but the greatest interest has been generated over last 40 years. Alston and Turner (1963), Swain (1966), Heywood (1973), Gibbs (1974), Cagnin *et al.* (1977) have put this promising discipline on a sound footing.

Allelochemic studies throw light on chemical co-evolution. The concept is based on realisation that plants have continuously evolved new defensive chemical mechanisms to save themselves from predators, and animals evolved methods to overcome these defenses. In the process, in course of time, some plant species have developed animal hormones, thus disturbing the hormonal levels, if ingested.

A wide range of chemical compounds are found in plants and to them are added various chemicals produced in various metabolic pathways. Chemical data of taxonomic value, include micromolecules (mol wt upto 1000), including both primary and secondary metabolites and macromolecules (mol wt 1000 or more), both non-semantide (not involved in information transfer, such as starches, celluloses etc.) and semantides (information carrying molecules, such as DNA, RNA and proteins). Primary metabolite micromolecules include compounds involved in vital metabolic pathways, *e.g.* citric acid, aconitic acid, protein, amino acids etc., whereas secondary metabolite micromolecules include the compounds which are the bye products of metabolism and often, perform non-vital functions, *e.g.* non-protein amino acids, phenolic compounds, alkaloids, glucosinolates, terpenes etc.

The chemical characters of plants can often find a wide application in taxonomic delimitations. The occurrence and the distribution of various types of chemical substances present in plants, form the taxonomic evidence. But all kinds of chemical substances do not reveal information useful to taxonomist. The chemical substances, mainly secondary metabolites, of greater use as taxonomic evidence include :

Non-protein Amino Acids

A large number (more than 300) of amino acids not associated with proteins are of great taxonomic value. Their distribution is not universal but specific to certain groups. They are present in high concentration. For example lathyrine is known only in *Lathyrus* (Fabaceae). On the basis of distribution of these amino acids, seven intrageneric groups have been recognised in the genus *Lathyrus*.

Phenolics

They form a loose class of compounds, based upon phenol, and are present in plants in considerable number, diversity and quantity. Simple phenols are widely distributed in plants and as such give them characteristic smell. The crushed leaves of *Anthoxanthum odoratum* can be identified with their characteristic smell due to presence of *coumarins*. Some specific problems could also be solved on the basis of phenolic compounds, *e.g.* Bate-Smith (1958) studied 5 phenolic characters in different sections in the genus *Iris*. The chemical evidence favoured the division into various sections but *Iris flavissima* (originally placed in the section *Pogoniris*) resembled species of the section *Regelia* on the basis of phenolic characteristics.

Flavonoids

These include the largest group of naturally occurring phenols that are more extensively studied in chemotaxonomy. Various types of flavonoids are flavones, flavonones, isoflavonoid, flavonols, anthocynidins etc. Flavonoids, from vegetative parts and seeds, provide more reliable taxonomic evidence than floral pigments. Anthocyanins and Anthoxanthins are important pigments, occurring in the cell sap of petals, providing red and blue (anthocyanin) and yellow (anthoxanthin) colour in a large number of angiosperm families. In some families, these pigments are replaced by betacyanins and betaxanthins (together known betalains) but functioning similarily. Betalains are concentrated in the traditional group Centrospermae (of Engler and Prantl), now recognised as order Caryophyllales. Of the 9 betalains containing families, 7 were included in Centrospermae, 8th, the Cactaceae, placed in Cactales or Opuntiales and the 9th was placed in Sapindales. The traditional Centrospermae also included Gyrostemonaceae, Caryophyllaceae and Molluginaceae lacking betalains and contain anthocyanins instead. Mabry, *et al.* (1963), suggested the placement of only betalain containing families in Centrospermae and thus advocated inclusion of Cactaceae and Didieraceae and exclusion of Gyrostemonaceae, Caryophyllaceae and Molluginaceae, whereas inclusion of Cactaceae and Didieraceae was readily accepted–thus bringing together all betalain containing families of the same order (the exclusion of Caryophyllaceae and Molluginaceae was strongly opposed on the basis of structural data).

Alkaloids

This heterogenous group of organic nitrogen containing bases, is distributed throughout the plant tissues and is present in the vacuoles in the form of salts. There are about 5,000 alkaloids reported from angiosperms, mostly from dicots. Their distribution has proved useful in taxonomic delimitations. The three genera of Fabaceae namely *Genista, Ammodendron* and *Adenocarpus* contain ammodendrine-hystrine alkaloids. Of these *Genista* and *Adenocarpus* were included in the Genistae, whereas *Ammodendron* was placed in Sophorae–a tribe characterised by the presence of matrine alkaloids. This suggests transfer of *Ammodendron* to the tribe Genistae.

Familiar examples of alkaloids include nicotine (*Nicotiana*) and ephedrine (*Ephedra*). Their distribution is often specific and thus of taxonomic importance. Morphine is present only in *Papaver somniferum.* Families Papaveraceae and Fumariaceae are closely related, as is supported by occurrence of alkaloid protopine in both. Some families rich in alkaloid species are Leguminosae, Berberidaceae, Ranunculaceae, Solanaceae etc.

Glucosinolates

They are mustard oil glucosides occurring in Capparales. Originally, Cruciferae (Brassicaceae), Capparaceae, Papaveraceae and Fumariaceae were placed in the same order Rhoeadales but chemical (presence of glucosinolate) and other evidences supported the placement of Cruciferae and Capparaceae in the order Capparales and Papaveraceae and Fumariaceae in the order Papaverales (or suborder Papaverineae of Ranunculales, Thorne, 2003) on the basis of the absence of glucosinolates.

Terpenes

These include a large group of compounds, mostly derived from polymerised isopropane, *e.g.* camphor (*Cinnamomum*), menthol (*Mentha*) and carotenoids which seem to have a definite role in the allelopathic effects of plants.

1. Terpenoids. A major contribution of terpenoid chemistry has been the use of sesquiterpene lactones in the family Compositae in establishing centres of origin of *Vernonia* and *Xanthium.*

2. Iridoids. They are present in over 50 families and their presence is correlated by sympetaly, unitgmic tenuinucellate ovules, cellular endosperm and endosperm haustoria. Dahlgren brought together all iridoid–producing families. *Budleia,* which contains aucubin, has been transferred from Loganiaceae to Budleiaceae.

Directly visible chemical characters, although restricted in number, have attracted the attention of taxonomists. These include starch grains (Solanaceae), raphides (Onagraceae), calcium oxalate crystals (Liliaceae), silica (Poaceae), gypsum (Tamaricaceae) etc.

Studies on secondary metabolites of Indian plants, are largely based on structural and synthetic aspects of phenolic compounds, quinones, essential oils, resins, saponins, alkaloids and flavonoids. The work has been carried out mostly by the school led by Late Professor Shesadri and his students (1943-69) at Delhi. Atal, *et al.* (1962-71) have investigated micromolecules of *Piper* and *Crotalaria.* Dakshini *et al.* (1970, 75), have performed preliminary screening of *Blumea* and *Bidens* for flavonoids. Central Drug Research Institute, Lucknow, is actively engaged in screening of Indian plants for antibacterial and antitumor active principles.

SEROTAXONOMY

In serology, the protein extract from plant antigen (whose relationship is to be determined), is injected into the body of an experimental animal, to form antibodies. After some time, the blood serum is taken out from the body of the animal and made to react *in vitro* with the antigenic proteins of another plant (whose relationships are to be ascertained). The amount of the precipitation indicates the degree of protein homology, and thus taken as phylogenetic marker and taxonomic character. If there is no precipitation there appears no relationship between the taxa in question.

This method was first suggested by KC Mez (1926, 36), a German botanist, calling it *Serum Diagnosis* and building up phylogenetic trees of families based on this technique. The trees were to a degree in agreement with the schemes proposed by Bessey and Hutchinson. The serological information gathered in last few decades of 20th century, by and large, tend to correlate recent systematic treatments. John and Fairbrothers (1965), used this technique and found that *Magnolia* and *Michelia*, displayed closest affinity within the Magnoliaceae whereas *Liriodendron* is quite distinct from other members of the family. Similar studies, made by Simon (1971), have also been helpful in establishing relationships of Nymphaeaceae and Nelumbonaceae and generic kinship in Caprifoliaceae and Rubiaceae with related groups. Serological studies also support subdivision of Umbelliferae into Hydrocotyledeae, Saniculoideae and Apioideae. Crispels and Gartner (1978), gave serological evidence to confirm assignment of *Phaseolus aureus* and *P. mungo* to the genus *Vigna*.

For derivation of taxonomic evidence at lower taxonomic levels, peptide finger printing and electrophoresis have proven to be quite useful. In peptide finger printing, carboxymethylated protein (hydrolysed into proteins by incubating with proteolytic enzymes), is separated by unidimensional chromatography on paper, and further into a second dimension by electrophoresis. Separated peptides are located by using chromogenic substances. The distribution map of the peptides is characteristic of each protein. Differences in the map help to detect difference in proteins. Finger printing patterns of globulins of five genera of Viceae showed that *Vicia* and *Lathyrus* had almost identical patterns and were increasingly different from *Lens, Pisum* and *Cicer* in that order. Electrophoretic studies have also supported the origin of hexaploid wheat (*Triticum aestivum*) from *Aegilops squarrosa* and *Triticum dicoccum*, to assess species relationships of *Chenopodium* (Crawford and Julian, 1976) and in establishing the fact that *Brassica campestris* and *B. oleracea* are closer to each other than *B. nigra*.

Antigens are mostly extracted from seeds and pollen grains. In works performed earlier, crude total comparison of precipitation reactions was done, but now more refined methods have been developed which can bring about individual antigen–antibody reactions. These methods include *Double-diffusion Serology, Immuno-electrophoresis,* Absorption, Radio-immuno- assay (RIA), *Enzyme–linked Immunosorbent Assay* (*ELISA*) etc.

PHOTOSYNTHETIC TAXONOMY OF ANGIOSPERMS

Plants being *photoautotrophs, photosynthesis* becomes the most important functional activity, and thus the occurrence of diversity in the mechanism of this process needs to be examined in relation to the taxonomic groupings. Three different kinds of photosynthesis are known to operate in plants namely *Calvin cycle* (C_3), *dicarboxylic acid cycle* (C_4) and *crassulacean acid metabolism* (CAM). C_3 photosynthesis is only mode found to occur in gymnosperms and great majority of families (about 300) of flowering plants. C_4 photosynthesis is known to occur in atleast 945 species of 196 genera belonging to 18 families. *CAM metabolism* has been identified in 18 families, including a succulent epiphytic fern *Drymoglossum* and a desert gymnosperm, *Welwitschia.* Both C_4-photosynthesis and CAM pathway exist in a diverse group of plants united more by adaptations to xerophytic environments than by phylogeny based on morphological considerations.

Although *C_4 plants* form only a minority (about 1/300th) of angiospermic species, they offer interesting taxonomic implications. In the Gramineae (Poaceae), the C_4 species are confined to only three subfamilies-Aurandinoideae, Eragrostroideae and Panicoideae, which are mostly tropical in their distribution. Agrosteae and Festuceae, on the other hand, with very few C_4 plants reported among them, occur predominantly in temperate regions.

The occurrence of C_4 pathway, which is being considered to be an advanced character, can be used to solve some of the controversies over the systematic position of the genera such as *Aristida* and *Eragrostis.* The separation of C_4 pathway supports the inclusion of *Aristida* in a separate tribe Aristideae (rather than in Stipeae) and suggests close affinity of *Eragrostis* with panicoid grasses (rather than with members of Festuceae). Similarily the C_4-nature of *Uniola* and *Leptochloopsis* justifies their placement in the subfamily Eragrostioideae (rather than the old tribe Unioleae). Majority of C_4 dicot plants belong to Centrospermae. The C_4 species of *Euphorbia* allow a taxonomic circumscription of subgenus *Chamaesyce* and warrant the treatment of *Chamaesyce* as distinct genus.

Apart from the taxonomic significance, the distribution of C_4 plants has also phylogenetic implications. Among the tribes of Gramineae (Poaceae), the Bambuseae, with no C_4 species, appear to be primitive whereas the subfamily Panicoideae, with all three tribes Andropogoneae, Maydeae and Paniceae, possessing a number of C_4 plants, is considered to be highly evolved. The wide diversity of families with C_4 genera, suggests that the C_4 pathway has evolved independently atleast more than twenty times in the angiosperm families.

MOLECULAR SYSTEMATICS

Macromolecular studies towards *DNA* and *RNA*, in last few years of the 20th century, gave a way for establishment of an emerging field of molecular systematics which commonly deals exclusively with utilisation of nucleic acid data. A molecular data gives an understanding of gene level changes reflecting true phylogeny, better than morphological data and data obtained from other sources. But at the same time, it was also realised that molecular data may also pose similar problems. Since there are more molecular characters available, they make the comparison easier. Despite, various taxa (especially species) have been characterised primarily on morphological differences and placed in various evolutionary trees based on data obtained from various other disciplines such as anatomy, embryology, physiology, palynology, biochemistry, cytology etc., but for the fact that evolution is based on genetic changes, require the data at the level of the gene. It is expected that species which are closely related, will have greater similarities in their genetic material than the distantly related species. During last few years, *molecular genetics* has played a significant role in understanding of speciation and evolution clearly. Taking the help of mathematical principles, using *computer programme*, the differences in nucleotide sequences, which are quantitative, can be analysed. To establish evolutionary relationships among different species, evolutionary changes at the level of DNA, can be objectively compared.

Molecular data are obtained from nucleus, chloroplast and mitochondria yielding nuclear, chloroplast and mitochondrial genome, respectively. Among these, chloroplast genome is the smallest (120-160 kbp) in higher plants, mitochondrial genome ranges between 200-2500 kbp and the nuclear genome (which is much larger) ranges between million to more than billion kbp. Mitochondrial and chloroplast genomes are inherited from maternal parent whereas *nuclear genome* is biparental. The *mitochondrial genome* undergoes lot of rearrangements and is of little importance in interpreting phylogenetic relationships, whereas chloroplast and nuclear genomes are highly stable, not only within the same cell but also within a species, and as such are used as taxonomic tools.

Taking into consideration, the early attempts to make use of *in vitro* DNA/DNA hybridisation in related plants, by McCarthy (1962) and Bolton (1966) found that only half of the nucleotide sequences in the DNA of *Vicia villosa* are similar (homologous) with those of *Pisum* while only 1/5th are similar (homologous) between *Pisum* and *Phaseolus*. Similarly, Mabry (1976), during DNA/RNA hybridisation in Centrospermae (Caryophyllales), reported that family Caryophyllaceae (although lacking betalins)) is quite close to betalin-containing families, but not as close as the latter are to each other.

One of the earliest important results using gene mapping, was obtained by Jansen and Palmer (1987). They reported, in Asteraceae (Compositae), a large single-copy region of the *chloroplast genome* and this unique order could be explained by a single inversion of DNA (a feature lacking in all other angiosperms), strongly confirming that Asteraceae are monophyletic. Similarly, Poaceae (Gramineae) have three inversions in chloroplast genome, of which one is unique to family Poaccae and confirms its monophyletic status and of the other two, one is shared with Joinvilleaceae and one with both families Joinvilleaceae and Restionaceae, suggesting that these two are the sister groups of Poaceae. Riesberg *et al.* (1996) reported that in nuclear genome mapping in *Helianthus annuus* and *H. petiolaris*, there were atleast seven translocations and three inversions. Some progress in this area has also been made in Solanaceae and grasses.

Using gene sequences, Belford and Thomson (1979), during side-copy sequence hybridisation in *Atriplex*, concluded that division of this genus into two subgenera is not logical. Bayer *et al.* (1999), based on sequence analysis of the plastid *atp*B and *rbc*L DNA, got support for an expanded order Malvales (including most of the genera previously included in Sterculiaceae, Tiliaceae, Bombacaceae and Malvaceae) and thus these authors propose to merge Sterculiaceae, Tiliaceae and Bombacaceae in Malvaceae and subdivide the enlarged family Malvaceae into nine subfamilies based on molecular, morphological and biogeographical data.

Molecular systematics provides powerful tools for constructing phylogenetic trees based on chloroplast DNA using restriction site polymorphism (cpRFLP), analysis of chloroplast gene for subunit F of NADP dehydrogenase (*ndh*F–in the small copy region), for 'a' and 'b' subunits of RNA polymerase II (*rpo*A and *rpo* C2 in the large single copy region), for 'b' subunit of ATP synthase (*atp*B), ITS region of ribosome, phytochrome B and granule bound starch synthase I. Except for tribe Stipeae of grasses, in other cases these studies did not prove of much help in constructing molecular ***phylograms***. The gene tree constructed from *rbc*L, is of great use in angiosperms. Using 499 *rbc*L, Chase *et al.* (1993), made an attempt to yield the phylogeny of all seed plants. Their analysis proved a few sequences to be ***pseudogenes***. The entire families were represented by single sequences. The data sets have been reanalysed (Rice *et al.* 1997) to yield parsimonious trees. The *rbc*L data has provided support for the union of family pairs Asclepiadaceae–Apocyanaceae, Araliaceae–Apiaceae and Brassicaceae–Capparaceae as also the phylogenetic nature of Saxifragaceae and Caprifoliaceae.

NUMERICAL TAXONOMY

The main aim of the systematics is to develop classifications based on various criteria and often a distinct methodology, is used for the analysis of data. While establishing relationships between the organisms, data handling is more commonly done with phenetic methods. Numerical taxonomy has emerged as a branch of taxonomy which has received a great impetus with the development and advancement of computers. Also known as *mathematical taxonomy* (Jardin and Sibson, 1971), *taxometrics* (Mayr, 1966) or *phenetics*, the numerical taxonomy marked its beginning from contributions of Sneath (1957), Michner and Sokal (1957) and Sokal and Michner (1958) which culminated in the *Principles of Numerical Taxonomy* (Sokal and Sneath, 1963) with an expanded and updated version *Numerical Taxonomy* (Sneath and Sokal, 1973). The last few decades have witnessed a forceful debate on whether (i) the empirical approach (*empirical taxonomy*), *i.e.*, classification on the basis of taxonomic judgment on the basis of observation of data and not the assumptions or (ii) operational, approach (*operational taxonomy*), *i.e.*, based on operational methods on the basis of experimentation to evaluate the observed data before a final classification, would be suitable. Numerical taxonomy finds balance between empirical approach and operational approach as it incorporates both the approaches. Numerical taxonomy in itself neither produces new data nor a new system of classification, but is rather a new method of organising data that may prove helpful in better understanding of relationships. Numerical taxonomy results a classification based on a greater number of characters from many sets of data in an effort to develop an entirely phenetic classification of maximum predictivity.

The lowest ranking taxa (basic units) in any particular study are called *Operational Taxonomic Units* (*OTU's*) which could be an individual or a supra-individual category such as species, genus or even higher ranked taxa. A sufficiently large number of suitable characters of OTU's are selected (not less than 50 characters). The value of similarity coefficient becomes more stable with increase in the number of characters sampled. It is recommended that the taxonomic characters should be selected from all parts and all stages of life cycle; and all characters varying within the group studied should be used. Characters (attributes) capable of expression as binary, qualitative multistate or quantitative multistate data are generally used in studies on numerical taxonomy. Binary characters possess two contrasting states, such as the presence or absence of some features (*e.g.* presence or absence of stipules) or contrasting alleles of a single character (*e.g.* fruit being dehiscent or indehiscent). Qualitative multistate characters possess three or more contrasting forms, each ranking equal (*e.g.* flower colour : red, white or yellow) whereas quantitative multistate characters represent measures of size of a continuous scale, such as weight or length (*e.g.* chemicals produced by bacteria, bacterium strain or length of spore). To estimate similarity between pairs of OTU's, the basic data are recorded in the next matrix consisting of OTU's classified by *n* characters. There are various ways to express the scores in a data matrix. The resemblance between a pair of OTU's can be expressed in the terms of similarity/dissimilarity, *i.e.*, the percentage of character in which they agree/disagree. There are several classes of similarity coefficient suggested for biological taxonomy. Then a *similarity matrix* (*txt* matrix) is prepared where computed similarities are arranged to fall on a scale between 0 and 1. Afterwards OTU's are rearranged to produce a *sortex matrix* where OTU's with high mutual similarity are arranged together to form a phenetic group.

Numerical taxonomy has many advantages, such as it has power to integrate the data from variety of sources, being mathematical and quantitative it provides a more exact basis for classification and permits classifications to be built on wide range of characters.

Application of Numerical Taxonomy in Angiosperms

Numerical taxonomy has been able to solve many taxonomic problems. Numerical analysis of four subclasses of monocots, namely Alismidae (Cronquist's-Alismatidae), Liliidae, Commelinidae and Arecidae, have been recognised by Takhtajan (1969) and Cronquist (1968). Clifford (1977), based on the results of cluster analysis and principal co-ordinate analysis, concluded that Alismidae (water lilies) form a distinct group among monocots. The studies made by him also concluded that Triuridales (placed by Cronquist in Alismatidae and by Takhtajan in Liliidae) belong to a distinct group among anemophilous (wind pollinated) families. Interestingly, Takhtajan (1987, 97) separated Triuridales under a distinct subclass Triuridinae (a placement reversed by him in 2003). The study also supported the placement of Typhales among Arecidae by Takhtajan but opposed the placement of Arales, stating that although Arales are not related to Arecidae, they fit uneasily even among Liliidae (lillies) to which they are closer. It is notable that Takhtajan (1997) separated Arales to a distinct subclass Aridae. The cluster analysis resulted a *phenogram* (*dendrogram*) with ten groups. The findings of principal co-ordinate analysis were presence or absence of endosperm in the seed. The squamulae and endosperm were found to be highly diagnostic of the primary division within monocots. The attributes found to be most diagnostic for separation of the anemophilous (wind pollinated) groups include : ovary position (superior, inferior or subinferior), nature of endosperm (starchy or not), number of subsidiary cells (0, 2 or more) and fruit type (achene, nut, berry, capsule or drupe).

Young and Watson (1971) undertook another significant work on dicots using 83 attributes in their analysis of 543 representative genera and proposed the division of dicots into *Crassinucellatae* and *Tenuinucellatae*. Studies on numerical analysis of Gramineae made by Hilu and Wright (1982), utilising morphological and microscopic characters separately, did not give satisfactory results as the results were incongruent. However, combination of characters, both morphological and microscopic, resulted in the establishment of eight clusters comparable to eight subfamilies recognised within Gramineae.

BIOSYSTEMATICS

Biosystematics deals with the study of populations of living organisms to establish the variation patterns arising from genetic sources within and between populations. According to Rollins (1958), this type of study provides a rich source of understanding of a given taxon, and when these patterns are properly evaluated in terms of their repetition and geographical spread, they are at the base of a taxon's proper definition. The development of taxonomy in relation to population sampling, known as biosystematics, has also been termed variously as **Experimental Taxonomy, Genenomy, Genecology, Modern Taxonomy, New systematics** and so on.

The aims of biosystematics, first stated by Camp and Gily (1943), included to delimit the natural biotic units and to apply to these units a system of nomenclature adequate to the task of conveying precise information regarding their defined limits, relationships, variability and dynamic structure. According to Clausen, Keck and Hiesey (1945), genetics, cytology, comparative morphology and ecology are furnishing the critical data which in combination, when applied to the organic evolution, make up biosystematics.

Taxonomists often arrive at divergent views in the relationsips of the species within a large genus, some believing the genus to be largely consisting of unrelated species or a large number of polymorphic species, whereas others considering the species to be showing continuous or overlapping variations, and thus the genus is taken to be composed of several sections or subsections, while still others take it to be a phylogenetic group. Such type of disputes are generally settled more or less satisfactorily with the help of cytogenetics or cytotaxonomic studies.

During the last few decades, biologists have been able to comprehend, to an increasing extent, some of the ways in which small-scale evolution of plants and animals takes place. Biosystematics appears to be modifying procedure in taxonomy and building systematics into a synthetic science. In general, biosystematic studies have been able in delimiting natural biotic units and the application of formal nomenclature to these units coveying precise information regarding their definite limits, relationships, variability and dynamic structure. Davis and Heywood (1963) regards biosystematics, essentially an expansion of classical taxonomy. It still employs comparative morphology as primary tool for describing, delimiting and defining taxa. For example, in genus *Nicotiana*, about 60 species have been studied in great details by Goodspeed (1945) but there are others who correlated the findings of distributional patterns and morphological differences with studies of chromosome counts and breeding behaviour. Similar studies on some genera of Cruciferae, *Oenothera, Crepis* etc., have been made by various other authors.

❑❑❑

Chapter **14**

Morphological Features

Taxonomic studies require a perfect knowledge of technical terms for description and identification of the plants. Following are some commonly used terms and their meanings.

HABITAT AND HABIT

Habitat. The immediate environment occupied by an organism (plant), *e.g.* aquatic or terrestrial.

Habit. The general external appearance of a plant, *e.g.* herbs, shrubs or trees.

Habitat form. A plant showing some atypical features but can be related to the place where it is growing, such as dwarfing under adverse environment.

Annual. A plant completing its life cycle in a single season, *i.e.*, from seed to germination, reproduction and death, *e.g. Anagalis.*

Ephemeral. A plant which completes its life cycle in a very short time, say within some weeks, *e.g. Chenopodium.*

Biennial. A plant which completes its life cycle in two seasons, *e.g.* radish (*Raphanus sativus*).

Perennial. A plant which grows for several years.

Herb. A plant which has no persistent parts above the ground, *e.g. Chenopodium, Ranunculus, Fumaria* etc.

Shrub. A woody short plant (less than thirty feet) in which side shoots are well developed, so that there is no trunk, *e.g. Lantana.*

Tree. A tall woody perennial plant with a well-marked trunk, *e.g. Eucalyptus, Mangifera, Tectona* etc.

Forms of Trees may be :

(i) Caudex–main trunk straight, unbranched and columnar, bearing crown of leaves, *e.g. Cocos nucifera* (fig. 14.1A).

(ii) Excurrent–main trunk keeps on growing but bears lateral branches in acropetal order, giving a conical shape to the tree, *e.g. Polyalthia* (fig. 14.1B).

Fig. 14.1 Types of trees. A–caudex, B–excurrent, C–deliquescent and D–culm.

(iii) Deliquescent–canopied spreading habit due to suppressed apical bud and dominant lateral buds, *e.g. Mangifera* and *Ficus* (fig. 14.1C). **Culms.** Grass-like strong stem with distinct nodes (solid) and internodes (hollow), may look-like tree, as in bamboo (fig. 14.1D).

Creeper. Plants with stem growing along the surface of the ground and rooting at its node, *e.g. Hydrocotyle asiatica* (fig. 14.2 A) and *Ipomoea batatas.*

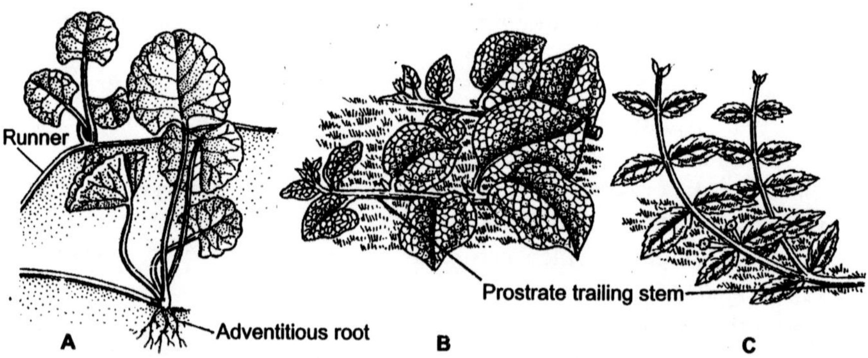

Fig. 14.2 Weak forms of plants. A-creeper (*Hydrocotyle*), B-procumbent (*Basella*) and C-decumbent (*Lindenbergia*).

Trailor. A plant which trails over the ground surface by its weak, prostrate and slender stem. A trailer may be : **(i) procumbent,** *i.e.,* trailing flat on the ground, *e.g. Basella* (fig. 14.2B), or **(ii) decumbent**, *i.e.,* trailing on the ground to some distance and then straightens upright, *e.g. Lindenbergia* (fig. 14.2C) and *Portulaca* or **(iii) diffuse**, *i.e.,* trailers growing in all directions, *e.g. Boerhaavia.*

Climber. A plant with weak stem which climbs by taking support of neighbouring objects. A climber may be :

(I) Stem Climbers.

(i) **Twiners,** *i.e.,* Climbing by twinning around its support, *e.g. Ipomoea quamaclit* (fig. 14.3A) or *Dolichos lablab* (fig. 14.3B).

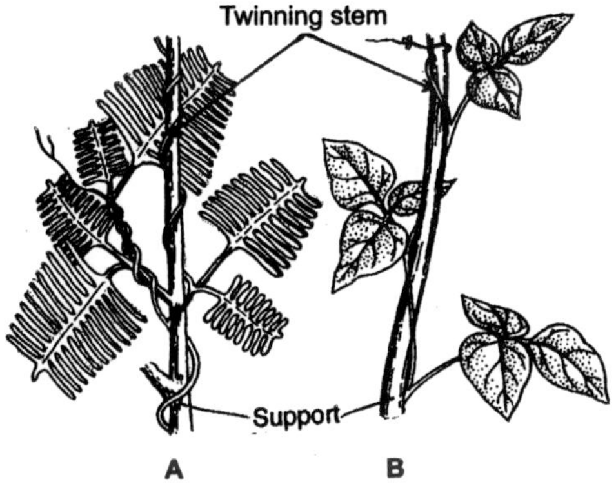

Fig. 14.3 Stem climbers, A-*Ipomoea* and B-*Dolichos.*

(II) Tendril Climbers. Climbing by formation of special wire-like structures twinning around to seek support. A tendril may be formed by : **(i) the leaf**, *e.g. Lathyrus* (fig. 14.4A), **(ii) the stem apex** or **axillary branch**, *e.g. Vitis* and *Passiflora* (fig. 14.4B, C),

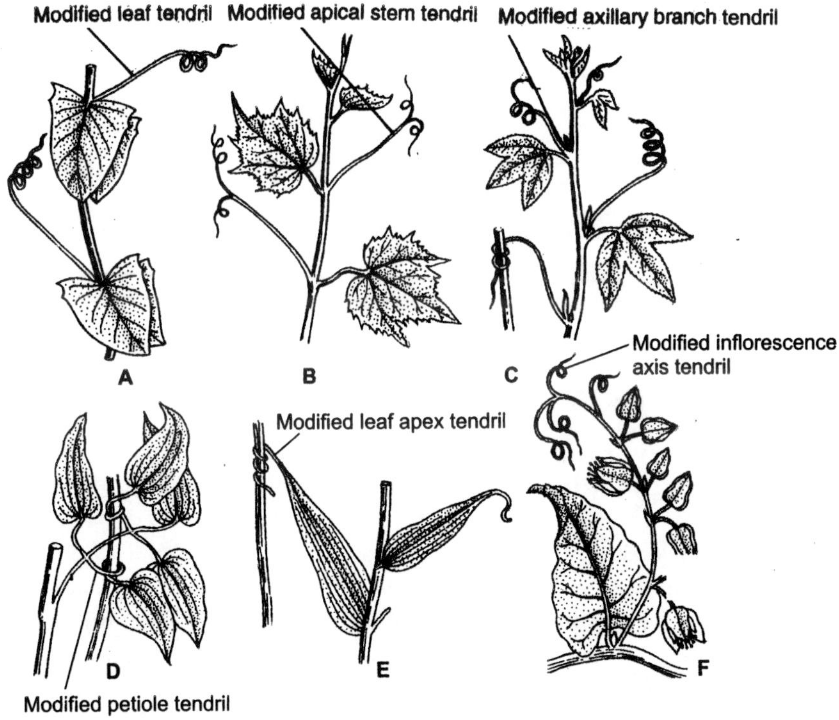

Fig. 14.4 Tendril climbers. A-*Lathyrus,* B-*Vitis,* C-*Passiflora,* D-*Clematis,* E-*Gloriosa* and F-*Antigonon.*

(iii) **the petiole**, *e.g. Clematis* (fig. 14.4D), **(iv) the leaf tip**, *e.g. Gloriosa* (fig. 14.4E) or by **(v) the inflorescence axis**, *e.g. Antigonon* (fig. 14.4F).

(III) Root Climbers. Climbing by the formation of adventitious roots arising from the nodes, *e.g. Piper betle* and *P. longum* (fig. 14.5).

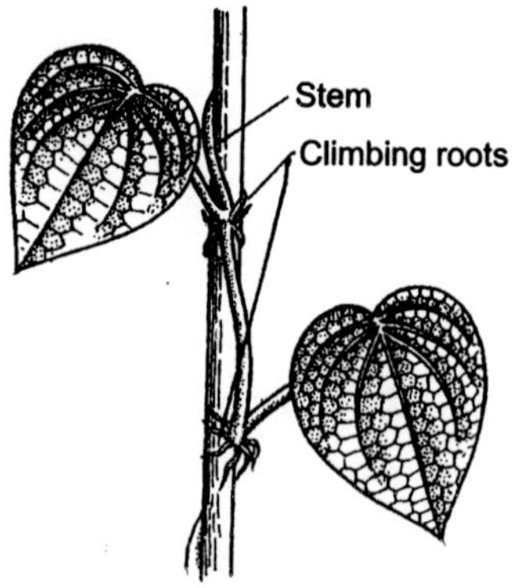

Fig. 14.5 *Piper betle*

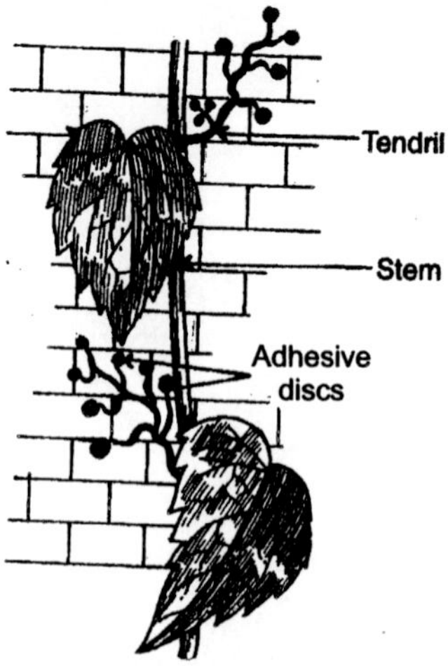

Fig. 14.6 *Ampelopsis*

(IV) Scramblers or **Hook Climbers.** Climbing by the formation of hooks (*e.g. Artabotrys*) (fig. 14.7A), or curved prickles (*e.g. Rosa*) (fig. 14.7B) and *Lantana* (fig. 14.7C).

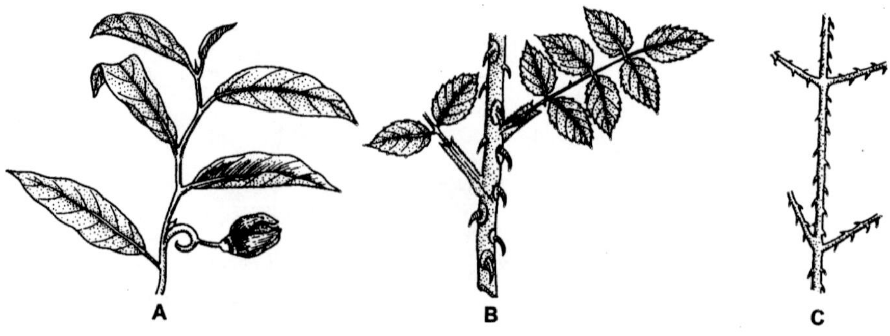

Fig. 14.7 A-*Artabotrys*, B-*Rosa* and C-*Lantana*

(V) Adhesive Climbers. Climbing by the formation of adhesive discs, either by the tendrils as in *Ampelopsis* (fig. 14.6) or by the climbing roots, *e.g. Hedera helix* (ivy).

(VI) Lianas. Climbing by twinning woody and stout stems which hang down by thick ropes, *e.g. Tinospora, Hiptage, Quisqualis* etc.

Epiphyte. A plant growing on another plant, but not deriving any food from it. It grows on other plant to get only mechanical support, *e.g.* most orchids (fig. 14.11A).

Parasite. A plant living on another organism (or plant) and extracts food from it (called host), *e.g. Cuscuta* (fig. 14.12A).

ROOT

A root is the positively geotropic leafless lower portion of the plant axis. Basically, it arises from radicle and is characterised by non-green smooth surface and not differentiated into nodes and internodes. Lateral branches arise endogenously. Main function of the root is the anchorage of the plant into the soil and absorption of water and dissolved minerals from the soil by means of root hairs. Three kinds of roots have been recognised on the basis of their nature (tap roots, seminal roots and fibrous roots), and into two types on the basis of their origin (normal roots and adventitious roots).

Tap root. A root system with a prominent main root, directed vertically downwards bearing small lateral roots. It is characteristic to dicot plants, *e.g.* carrot, radish, rose, trees etc. (fig. 14.8A). Tap root is primary in nature as it is formed from the radicle.

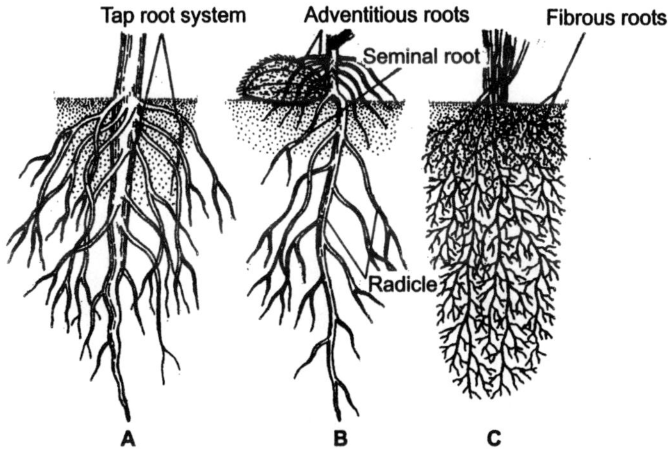

Fig. 14.8 Types of roots. A-tap root system, B-seminal roots in rice seedlings and C-fibrous root system.

Seminal root. In monocots, such roots arise from near the base of the radicle, as in rice (fig. 14.8B) and barley seedlings.

Fibrous root. A root system that develops from the base of the plumule (hypocotyl) in which all roots are equally prominent, as characteristically found in monocots (grasses) (fig. 14.8C).

Normal root. A root arising from the radicle.

Adventitious root. A root developing from any other plant part, except the radicle. It may arise from the underground or aerial stem or from leaf margins.

Modified Roots. Some roots are modified to carry out specialised functions of mechanical and physiological nature.

Modified tap roots. Some tap roots are modified for food storage. These are classified on the basis of their shape, as follows :

(i) **Fusiform roots**–swollen in the middle and tapering at the two ends, as in radish (*Raphanus sativus*) (fig. 14.9A).

(ii) **Conical roots**–broad at the upper end and gradually tapering towards the lower end, as in carrot (*Daucus carota*) (fig. 14.9B).

(iii) **Napiform roots**–almost spherical at the upper end which abruptly tapers at the lower end, *e.g.* beet root (*Beta vulgaris*) and turnip (*Brassica oleracea*) (fig. 14.9C).

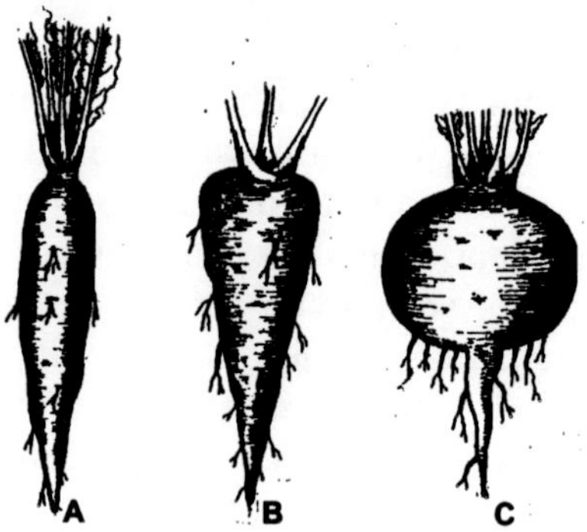

Fig. 14.9 Modification of tap roots. A-fusiform (radish), B-conical (carrot) and C-napiform (turnip).

Modified adventitious roots. Some adventitious roots are modified for various purposes, as follows :

(i) **Root tubers**–arising from the node. These swell like tubers for food storage, *e.g.* sweet potato (*Ipomoea batatas*) (fig. 14.10A).

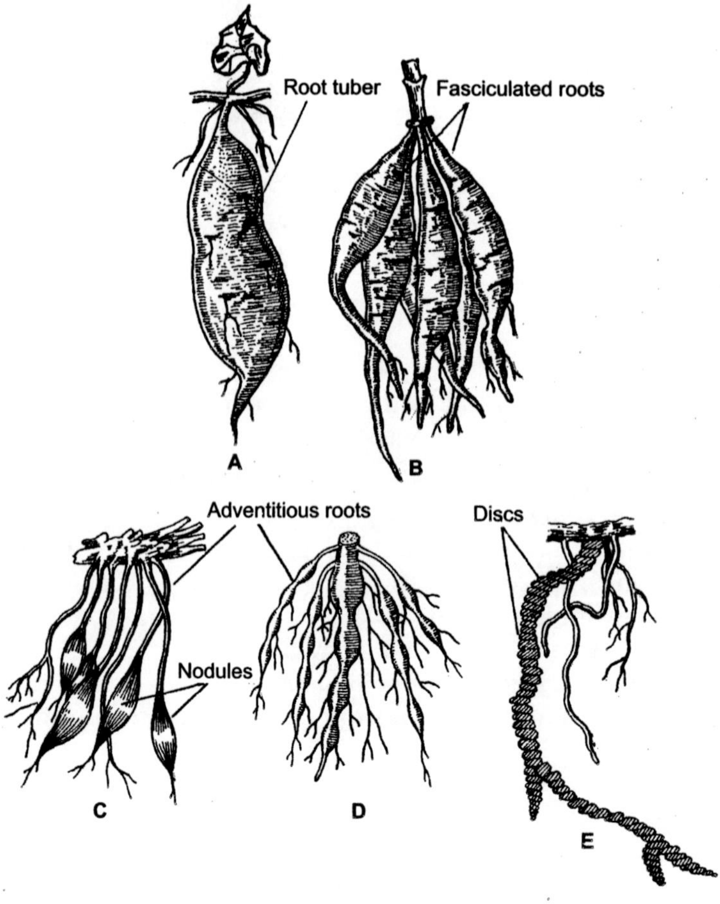

Fig. 14.10 Modifications of adventitious roots. A-root tuber (sweet potato), B-fasciculated (dahlia), C-nodulose (mango-ginger), D-moniliform (grasses) and E-annulated (ipecac).

(ii) **Fasciculated roots**–the cluster of adventitious tuberous roots, as found in *Asparagus* and *Dahlia* (fig. 14.10B).

(iii) **Nodulose roots**–apices of the roots swell up to form bead-like structures, *e.g.* mango ginger (*Curcuma amada*) (fig. 14.10C).

(iv) **Moniliform roots**–roots develop alternate swellings and constrictions to give a beaded appearance, as found in *Vitis*, *Dioscorea* and some grasses (fig. 14.10D).

(v) **Annulated roots**–roots appear as if a number of discs are placed one above the other, as in ipecac (*Cephalis ipecacuanha*) (fig. 14.10E)

Aerial roots. Orchids grow as epiphytes on trees and bear hanging aerial roots to absorb atmospheric moisture with the help of spongy velamen tissues (fig. 14.11A).

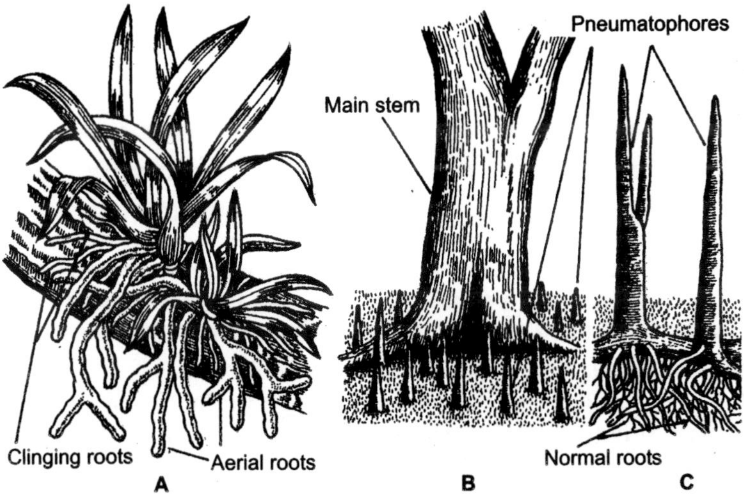

Fig. 14.11 A-aerial roots of orchids and B & C-respiratory roots (pneumatophores) of mangrove plants.

Assimilatory roots. These are hanging green aerial roots which have the power of assimilation (photosynthesis), as found in *Tinospora cordifolia.*

Respiratory roots. These are negatively geotropic vertically growing roots into the air for breathing or oxygen requirement found in mangrove plants growing in saline waters, as in *Rhizophora, Heritiera, Avicennia* etc. These are also called **pneumatophores** (fig. 14.11B and C).

Parasitic or **haustorial roots.** These absorb nutrition by sending out haustoria into the tissue of hosts, as found in parasitic plants like *Loranthus* and *Cuscuta* (fig. 14.12A and B).

Reproductive roots. Roots that bear adventitious buds which grow into plantlets as those of sweet potato, pointed gourd (*Trichosanthes dioica*) and wood apple (*Feronia elephantum*).

Mycorrhizal roots. Roots of higher plants infested with fungal mycelia which absorb food from the soil. Mycorrhiza facilitates nutrient absorption specially N, P and K by the higher plants. These are found in *Monotropa* (a saprophytic angiosperm), *Sarcodes* and *Pinus* (fig. 14.12C and D).

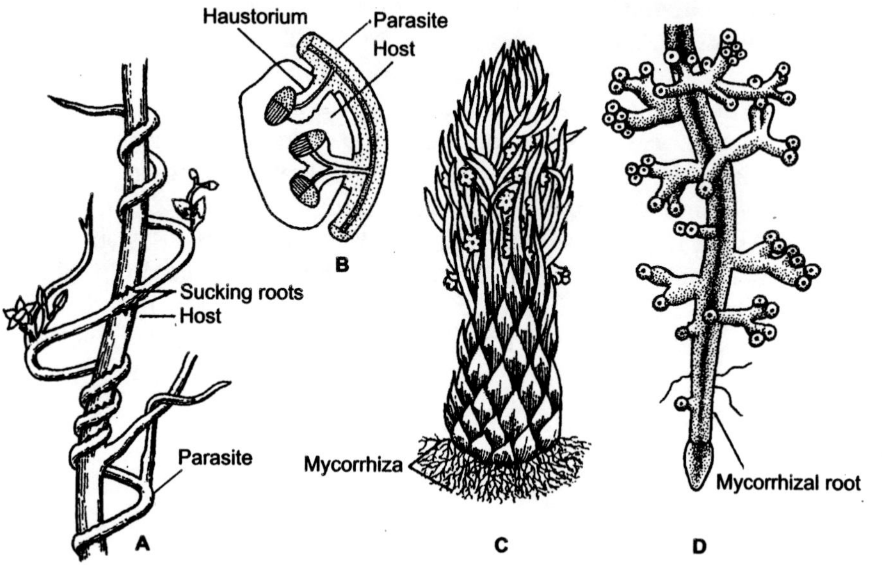

Fig. 14.12 A-haustorial roots of *Cuscuta*, B-T.s. of host plant showing haustoria of *Cuscuta*, C-mycorrhizal roots of *Sarcodes* and D-mycorrhizal root of *Pinus*.

Food storing roots. These accumulate reserve food material to carry out physiological functions, *e.g.* fusiform, conical and napiform roots (tap roots) or tuberous, fasciculated, nodulose, moniliform or annulated roots (adventitious roots).

Roots modified for mechanical support. Roots may be modified variously to provide mechanical support to the plants and are named accordingly, *e.g.*

(i) **Prop roots**–strong, woody and sunken in the soil, these aerial roots arise from horizontal branches (stem) to provide stem-like support, as found in banyan tree (*Ficus benghalensis*) (fig. 14.13A).

(ii) **Stilt roots**–roots arising from lower nodes of the stem and growing obliquely downwards to anchor the plant into the soil, as

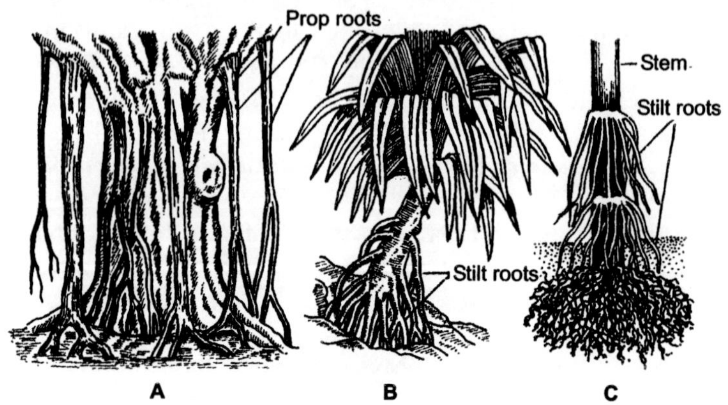

Fig. 14.13 A-prop roots of banyan, B-stilt roots of screw pine and C-stilt roots of maize.

found in maize (*Zea mays*), sugar cane (*Saccharum officinarum*), screw pine (*Pandanus foetidus*) and in some mangrove plants (*Rhizophora*) (fig. 14.13B and C).

(iii) **Clinging roots**–found in epithytic plants (orchids) to cling the plant on the stem of other plant (fig. 14.11A).

(iv) **Climbing roots**–adventitious roots arising from the node of the stem that help in climbing or twinning of the plant, as found in betel vine (*Piper betle*) (fig. 14.5) and money plant (*Pothos*).

(v) **Root buttresses**–found in some large trees to provide additional support by producing plank-like roots arising from the base of trunk, as found in *Salmalia, Ficus* and *Terminalia* (fig. 14.14).

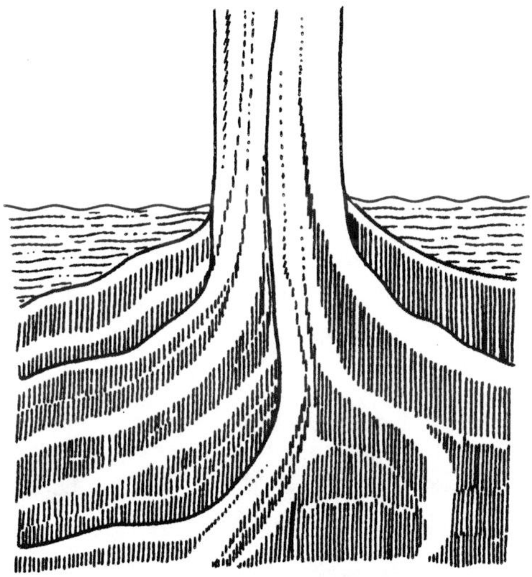

Fig. 14.14 Root buttresses.

(vi) **Floating roots**–found in aquatic floating plants–special adventitious spongy and whitish roots filled with air to facilitate

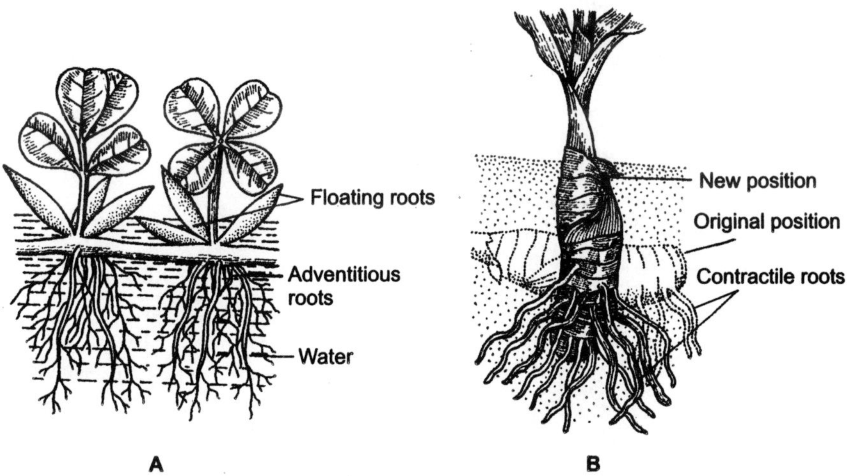

Fig. 14.15 A-floating roots of *Jussiaea* and B-contractile roots of *Canna*.

plant floating on the water. These also help in respiration, as found in *Jussiaea repens* and *Pistia stratiotes* (fig. 14.15A).

(vii) Contractile roots–special type of roots which can contract or pull the stem deep into the soil, as found in *Lilium, Allium* and *Canna* (fig. 14.15B).

STEM

Stem is the axis of a plant bearing leaves with buds in their axils. It is formed by the plumule of the seed.

Acaulescent. A plant which is stemless or without visible stem.

Arborescent. A plant almost tree-like with a short main trunk.

Arboreous. A tree with well-developed trunk.

Culm. A jointed stem with distinct nodes and internodes, as in grasses and sedges (fig. 14.1D).

Branching of the Stem. Branching may be :

(i) **Racemose** or **monopodial branching**–main stem (terminal bud) grows indefinitely and lateral branches are produced in an acropetal order (*i.e.*, older towards base), as in *Polyalthia.*

(ii) **Cymose**–the terminal (apical) bud loses its activity and axillary buds grow vigorously to form a large spreading tree as in mango (*Mangifera indica*), neem (*Azadirachta indica*), banyan (*Ficus benghalensis*) etc. Cymose branching may be : ***(a) Uniparous*** (only one branch at each node), ***(b) Biparous*** (two branches at each node) or ***(c) Multiparous*** (more than two branches at each node).

Modified Stems

(I) Underground Modifications

(1) Rhizome–underground, dorsiventral, horizontally growing below the soil surface with distinct nodes and internodes. Nodes bear scale leaves on the upper surface and adventitious roots on the lower surface. They are usually fleshy due to the storage of food, as found in ginger (*Zingiber officinale*) and turmeric (*Curcuma longa*) (fig. 14.16A).

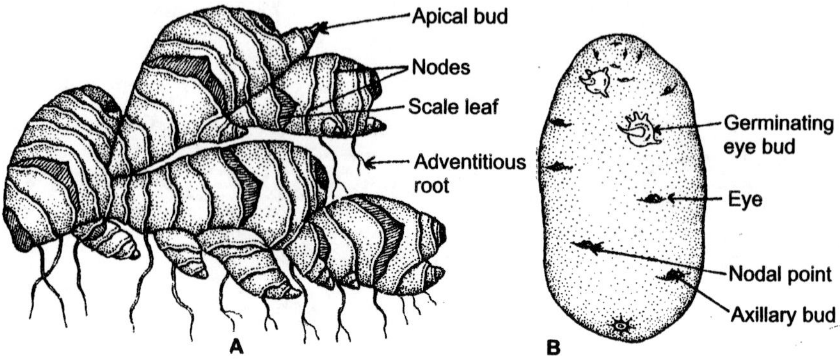

Fig. 14.16 A-rhizome of ginger and B-tuber of potato.

(2) Tubers–fleshy, swollen, underground stem bearing scale leaves with buds (eyes) in their axils. The eyes are spirally arranged, but more crowded towards the distal end (called *rose end*), each eye develops into a small shoot. Stem tubers are commonly found in potato (*Solanum tuberosum*) (fig. 14.16B), Jerusalem artichoke (*Helianthus tuberosus*) and *Cyperus rotundus*.

(3) Bulb–a rounded, modified, underground bud bearing highly reduced (disc-like) stem and fleshy scale leaves which make up the bulk of the organ. According to the arrangement of the scale leaves, the bulbs are called ***(i) tunicated***–scale-leaves arranged in a concentric manner and the whole bulb is enclosed by dry membranous scale leaves, called *tunic*, as seen in onion (*Allium cepa*), tulips (*Tulipa*) and tuberose (*Polianthes*) (fig. 14.17 A, B),

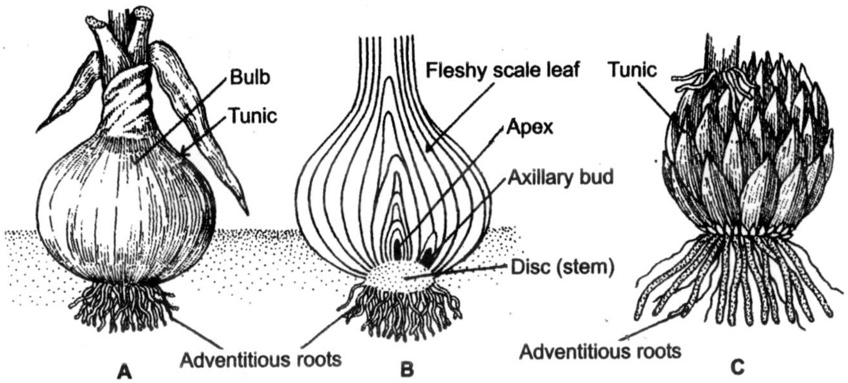

Fig. 14.17 Bulbs. A-tunicated bulb of onion, B-L.S. of the bulb and C-scaly bulb of lily.

and ***(ii) scaly*** or ***imbricated***–scale-leaves arranged loosely with overlapping margins and are not covered by any common tunic, as in garlic (*Allium sativum*) and lily (*Lilium*) (fig. 14.17C).

(4) Corm–a solid, highly condensed, vertical root-stock with a large apical bud. Some scale leaves are present on its body and adventitious roots grow either from the base or all over its body. Corms are found in *Colocasia, Crocus, Colchicum, Gladiolus* and *Amorphophallus*. In *Amorphophallus complanatus*, the corm is single internode with numerous adventitious buds and roots all over its body (fig. 14.18).

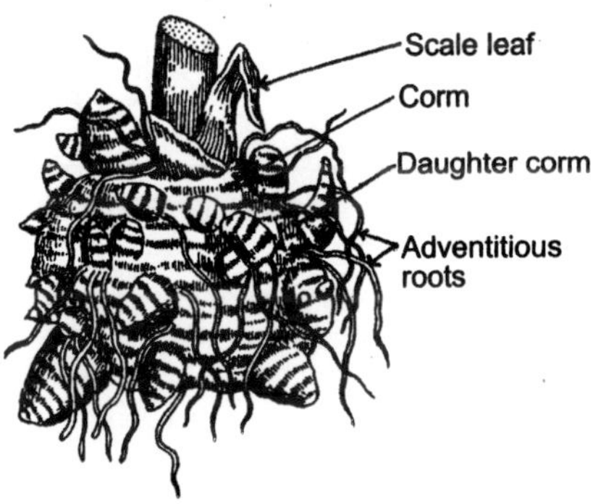

Fig. 14.18 Corm of *Amorphophallus complanatus*.

(II) Sub-aerial Modifications

(1) Runners–develop from the basal internode of the axillary buds of the lowest leaves and which elongate horizontally along the ground carrying the buds, and after trailing for a short distance along the ground, each runner attaches itself to the soil by means of adventitious roots and develops into a new daughter plant, as in *Cynodon, Oxalis* (fig. 14.19A) and *Hydrocotyle.* **Sobole** is a special type of underground runner found in many grasses. The grasses send out soboles in all directions and develop numerous offsprings (fig. 14.19B). It is a more effective method of propagation, as in *Saccharum spontaneum, Agropyron* and *Imperata arundinacea.*

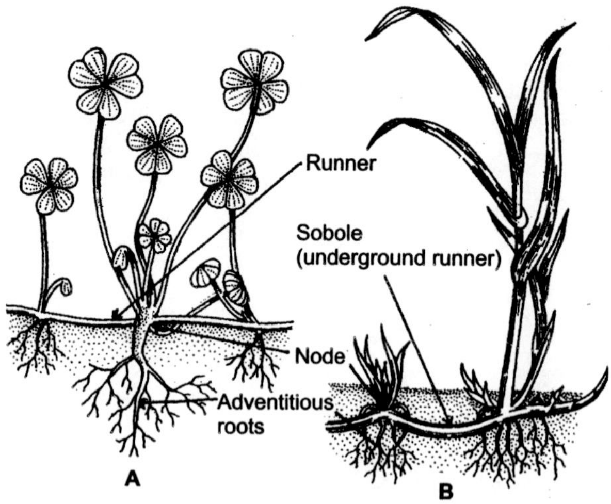

Fig. 14.19 A-runner (*Oxalis*) and B-sobole (*Agropyron*).

(2) Stolon–a special type of runner which grows upward like ordinary branch and then archs down to strike roots in the soil and forms a daughter plant, as in *Mentha* (fig. 14.20A).

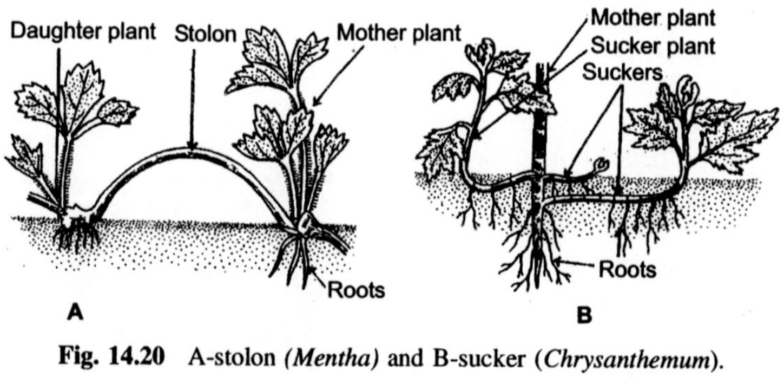

Fig. 14.20 A-stolon *(Mentha)* and B-sucker (*Chrysanthemum*).

(3) Sucker–an underground runner which after travelling a short distance grows up and forms a daughter plant. These may also develop from the axillary buds. Suckers are commonly found in *Mentha arvensis, Musa* and *Chysanthemum* (fig. 14.20B).

(4) Offset–similar to a runner but shorter and thicker and found in aquatic plants. It runs for sometimes under-water

Fig. 14.21 Offset of *Eichhornia.*

and then surfces to bear a bunch of leaves and adventitious roots, as in *Pistia stratiotes* and *Eichhornia crassipes* (fig. 14.21).

(III) Aerial Modifications

(1) Phylloclade or **Cladophyll**–a stem in which several nodes and internodes become flattened or swollen, fleshy and green leaf-like. Such stems carry out the photosynthesis and also store the food. Leaves, present on the nodes, are reduced to scales or modified into spines. Phylloclade is a xerophytic adaptation to reduce transpiration, as found in cocoloba (*Muehlenbeckia platyclada*), cactus (*Opuntia*) and ruscus (*Ruscus*) (fig. 14.22 A, B and C). In *Ruscus*, phylloclades look exactly like leaves, but occurrence of flower at the nodal points, justify their stem nature.

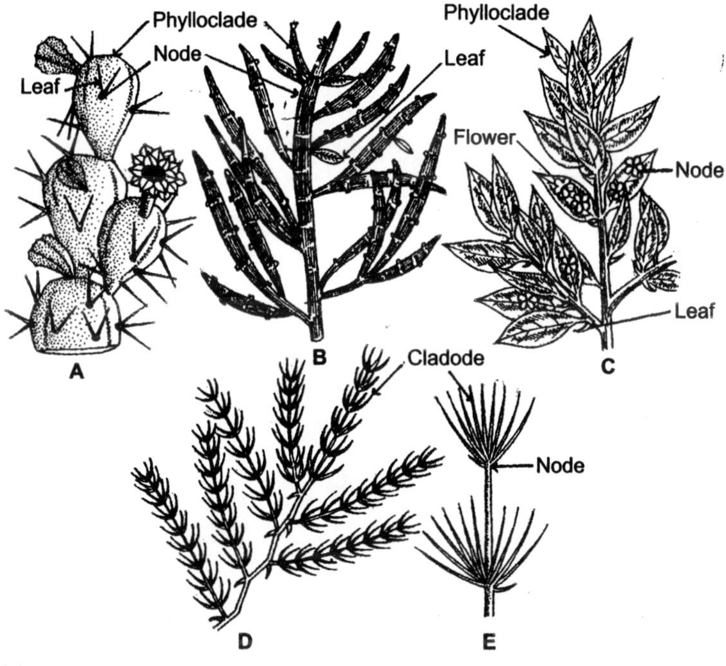

Fig. 14.22 A-phylloclade (*Opuntia*), B-phylloclade of *Muehlenbeckia*, C-phylloclade of *Ruscus* and D and E-cladodes of *Asparagus*.

(2) Cladode–a special type phylloclade consisting of one or two internodes only. In *Asparagus*, the ultimate branchlets are of one internode only and modified into cylindrical leaf-like structures, whereas leaves are reduced to scales or spines (fig. 14.22 D, E). It is a xerophytic adaptation to reduce transpiration.

(3) Thorn–a modified straight, hard and pointed axillary branch. Thorns are deep seated, direct prolongations of the stem and have vascular connections. These are xerophytic adaptations. Thorns of *Duranta* bear leaves (fig. 14.23) and those of *Prunus* flowers.

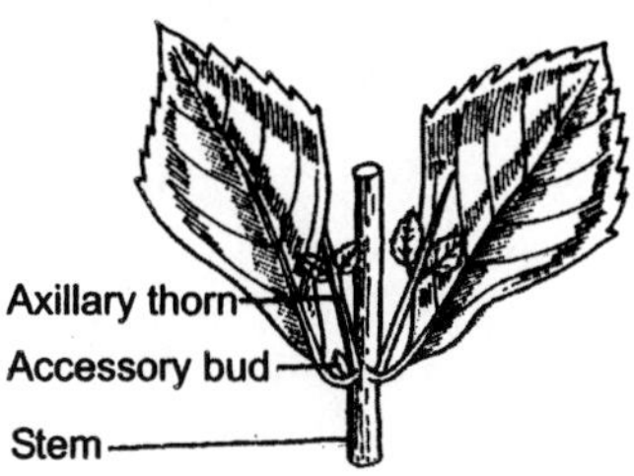

Fig. 14.23 Axillary branch modified into thorn (*Duranta*).

(4) **Stem-tendril**–a modified axillary or terminal bud into tendril for climbing. Tendrils of *Passiflora* (fig. 14.4C) are modified axillary branches whereas those of *Vitis* are modified tips of sympodial branches.

LEAF

Leaf is a lateral appendage of the stem borne at the node and bears an axillary bud in its axil. It is usually expanded and concerned with the manufacture of food (photosynthesis).

Kinds of Leaves

(1) **Cotyledonary leaf**–cotyledon green and appearing like foliage leaf as in *Tilia* and *Geranium.*
(2) **Cataphylls** or **Scale leaves**–dry, papery or fleshy scale leaves or bud scales, as found on various parts of *Casuarina, Asparagus* and bamboos.
(3) **Hypsophylls** or **bract leaves**–special leaves which bear floral buds at their axils.
(4) **Prophylls**–first few leaves of a branch which differ from other leaves as found in wood apple (*Aegle marmelos*) where these are represented by spines.
(5) **Floral leaves** or **sporophylls**–stamens and carpels are called sporophylls in common with lower plants.
(6) **Foliage leaves**–common green leaves (discussed in detail here).

Duration of Leaves

(1) **Caducous** or **fugacious**–leaves fall off as soon as they are formed as in *Opuntia.*
(2) **Deciduous** or **annual**–leaves fall off at the end of a particular season as in *Acer.*
(3) **Persistent**–leaves last for a long time as in most tropical trees, as in *Polyalthia.*

Phyllotaxy

It is the mode of arrangement of leaves on the stem. Leaves borne directly on the main trunk are called **cauline** as in *Dracaena* and other monocots and borne on the branches are called **ramal** as in most dicot trees. Leaves are called **radical** if these arise from reduced stem and appear as if arising from the root as in radish.

In **cauline** or **ramal** leaves, there may be one, two, three or more leaves at each node. When there is only one leaf on each node, the arrangement is called **spiral, alternate** or **acyclic** and if there are, two or more leaves on each node, it is called **cyclic.**

Spiral or **alternate phyllotaxy.** Leaves are produced alternately or spirally on the long axis or stem. If these are found to lie in a fixed number of vertical rows, the arrangement is called **orthostichous** and is variously termed :

(1) **Distichous** or $\frac{1}{2}$ **phyllotaxy**–the leaf of second node is just opposite the leaf of the first node, as in grasses.

(2) **Tristichous** or $\frac{1}{3}$ **phyllotaxy**–leaves are three ranked and the fourth is just above the first leaf as in *Cyperus rotundus.*

(3) **Pentastichous** or $\frac{2}{5}$ **phyllotaxy**–the sixth leaf is found above the first leaf and seventh above the second leaf as in china-rose (*Hibiscus rosa-sinensis*) (fig. 14.24).

(4) **Octastichous** or $\frac{3}{8}$ **phyllotaxy**–the ninth leaf is found above first leaf and so on, as in *Carica papaya.*

Parastichous term is used when it is not possible to establish orthostichy, as on date palm trunk.

Cyclic phyllotaxy. The leaves at each node form a whorl with the leaves placed on a circle in which the angles between adjacent leaves are same. It can broadly be of two types :

(1) **Opposite phyllotaxy**–two leaves on each node and opposite to one another. It can be of two kinds : ***(i) Opposite decussate***–successive pairs of leaves placed at right angles to one another as in *Calotropis* (fig. 14.24) or ***(ii) Opposite superimposed***–successive pairs of leaves exactly on top of one another so that all leaves lie in one plain as in *Psidium guajava* and *Quisqualis* (fig. 14.24).

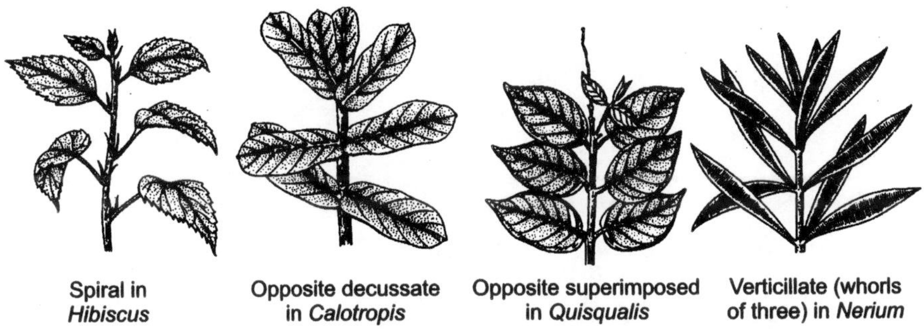

Fig. 14.24 Phyllotaxy-spiral, opposite decussate, opposite superimposed and verticillate.

(2) **Verticillate phyllotaxy**–also called **whorled** phyllotaxy in which three to five leaves form a whorl in each node as three leaves in *Nerium odorum* (fig. 14.24) and five or more in *Alstonia scholaris.*

Heterophylly. Presence of different types of foliage leaves as in *Ranunculus aquatilis* and *Limnophila heterophylla* where submerged leaves (in water) are dissected and aerial leaves are more or less entire.

Anisophylly. In some opposite phyllotaxy, leaves are of unequal size, as in *Pellionia daveanana.*

Leaf Parts (of Foliage Leaves)

A typical leaf is made up of three parts, *(i)* the leaf base or *hypopodium*, *(ii)* the stalk or petiole or *mesopodium* and *(iii)* the leaf blade or lamina or *epipodium* (fig. 14.25).

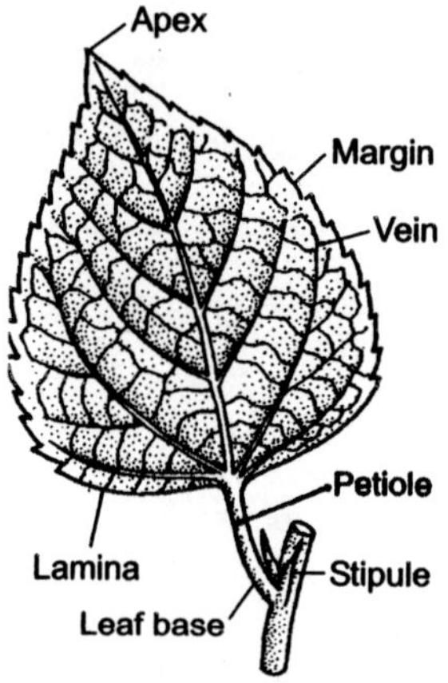

Fig. 14.25 Parts of a typical leaf.

Leaf base. The point of attachment of the leaf to the stem. It may be variously modified, *e.g.* ***(a) pulvinus***–a swollen leaf base, as in mango (*Mangifera*) and *Mimosa pudica* (fig. 14.26A), ***(b) sheathing***–clasping or forming sheath around the stem, as in banana (*Musa paradisiaca*) (fig. 14.26B), ***(c) amplexicaul***–a sheathing leaf base which surrounds stem completely, as in grasses and *Aethusa cynapium*, (fig. 14.26C) and ***(d) decurrent***–a winged leaf base which extends down to the stem, as in *Symphytum* (fig. 14.26D).

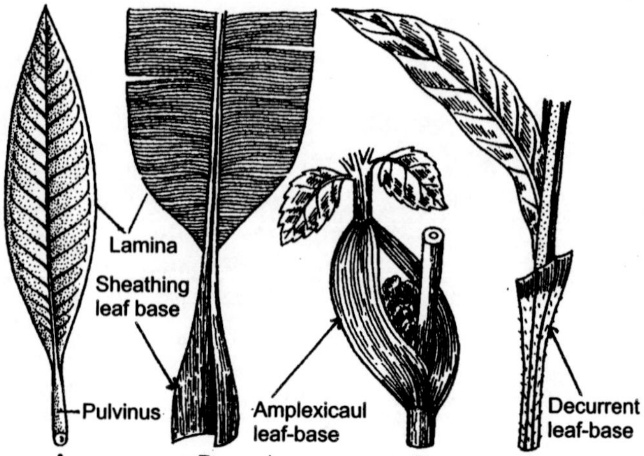

Fig. 14.26 Variations in leaf bases. A-pulvinus (*Mangifera indica*), B-sheathing leaf base (*Musa paradisiaca*), C-amplexicaul (*Aethusa cynapium*) and D-decurrent (*Symphytum officinale*).

Stipules. A pair of lateral outgrowth present at the base of the leaf. Leaves possessing stipules are called **stipulate** and lacking are **exstipulate** (as in mango). Stipules may be of various types, *e.g.* **(i) Adnate**–adhering to the petiole upto a certain height and making it somewhat winged as in rose (*Rosa*) (fig. 14.27A), **(ii) Free lateral**-free on the two sides of the leaf base as in china rose (*Hibiscus rosa-sinensis*), (fig. 14.27B), **(iii) Intrapetiolar**–present between the leaf and the axis as in *Gardenia* (fig. 14.27C), **(iv) Interpetiolar**–two stipules lying between the two leaves, as in *Ixora* (fig. 14.27D), **(v) Ochreate**–stipules which form a hollow tube encircling the stem from the node upto a certain height, as in *Polygonum* (fig. 14.27E), **(vi) Tendrillar**–stipules modified into tendril, as in *Smilax* (fig. 14.27F), **(vii) Foliaceous**–leafy stipules, as in *Pisum* (fig. 14.27G), **(viii) Scaly**–small and dry as in *Artocarpus*, and **(ix) Spinous**–modified into spines, as in *Acacia* and *Zizyphus* (fig. 14.27H).

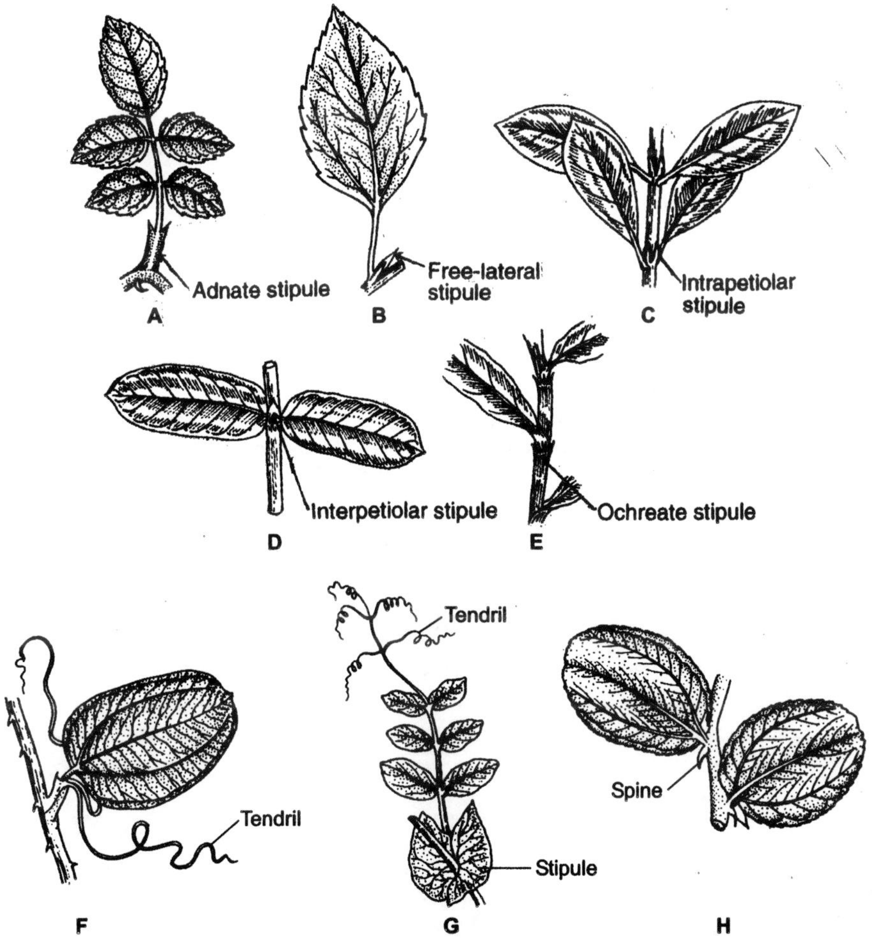

Fig. 14.27 Various types of stipules. A-adnate (*Rosa*), B-free lateral (*Hibiscus rosa-sinensis*), C-intrapetiolar (*Gardenia*), D-interpetiolar (*Ixora*), E-ochreate (*Polygonum*), F-tendrillar (*Smilax*), G-foliaceous (*Pisum*) and H-spinous (*Zizyphus*).

The **duration** of the stipules may be **(i) caducous**–shed very early as in *Michelia champaca*, **(ii) Deciduous**–lasting for one season, as in *Cassia tora* or **(iii) Persistent**–as in foliage leaves, *e.g. Pisum* and *Rosa*.

Stipels. Stipule-like appendages at the base of leaflets, as in *Dolichos lablab*, *Vicia faba* and *Desmodium gyrans* (all leguminous plants).

Petiole. Stalk of the leaf, and also called the **mesopodium.** A leaf having a petiole is called *petiolate* as in mango (*Mangifera indica*) and without petiole is called sessile, as in *Calotropis*. The petiole shows, sometimes modifications, as it is **winged** in *Citrus* (fig. 14.28A), swollen and **spongy** in *Eichhornia* (fig. 14.28B) and *Trapa*, **flattened** or leaf-like called phyllode in australian acacia (*Acacia auriculaeformis*) (fig. 14.28C and D), **tendrillar** in *Smilax*, partly **wing-like** and partly **tendrillar** in *Nepenthes* and **spinous** in *Quisqualis*.

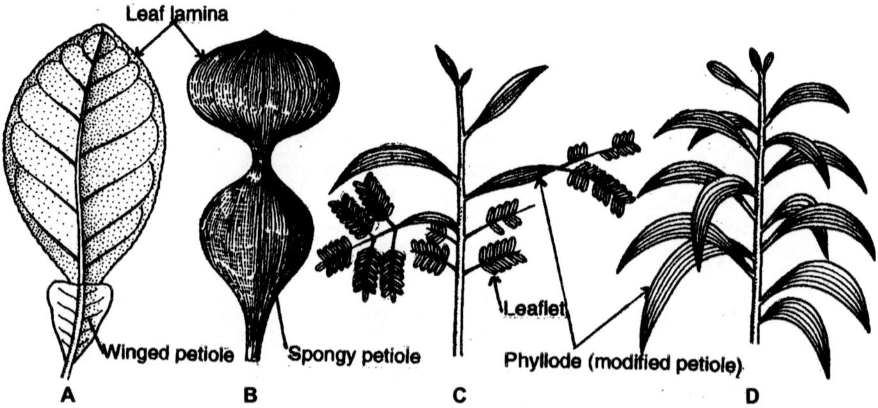

Fig. 14.28 Modifications of petiole. A-winged (*Citrus*), B-spongy (*Eichhornia*) and C-phyllode (*Acacia*). (C-young seedling part and D-mature leaf (phyllode)).

Lamina (Epipodium). The most important part of the leaf and a seat of gaseous exchange for photosynthesis, respiration, transpiration etc. If the upper surface (*adaxial*) differs in structure from the lower (*abaxial*) surface, the leaves are called **dorsiventral** as in most dicots and if both surfaces receive equal light and there is no difference between the two surfaces, they are called **isobilateral**, as in maize and other grasses. Sometimes, leaves are cylindrical or centric (no distinct two surfaces), *e.g. Allium cepa*.

Several features of leaf are considered regarding appearance of lamina, such as shape, base, margin, apex, surface, texture, venation, incision and simple or compound etc.

(I) Shape

Important shapes of lamina are :

(1) **Acicular**–needle-shaped (as in *Pinus*).

(2) **Linear**–longer and slightly broader as in grasses and tube roses (*Polianthes tuberosa*).

(3) **Lanceolate**–lance-shaped as in *Nerium* and *Polyalthia*.

(4) **Oblong**–more or less rectangular as in banana (*Musa*).
(5) **Ovate**–egg-shaped as in china-rose (*Hibiscus rosa-sinensis*) and banyan.
(6) **Cordate**–heart-shaped as in betel vine (*Piper betle*).
(7) **Sagittate**–shape like arrow head with two basal lobes as in *Sagittaria sagittifolia.*
(8) **Hastate**–like sagittate but the two basal lobes are directed outwards as in some *Ipomoea.*
(9) **Reniform**–kidney-shaped as in *Hydrocotyle.*
(10) **Lunate**–half moon-shaped as in *Passiflora lunata.*
(11) **Obovate**–reverse of ovate as in jackfruit (*Artocarpus*).
(12) **Obcordate**–reverse of cordate with an apical notch as in *Bauhinia.*
(13) **Spathulate**–spathula-shaped as in *Lippia nodiflora* and *Euphorbia nerifolia.*
(14) **Cuneate**–wedge-shaped as in *Pistia stratiotes.*
(15) **Elliptical**–like an ellipse as in *Vinca rosea* and guava (*Psidium guajava*).
(16) **Rotund**–orbicular or circular as in water lily and lotus (*Nelumbo*).

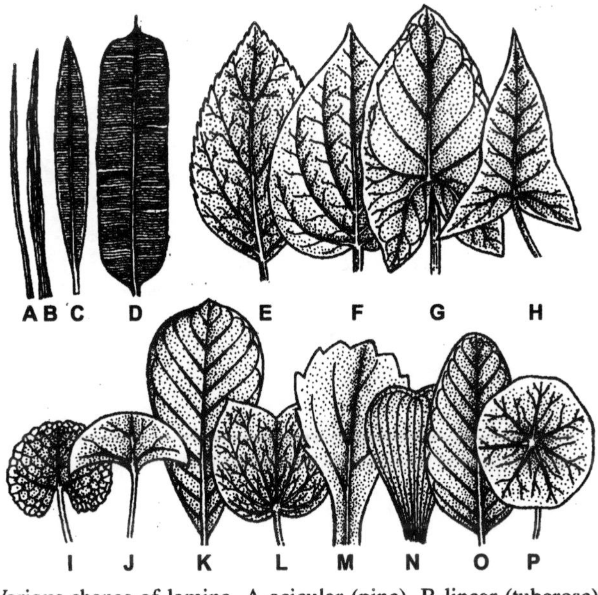

Fig. 14.29 Various shapes of lamina. A-acicular (pine), B-linear (tuberose), C-lanceolate (*Nerium*), D-oblong (banana), E-ovate (china-rose), F-cordate (betel vine), G-sagittate (*Sagittaria*), H-hastate (*Ipomoea*), I-reniform (*Hydrocotyle*), J-lunate (*Passiflora*), K-obovate (jackfruit), L-obcordate (*Bauhinia*), M-spathulate (*Lippia*), N-cuneate (*Pistia*), O-elliptical (guava) and P-rotund (water-lily).

(II) Margin

The outermost boundary of the lamina is known as margin. Margins show variation, *e.g.*

(1) **Entire**–smooth margin as in mango (*Mangifera*).
(2) **Repand**–wavy margin as in *Polyalthia.*

(3) **Serrate**–margin with teeth pointed upwards as in saw, *e.g.* china-rose (*Hibiscus rosa-sinensis*) and rose.
(4) **Biserrate**–teeth again serrated as in elm (*Maple*).
(5) **Retroserrate**–teeth pointed downwards.
(6) **Dentate**–margin toothed, but teeth pointed outwards (at right angles) as in water-lily or *Nymphaea.*
(7) **Bidentate**–teeth again dentate.
(8) **Crenate**–margin with rounded teeth as in *Bryophyllum* and *Hydrocotyle.*
(9) **Bicrenate**–margin with tooth and teeth again crenate.
(10) **Spiny**–teeth pointed to form spines as in *Ananas* (pineapple) and *Argemone* (Mexican poppy).
(11) **Lobed** or **incised**–margin much dissected or incised as in *Raphanus.*

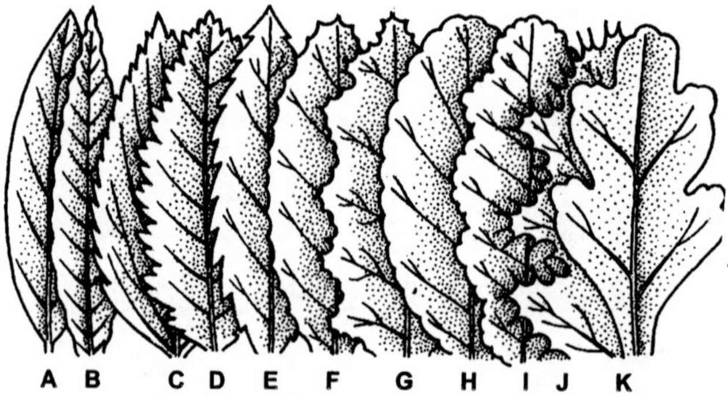

Fig. 14.30 Leaf margin. A-entire, B-repand, C-serrate, D-biserrate, E-retroserrate, F-dentate, G-bidentate, H-crenate, I- biserrate, J-spiny and K-lobed.

(III) Apex

The anterior tip of the lamina is called apex. It may be of various shapes, *e.g.*

(1) **Acute**–pointed and narrow as in mango.
(2) **Acuminate**–apex drawn out into long tapering tail as in peepal (*Ficus religiosa*).
(3) **Obtuse**–apex blunt or broad angled as in banyan (*Ficus benghalensis*).
(4) **Mucronate**–apex broad but with a sharp point as in periwinkle (*Vinca rosea*).
(5) **Cuspidate**–spiny as in pineapple (*Ananas*), date palm (*Phoenix*), screw-pine (*Pandanus*) etc.
(6) **Tendrillar**–apex forms tendril for climbing as in *Gloriosa.*
(7) **Cirrhose**–mucronate apex with a fine thread-like structure as in banana (*Musa*).
(8) **Truncate**–apex abruptly cut across as in *Paris polyphylla.*
(9) **Retuse**–slightly notched obtuse apex as in *Pistia* and *Clitoria.*
(10) **Emarginate**–deeply notched obtuse apex as in *Bauhinia.*

Fig. 14.31 Leaf apices. A-acute, B-acuminate, C-obtuse, D-mucronate, E-cuspidate, F-tendrillar, G-cirrhose, H-truncate, I-retuse and J-emarginate.

(IV) Surface

The external surface of lamina may show variation, *e.g.*

(1) **Glabrous**–smooth and without any hair, as in mango.

(2) **Glaucus**–shiny and covered with waxy coating as in *Calotropis* and lotus.

(3) **Scabrous**–rough because of the presence of short rigid points, *e.g.* fig leaves.

(4) **Viscose**–sticky because of the some secretion as in leaves of *Cleome* and *Nicotiana tabacum*, sometime called **glutinous**.

(5) **Rugose**–wrinkled surface as in *Rubus rugosus*.

(6) **Gland dotted**–glands found on the leaf surface as in lemon (*Citrus*).

(7) **Hairy**–surface covered with hairs. Hairs may be of various types and accordingly called :

(a) **Pubescent**–hairs soft and woolly as in tomato.

(b) **Pilose**–hairs long, distinct and scattered as in *Grewia pilosa*.

(c) **Villose**–hairs long, soft, closely set as in *Leucas aspara*.

(d) **Tomentose**–hairs short, dense, cottony as in *Terminalia tomentosa* and *Calotropis procera*.

(e) **Hirsute**–hairs stiff, fine, scattered as in *Eclipta alba*.

(f) **Hispid**–hairs long and rigid as in *Cucurbita*.

(8) **Spinose**–surface covered with small prickles as in brinjal (*Solanum melongena*).

(V) Venation

The pattern of arrangement of vascular system in the lamina is called venation. It is mainly of two types :

(i) Reticulate venation. The main vein branches repeatedly to form a network or reticulum covering the entire leaf. It is a characteristic of dicotyledons except in *Callophyllum*. It shows two variations :

(1) ***Unicostate*** or ***Pinnate***–almost racemose type. One prominent mid-rib runs from the base to the apex with several secondary and tertiary and so on anastomosing veinlets forming a net-like

framework, as in peepal (*Ficus religiosa*), banyan (*F. benghalensis*), *Eugenia jambolina* and most dicots (fig. 14.32 A and B).

(2) ***Multicostate*** or ***Palmate***–from the base of the lamina arises a number of equally strong veins (or costas) which spread out like fan, hence called palmate. Palmate veins may be :

(a) **Convergent**–veins converging towards the tip as in *Zizyphus jujuba* and *Cinnamomum tamala* (fig. 14.32C).

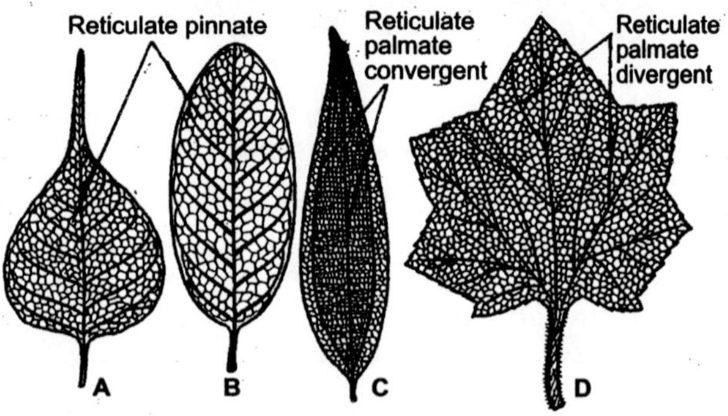

Fig. 14.32 Reticulate venation. A-unicostate (peepal), B-unicostate showing submarginal veins (*Eugenia jambolana*), C-palmate convergent (*Cinnamomum*), and D-palmate divergent (*Cucurbita*).

(b) **Divergent**–veins spread out from one point to different directions as in *Cucurbita maxima* (fig. 14.32D).

(ii) Parallel or **Striate venation.** Main veins run parallel to one another with laterally connected unbranched veinlets.

Parallel venation may be :

(a) ***Unicostate*** or ***Pinnate***–presence of single strong mid-rib which gives rise to parallel branches, joined by parallel transverse veinlets, as in banana (*Musa*) (fig. 14.33A) and *Canna*.

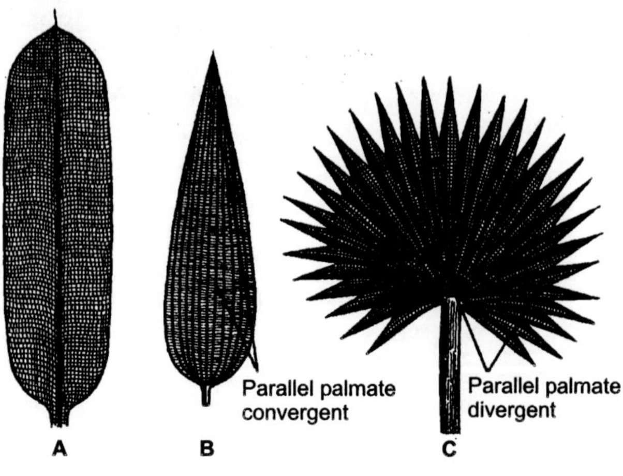

Fig. 14.33 Parallel venation. A-unicostate (banana), B-palmate convergent (bamboos) and C-palmate divergent (fan palm).

(b) ***Multicostate*** or ***Palmate***–presence of a number of strong veins arising from the base, which may be :

Convergent–as in bamboos (fig. 14.33B) and grasses or

Divergent–as in fan palm (*Borassus flabellifer*) (fig. 14.33C)

(VI) Leaf Incision (Simple and Compound Leaves)

Indentation of the lamina margin is called leaf incision. Accordingly two main categories are recognised, *i.e.*, simple leaf and compound leaf.

Simple Leaf

A leaf having entire margin (as in mango) (fig. 14.34A) or, if dissected, presenting a single appearance due to presence of some apparent laminar connection as in radish and mustard (*Raphanus sativus* and *Brassica campestris*). Leaf-shape of radish is called **lyrate.**

Fig. 14.34 Types of leaf. A-simple (mango), B-pinnate-paripinnate compound (*Tamarindus*), C-pinnate-imparipinnate compound (rose), D-bipinnate compound (*Mimosa*), E-tripinnate compound (*Moringa*) and F-decompound (*Foeniculum*).

Compound Leaf

The leaf has incised margin, and the incision goes so deep, touching the mid-rib, so that lamina breaks up into lobes or segments called leaflets.

A compound leaf may be (A) Pinnate or (B) Palmate.

(A) Pinnate Compound

The midrib (or rachis) directly bears the leaflets or pinnae along both the sides. It may be simple pinnate, bipinnate, tripinnate and decompound.

(i) **Simple pinnate** or **unipinnate**–leaflets borne directly on rachis, and may be.

(a) ***Paripinnate***–leaflets present in pairs as in *Tamarindus indica* (fig. 14.34B), *Cassia, Swietinia* etc.

(b) ***Imparipinnate***–rachis is terminated by an odd leaflet as in rose (*Rosa indica*) (fig. 14.34C), *Clitoria, Murraya, Melia azadirachta* etc.

(ii) **Bipinnate**–a compound leaf with main segments again pinnately divided into *pinnules, i.e.*, leaflets (pinnules) borne on the rachis but on its branches as in *Acacia arabica, Mimosa pudica* (fig. 14.34 D), *Caesalpinia pulcherrima* etc.

(iii) **Tripinnate**–incision goes to the third order so that leaflets are found on secondary branches of the rachis, as in *Moringa pterygosperma* (fig. 14.34E).

(iv) **Decompound**–incision high in number to give dissected appearance of the leaf as in fennel (*Foeniculum vulgare*) (fig. 14.34F) and other plants of Umbelliferae.

(B) Palmate Compound

In such leaves, rachis does not differentiate and the lamina appears articulated or attached to a point on the top of the petiole. Such compound leaves may be of following types :

(1) **Unifoliate**–single leaflet attached to the top of the petiole, as in *Citrus* (fig. 14.35A).

(2) **Bifoliate**–only two leaflets articulated to the rachis, as in *Hardwickia binata* and *Bignonia grandiflora* (fig. 14.35B).

Fig. 14.35 Palmate compound leaves. A-unifoliate (*Citrus*), B-bifoliate (*Bignonia*), C-trifoliate (*Aegle*), D-quadrifoliate (*Marsilea*) and E-multifoliate (*Bombax*).

(3) **Trifoliate**–three leaflets articulated at the top of the rachis, as in *Trifolium, Aegle marmelos, Oxalis corniculata, Dolichos lablab* (fig. 14.35C).

(4) **Quadrifoliate**–four leaflets articulated to the top of the rachis, as in *Paris quadrifolia* and in a pteridophyte, *Marsilea quadrifoliata* (fig. 14.35D). It is a rare type.

(5) **Multifoliate**–more than four leaflets attached at the top of rachis (petiole), as in *Cleome viscosa, Gynandropsis pentaphylla, Bombax* (fig. 14.35E) or *Salmalia malabarica*.

Modifications of the Leaf Lamina

The lamina or sometimes the entire leaf undergoes modification to suit various purposes. Some important modifications are :

(1) **Leaf tendrils**–tendrils formed by climbing plants which may be modification of entire leaf or its parts, *e.g.* petiole tendril (as in *Clematis*), entire lamina (in *Lathyrus*), terminal leaflets (*Bignonia vensuta* and *Pisum*) and leaf apex (*Gloriosa superba*) (fig. 14.4E).

(2) **Hooks**–leaf modified into hooks for climbing as in *Bignonia unguis-cati*, three terminal leaflets become claw-like hooks.

(3) **Spines**–entire or part of leaf modified into spines (pointed hard structure), as in *Opuntia*, whole leaf is modified into a spine; in pineapple (*Ananas*), *Agave* and *Argemone mexicana*, leaf margin and in *Phoenix*, the apex is modified into a hard spine. In *Citrus*, prophyll is a spine.

(4) **Pitcher**–a pitcher-like structure formed either by entire leaf (as in *Dischidia rafflesiana*) or by part of it such as by lamina (as in *Nepenthes*, the pitcher plant). In *Nepenthes*, pitchers are modified to catch insects to meet the need of nitrogenous substances (fig. 14.36B) and in *Dischidia* (an epiphytic climber), the pitcher accumulates the rain water as well as debris which is absorbed by the adventitious roots formed within.

(5) **Bladder**–in *Utricularia* (aquatic insectivorous plant), the leaf segments modify into bladder to catch the insects (fig. 14.36D).

(6) **Fleshy leaves**–in some xerophytes and halophytes leaves become fleshy or succulent to store water due to presence of special storage tissue. Such plants are commonly called succulents, *e.g. Portulaca oleracea, Basella rubra, Swaeda maritima* and species of *Aloe, Agave, Bryophyllum, Sedum* etc.

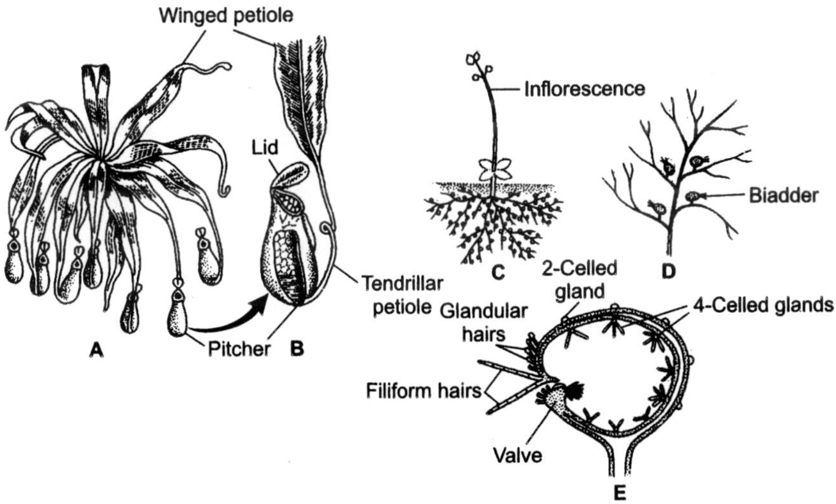

Fig. 14.36 Leaves in insectivorous plants. A-a plant of *Nepenthes*, B-a single leaf with pitcher, C-a plant of *Utricularia*, D-a part of leaf bearing bladders and E-c.s. of the bladder.

(7) **Absorbing organs**–submerged leaves of aquatic plants absorb water by their general surface. Such plants are often rootless, *e.g. Ceratophyllum* and *Utricularia.*

THE REPRODUCTIVE ORGANS AND FRUITS
The Essential, Valuable and Regular Features for Classification

---

BRACT AND BRACTEOLES

Axillary floral buds are usually borne in the axils of the specialised leaves called **bracts.** Such flowers are called **bracteate**, as in most members of Acanthaceae. In some cases, bracts are absent where flowers are called **ebracteate**, as in Brassicaceae and Boraginaceae. Sometimes additional bract-like small and thin structures are borne on the pedicel (or peduncle) between the flower and bracts. These are called **bracteoles** and such flowers are called **bracteolate.**

Bracts and bracteoles may be variously modified, such as :

(1) **Foliaceous** or **Leafy bract**–bract appears like ordinary green leaf, as in china rose (*Hibiscus rosa-sinensis*).

(2) **Petalloid bract**–bract becomes brightly coloured and appears like a petal as in *Bougainvillea spectabilis* (fig. 14.37A) in which three small flowers are enclosed within three coloured bracts. Similarly, bright red or orange bracts are found in *Poinsettia pulcherrima.*

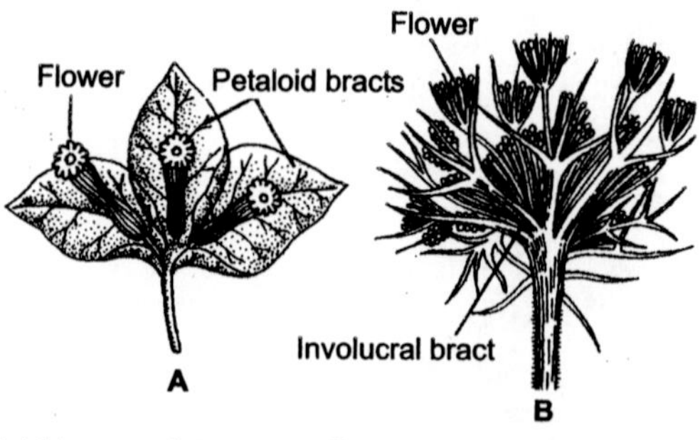

Fig. 14.37 A-petaloid bracts of *Bougainvillea* and B-involucral bracts of *Daucus carota.*

(3) **Spathy bract**–large, thick, boat-shaped bract and called *spathe* covering full or a part of a *Spadix* inflorescence, as in *Colocasia antiquorum* (fig. 14.40D) and *Amorphophallus titanum.* In some palms, the spathe may be several feet (upto 20) long. Spathes in maize and banana are well developed and prominent.

(4) **Involucral bract**–whorls of green sepaloid bracts occurring around the base of the inflorescence. These form an involucre, as in family Compositae. Involucre of bracts is also found in Umbelliferae (as in *Daucus carota,* fig. 14.37B). In addition to bracts, there may be smaller involucres round the base of each branch of the

inflorescence as in *Daucus* and these are called **involucels** or **partial involucres.**

(5) **Scaly bract**–additional scaly bracts found with each individual floret as in *capitulum* (fig. 14.38), *catkin* or *cyathium* inflorescences.

(6) **Cupule**–hard and woody bracts and bracteoles found at the base of flower, somewhat like an involucre, as in Cupulifereae. These fuse together to form a cup-like structure (cupule) as the fruit matures.

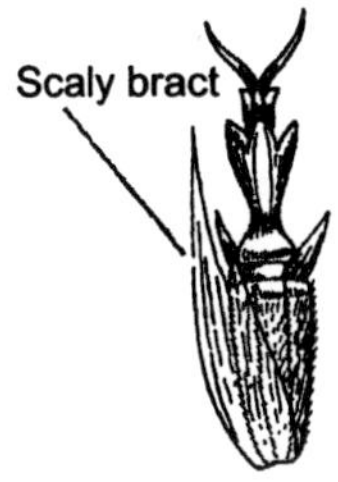

Fig. 14.38 Scaly bract of a floret of sunflower (*Helianthus*).

(7) **Epicalyx**–a whorl of sepaloid green bracteoles found below the calyx, as in china rose.

(8) **Glumes**–two scaly bracts present at the base of each spikelet as in Poaceae, also called *empty glumes* (fig. 14.41C).

(9) **Lemma and Palea**–in the spikelet of Poaceae, each flower bears a bract called *flowering glume* or *lemma* and a bracteole called *palea* which fits within the lemma (fig. 14.41C).

INFLORESCENCE

The arrangement of flowers and mode of distribution of flowers on the shoot system of a plant is called inflorescence. Sometimes, flowers are found singly and such a condition is called solitary. Solitary flowers may be terminal as in *Papaver* or axillary as in *Hibiscus rosa-sinensis* (fig. 14.39).

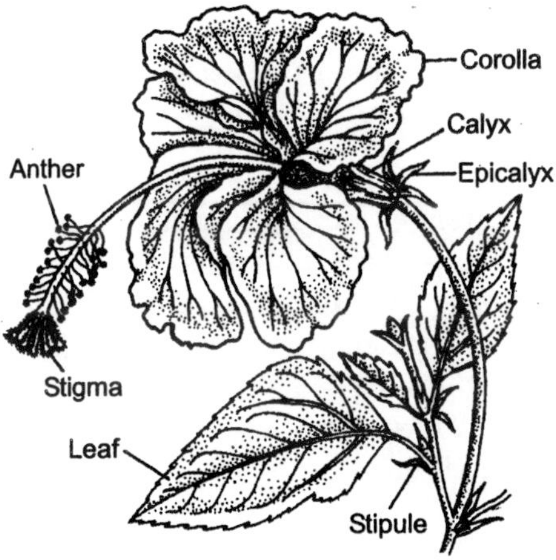

Fig. 14.39 Solitary axillary flower of *Hibiscus.*

Kinds of Inflorescence

Different types of inflorescences are grouped into three categories, *e.g.*

(i) Simple inflorescence (ii) Compound inflorescence

(iii) Special inflorescence

(I) SIMPLE INFLORESCENCE

In such inflorescence, flowers are arranged directly on the main axis. These are of two types :

I. Racemose or indefinite inflorescence

II. Cymose or definite inflorescence

I. Racemose inflorescence. It has indeterminate growth of main axis, all flowers are lateral and arranged in acropetal succession, *i.e.*, oldest towards base and youngest towards apex. These are of following types :

1. **Raceme.** Peduncle is elongated and flowers are pedicillate (stalked), *e.g.* mustard (fig. 14.40A) and larkspur (*Delphinium*).
2. **Spike.** It is similar to raceme but flowers are sessile, *e.g. Achyranthes* (fig. 14.40C) and *Amaranthus.*

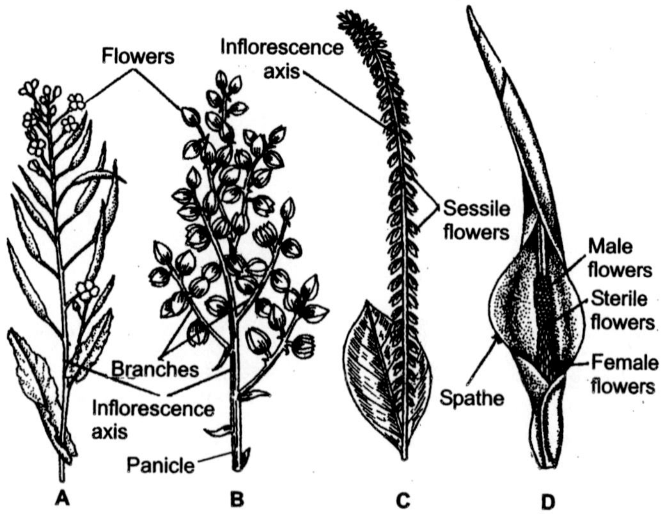

Fig. 14.40 Different types of racemose inflorescences. A-raceme of mustard, B-panicle of *Yucca,* C-spike of *Achyranthes* and D-spadix of *Colocasia.*

3. **Spikelet.** It is a secondary small spike with one or a few flowers. Spikelets are arranged in spike, raceme or panicle (fig. 14.40B). Each spikelet bears at its base two minute bracts called **empty glumes.** Above this is a **fertile glume,** a third bract or bracteole. The fertile bracteole may be called *palea* or *lemma.* Opposite to the lemma, it bears a small glume called **upper palea** (or superior palea). Each flower of the spikelet remains enclosed by a lemma and palea. Such inflorescence is characteristic to family Gramineae (Wheat and rice) (fig. 14.41B, C).
4. **Catkin (Ament).** It has a delicate and pendulous floral axis which bears unisexual flowers. All flowers are sessile and densely set. Entire inflorescence appears like one unit, *e.g.* mulberry (*Morus*) (fig. 14.41A) and *Acalypha.*
5. **Spadix.** It has a fleshy floral axis which remains covered with large and coloured bract, called **spathe.** Flowers are sessile and unisexual.

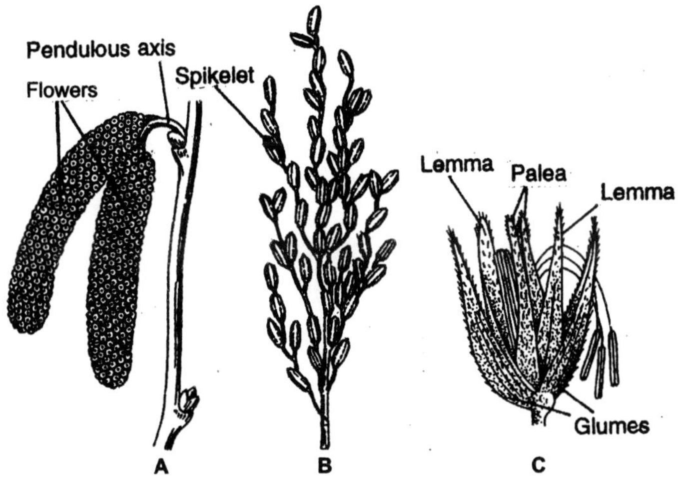

Fig. 14.41 Inflorescence. A-catkin (mulberry), B-spikelet (rice) and C-spikelet with two florets.

Male flowers are arranged on the upper part of the axis while female flowers on the lower part of the axis. Space on axis left between two types of flowers remain covered with sterile hairs. Spadix is found in *Arum, Colocasia* (fig. 14.40D) and *Musa.*

6. **Corymb.** It has a comparatively short main axis and the lower flowers have much longer pedicels than the upper ones so that all flowers are brought more or less to the same level, *e.g.* candytuft (*Iberis*) and cherry (fig. 14.42A).
7. **Umbel.** It has a very much shortened main axis and all flowers appear to be arising from the same point. Flowers are bracteate. Younger flowers are arranged towards centre while older flowers towards periphery. The whorl of bracts is called involucre. Such inflorescence is characteristic to family Umbelliferae (*Coriandrum, Foeniculum* (or fennel), *Daucus* etc. (fig. 14.42B, C).

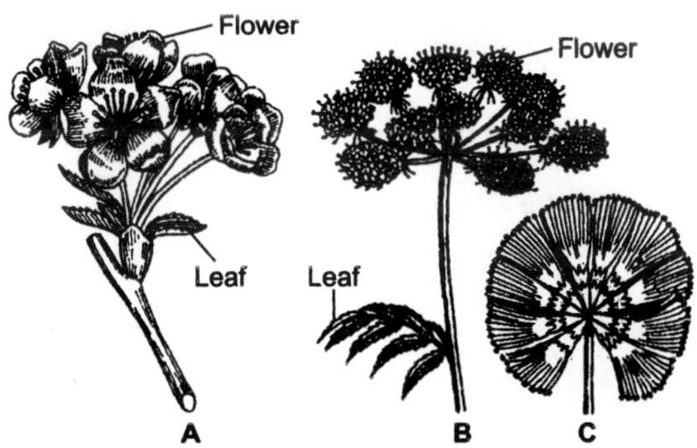

Fig. 14.42 A-corymb of cherry, B-compound umbel, and C-a single umbel (enlarged).

8. **Capitulum** or **Head.** It has a flattened receptacle on which several sessile small florets are arranged in a centripetal manner, *i.e.,* youngest in the centre and oldest towards the periphery. Individual

florets are bracteate. The whorl of bracts present is called involucre. On the receptacle, two kinds of florets are usually found :

(i) **Ray floret.** These are found at the rim of receptacle (periphery) and have ligulate or strap-shaped showy corolla. Flowers are either sterile (neuter) or female (pistillate) and are zygomorphic.

(ii) **Disc florets.** These are grouped in the centre and are actinomorphic, tubular and bisexual.

Capitulum of sunflower has both types of flowers (fig. 14.43) but that of *Ageratum* has only disc florets. Capitulum inflorescence is characteristic to family Compositae.

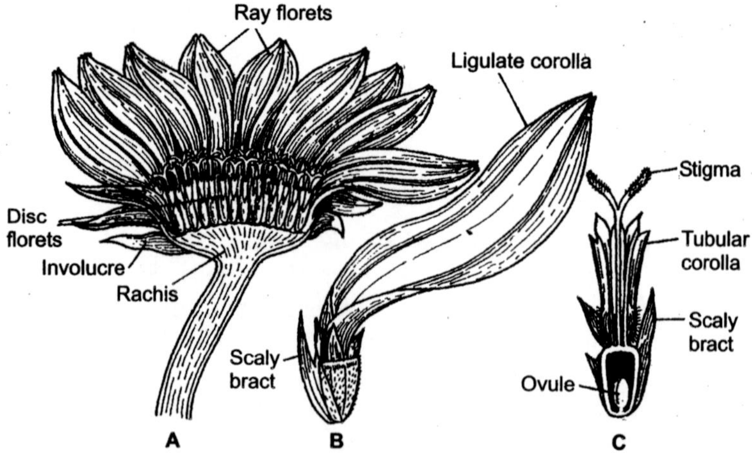

Fig. 14.43 Capitulum inflorescence. A-L.S. of inflorescence of sunflower, B-a ray floret and C-a disc floret.

II. Cymose inflorescence. Such inflorescence has determinate (definite) growth of the main axis, as the apical bud forms the first flower. Other flowers are produced on lateral branches which also bear terminal flower. Thus, every time new lateral branch is formed to produce new flower. Flowers are arranged centrifugally.

Cymose inflorescences are of following types :

1. **Monochasial** or **Uniparous cyme.** A single lateral branch arises from the peduncle of old flower which terminates in a flower. The lateral branch also terminates in a flower. Monochasial cyme can be of two types :

   (i) **Helicoid cyme (Bostryx).** In this case, all lateral branches arise only on one side of main axis, *e.g. Heliotropium* and *Drosera* (fig. 14.44B).

   (ii) **Scorpoid cyme.** In this case, lateral branches arise alternately on left and right sides, as in *Ranunculus* and *Gossypium* (fig. 14.44A).

2. **Dichasial** or **Biparous cyme.** The peduncle bears a terminal flower and stops growing. From the main peduncle arise two lateral branches, each of which also terminates in a flower. From the pedicel of each lateral flower further branching may arise repeatedly, as in *Dianthus* (fig. 14.44C).

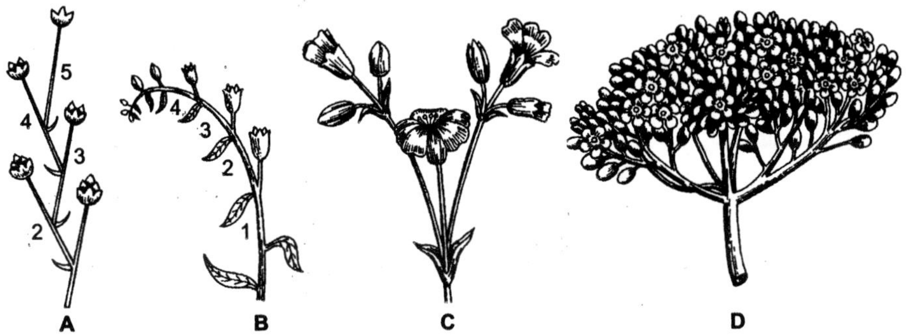

Fig. 14.44 A, B-monochasial cyme (A-scorpoid cyme and B-helicoid cyme), C-dichasial cyme and D-polychasial cyme.

3. **Polychasial, Polychasium** or **Multiparous cyme.** The peduncle bears a single terminal flower and below it more than two branches arise. These also bear terminal flowers. In *Calotropis,* the inflorescence is a polychasial umbellate cyme, *i.e.,* groups of multiparous cymes in which main peduncle becomes much short (fig. 14.44D).

(II) COMPOUND INFLORESCENCE

In a compound inflorescence, the peduncle (main axis) branches repeatedly once or twice in racemose or cymose manner and ultimate branches bear flowers in a racemose or cymose manner. These can be of following types :

1. **Compound raceme** or **panicle**, *e.g.* gold mohur (*Delonix regia*) and *Yucca.*
2. **Compound spike** or **spike of spikelets,** *e.g.* wheat
3. **Compoud corymb,** *e.g.* cauliflower
4. **Compound umbel,** *e.g. Coriandrum* and *Foeniculum* (fig. 14.42B)

(III) SPECIAL INFLORESCENCES

These are of following types :

1. **Cyathium.** The bracts or involucre becomes fused to form a cup-shaped structure. Few nectaries are found in the involucral cup. This cup completely encloses a centrally placed female flower surrounded by a large number of male flowers. Each male flower is bracteate and achlamydeous and is represented by a single stalked stamen. The female flower is bracteate and achlamydeous and is represented by stalked (gynophore), tricarpellary and syncarpous gynoecium. This type of inflorescence is characteristic to genus *Euphorbia.* The inflorescence is a special cymose type in which male flowers are arranged basipetally (fig. 14.45).
2. **Verticillaster.** In this type, leaves are borne on opposite side on a node, each node bears flowers arranged in dichasial cyme. After the formation of three flowers, each dichasial cyme passes on to

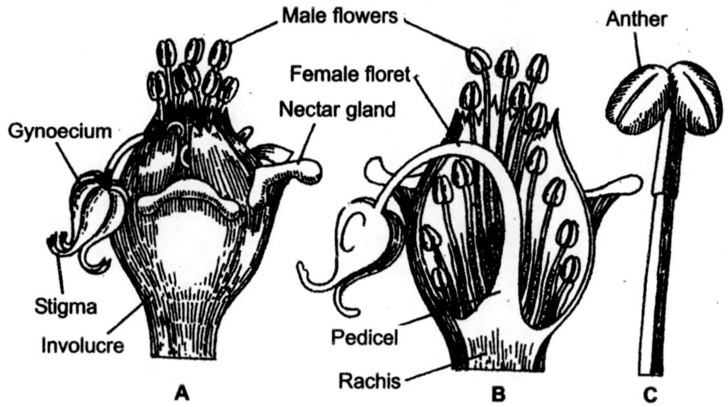

Fig. 14.45 Cyathium inflorescence in *Poinsettia* or *Euphorbia pulcherrima*. A-external view. B-L.S. of inflorescence and C-a male (staminate) flower.

monochasial cyme of scorpoid type. Flowers are condensed, sessile, bisexual and bilipped. The inflorescence is characteristic to family Labiatae (*e.g. Ocimum*) (fig. 14.46).

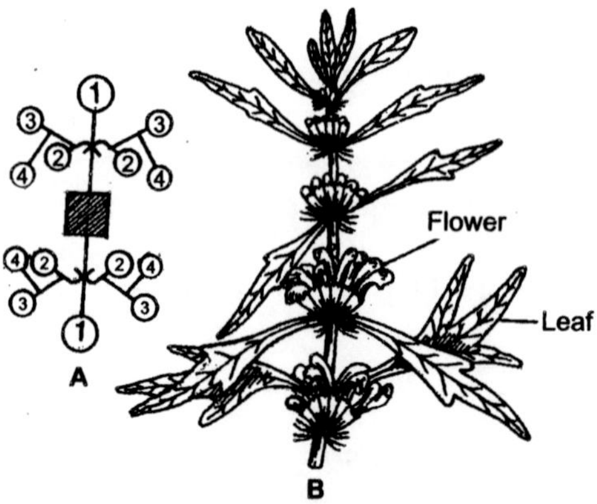

Fig. 14.46 Verticillaster inflorescence. A-flowering sequence and B-a twig with inflorescence.

3. **Hypanthodium.** The fleshy receptacle forms a hollow pear-shaped cavity with a narrow small opening. The flowers are borne on the

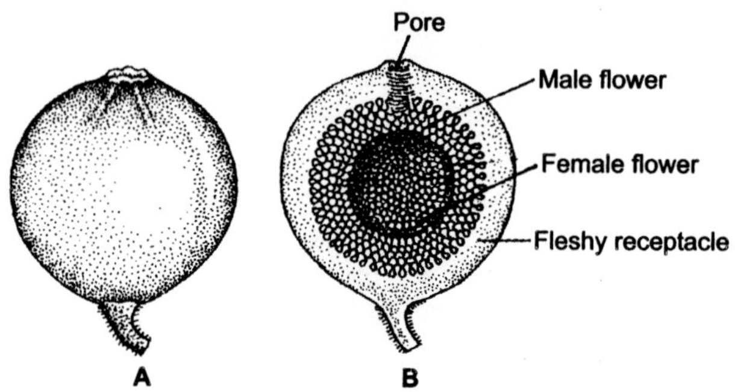

Fig. 14.47 Hypanthodium inflorescence. A-external view and B-L.S. of inflorescence.

inner wall of the cavity and are unisexual. Male flowers are borne on upper side, *i.e.*, towards mouth and female flowers are borne at the base of cavity. Such inflorescence is characteristic to genus *Ficus* (fig. 14.47).

Some Special Terms

Peduncle. A stalk which bears inflorescence.

Scape. A leafless inflorescence stalk arising from the middle of a rosette of leaves. It bears a flower, several flowers, or a crowded infloresence (as in onion).

Rachila. The axis in the centre of a grass spikelet.

Strobile. A spike in which each flower is borne in the axil of a persistent membranous bract.

Umbellule. A unit of a compound umbel.

Cymule. A diminutive of cyme inflorescence.

Cincinus (or **Cincinnus**). A cymose inflorescence with the lateral axes arising on alternate sides of the relatively main axis.

Drepanium. A monochasial cyme in which all branches arise from the same side of the relatively main axis.

Rhipidium. A monchasial cyme in which the branches lie in one plane, so that it becomes fan-shaped.

Glomerule. A cluster of short-stalked cyme. It is distinguished from a head inflorescence by blooming of the central flower first.

Tassel. Male inflorescence in maize.

Thyrse (or **Thyrsus**). A densely branched inflorescence with the main branching racemose (indeterminate), but the lateral branches cymose (determinate), as in grapes.

Secund. Having lateral members all turned to one side, *i.e.*, appearing as if arising from one side.

FLOWER

Flower is considered as a modified shoot in which every appendage represents a single leaf. Sepals and petals are sterile leaves while stamens and carpels are fertile leaves. The different floral appendages are arranged in whorls. Different parts of flower are shown in fig. 14.48.

1. Calyx. It is a collective name for sepals which are present in the outermost whorl. Sometimes a whorl of epicalyx may be present outer to it as in china rose.

2. Corolla. It is a collective name of petals found next to calyx. Petals are variously coloured.

3. Androecium. It is third inner whorl of a flower that consists of male reproductive structures called stamens.

4. Gynoecium. It is a collective name of female reproductive structures called carpels which are centrally placed in a flower.

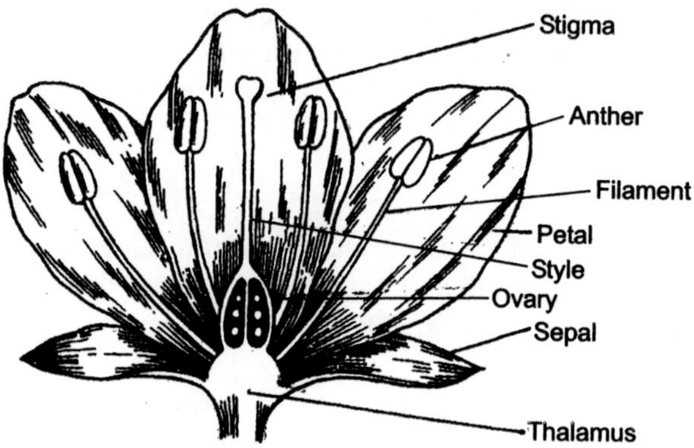

Fig. 14.48 A-complete flower, (L.S.).

These structures give a fundamental organisation to a flower, however, there may be several variations. These variations are described under the heading Terminology.

Terminology

Bracteate. A flower is said to be bracteate when bract is present.

Ebracteate. Bracts are not found.

Pedicillate. Flower is produced on a stalk or pedicel.

Sessile. Stalk is not found.

Complete. A flower when it contains all the four whorls (calyx, corolla, androecium and gynoecium) is called complete.

Incomplete. Absence of either one of the whorls.

Hermaphrodite. Bisexual, bearing both androecium and gynoecium.

Unisexual. Presence of only one type of sex organs.

Staminate. Presence of only male reproductive structures (stamens).

Pistillate. Presence of gynoecium (stamens are absent).

Neuter. Sex organs are altogether absent (no androecium and gynoecium).

Actinomorphic. A flower with radial symmetry and the parts with similar size and shape (fig. 14.49 A, B).

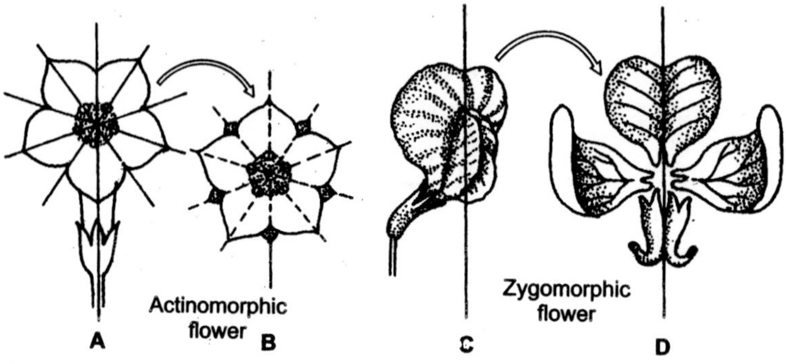

Fig. 14.49 A&B-actinomorphic flower (B-surface view) and C&D-zygomorphic flower (D-petals separated).

Zygomorphic. A flower capable of equal division by only one plane of symmetry (fig. 14.49 C,D).

Asymmetric. A flower which cannot be divided equally from any plane.

Dioecious. When male and female flowers are produced on separate plants.

Monoecious. When male and female flowers are present separately on the same plant.

Polygamous. A plant in which both bisexual and unisexual flowers are present.

Trimerous. The floral whorls contain units in multiple of three (as in Liliaceae).

Tetramerous. The floral whorls contain units four in number or in multiple of four (as in Cruciferae).

Pentamerous. The floral whorls contain units five in number or in multiple of five (as in Malvaceae).

Hypogynous. A flower with superior ovary (as in Cruciferae).

Perigynous. A flower with half embedded or sunken ovary (as in Rosaceae) (fig. 14.50B).

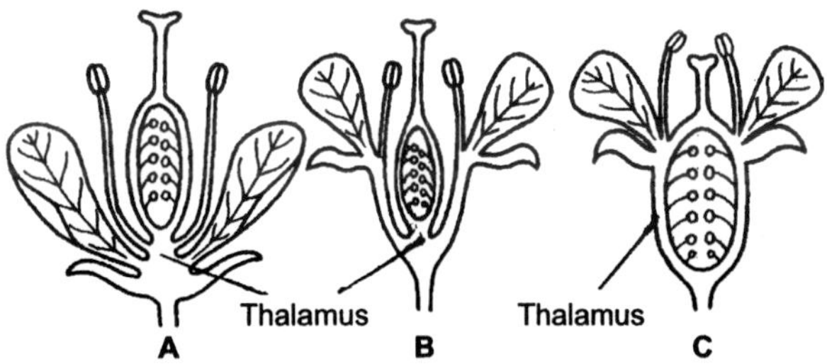

Fig. 14.50 A-hypogynous flower, B-perigynous flower and C-epigynous flower.

Epigynous. A flower with inferior (completely sunken) ovary (as in Compositae) (fig. 14.50C).

Terms Used in Description of Calyx

Polysepalous. Sepals are free, *i.e.*, not united with each other (as in *Brassica*).

Gamosepalous. Sepals are fused with each other.

Asepalous. Sepals are absent.

Caducous. Sepals fall off after fertilization (as in *Brassica*).

Persistent. Sepals do not fall off rather these remain with the developing fruits (as in brinjal and tomato).

Petaloid. Sepals are coloured.

Tubular. Sepals fuse to form a tubular structure (fig. 14.51B).

Cupulate. Sepals fuse to form a cup-like structure (fig. 14.51E).

Urceolate. Sepals fuse to form urn-shaped structure (fig. 14.51C)

Campanulate. Sepals fuse and form a bell-shaped structure (fig. 14.51D).

Bilabiate. Sepals fuse and form a two lipped structure (fig. 14.51G).

Infundibuliform. Sepals form a funnel-shaped structure (fig. 14.51F).

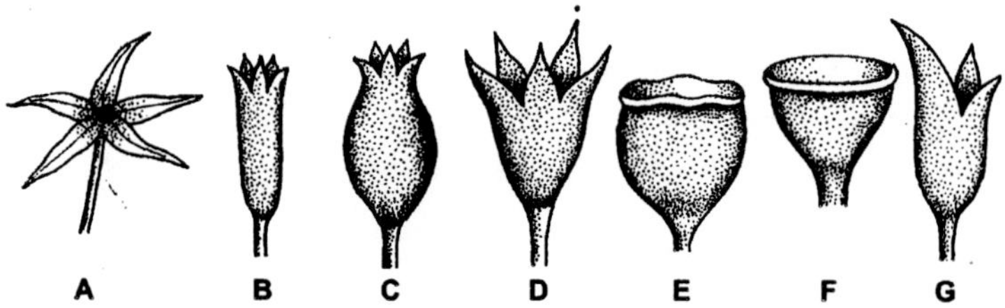

Fig. 14.51 Various forms of calyx. A-free; B-tubular, C-urceolate, D-campanulate, E-cupulate, F-infundibuliform and G-bilabiate.

Spurred. One or more sepals form a long, pointed, spine-like structure, which is called spur (fig. 14.52A).

Pappus. Sepals are modified into hair-like structures (fig. 14.52B).

Spinous. Sepals are modified into spines as in water chestnut (fig. 14.52C).

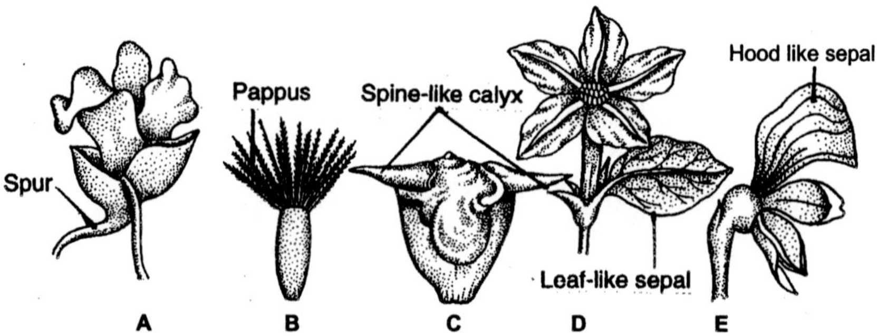

Fig. 14.52 Different forms of calyx. A-spurred, B-pappus, C-spinous, D-leaf-like and E-hooded.

Leaf-like. Sepals form a leaf-like structure (fig. 14.52D).

Hooded. Sepals enlarge to form a hood-like structure (fig. 14.52E).

Aestivation of calyx. Arrangement of sepals in the bud. It may be of following types :

(i) **Valvate.** Sepals are closely arranged in a circle without overlapping to each other (as in *Mimosa*) (fig. 14.53).

(ii) **Twisted.** With parts rolled up in such a way that the outer part of each covers the inner part of the one in front of it, while in turn its inner part is covered by the one behind it (fig. 14.53).

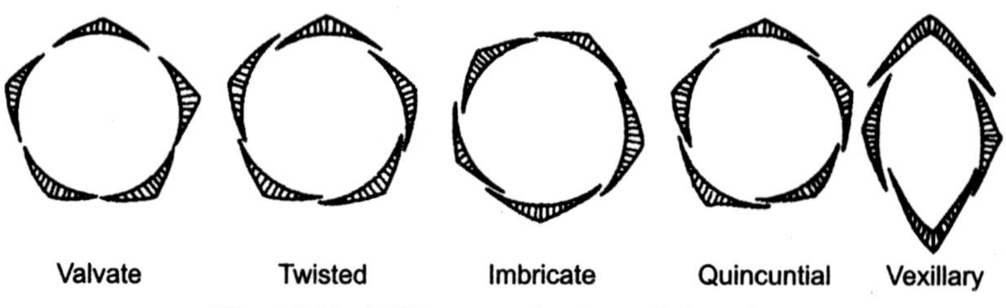

Fig. 14.53 Different aestivations of the calyx.

(iii) **Imbricate.** Both ends of one sepal are outer and of another sepal are inner while other three sepals have one end outer and another one inner (fig. 14.53).

(iv) **Quincuncial.** Two sepals are outer and two sepals are inner while one sepal has one overlapped and one exposed margin (fig. 14.53).

(v) **Vexillary.** See description under corolla (fig. 14.53).

Terms Used in Description of Corolla

Polypetalous. Petals are free (not united) as in Cruciferae.

Gamopetalous. Petals are fused as in Solanaceae.

Apetalous. Petals are absent.

Sepaloid. Petals are green in colour or similar to sepals.

Caducous. Petals fall off after fertilization.

Persistent. Petals do not fall off, rather remain attached with fruits.

Deciduous. Petals fall off after drying.

Cruciform. Petals arranged in a cross-like fashion as in *Brassica* (fig. 14.54A).

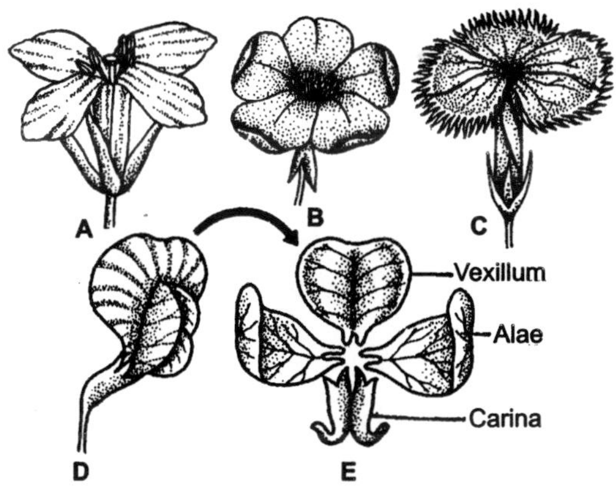

Fig. 14.54 Various forms of polypetalous corolla. A-cruciform, B-rosaceous, C-caryophyllaceous and D-papilionaceous (E-exposed petals of D).

Rosaceous. Petals arranged as found in rose (fig. 14.54B).

Caryophyllaceous. Petals with comparatively long claws and with limbs placed at right angles to the claws (fig. 14.54C).

Papilionaceous. Petals appearing butterfly like; a zygomorphic corolla with one large posterior *standard*, two lateral *wings* and two inner most fused petals, the *keel* (fig. 14.54D), as in pea.

Tubular. Petals fused forming tube-like structure (fig. 14.55A).

Infundibuliform. Petals fused forming funnel-shaped structure (fig. 14.55B).

Campanulate. Petals fused forming bell-shaped structure (fig. 14.55C).

Hypocrateriform. Petals fused with basal portion narrow and tube-like and upper portion bowl-shaped (fig. 14.55D).

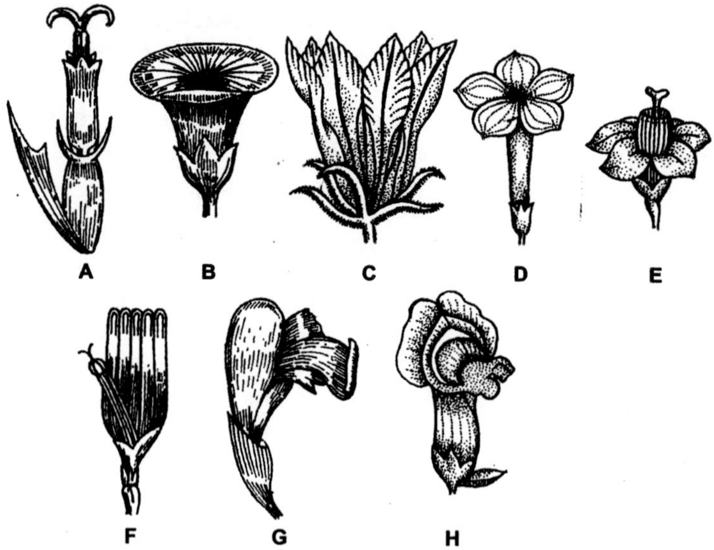

Fig. 14.55 Various forms of gamopetalous corolla. A-tubular, B-infundibuliform, C-campanulate, D-hypocrateriform, E-rotate, F-ligulate, G-bilabiate ringent and H-bilabiate personate.

Trumpet-shaped. Petals fused with basal portion tube-like and upper funnel-like appearance.

Urceolate. Petals fuse to form a pitcher like structure (swollen in middle and narrow at the apex).

Rotate. Petals fuse to form a saucer-shaped structure at their uper part (fig. 14.55E).

Bilabiate ringent. Corolla irregular with petals used to form a bilipped structure but the slit in between two remains open or broader (fig. 14.55G).

Bilabiate personate. Corolla irregular with petals fused to form a bilipped structure having compactly closed slit (fig. 14.55H).

Ligulate. Petals fused to form a tube-like structure at the base and broad flap-like structure in the upper part (fig. 14.55F), as in ray florets of Compositae.

Aestivation. Arrangement of petals in the bud which may be valvate (fig. 14.56A), twisted (fig. 14.56B), imbricate (fig. 14.56C), quincuncial (fig. 14.56D) or vexillary (fig. 14.56E). The details are described under the aestivation of calyx.

Perianth. Petals and sepals are of similar appearance and the two can not be distinguished. The unit of perianth is called **tepal.** If tepals are fused, the condition is called **gamotepalous** and if free called **polytepalous.** If tepals are green in colour, these are called **sepaloid** and if coloured like petals called **petaloid.**

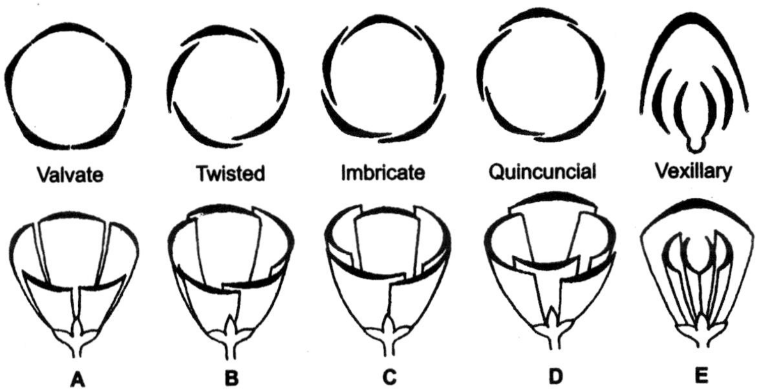

Fig. 14.56 Aestivation of corolla. A-valvate, B-twisted, C- imbricate, D-quincuncial and E-vexillary in two different views.

Terms Used in Description of Androecium

Androecium is a collective name given to male reproductive organs, the unit of which is called **stamen.** A stamen consists of **filament, connective** and **anther** (fig. 14.57). The two lobes of anther remain connected by a connective.

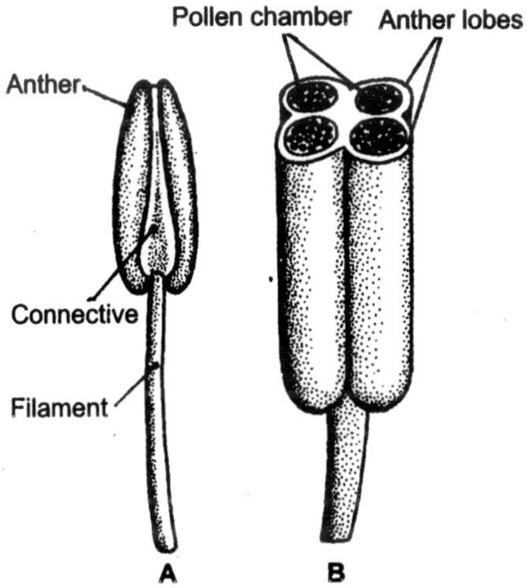

Fig. 14.57 A-different parts of a stamen and B-anther showing internal structure in T.S.

Other terms used to describe androecium are as follows :

Number of Stamens

Monandrous. Single stamen in each flower as in *Euphorbia.*
Diandrous. Two stamens in each flower as in *Coronopus.*
Triandrous. Three stamens in a flower as in *Triticum.*
Tetrandrous. Four stamens in a flower as in *Scoparia.*
Pentandrous. Five stamens in a flower as in *Solanum.*
Hexandrous. Six stamens in a flower as in *Gynandropsis.*
Polyandrous. Indefinite number of stamens as in *Nymphaea.*

Cohesion of Stamens

Monadelphous. Filaments are fused but anther lobes are free as in Malvaceae (fig. 14.58A).

Diadelphous. Stamens with fused filaments arranged in two groups (9 + 1) (fig. 14.58B), as in *Pisum* (Fabaceae).

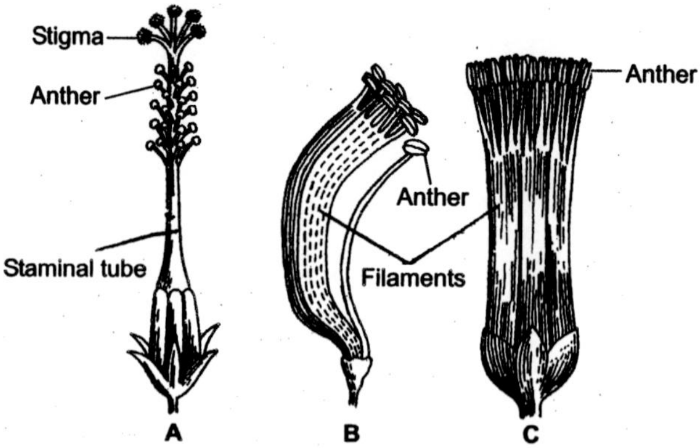

Fig. 14.58 Stamen types. A-monadelphous, B-diadelphous and C-polyadelphous.

Polyadelphous. Stamens with fused filaments arranged in several groups (fig. 14.58C), as in *Salmalia.*

Syngenesious. Anthers are fused with each other but filaments remain free as in Compositae (fig. 14.59A).

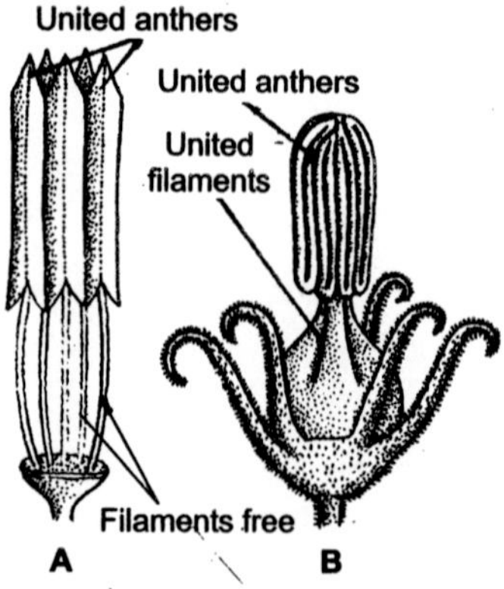

Fig. 14.59 Stamen types : A-syngenesious and B-synandrous

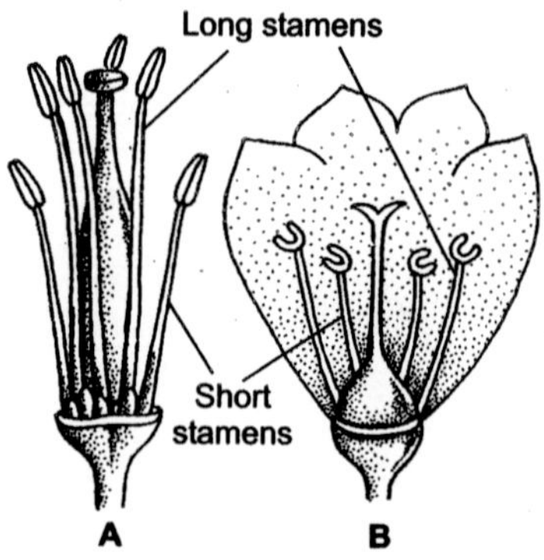

Fig. 14.60 A-tetradynamous and B-didynamous.

Synandrous. Both anthers and filaments remain fused as in Cucurbitaceae (fig. 14.59B).

Adhesion of Stamens

Epipetalous. Stamens are found attached to petals, as in *Solanum.*

Epiphyllous. Stamens are found attached to tepals, as in *Asparagus.* It may also be called epitepalous.

Gynandrous. Stamens are fused with the gynoecium, as in *Calotropis* (14.61B).

Length of Filaments

Didynamous. Four stamens, of which two have long filaments and two short filaments (fig. 14.60B), as in Labiatae and Scrophulariaceae.

Tetradynamous. Six stamens with outer two short and inner four long as in *Brassica* (fig. 14.60A).

Inserted. Length of stamens shorter than petals so these remain within corolla tube.

Exerted. Length of stamens longer than petals so these come out of corolla tube.

Branched stamens. Filaments of the stamens found branched as in castor (*Ricinus communis*) (fig. 14.61A).

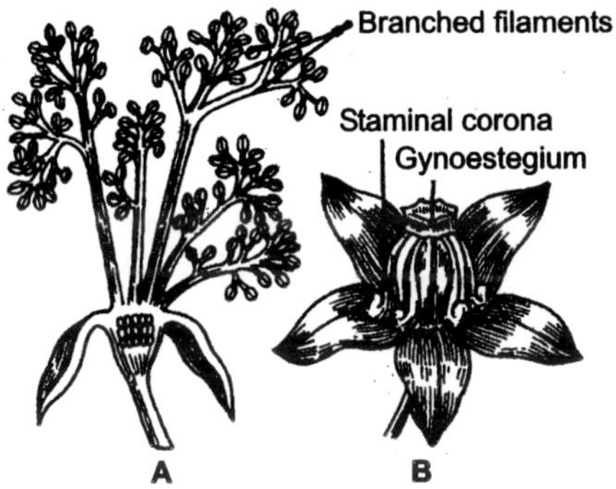

Fig. 14.61 A-branched stamens and B-gynostegium.

Gynostegium. Anthers crowning the column are adnate by the connective to the pentagonal stigma and form gynostegium (fig. 14.61B) as in *Calotropis*.

Insertion on Thalamus

Diplostemonous. Stamens arranged in two whorls on the thalamus and stamens of outer whorl alternate with petals and of inner whorl opposite the petals, as in Rhamnaceae and Portulacaceae.

Obdiplostemonous. Stamens are arranged in two whorls and stamens of outer whorl opposite and inner whorl alternate with the petals, as in Rutaceae.

Attachment of Filament with Anther Lobe

Basifixed. Filament attached with the base of anther lobe (fig. 14.62A), as in *Brassica*.

Dorsifixed. Filament attached to the dorsal surface of anther (fig. 14.62B), as in *Passiflora*.

Adnate. Filament running the whole length of the anther from the base to the apex (fig. 14.62C), as in *Magnolia*.

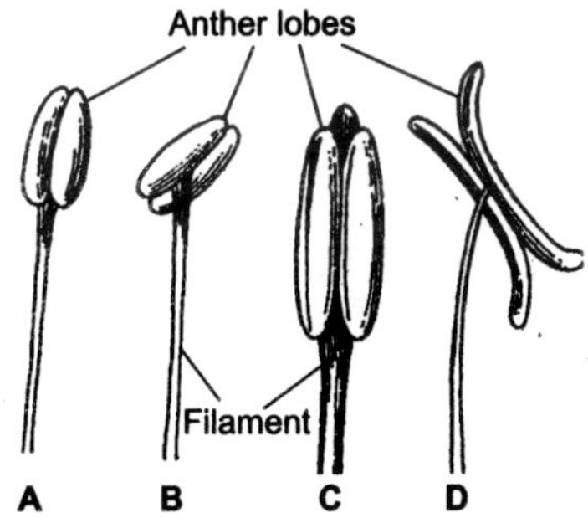

Fig. 14.62 Fixation of anthers with filament, A-basifixed, B-dorsifixed, C-adnate and D-versatile.

Versatile. Filament attached to the back of anther at a point only, so that anther can swing freely, as in grasses (fig. 14.62D).

Dehiscence

Extrorse. An anther dehiscing towards the periphery of the flower.

Introrse. An anther dehiscing towards the centre of the flower.

Longitudinal. Anther lobe bursts along longitudinal suture (fig. 14.63A). It is common type, as in *Datura.*

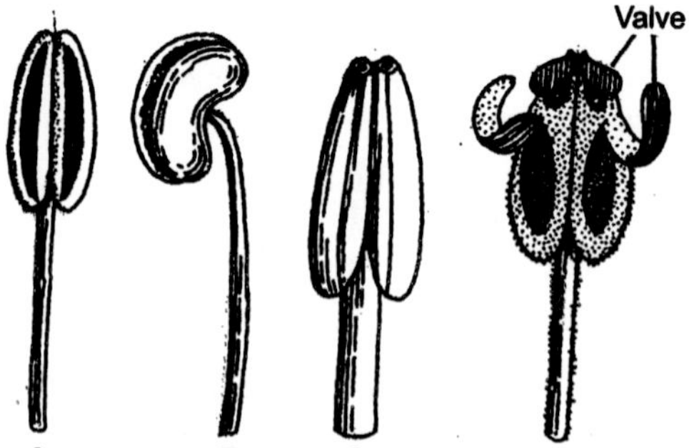

Fig. 14.63 Dehiscence of anthers. A-longitudinal, B-transverse, C-porous and D-valvular.

Transverse. Found in unilocular anthers, it appears to be transverse because suture is placed that way, as in Malvaceae (fig. 14.63B).

Porous or **Apical.** Discharge of pollens through apical pore, as in potato (fig. 14.63C).

Valvular. Wall of the anther opens like trap door, as in *Barberis* (fig. 14.63D).

Terms Used in Description of Gynoecium

Gynoecium is the collective name of female reproductive organs and the unit of which is carpel. A carpel consists of ovary, style and stigma (fig. 14.64). Ovules are contained in the ovary.

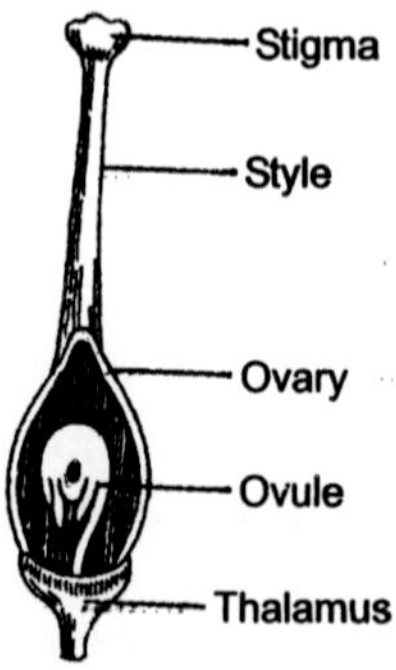

Fig. 14.64 A carpel showing different parts.

Fusion of Carpels

Apocarpous. Carpels are free (not united), as in *Nymphaea* and *Ranunculus* (fig. 14.65A).

Syncarpous. Carpels are fused, as in *Brassica* (fig. 14.65B).

Number of Carpels

Monocarpellary. Presence of a single carpel in a flower, as in Leguminosae.

Bicarpellary. Gynoecium formed by fusion of two carpels, as in Cruciferae.

Tricarpellary. Gynoecium that consists of three carpels, as in Cucurbitaceae.

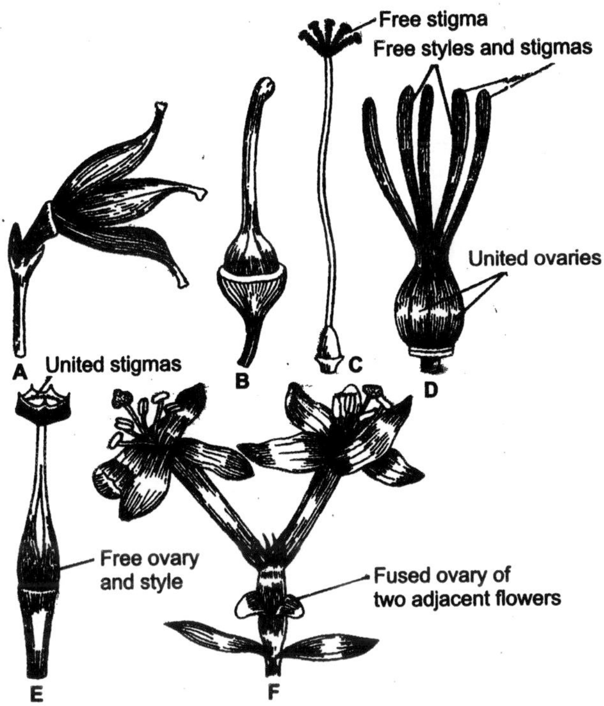

Fig. 14.65 Different types of pistils A-apocarpous, B-syncarpous, C-pistil with free stigmas (Malvaceae), D-free style and stigma (*Linum*), E-ovary and style free (*Calotropis*) and F-fused ovary of two adjacent flowers, *i.e.*, syngynia (*Lonicera*).

Pentacarpellary. Gynoecium that consists of five carpels. It may also be called multicarpellary as in Malvaceae.

Number of Locules

Unilocular. Ovary with single chamber (locule), as in Leguminosae and Compositae (fig. 14.66 A,B).

Bilocular. Ovary with two chambers, as in Cruciferae and Solanaceae (fig. 14.66C).

Trilocular. Ovary with three chambers, as in Liliaceae (fig. 14.66D).

Tetralocular. Ovary with four chambers, as in Labiatae or *Ocimum* (fig. 14.66E).

Penta or **polylocular.** Ovary with 5 or more chambers as in Malvaceae (fig. 14.66F).

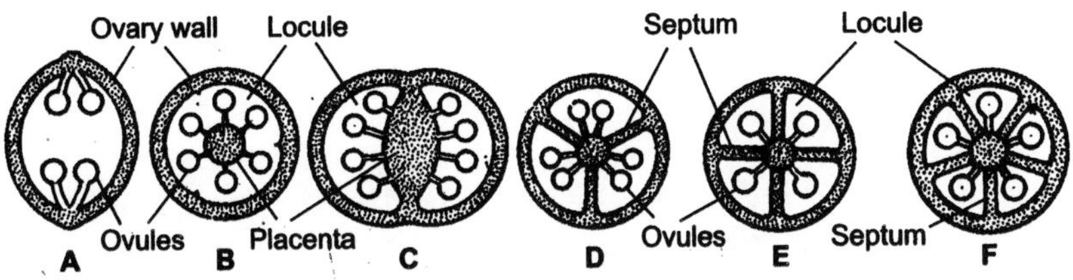

Fig. 14.66 Locules of ovary, A-B-unilocular, C-bilocular, D-trilocular, E-tetralocular and F-pentalocular.

Placentation *(arrangement of ovules within the ovary)* :

Marginal. Placenta present along the junction of two margins of the carpel in one chambered ovary (fig. 14.67 A,B), as in Leguminosae.

Parietal. Placentae bearing ovules developed on the inner wall of the ovary and their position corresponds to the confluent margins of carpel in which ovary is unilocular (fig. 14.67 C,D), as in Cruciferae and Cucurbitaceae.

Axile. Ovules present at or near the centre of ovary on the axis formed by the fusion of septa and usually in vertical rows (ovules are attached on the axis of a compound ovary) (fig. 14.67 E,F), as in Malvaceae, Solanaceae and Liliaceae.

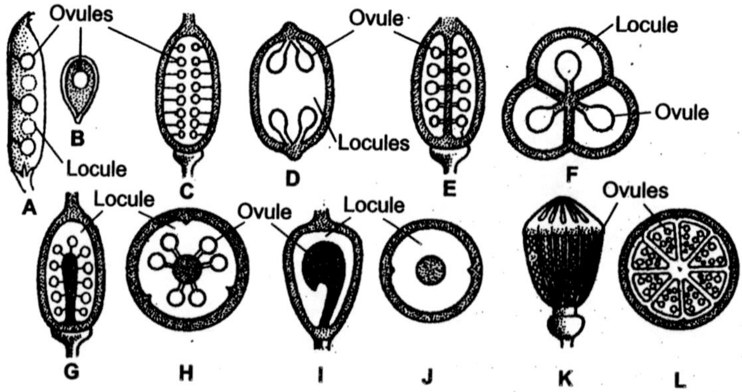

Fig. 14.67 Different types of ovaries with placentation. A, B-monocarpellary with marginal placentation, C, D-bicarpellary with parietal placentation, E, F-tricarpellary with axile placentation, G, H-tricarpellary with free central placentation, I, J-bicarpellary with basal placentation and K, L-multicarpellary with superficial placentation.

Free central. The ovules are borne on a central column with no septa present, the ovary remains unilocular (fig. 14.67 G,H), as in Caryophyllaceae.

Superficial. The placenta develop all round the inner surface of a multilocular and multicarpellary gynoecium, as in *Nymphaea* (fig. 14.67 K,L).

Basal. A single ovule borne at the base of a unilocular ovary, as in Compositae (fig. 14.67 I,J).

Position on Thalamus

Superior. Ovary placed above the point of attachment of corolla and stamens.

Inferior. Ovary borne below attachment of all other floral parts and adnate to them.

Semi-inferior. A condition intermediate to superior and inferior.

Style

Part between ovary and stigma is called style. Commonly, it arises from the top (apex) of ovary. However, style may arise laterally as in mango (fig. 14.68A), or from the central base of the ovary, *i.e., gynobasic* (as in Labiatae) (fig. 14.68 B). Sometimes, style may be branched (fig. 14.68C), as in some plants of Euphorbiaceae.

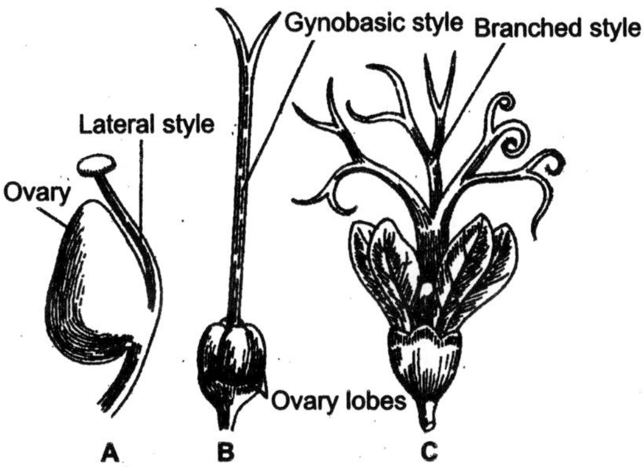

Fig. 14.68 A-lateral style, B-gynobasic style and C-branched style.

Stigma

The apex of carpel is called stigma on which pollen grains are collected. It may be bilobed, trilobed, bifid, capitate, discoid, dumbell-shaped, feathery, linear, highly branched or striate as illustrated in fig. 14.69.

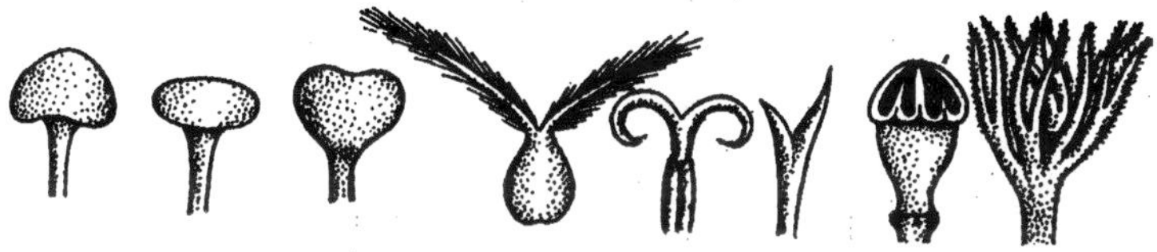

Fig. 14.69 Stigma types. A-capitate, B-discoid, C-lobed, D-bifid, E-bifid curved, F-bifid straight, G-sessile and straight and H-highly branched.

OVULE

Ovules are found in the basal part of the carpel and are :

(i) integumented megasporangia,

(ii) the nucellus containing the embryo sac and enclosed by one or two integuments, which after fertilisation, and subsequent development, becomes a seed, or

(iii) young seed in the course of development.

Ovule arises from the placenta and the arrangement of ovules in an ovary is called **placentation.**

The various parts of the ovule are—**Funicle** or Funiculus (short stalk through which it remains attached), **Raphe** (a ridge-like structure formed by fusion of funicle and body of the ovule), **Integument** (the membrane enclosing the nucellus), **Nucellus** (the central tissue of the ovule, containing embryo-sac), **Micropyle** (a canal-like pore or structure formed by an upgrowth of the integuments), **Chalaza** (the base of the nucellus of the ovule), **Embryo-sac** (the female gametophyte of the angiosperm comprising antipodal cells, polar nuclei/secondary nucleus, egg and synergid cells) etc.

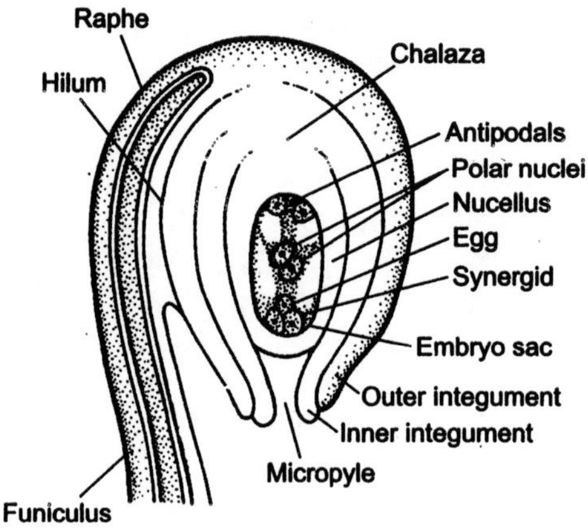

Fig. 14.70 L.S. of an ovule.

Ategmic ovule. The ovule without any integument as found in *Loranthus* and *Santalum.*

Tegmic ovule. The ovule with integument. It may be **unitegmic** (with a single integument as generally found in sympetalae / gamopetalae, *e.g.* Compositae / Asteraceae), or **bitegmic** (with two integuments found generally in polypetalae and monocots).

Types of Ovules

The mature ovule may be classified into six types, based on the position of micropyle in relation to the funiculus (fig. 14.71).

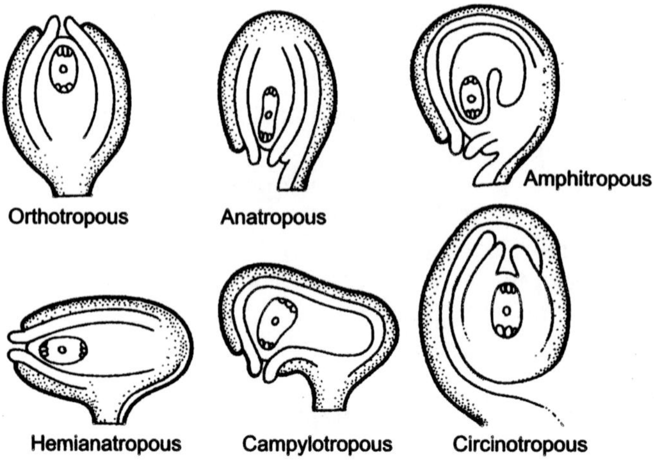

Fig. 14.71 Different types of ovules.

1. **Orthotropous type** (also called *atropous*). The mature ovule is straight and the micropyle and funiculus are in a straight line and above the hilum, as found among members of Polygonaceae and Piperaceae.

2. **Anatropous type** (also called *inverted*). It is the most common type of ovule found in angiosperms and is characteristic of sympetalous families. The ovule bends at 180° and gets inverted, so that the micropyle comes to lie close to the funicle.
3. **Campylotropous type.** In this type of ovule, the curvature is incomplete, and the body of the ovule lies at right angle to the funiculus, and is somewhat curved, as in Capparaceae/ Capparidaceae and Leguminosae.
4. **Amphitropous type.** When the curvature of the body of the ovule becomes more than that of campylotropous and the embryo sac assumes a horse-shoe shape, the ovule is said to be amphitropous type. It is commonly found in Alismaceae.
5. **Hemianatropous type** (also called *hemitropous*). It is a modified type of anatropous ovule in which the body of ovule lies at right angle to the funicle and micropyle and chalaza lie in a straight line. The micropyle is not found near the funicle as in *Ranunculus* and in members of Primulaceae.
6. **Circinotropous type.** It is a special type of ovule formed due to unilateral growth. Initially, the nucellar mass is in line with the funicle and appears orthotropous, then unilateral growth of ovule takes place and it gets inverted, thus becoming anatropous. The curvature continues till the ovule takes a complete 180° turn and the micropyle again faces upwards as in *Opuntia* (Cactaceae).

POLLINATION

Pollination. The transfer and deposition of pollen on the stigma of the flower.

Self pollination. The pollination between the different flowers or same flower of the same plant, as in groundnut, tobacco, barley, oat, wheat etc.

Autogamy. The transfer of pollen grains from the anther to the stigma of the same flower.

Geitonogamy. The transfer of pollen grains from the anther to the stigma of another flower of the same plant.

Homogamy. Simultaneous maturation of sex organs, as in *Mirabilis jalapa* and *Vinca rosea.*

Cleistogamy. Flowers do not open at the time of the fertilisation and do not expose their sex organs, as in *Commelina benghalensis.*

Cross pollination (Allogamy). The transfer of pollen grains from the anther of a flower to the stigma of another flower (*i.e.*, pollination between the flowers of two different plants).

Xenogamy. Pollination between flowers of different genetic constitution.

Dicliny or **Unisexuality.** Plant bearing unisexual flowers, either only male or only female flowers, as in *Carica papaya, Phoenix sylvestris, Morus alba* etc.

Dichogamy. Maturation of sex organs at different time periods. If anthers mature first, flowers are called **protandrous,** and if stigma (female sex organ) mature first, they are called **protogynous.**

Protandrous. A dichogamous flower in which anthers mature first, as in Lamiaceae (Labiatae), Malvaceae and Rubiaceae.

Protogynous. A dichogamous flower in which stigma matures before the anthers, as in *Magnolia, Michelia* and *Adhatoda.*

Heteromorphy. The phenomenon of bearing two types of morphologically dissimilar flowers, as in *Primula vulgaris.*

Herkogamy. Presence of some morphological barriers which make self pollination impossible, as in *Calotropis, Asclepias, Pergularia, Coelogyne* etc.

Entomophily. Pollination brought about by insects, as in *Salvia, Yucca, Ficus, Calotropis* etc.

Ornithophily. Pollination carried out by birds, as in *Salmalia, Callistemon, Erythrina* etc.

Cheiropterophily. Pollination carried out by bats, as in *Kigelia pinnata, Bauhinia megalandra, Eperua falcata* etc.

Malacophily. Pollination carried out by snails and slugs, may be found in subterranean runners.

Anemophily. Pollination by wind as in wheat, maize, coconut, bhang etc.

Hydrophily. Pollination by water as found in *Elodea, Hydrilla* and *Vallisnaria.*

FRUIT

A ripened ovary is called **fruit,** and study of fruit and fruit bearing plants is called **pomology.** Fruit is an important constituent of a balanced diet and is rich in carbohydrates, minerals, vitamins etc.

On the basis of climate, three types of fruits are recognised :

1. **Temperate fruits.** These plants grow in cold climate, *e.g.* apple, pear, khumani, cherry, musk melon, water melon, mulberry, strawberry etc.
2. **Tropical fruits.** These plants grow in warm climate, *e.g.* pineapple, custard apple, jack fruit, cheeku, mango, lemon, madrin, banana, guava, papaya etc.
3. **Sub-tropical fruits.** These plants grow in moderate climate, *e.g.* litchi, ber, pome granate, loquat, fig, date, phalsa, jamun, grapes, bel etc.

Classification of Fruits

There are three main types of fruits :

I. **Simple fruits.** These develop from monocarpellary ovary or multicarpellary syncarpous ovary.

II. **Aggregate fruits (Etaerio).** These develop from multicarpellary, apocarpous ovary.

III. **Composite fruits (Multiple).** These develop from the complete inflorescence.

FRUIT

- **Simple**
  - Dry
    - Indehiscent or Achenial
      1. **Achene** *e.g., Clematis,* Rose
      2. **Caryopsis** *e.g.,* Maize, Wheat
      3. **Cypsela** *e.g.,* Sunflower, *Dahlia Chrysanthemum*
      4. **Samara** *e.g.,* Chilbil
      5. **Nut** *e.g.,* Cashew nut, *Trapa,* Litchi
    - Dehiscent or Capsular
      1. **Legume** or **pod** *e.g.,* Pea, Bean, Gram
      2. **Follicle** *e.g., Calotropis*
      3. **Siliqua** *e.g.,* Mustard, Radish
      4. **Silicula** *e.g.,* Candytuft
      5. **Capsule** *e.g.,* Poppy, *Antirrhinum* (porous), Cotton, Ladies finger (loculicidal) *Aristolochia* (septicidal), *Datura* (septifragal) *Celosia* (Pyxidium or transverse)
    - Splitting or Schizocarpic
      1. **Lomentum** *e.g., Acacia, Mimosa*
      2. **Regma** *e.g.,* Castor
      3. **Cremocarp** *e.g., Cariander,* Fennel
      4. **Carcerulus** *e.g.,* Hollyhock, *Abutilon*
      5. **Double samara** *e.g.,* Maple
      6. **Triple samara** *e.g.,* Hiptage
      7. **Samaroid** *e.g.,* Sal
  - Succulent
    1. **Drupe** *e.g.,* Mango, Coconut, *Zizyphus*
    2. **Berry** *e.g.,* Tomato, Guava, Brinjal, Grape, Banana
    3. **Balausta** *e.g.,* Pome granate
    4. **Hesperidium** *e.g.,* Lemon, Orange
    5. **Pome** *e.g.,* Pear, Apple
    6. **Pepo** *e.g., Cucurbita,* Muskmelon
    7. **Amphisarca** *e.g.,* Wood apple, Elephant apple
- **Aggregate**
  - **Etaerio of achenes** *e.g., Lotus, Ranunculus*
  - **Etaerio of follicles** *e.g.,* Champa
  - **Etaerio of berries** *e.g.,* Custard apple
  - **Etaerio of drupes** *e.g.,* Raspberry
- **Composite**
  - **Sorosis** *e.g.,* Mulberry, Pineapple
  - **Syconus** *e.g., Ficus,* Fig

I. SIMPLE FRUITS

On the basis of **nature of pericarp**, simple fruits are divided into two types :

1. **Dry fruits.** Pericarp remains dry and undifferentiated into three layers.
2. **Succulent fruits.** Pericarp is fleshy or fibrous and remains distinguishable into three layers. These are indehiscent.

1. DRY FRUITS

According to mode of dehiscence, these fruits are divided into three groups :

(a) **Dehiscent** or **Capsular fruits.** Pericarp is ruptured after ripening and seeds are dispersed.

(b) **Indehiscent** or **Achenial fruits.** These fruits do not dehisce.

(c) **Schizocarpic** or **Splitting fruits.** After ripening, these fruits split (divided) into one seeded segments, called mericarps. Mericarps do not rupture further.

(a) ***Dehiscent Fruits*** **(Capsular Fruits).** According to the mode of dehiscence, these are broadly divided into five groups :

(1) **Legume** or **Pod.** Such fruits develop from monocarpellary superior ovary with marginal placentation. Ovary is unilocular with many ovules. The fruit dehisces by both sutures, as in pea, gram, red gram etc. (fig. 14.72A).

(2) **Follicle.** Such fruit develops from superior and unilocular ovary. These are usually found in pairs or groups. The fruit dehisces by one suture as in larkspur, *Calotropis, Michelia, Vinca* etc. (fig. 14.72B).

(3) **Siliqua.** The fruit develops from bicarpellary, syncarpous, superior ovary with parietal placentation. The ovary remains unilocular in the beginning but becomes bilocular due to formation of false septum, called **replum.** Each locule has many seeds. On maturity

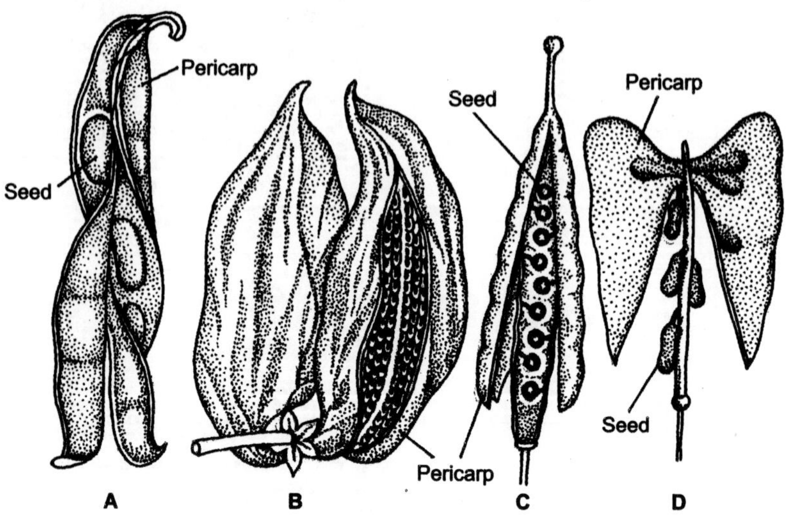

Fig. 14.72 Different types of dehiscent fruits. A-legume (Pea), B-a pair of follicle (*Calotropis*), C-siliqua (*Brassica*) and D-silicula (*Capsella*).

fruit dehisces from below to upwards. Such fruits are characteristic of family Cruciferae, *e.g.* mustard, radish, turnip etc. (fig. 14.72C).

(4) **Silicula.** It resembles with siliqua but its breadth and length are equal. It is wide and flat, as in candytuft (*Iberis amara*) and shephard's purse (*Capsella bursa-pastoris*) (fig. 14.72D).

(5) **Capsule.** It develops from multicarpellary, syncarpous, superior ovary with several locules and axile placentation. According to mode of dehiscence, these fruits are further classified as follows :

(i) **Porous dehiscence.** Fruit dehisces through the pores developed in the terminal part. Seeds are slowly dispersed by wind through these pores, as in poppy (*Papaver*).

(ii) **Loculicidal dehiscence.** In these fruits pericarp opens through the middle of locules and placenta is divided, as in cotton (*Gossypium*) and okra (*Hibiscus esculentus*).

(iii) **Septicidal dehiscence.** In such fruits, during dehiscence septa are ruptured and locules are separated. Pericarp is also separated in the form of valves, as in linseed and *Aristolochia*.

(iv) **Septifragal dehiscence.** In such fruits, during dehiscence placenta remains attached in the centre of undivided fruit and pericarp's segments become separated. Dehiscence can be either septicidal or septifragal, as in *Datura* (fig. 14.73B).

(v) **Transverse dehiscence.** In such fruits, pericarp dehisces transversely into two parts. The upper part opens out like a lid and exposes the seed to the wind for dispersal, as in *Celosia*. Such fruits are also called *pyxidium* (fig. 14.73D).

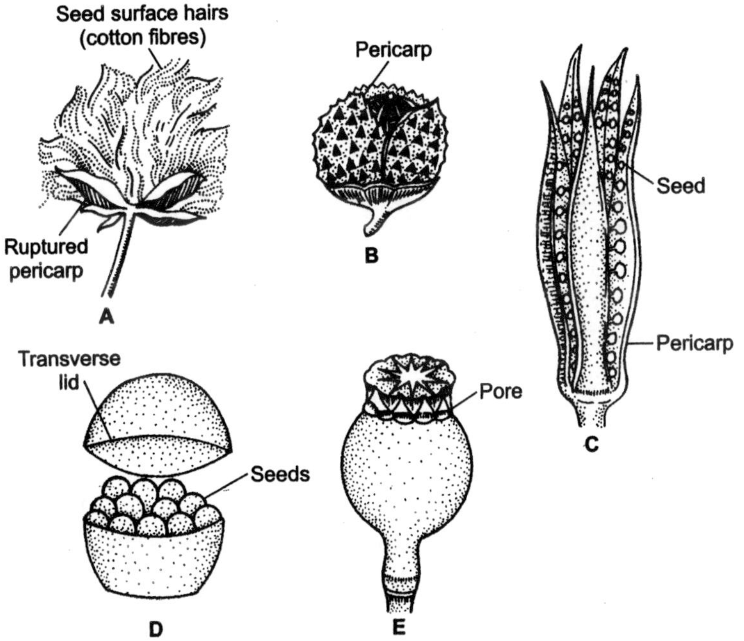

Fig. 14.73 Different modes of dehiscence in capsule fruits. A and C-loculicidal, B-septifragal, D-pyxidium and E-porous.

(b) ***Indehiscent*** or ***Achenial Fruits.*** In such fruits, pericarp does not rupture and seeds remain enclosed within it. Achenial fruits are further divided into :

(1) **Achene.** The fruit develops from monocarpellary superior ovary. It is unilocular and single seeded. The pericarp does not fuse with seed coat, as in *Clematis* and *Naravelia* (fig. 14.74A).

(2) **Caryopsis.** The fruit develops from monocarpellary, superior ovary and it remains unilocular and single seeded. The **pericarp remains fused with testa.** Such fruits are characteristic to family Gramineae, *e.g.* wheat, rice, maize etc.

(3) **Cypsella.** The fruit develops from bicarpellary, syncarpous, inferior ovary with basal placentation. Fruits are unilocular and single seeded. Persistent hairy calyx (pappus) are found at the apex of fruit. These are characteristic to family Compositae, *e.g. Taraxacum* and *Cosmos* (fig. 14.74B).

(4) **Nut.** Fruit develops from unilocular, syncarpous, multicarpellary, superior ovary. The fruit is single seeded. Pericarp becomes hard and stoney as in cashew nut (*Anacardium*), litchi, water chestnut (*Trapa*) (fig. 14.74C).

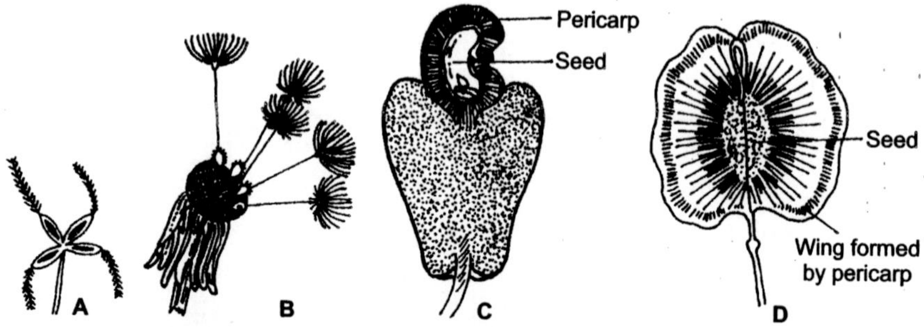

Fig. 14.74 A-achene (*Clematis*), B-cypsella (*Taraxacum*), C-nut (cashew), D-samara (*Holoptelea*).

(5) **Samara.** The fruit develops from bicarpellary, syncarpous, superior ovary. It is single seeded. The **pericarp becomes flat like wings,** *e.g.* chilbil (*Holoptelea*) (fig. 14.74D).

(c) ***Schizocarpic or Splitting Fruits.*** These fruits are dry and multiseeded, and after ripening are divided into one seeded segments or **mericarps.** Mericarps do not rupture further. Such fruits are further divided into :

(1) **Lomentum.** The fruit develops from monocarpellary, unilocular, superior ovary. It is a modification of legume. These are bisutural fruits which are constricted or divided into one seeded mericarps as in groundnut, *Mimosa, Tamarindus* etc. (fig. 14.75A).

(2) **Cremocarp.** The fruit develops from bicarpellary, syncarpous, inferior ovary. On maturation, these divide along with carpophore into two mericarps, each single seeded (fig. 14.75B). The fruit is characteristic to family Umbelliferae, as in coriander, carrot, fennel etc.

(3) **Regma.** The fruit develops from tricarpellary (multicarpellary), syncarpous, superior ovary. The fruit is multilocular. On maturation, after splitting, these divide into as many parts as the number of carpels. Each part is known as **coccus** and has one seed, as in castor (*Ricinus*) and *Geranium* (fig. 14.75C).

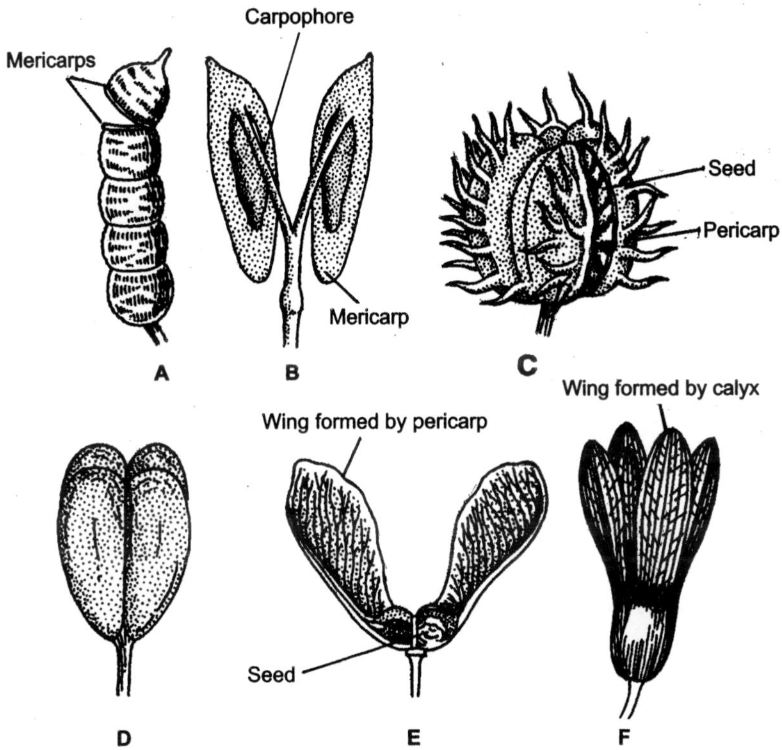

Fig. 14.75 Different types of fruits. A-lomentum (*Mimosa*), B-cremocarp, C-regma, D-carcerule (*Labiatae*), E-double samara and F-samaroid.

(4) **Carcerulus.** The fruit develops from bi- or polycarpellary, syncarpous, superior ovary and is mutilocular. Number of locules may increase due to false septation. On maturity, single seeded mericarp splits away, as in *Althaea, Ocimum* and *Malva* (fig. 14.75D).

(5) **Double samara.** The fruit develops from bicarpellary, syncarpous, superior ovary. Pericarp develops into two wings. On maturation it is divided into two one seeded parts, as in *Acer* (maple) (fig. 14.75E).

Samaroid. In sal (*Shorea robusta*), persistent calyx forms wing, hence look-like samara fruit.

2. SUCCULENT OR FLESHY FRUITS

In these fruits, the pericarp is distinguished into three layers - epicarp, mesocarp and endocarp. Mesocarp is fleshy or fibrous. These are of following types :

(1) **Drupe.** The fruit develops from mono- or polycarpellary, syncarpous, superior ovary. Fruits are usually single seeded, rarely many seeded. Pericarp comprises three layers. The epicarp forms

the skin of the fruit. Mesocarp is fleshy or fibrous and endocarp is hard and stoney, as in mango, coconut, peach, walnul, cherry, almond etc. (fig. 14.76A).

(2) **Berry.** The fruit develops from mono- or multicarpellary, syncarpous, superior or inferior ovary with axile or parietal placentation. Epicarp forms the rind of the fruit, mesocarp becomes fleshy and endocarp remains thin or membranous, as in tomato, brinjal, guava, date, papaya, chiku, areca nut etc. (fig. 14.76B).

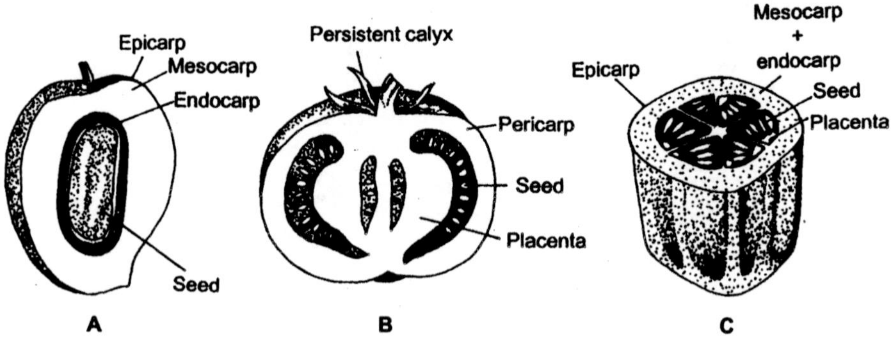

Fig. 14.76 Different types of fruits. A-drupe (mango), B-berry (tomato) and C-pepo (cucumber).

(3) **Pepo.** The fruit develops from tricarpellary, syncarpous, unilocular, inferior ovary with parietal placentation. These fruits are full of swollen placenta and are many seeded. Epicarp makes the hard rind. Mesocarp and endocarp are fleshy. Such fruits are characteristic of family Cucurbitaceae as in bottle gourd, cucumber, muskmelon etc. (fig. 14.76C).

(4) **Hesperidium.** The fruit develops from multicarpellary, syncarpous, superior ovary with axile placentation. These are multilocular. Epicarp is firm and leathery and has several oil glands. Mesocarp is in the form of white, fibrous part fused with epicarp. Membranous endocarp projects inwards forming distinct chambers. Many juicy unicellular hairs are found on the inner side of endocarp, as in lemon and orange (fig. 14.77A).

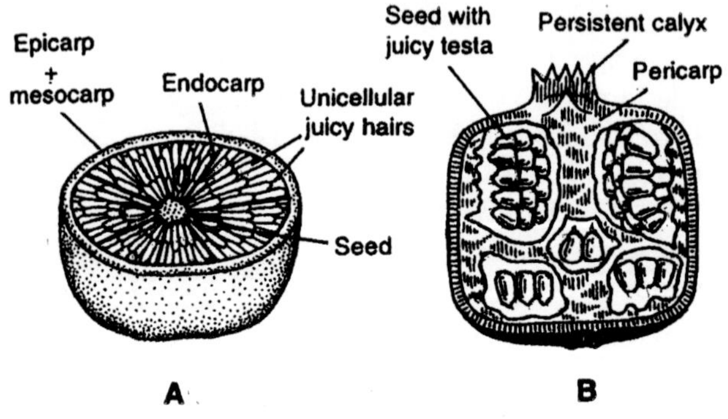

Fig. 14.77 Different types of fruits. A-hesperidium fruit (T.S.) and B-balausta (L.S.).

(5) **Balausta.** The fruit develops from multicarpellary, syncarpous, multilocular superior ovary. It is a many seeded fruit. The epicarp is rough and leathery, mesocarp is papery and thin and endocarp is hard and it forms chambers to enclose seeds irregularly. The fruit has persistent calyx. Testa is red and fleshy whereas tegmen becomes hard, as in pome granate (fig. 14.77B).

(6) **Amphisarca.** The fruit develops from multicarpellary, syncarpous, multichambered superior ovary. The fruit is many seeded. Epicarp is hard. Mesocarp and endocarp are fleshy on which are found scattered seeds, as in wood apple (*Aegle marmelos*) (fig. 14.78A).

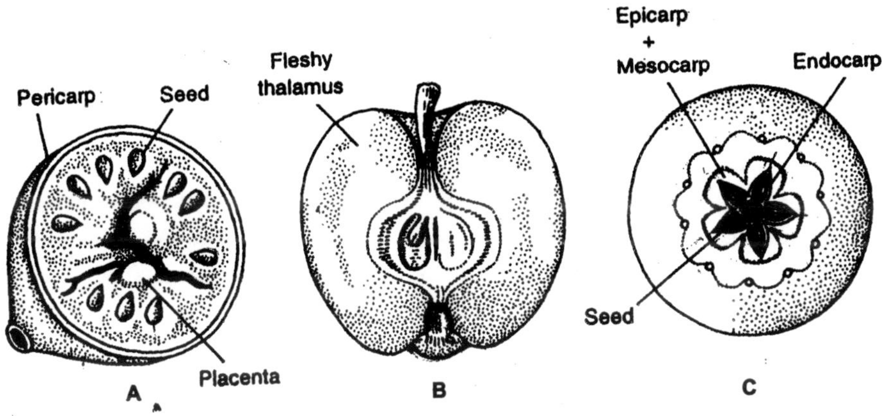

Fig. 14.78 Different types of fruits. A-amphisarca fruit (T.S.), B and C-pome (L.S. and T.S.).

(7) **Pome.** The fruit develops from bi- or polycarpellary, syncarpous, inferior ovary. The thalamus becomes fleshy and swollen and surrounds the fruit. So it is a false fruit. The pericarp is thin and papery. Fleshy swollen thalamus forms the edible part of the fruit, as in apple and pear (fig. 14.78B and C).

II. AGGREGATE FRUITS

Infact, these fruits are the group of fruitlets developed from multicarpellary, apocarpous ovary of a single flower. Such type of aggregate fruits is called **etaerio,** *i.e.*, aggregate of fruitlets.

These are of following types :

1. **Etaerio of follicles.** In such fruits, each free (separate) carpel develops into a fruitlet which is known as follicle. Many follicles of a flower make it etaerio, *i.e.*, etaerio of follicles, as in *Calotropis* and *Catharanthus* (aggregate of two follicles) and *Michelia* (aggregate of several follicles) (fig. 14.79A).
2. **Etaerio of achenes.** It is an aggregate of achene fruitlets developed from a single flower, as in rose, strawberry, *Clematis* and *Naravelia* (fig. 14.79C).

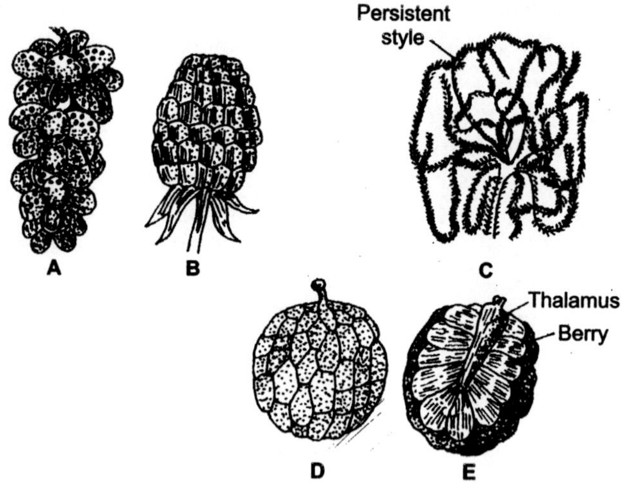

Fig. 14.79 Different types of aggregate fruits. A-etaerio of follicles (*Michelia*), B-etaerio of drupes, C-etaerio of achenes (*Naravelia*), D-etaerio of berries (*Annona*) (E = L.S.).

3. **Etaerio of berries.** It is an aggregate fruitlets of berries developed from a single flower, as in custard apple (*Annona*) (fig. 14.79D and E).
4. **Etaerio of drupes.** It is an aggregate of fruitlets of drupes developed from apocarpous ovaries of a single flower, as in raspberry (*Rubus*) (fig. 14.79B).

III. COMPOSITE OR MULTIPLE FRUITS

These fruits develop from the complete inflorescence, and are known as **infructescence.** These are of two types :

1. **Syconus.** Such fruit develops from hypanthodium inflorescence. The receptacle becomes fleshy and hollow cup-shaped with a narrow apical opening. Unisexual flowers are found on its inner surface, male flowers towards upper side and female (pistillate) flowers towards lower side. Larvae of insects live in gall flowers (female flowers). The fruit is achene. Fleshy receptacles form the edible part, as in figs (*Ficus*) (fig. 14.80D).

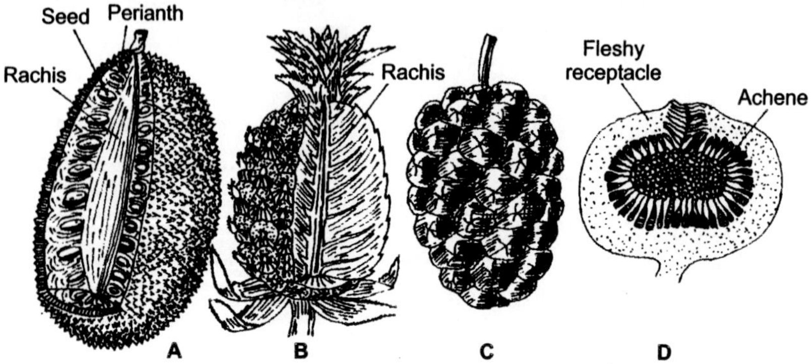

Fig. 14.80 Different types of composite fruits. A-jackfruit (sorosis), B-pineapple (sorosis), C-mulberry (sorosis), and D-fig (syconus).

2. **Sorosis.** This type of fruit develops from spike, spadix or catkin inflorescence, as in jackfruit, mulberry and pineapple. The fruits are so compactly set that entire inflorescence appears as one fruit. In jackfruit, stigmas fuse with each other to make rough and spiny rind. Bracts, perianth and seeds become edible in jackfruits. In mulberry, perianth present around the dry achenes is eaten up (fig. 14.80A, B and C).

Table 14.2. Some Common Fruits and their Edible Parts

Name	*Type of fruit*	*Edible parts*
I. Simple Fruits		
1. Pea (*Pisum sativum*)	Legume	Seeds
2. Okra or ladies finger (*Abelmoschus esculentus*)	Capsule	Whole fruit
3. Wheat (*Triticum aestivum*)	Caryopsis	Endosperm and embryo
4. Cashew nut (*Anacardium occidentale*)	Nut	Cotyledons and fleshy peduncles
5. Litchi (*Nephelium litchi*)	Nut	Fleshy aril
6. Water chestnut (*Trapa bispinosa*)	Nut	Cotyledon
7. Tamarind (*Tamarindus indica*)	Lomentum	Mesocarp
8. Groundnut (*Arachis hypogea*)	Lomentum	Embryo
9. Coconut (*Cocos nucifera*)	Drupe	Endosperm
10. Walnut (*Juglans regia*)	Drupe	Lobed cotyledons
11. Mango (*Mangifera indica*)	Drupe	Mesocarp
12. Grape (*Vitis vinifera*)	Berry	Pericarp and placenta
13. Areca nut (*Areca catechu*)	Berry	Endosperm
14. Guava (*Psidium guajava*)	Berry	Whole fruit
15. Banana (*Musa sapientum*)	Berry	Mesocarp and endocarp
16. Tomato (*Lycopersicon lycopersicum*)	Berry	Pericarp and placenta
17. Cucumber (*Cucumis sativus*)	Pepo	Mesocarp, endocarp and placenta
18. Muskmelon (*Cucumis melo*)	Pepo	Mesocarp and endocarp

19. Bottle gourd (*Lagenaria vulgaris*)	Pepo	Mesocarp and endocarp
20. Pome granate (*Punica granatum*)	Balausta	Fleshy testa
21. Madarin orange (*Citrus reticulata*)	Hesperidium	Fleshy unicellular hairs of endocarp
22. Sweet orange (*Citrus sinensis*)	Hesperidium	Fleshy unicellular hairs of endocarp
23. Lemon (*Citrus lemon*)	Hesperidium	Fleshy unicellular hairs of endocarp
24. Lime (*Citrus aurantifolia*)	Hesperidium	Fleshy unicellular hairs of endocarp
25. Apple (*Pyrus malus*)	Pome	Fleshy thalamus
26. Pear (*Pyrus communis*)	Pome	Fleshy thalamus

II. Aggregate Fruits

1. Lotus (*Nelumbium speciosum*)	Etaerio of achenes	Thalamus and seeds
2. Strawberry (*Fragaria indica*)	Etaerio of achenes	Fleshy receptacle
3. Custard apple (*Annona squamosa*)	Etaerio of berries	Pericarp (except apical part)

III. Composite Fruits

1. Fig (*Ficus carica*)	Syconus	Fleshy receptacle and seeds
2. Pineapple (*Ananas sativus*)	Sorosis	Outer rachis, bracts, perianth and pericarp
3. Jackfruit (*Artocarpus integrifolium*)	Sorosis	Bracts, perianth, seeds
4. Mulberry (*Morus indica*)	Sorosis	Perianth

DISTINGUISHING CHARACTERS OF THE FRUITS

Table 14.3 Differences between Achene, Caryopsis and Cypsella

	Achene	Caryopsis	Cypsella
1.	The fruit develops from monocarpellary and superior ovary.	The fruit develops from monocarpellary and syncarpous ovary.	The fruit develops from bicarpellary, syncarpous and inferior ovary.
2.	Pericarp and testa (seed coat) are separate.	Pericarp and testa are fused and can not be separated.	Pericarp and testa are separate.
3.	The fruit is etaerio of achenes. **Examples.** *Clematis*, *Naravelia* and *Ranunculus*.	Fruit is not found in groups. **Examples.** *Triticum*, *Oryza* and *Zea*.	Fruit is not developed in groups. **Examples.** *Helianthus* and *Cosmos*.

Table 14.4 Differences between Legume and Follicle

	Legume or Pod	Follicle
1.	The fruit develops from monocarpellary, unilocular and superior ovary	It also develops from monocarpellary, unilocular and superior ovary.
2.	Fruit is found singly.	Fruit is found usually in pairs, two or more in number.
3.	It dehisces from both the sutures and from top towards base.	It dehisces by a single suture.
4.	When fruit dehisces, all seeds come out.	When fruit dehisces, seeds are exposed and are released slowly with drying.
	Examples. Fruit is characteristic of family **Papilionaceae**, *e.g.* *Pisum*, *Dolichos*, *Vigna* etc.	**Examples.** *Calotropis* and *Michelia*.

Table 14.5 Differences between Siliqua and Silicula

	Siliqua	Silicula
	Both types of fruits are found in Cruciferae.	
1.	The fruit is long and cylindrical.	The fruit is broad and flat.
2.	Fruit has a large number of seeds.	Fruit contains a few seeds.
	Examples. *Brassica* and *Raphanus*.	**Examples.** *Iberis* and *Capsella*.

Table 14.6 Differences between Legume and Lomentum

	Legume	Lomentum
1.	It is a multiseeded, dry and dehiscent fruit.	It is a multiseeded schizocarpic fruit.
2.	Mericarps are not formed.	Mature fruit breaks transversely into single seeded parts, called mericarp.
3.	The fruit has two sutures and dehisces by both the sutures.	Mericarps are indehiscent.
	Examples. *Pisum*, *Cicer*, *Phaseolus* etc.	**Examples.** *Arachis*, *Tamarindus*, *Mimosa* etc.

Table 14.7 Differences between Aggregate and Composite Fruits

	Aggregate Fruit	Composite Fruit
1.	It is a group of fruitlets which develops from multicarpellary and apocarpous gynoecium of a single flower.	It develops from an entire inflorescence.
2.	The fruit develops from carpels of a single flower.	In the formation of a fruit, ovaries, perianths and thalamus participate.
	Examples. *Michelia*, *Calotropis*, *Annona* etc.	**Examples.** *Morus* (mulberry), *Ananas*, *Ficus* etc.

Table 14.8 Differences between Follicle and Capsule

	Follicle	Capsule
1.	It develops from a single carpel with unilocular and superior ovary.	It develops from multicarpellary, syncarpous and superior ovary.
2.	The fruit dehisces by the single suture.	On maturity, the fruit shows longitudinal, transverse and porous dehiscence.
3.	The fruits are formed in pairs or in groups.	The fruit is formed singly.
4.	The fruit is dehiscent.	Fruit is schizocarpic or splitting.
	Examples. *Calotropis* and *Michelia.*	**Examples.** *Papaver, Gossypium, Datura* etc.

Table 14.9 Differences between Caryopsis and Regma Fruits

	Caryopsis	Regma
1.	The fruit is dry and indehiscent.	The fruit is dry and schizocarpic.
2.	The fruit develops from monocarpellary and superior ovary.	The fruit develops from tri-carpellary, syncarpous and superior ovary.
3.	Fruit is single chambered and single seeded.	The fruit is multichambered, each chamber with a single seed.
4.	Pericarp and testa are fused and can not be separated. The seed remains completely enclosed in the fruit.	Fruit on maturity splits into as number of parts as the number of carpels. Each part is called coccus. Cocci split to release seed.
	Examples. *Triticum, Oryza, Zea mays* etc.	**Examples.** *Ricinus* and *Geranium.*

Table 14.10 Differences between Cremocarp and Regma Fruits

	Cremocarp	Regma
1.	The fruit develops from bicarpellary, syncarpous and inferior ovary.	The fruit develops from tricarpellary, syncarpous and superior ovary.
2.	The fruit is bilocular and two seeded.	The fruit is multilocular and many seeded.
3.	The fruit on maturity splits into two mericarps, each placed at the top of carpophore. The mericarp is single seeded.	The fruit splits into as many number of cocci as the number of carpels. The coccus is single seeded.
4.	Fruit wall has ridges and furrows.	On the outer surface of the fruit wall, spiny structures are found.
	Examples. *Foeniculum* and *Coriandrum.*	**Example.** *Ricinus communis.*

Table 14.11 Differences between Drupe and Schizocarpic Fruits

	Drupe	Schizocarpic Fruit
1.	It is a succulent type of fruit.	It is a dry type of fruit.
2.	Fruit develops from mono- or polycarpellary, syncarpous, superior ovary.	It develops from monocarpellary or polycarpellary, syncarpous, superior ovary.
3.	Pericarp is clearly distinguishable into three layers : epicarp, mesocarp and endocarp.	Such differentiation in pericarp is not found.
4.	The epicarp forms skin of the fruit, mesocarp is fleshy or fibrous and endocarp is hard and woody.	No such structure is found.
5.	The fruit is usually single seeded.	The fruit is usually multiseeded.
6.	Fruit does not split open at maturity.	Fruit breaks into number of indehiscent mericarps.
	Examples. *Mangifera* (mango) and *Cocos* (coconut).	**Examples.** *Arachis* (groundnut) and *Tamarindus* (tamarind).

Table 14.12 Differences between Pome and Drupe Fruits

	Pome	Drupe
1.	It is a kind of false fruit as the fleshy fruit-like structure develops from the thalamus.	It is a true fruit that develops from ovary wall.
2.	The fruit develops from bi- or polycarpellary, syncarpous, inferior ovary and it remains surrounded by enlarged fleshy thalamus.	The fruit develops from monocarpellary or polycarpellary superior ovary.
3.	The pericarp is thin, papery and inedible.	The pericarp is well developed and differentiated into epicarp, mesocarp and endocarp.
4.	The edible part is actually a fleshy thalamus.	The edible part is mesocarp.
5.	The fruit has 5-6 seeds.	The fruit has a single seed.
	Examples. *Pyrus* (apple and pear).	**Examples.** *Mangifera* (mango) and *Cocos* (coconut).

Table 14.13 Differences between True and False Fruits

	True Fruit	False Fruit
1.	It develops from ovary after fertilization.	In its formation, thalamus, perianth or rachis participate along with ovary.
	Examples. *Mangifera* (mango) and *Carica* (papaya).	**Examples.** *Pyrus* (apple) and *Ficus* (fig).

Table 14.14 Differences between Pepo and Pome Fruits

	Pepo	Pome
1.	It is a true fruit that develops from ovary after fertilization.	It is a false fruit in which fruit like structure is formed by the thalamus.
2.	It develops from tricarpellary, syncarpous, inferior ovary with parietal placentation.	It develops from bi- or polycarpellary, syncarpous inferior ovary and remains surrounded by fleshy thalamus.
3.	The pericarp is well developed and differentiated into epicarp, mesocarp and endocarp.	The pericarp is thin and papery and indistinguishable into three layers.
4.	Seeds are produced in a large number.	Only 5-6 seeds are formed.
5.	The edible part is fleshy mesocarp and endocarp.	The edible part is thalamus.
	Examples. *Cucurbita*, *Citrullus*, *Momordica* etc.	**Example.** *Pyrus malus* (apple).

Table 14.15 Differences between Lomentum and Caryopsis Fruits

	Lomentum	Caryopsis
1.	The fruit develops from monocarpellary and superior ovary.	It develops from monocarpellary superior ovary.
2.	The fruit is unilocular, multiseeded and bisutural. It is divided into several mericarps.	The fruit is unilocular and single seeded.
3.	Pericarp and testa remain separate.	Pericarp and testa are united or fused.
	Examples. *Arachis* (groundnut) and *Tamarindus* (tamarind).	**Examples.** *Triticum* (wheat) and *Oryza* (rice).

Table 14.16 Differences between Drupe and Berry Fruits

	Drupe	Berry
1.	The fruit develops from mono- or polycarpellary, syncarpous and superior ovary.	The fruit develops from mono- or polycarpellary, syncarpous, superior or inferior ovary.
2.	Pericarp is divided into three layers, *i.e.*, epicarp, mesocarp and endocarp. The endocarp is hard and woody.	Pericarp is divided into epicarp, mesocarp and endocarp. The endocarp is thin and membranous.
3.	The fruit is usually single seeded but sometimes multiseeded.	The fruit is usually multiseeded.
	Examples. *Mangifera* (mango) and *Cocos* (coconut).	**Examples.** *Lycopersicum* (tomato) and *Vitis* (grapes).

Table 14.17 Differences between Hesperidium and Syconus Fruits

	Hesperidium	Syconus
1.	Fruit develops from the ovary of a single flower.	The fruit develops from the entire inflorescence, thus a type of false fruit. It develops from hypanthodium inflorescence and each fruit is achene.
2.	Pericarp is distinguishable into three layers, the epicarp, the mesocarp and the endocarp.	Pericarp is indistinct.
3.	The edible part of the fruit is the fleshy unicellular juicy hairs developed from endocarp.	The edible part comprises pericarp, thalamus, perianth and rachis.
4.	The epicarp has a large number of oil glands.	Oil glands are not found
	Example. *Citrus* species.	**Example.** *Ficus* species (figs).

SEED

Seed is a product of a fertilised ovule, consisting of an embryo, enclosed in the testa which is derived from the integument(s). Food reserves may or may not be contained in the endosperm.

Testa. Also called seed coat, and derived from the outer integument and is protective in nature.

Aril. A fleshy covering on the seed which arises as an outgrowth of the funicle or base of the ovule. It may be a tuft of hairs. A fleshy white aril is distinctive and is the edible part of litchi.

Caruncle. An outgrowth near the micropyle and hilum of the seed (as in castor).

Cotyledon. The leaf formed directly from the embryo of an angiosperm (also in gymnosperms). In angiosperms, monocots have one cotyledon (*i.e.* Monocotyledons) or dicots have two cotyledons (*i.e.* Dicotyledons.)

Kernel. The seed inside the stoney endocarp of a drupe.

Perisperm. A nutritive tissue derived from the nucellus of the ovule (present in some seeds) or may be called persistent nutritive nucellus (as in members of Nymphaeaceae). In coffee seed, it is the perisperm that is the source of coffee powder.

Straphiole. An appendage of hilum (found in some seeds).

Methodical Vegetative and Floral Description (Taxonomic Description)

As far as possible fresh flower must be taken for description. The minute vegetative and floral structures should be viewed under dissecting microscope. For the study of carpels and locules a transverse section through immature ovary should be examined.

The taxonomic description should be written adopting following sequence :

1. Habitat
(a) Hydrophyte, mesophyte or xerophyte.
(b) Cultivated, ornamental, food crop or wild.
(c) Tropical, temperate or arctic.

2. Habit
(a) Herb (whether prostrate, erect, diffuse or weak-stemmed), shrub or tree.
(b) Annual, biennial or perennial.

3. Root
(a) Tap or adventitious.
(b) Branched or unbranched.
(c) Any modification (both for tap and adventitious types of roots).

4. Stem
(a) Erect, prostrate or climbing.
(b) Modification-bulb, tuber, corm, rhizome etc.
(c) Branching : monopodial or sympodial.
(d) Herbaceous (solid or fistular) or woody.
(e) Angular, cylindrical or flattened.
(f) Hairy, spiny, glabrous or glaucous.
(g) Colour : green or brown.

5. Leaves
(a) Radical, cauline or ramal.
(b) Alternate, opposite (superposed or decussate) or whorled.
(c) Sessile or petiolate.
(d) Stipulated or exstipulated, type of stipule (if present).
(e) Leaf base sheathing connate, perfoliate or ligulate.
(f) Simple or compound.
(g) If compound, pinnate or palmate, number of leaflets.
(h) Venation-uni or multicostate, parallel or reticulate.
(i) Leaf margin : entire, serrate or reticulate.
(j) Surface : hairy, glabrous or glaucous.
(k) Apex : acute, acuminate or obtuse.

6. Inflorescence
(a) Racemose, cymose or any special type.

7. Flower
(a) Pedicellate or sessile.
(b) Bracteate or ebracteate.
(c) Complete or incomplete.
(d) Unisexual or bisexual.
(e) Regular or irregular.
(f) Actinomorphic or zygomorphic.
(g) Hypo, peri- or epigynous.
(h) Colour.
(i) Nectaries present or absent.

8. Calyx
(a) Number of sepals.
(b) Gamosepalous (fused) or polysepalous (free).
(c) Shape and form.
(d) Green or petaloid.
(e) Aestivation : valvate, twisted or imbricate (ascending, descending or quincuncial).

9. Corolla
(a) Number of petals.
(b) Gamopetalous (fused) or polypetalous (free).
(c) Shape or form.
(d) Coloured or sepaloid.
(e) Aestivation.

10. Perianth When there is no distinction between calyx and corolla it is called perianth.
(a) Number of tepals.
(b) Gamophyllous (fused) or polyphyllous (free).

11. Androecium
(a) Number of stamens.
(b) Polyandrous (free), adelphous (fused to form a tube) or syngenesious (anthers fused).
(c) Epipetalous or free.
(d) Filament long or short.
(e) Anthers basifixed, dorsifixed or versatile.
(f) Anthers monothecous or dithecous, introrse or extrorse.

12. Gynoecium
(a) Number of carpels.
(b) Syncarpous (fused) or apocarpous (free).
(c) Ovary superior or inferior.
(d) Number of locules.
(e) Placentation (marginal, axile, parietal, free central or basal).
(f) Number of ovules in each locule.
(g) Styles free or united.
(h) Stigma simple, lobed, capitate etc.

13. Seeds
(a) Endospermic or non-endospermic.
(b) Number of cotyledons.

14. Fruits
(a) Kinds of fruits-dry, succulent or of any special type.

Floral Formula. It is written the help of the following symbols :

Br	Bracteate
brl	bracteolate
%	Zygomorphic
⊕	Actinomorphic
♂	Staminate
♀	Pistillate
⚥	Bisexual or Hermaphrodite
☉	Neuter
Ep. K	Epicalyx
K	Calyx free
(K)	Calyx fused
C	Corolla free
(C)	Corolla fused
P	Perianth

A	Stamens free
(A)	Stamens fused
G	Gynoecium free
(G)	Gynoecium fused
$\overline{G}$	Ovary inferior
$\underline{G}$	Ovary superior
∝	Indefinite
$\widehat{CA}$	Epipetalous stamens
$\widehat{PA}$	Epiphyllous stamens

For example floral formula of shoe flower (*Hibiscus rosa-sinensis*) is :

$$\oplus ⚥ \text{ Epi } K_{5-8}\ K_{(5)}\ \widehat{C_5\ A_{(\alpha)}}\ \underline{G}_{(5)}$$

Floral diagram. Arrangement of floral parts and method of drawing the floral diagram is illustrated in fig. 14.81.

The procedure of drawing floral diagram should be taken up carefully, starting from outermost whorl to central whorl (gynoecium). The position of each floral part should be ascertained with its relative position in respect to mother axis (hypothetical).

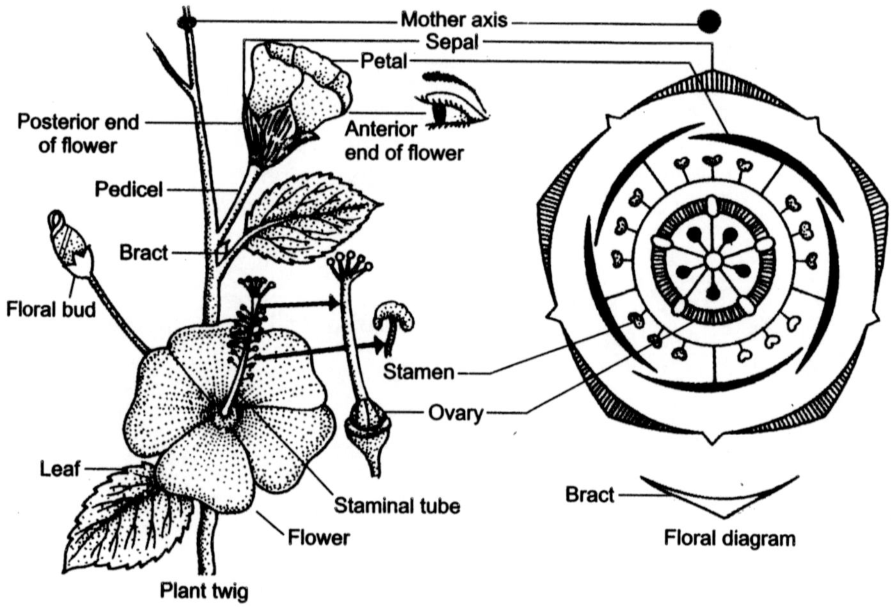

Fig. 14.81 Depicting method for sketching floral diagram.

□□□

Chapter **15**

Plant Identification : Construction and Use of Keys

Identification of an unknown plant is one of the basic components of plant taxonomy. A plant specimen can be identified with already known herbarium specimen, utilising the available literature and comparing the description of unknown plant with published description or taking the help of botanical gardens or some organisations like Botanical Survey of India or similar other institutions. Taxonomic keys prove an ideal aid for rapid identification of unknown plants.

Construction and Use of Keys

First step for the identification of a plant is to determine whether the plant is a dicotyledon or monocotyledon. If a plant is woody or herbaceous, has tap root system, leaves with reticulate variation, floral parts in multiple of 5 whorls (or 4 or 2 whorls), it is assigned as a **dicotyledonous plant**. Of course, anatomical characters like vascular bundles arranged in a ring, secondary growth present and two cotyledons also provide supportive evidences. And if a plant is herbaceous, has adventitious root system, leaves with parallel venation, floral parts in multiple of 3 whorls (rarely less or more), it is assigned as a **monocotyledonous plant**. Anatomical characters like scattered vascular bundle in the stem, absence of secondary growth and cotyledon one, further support the assignment.

Then comes assignment of the plant to its specific category. Initial identification has to be at the **family** level and subsequently upto **genus, species, sub-species, variety, forma** etc. Keys based on comparative characters, are quite helpful in plant identification.

Specimens from newly explored areas are usually identified with the help of keys provided in the floras or manuals. Special keys have been prepared for this purpose (Hutchinson, 1967; Davis and Cullen, 1965;

Saldanha and Rao, 1974; Naik, 1977 etc.). Based on comparative characters, such type of keys play a significant role in determining the identity of the plants by a sequence of choice between two (or more) statements and selecting one of them.

Need for Accurate Identification

It has often been noticed that scholars, other than taxonomists, undertaking studies on plants, do not care much for the accurate identification of a taxon with which they work. This obviously leads to misinterpretation of the data and misuse in disciplines, such as pharmacology, pharmacognosy, biochemistry etc. The success of Ayurvedic medicines largely depends upon the accurate identification of plant wealth. And for this, one has to rely on keys for identification, apart from his own judgement acquired through long experience and critical observations.

Construction of Keys

For identification of unknown plants, keys are very convinient, reliable and quick to use. The keys are prepared in such a way that any one possessing working knowledge of morphological terms may be able to use them correctly. Obviously, it is often difficult to arrange the taxa according to their natural relationships, and thus, in order to make a key convenient, usually artificial separation of the groups, using one or two easily observable characters, is employed. According to Davis and Heywood (1963), the first aim of the key is to provide ease and certainity of identification.

Keys are constructed using statements regarding contrasting characters to divide possible names in the keys into smaller and smaller groups, each division employing a choice and eliminating one or more taxa, *e.g.* separating a taxa using choices such as herbaceous v/s woody, if herbaceous other choice woody may be eliminated; zygomorphic flowers v/s actinomorphic flowers, if zygomorphic, other choice actinomorphic may be eliminated and so on. Each time a choice is made and the number of taxa that remains is reduced by use of the constrasting characters. If sufficient number of characters are contrasted, the number of possibilities is eventually reduced to one.

In recent years, the keys are prepared based on cards, tables and computer programmes. Computer programmed keys are usually designed for identification by non-professionals, such as persons engaged in pharmaceutical industries. Such type of keys are based on characters that are stable and reliable. The keys facilitate faster preliminary identification but it should be followed by confirmation through comparison with the detailed description of the taxon provisionally identified. Before identifying an unknown plant, it should be carefully studied, described and a list of its characters be prepared.

Types of Keys

Based on the arrangement of characters and their utilisation, keys are mainly of two types : **single-access (dichotomous, diagnostic** or **sequential)** keys or **multiaccess (polyclaves)** keys.

I. Single-Access, (Dichotomous, Sequential or **Diagnostic) Keys.**

They are usual component of floras, manuals, monographs and such other literature meant for identification. These are called **single - access keys** because they have a single entry character included in a **couplet** (a pair of contrasting character such as woody v/s herbaceous) with two **leads** (two choices of contrasting characters such as woody and herbaceous). These are called **dichotomous keys** because they consist of contrasting characters (couplet), each statement of which is a lead. The leads are numbered. Commonly both leads begin with the same word. These are called **diagnostic keys** because they are based on important and conspicous key words. The characters to be used for construction of such keys are easily observable, constant and thus dependable. Preference is given to quantitative characters over qualitative ones. These are called **sequential keys** because they are constructed as a sequence of alternative choices, each pair forming a character couplet, *e.g.* habit woody v/s herbaceous, in which woody and herbaceous are the leads. For characters having more than two available choices, they can be split to make it dichotomous, *e.g.* if flowers in a taxon are red, yellow or white, the first couplet would constitute flowers red v/s non-red and the second couplet-flowers yellow v/s white. In order to identify closely related taxa with overlapping characters, usually more than one contrasting characters are employed so that the combination of characters is taken to decide the similarity or difference between taxa.

The device involved in construction of keys is known as **determinator.** Since these keys are based on artificial characters or conventional phenotypic characters, they are frequently termed as **artificial keys.**

Based on the arrangement of couplets (*e.g.* woody v/s herbaceous) and their leads (*e.g.* woody and herbaceous), the keys are mainly of three types : **indented (yoked)** keys, **bracketed (parallel)** keys and **serial (numbered)** keys.

(1) Indented or **Yoked Keys.** These are the most commonly used in floras and manuals, especially when they are of smaller size. In them, the statements (leads) and the taxa identified from them are arranged in usual groups or yokes. Additionally, the subordinate couplets are indented below the primary one at a fixed distance from the margin, the distance increasing with each subordinate couplet. An example of indented or yoked key for taxa under consideration is given :

1. Fruit achene
    2. Calyx differentiated from corolla
        3. Petals with basal nectary — 1. *Ranunculus*
        3. Petals without basal nectary — 2. *Adonis*
    2. Calyx not differentiated from corolla
        4. Plants woody — 4. *Clematis*
        4. Plants herbaceous — 3. *Anemone*
1. Fruit follicle
    5. Spur present
        6. Number of spurs 1........... 6. *Delphinium*
        6. Number of spurs 5........... 7. *Aquilegia*
    5. Spur absent .... 5. *Caltha*

The shorter indented keys have an advanage of giving a visual presentation of the group so that the user can readily obtain a picture of the taxon but if these are far large (as in *Flora Europea*) the individual leads in the couplets get widely separated from each other while these occur in pairs in the bracket keys.

(2) Bracketed or **Parallel Keys.** These have been used in larger floras (as in *Flora of USSR, Plants of Central Asia* and *Flora of British Isles*). In them, the two leads of a couplet are always together and the distance from the margin is always the same, thus saving the page space. Athough they are constructed on the basis, similar to indented or yoked keys on contrasting characters for the comparison, the placement of these comparative characters (two leads of each couplet) is adjacent to each other and leads are arranged parallely. These characters are numbered as in *Flora of Siberia.* An example of bracketed or parallel key for taxa under consideration is given below :

1. Fruit achene................ 2
1. Fruit follicle................ 5
2. Calyx differetiated from corolla.... 3
2. Calyx not differentiated from corolla.... 4
3. Petal with basal necetary..... 1. *Ranunculus*
3. Petals without basal nectary..... 2. *Adonis*
4. Plants woody........... 4. *Clematis*
4. Plants herbaceous.......... 3. *Anemone*
5. Spur present.................... 6
5. Spur absent.................... 5. *Caltha*
6. Number of spurs 1............ 6. *Delphinium*
6. Number of spurs 5............. 7. *Aquilegia*

The positive features of parallel (bracketed) key and visual groups of yoked (indented) key have been retained in the serial key.

(3) Serial Keys. Although these have been used for the identification of animals but they have also been adopted in some botanical works. In them, yoked key arrangement is retained but there is no indentation so that the distance from the margin remains the same. The location of alternate leads is done by serial numbering of couplets (or leads when separated) and indicating the serial numbers of the alternate lead within paranthesis the, key appearing like :

1. (6) Fruit achene.
2. (4) Calyx differentiated from corolla.
3. Petal with basal nectary........ 1. *Ranunculus*
3. Petal without basal nectary....... 2. *Adonis*
4. (2) Calyx not differentiated from corolla
5. Plants woody.............. 4. *Clematis*
5. Plant herbaceous.......... 3. *Anemone*
6. (1) Fruit follicle.
7. (9) Spur present.
8. Number of spurs 1........ 6. *Delphinium*
8. Number of spurs 5......... 7. *Aquilegia*
9. (7) Spur absent......... 5. *Caltha*

In these keys, the visual groups of statements and taxa are retained. Alternate leads, even though separated, are easily located. Morever, there is no wastage of page space.

Dichotomous key has an inherent draw-back in that the user has a single fix choice of the sequence of the characters fixed by constructor of the key and information about one of the successive characters is not available and thus, it is not possible to go beyond the first couplet.

Guidelines for Construction of Dichotomous Keys

An ideal dichotomous key should be based on certain basic considerations, such as :

1. The keys should be dichotomous with couplets having only two possible leads.
2. The two leads of the couplet should be mutually exclusive so that when one lead is accepted the other should automatically be rejected.
3. The statements of the leads should not be overlapping.
4. The first word of each lead of the couplet should be identical and start with the same capitalised initial word, *e.g.* if the first lead of a couplet begins with the word Stamen, the second lead of the same couplet must begin with word Stamen.
5. The leads of two successive couplets should not start with same initial letter.
6. For identification of trees, to facilitate identification round the year, two separate keys should be prepared on the basis of vegetative and reproductive characters. This is in view that trees commonly have leaves throughout or the most part of the year but flowers appear commonly briefly when in many trees by this time the leaves have not yet reappeared.
7. Use of vague statements such as large, small, narrow and broad, which often create confusion, should be avoided.

8. For dioecious plants, based on male and female flowers, two separate keys should be constructed.
9. The leads, especially in longer keys, should be prefixed by numbers (1, 2, 3) or letter (A, B, C) to make the location of the lead easier.
10. Terms used should be easy to understand; measurements should be definite; lens should be used for keen observations, especially for floral parts; specimen should be studied comprehensively; only observable features should be used etc., are the other general recommendations for construction and use of dichotomous keys.

Since these keys have a single character included in a couplet, with a pair of leads with contrasting characters, they are kown as **monothetic sequential keys.**

Polythetic sequential keys (synoptic keys) are the commonest forms of keys used in floras. However, they have atleast some couplets with several statements about different characters in each lead. These keys are specially useful for identification of higher categories of plants. An example of synoptic key is given below :

Plants rarely woody, mostly herbaceous, petals coalescent, stamens adnate to the petals, corolla bilaterally symmetrical, stamens 5 and carpels 2-3 .... order Campanulales

(i) Flowers not in a dense head like spike :

2. No succulent, stigma is surrounded by cup-like structure....... Campanulaceae

2. Plants succulent, stigma is surrounded by cup-like structure...... Goodeniaceae

(ii) Flowers arranged in a dense head or head like spike inflorescence....... Calyceraceae

Polythetic sequential keys or synoptic keys are advantageous over monothetic sequential keys because :

1. A monothetic key becomes useless if one or more characters may be unobservable due to damage or non-occurrence of requisite stage in the specimen.
2. The mistake caused by user in deciding about a single character gets minimised, if more than one character is used.
3. Using more than one character may reduce the chances, if the single character used in the couplet is exceptional.

II. Multi-Access Keys (Polyclaves)

They are multientry order-free user oriented keys with many choices of sequence of characters. The sequence in which to use characters, depends upon the choice of the user and if the information about few characters is not available, the user can go ahead with identification. The identification may often be achieved without having to use all characters available to the user. Such identification methods employ card placed on the top of one another to eliminate taxa that disagree with the specimen to be identified.

Another form is computer-stored multientry key. Still other form is a printed table or matrix giving the status of various taxa and characters useful for separating the texa.

Punch Cards

Two basic types of punch cards are commonly used.

Body punched cards. These cards, also named as **window cards** or **peak-a-boo cards**, make use of cards with appropriate holes in the body of the card. In the process, one card is used for one attribute (character-state) for world angiosperm families. Round holes are punched besides a family

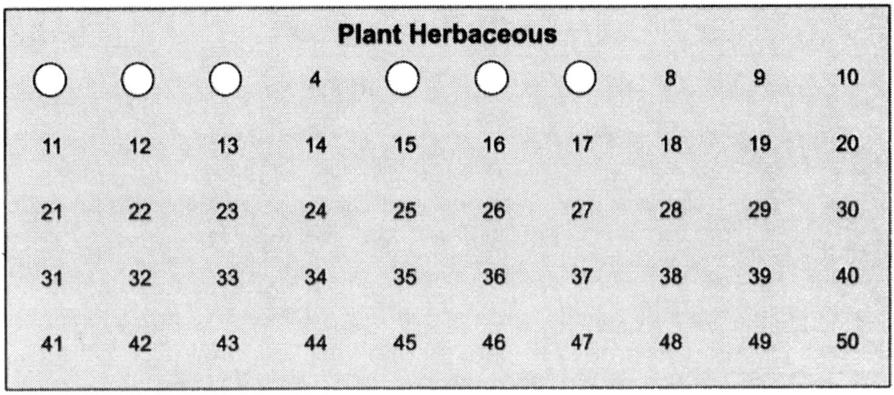

Fig. 15.1 A body-punched card for herbaceous habit for the seven representative genera of Ranunculaceae : 1. *Ranunculus*, 2. *Adonis*, 3. *Anemone*, 4. *Clematis*, 5. *Caltha*, 6. *Delphinium*, 7. *Aquilegia*.

number for the families that have that character. The plant to be identified is examined and its characters noted. A good specimen often fetches twenty to twenty five characters. One selects those cards corresponding to the characters found in the plant. The cards are put on the top of one another in a stack. Families having all observed characters are indicated by holes or windows which are easily seen when the cards are held in front of light. The logic is simple. Each time a card is added, families not perforated in this new card are excluded. With 20 to 25 cards, it is likely that a family may be determined or search narrowed to limited families.

Edge-Punched Cards. It differs from body-puched card in having one card for each taxon and holes punched all along the edge of the card, one for each attribute. These holes are normally closed along the edge. For each attribute, present in the taxon, the hole is clipped out to form an open notch instead of a circular hole along the edge. For identification, all cards are held together as a pack. Thereafter a needle is inserted in the hole corresponding to the first attribute of the unknown plant. As this needle is lifted up, the taxa containing this attribute would fall down and those lacking that attribute would remain in the peak lifted by the needle. The latter are

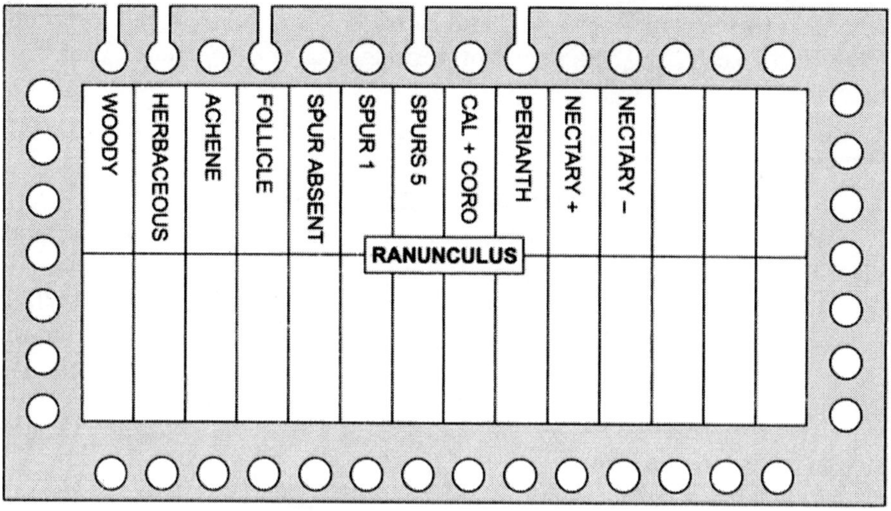

Fig. 15.2 Edge-punched card for genus *Ranunculus*. Only the attributes represented in the example above are pictured. Many more attributes could be added along the vacant holes to make the identification process more versatile.

rejected. The cards falling down are again rearranged in a pack and the needle is inserted in the hole corresponding to the next attribute of the unknown plant. This process is repeated untill finally a single card falls down and the taxon, which this card represents, is the identification of unknown plant. One may not have to explore all attributes of the unknown plant. The identification may be achieved much before reaching the end of the list of attributes of the unknown plant.

Tabular Keys

Tabular keys, essentially similar to polyclaves and meant for taking care of exhaustive list of attributes, are easier to use. The data are incorporated in the taxon-character frequency table and not on the cards. The taxa are arranged along the rows and attributes along the columns. The attributes are depicted as symbols or drawings. As such the table will have as many rows as taxa and as many columns as the number of attributes for which information is available. Identification is made using a strip of paper whose width is equal to each row and vertical lines separated by the width of the columns. This strip of paper pictured with the attributes of the plant to be identified, is next placed towards the top of the table and slowly lowered and compared with each row, the row with which the entries match represents the identification of plant in question.

Taxonomic Formulae

A taxonomic formula is an alphabetic formula based on specific combination of alphabets that are loaded with various attributes. Each taxon thus gets a specific alphabetic formula that next are arranged on alphabetic manner as words in a dictionary so as to locate them easily. The formula is located in the alphabetic list. It is decoded and identification is read against the formula. The examples of Ranunculaceae as dipicted in yoked and parallel keys is being extended here by assigning alphabets to attributes as A : Woody; B : Herbaceous; C : Achene; D : Follicle; E : Spur absent; F : Spur 1; G : Spurs 5; H : Calyx differentiated from corolla; I : Calyx not differentiated from corolla, only perianth present; J : Nectary present; K : Nectary absent, and the seven representative genera would be assigned the formulae as ACEIF (*Clematis*), BCEHJ (*Ranunculus*), BCEHK (*Adonis*), BCEIK (*Caltha*), BDFIJ (*Delphinium*) and BDGIK (*Aquilegia*). Such formulae are of great use in identification process. Hedge and Lamond (1972) have incorporated in written version of the multi-access key to the Genera of Apiaceae in the *Flora of Turkey*.

Computers in Identification

Rapid use of computers in data collection, over the years, has made plant identification online, an easier and quicker method, even without the help of a taxonomist. For the purpose, appropriate programme may be developed which can construct a key based on taxonomic information about the taxa. Constructed dichotomous keys are fed into a computer and run using an appropriate programme. Programming starts with the first couplet of the key, enquiries about the attribute in plant in question, and on the basis of information provided, handles the key asking relevant questions until finally the actual identification is achieved. Such keys are permanently stored in a computer and can be handled easily for the step-wise process of identification. Various methodologies involved in computer identification are **simultaneous character set identification methods, automated pattern-recognition systems, DELTA system (DELTA program key), setting up a simple DELTA identification etc.**

In recent years, information exchange through internet, providing illustrations of the plants from various parts of the world, has made the plant identification online. A number of electronic lists, *e.g.* **Taxacom**, are maintained by list servers and taxonomists of various parts of the world. In last few years, a number of interactive online identification programmes have been raised. Based on **Delta Intkey**, a programme for identification of vascular flora of Western Australia available in Flora Base.http://florabase.com.wa. gov.au has been developed. Web-based interactive identification keys are now also available in a program '**Actkey**' maintained by the Harvard University Herbaria Editorial Centre, http://flora.huh.harvard.edu : 8080/ actkey/index.jsp. for the floras of China, North America, Madagascar, Borneo, and the world.

❑❑❑

Chapter **16**

Magnoliidae
(Magnoliaceae, Annonaceae, Nymphaeaceae, Ranunculaceae, Papaveraceae, Fumariaceae)

MAGNOLIACEAE (Magnolia Family)

Bentham & Hooker	Hutchinson	Cronquist	APG
Dicotyledons	Dicotyledones	Magnoliophyta	II. APG Group
Polypetalae	Lignosae	Magnoliopsida	Neither Monocots
Thalamiflorae	Magnoliales	Magnoliidae	Nor Dicots
Ranales		Magnoliales	Magnoliales

The family is regarded as the primitivemost family among the living angiosperms. It comprises **12 genera** and about **230 species** distributed mainly in Indo-Malayan region, South China, Japan, Australia, North East USA, Brazil and West Indies. In India, it is distributed in NE and SE regions, mainly in evergreen climax forests of West Bengal, Assam, Naga and Khasi hills, Uttarakhand, UP etc. with **8 genera** and about **27 species.** Common Indian representatives are *Magnolia, Michelia, Manglietia, Schizandra, Talauma* and *Kadsura. Magnolia* (80 species) and *Michelia* (40 species) are major genera of the family.

Vegetative Characters

Plants are shrubs or trees–deciduous or evergreen. Climbing or scandent plants are also met with, *e.g. Schizandra* and *Kadsura*. Plants have erect, woody and branched stem with oil glands (aromatic glands) present in the parenchyma. Wood is vessel-less in *Drimys, Exospermum* and *Trochodendron.* **Leaves** are ramal, usually large (upto 70 cm long and 30 cm wide), simple, alternate or opposite (decussate), with entire or toothed margin. Stipules are large, foliaceous, bud protecting and deciduous. Lamina has unicostate reticulate venation and has rubaceous type of stomata and oil glands. Stipules are absent in *Drimys, Schizandra* and *Illicium.*

Inflorescence and Flowers

Flowers are usually solitary and terminal or axillary but in *Illicium*, these are crowded near the tips of branches.

Flowers are large (upto 20 cm in dia), showy, fragrant, often subtended by a spathaceous caducous bract, actinomorphic, complete, hypogynous and usually spirocyclic with an elongated or convex receptacle. However, these are unisexual in *Kmeria, Drimys* and *Schizandra.*

Perianth is represented by many segments, differentiated into calyx and corolla or all petaloid, arranged spirally on receptacle or in whorls and imbricate. In *Michelia*, 6-15 segments remain set in 3-5 series.

Androecium consists of numerous stamens. Stamens are spirally arranged, free or monadelphous on the basal part of the cone-shaped axis (*Magnolia*). Filaments show a transition from laminar type of terete filaments and proper anthers. Anthers are long, bicelled (dithecous), longitudinally dehiscing, introrse (*Mangolia*) or extrorse (*Liriodendron* and *Michelia*). Connective is prolonged, sometimes extending beyond the anther lobe as in *Illicium.*

Gynoecium comprises indefinite number of carpels. Carpels are free (apocarpous), often spirally arranged on an elongated torus (called gynophore). In *Illicium* and *Drimys*, 5-20 free carpels remain arranged in a single whorl. Ovary is superior and unilocular with one to several anatropous or amphitropous ovules in parietal placentation. In *Pachylarnax*

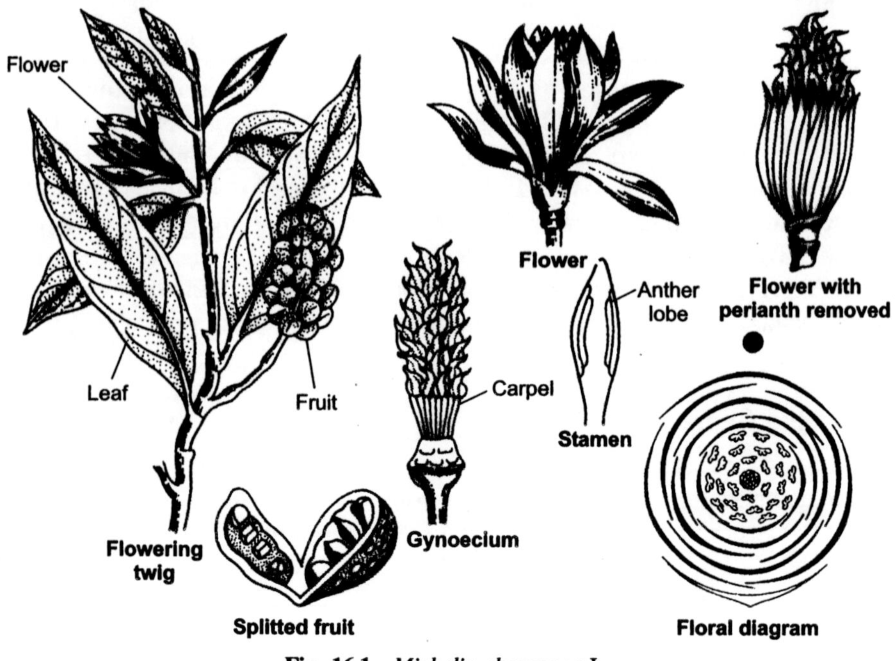

Fig. 16.1 *Michelia champaca* L.

and *Zygogynum*, carpels are fused while in *Talauma*, these fuse in fruiting. Style is short and distinct. Stigma one.

Flowers are protandrous and entomophilous.

Fruits and Seeds

Fruits are mostly aggregate, etaerio of follicles. Individual fruit in *Liriodendron* and *Euptelea* is samara, in *Schizandra* berry and in *Magnolia* follicle. **Seeds** are large with abundant oily endosperm and brightly orange decorative testa. In *Magnolia*, carpels dehisce and seeds appear suspended by fine silky threads (funiculus).

Floral Formula $\oplus$ ⚥ $P_{3-\alpha}$ or $K_3\ C_{6-\alpha}\ A_\alpha\ \underline{G}_\alpha$

Economic Importance

Family has a great economic value.

Timber values. An useful wood, for making boxes, musical instruments, toys etc., is obtained from some species of *Michelia* (*M. champaca* L., *M. nilagirica* Zenker and *M. exelsa* Blume), *Talauma* (*T. phellocarpa* King, *T. rabaniana* Hook. f. & Thoms.), *Magnolia* (*M. pterocarpa* Roxb.) and *Alchimandra* (*A. cathcartii* Hook. f. & Thoms.) Dandy syn. *M. cathcartii* Hook. f. & Thoms.). *Liriodendron tulipifera* L. (Tulip tree) yields an excellent commercial timber, called white wood or yellow poplar, valuable for furnitures and cabinets.

Ornamental values. Several plants are grown for their large and sweet smelling flowers and glossy leaves, *e.g. Magnolia grandiflora* L., *M. pterocarpa* Roxb., *M. denudata* Desv., *Michelia champaca* L., *M. fuscata* Blume, *Liriodendron tulipifera* L. (Tulip tree) etc.

Medicinal values. *Michelia champaca* L. is a valuable medicinal plant. Its roots and root bark are used as purgative, the flowers and fruits as carminative and in renal and veneral diseases like gonorrhoea and the oil in gout. *Drimys winteri* Forst. is the source of winter's bark used as an astringent and stimulant. The fruits of *Illicium griffithii* Hook. f. & Thoms. and *I. verum* Hook. f. yield an essential oil which is stimulant, stomachic and carminative. It also relieves colic.

Other uses. Flowers of *Magnolia grandiflora* L. and *Michelia champaca* L. are the source of a volatile oil which is highly valued in perfumery and cosmetics. Flowers of *M. champaca* L. yield yellow dye used in textiles. Fruits of *M. champaca* L., *Schizandra grandiflora* Hook. f. & Thoms. and *Kadsura roxburghiana* Arn. are edible.

Systematics and Phylogeny (Evolutionary Correlations of Magnoliaceae)

Magnoliaceae, in angiosperms, hold an important place as it shows several primitive characters, both in vegetative and reproductive organisations. The elongated floral axis bearing several free perianth segments, usually spirally arranged, the numerous stamens and carpels, also often acyclic in arrangement, monosulcate pollen grains and fruit a follicle, indicate the primitive composition of a flower. Presence of a copious endosperm in the seed is also a primitive character. The scalariform pitting of the vessels resembles the xylem of Mesozoic Cycadeoideae. Presence of vessel-less wood in some genera also points out its primitiveness.

Hallier (1905) regarded *Magnoliaceae* as most primitive family of existing angiosperms. He compared floral axis of *Magnolia*, bearing numerous spirally arranged free sporophylls, with sporophyll bearing axis of Bennettitales (order of fossil Gymnosperms). Bentham and Hooker (1862-83) placed it in the order **Ranales**, series **Thalamiflorae**, sub-class **Polypetalae** of **Dicotyledons**. They kept eight families in the order **Ranales** (*Ranunculaceae, Dilleniaceae, Calycanthaceae, Magnoliaceae, Annonaceae, Menispermaceae, Berberidaceae* and *Nymphaeaceae*).

Hutchinson (1959) included the family *Magnoliaceae* in the order **Magnoliales** of Division **Lignosae** of Sub-phylum **Dicotyledones**. He regards *Magnoliaceae* as the basic and most primitive family of Woody Dicotyledons (**Lignosae**) of Angiosperms. He considered it to be closely related to *Illiciaceae* and regards *Ranunculaceae* showing parallelism to *Magnoliaceae.* Takhtajan (1969) regards *Magnoliaceae* as closely related to *Degeneriaceae* and *Himatandraceae*, the latter forming a link between *Magnoliaceae* and *Degeneriaceae* on one hand and *Annonaceae* on the other. Cronquist (1968) stated that "Within the **Magnoliales** as here defined, the *Magnoliaceae, Winteraceae, Degeneriaceae, Himatandraceae* and *Annonaceae* form a cluster which has been evident to most recent studies of the group". These families have hypogynous, nearly always apocarpous flowers with mostly numerous stamens undivided into filaments and anthers, uniaperturate pollens and seeds with copious endosperms.

All Ranalian families share a common feature of release of pollen grains from the pollen sac at a two nucleate stage. Brewbaker's (1967) proposition that phylogenetically primitive taxa have binucleate pollen grains, further supports primitive status of these forms. *Magnoliaceae* further resemble *Degeneriaceae* and *Eupomatiaceae* in possessing multilacunar and heterogenous rays and open strand pattern of protoxylem (Benzing, 1967). Earlier, Eames (1961) pointed out resemblances between *Magnoliaceae* and *Annonaceae* in the presence of similar pattern of cortical vascular system in their floral axis.

The status of *Magnoliaceae* as the most primitive, has been challenged by Carlquist (1969), Gottsberger (1974) and Thorne (1976) who believe that *Winteraceae* is the most primitive family.

Qui and coworkers (1993, 99, 2000) studied molecular phylogenetic aspects of **Magnoliidae**. Their cladistic analyses of nucleotide sequences of plastid genes, *rbc*L and *ndh*F support monophyletic origin of the family *Magnoliaceae.* Judd and coworkers (2002), recognised only two genera–*Magnolia* and *Liriodendron* within the family. However, Thorne (2003) reviewed the status of sub-classes **Magnoliidae**, **Ranunculidae**, **Caryophyllidae**, **Dilleniidae**, **Rosidae**, **Asteridae** and **Lamiidae** and placed *Magnolia* and other 5 genera in the sub-family *Magnolioideae* but separated *Liriodendron* in monogeneric *Liriodendroideae.*

Phylogenctic interrelationships of the orders of flowering plants, compiled from recent cladistic analysis (based on 545 sequence of *rbc*L and 18S rDNA genes and 2538 *rbc*L sequences) place family *Magnoliaceae* in the order **Magnoliales** in II APG Group : Neither Monocots nor Dicots (Kallersjo, *et al.* 1998).

ANNONACEAE (Custard-apple Family)

Bentham & Hooker	Hutchinson	Cronquist	APG
Dicotyledons	Dicotyledones	Magnoliophyta	II. APG Group
Polypetalae	Lignosae	Magnoliopsida	Neither Monocots
Thalamiflorae	Annonales	Magnoliidae	Nor Dicots
Ranales		Magnoliales	Magnoliales

The family comprises **128 genera** and about **2300 species** widely distributed in tropics. In India, **26 genera** and about **200 species** are found with common examples like *Polyalthia, Annona, Artabotrys, Miliusa, Cananga, Goniothalamus* etc. Major genera of the family are *Guateria* (250 species), *Xylopia* (150 species), *Annona* (100 species) and *Polyalthia* (100 species).

Vegetative Characters

Plants are usualy aromatic trees or shrubs or often scandent or lianas. **Leaves** are simple, entire, exstipulate and alternate. Oil passages are found in the stem and leaves.

Inflorescence and Flowers

Flowers are often solitary, axillary or terminal, sometimes fascicled. *Polyalthia fragrans* is cauliflorus (flowers borne on old axis).

Flowers are usually ebracteate, pedicellate, complete, actinomorphic, bisexual and hypogynous. In *Stelechocarpus,* unisexual flowers are found. **Calyx** consists of 3 sepals, free or united (connate) at the base and valvate. **Corolla** comprises 6 petals in two whorls. Petals are free or basally connate and show imbricate or valvate aestivation. In *Unona*, inner whorl is absent.

Androecium consists of numerous free stamens, spirally arranged on a large convex receptacle. Anthers are adnate, dithecous and extrorse. Filaments are short. Connective is usually truncate and prominent. In *Xylopia,* anthers are transversely septate.

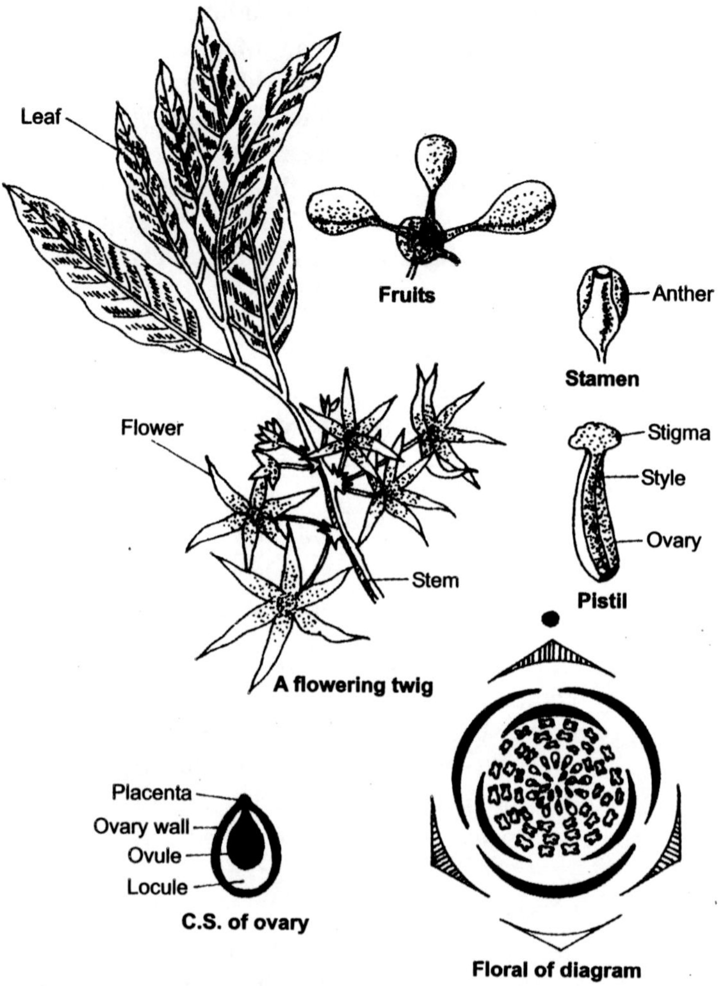

Fig. 16.2 *Polyalthia longifolia* (Sonner) Thaw.

Gynoecium has one to many carpels, spirally arranged on the receptacle above the stamens. Carpels are 1-3 in *Cyathocalyx*, few to many in *Artabotrys* and indefinite in *Polyalthia*. In *Annona*, carpels are subconnate with distinct stigmas. Ovary is superior, unilocular with many anatropous ovules in two rows on the ventral suture of the carpel. Placentation is marginal, basal or sub-basal with 1 or 2 ovuloes as in *Polyalthia*. Style is short or stigma sessile.

Fruits and Seeds

Fruit is usually an aggregate of berries. Berries in *Annona* become confluent with each other and with the receptacle to form a globose or ovoid, many seeded fleshy fruit (etaerio of berries). *Anaxogorea* has follicular fruits. **Seeds** are large with a small embryo and dense ruminate endosperm, often arillate.

Pollination and Dispersal

Flowers are often fragrant and insect pollinated. Seeds are dispersed by birds and bats. Fleshy fruits of *Annona* are usually dispersed by mammals.

Floral Formula $\oplus$ ⚥ $K_3 C_{3+3} A_\alpha \underline{G}_\alpha$

Two subfamilies and four tribes have been recognised in the family.

Sub-family-Annonoideae. Characterised by spirally set free carpels, comprises three tribes.

(i) ***Uvarieae.*** Leaves distichous, petals imbricate (rarely valvate), *e.g. Uvaria* and *Unona*.

(ii) ***Annoneae.*** Leaves distichous, petals valvate, rarely imbricate, *e.g. Annona* and *Polyalthia*.

(iii) ***Tetramerantheae.*** Leaves spirally arranged, sepals imbricate, stigma trilobed, *e.g. Tetrameranthus* (one genus).

Sub-family – Monodoroideae. Carpels united into a unilocular ovary, cyclic arrangement, comprises a single tribe.

(iv) ***Monodoreae.*** Placentation parietal, includes two genera-*Monodora* and *Isoloma*.

Economic Importance

Several plants of the family have great economic value.

Edible fruits. Fruits of several species of *Annona, e.g. A. cherimolia* Mill. (Hanumanphal), *A. reticulata* L. (Ramphal) and *A. squamosa* L. (Sitaphal, Sharifa) and *Asimina triloba* Dun. are edible.

Ornamentals. Several plants are grown for their fragrant flowers and attractive foliage, *e.g. Artabotrys odoratissimus* R.Br., *Cananga odorata* (Lamm.) Hook. f. Thoms. and *Polyalthia longifolia* (Sonner) Thaw.

Other uses. Flowers of *Cananga odorata* (Lamm.) Hook. f. Thoms. and *Artabotrys odoratissimus* R. Br. yield high class essential oil. *Bocagea longifolia* Bth. & Hk.f., *Miliusa indica* Loschen. and *Polyalthia longifolia* (Sonner) Thaw. produce valuable timber. Seeds of *Annona squamosa* L. are used for killing insects and lice (insecticidal properties).

Systematics and Phylogeny

Annonaceae are allied to *Magnoliaceae*, is supported by the presence of similar characters like free floral parts, numerous stamens and carpels, superior gynoecium etc. But it is more advanced than *Magnoliaceae* on the basis of sepals and petals. Hutchinson (1959) proposed that the family has been evolved from the Magnolaceous stock and that there are two different Orders, the **Magnoliales** and the **Annonales**, derived from the common stock. Takhtajan (1980, 91), stated that the *Annonaceae* are related to *Magnoliaceae* but noticeably more advanced. Cronquist (1981) relates *Annonaceae* with *Magnoliaceae* and regards these families to form a cluster of closely related families in the order **Magnoliales**. He, however, does not agree with Hutchinson to regard *Annonaceae* as belonging to a distinct order

Annonales. Takhtajan (1969, 1980) has also kept *Annonaceae* in the order **Magnoliales.**

The *Annonaceae* differ from the *Magnoliaceae* in reduced and whorled outer whorls (sepals and petals), exstipulate leaves and the presence of ruminate endosperm. The family resembles the **Monocotyledonae** by its typically trimerous flowers, and a small embryo with copious endosperm.

Phylogenetic interrelationships of flowering plants, compiled from recent cladistic analyses, place the family in the order **Magnoliales** in the II. APG Group : Neither Monocots nor Dicots (Kollersjo *et al.*, 1998). Such studies are based on sequences of *rbc*L, 18S rDNA, *atp* B genes etc.

NYMPHAEACEAE (Water-lily Family)

Bentham & Hooker	Hutchinson	Cronquist	APG
Dictoyledons	Dicotyledones	Magnoliophyta	II. APG Group
Polypetalae	Herbaceae	Magnoliopsida	Neither Monocots
Thalamiflorae	Ranales	Magnoliidae	Nor Dicots
Ranales		Nymphaeales	Not assigned to order

The family comprises **6 genera** and about **62 species** widely distributed in tropical and temperate regions of the world. Common genera found in India are *Nymphaea, Euryale* and *Victoria.* Major genera of the family are *Nymphaea* (40 species) and *Nuphar* (15 species).

Vegetative Characters

Plants are perennial aquatic herbs with creeping (rhizome) or small erect stem. Stems have peculiar anatomical structures such as scattered vascular bundles, lack of cambium, reduced xylem, occurrence of laticiferous canals and stellate hairs. **Leaves** are large, sometimes attaining a diameter upto 2 meters (*Victoria*), simple and long petioled and often floating on the surface of water.

Inflorescence and Flowers

Flowers are solitary axillary, usually on a long scape.

Flowers are situated on long pedicels and are large, showy, beautifully coloured, often sweet scented, bisexual, actinomorphic, complete, hypogynous or more or less epigynous or perigyous and spirocyclic or cyclic. Perianth often shows gradual transition from sepals to petals and petals to stamens. **Calyx** consists of 3 to many green sepals, often becoming petaloid and even larger than petals (*Nuphar*). **Corolla** comprises numerous petals, set in numerous whorls of 4-5 each and frequently grading into stamens. Petals are white or variously coloured and showy, and are spirally arranged in *Nymphaea.* In *Nuphar,* petals become minute and scale-like.

Androecium consists of many distinct stamens. Anthers are dithecous and introrse and filaments often extend as sterile appendages beyond the anther sacs. In *Nymphaea* and *Victoria*, stamens are laminar, flat and broad and without distinction between anthers and filaments and anther cells are deeply sunk in the lamina.

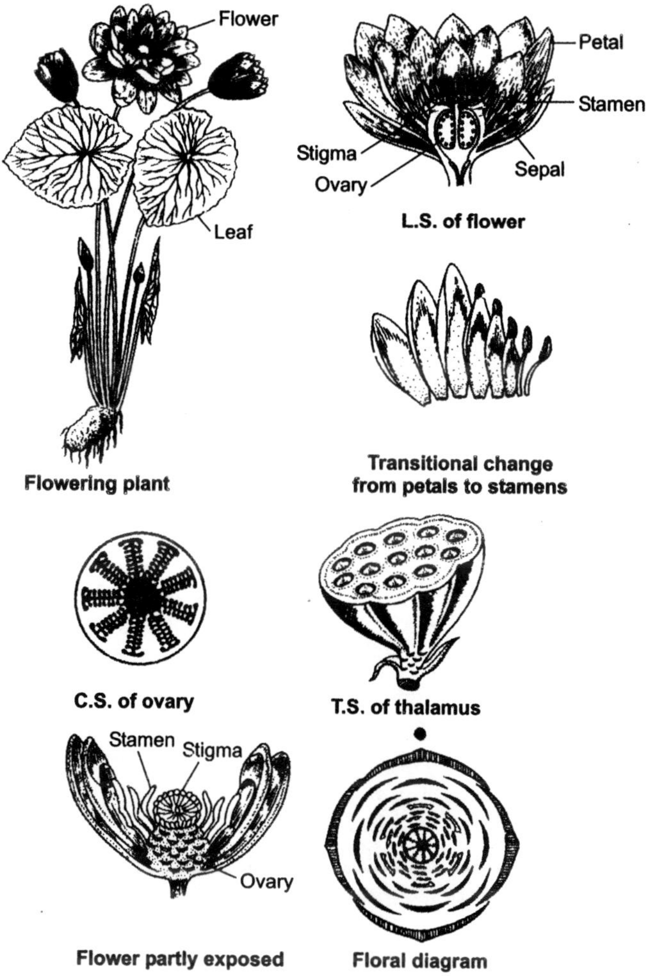

Fig. 16.3 *Nymphaea nouchali* Burm. f.

Gynoecium consists of numerous (5-35) united carpels (syncarpous). Ovary is superior (*Victoria*) to inferior (*Euryale*), unilocular with numerous anatropous ovules in lamellar or marginal placentation. Carpels remain enveloped by the thalamus and ovary is crowned with sessile and radiate stigmas, the number being equal to the number of carpels.

Fruits and Seeds

Fruit is an etaerio of indehiscent nutlets or a spongy berry. **Seeds** have small embryo and endosperm but with abundant perisperm.

Pollination and Dispersal

Flowers are both self and cross pollinated (entomophilous). Fruits and seeds are dispersed by water and aquatic animals, like ducks which eat them.

Floral Formula $\oplus$ ⚥ $K_{3-6}\, C_{\alpha}\, A_{\alpha}\, \underline{G}_{(5-35)}$

Economic Importance

The family has a considerable economic value.

Edibles. Seeds of *Euryale ferox* Salisb. (gorgan nut, makhana) are roasted and eaten. Seed flour is easily digestible.

Medicinal values. Rhizomes of *Nymphaea nouchali* Burm. f. and *N. stellata* Willd. are used in the treatment of dysentry and dyspepsia.

Ornamentals. Several species have high ornamental value due to their beautifully coloured, elegant and sweet scented flowers, *e.g. Nymphaea nouchali* Burm. f. syn. *Nymphaea lotus* L. (Indian red water lily), *N. stellata* Willd. (Indian blue water lily), *N. alba* L. (European white water lily) and *Victoria amazonica* Mart. & Zucc. syn. *Victoria regia* Lindl. (royal water lily) etc.

Systematics and Phylogeny

Nymphaeaceae are considered as a primitive family of **palaeoherb complex** as characterised by scattered vascular bundles, absence of vascular cambium, leaves alternate, usually palmately veined, adventitious root system and absence of etherial cells (cells producing etherial oils). Its status has been assigned in varied ways. Many present taxonomists believe it as not a single family. While Hutchinson (1959) kept it within the Order **Ranales** and splitted it into two families, *Nymphaeaceae* and *Cabombaceae*, Cronquist (1981, 88) splitted it into *Nymphaeaceae* and *Nelumbonaceae* and placed them into a new Order **Nymphaeales** along with the family *Ceratophyllaceae*, Takhtajan (1987, 91) maintained two orders **Nymphaeales** (includes *Nymphaeaceae, Cabombaceae* and *Barclayaceae*) and **Nelumbonales**. The **Nelumbonales** is a monotypic order with a single family, the *Nelumbonaceae*. It is separated because of distinct tricolpate pollen grains and lack of laticifers.

The affinities of *Nymphaeaceae* can be traced to the *Ranunculaceae* through **Cabomboideae**. The family resembles *Berberidaceae* in having scattered vascular bundles and arillate seeds, and with *Papaveraceae* in the presence of latex in tissues, large peltate-rayed stigma and lamellar placentation. Some investigators believe that aquatic habit and structure of ovary recalls a monocotyledonous family, the *Alismataceae*.

Recent trends in classification do not recommend separation of *Cabombaceae* as a separate family. APG classification and Judd and coworkers (2002) include it as a subfamily of *Nymphaeaceae* to avoid its polyphyletic origin. The molecular genetic studies strongly support to retain *Nymphaeaceae* as arising from the basal angiosperm clades and refute the proposition of Thorne (2000, 2003) to place the family *Nymphaeaceae* after the usual families of Magnoloid complex.

Poylogenetic interrelationships of flowering plants, compiled from recent cladistic analyses, did not assign any order to the family and placed it in II. APG Group : Neither Monocots nor Dicots and in between Dicots and Monocots, and with the ordinal position uncertain.

RANUNCULACEAE (Buttercup Family)

Bentham & Hooker	Hutchinson	Cronquist	APG
Dicotyledons Polypetalae Thalamiflorae Ranales	Dicotyledones Herbaceae Ranales	Magnoliophyta Magnoliopsida Magnoliidae Ranunculales	I. APG Group Eudicots 1. APG Group Eudicot-Unassigned Ranunculales

The family comprises **58 genera** and about **2505 species** found mainly in temperate and boreal regions of Northern Hemisphere. In India, some **20 genera** with about **154 species** are found, and common examples are *Ranunculus, Delphinium, Clematis, Anemone, Aconitum, Nigella* etc. The major genera of the family are *Ranunculus* (400 species), *Delphinium* (250 species), *Aconitum* (245 species) and *Clematis* (200 species).

Vegetative Characters

Plants are mainly annual or perennial herbs, rarely woody climbers (*Clematis* and *Naravelia*) or tall shrubs (*Paeonia*). Some species of *Ranunculus* are aquatic (*R. aquatilis*). Perennial forms develop rhizome. **Leaves** are alternate (*Ranunculus* and *Delphinium*) or opposite (*Clematis*) and exstipulate with sheathing bases. Lamina may be simple or pinnately or palmately compound or decompound. Aquatic species of *Ranunculus* show heterophylly. In climbing species, petiole coils around the support.

Inflorescence and Flowers

Inflorescence shows variation from solitary terminal (*Anemone* and *Nigella*), long raceme (*Aconitum* and *Delphinium*) to much branched panicles (*Clematis*) and dichasial cyme (*Ranunculus*).

Flowers are bracteate, hermaphrodite, actinomorphic, pentamerous and hypogynous. Flowers are zygmorphic in *Delphinium* and *Aconitum* and unisexual in *Thalictrum*. Involucre of 3-bracts is found in *Anemone* and *Nigella*. **Calyx** has 5-8 distinct and deciduous sepals. Sepals in *Delphinium, Clematis, Anemone* and *Aconitum* are petaloid (tepals) and the posterior sepal is spurred. Aestivation is imbricate or valvate. **Corolla** consists of 5-free (polypetalous), regular (*Ranunculus*) or irregular (*Aconitum* and *Delphinium*) petals. In *Delphinium*, posterior pair of petals forms spur. Anterior and lateral petals are small. In *Ranunculus*, nectary is found at the base of each petal. Petals are absent in *Clematis* and *Anemone*. Sometimes, petals turn into nectaries (*Delphinium* species). Aestivation is mainly imbricate.

Androecium consists of several free (polyandrous) and spirally arranged stamens. In *Nigella*, stamens are arranged in definite rings. Anthers are adnate, dithecous, extrorse and dehiscing longitudinally. The outer whorl of stamens, in most cases, is modified into nectaries.

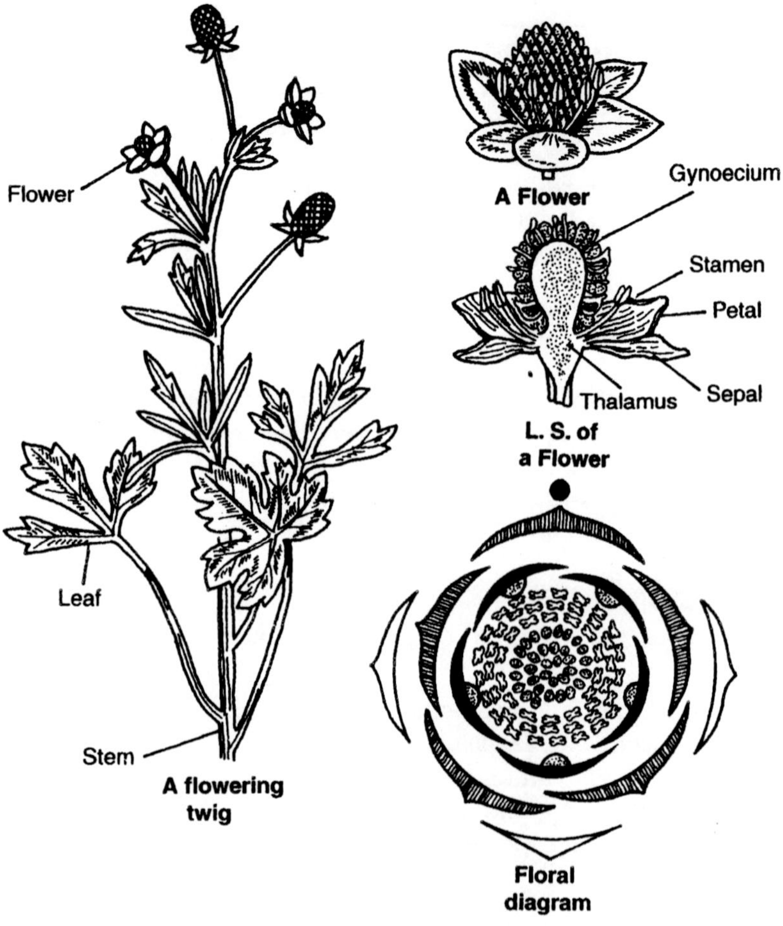

Fig. 16.4 *Ranunculus sceleratus* L.

Gynoecium comprises numerous free carpels arranged spirally on the distinct thalamus. However, in some cases, number of carpels varies, *e.g.* 1 (*Actaea*), 1-3 (*Delphinium*), 5-8 (*Nigella*) or indefinite (*Ranunculus*). In *Nigella*, carpels are united. Ovary is superior, unilocular with 1-many anatropous ovules. A variation is found in placentation, *e.g.* basal in *Ranunculus*, marginal in *Delphinium* and axile in *Nigella*. Styles are short or long with simple stigma.

Fruits and Seeds

A group of one seeded achene **fruit** is found in *Clematis* and *Ranunculus*, capsule in *Nigella* and berry in *Actaea*. A long persistent feathery style is found in *Naravelia* and *Clematis*. **Seeds** are endospermic.

Pollination and Dispersal

Flowers are normally protandrous and insect pollinated (entomophilous). In *Anemone, Clematis* and *Naravelia*, fruits are wind dispersed (anemochory) due to presence of persistent, long, feathery style. Seeds of *Ranunculus* are disseminated by animals or birds.

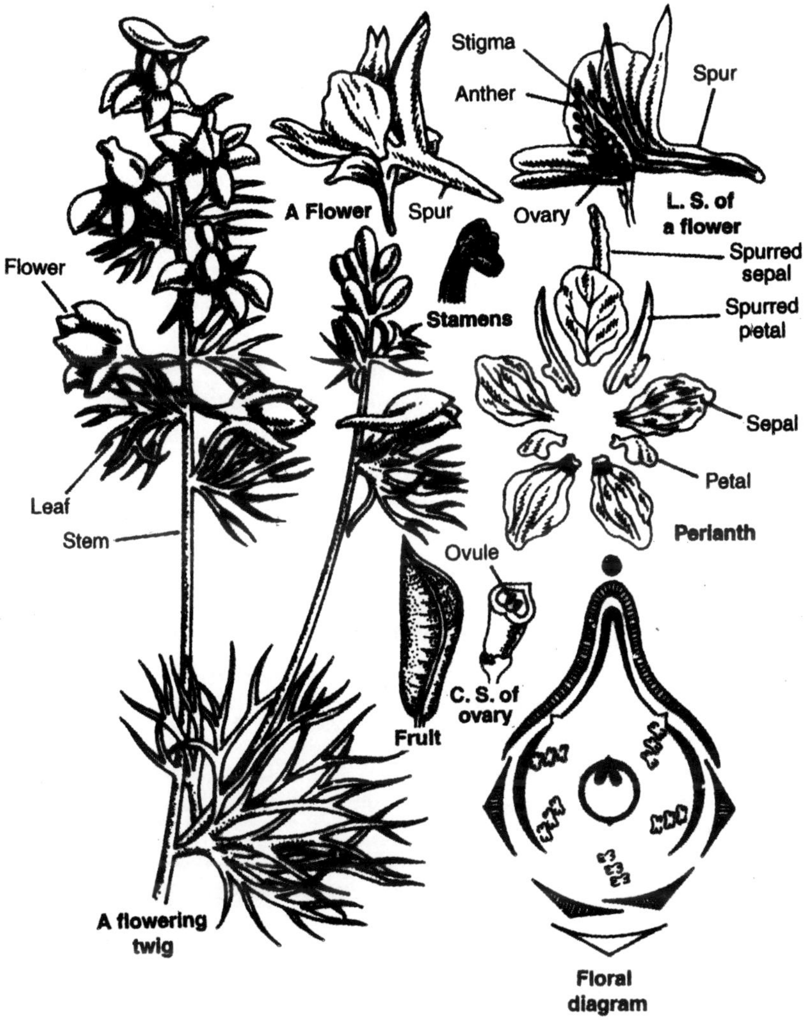

Fig. 16.5 *Delphinium ajacis* L.

Floral Formulae

Ranunculus	Br brl ⊕ ⚥ $K_5\ C_5 A_{\alpha}\ \underline{G}_{\alpha}$
Delphinium	Br brl % ⚥ $K_5\ C_{(4)}\ A_{\alpha}\ \underline{G}_1$
Nigella	Ebr ⊕ ⚥ $K_5\ C_0\ A_{\alpha}\ \underline{G}_{(5-8)}$
Clematis	Br brl ⊕ ⚥ $K_4\ C_0\ A_{\alpha}\ \underline{G}_{4-6}$

Economic Importance

Family has cosiderable economic value.

Medicinal plants. Roots of *Aconitum napellus* L. yield drug **aconite** used in the treatment of rheumatism and as nerve sedative. Plant extracts of *Ranunculus arvensis* L. and *R. muricatus* L. are used to cure intermittent fever. *Delphinium brunonianum* Royle is used as cardiac and respiratory depressant. Roots of *Hydrastis canadensis* L. (orange root) are used as an antidote for snake bite. Seeds of *Nigella sativa* L. (kala jeera) are used as condiment and in the treatment of fever, cough and bronchial asthma. *Anemone pulsatilla* L. yields valuable medicine **pulsatilla** used to cure nervous exhaustion.

Ornamental plants. Several plants are grown for ornamental purposes, *e.g. Ranunculus asiaticus* L. and *R. repens* L., *Clematis vitalba* L., *Anemone parviflora* Michx., *A. sylvestris* L., *Delphinium ajacis* L., *Aconitum napellus* L. and *A. bicolor* Schult, *Aquilegia vulgaris* L., *Nigella damascena* L., *Paeonia emodi* Wall. etc.

Systematics and Phylogeny

Most phylogenetists regard *Ranunculaceae* as a primitive family of angiosperms which shows affinity and relationship with *Papaveraceae* and *Magnoliaceae* of dicotyledons and *Alismaceae* and *Hydrocharitaceae* of monocotyledons. The primitive features met in the family are : **(1)** predominant herbaceous habit, **(2)** exstipulate leaves, **(3)** xylem vessels with simple pits, **(4)** stomata without subsidiary cells (ranunculaceous type), **(5)** bisexual and actinomorphic flowers, **(6)** uniform sepals and petals and their imbricate aestivation, **(7)** numerous free and spirally arranged stamens and carpels, **(8)** laminate stamens, **(9)** uniaperturate pollen grains, **(10)** unsealed carpels, **(11)** follicular fruits, **(12)** anatropous ovules, **(13)** endospermic seeds etc.

The floral parts of the family show similarities with *Magnoliaceae* but latter is separated from it in presence of etherial oil cells, absence of laminate stamens, woody habit and distinct stipules. The family also resembles *Papaveraceae* on the basis of stamens and carpels but latter is characterised by the presence of latex, dimerous calyx and syncarpous ovary. *Rosaceae*, although resemble it but stand out due to perigynous flowers and cyclic arrangement of stamens.

Hutchinson (1964) considered *Ranunculaceae* showing *parallel development* with the *Magnoliaceae* and suggested its evolution from common but remote ancestors. Further, he has gone to the extent of linking *Ranunculaceae* with some of the primitive monocotyledons (*e.g. Alismaceae*) based on similarities and suggested *Ranunculaceae* as probable ancestors of *Alismaceae*.

Rendle divided *Ranunculaceae* into three tribes, **(i)** *Helleboreae*, **(ii)** *Anemoneae* and **(iii)** *Paeonieae*. Hutchinson segregated tribe *Paeonieae* and raised it as a separate family *Paeoniaceae*, similarly *Helleboreae* into *Helleboraceae*. Takhtajan (1980, 91) and Cronquist (1981) have removed family *Paeoniaceae* from **Ranunculales** (or **Ranales**) and placed it close to *Dilleniaceae*.

Takhtajan (1997) removed *Hydrastasis* (having trimerous perianth) and *Glaucidium* and kept them under *Hydrastidaceae* and *Glaucidiaceae* under **Hydrostidales** and **Glaucidiales**, respectively. Thorne (2003) shifted *Glucidiaceae* under **Paeoniales** and *Hydrastidaceae* near *Ranunculaceae* under **Ranunculales**. Hoot (1995) studied restriction sites and sequence data of chloroplast DNA and suggested that these two genera should be retained within the family *Ranunculaceae*. Thorne (2003) divided *Ranunculaceae* into three subfamilies (*Coptidoideae, Isopyroideae* and *Ranunculoideae*) but Stevens (2003) has recognised five subfamilies adding *Hydrastidoideae* and *Glaucidioideae* (APweb Classification).

Despite all these facts, the family is accepted typically primitive with several internal displacements and interactions. APG classification places the family in 1 APG group : **Eudicots**, APG group **Eudicot - Unassigned** and in the order **Ranunculales.**

Table 16.1 Differences between Ranunculaceae and Papaveraceae

Ranunculaceae	Papaveraceae
1. **Flowers** are bracteate.	**Flowers** are usually ebracteate.
2. **Calyx** and **corolla** are distinguishable, sometimes perianth (as in *Delphinium*) is found.	Both **calyx** and **corolla** are distinguishable.
3. **Calyx** or outer whorl of perianth has 3-5 sepals or tepals.	**Calyx** has usually two sepals.
4. **Corolla** consists of five petals.	**Corolla** consists of 4 or 6 petals.
5. **Stamens** are arranged spirally.	**Stamens** are arranged in several whorls.
6. **Gynoecium** is apocarpous and carpels are spirally arranged on receptacle. Each carpel is unilocular.	**Gynoecium** is syncarpous and unilocular.
7. **Placentation** may be axile, basal or marginal.	**Placentation** is usually parietal.
8. **Fruit** is achene or several seeded follicle.	**Fruit** is several seeded capsule.
9. **Floral formula** $Br \oplus ⚥ K_5 C_5 A_\alpha \underline{G}_\alpha$	**Floral formula** $Br \oplus ⚥ K_2 C_{2+2} A_\alpha G_{(\alpha)}$

Table 16.2 Differences between Ranunculaceae and Capparaceae

Ranunculaceae	Capparaceae
1. **Leaves** are exstipulate and with sheathing leaf bases.	**Leaves** are stipulate.
2. **Inflorescence** may be solitary terminal or dichasial cyme.	**Inflorescence** may be solitary or racemose type.
3. **Flowers** are pentamerous.	**Flowers** are tetramerous.
4. **Calyx** consists of 5-8 sepals.	**Calyx** consists of four sepals arranged in two shorls.
5. **Corolla** consists of five petals.	**Corolla** consists of four petals.

6. **Androecium** consists of spirally arranged indefinite number of stamens.	**Androecium** consists of usually 4-6 stamens and androphores are present.
7. **Gynoecium** consists of a large number of free carpels **(apocarpous)**.	**Gynoecium** is bicarpellary and syncarpous. Usually, gynophore is found.
8. **Placentation** may be basal, (*Ranunculus*), apical (*Clematis*) or marginal (*Delphinium*).	**Placentation** is parietal.
9. **Fruit** is an achene.	**Fruit** is simple, berry or drupe.
10. **Floral formula**	**Floral formula**
Br brl ⊕ ⚥ $K_5\ C_5\ A_\alpha\ \underline{G}_\alpha$	Br ⊕ ⚥ $K_{2+2}\ C_5\ A_{4-\alpha}\ \underline{G}_{(2)}$

PAPAVERACEAE (Poppy Family)

Bentham & Hooker	Hutchinson	Cronquist	APG
Dicotyledons Polypetalae Thalamiflorae Parietales	Dicotyledones Herbaceae Rhoeadales	Magnoliophyta Magnoliopsida Magnoliidae Papaverales	1. APG Group : Eudicots APG Group : Eudicots-unassigned Ranunculales

Papaveraceae comprise **35 genera** and about **700 species** distributed worldwide, preferably in North Temperate regions. In India, about **5 genera** spread in **20 species** are found, mostly in Himalayas. Common examples met in the country are : *Papaver, Argemone, Mecanopsis* etc. *A. mexicana* is a common roadside and barren area's weed.

Vegetative Characters

Plants are mostly annual or perennial herbs containing usually milky or coloured latex. A few forms are climbers or scandent shrubs *e.g. Adlumia* and *Dactylocapnos. Bocconia arborea* is a small tree. **Leaves** are alternate with entire or lobed or cut margins with spiny teeth. These are exstipulate. *Dactylocapnos* and some species of *Corydalis* have compound or decompound leaves.

Inflorescence and Flowers

Flowers are mostly solitary but in certain species racemose or cymose inflorescence may be found. In *Chelodonium,* the small flowers are in umbels and in *Macleya,* inflorescence is paniculate.

Flowers are showy and attractive, bisexual, regular, complete and hypogynous. Perianth is biseriate or triseriate. **Calyx** consists of 2-3 free caducous and imbricate sepals. In *Eschscholzia,* sepals are united to form a

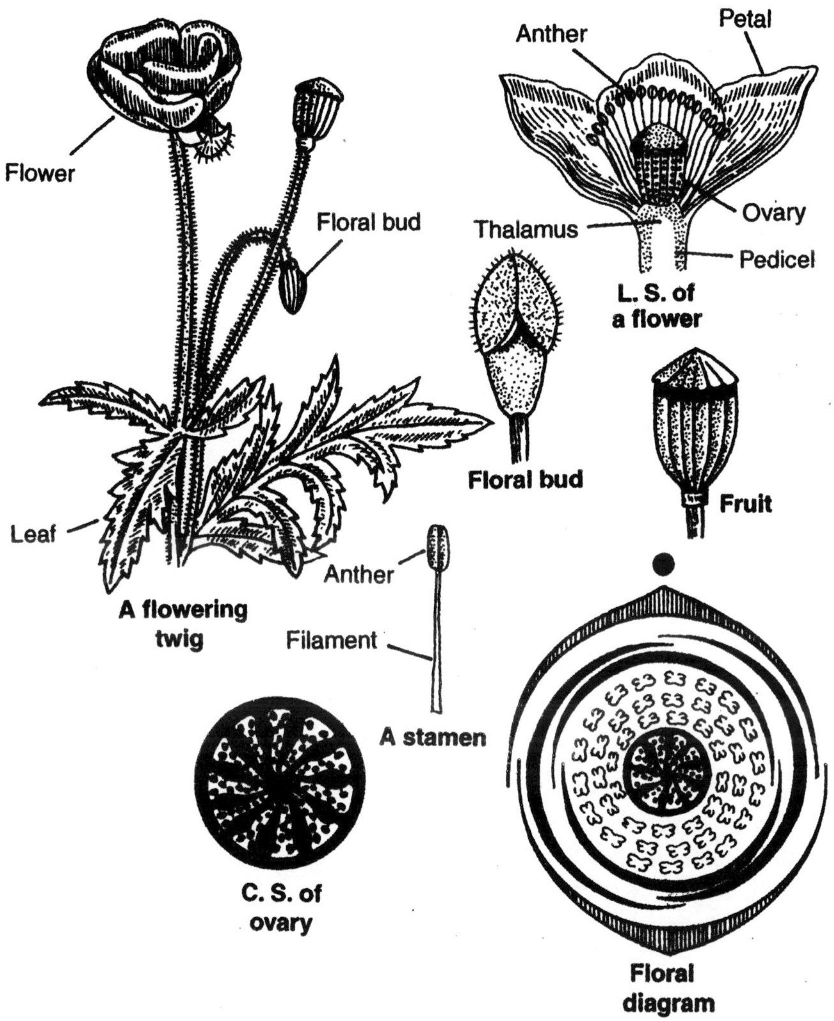

Fig. 16.6 *Papaver somniferum* L.

calyptra which falls off as the petals grow. **Corolla** consists of 4-6 petals (usually double the number of sepals), rarely 8-12. These are arranged in two whorls and are crumpled or rolled in bud. Petals are large, free (polypetalous) and variously coloured and imbricate. In *Bocconia* and *Macleya*, corolla is absent.

Androecium consists of indefinite number of stamens arranged in several alternating whorls. Anthers are dithecous, extrorse and dehiscing longitudinally.

Gynoecium has 2-many united carpels (syncarpous). Ovary is superior and

Fig. 16.7 Floral diagram of *Argemone mexicana* L.

unilocular with parietal placentation. Number of placenta is equal to the number of carpels. In *Papaver*, ovary becomes multilocular by connation of intruding placentae. Stigma are as many as the carpels. Style is very short or absent. In *Glaucium*, ovary becomes bilocular due to the formation of a replum, uniting the two opposite placentae. *Bocconia* has a unilcular ovary with a single basal ovule.

Fruits and Seeds

Fruit is many seeded capsule opening by valves or pores. Pod-like fruit of *Hypecoum* breaks up into several one-seeded parts by developing transverse septa. **Seeds** are small and endospermic and arillate in *Bocconia*.

Pollination and Dispersal

Flowers are both self- and cross pollinated. Seeds are dispersed by wind, birds or animals.

Floral Formulae

Papaver Ebr $\oplus$ ⚥ $K_2\ C_{2+2}\ A_\alpha\ \underline{G}_{(\alpha)}$

Argemone Ebr $\oplus$ ⚥ $K_3\ C_{3+3}\ A_\alpha\ \underline{G}_{(4-6)}$

The family has been divided into 3 sub-families.

(i) Papaveroideae. Latex present, flowers regular, petals not spurred, stamens many, free, fruit-capsular.

(ii) Hypecoideae. Latex absent, flowers regular, outer petals trilobed and inner tripartite, petals not spurred, stamens 4, fruit a lomentum.

(iii) Fumarioideae. Latex absent, oil sacs present, leaves much dissected, flowers zygomorphic, outer petals spurred, inner petals hooded, stamens 6, tripartite, fruit dehiscent or indehiscent.

Economic Importance

Economically, most important plant of the family is opium poppy (*Papaver somniferum* L.). It is widely cultivated in warm countries like India and Afganistan. The milky latex collected by the incision of unripen capsule wall is the source of **opium** which contains several important alkaloids, *e.g. morphine, codeine, papaverine, narcotine* etc. Opium is used medicinally as sedative to relieve pain, anxiety and insomnia (sleeplessness). However, opium is quite habit forming and its toxic doses produce sleep, respiratory depression or paralysis, coma and even death. Seeds of *P. somniferum* L., the *postadana*, are edible and source of oil. Seeds of *Argemone mexicana* L. yield a bitter, non-edible oil used as an illuminant or lubricant. The oil is toxic and its consumption produces dropsy.

Several species are grown as ornamental plants, *e.g. Papaver rhoeas* L. (corn poppy, garden poppy), *P. orientale* L. (oriental poppy), *P. nudicaule* L. (iceland poppy), *Eschscholzia californica* Cham. (californian poppy), *Mecanopsis* species (blue poppy) etc.

Systematics and Phylogeny

The floral characters of *Papaveraceae* are somewhat different from those of the *Ranunculaceae* as the family shows a great divergence in the occurrence of syncarpy (fused carpels) and capsular fruit. *Papaveraceae* resemble with *Capparidaceae* (*Capparaceae*) and *Brassicaceae* in presence of dimerous calyx and corolla, syncarpous, superior, unilocular ovary with ovules on parietal placentation and pod-shaped ovary but it differs from both in having albuminous seeds. The sessile stigma and projecting placentae resemble the family *Nymphaeaceae* to some extent.

Rendle kept *Papaveraceae* along with *Capparidaceae, Cruciferae* and *Resedaceae* under the order **Rhoeadales.** *Fumariaceae* was not recognised. Engler kept 7 families in the order **Rhoeadales** which have now been distributed in three orders by Takhtajan (1980, 97)—*Papaveraceae* and *Fumariaceae* in **Papaverales;** *Capparaceae, Tovariaceae, Moringaceae, Résedaceae* and *Brassicaceae* in **Capparales** and *Bretschneideraceae* in **Sapindales.** Hutchinson (1959) and several others have treated *Fumariaceae* as a separate family parallel to *Papaveraceae* under the order **Rhoeadales,** an order derived from **Ranales.** Cronquist (1981, 88) and Takhtajan (1980, 97) placed the family in a separate order the **Papaverales.** APG classification places the family under the APG Group—**Eudicots-unassigned** and order **Ranunculales,** as monophyletic familial group not aliied to any subgroup of Eudicots.

Table 16.3 Differences between Cruciferae and Papaveraceae

	Cruciferae (Brassicaceae)	Papaveraceae
1.	**Leaves** are simple and dissected.	**Leaves** are simple, divided, radicle or cauline.
2.	**Inflorescence** is raceme type.	**Flowers** are borne solitary.
3.	**Flowers** are tetramerous.	**Flowers** are bi- or trimerous.
4.	**Calyx** consists of four sepals showing 2 + 2 arrangement.	**Calyx** consists of two sepals.
5.	**Corolla** consists of four free petals, arranged in cruciform with valvate aestivation.	**Corolla** consists of 4-6 free petals with imbricate aestivation.
6.	**Androecium** consists of six stamens showing **tetradynamous** condition.	**Androecium** consists of indefinite number of stamens.
7.	**Gynoecium** consists of two carpels. **Ovary** becomes bilocular due to formation of false septum, called replum.	**Gynoecium** consists of two to many carpels. Ovary is unilocular.
8.	**Fruit** is **siliqua** or **silicula**.	**Fruit** is a multiseeded capsule showing porous dehiscence.
9.	**Floral formula** Ebr ⊕ ⚥ $K_{2+2}\, C_4\, A_{2+4}\, \underline{G}_{(2)}$	**Floral formula** Ebr ⊕ ⚥ $K_2\, C_{2+2}\, A_{\alpha}\, \underline{G}_{(\alpha)}$

Table 16.4 Differences between Papaveraceae and Capparaceae

	Papaveraceae	Capparaceae
1.	**Leaves** are simple, entire and exstipulate.	**Leaves** are simple or palmately lobed and stipulate.
2.	**Flowers** are ebracteate.	**Flowers** are bracteate.
3.	**Calyx** usually consists of two sepals.	**Calyx** consists of four sepals.
4.	**Corolla** consists of four free petals, arranged in two whorls.	**Corolla** consists of four free petals.
5.	Floral appendages are sessile, *i.e.* androphore and gynophore are not found.	Both androphore and gynophore may be present.
6.	**Androecium** consists of indefinite number of stamens.	**Androecium** consists of four to indefinite number of stamens.
7.	**Gynoecium** consists of two to many carpels.	**Gynoecium** consists of two carpels.
8.	**Floral formula** Ebr ⊕ ⚥ $K_2\ C_{2+2}\ A_\alpha\ \underline{G}_{(\alpha)}$	**Floral formula** Br ⊕ ⚥ $K_{2+2}\ C_4\ A_{4-\alpha}\ \underline{G}_{(2)}$

FUMARIACEAE (Fumaria Family)

Hutchinson	Cronquist	APG
Dicotyledons	Magnoliophyta	1. APG Group : Eudicots
Herbaceae	Magnoliopsida	1. APG Group :
Rhoeadales	Magnoliidae	Eudicots-unassigned
	Papaverales	Ranunculales

The family consists of **19 genera** and about **425 species** found mostly in temperate countries. In India, it is fairly represented by **4 genera** and about **40 species** found distributed in plains of North India, Nilgiri and Himalayas. Common examples are species of *Dactylocapnos, Dicentra* and *Corydalis* growing in Himalayas (temperate region) and *Fumaria indica* as weed in Indo-Gangetic plains, Nilgiris and foot hills of Himalayas.

Vegetative Characters

Plants are delicate, smooth, annual or perennial herbs with a watery sap. Some forms are climbers or scandent shrubs. Latex absent. **Leaves** are alternate, exstipulate and much divided (repeatedly pinnate).

Inflorescence and Flowers

Flowers are set in racemes subtended by bracts. In some *Corydalis* species, subumbellate condition is found.

Flowers are bracteate, complete, hermaphrodite, transversely zygomorphic (irregular), dimerous and hypogynous. **Calyx** consists of two small, scale-like caducous sepals. **Corolla** consists of 4 petals, set in more or less connate, in two whorls in usually very dissimilar pairs, the outer are large and one of them (in *Fumaria* and *Corydalis*) or both (in *Dicentra*) being basally saccate or spurred. Inner petals are narrower and coherent. In *Hypecoum*, the outer petals are trilobed and the inner tripartite.

Androecium consists of 6 stamens united into two lateral groups (series). Each stamen is branched into 3 near the upper end, the middle branch is bithecal and the lateral branches are monothecous or monothecal.

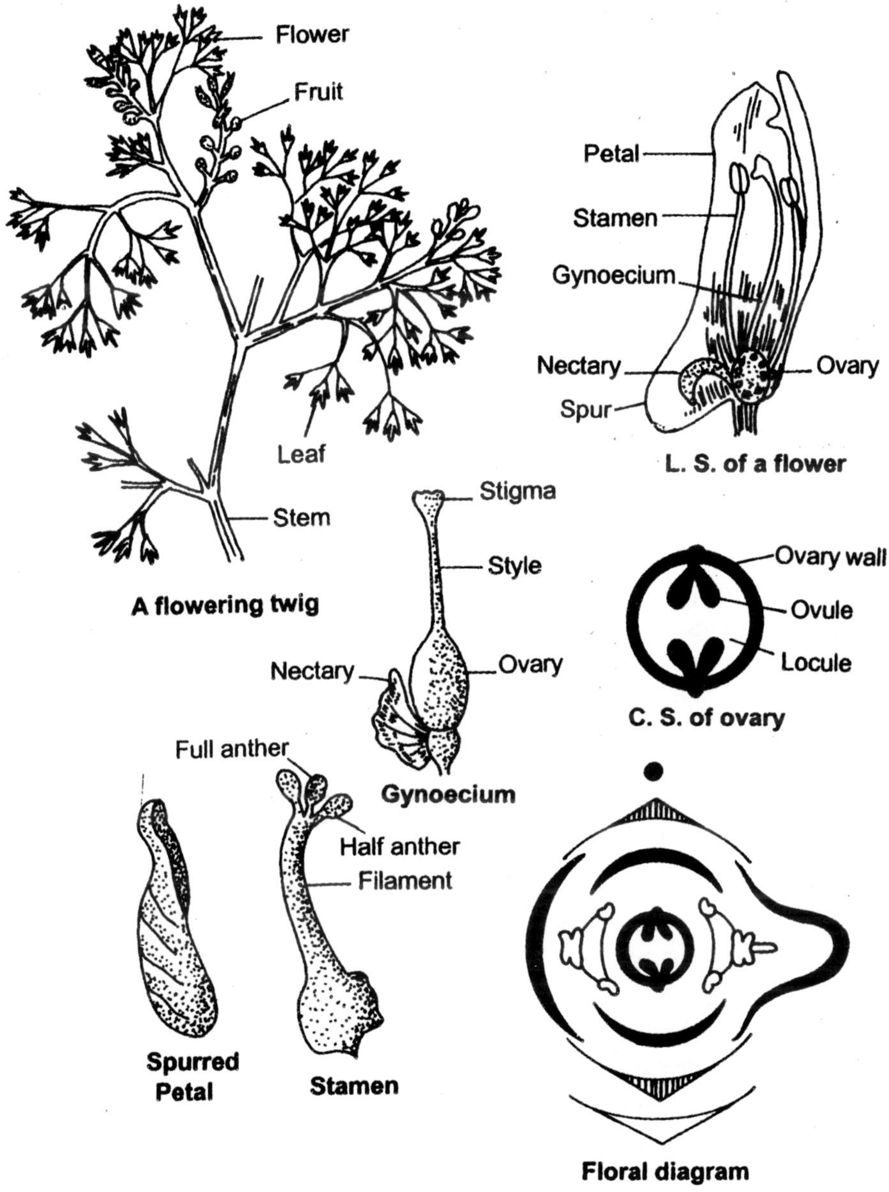

Fig. 16.8 *Fumaria indica* Pugsley.

One (as in *Fumaria*) or both (as in *Dicentra*) lateral stamen groups are with a basal nectary, enclosed in the petal spur. In *Hypecoum*, stamens are 4 and opposite the petals.

Gynoecium is bicarpellary and syncarpous with superior unilocular ovary. Ovules one (in *Fumaria*) to many set in parietal placentation. Style is slender and filiform and the stigma is capitate or shortly bilobed.

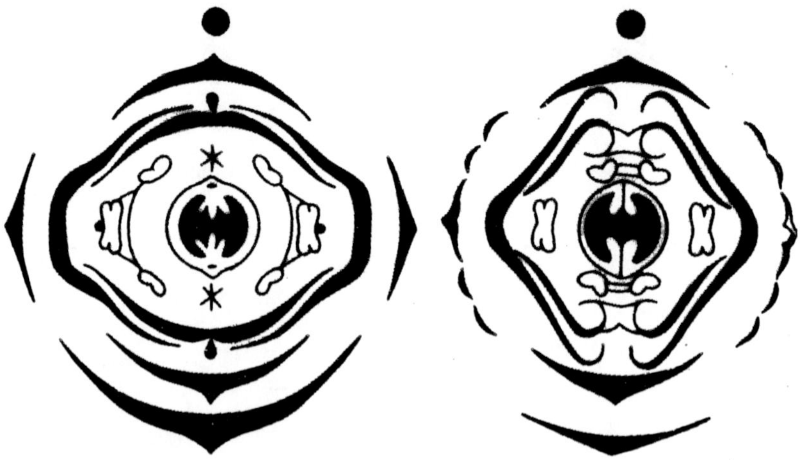

Floral diagram of *Dicentra* Floral diagram of *Hypecoum*

Fig. 16.9 Floral diagrams.

Fruits and Seeds

Fruit is a bi-valved capsule dehiscing transversely or an indehiscent, globose, one seeded nut (as in *Fumaria*). **Seeds** have a minute embryo and a fleshy endosperm.

Pollination and Dispersal

Flowers are entomophilous and seeds are dispersed by ants, birds and human beings.

Floral Formula

Fumaria Br % ⚥ $K_2\, C_{2+2}\, A_{(\frac{1}{2}+1+\frac{1}{2})+(\frac{1}{2}+1+\frac{1}{2})}\, \underline{G}_{(2)}$

Economic Importance

Some species of *Dicentra* and *Corydalis* are grown as ornamentals. *Hypecoum leptocarpum* Hook. f. & Thoms. and *H. procumbens* L. yield alkaloid **protopine** (a narcotic) used in cardiac and blood pressure problems. *Fumaria officinalis* L. (pitt-papara) is an efficient liver tonic for hepatic ailments.

Sytematics and Phylogeny

Bentham and Hooker, Rendle, Engler etc. considered the family as a component of family *Papaveraceae*. However, Hutchinson (1959), Cronquist (1981, 88) and Takhtajan (1980, 97) maintained it as a separate family under the order **Papaverales**. *Fumariaceae* are considered as advanced than *Papaveraceae* due to absence of latex, presence of zygomorphic flowers with spurred or saccate corolla, two stamen bundles in transverse plane, each with three stamens. *Hypecoum,* which was considered as a link between *Fumariaceae* and *Papaveraceae,* has been assigned a rank of a distinct family by Takhtajan (1980).

The family comes close to monocotyledons due to presence of a single active cotyledon, subterranean tubers and rhizomes bearing scales in several species of *Corydalis* and *Dicentra.* In APG classification, the family is considered as a small family that may be reduced to synonymy of their sister group. It has been placed in I. APG Group : **Eudicots - unassigned** and in the order **Ranunculales.**

❑❑❑

Chapter 17

Hamamelidae
(Moraceae and Casuarinaceae)

MORACEAE (Mulberry Family)

Hutchinson	Cronquist	APG
Dicotyledons Lignosae Urticales	Magnoliophyta Magnoliopsida Hamamelidae Urticales	3. APG Group Eudicots-Rosids Subgroup 1. Eurosids I Rosales

The family comprises **37 genera** and about **1,100 species** found mainly in tropical and sub-tropical regions of the world. In India, **15 genera** and about **150 species** are found. Common examples found in India are species of *Morus* (mulberry), *Ficus* (*F. benghalensis* and *F. religiosa*) and *Artocarpus heterophyllus* (jackfruit). *Ficus* is the largest genus having about 600 species.

Vegetative Characters

Plants are shrubs or trees with milky sap. Some species of *Ficus* (*F. foveolata* and *F. banjamina*) climb up on other trees with the help of aerial clasping roots. *Ficus benghalensis* produces characteristic prop roots.

Leaves are alternate or opposite, simple, entire and stipulate. Stipules, often caducous, are small and lateral in *Morus* or intrapetiolar in *Ficus* forming a cup covering the young bud. Leaves of *F. krishnae* are curiously somewhat cup-shaped. Often cystoliths are present in leaf epidermis.

Inflorescence and Flowers

Inflorescence shows a large variation. These may be heads, racemes, spikes, catkin etc. In *Morus*, male flowers are set in catkins and the female into pseudospikes. In *Artocarpus*, male inflorescence is pseudocatkin and female is pseudohead. In *Ficus*, it is hypanthodium in which peduncles of cymes coalesce to develop into a fleshy shallow axis bearing three types of flowers (male, female and gall flowers) on the interior surface. Species of *Ficus* may be monoecious or dioecious. *Streblus* is usually dioecious with male flowers in spikes while female ones solitary or 2-4 in groups. In *Dorstenia*, peduncles of several cymes unite to form a flat saucer-shaped structure with both types of flowers in the central portion. Cauliflory is found in *Artocarpus* and in some species of *Ficus*. Some species of *Ficus* are geocarpic, fruiting on underground stolons.

Flowers are unisexual and plants may be monoecious or dioecious. These are actinomorphic, tetramerous and hypogynous to epigynous. Bracts are often present. **Perianth** consists of usually 4 sepaloid tepals, free or united with valvate or imbricate aestivation. Female flowers of *Ficus* lack perianth. Tepals are rarely 2 or 1 and persistent.

Androecium comprises stamens equal in number and opposite to the perianth segments. In *Ficus, Artocarpus, Morus* and *Dorstenia*, the number is

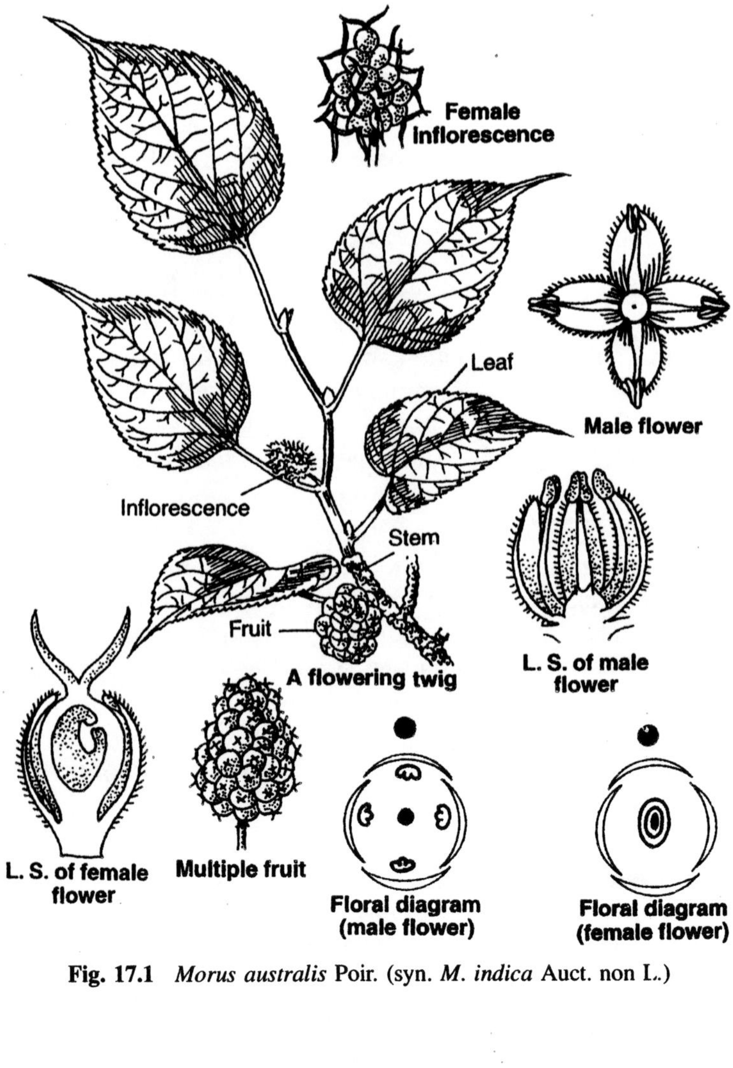

Fig. 17.1 *Morus australis* Poir. (syn. *M. indica* Auct. non L.)

sometimes reduced to 1 or 2. Anthers are dithecous, versatile and longitudinally dehiscing. Filaments remain inflexed in bud in *Morus, Streblus* and *Broussonetia* but erect in *Balanostreblus, Artocarpus, Ficus* and *Cudrania.*

A rudimentary ovary is sometimes found in the male flowers of *Morus.*

Gynoecium consists of two united (syncarpous) carpels, often one is abortive. Ovary is superior (as in *Morus*) or inferior (as in *Dorstenia*), unilocular with a solitary ovule, pendulous from the apex. Styles are two, rarely one, short with bifid stigma.

Fruits and Seeds

Individual **fruit** is achene (as in *Ficus*) or drupe (as in *Morus*). In *Ficus,* whole *hypanthodium* (inflorescence) matures in a fruit, the **syconus** and in *Artocarpus,* whole female inflorescence including receptacle forms a fruit, the **sorosis. Seeds** may be non-endospermic or endospermic.

Pollination and Dispersal

Flowers of *Morus* are wind pollinated (anemophilous). *Ficus* shows specialised mode of entomophily being carried out by gall-wasps (*Blastophaga*). Seeds are dispersed by birds and animals.

Floral Formulae

Male flower ⊕ ♂ P_{2+2} A_4 G_0

Female flower ⊕ ♀ P_{2+2} A_0 $\underline{G}_{(2)}$

Economic Importance

Edibles. *Morus alba* L. (white mulberry), *M. nigra* L. (black mulberry, shahtut), *M. serrata* Roxb. (Himalayan mulberry, tut) and *M. macroura* Miq. produce edible fruits which are laxative and refrigerant.

Ficus carica L. (common fig, anjir), *F. palmata* Forsk. (anjiri) produce edible figs. Fruits of *F. glomerata* Roxb. (gular) are cooked as vegetable and eaten raw when ripe.

Several species of *Artocarpus* (*A. integrifolia* L.f. (jackfruit, kathal), *A. lakoocha* Roxb. (monkey fruit, barhal) and *A. incisia* (Thunb.) L.f. (bread fruit etc.) are grown for edible fruits. People of Venezeula use copious latex of *Brosimum galactodendron* H.B. & K. (cow tree) as milk for its sweet and nutritious nature.

Medicinal Plants. Bark of *Morus alba* L. and *M. nigra* L. is used as purgative and vermifuge. Latex of *Ficus benghalensis* L. and *F. rumphii* Blume is anthelmintic, and of *Streblus asper* Lour. is astringent and antiseptic. Latex of *Antiaris toxicaria* Leschen. (upas tree) is poisonous but stem juice is mild cardiac and circulatory stimulant.

Other uses. The wood of *Artocarpus hirsuta* Lamk., *A. chaplasha* Roxb., *A. altilis* (Park.) Fosberg and *A. lakoocha* Roxb. is used for construction purposes. Hockey sticks and some other sport goods are made from the wood of *Morus alba* L.

The latex of *Castilla elastica* Cerv. forms "Panama rubber", *C. ulei* Warb. "Caucho rubber" and of *F. elastica* Roxb. produce "India rubber".

Ficus benghalensis L. is a sacred plant of Hindus and is grown closeby temples. *F. krishnae* DC is planted in gardens for its peculiar cup-shaped leaves.

Several spcies of *Ficus* and *Morus* are hosts of Indian lac insects. Their leaves are used as fodder.

Bark fibres of *Broussonetia papyrifera* Vent. (paper mulberry) are woven into coarse cloth and used to produce paper and cardboards.

Systematics and Phylogeny

Bentham and Hooker recognised tribe *Moreae* under the family *Urticaceae* of order **Unisexuales**, but Hutchinson (1959) believed that their inconspicuous unisexual flowers and tree habit indicate advancement and not primitiveness. He considered family *Moraceae* (a separate family) as a recent one and placed it in **Lignosae** in the order **Urticales** after the orders **Fagales, Juglandales** and **Casuarinales. Urticales** is related to **Juglandales** and **Fagales** and has been considered to be derived from the order **Hamamelidales.** Cronquist (1981) and Takhtajan (1980) placed the family *Moraceae* within the order **Urticales** in the sub-class, **Hamamelidae**, *i.e.*, under *Hamamelid complex.* Dahlgren (1980, 89) and Thorne (1992) have, however, placed them within *Malvanean groups.* APG classifications (1998) stretch them close to *Rosaceae* and *Rhamnaceae* under the order **Rosales.** Sytsma and Baum (1996) emphasising molecular phylogenies, proposed it as monophyletic in nature. They believe that the reduction of one carpel is also indicated (strongly unequal styles) in *Artocarpus, Dorstenia* and *Ficus.* A complete loss of one style probably occurred in common ancestors of *Cecropiaceae* + *Urticaceae* clade. APG II (2003) has recommended to merge *Cecropiaceae* with *Urticaceae.*

CASUARINACEAE (Casuarina Family)

Bentham & Hooker	Hutchinson	Cronquist	APG
Dicotyledons	Dicotyledones	Magnoliophyta	3. APG Group
Monochlamydeae	Lignosae	Magnoliopsida	Eudicots-Rosids
Unisexuales	Casuarinales	Hamamelidae	Subgroup 1. Eurosids I
		Casuarinales	Fagales

The family is represented by **a single genus**, *Casuarina* (the horse-tail tree) with about **40 species** found mainly in the Southern Hemisphere, particularly in Australia. In India, *C. equisetifolia* is most common throughout the coastal region and is now often grown in parks and along the roadsides. It grows well on sea shores. *C. suberosa* is commonly grown in Nilgiris.

Vegetative Characters

Casuarina plants are perennial, evergreen, profusely branched, woody shrubs or trees of xerophytic habit. It has jointed stem and branches and a whorl of slender branches at each joint (node). Branchlets are grooved, green in colour and resemble the pine needles to some extent, and are usually drooping. Internodes have striate grooves as many as leaves on the node.

Leaves are scale-like, whorled–whorls of 4-16, usually linear to lanceolate and always connate basally forming a sheath surrounding the node. Leaf tips appear only as teeth.

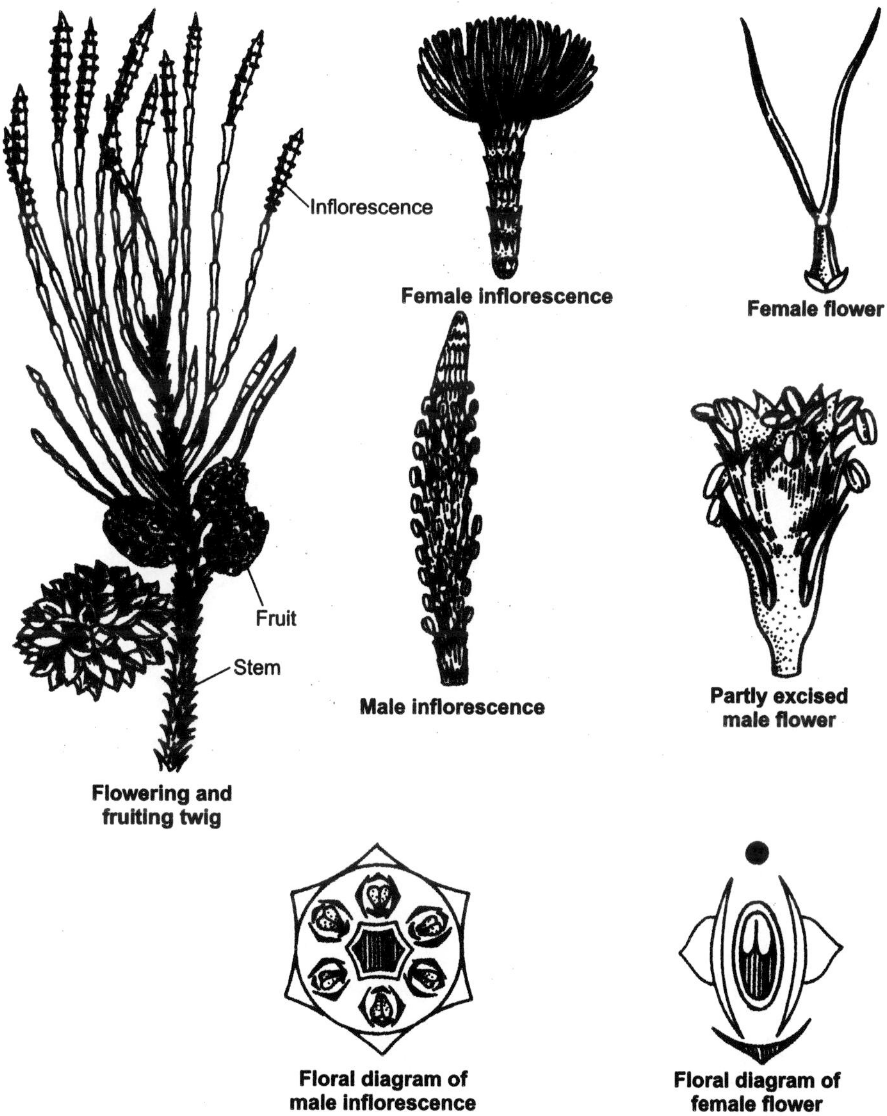

Fig. 17.2 *Casuarina equisetifolia* Forst.

Stem takes the function of leaf (phylloclade) with palisade cells below the ridges. The cortical vascular bundles are leaftraces. Stomata are sunken and situated in the furrows (a xerophytic adaptation).

Inflorescence and Flowers

Plants are monoecious or dioecious. Male flowers are produced in catkin-like inflorescences at the end of branchlets. These are formed in many whorls, each whorl in a sheath formed by the fusion of bracts. The female flowers are borne in a small capitate spike, each flower in the axil of a bract and subtended by a pair of bracteoles which enlarge and form a hard covering of the fruit.

Male flower has a rudimentary perianth represented by an anterior and a posterior scale, often one or both absent, and a single stamen, the **Androecium**, having slender filament and a bilobed anther. Anther lobes dehisce longitudinally. Stamen hangs out over the edge of the sheath.

Female flower is without perianth (*i.e.* naked). **Gynoecium** is bicarpellary, syncarpous and superior. The posterior carpel is abortive with the result ovary is unilocular. The anterior locule contains two parietal ovules, of which only one matures in the seed and the other one aborts. Ovule is ascending orthotropous. The style is short and terminal with two long filamentous stigmas.

Pollination. Flowers are wind pollinated.

Fruits and Seeds

Fruit is a winged nut enclosed by the two woody bracteoles which separate on maturity. Nuts are aggregated into a dry cone-like structure and form a multiple fruit.

Seed is non-endospermic with two large cotyledons.

Floral Formulae

Male flower $\oplus$ ♂ P_2 or $_0$ A_1

Female flower $\oplus$ ♀ P_0 $\underline{G}_{(2)}$

Economic Importance

Casuarina equisetifolia Forst. (jangalisaru or jan) yields good fuelwood. The wood is hard but not suitable for planking. Trees are planted on roadsides and as ornamental plants in plains and parks. These have been found very useful in land reclamation near sea and are now grown all along Indian coastal regions. Bark is sometimes used for tanning and has antidysentry and antidiarrhoeal properties.

Systematics and Phylogeny

Floral structure similarities between *Casuarina* of Angiosperms and *Ephedra* of Gymnosperms, such as simple flowers, presence of rudimentary perianth and a single stamen in male flower and presence of hard and protective covering of the seed coupled with their certain vegetative structures such as the presence of scaly leaves, phylloclades, distinct nodes and internodes in young branches, *i.e.*, general appearance, led earlier taxonomists like Engler and Rendle to conclude that *Casuarinaceae* are most primitive among Dicotyledons and that they have originated from *Ephedra*–like ancestors of Gymnosperms, *i.e.*, from Gnetales. They were of the opinion that **Amentiferae** with reduced unisexual flowers in catkins (or aments) constitute the most primitive living angiosperms.

However, modern taxonomists like Hutchinson, Cronquist, Thorne etc. believe that the simplicity of flowers in *Casuarina* is due to reduction, and features like unisexual flower, single terminal stamen, perianth rudimentary or absent, bicarpellate ovary with one carpel abortive and only one of the two ovules developing into a seed, point out to the advancement to some degree.

Hutchinson (1959) has placed it at the top of the **Amentiferae** and derived the entire group from **Hamamelidales.** He includes **Casuarinales** in the division **Lignosae** and places it after **Juglandales**. Cronquist (1980) classifies it as 11th order of the sub-class **Hamamelidae**, and considers it as the most evolved order of the sub-class which has developed advanced characters by the reduction in the course of evolution. Most taxonomists consider the **Casuarinales** to be a **monotypic order** which comprises a single family *Casuarinaceae.*

Takhtajan (1980) places this order between orders **Barbeyales** and **Fagales** in the sub-class **Hamamelididae.** Some other advanced characters met in the order are the presence of tricolpate pollen grains, trilacunar nodes (wood anatomy) and syncarpous pistil. It is believed that wind pollination is secondarily achieved. APG classification (2003) places the family in the order **Fagales** of subgroup 1. **Eurosids** I of 3. APG Group : **Eudicots-Rosids** (monophyletic familial group).

❑❑❑

Chapter **18**

Caryophyllidae
(Cactaceae, Chenopodiaceae, Amaranthaceae, Portulacaceae, Caryophyllaceae, Polygonaceae)

CACTACEAE (Opuntia Family)

Bentham & Hooker	Hutchinson	Cronquist	APG
Dicotyledons	Dicotyledones	Magnoliophyta	2. APG Group :
Polypetalae	Lignosae	Magnoliopsida	Core Eudicots
Calyciflorae	Cactales	Caryophyllidae	
Ficoidales		Caryophyllales	Caryophyllales

Family is fairly large, comprising about **50** distinctive **genera** and **2000 species.** There is dispute about the number of genera because some cactus experts record upto **150 genera** or so. Forms are naturally spread in tropical and drier parts of America, most abundantly in Mexico and Central America. In India, about **6 species** are indigenous but a large number of species have become localised and are grown as ornamental forms. The familiar examples are species of *Opuntia, Cereus, Pereskia, Mammillaria, Echinocactus, Echinopsis, Eriocereus* etc.

Vegetative Characters

Plants are commonly xerophytic, succulent, perennials, shrubs (*Opuntia*) or rarely trees (*Pereskia grandiflora* and *Carnegia gigantea*) growing upto 20 meters high. Some of them are woody climbers (*Pereskia aculeata* and *Rhipsalis cassytha*). *Cereus* is epiphytic. *Mammillaria* and *Notocactus* have condensed short stem that looks like a ball studded with spines.

Cacti are popular among plant lovers due to their great range of variation in vegetative characters. Roots are well developed, slender, fleshy or tuberosus.

Stem is peculiar in appearance, usually fleshy, erect (*Opuntia*), condensed (*Mammillaria*), climbing (*Pereskia*), epiphytic and hanging (*Rhipsalis*), joints frequently constricted and armed with areoles (spine-bearing areas), columnar or flattened into phylloclades (*Opuntia*).

Leaf spines are regarded as modified leaves of an axillary branch. These grow from rounded cushion-like tubercles and together are called areoles. In *Opuntia*, the areoles are often furnished with glochidia or barbed bristles. Areoles are often arranged in a definite phyllotaxy increasing the beauty of plants. Spiny leaves have several advantages, as these retard transpiration, protect from animals, promote dew formation at their tips etc. Leaves are broad, flat and succulent in *Pereskia* and *Pereskiopsis*.

Cell sap is mucilaginous which helps in checking transpiration. Epidermis is thick and highly cuticularised. Stomata are present in furrows. Internal tissues are water storing and parenchymatous. Chlorophyllous cells are present in the cortical region. In the ridge region, mechanical tissues are present. Most cacti are succulent-xerophytes.

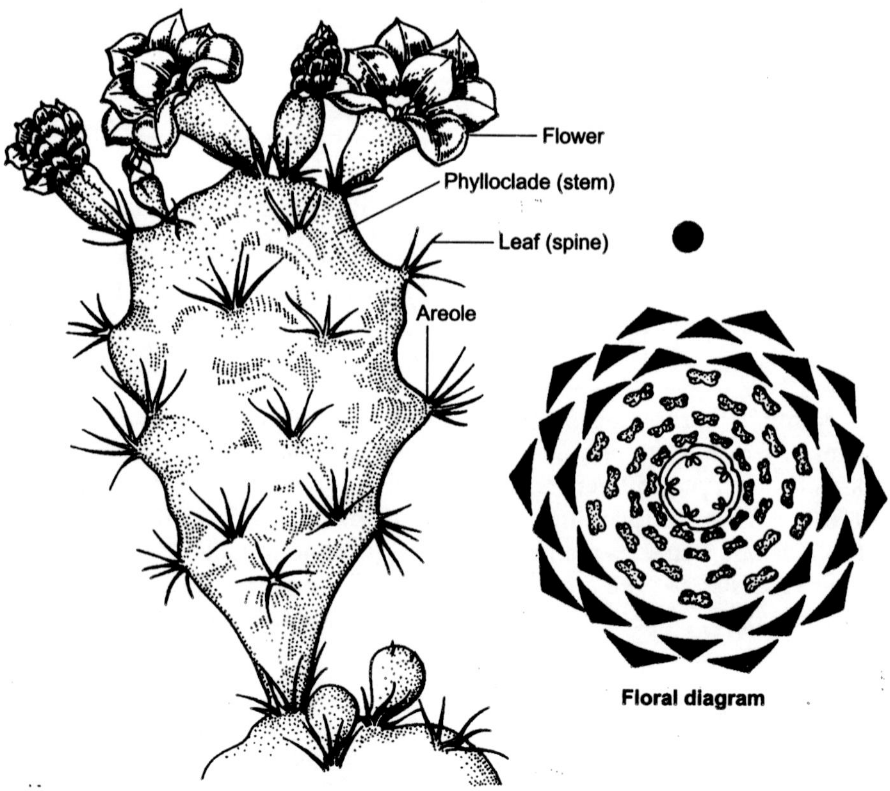

Flowering twig

Fig. 18.1 *Opuntia dillenii* Haw.

Inflorescence and Flowers

In most forms, flowers are solitary–borne upon or near the areole or in the axils of tubercles, or in clusters, rarely paniculate (*Pereskia*) or spikate. In *Rhipsalis*, several flowers develop from an areole in succession.

Flowers are usually sessile, large, showy, hermaphrodite, actinomorphic or zygomorphic, epigynous, often clothed with stiff hairs or spinous structures. Receptacle is cup-shaped on which numerous spirally set perianth segments are found. **Perianth** exhibits a gradual transition from sepals to petals and often merges into numerous bracts, free or united to form tube. All these are often spinous at the tip. The receptacle with bases of perianth segments forms the *hypanthium.*

Androecium comprises large number of stamens which are arranged spirally. Filaments are free, slender, sometimes much exerted, often united to form a column or are flat. Anthers are dithecous (2-celled), basi- or dorsifixed and dehiscing longitudinally. Pollens are minute, globular with finely warted exine.

Gynoecium comprises 2–α number of carpels. Carpels are syncarpous. Ovary is inferior, unilocular with numerous ovules set on parietal placentae, the number of placentae is equal to the number of carpels. Ovules are anatropous or campylotropous. Style is simple, rarely branched and stigma is capitate, rarely lobed.

Pollination

Flowers being large, brightly coloured and sweet scented, are well adapted for pollination by insects (entomophilous).

Fruits and Seeds

Fruit is a berry, spiny or bristly or glochidiate, often dry (*Echinocactus*). **Seeds** are numerous, often finely pitted or with smooth testa, rarely winged and with corky aril. Endosperm is scanty or absent. Fruits are often sweet and pleasant to taste and are distributed by animals (zoochory).

Floral Formula $\oplus$ or % ⚥ $P_{\alpha}\ A_{\alpha}\ \overline{G}_{(\alpha)}$

Economic Importance

A large number of cacti are grown in green houses as ornamental plants and used in developing rock gardens due to their peculiar plant body, large showy flowers and xerophytic habits, *e.g. Cereus, Echinocactus, Epiphyllum, Mammillaria, Echinocereus* etc. Grafting is widely practiced. Species of *Opuntia* and *Nopalaea* are grown as hedges along the fields.

Fruits of some species of *Opuntia* are sweet, tasty and edible. Spineless fruits are used as fodder. The acidic juice of *Opuntia* fruits is used in the treatment of whooping cough and gonorrhoea. Some species of *Anhalonium*, are highly poisonous containing saponine and other alkaloids. Species of *Lophophora* and *Ariocarpus* are used for preparing an intoxicating drink by American Indians on religious festivals for hallucinogenic purposes. A cochineal dye, carmine, is obtained from insects reared on different cactii, *e.g. Nopalaea coccinellifera* Solm-Dyke and *Opuntia tuna* Mill.

Systematics and Phylogeny

The systematic and phylogenetic position of family *Cactaceae* remains controversial owing to diverse opinions of the taxonomists. Bentham and Hooker included the family in the order **Ficoidales** after **Passiflorales** and before **Umbellales**. Engler considers that *Cactaceae* show affinity with *Aizoaceae* and placed it in the order **Opuntiales** near **Myrtiflorae**. Hutchinson (1973) included the family in an order of its own, the **Cactales** and placed it between **Cucurbitales** and **Tiliales**. Maheshwari (1945), on the basis of embryological, morphological and anatomical characters, suggested that the family should be assigned a position near or within **Centrospermae**. On the basis of *Anemarrhena* type of seedling, spirocyclic flowers, transition from sepals to petals, fasciculate androecium and large persisting antipodals, many taxonomists supported Hutchinson to develop a monotypic order **Cactales** to include a single family *Cactaceae*. However, Cronquist (1981) placed family *Cactaceae* in the order **Caryophyllales** (*i.e.* **Centrospermae)** of the sub-class **Caryophyllidae** and did not develop a new order **Cactales** or **Opuntiales.** Affinity of *Cactaceae* with *Aizoaceae, Portulacaceae* and *Phytolaccaceae* has been supported by morphological and chemotaxonomic investigations and it is now believed that the family has been derived from the most primitive apocarpous *Phytolaccaceae*. Takhtajan (1991) and Thorne (1999, 2001) have also placed the family *Cactaceae* in the order **Caryophyllales.** APG classification (1998, 2003) places the family in the order **Caryophyllales** of 2 APG Group : **Core Eudicots.**

CHENOPODIACEAE (Chenopodium Family)

Bentham & Hooker	Hutchinson	Cronquist	APG
Dicotyledons Monochlamydeae Curvembryae	Dicotyledones Herbaceae Chenopodiales	Magnoliophyta Magnoliopsida Caryophyllidae Caryophyllales	2. APG Group : Core Eudicots Caryophyllales

The family is represented by **105 genera** and **1500 species.** Plants are found worldwide, especially in xerophytic and halophytic areas. A fairly large number of species are found in Prairies, North America, Mediterranean coasts, shores of Red Sea, salt plains of Australia, Central Asia etc. In India, **20 genera** and about **50 species** are found spread from coastal region to Himalayan ranges. Common examples met in India are *Beta vulgaris, Spinacia oleracea, Chenopodium album, C. maritima, Basella rubra, Salicornia brachiata* etc. *Acroglochin chenopodiodes* and *C. botrytis* occur in Himalayas upto 7000 ft. *Kochia indica* is a tall erect annual, upto 2 m. high.

Vegetative Characters

Plants are usually annual or perennial herbs, rarely shrubs (some species of *Suaeda* and *Salicornia*) or trees (*Haloxylon*) reaching upto the height of 20 ft or twiner (*Basella*). Plants are mostly halophytes, *e.g. Salsola kali* and *Salicornia herbacea.* Plants are glabrous or mealy (covered with small hairs). Stem is often succulent or jointed as in *Salicornia.* **Leaves** are simple, usually alternate, rarely opposite (*Salicornia* and *Nitrophila*) and exstipulate and are often covered with hairy growth. In *Salicornia,* leaves are highly reduced and stem becomes succulent and green cactus-like.

Stem is characterised by the presence of several concentric rings of vascular bundles. Anomaly is found due to appearance of pericyclic rings or strips of cambium, which originate and cease their activity successively forming secondary bundles and conjunctive tissue.

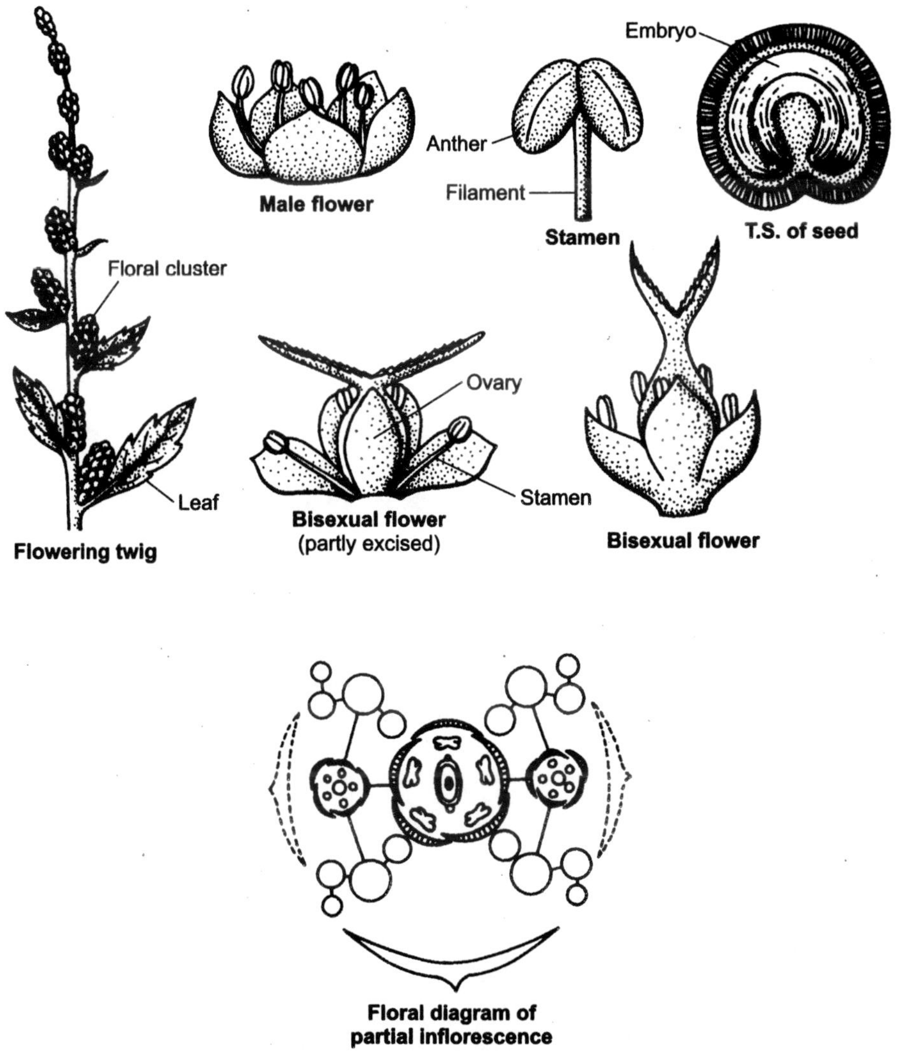

Fig. 18.2 *Chenopodium album* L.

Inflorescence and Flowers

Flowers are set in a mixed panicle, small cymes arranged in racemes or spikes or often a thyrsus.

Flowers are minute or inconspicuous, green, bracteate or bracteolate or without bracts and bracteoles. These are usually regular (actinomorphic) and bisexual, rarely unisexual and hypogynous, except in *Beta*. When unisexual, plants may be monoecious (*Sarcobatus* and *Spinacia*) or dioecious (*Grayia* and some species of *Atriplex*).

Perianth consists of 5 segments, uniseriate and sepaloid. However, in *Salicornia*, 3 or 4 and in *Atriplex* only two perianth segments are found. Tepals are persistent, free or connate. Aestivation is of imbricate type. Sometimes, perianth is absent in male flowers.

Androecium consists of stamens as many as the perianth segments and opposite to them. They are inserted on the floral axis (as in *Salsola*) or on a disc (as in *Haloxylon*). Number of perianth segments and stamens may be variable in the same genus or even in one and the same plant (*Chenopodium* species). Anthers are bicelled and introrse, dehiscing longitudinally. Pollen grains are spherical and with numerous pores. In *Beta* and *Haloxylon*, a nectariferous disc may be present below stamens.

Gynoecium is bi- or tricarpellary, syncarpous with a superior ovary, except in *Beta*, where it is either half inferior or inferior. Ovary is unilocular with one basal, rarely lateral, campylotropous ovule. Style is usually one, rarely 2 or 3 and free and stigma is capitate or 2-3 lobed.

Pollination

It occurs usually through the insects (entomophily).

Fruits and Seeds

Fruit is usually a roundish nut or a utricle, often enclosed by the persistent perianth. In *Basella*, fruit is a fleshy berry. Fruits are dispersed by water or wind. **Seeds** of *Haloxylon* and *Atriplex* are winged. Seeds are endospermic with a bent or spirally twisted embryo.

Floral Formula

$$\oplus\ ⚥ \text{ or } ♂ \text{ or } ♀\ Br\ brl\ P_{5 \text{ or } (5)} \text{ or less } A_5\ \underline{G}_{(2-3)}$$

Economic Importance

The family has great economic value, especially for some food plants.

Edibles. Common edible plants are *Chenopodium album* L. (chenopod or bathua), *Beta vulgaris* L. (sugar beet or chukandar), *B. benghalensis* (gubru palak) and *Spinacia oleracea* L. (spinach or chirva palak). *C. album* L., a common winter weed of north and central India is popularly eaten as vegetable or salad. Plants are laxative, and extremely rich in potash salts. Seeds of *C. quinoa* are boiled and eaten like rice.

Beta vulgaris L., economically most important, is cultivated extensively for its nourishing valued roots being eaten as salad or mixed in vegetables. Roots are the most important source of sugar in European countries. Some improved varieties have sugar content more than 20 per cent of the total weight.

Plants of *Spinacia oleracea* L. are largely cultivated for their nourishing leaves which are eaten raw as salad or mixed with various vegetables.

Ornamental plants. Several plants are grown for their ornamental value, *e.g. Beta vulgaris* L., *Spinacia oleracea* L., *Kochia indica* Wight., *K. scoparia* (L.) Schrad (syn. *S. trichophylla* Voss.), *Chenopodium amaranticolor* (L.) Coste & Reyn. etc.

Medicinal plants. Some plants have medicinal importance, *e.g.* leaves of *C. album* L. are laxative; leaves of *B. vulgaris* L. are applied to burn and bruises and its seeds are diaphoretic and cooling; leaves of *S. oleracea* L. are cooling and are useful in lung inflamation and seeds are laxative, cooling and used in difficult breathing, inflamation of liver and in jaundice; plants of *Kochia indica* Wight. are cardiac stimulant and are used in weak and irregular heart beating; *C. anthelminticum* L. yields "Oil of Chenopod" of considerable medicinal value etc.

Systematics and Phylogeny

Bentham and Hooker have placed family *Chenopodiaceae* in the series **Curvembryae** of the class **Monochlamydeae** along with *Amarantaceae, Phytolaccaceae* etc. Hutchinson (1959, 80) kept *Chenopodiaceae* under a separate order **Chenopodiales** and believes that *Chenopodiaceae, Amaranthaceae, Phytolaccaceae, Gyrostemonaceae* and *Basellaceae* are more advanced families. They are considered derived from the order **Caryophyllales** whereas the families included in **Caryophyllales** and *Polygoniaceae* are derived from **Ranales.** According to Cronquist (1981), family *Chenopodiaceae* stands between *Amaranthaceae* and *Phytolaccaceae,* and has been derived through *Phytolaccaceae.* He believes that *Phytolaccaceae* is most primitive among other families of order **Caryophyllales** from which families like *Cactaceae, Chenopodiaceae, Amaranthaceae, Portulacaceae* and *Caryophyllaceae* have been derived. *Chenopodiaceae* are closely related to *Amaranthaceae* from which it differs chiefly in green and fleshy perianth. Cronquist (1981) and Takhtajan (1997) included *Chenopodiaceae* and *Amaranthaceae* in the order **Caryophyllales**.

Family is distinguished from *Amaranthaceae* in its herbaceous perianth, all fertile stamens equal in length and free filaments. In APG classification, it has been merged with *Amaranthaceae,* because separation of *Chenopodiaceae* leads to paraphyletic affinity group (other than **core eudicots**) owing to inclusion of genus like *Sarcobatus* in it which has been kept by Bentham and Hooker in a separate monogeneric tribe. Thorne (2003) has placed *Sarcobataceae* and *Nyctaginaceae* in a separate sub-order **Phytolaccineae**.

AMARANTHACEAE (Amaranth Family)

Bentham & Hooker	Hutchinson	Cronquist	APG
Dicotyledons	Dicotyledones	Magnoliophyta	2. APG Group :
Monochlamydeae	Herbaceae	Magnoliopsida	Core Eudicots
Curvembryae	Chenopodiales	Caryophyllidae	Caryophyllales
		Caryophyllales	

The family contains **69 genera** and about **1000 species** distributed in tropical and temperate regions of the world. In India, the family is represented by **17 genera** and over **50 species** growing in warmer parts of the country. Common examples met in the country are *Amaranthus, Achyranthes, Aerva, Celosia, Gomphrena, Alternanthera* etc. *Gomphrena* (120 species) and *Alternanthera* (100 species) are some large genera.

Vegetative Characters

Plants are herbs or undershrubs. *Deeringia* is a climbing shrub with long hanging branches. Branching is often trichotomous in *Pupalia* and *Psilotrichum.* **Stem** shows abnormal secondary growth and primary anamolous medullary vascular bundles (similar to *Chenopodiaceae*). **Leaves** are alternate or opposite, simple, exstipulate and often coloured or molted. Spines are rarely present in leaf axils.

Inflorescence and Flowers

Small flowers are aggregated into simple or branched panicles, spikes, heads or cymes. Inflorescence of *Celosia cristata* is very dense and showy.

Flowers are bracteate, bracteolate, actinomorphic, monochlamydeous, bisexual or rarely unisexual, pentamerous and hypogynous. Each flower is subtended by a persistent bract and a pair of similar bracteoles which are often hooked.

Perianth comprises 3-5 tepals, set in a single whorl, free or basally connate, rigid or scarious, more or less membranous often coloured and persistent with imbricate aestivation.

Androecium consists of 5 stamens which lie opposite the perianth segments. Filaments are free (*Digera* and *Amaranthus*) or connate in a cup or tube (*Cyathula* and *Celosia*). Anthers are ditchecous and introrse opening by longitudinal slits. These are monothecous in *Gomphrena.* Often staminodes are present. In *Nothosaerva*, there is only one or sometimes two stamens.

Gynoecium is bi- or tricarpellary and syncarpous. Ovary is superior and unilocular with a solitary pendulous or erect campylotropous ovule arising from the base of the locule. Styles are one to three with as many stigmas.

Fruits and Seeds

Fruit is a nutlet or utricle. It is berry in *Deeringia*, circumsessile capsule in *Celosia* and nut in *Digera*. **Seeds** are lenticular with a rough or polished testa, embryo annular or horse-shoe shaped and surrounded with copious (mealy) endosperm.

Pollination and Dispersal

Flowers are insect pollinated. The winged fruits of *Gomphrena* are wind dispersed while those of *Achyranthes*, *Cyathula* and *Pupalia* are dispersed by animals due to presence adhering spinous hooks on their segments.

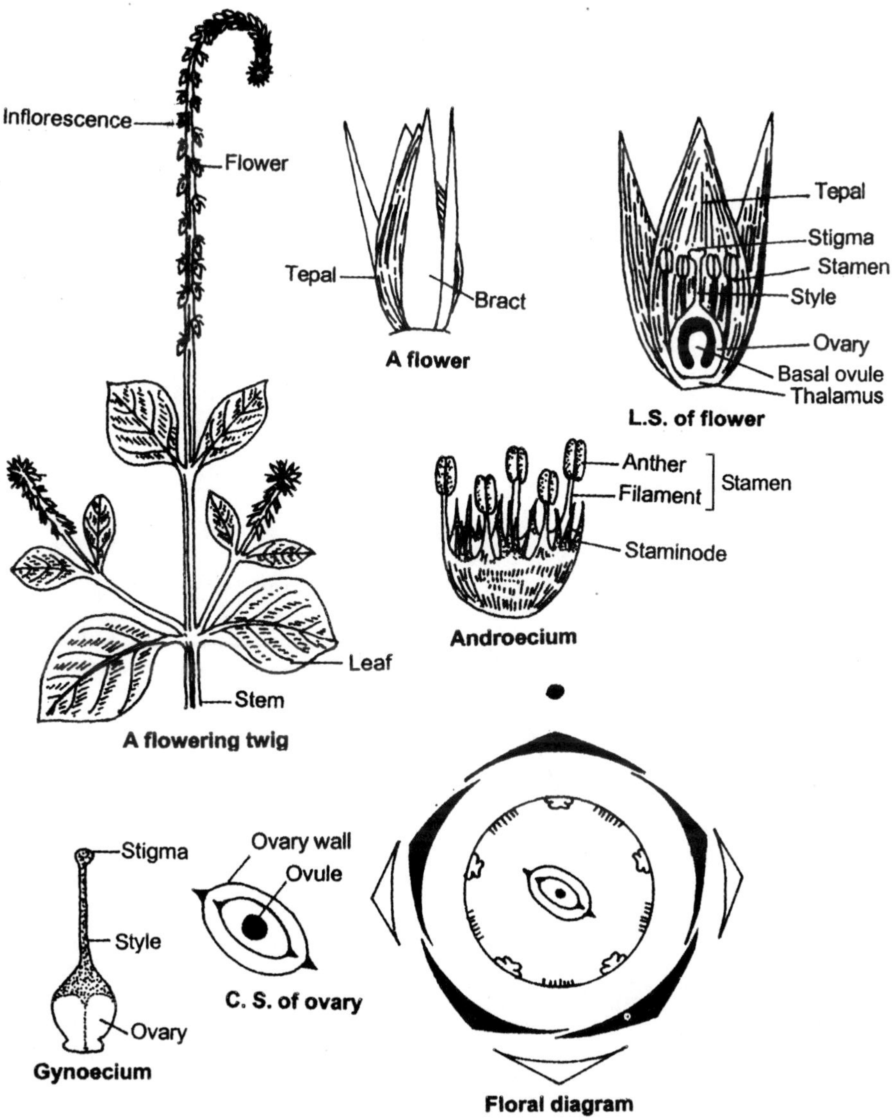

Fig. 18.3 *Achyranthes aspera* L.

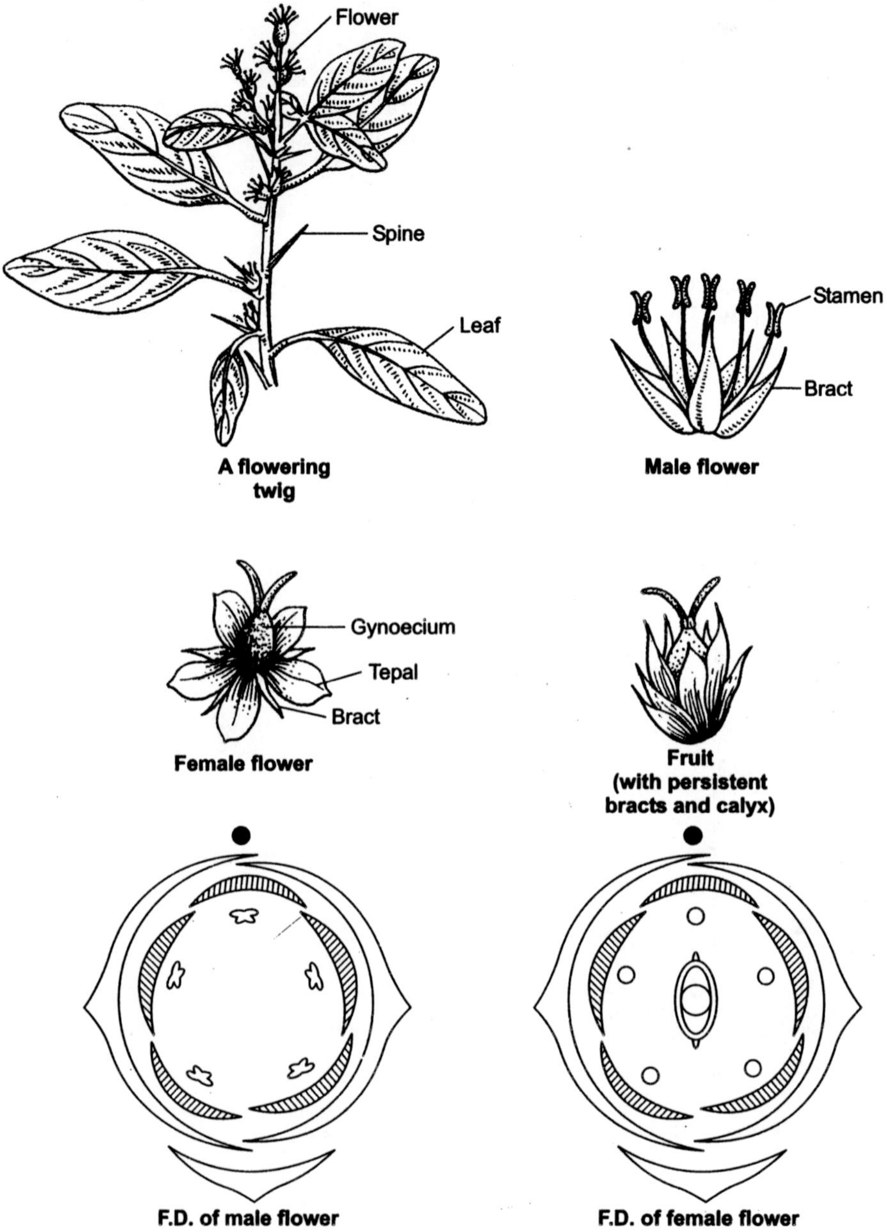

Fig. 18.4 *Amaranthus spinosus* L.

Floral Formulae

Br brl ⊕ ⚥ $P_{4-5 \text{ or } (4-5)}$ A_{4-5} $\underline{G}_{(2-3)}$

Achyranthes aspera	Br brl ⊕ ⚥ P_5 A_5 $\underline{G}_{(3)}$
Amaranthus spinosus	
Male flower	Br brl ⊕ ♂ P_5 A_5 G_0
Female flower	Br brl ⊕ ♀ P_5 A_0 $\underline{G}_{(2)}$

Economic Importance

Ornamental plants. Several forms are grown as ornamental plants, such as *Amaranthus salicifolius* Hort., *A. tricolor* L., *Celosia argentea* L. (Cock's comb), *Iresine herbtsii* Hook. f., *Gomphrena globosa* L. (globe amaranth) etc. These are often grown for their coloured leaves or spikes.

Medicinal plants. *Achyranthes aspera* L. (Chirchita) is used in cough and its decoction is given in renal dropsy and bronchial infection. Leaves of *Deeringia amaranthoides* (Lamk.) Merrill. are applied to cure sores.

Edibles. *Amaranthus caudatus* L. yields edible grains while *A. cruentus* L. (Ramdana) and *A. frumentaceus* are raised as cereals by primitive tribes in Tropical Asia. Leaves of *A. gangeticus* L. and *A. blitum* L. var. *oleracea* Duthie (Chaulai) are used as vegetables.

Phylogeny and Affinities

The family shows relationship with *Chenopodiaceae* in having monochlamydeous flowers, uniseriate stamens and a single basal ovule in unilocular ovary but it differs in the congested inflorescence, scarious bracts, and membranous perianth.

Many phylogenists believe the family as primitive and belonging to **Centrospermales.** Recent studies indicate that the basic inflorescence is a dichasium of three flowers, of which two are aborted but are represented by bracteoles. Each bracteate flower is supposed to represent a relic of an ancestral dichasium. Accepting this view, the family is not primitive and is considered derived from caryophyllaceous ancestors. In APG classification, it has been kept in 2 APG Group : **Core Eudicots** and **Caryophyllales.**

(*For more discussion, see family Chenopodiaceae*)

PORTULACACEAE

Bentham & Hooker	Hutchinson	Cronquist	APG
Dicotyledons	Dicotyledones	Magnoliophyta	2. APG Group :
Polypetalae	Herbaceae	Magnoliopsida	Core Eudicots
Thalamiflorae	Caryophyllales	Caryophyllidae	Caryophyllales
Caryophyllinae		Caryophyllales	

The family comprises **19 genera** and **580 species** found in tropical and temperate regions of the world with preference to America and South Africa. In India, *Portulaca oleracea* and *P. quadrifida* are common weeds of plains. *Talinum* and *Calandrina* are cultivated in gardens for their showy flowers.

Vegetative Structures

Plants are usually succulent herbs or undershrubs. Calcium oxalate crystals are commonly found in leaf cells. **Leaves** are simple, opposite or alternate, usually stipulate (except *Claytonia*). Stipules are scarious or bristle-like.

Inflorescence and Flowers

Flowers are terminal, solitary or grouped, surrounded by a whorl of leaves. Flowers are set in raceme or cymose inflorescence. In *Talinum*, panicle is found. **Flowers** are regular, bisexual and hypogynous or perigynous (*Portulaca*). **Calyx** is represented by two anterio-posterior imbricate sepals which may be free or basally connate. **Corolla** consists of usually 4 to 6 (rarely 2 or 3), free or united, often caducous, imbricate petals.

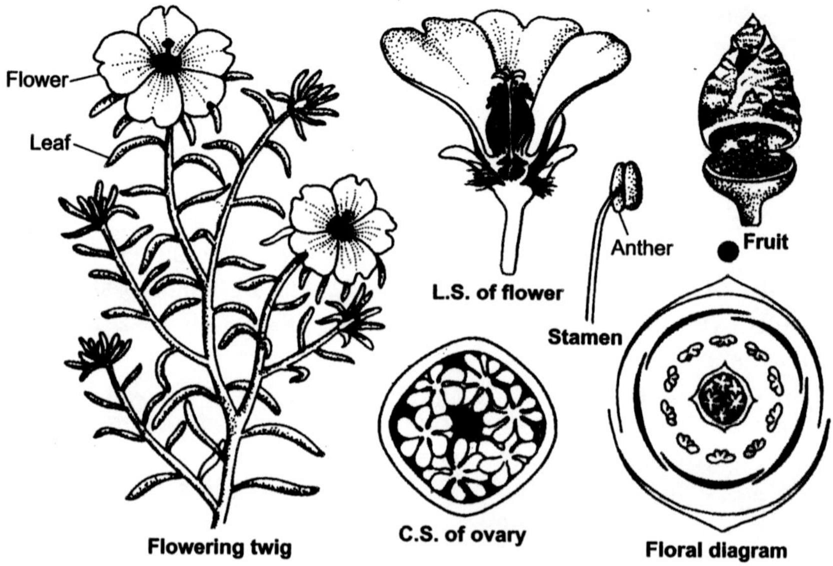

Fig. 18.5 *Portulaca grandiflora* L.

Androecium has stamens as many as petals, rarely 3 or double the number of petals (as a result of splitting). These are free or inserted on the petals. Anthers are dithecous and introrse, dehiscing longitudinally.

Gynoecium consists of 2-5 united carpels (syncarpous). Ovary is superior or semi-inferior (*Portulaca*) and unilocular with 2 to α-campylotropous ovules set in free central placentation. Styles and stigma correspond to number of carpels.

Flowers secrete nectar and are insect pollinated (entomophilous).

Fruits and Seeds

Fruit is a capsule dehiscing circumscissilely, called pyxidium or by 2-5 apical valves. Rarely, fruit is a nut. **Seeds** are compressed with a much curved embryo surrounding a mealy endosperm. These are dispersed either by rain wash or through animals.

Floral Formula $\oplus$ ⚥ $K_2\, C_{4-6}\, A_{4-6}\, \underline{G}_{(2-5)}$

Economic Importance

Species of *Portulaca* : *P. oleracea* L., *P. quadrifida* L. and *P. grandiflora* Hook., are grown as ornamentals for their beautifully coloured flowers. Other plants grown in the gardens are *Talinum portulacifolium* (Forsk.) Aschers. ex Schwf. (syn. *T. cuneifolium* Willd.) and *Portulacaria afra* Jacq.

Systematics and Phylogeny

Portulacaceae show resemblances with other families of **Centrospermae** or **Caryophyllales** but more closeness towards *Basellaceae.* Rendle believes its origin from *Phytolaccaceae* as a separate diversion from other families. Cronquist (1981) believes that it has been derived from *Phytolaccaceae* in the line of *Basellaceae* and has placed the family *Portulacaceae* under **Caryophyllales** of sub-class **Caryophyllidae**. This placement has been supported by Takhtajan (1980, 91). However, Thorne (1992, 2001) raised a new group **Caryophyllana** which includes order **Caryophyllales** and family *Portulacaceae.* APG classification (1998) places the family in order **Caryophyllales** of APG Group : **Core Eudicots.**

CARYOPHYLLACEAE (Pink Family)

Bentham & Hooker	Hutchinson	Cronquist	APG
Dicotyledons	Dicotyledones	Magnoliophyta	2. APG Group :
Polypetalae	Herbaceae	Magnoliopsida	Core Eudicots
Thalamiflorae	Caryophyllales	Caryophyllidae	Caryophyllales
Caryophyllineae		Caryophyllales	

The family comprises **96 genera** and **2415 species** found worldwide, preferably in temperate regions. In India, some **20 genera** and **105 species** are found, mostly in temperate parts with common examples *Stellaria, Saponaria, Spergula, Dianthus* etc. *Silene* (700 species) and *Dianthus* (300 species) are some large genera. *Stellaria aquatica* is an aquatic herb.

Vegetative Characters

Plants are mostly either annual or perennial herbs, rarely undershrubs. **Stem** is herbaceous, erect, branched with swollen nodes. **Leaves** are simple, linear to lanceolate, sessile and opposite decussately arranged. In some species of *Spergula,* leaves are dissected from the base, hence appear whorled. Stipules are either absent or when present, these are thin, dry and membranous (scarious).

Inflorescence and Flowers

Inflorescence is typically a dichasial cyme, but one of the two branches usually outgrow the other, and thus it becomes monopodial cyme. This characteristic inflorescence is called *cincinus* or *caryophyllous* type. **Flowers** are solitary in *Arenaria* and *Githago.* Bracts are present subtending the branches of the inflorescence.

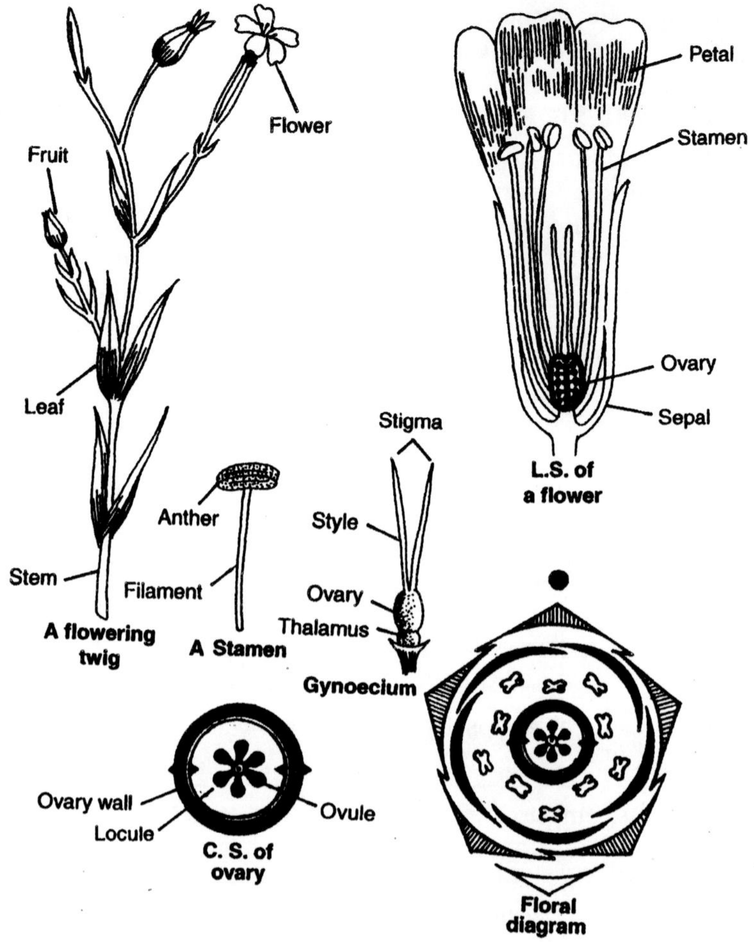

Fig. 18.6 *Saponaria vaccaria* L.

Flowers are hermaphrodite, actinomorphic, pentamerous and hypogynous. Unisexual flowers are found in *Lychnis* and tetramerous flowers in *Sagina*. **Calyx** consists of 5 sepals, which may be poly - or gamosepalous, persistent and imbricate. In *Sagina*, 4 sepals are found. **Corolla** consists of 5, rarely 4 petals. Petals are free and are usually differentiated into a limb and a claw. In *Stellaria media*, petals are deeply bifid and in *Dianthus*, ligulate outgrowth called **corona** is found on the dorsal surface. In *Sagina apetala*, petals are absent or minute.

Androecium consists typically of ten polyandrous stamens, rarely 8, 5 or 4. Stamens are usually arranged in two equal or alternate whorls, outer whorl being opposite the petals. Such arrangement is called **obdiplostemonous** type. Floral anatomy reveals that obdiplostemony is not true. Filaments may be distinct or connate type. Anthers are dithecous, extrorse and dehiscing longitudinally. Sometimes, petaloid staminodes are present. In *Lyallia*, stamens are only 3 and in *Drymaria*, *Polycarpon* and *Polycarpaea*, stamens are found in a single whorl, 5 or less. Nectaries are found at the base of calyx tube.

Floral diagram of *Dianthus*

Floral diagram of *Spergula*

Fig. 18.7 Floral diagrams.

Gynoecium consists of 5 carpels, but number may vary from 2-5. Ovary is syncarpous, superior, unilocular with many ovules set in free central placentation. Styles are free.

Fruits and Seeds

Fruit is a nut or achene or capsule opening by apical valves or teeth as many as or double the number of carpels. Fruit is rarely berry. **Seed** has curved embryo with mealy endosperm.

Pollination and Dispersal

Flowers often show dichogamy and protandry, hence are entomophilous. Seeds are dispersed by water, wind or animals.

Floral Formula $\oplus$ ⚥ $K_5\ C_5\ A_{5+5}\ \underline{G}_{(2-5)}$

Economic Importance

Plants are important as ornamentals and the source of medicines.

Ornamental plants. Important ornamental plants are species of *Dianthus* (about 75 species grown in the gardens), *Vaccaria, Silene, Gypsophila, Lychnis* etc. Commonly grown species of *Dianthus* are *D. caryophyllus* L. (the carnation), *D. chinensis* L., *D. barbatus* L., *D. latifolius* Willd. etc.

Medicinal plants. Plant extract of *Stellaria semivestita* Edgew. is used in anticancer therapy, and decoction of *S. vestila* Kurz. is used in bone-ache and rheumatic pains. *Stellaria media* L. is used in inflammation of digestive, renal and respiratory tracts. *Dianthus anatolicus* is antiperiodic in intermittent fever.

Other uses. Roots of *Saponaria, Vaccaria, Lychnis* and *Gypsophila* contain **saponin** and are used for washing silk and wool.

Division of Caryophyllaceae

Caryophyllaceae have been divided into three sub-families :

(1) Alsinoideae : Sepals distinct, petals present, fruit a capsule. Examples : *Arenaria* and *Cerastium.*

(2) Paronychioideae. Sepals distinct or united, petals absent, fruit an utricle. Examples - *Paronychia* and *Scleranthus.*

(3) Silenoideae. Sepals united, petals present, fruit a capsule. Examples : *Dianthus* and *Lychnis.*

Systematics and Phylogeny

A controversy exists regarding origin of the family. Eichler believes that it has originated from *Phytolaccaceae* by the evolution of petals from the outer whorl of stamens and by the evolution of stamens from the outer whorl of carpels. Others believe that the family has been derived either from the **Ranales** or from *Primulaceae* (Hutchinson, 1964). The family is related to *Portulacaceae* and *Geraniaceae.* Resemblances with *Portulacaceae* are supported by the presence of unilocular ovary with free central placentation while with *Geraniaceae* on account of the pentamerous flowers and nature of stamens.

The family is related to other families of order **Centrospermae**, all having curved embryo and perisperm in the seed. Cronquist (1981) suggested that the family has been derived from *Phytolaccaceae* in the line of *Polygonaceae* and has kept it in the order **Caryophyllales** of sub-class **Caryophyllidae.** Takhtajan (1980, 91) has supported Hutchinson in its placement. Thorne (1999, 2003) recognised 5 suborders within **Caryophyllales** and has placed *Caryophyllaceae* under the monotypic suborder **Caryophyllineae.** Morphological studies and *rbc*L sequence suggest that the family forms a well defined monophyletic clade. APG classification (1998, 2003) has placed the family under the order **Caryophyllales** of APG Group : **Core Eudicots**, a monophyletic familial group.

POLYGONACEAE (Buckwheat Family)

Bentham & Hooker	Hutchinson	Cronquist	APG
Dicotyledons	Dicotyledones	Magnoliophyta	2. APG Group :
Monochlamydeae	Herbaceae	Magnoliopsida	Core Eudicots
Curvembryae	Polygonales	Caryophyllidae	Caryophyllales
		Polygonales	

It is a small family comprising **40 genera** and **800 species**, mainly distributed in temperate regions of the world. In India, **8 genera** and about **110 species** are found. Common members found in the country are *Polygonum, Fagopyrum, Rumex, Rheum, Antigonon, Calligonum, Coccoloba* etc. Family shows a great variation in vegetative as well as floral characters.

Vegetative Characters

Plants are mostly herbs, sometimes shrubs or trees. *Antigonon* is a climber and *Coccoloba* may be a tree. **Stem** is usually herbaceous with swollen nodes. In *Coccoloba platyclada,* stem is modified into phylloclade (flat and green) and leaves are suppressed or membranous. **Leaves** are usually alternate or whorled, stipulate, simple and generally entire. The most important and a constant character is the presence of a sheathing stipular growth around the stem above the leafbase, called *ochrea.* Ochreate stipules are not found in *Antigonon* and *Erigonum.*

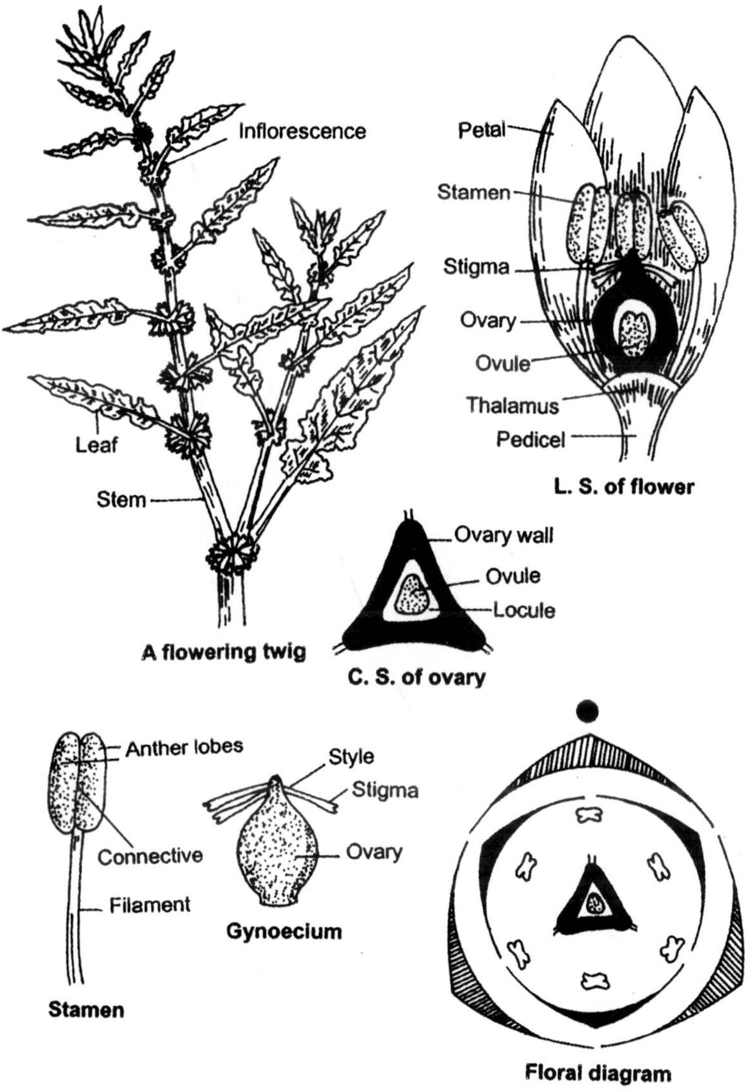

Fig. 18.8 *Rumex dentatus* L.

Inflorescence and Flowers

Inflorescence is primarily racemose, often forming panicles, rarely cymose or the flowers may be in clusters.

Flowers are usually bracteate, bisexual, actinomorphic, cyclic or acyclic and hypogynous, sometimes perigynous. In some species of *Rumex* and in *Triplaris*, flowers are unisexual. Flowers are usually trimerous but dimerous in *Oxyria* and *Rumex*. In *Fagopyrum*, flowers are dimorphic (long stamens with short style or short stamens with long style).

Perianth in cyclic flowers, consists of usually six homochlamydeous segments (tepals) arranged in two whorls. In acyclic flowers, there are five perianth leaves arranged according to 2/5 phyllotaxy (as in *Polygonum*). In *Oxyria*, four perianth leaves are found in two dimerous whorls. The perianth leaves are either entirely free or more or less connate. Aestivation is imbricate.

Androecium consists of 6-9 stamens. The filaments are free or united at the base. Anthers are dithecous, dehiscing longitudinally. Doubling of the

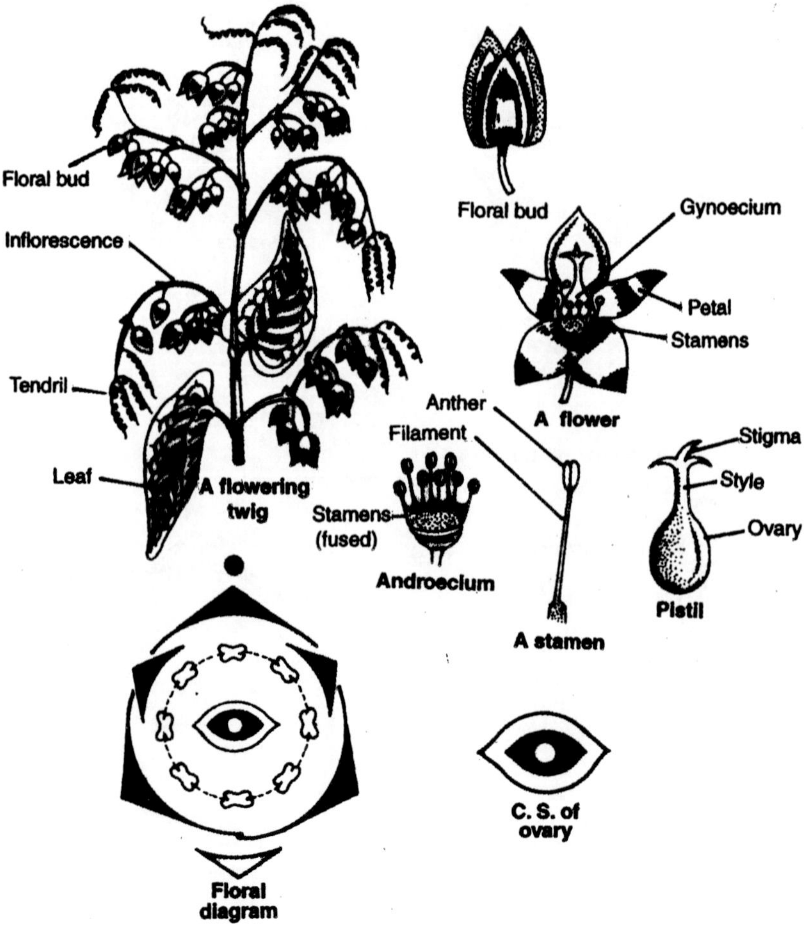

Fig. 18.9 *Antigonon leptopus* Hook. & Arn.

outer whorl of stamens is found in several species while partial or complete suppression of inner whorl stamens is also met with.

Gynoecium is usually tricarpellary (rarely bi- or tetracarpellary) and syncarpous. The ovary is superior and unilocular with a solitary orthotropous basal ovule. Style is 2 to 4 clefted and stigma is capitate (*Polygonum*) or brush-like (*Rumex*).

Fruits and Seeds

Fruit is an angled (trigonous or ovoid) nutlet usually enclosed in a persistent, enlarged perianth (calyx). In *Coccoloba*, the perianth becomes fleshy in the fruit. **Seeds** are endospermic, ruminate endosperm in *Coccoloba*.

Pollination and Dispersal

Flowers may be anemo - or entomophilous. Seeds are dispersed by wind (anemochory), birds (ornithochory) or animals (zoochory).

Floral Formulae

Polygonum $Br \oplus ⚥ P_5 A_{5+3 \text{ or } 5+2} \underline{G}_{(3)}$

Rumex $Ebr \oplus ⚥ P_{3+3} A_{3+3} \underline{G}_{(3)}$

Economic Importance

Some plants are of considerable economic value.

Edibles. Seeds of *Fagopyrum esculentum* Monech. (buckwheat, kuttu) are used as cereal because of starch rich endosperm. For this, plants are commonly grown throughout the Himalayan track. It's flowers are used to increase fragility associated with hypertension. Plants are used as green fodder, being enriched with nectar. *F. tataricum* Gaertn. (duckwheat) is grown for the same reasons as buckwheat. *Coccoloba uvifera* L. produces edible fruits.

Medicinal plants. *Rheum* (Rhubarb) is an important medicinal plant. The dried rhizomes and roots of *R. officinale* Baill., *R. emodi* Wall., *R. webbiana* Royle and *R. palmatum* L. yield a well known drug **Rhubarb** used as laxative, purgative and in stomachic tonics. The long succulent petioles of *Rheum* species are eaten as salad. The young shoots of *Rumex acetosella* L. are used in cancer and tumour treatments. *R. acetosa* L. (garden sorrel) is used as blood purifier and in bronchial diseases. *Polygonum aviculare* L. is an important species used as **Homerian tea** in the treatment of diabetes, rheumatism and external and internal ulcers. Plants of *P. alpinum* are eaten raw or cooked.

Ornamental plants. Several members of the family are grown as ornamental plants, *e.g.* *Antigonon letopus* Hook. & Arn., *Coccoloba platyclada* F. Muell. (syn. *Muehlenbeckia platyclada* Meissn.), *Polygonum orientale* L., *P. alpinum* All., *P. paniculatum* Blume, *P. aubertii* L., *P. vaccinifolium* Wall. etc.

Systematics and Phylogeny

The family *Polygonaceae* is related to the *Amaranthaceae, Chenopodiaceae* and *Nyctaginaceae* in the plan of floral structure but differs from them in the presence of ochreate stipules, triangular ovary, solitary erect ovule and S-shaped embryo. Its unique feature is the multiplication of the stamens and the relation between the two-whorled trimerous and the cyclic perianth. Hutchinson (1959, 64) included family *Polygonaceae* in the order **Polygonales** of the sub-class **Herbaceae** after the order **Caryophyllales,** or in between **Caryophyllales** and **Chenopodiales.**

Cronquist (1981) kept it in the order **Polygonales** of the sub-class **Caryophyllidae** and suggested its derivation from *Phytolaccaceae* through *Caryophyllaceae.* Takhtajan (1980, 91) has supported Cronquist's proposition. Thorne (1999, 2001), however, has placed it in **Dillenianae** and **Caryophyllales.** APG classification (1998, 2003) places it in monophyletic group-**Core Eudicots** (APG II). Inclusion of the family *Polygonaceae* into in its own order **Polygonales** has been proposed by Cronquist (1981), Takhtajan (1980, 91) and Thorne (2001).

❑❑❑

Chapter **19**

Dilleniidae
(Tiliaceae, Sterculiaceae, Malvaceae, Violaceae, Cucurbitaceae, Capparaceae, Brassicaceae, Sapotaceae)

TILIACEAE (Linden Family)

Bentham & Hooker	Hutchinson	Cronquist	APG
Dicotyledons	Dicotyledones	Magnoliophyta	3. APG Group :
Polypetalae	Lignosae	Magnoliopsida	Eudicot-Rosids
Thalamiflorae	Tiliales	Dilleniidae	Eurosids II
Malvales		Malvales	Malvales

The family comprises **41 genera** and **400 species** distributed in tropical and temperate parts of the world. Common examples like *Grewia, Tilia, Muntigera* and *Berria* grow in tropical parts of India. *Corchorus* species are commercially grown in north-eastern part of the country.

Vegetative Characters

Plants are mostly trees or shrubs, rarely herbs (*Corchorus* species). **Leaves** are usually alternate, rarely opposite, simple, entire or toothed, stipulate or ex-stipulate. Stipules are free and often caducous.

Inflorescence and Flowers

Flowers are often arranged in axillary or terminal cyme, rarely axillary raceme as in *Elaeocarpus*.

Flowers are complete, bisexual rarely unisexual, actinomorphic (regular) and hypogynous. **Calyx** consists of 5 sepals (rarely 3 or 4), free or connate with valvate aestivation and caducous. **Corolla** consists of 5 petals or equal the number of sepals, free with imbricate, contorted or valvate aestivation. Petals are sepaloid in *Elaeocarpus* and absent in *Chartacalyx*.

Androecium comprises 10 to numerous stamens inserted on a disc, free or polyadelphous (slightly connate at the base forming 5-10 bundles). Anthers are dithecous, introrse and opening by a longitudinal (lateral) slit or apical pore. Often a few stamens are reduced to staminodes. A gynandrophore is sometimes present.

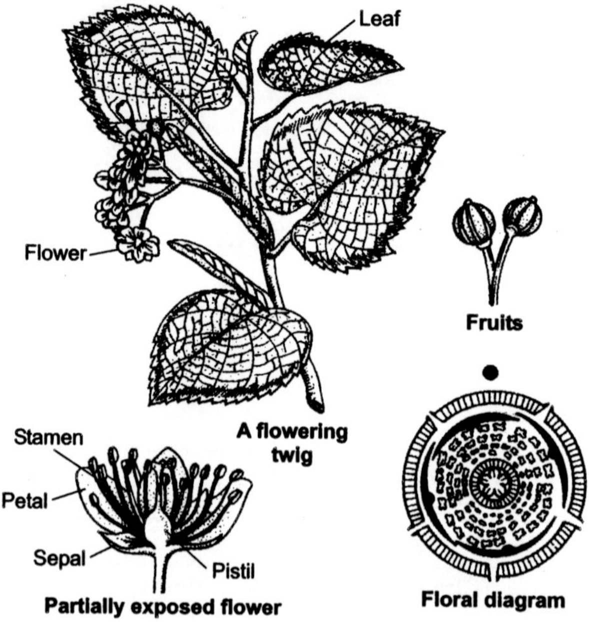

Fig. 19.1 *Tilia cordata* Mile.

Gynoecium is made up of 2-10 carpels with superior ovary and axile placentation. Number of locules corresponds to the number of carpels. In *Elaeocarpus*, ovary becomes unilocular due to dissolution of septa. Number of ovules per locule varies from one to many. Style is one but number of stigma is equal to the number of carpels.

Fruits and Seeds

Fruit is a capsule or drupe. **Seeds** are albuminous (copious endosperm) with a curved embryo having leafy cotyledons.

Pollination and Dispersal

Flowers are insect pollinated (entomophilous) and are usually protandrous. Fruits and seeds are dispersed by animals (zoochory).

Floral Formula Br brl $\oplus$ ⚥ $K_5\ C_5\ A_{10}\ \underline{G}_{(2-10)}$

Economic Importance

Plants are economically important in several ways.

Ornamental plants. Several forms are grown as ornamental plants. The common examples are : *Tilia europaea* L. (Linden tree of Europe), *Muntingia* *calabura* L., *Elaeocarpus ganitrus* Roxb., *Grewia tiliaefolia* Vahl., *G. elastica* Royle etc. Most of them are grown for their spreading habit with large foliage and white flowers.

Medicinal plants. The bark of *Grewia tiliaefolia* Vahl. is used in dysentry and its wood as antidote to opium poisoning. Roots of *G. villosa* Willd. are used in diarrhoea and of *G. sclerophylla* Roxb. ex Don. in cough. Plant mucilage of *Triumfetta rhomboidea* Jacq. is used in the treatment of gonorrhoea.

Timber. A good quality of timber is produced by *Tilia europaea* L., *Grewia elastica* Royle, *G. sclerophylla* Roxb. ex Don, *Elaeocarpus ganitrus* Roxb. etc.

Fibres. Valuable bast fibres (jute of commerce) are derived from *Corchorus capsularis* L. and *C. olitorius* L. These fibres are used for making gunny bags and in linoleum backing.

Food. Fruits (berries) of *Grewia subinaequalis* DC syn. *G. asiatica* L. are eaten with taste and are commonly called 'phalsa'.

Systematics and Phylogeny

Tiliaceae show resemblances with *Malvaceae* and *Sterculiaceae* in habit, leaves and stipules, and is allied to some members of *Euphorbiaceae* which have free stamens inserted upon gynophore. However, it can easily be distinguished from them by free and polyadelphous stamens, dithecous anthers, typical cymose inflorescence and the presence of a disc. Bentham and Hooker, Cronquist (1981), Takhtajan (1991) and Dahlgren (1980, 83) kept the family under the order **Malvales** but Hutchinson developed a new order **Tiliales** to include *Tiliaceae, Bombacaceae* and *Sterculiaceae.* Engler separated *Elaeocarpus* with some genera and raised a separate family *Elaeocarpaceae.* APG classification (Angiosperm phylogeny group II and AP web) (2003) has merged it into *Malvaceae* as sub-family *Tilioideae* with genus *Tilia.* It shows elevation of sub-family *Grewioideae* of *Malvaceae* as a new family *Grewiaceae.* APG classification placed **Malvales** in 3 APG : **Eudicots-Rosids**, sub-group **Eurosids II**.

Table 19.1 Comparison between Tiliaceae and Sterculiaceae

Tiliaceae	Sterculiaceae
1. **Leaves** are simple, alternate and stipulate.	**Leaves** are simple, stipulate, alternate or rarely digitate.
2. **Flowers** are regular, hypogynous, bisexual or unisexual.	**Flowers** are regular or irregular, bi- or unisexual and hypogynous.
3. **Epicalyx** may or may not be present.	**Epicalyx** are absent.
4. **Corolla** consists of 3–5 petals, twisted, often absent.	**Petals** 5, twisted, often reduced or absent.
5. **Androecium** consists of indefinite number of stamens, free or slightly connate at base, or in 5–10 bundles (polyadelphous).	**Stamens** indefinite in number, in two whorls, outer sterile or absent, inner fertile and monadelphous.

6. **Androgynophore** is not found.	**Androgynophore** may be present or absent.
7. **Carpels** are 2–α and syncarpous.	**Carpels** 2–5 and syncarpous.
8. **Ovary** bi- or multilocular with one to many ovules in each locule.	**Ovary** 4–5 locular with 2–α ovules in each locule.
9. **Fruit** is baccate or capsular.	**Fruit** is capsule or follicle.
10. **Floral formula**	**Floral formula**
Br brl ⊕ ⚥ $K_5\ C_5\ A_{10}\ \underline{G}_{(2-10)}$	Br brl ⊕ ⚥ $K_5\ C_5\ A_{\alpha}\ \underline{G}_{(2-5)}$

STERCULIACEAE (Sterculia Family)

Bentham & Hooker	Hutchinson	Cronquist	APG
Dicotyledons	Dicotyledones	Magnoliophyta	3. APG Group :
Polypetalae	Lignosae	Magnoliopsida	Eudicot-Rosids
Thalamiflorae	Tiliales	Dilleniidae	Eurosids II
Malvales		Malvales	Malvales

The family is represented by **50 genera** and **750 species** distributed mainly in tropical parts of the world. In India, some **18 genera** and over **90 species** are found in tropical parts and Himalayas. Common Indian examples are *Heritiera littoralis* (Sundari, the Sunderban is named after this tree), *Pterospermum acerifolium, Guazuma ulmifolia, Helicteres isora, Dombeya mastersii* etc.

Vegetative Characters

Plants are usually soft wooded trees or shrubs, rarely herbs (*Pentapetes*) or climbers (*Buettneria* species). Young parts are often stellate-tomentose. **Leaves** are alternate, simple, entire or palmately lobed or digitate (*e.g. Sterculia*). Stipules are present and caducous.

Inflorescence and Flowers

Flowers are found arranged in axillary complex cymes. Sometimes, cauliflory is met, as in *Theobroma*.

Flowers are bracteate, hermaphrodite, actinomorphic or zygomorphic, pentamerous and hypogynous. Flowers are unisexual in *Sterculia*. **Calyx** consists of 5 sepals, somewhat basally connate with valvate aestivation. Sepals are often coloured in *Sterculia*. **Corolla** comprises 5 petals, often minute and deciduous, occasionally absent as in *Sterculia*. Petals are free or adnate to the base of staminal tube. In *Dombeya*, petals are persistent.

Androecium consists of few to many stamens, free or often connate into a tube and antisepalous. In *Melochia*, 5 stamens opposite the petals and connate below, are found, and in *Pentapetes* and *Pterospermum*, there are 20 stamens : 15 fertile in five groups of three each alternating with 5 antipetalous staminodes. In *Sterculia*, numerous stamens with sessile anthers on a staminal column, are found while in *Melhania*, ten stamens connate below to form a short tube, are found. Staminodes are often present alternating with the fertile stamens and are antipetalous. In *Guazuma*, stamens and staminodes connate together in a monadelphous cup. Anthers are dithecous, parallel or divergent.

Gynoecium consists of 2-5 carpels. Ovary is syncarpous and superior. Number of locules corresponds to the number of carpels, rarely 10-12 locules are found. Ovary is sessile, sometimes androgynophore may be found. Placentation is axile with few to many anatropous ovules. Style is simple or lobed or rarely divided upto the base, the number being equal to the number of carpels.

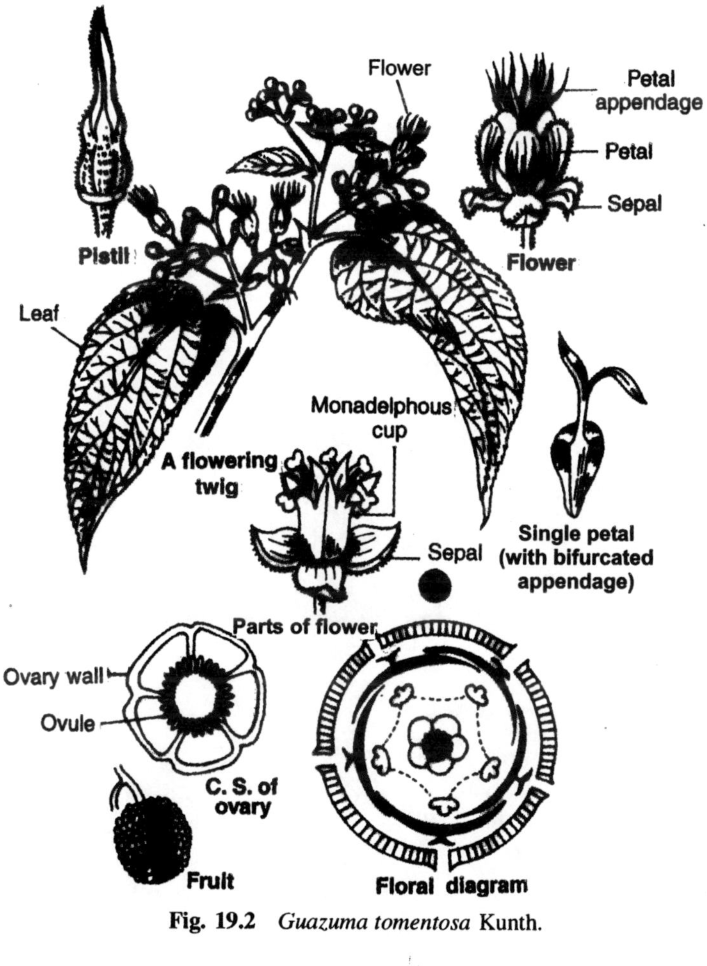

Fig. 19.2 *Guazuma tomentosa* Kunth.

Fruits and Seeds

Fruit is indehiscent or dehiscent capsule, sometimes separating into cocci. Capsule is woody in *Pterospermum*. **Seeds** may be endospermic or non-endospermic. Sometimes, seeds are winged (as in *Pterospermum*) or arillate (*Sterculia*).

Pollination and Dispersal

Flowers show entomophily and seeds show anemochory (winged seeds) or zoochory.

Floral Formula Br brl ⊕ ⚥ $K_5\ C_5\ A_\alpha\ \underline{G}_{(2-5)}$

Economic Importance

Plants are economically important in various ways.

Ornamental plants. Several species are grown as ornamentals, *e.g.* *Dombeya spectabilis* Bojer, *Pentapetes phoenicea* L. (Dopharia), *Guazuma ulmifolia* Lam., *Pterospermum acerifolium* Willd. (the Muchkundo Champa), and many species of *Sterculia*.

Medicinal plants. Root extracts of *Helicteres isora* L. (marorphali) are used in diabetes and seed decoction of *H. littoralis* Dryand. is given for the treatment of diarrhoea and dysentry. Roasted seeds of *Guazuma ulmifolia* Lam. are stringent and are used in stomach disorders. Fruits of *Sterculia balanghas* L. are laxative.

Other uses. Several species of *Pterospermum, Sterculia, Pterygota* and *Heritiera* yield useful timber for furniture, cabinet and construction work. Bark of *Abroma augusta* L., *Pterygota alata* R.Br. and several species of *Sterculia* (*S. balanghas* L., *S. guttata* Roxb.) produces good coarse fibres used for cordages.

Seeds of *Theobroma cacao* L. (cocoa tree) after roasting yield cocoa powder (a nutritious beverage and chocolate) used as food world over. It is now commercially cultivated in South India. Seeds of *Cola acuminata* Schott. & Endl. (kola nut) contain caffeine and theobromine and are used as soft drinks or beverages.

Sterculia urens Roxb. is the source of **Karaya gum** used in textile industry.

Systematics and Phylogeny

Family shares several characters with *Malvaceae*, but it can be distinguished by the absence of epicalyx, dithecous anthers, an outer whorl of staminodes and usually stalked ovary. Being intimately related with *Tiliaceae* and *Malvaceae*, the family is considered as derived from the same stock. Although, Hutchinson has kept it under the order **Tiliales** along with *Tiliaceae* and *Bombacaceae*, most botanists follow Bentham and Hooker, who considered this as a member of order **Malvales.** APG classification also supports this proposition, but it is included in the family *Malvaceae*.

Table 19.2 Comparison between Sterculiaceae and Malvaceae

	Sterculiaceae	Malvaceae
1.	**Members** are trees or shrubs, rarely herbs.	**Members** are herbs or shrubs, rarely trees.
2.	**Leaves** are simple, rarely digitate.	**Leaves** are simple.
3.	**Flowers** are regular or irregular, bisexual or unisexual and hypogynous.	**Flowers** are regular, bisexual and hypogynous.
4.	**Epicalyx** are absent.	**Epicalyx** are usually present.
5.	**Sepals** are 3–5, free or slightly united at the base.	**Sepals** are 5 and united.
6.	**Petals** are 5, often reduced or absent and twisted.	**Petals** are 5, slightly united at the base and twisted.
7.	**Stamens** are α, in 2 whorls-outer whorl sterile or absent and inner whorl fertile and monadelphous.	**Stamens** are α and monadelphous and epipetalous.
8.	**Anthers** are dithecous and **pollens** are smooth.	**Anthers** are monothecous and pollens are spiny.
9.	**Androgynophore** may be present or absent.	**Androgynophore** is absent.
10.	**Carpels** are 2–5 and syncarpous.	**Carpels** are 2–α and syncarpous.
11.	**Ovary** is 4–5 locular with 2–α ovules in each locule.	**Ovary** is 1–many locular with 1 or few ovules in each locule.
12.	**Fruit** is capsule or follicle.	**Fruit** is capsule or schizocarp.
13.	**Floral formula**	**Floral formula**
	Br brl ⊕ ⚥ $K_5\ C_5\ A_{\alpha}\ \underline{G}_{(2-5)}$	Br brl ⊕ ⚥ $Epi_3\ K_{(5)}\ \overbrace{C_5\ A_{(\alpha)}}\ \underline{G}_{(\alpha)}$

MALVACEAE (Mallow Family)

Bentham & Hooker	Hutchinson	Cronquist	APG
Dicotyledons	Dicotyledones	Magnoliophyta	3. APG Group :
Polypetalae	Lignosae	Magnoliopsida	Eudicot-Rosids
Thalamiflorae	Malvales	Dilleniidae	Eurosids II
Malvales		Malvales	Malvales

The family comprises **197 genera** and about **2865 species** spread over tropical and subtropical regions of the world. Some **22 genera** and **110 species** are found in India. Common forms met in the country are—*Hibiscus, Malva, Althaea, Gossypium, Abutilon, Sida, Malachra, Malvastrum* etc. *Hibiscus* (300 species) and *Sida* (200 species) are some large genera.

Vegetative Characters

Plants are mainly herbs, rarely shrubs (*Hibiscus*) or woody trees (*Kydia* and *Thespesia*). **Stem** is erect, jointed, cylindrical, branched, solid and hairy. **Leaves** are alternate, stipulate, simple with multicostate reticulate venation. Stipules are free and often caducous.

Inflorescence and Flowers

Flowers may be solitary, axillary (*Hibiscus*) or terminal or in some cases set in racemose or cymose inflorescence.

Flowers are bracteate, bracteolate, pedicillate, complete, hermaphrodite, actinomorphic, hypogynous and pentamerous. **Epicalyx** may be present, 3-8 in number or absent (*Sida*). **Calyx** consists of 5, free or connate sepals with valvate aestivation. **Corolla** consists of 5 petals, often large and showy, free or basally connate with staminal column and with twisted or imbricate aestivation.

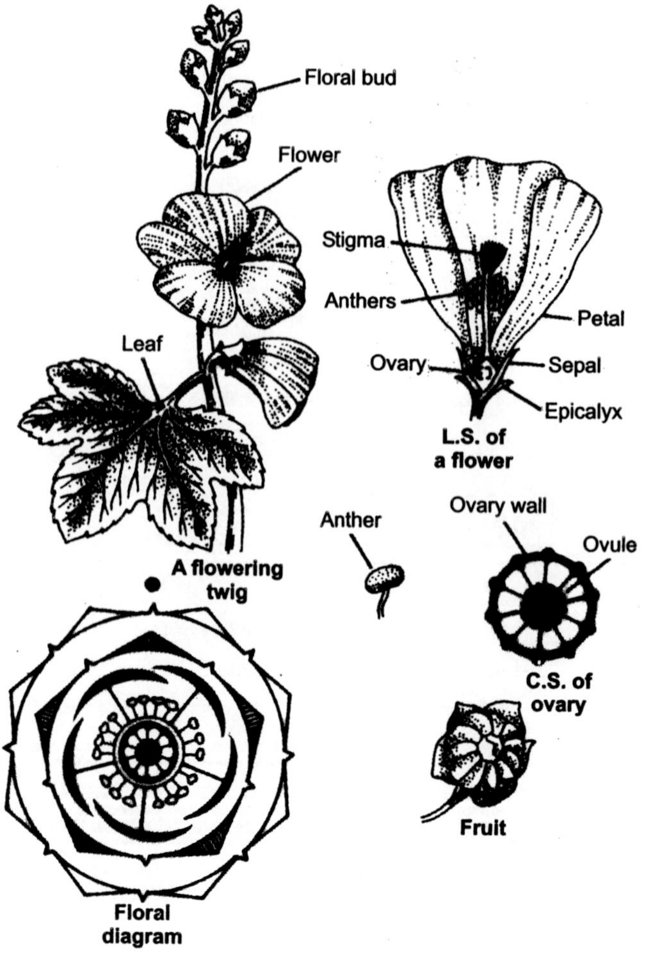

Fig. 19.3 *Althaea rosea* Cav.

Androecium consists of numerous stamens, **monadelphous**, all of them being united below into a tube and the upper parts of the filaments remaining distinct. Anthers are transversely attached to the filaments, monothecous, reniform, extrorse and dehiscing by longitudinal slits. Pollen grains are characteristically spiny.

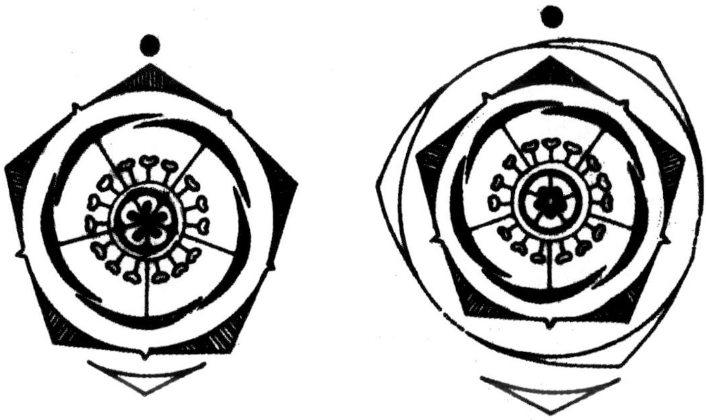

F.D. of *Sida* F.D. of *Malva*

Fig. 19.4 Floral diagrams of *Sida* and *Malva*.

Gynoecium consists of 5 to many fused carpels (syncarpous). Ovary is superior, multilocular with axile placentation. Style is connate below or throughout its length and passes through the staminal tube but branches terminally. Stigmas are free and their number corresponds to the number of carpels, often velvety.

Although the number of carpels in most genera is 5, these are only 1 or 2 in *Plagianthus*, while 8-10 in *Malvastrum*, 15-20 in *Malva* and numerous in *Malope trifida*. In *M. trifida*, carpels are arranged vertically on an elongated thalamus. Number of style branches are double the number of carpels in *Urena*, *Pavonia* and *Malachra*.

Fruits and Seeds

Fruits show variation, *e.g.* loculicidal capsule in *Hibiscus* and *Gossypium*, schizocarpic in *Malva* or fleshy berry in *Malvaviscus*. Carpels commonly separate into mericarps. **Seeds** are usually with endosperm and folded cotyledons, and reniform or ovovoid.

Pollination and Dispersal

Flowers are mostly protandrous and insect pollinated. Seeds are dispersed by air (*Gossypium*), animals (*Abutilon*) or water.

Floral Formulae

Hibiscus Br $\oplus$ ⚥ $Epi_{3-8}\ K_{(5)}\ \widehat{C_5\ A_{(\alpha)}}\ \underline{G}_{(5)}$

Malva Br brl $\oplus$ ⚥ $Epi_3\ K_{(5)}\ \widehat{C_5\ A_{(\alpha)}}\ \underline{G}_{(\alpha)}$

Economic Importance

The family is of great economic value.

Vegetables. *Abelmoschus esculentus* Moench. (lady's finger) is cultivated in India for its fruits used as vegetables.

Fibres. Cotton fibre of commerce is obtained from the epidermal hairs of seeds of *Gossypium* species. *G. herbaceum* L. and *G. arboreum* L. are commonly called old world cotton (diploid plants) while *G. hirsutum* L. and *G. barbadense* L. are new world cotton (tetraploid plants). These yield most important and valuable fibres for textile industries.

Hibiscus cannabinus L. (patsan) is grown as a fibre crop in India to procure bast fibre used for ropes, cordages, coarse sacking etc. Inferior quality of fibre is also obtained from *Abutilon indicum* Sweet., *A. polyandrum* G. Don., *Urena lobata* L., *Sida acuta* Burm. f., *S. cordifolia* L., *Kydia calycina* Roxb., *Malachra capitata* L. etc.

Oil. Oil obtained from the seeds of *Gossypium* is used in the manufacture of soaps and lubricants. The oil cake is an excellent protein rich cattle feed. Seeds of *Hibiscus cannabinus* L. yield a fatty oil used as a lubricant and in the manufacture of linoleum, paints and varnishes.

Medicinal plants. Seeds and bark of *Urena lobata* L. (bachital) are used to cure hydrophobia. Roots of *Malva verticillata* L. are used against whooping cough and ash of dried leaves in scabies. Roots of *Pavonia odorata* Willd. are used in Indian perfume industry, particularly in Hina perfume. Roots of *Malachra capitata* L. are used in rheumatism and lumbago.

Ornamentals. Fairly large number of plants are grown for their ornamental values, *e.g. Hibiscus rosa-sinensis* L. (shoe-flower) for showy large red flowers; *H. sabdariffa* L. for yellow showy flowers; *H. mutabilis* L. (cotton rose) for double flowers, white, pink or red; *Malvaviscus arboreus* Cav. for pendulous scarlet flowers; *Althaea rosea* Cav. (hollyhock) for sessile brightly coloured-red, pink or white flowers; *Malva sylvestris* L. for purple-rose flowers etc. Some beautiful trees grown in gardens as avenues belong to Malvaceae, such as *Kydia calycina* Roxb. (White or pink flowers and branched inflorescence), *Thespesia populnea* Soland. (Indian tulip tree, yellow to purple flowers) and *Adansonia digitata* L. (monkey bread tree, white flowers and purplish stamens) etc.

Systematics and Phylogeny

Malvaceae resemble *Ranunculaceae* in the presence of distinct calyx and corolla, numerous stamens and numerous carpels but it is remarkably distinct in the nature of androecium–monadelphous and monothecous anthers. It shows closeness with *Tiliaceae* and *Sterculiaceae* in the presence of mucilage, stipules and valvate aestivation of calyx. *Malvaceae* differ from them due to monadelphous stamens and monothecous anthers.

Engler developed the order **Malvales** which included eight families. Bentham and Hooker (in *Genera plantarum*) included *Malvaceae, Sterculiaceae* and *Tiliaceae* in the cohort **Malvales**. Hutchinson (1959) restricted **Malvales** to include only *Malvaceae* and his **Tiliales** included *Tiliaceae, Bombacaceae* and *Sterculiaceae*. Cronquist (1981) kept the family in order **Malvales** of the sub-class **Dilleniidae**. Such proposition has been supported by Thorne (1992, 2001). Takhtajan (1991) has assigned **Malvales** in the subclass **Malvanae**. APG classification (1998, 2003) placed it in monophyletic familial group 3 (APG III)—**Eudicots-Rosids (Eurosids II)** (subgroup 2) and order 2—**Malvales**. Thorne (2003) earlier maintained *Tiliaceae* as a separate family but now shifted *Tilia* and *Craigia* to *Malvaceae* (abolishing *Tiliaceae*) and developed a new family *Grewiaceae* for other genera.

APweb has recognised 9 subfamilies under *Malvaceae* (Malvoideae, Bombacoideae, Sterculioideae, Tilioideae, Dombeyoideae, Brownlowioideae, Helicteroideae, Grewioideae and Byttnerioideae). Thorne (2003) removed last two subfamilies and projected them as two separate families.

VIOLACEAE (Violet Family)

Bentham & Hooker	Hutchinson	Cronquist	APG
Dicotyledons	Dicotyledones	Magnoliophyta	3. APG Group :
Polypetalae	Lignosae	Magnoliopsida	Eudicots-Rosids
Thalamiflorae	Violales	Dilleniidae	Eurosids I
Parietales		Violales	Malpighiales

The family consists of **18 genera** and about **1000 species** found in tropical and temperate parts of the world. In India, **3 genera** and **24 species** occur mostly in Himalayas and in hilly regions of western and southern parts. *Viola* is the largest genus of family having widely distributed about 500 species, of which 15 grow in India. *Hybanthus enneaspermus* is found in the plains. Third genus met in the country is *Vahila.*

Vegetative Characters

Plants are annual or perennial herbs, rarely erect shrubs or climbers, very rarely trees. *Hymenanthera* is an erect shrub while *Anhietia* and *Agatea* are climbers and *Rinorea* and *Paypayrola* are trees. **Leaves** are alternate or opposite (*Hybanthus* and *Alsodeia*) or basal (some species of *Viola*), petiolate and stipulate. Stipules are often large and foliaceous (*e.g. Viola*). Leaves are simple, sometimes lobed or divided, margin entire or toothed and cordate in *Viola.*

Inflorescence and Flowers

Flowers are set in an axillary spike or raceme (*Alsodeia*) or panicle, or sometimes solitary (*Viola*). **Flowers** are pedicillate, bracteate, actinomorphic (*Alsodeia*) or zygomorphic (*Viola*), complete, hypogynous and pentamerous. Dimorphic flowers are found in some species of *Viola* (winter flowers with

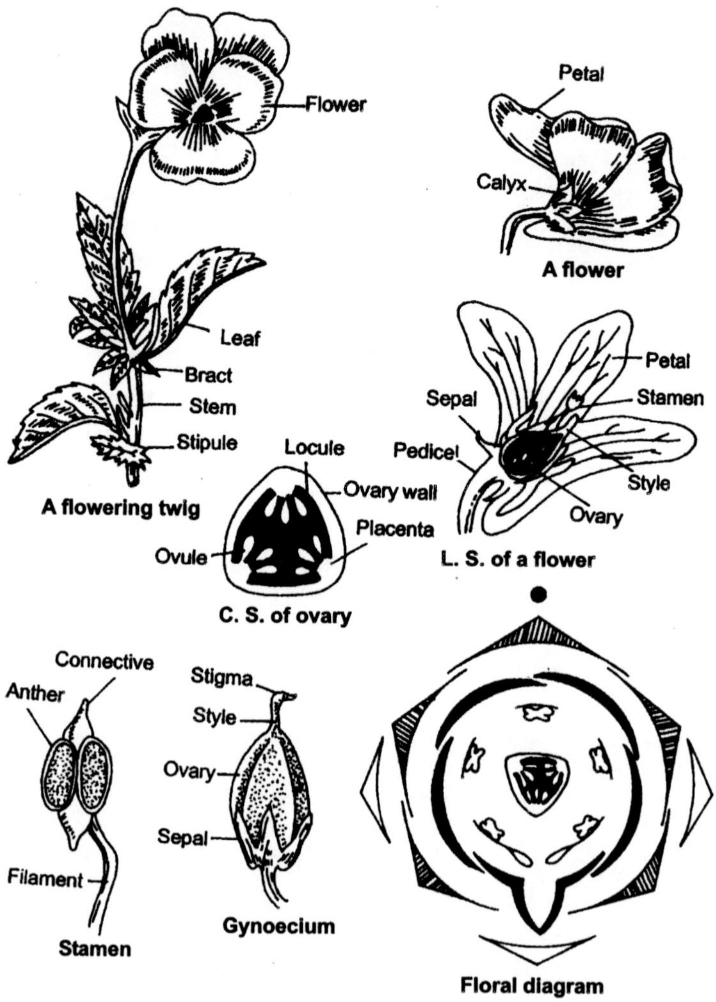

Fig. 19.5 *Viola tricolor* L.

showy petals and summer flowers apetalous and cleistogamous). In *Viola,* nectar is stored in the hollow spur of anterior petal and is secreted by the spurs of the anterior anther (an specific adaptation for pollination). **Calyx** consists of five free sepals, persistent, or connate at the base, imbricate in the bud. The posterior (5th) sepal is often caducous. **Corolla** comprises 5 free petals, variously coloured, of equal and unequal size (the anterior petal of *Viola* is spurred or gibbous) and imbricate or contorted in bud. In *Paypayrola,* corolla is gamopetalous.

Androecium consists of 5 stamens, alternate with petals with short and broad filaments. Anthers are free or connate into a ring around the ovary, dithecous, introrse and dehiscing lengthwise. Very rarely diadelphy is found (*Anchietia*). In *Viola,* the two anterior stamens are spurred at the base, and in *Agatea* and *Legnia,* stamens are monadelphous.

Gynoecium comprises usually 3 carpels (sometimes 2-5 in *Melicytus* and 4-5 in *Leonia*). Ovary is superior, syncarpous and unilocular with parietal placentation, 3 to many ovules on each placenta. Ovules are bitegmic and anatropous. Style is terminal, simple and thickened upwards or sigmoid in zygomorphic flowers. Stigmas are dilated (capitate, cupular or truncate).

Pollination. Flowers are insect pollinated.

Fruits and Seeds

Fruit is a capsule splitting into 3 boat-shaped valves, rarely berry. **Seeds** have hard and shining testa, often with an elaiosome in the form of an appendage of the raphe. Endosperm is copious and fleshy. In climbing genera like *Agatea* and *Anchietia*, seeds are winged. Several species of *Viola* and *Alsodeia* possess explosive capsules.

Floral Formula

$$Br \oplus ⚥ K_5 C_5 A_5 \underline{G}_{(3)}$$

or

$$Br \% ⚥ K_5 C_5 A_5 \underline{G}_{(3)}$$

Economic Importance

Some plants are economically important.

Ornamental plants. The family is mainly valued for the ornamental plants producing beautiful showy flowers. The common cultivated violet is *Viola odorata* L. and pansy is *V. tricolor* L. Hundred of varieties of these two species and of *V. palustris* L. have been produced by hybridisation.

Medicinal plants. Some forms have medicinal value, *e.g.* flowers of *V. odorata* L. are used as poultice in inflammatory affections of throat and other parts of the body. It is a source of Banafsha which is used as purgative in bilious affections. *V. tricolor* L. is used in dysentery. Some species of *Hybanthus* are good substitute of Ipecacuanha plant. Roots of *Anchietia salutaris* are emetic and are used to treat sore throat and lymphatic tuberculosis, and those of *Hybanthus enneaspermus* (L.) F. Muell. (Ratan-purus) in gonorrhoea, urinary affections, and bowel complaints of children.

Flowers of *V. odorata* L. are sweet scented and yield oil.

Systematics and Phylogeny

Bentham and Hooker placed the family in order **Parietales.** Hutchinson (1959), evaluating its closeness to the family *Flacourtiaceae*, placed it in order of its own, *i.e.*, **Violales** next to his **Tamaricales.** Takhtajan (1980, 91) included families *Flacourtiaceae*, *Bixaceae* and *Stachyuraceae* along with *Violaceae* within the order **Violales.** Cronquist (1981) also placed the family *Violaceae* under the order **Violales** of sub-class **Dilleniidae.** APG classification has assigned the family in the order 4. **Malpighiales** of Subgroup 1. **Eurosids** I of the monophyletic familial group-APG Group III—**Eudicots Rosids.**

CUCURBITACEAE (Gourd Family)

Bentham & Hooker	Hutchinson	Cronquist	APG
Dicotyledons	Dicotyledones	Magnoliophyta	3. APG Group :
Polypetalae	Lignosae	Magnoliopsida	Eudicot-Rosids
Calyciflorae	Cucurbitales	Dilleniidae	Eurosids I
Passiflorales		Violales	Cucurbitales

The family consists of **118 genera** and **760 species** distributed in sub-tropical and tropical countries. In India, **37 genera** and about **90 species** are found, several of them are being cultivated. Of the wild forms, common in India, are *Trichosanthes bracteata, T. cucumerina, Coccinia grandis, Momordica cochinchinensis, Zehneria umbellata* etc. *Cayaponia* (60 species) and *Momordica* (45 species) are some large genera of the family.

Vegetative Characters

Plants are mostly climbing or prostrate, annual or perennial herbs. Species of *Dendrosicyos* are soft wooded trees. *Ecballium* is a trailing herb without tendril. **Stem** is weak, herbaceous, often pentangular and usually fistular with two alternate rings of 5 bicollateral vascular bundles. Tendrils are extra- axillary, simple or branched. **Leaves** are alternate, simple, usually palmately lobed, but show considerable variations in form.

Tendrils found in some plants are controversial in nature, as these have been considered as modification of stem or leaf or organs *sui-generis*. Engler considered them as a modification of stipule wherein one stipule is suppressed and other one is transformed as tendril. Naudin believed that a tendril is an axillary branch, the axis of which gets united with the main stem and the tendrillar part represents the midrib of a reduced leaf. Muller's view has been widely accepted. He stated that the tendril is a metamorphosed branch, the lower thick portion representing the axis and the upper coiled portion representing the leaf. The occurrence of various transitional forms of tendril found in *Cucurbita* supports his proposition.

Inflorescence and Flowers

A considerable variation is found as regards to the inflorescence. Flowers are usually solitary axillary (as in *Cucurbita*) or arranged in different types of cymes (male flowers of *Luffa*). Plants may be **monoecious** (*Cucurbita*) or **dioecious** (*Coccinia* and *Momordica*).

Flowers are unisexual, actinomorphic, incomplete, pentamerous and epigynous. However, flowers are bisexual in *Schizopepon*.

Male Flowers

Calyx consists of 5 united sepals (gamosepalous), more or less tubular. These are often narrow and pointed with odd sepal posterior. Aestivation is usually imbricate or valvate, rarely quincuntial.

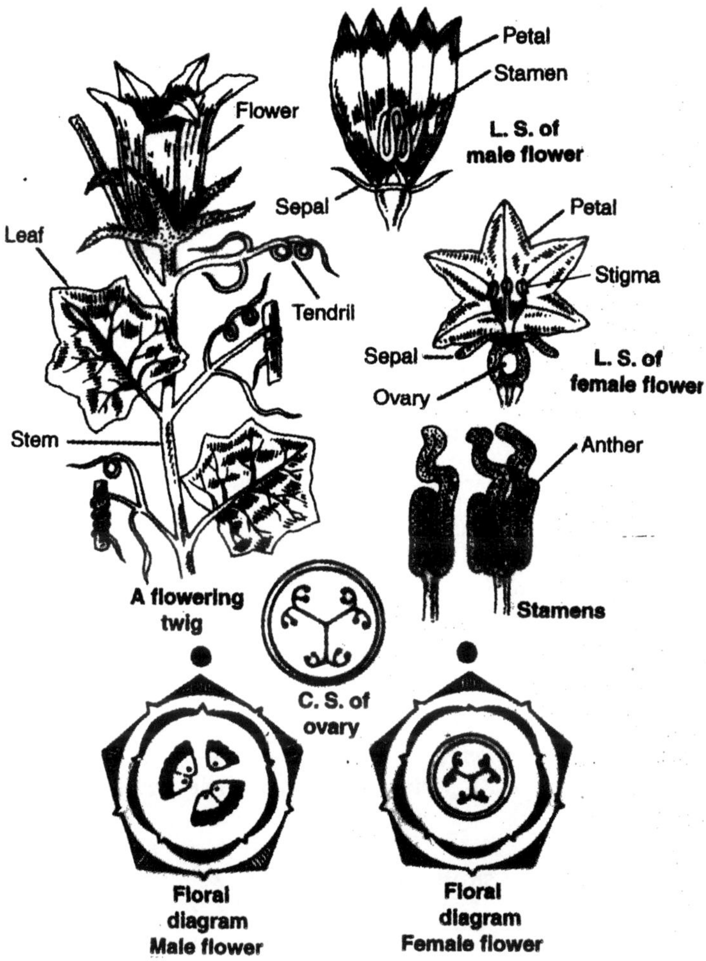

Fig. 19.6 *Cucurbita maxima* Duch.

Corolla consists of 5 united yellow or white petals (gamopetalous), inserted on the calyx tube. In *Trichosanthes*, petals are fimbriate and nearly quite free and in *Fevillia, Luffa* and *Lagenaria*, these are quite free (polypetalous). Corolla is campanulate, rotate or salverform. Aestivation may be valvate, imbricate or quincuncial.

Androecium has a great variation in the family. The number of stamens is typically five but the variations occur on account of their cohesion or fusion. Filaments are often fused and monadelphous and anthers unithecal and **synandrous**. In *Fevillia*, five stamens are free with dithecous anthers and in *Luffa cylindrica*, 5 free stamens are monothecous.

Androecium met in Cucurbitaceae can broadly be grouped into four types :

(i) filaments of 4 stamens coherent below in 2 pairs, the fifth remaining free, as in *Thladiantha*,

(ii) the coherent filaments unite completely so that the androecium seems to be made up of 3 stamens, 2 with dithecous anthers and one with monothecous anther, as in *Bryonia, Citrullus* and *Momordica,*

(iii) filaments free below and united above or completely united and all anthers united and spirally twisted, as in *Cucurbita,* and

(iv) stamens completely united with the lobes of the anthers forming two horizontal rings surrounding a peltate mass of connectives, as in *Cyclanthera.*

Fig. 19.7 Different types of androecium in Cucurbitaceae. A-*Lagenaria*, B-*Cucurbita* (anthers united into a column), C-*Cyclanthera* (stamens are all united into a column) and D-*Sicyos* (filaments as well as anthers united).

In *Sicyos,* all filaments are united into a column bearing the anthers, which are characteristically curved. Anthers are extrorse, dehiscing by a longitudinal slit. Anther lobes are sigmoid or convolute or conduplicate. An aborted or rudimentary pistil is often present in male flowers.

Female Flowers

Calyx and corolla are similar to those of male flowers.

Gynoecium consists of three united carpels (syncarpous). Ovary is inferior, unilocular with three intruding bifurcating parietal placentae which often become so close in the centre that ovary becomes spuriously (falsely) trilocular. In *Sechium,* only a single ovule is present. Style is one with three stigmas. In *Actinostemma,* the ovary is half inferior.

A **controversy** is noted regarding the *true nature of placentation* and ovary being *unilocular* or *trilocular.* Engler, Eichler, Wettstein etc., hold the opinion that the ovary is trilocular, the carpels intrude and meet at the centre and placentation is axile. Rendle has different view that the ovary is unilocular and the placentation is parietal, the placentae growing inside, meet at the centre and often fill the cavity of the ovary.

Pollination. Flowers are insect pollinated.

Fruits and Seeds

Fruit is a pepo, *i.e.,* special kind of berry developed from the inferior ovary. Fruits of *Luffa* opens by an aperture formed at the apex, and that of *Momordica* splits into three parts and seeds are ejected. Ripen fruit of *Ecballium* detaches and discharges its seeds violently (autochory). Fruits of *Lagenaria* break due to some external pressure. In other cases seeds are released when fruits rot.

Seeds are non endospermic and smooth. In *Zanonia*, seeds are winged and in *Sechium*, viviparous germination is found.

Floral Formulae

Male flower Br or Ebr ⊕ ♂ $K_{(5)}$ $C_{(5)}$ $A_{5 \text{ or } (2)+(2)+1}$ G_0

Female flower Br or Ebr ⊕ ♀ $K_{(5)}$ $C_{(5)}$ A_0 $\overline{G}_{(3)}$

Economic Importance

Family is often referred to as a **vegetable family** as fruits of several plants are used as vegetables. Important examples are *Benincasa hispida* (Thunb.) Cogn. (white gourd or petha), *Citrullus lanatus* (Thunb.) Mans. var. *fistulous* (Stokes) V.J. Nair (tinda), *Coccinia indica* W. & A. (ivy gourd, kanduri), *Cucurbita moschata* Duch. (pumpkin, sitaphal), *C. maxima* Duch. (red pumpkin), *C. pepo* L. (field pumpkin), *Lagenaria vulgaris* Ser. and *L. leucantha* (Duch.) Rusby (bottle gourd, lauki), *Luffa cylindrica* (L.) Roem. syn. *L. aegyptiaca* Mill. (gourd, bath sponge, ghiyatori), *L. acutangula* Roxb. (gourd, kalitori), *Momordica charantia* L. (bitter gourd, karela), *Trichosanthes anguina* L. (snake gourd, chachinda), *T. dioica* Roxb. (pointed gourd, parwal) etc.

Many plants are cultivated for the luscious **fruits,** *e.g. Citrullus lanatus* (Thunb.) Mak. (turbuz, watermelon), *Cucumis melo* L. (melon, kharbuz), *C. sativus* L. (cucumber, kheera), *C. melo* L. var. *momordica* Field & Gard. (phut), *C. melo* L. var. *utilissimus* Field & Gard. (kakri) etc.

Some plants have **medicinal value,** *e.g. Trichosanthes dioica* Roxb. is used as febrifuge, in stomachic and for killing worms. Fruit pulp of *Citrullus colocynthis* Scharad is purgative and is given in dropsy. Drug colocynthin obtained from it is used to cure rheumatism and urinary diseases. *Bryonia* is used in homoeopathy.

Some species of *Ecballium, Cyclanthera, Coccinia* and *Sechium* are grown as ornamentals.

Systematics and Phylogeny

Systematic position of the family remains quite controversial. Morphology of stamens and fruit is so unique that a difficulty arises in placing *Cucurbitaceae* among other epigynous groups with parietal placentation. Moreover, the family is remarkably distinct in anther morphology from any other family of angiosperm. Anther cells show several fantastic configuration, varying from simple to the curved, variously convoluted and annular patterns. Stamens may be monadelphous, syngenesious and variously disposed upon the filament, *i.e.,* vertical, horizontal or declined.

Robert Brown, de Condolles and Naudin believed that *Cucurbitaceae* are related to the *Passifloraceae.* They kept the family along with other members of the perigynous **Polypetalae** in the **Calyciflorae**. This has been supported by Bentham and Hooker who placed the family in the cohort **Passiflorales.**

Engler, supporting the view of Eichler, placed the family in the series **Cucurbitales,** next to **Campanulales**. He believed that *Cucurbitaceae* are more close to *Campanulaceae* than to the *Passifloraceae* on account of serodiagnostic tests of Mez and features like structure of ovule with a large persistent nucellus, extensive tapetal tissue and distinct integuments. However, the occurrence of tendrils, unisexual flowers and inferior ovary brings the family close to *Passifloraceae.* The family *Cucurbitaceae* is a highly evolutionary group which has undergone onward progress with structural metamorphosis coupled with reduction, amalgamation, connation, adnation and sterilisation of sex organs. It finds a close link with the *Passifloraceae* in having the placenta retained on the wall, unilocular ovary, fleshy fruit, tendril structure and extra-floral nectaries. Infact, it deserves a separate rank by itself due to type of male sex organs and typical fruit.

Hutchinson (1959) has kept the family under a separate order **Cucurbitales** in the **Lignosae** which is derived from **Passiflorales**. Takhtajan (1980, 91) places only family *Cucurbitaceae* in the order **Cucurbitales** in between **Passiflorales** and **Begoniales**. He believes that *Cucurbitaceae* are allied to *Passifloraceae* on one hand and to *Begoniaceae* on the another and that its derivation is through formation of unisexual flowers, parietal placentation, inferior ovary and stamen modification.

Cronquist (1981) included *Cucurbitaceae* in the large order **Violales** in the sub-class **Dilleniidae** along with *Passifloraceae, Loasaceae, Turneraceae, Begoniaceae, Datiscaceae* etc. (24 families). He believes that **Violales** is a natural group derived from **Theales**, but a heterogenous assemblage. Takhtajan (1991) also supported him. APG classification places the family *Cucurbitaceae* with *Begoniaceae* in the order **Cucurbitales** of sub-group **1. Eurosids I** in monophyletic familial group (clade) APG III : **Eudicots-Rosids.** Monophyly of **Cucurbitales** is supported by serological data and *rbc*L sequences. *Cucurbitaceae* and *Begoniaceae* share apomorphies of inferior ovary, strongly intruded placentae and imperfect flowers.

CAPPARACEAE (Capper Family)

Bentham & Hooker	Hutchinson	Cronquist	APG
Dicotyledons	Dicotyledones	Magnoliophyta	Family not
Polypetalae	Lignosae	Magnoliopsida	recognised, rather
Thalamiflorae	Capparidales	Dilleniidae	merged with
Parietales		Capparales	Brassicaceae

The family, with its **13 genera** and about **450 species**, is distributed in tropical and sub-tropical parts of the world. Some **7 genera** and about **53 species** are found in India, of which *Cleome, Capparis, Gynandropsis, Crataeva, Polanisia* etc. are common. *Capparis* (350 species) and *Maerua* (100 species) are some large genera of the family.

Vegetative Characters

Plants may be herbs (*Cleome*), shrubs (*Capparis*), trees (*Crataeva*) or woody climbers (*Maerua*). Plants have thorns or spines and small leaves provided with trichomes. **Leaves** are alternate, stipulate or ex-stipulate, simple or palmately foliate. Stipules are foliaceous in *Cleome* and spinose in *Capparis.*

Inflorescence and Flowers

Flowers are solitary or in groups of 3 or 4, but more commonly racemose setting is found.

Flowers are bracteate, complete, usually hermaphrodite, actinomorphic, often tetramerous and hypogynous. Flowers are zygomorphic in *Capparis* due to unequal development of sepals and petals. Unisexual flowers are found in *Podandrogyne.* Androphore and gynophore are often present.

Calyx consists of 4 free sepals, usually arranged in two whorls. In *Capparis,* the posterior sepal forms a hood-like structure. Aestivation is imbricate, rarely valvate.

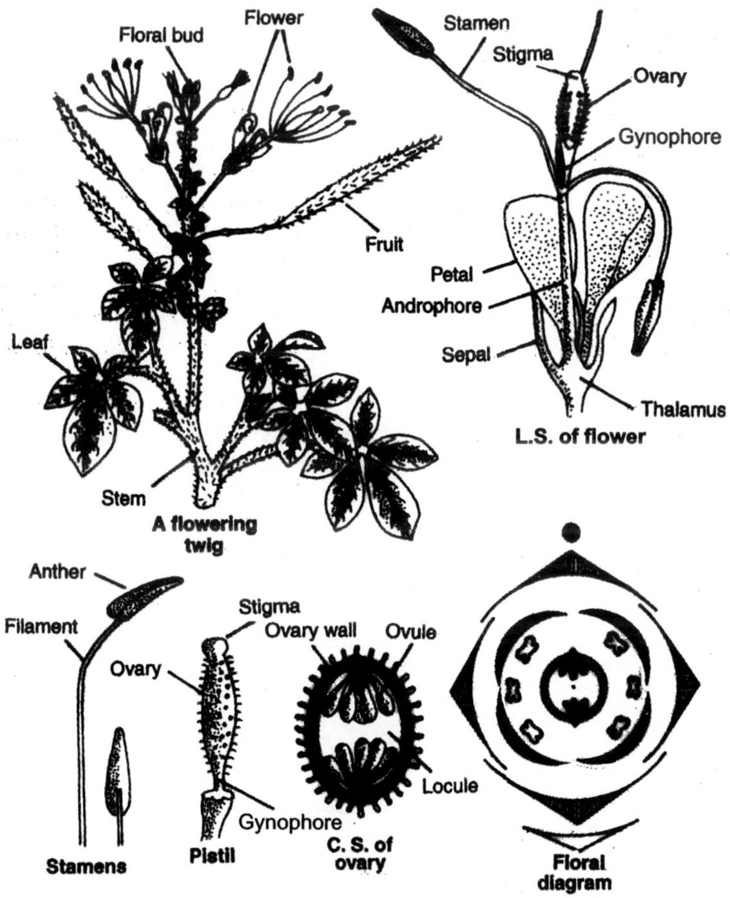

Fig. 19.8 *Cleome gynandra* L.

Corolla consists of 4 free petals, spreading with long claws and mostly set in diagonal fashion similar to that of family Brassicaceae (Cruciferae). Aestivation is valvate or imbricate. Rarely petals are absent. Sepals and petals are united in *Emblingia* where calyx is four-lobed and petals are also united and the posterior petals form a large hood protecting stamens.

Androecium shows a considerable variation in number of stamens, even in different species of the same genus, *e.g. Cleome*. Numerous stamens arranged in several whorls are found in *Capparis* and *Crataeva*. In *Cleome gynandra*, stamens are raised on an androphore. Filaments are long. Anthers are dithecous, introrse, dehiscing longitudinally. Staminodes are found in some forms.

Gynoecium is generally raised upon a long gynophore, *e.g. Cleome gynandra*. It is usually bicarpellary, syncarpous and transversely placed on thalamus. Ovary is superior, unilocular with many campylotropous ovules set in parietal placentation. Ovary is sometimes divided into two or more loculi by ingrowth of placentae. Style is short and stigma is capitate or lobed.

Fruits and Seeds

Fruit is a siliqua, capsule, berry or samara, rarely a nut. **Seeds** are usually reniform and exalbuminous. Embryo is large and variously folded.

Pollination and Dispersal

Flowers are insect pollinated and seeds are dispersed by cattles or water current.

Floral formula Br ⊕ ⚥ K_{2+2} C_4 $A_{6-\alpha}$ $\underline{G}_{(2)}$

Economic Importance

Economically the family is less important.

Medicinal plants. Seeds of *Cleome gynandra* L. (syn. *Gynandropsis pentaphylla* L.) are used to expel roundworms. Bark and leaves of *Crataeva religiosa* Forst. f. *nurvala* Buch-Ham. is diuretic and has antiseptic properties. Fruits of *Capparis decidua* (Forsk.) Pax. syn. *C. aphylla* Roth. are given to heart patients.

Ornamental plants. *Cleome spinosa* L. (giant spider flower) and *Capparis spinosa* L. are grown as ornamental plants.

Others uses. Floral buds of *Capparis spinosa* L. are the cappers used in flavouring food and making pickles.

Systematics and Phylogeny

Family *Capparaceae* resembles *Brassicaceae* in the presence of bicarpellary, syncarpous ovary with parietal placentation and similar arrangement of sepals and petals. It also resembles *Papaveraceae* to some extent. As such many workers believe that *Capparaceae* are intermediate between these two families. Bentham and Hooker designated it *Capparidaceae* and placed in **Parietales** but

Cronquist (1981), Dahlgren (1983) and Takhtajan (1991) designated it as family *Capparaceae* and placed in a separate order **Capparales**. APG II and APweb do not recognise it as a separate family, rather merges it with *Brassicaceae.*

BRASSICACEAE (CRUCIFERAE) (Mustard Family)

Bentham & Hooker	Hutchinson	Cronquist	APG
Dicotyledons	Dicotyledones	Magnoliophyta	3. APG Group :
Polypetalae	Herbaceae	Magnoliopsida	Eudicots-Rosids
Thalamiflorae	Cruciales	Dilleniidae	Eurosids II
Parietales		Capparales	Brassicales

It is a fairly large family comprising widely distributed **340 genera** and **3350 species.** In India, about **51 genera** and **138 species** are found, mostly growing in temperate and alpine Himalayas and in the plains of north India. The common Indian forms are *Brassica, Raphanus, Nasturtium, Coronopus, Iberis, Lepidium* etc. *Draba* (350 species) and *Lepidium* (170 species) are some large genera of the family.

Vegetative Characters

Plants are mostly herbs, may be annual (*Brassica campestris*), biennial (*Raphanus sativus*) or perennial (*Cherianthus*). A few are undershrubs (*Farsetia*). The tap roots become fleshy in *Raphanus sativus* (radish) and *Brassica rapa* (turnip). **Stem** is usually herbaceous, solid and erect, but extremely reduced in *Raphanus sativus* and trailing in *Coronopus.* **Leaves** are mostly simple, alternate, ex-stipulate and with unicellular branched or unbranched hairs. In *R. sativus,* leaves are radical. Lamina may be simple, entire, pinnatifid or pinnatisect or lyrate.

Inflorescence and Flowers

Flowers are mostly set in raceme, sometimes corymb (*Iberis*) or corymbose raceme (*Raphanus sativus*).

Flowers are complete, regular or actinomorphic, hermaphrodite, tetramerous and hypogynous. These are zygomorphic in *Iberis* and *Teesdalia* due to two outer enlarged petals.

Calyx consists of 4, free and caducous sepals arranged in two whorls of two each. The two outer lateral sepals are often large and saccate at the base, and the inner two are narrower. Aestivation is imbricate.

Corolla is made up of 4, free, distinct petals arranged in one whorl. Petals are often with long claws and are spread out in the form of Greek cross. The **cruciform arrangement** of petals is characteristic to the family. Aestivation is valvate. Petals are much reduced or almost absent in *Coronopus, Lepidium* and *Roripa.* In *Capsella,* petals are sometimes modified into stamens.

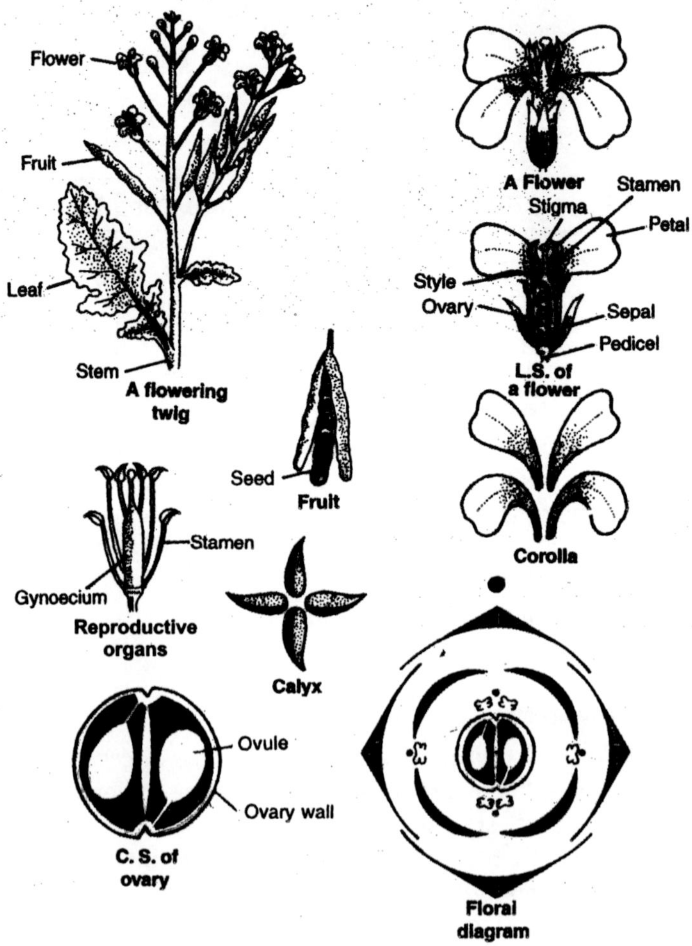

Fig. 19.9 *Brassica campestris* L.

Androecium consists of usually six stamens arranged in two whorls, outer two stamens short and inner four long. This arrangement of stamens is called **tetradynamous type** and is characteristic to the family. Stamens may be 2 (*Coronopus didymus*), 4 (some species of *Nasturtium*) or as many as 16 (*Megacarpaea*). Anthers are dithecous, introrse, basifixed, dehiscing longitudinally.

Gynoecium is bicarpellary, syncarpous and placed transversely in the flower. Ovary is superior, unilocular but becomes bilocular due to the formation of false septum, called *replum*. It bears two parietal placentae, each bearing one to many ovules. Style is short and stigma is bilobed or discoid. However, carpels are three in *Lepidium sativum* and four or more in *Tetrapoma* and *Holargidium*.

Fig. 19.10 Floral diagram of *Iberis amara* L.

Fruits and Seeds

Fruit is siliqua (*Brassica*) or silicula (*Iberis*). **Seeds** are small and non-endospermic with large cotyledons and folded embryo.

Pollination and Dispersal

Flowers are both, self or insect pollinated. Seeds are dispersed by wind, birds and cattles.

Foral formula $\oplus$ ⚥ $K_{2+2}\, C_4\, A_{2+4}\, \underline{G}_{(2)}$

Economic Importance

The family is of great economic value as it provides valuable vegetables, oils, medicines, condiments and ornamentals.

Vegetables. Important vegetable yielding plants and parts are—*Brassica rapa* L. (turnip, shaljam) fleshy roots; *B. oleracea* L. var. *botrytis* L. (cauliflower, phool gobhi) - underdeveloped inflorescence; *B. oleracea* L. var. *capitata* (cabbage, gobhi) - leafy buds; *B. oleracea* L. var. *gemmifera* Zenker. (Brussels sprouts) - leafy buds; *B. oleracea* L. var. *gongylodes* L. (knol-khol, ghant gobhi) - thickened fleshy stem; *B. campestris* L. var. *sarson* Prain (sarson) - leaves; *Raphanus sativus* L. (radish, muli) - fleshy roots, pods and leaves etc.

Oils. Important seed-oil yielding plants are - *Brassica campestris* L. var. *sarson* Prain (mustard or sarson), *B. alba* Boiss. (White mustard, safed sarson), *B. juncea* Hook. f. Thoms. (Indian mustard, rai), *B. nigra* Koch. (black mustard, kali sarson), *B. napus* L. (rape seed, toria), *Eruca sativa* Koch. (taramir) etc.

Condiments. Seeds of *Brassica nigra* Koch. and *B. alba* Boiss are used as condiments.

Ornamentals. Plants grown in the garden for their ornamental value are - *Iberis amara* L. (candytuft), *Alyssum maritimum* Lam., *Cherianthus* (wall flower), *Arabis* (rock cress), *Mathiola* (stocks), *Nasturtium* etc.

Medicinal plants. Some plants are medically important *viz.*, *Capsella bursa-pastoris* (shephard's purse-used to cure dropsy) and *Nasturtium indicum* (seeds used to cure asthma).

Systematics and Phylogeny

The papaverous ancestry of *Cruciferae* is supported by the morphology of stamens and carpels and anatomical features but for the presence of tetradynamous stamens. The family shows similarity with *Capparaceae* in the basic plan of floral setting but latter remarkably differ in the presence of gynophore, androphore and number and arrangement of stamens.

The three families have in common the features of tetramerous perianth, syncarpous pistil and parietal placentation. This poses a problem as to whether *Cruciferae* have been derived from the *Papaveraceae* or from *Capparaceae*.

Bentham and Hooker kept Cruciferae in the order **Parietales** and believed it to have been derived from *Papaveraceae*. However, Cronquist (1981), Takhtajan (1980, 91) and Dahlgren (1989) kept the family *Brassicaceae* in the order **Capparales** believing its origin from *Capparaceae*. Thorne (1999, 2003) retained it in the order **Brassicales.** APG classification retains the family *Brassicaceae* in the order **Brassicales** of subgroup **2. Eurosids II** of monophyletic APG Group III-**Eudicots Rosids.**

SAPOTACEAE (Sapodilla Family)

Bentham & Hooker	Hutchinson	Cronquist	APG
Dicotyledons	Dicotyledones	Magnoliophyta	4. APG Group :
Gamopetalae	Lignosae	Magnoliopsida	Eudicots
Heteromerae	Ebenales	Dilleniidae	Asterids
Ebenales		Ebenales	Ericales

The family includes **40 genera** and **600 species** widely distributed throughout the tropics. In India, **10 genera** and about **52 species** are found in the plains. Common examples are *Achras sapota, Madhuca indica, Manilkara hexandra, Mimusops elengi, Sideroxylon tomentosum* etc.

Vegetative Characters

Plants are shrubs or trees with laticiferous ducts (milky). Younger portions are often rusty and tomentose. **Leaves** are alternate or rarely sub-opposite (*Sacrosperma*), simple, entire, coriaceous or leathery. Stipules are usually absent except in *Sarcosperma* which has caducous stipules.

Inflorescence and Flowers

Flowers are solitary or in cymose clusters or in simple or compound panicle as in *Sarcosperma*. Sometimes, cauliflory is found in *Madhuca* and *Dichopsis*.

Flowers are bisexual, actinomorphic and hypogynous and bracteate. These are trimerous in *Palaquium* and *Manilkara* and pentamerous in *Sideroxylon*. **Calyx** consists of 4-8 sepals, sometimes in two whorls, free or slightly united. Calyx lobes are imbricate but in biseriate condition, outer series is valvate and persistent. In *Sarcosperma* and *Pouteria*, there are 5 sepals in one whorl. **Corolla** has petals as many as or 2-4 times as many as sepals and somewhat united (gamopetalous) to form a rotate or urceolate or funnel-shaped and often with dorsal appendage. Petals are caducous in *Madhuca* and *Mimusops*. In *Mimusops*, petals have a pair of appendages and corolla appears to be biseriate with 18-24 lobes. Aestivation is imbricate.

Androecium comprises stamens in 2-3 whorls of 4-5 each, epipetalous and lie opposite to corolla lobes. In *Madhuca*, stamens are 2-3 times as many as corolla lobes set in 2-3 whorls. Outer stamens are sometimes reduced to staminodes as in *Mimusops*. Anthers are oblong-lanceolate, dithecous, commonly introrse dehiscing longitudinally. Filaments are short.

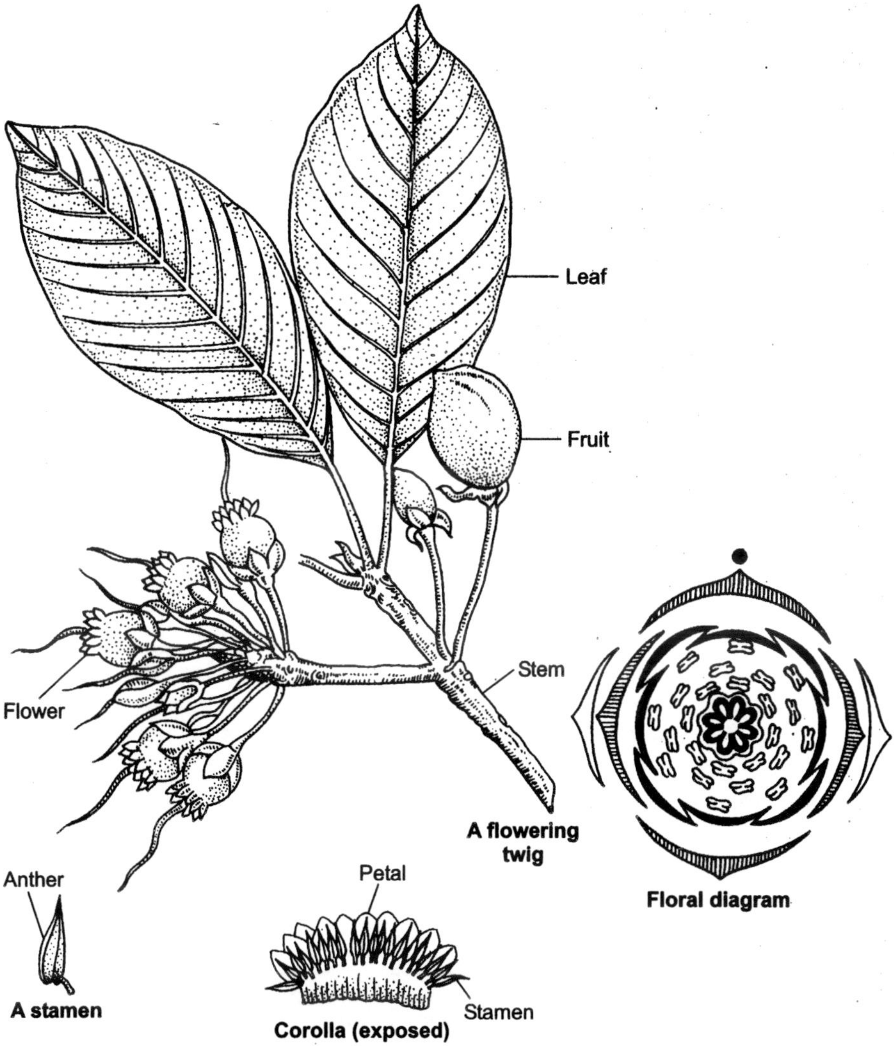

Fig. 19.11 *Madhuca indica* Gmel.

Gynoecium consists of 4-5 carpels, syncarpous, forming a superior 4-5 celled (loculed) ovary. Number of locules corresponds to the number of carpels. Each locule has a single, unitegmic and anatropous ovule ascending with a micropyle. Placentation is axile. Style is one and often apically lobed.

Fruits and Seeds

Fruit is 1-8 seeded berry with laticiferous sacs in the inner pulp, rarely one-seeded and drupe-like. **Seeds** are exalbuminous, embryo with a small radicle and broad foliaceous cotyledons.

Pollination and Dispersal

Flowers are insect pollinated and fruits are dispersed by birds, animals or water currents.

Floral Formula $Br \oplus ⚥ K_{(4-8)} \overbrace{C_{(4-8)} A_{8-16}} \underline{G}_{(2-8)}$

Economic Importance

The family has several plants of economic importance, *e.g. Calocarpum sapota* (Jacq.) Merr. (marmalade plum) syn. *Lucuma mammosa* Gaertn. f., *Chrysophyllum cainito* L., *Vittelaria mammosa* Radlk. f., *Manilkara achras* (Mill.) Fosb., *M. hexandra* (Roxb.) Dub. syn. *Mimusops hexandra* Roxb. etc. yield delicious edible fruits.

Seeds of *Madhuca butyracea* (Roxb.) Macbride, produce vegetable butter used as a cold cream and luminant. Flowers of *M. indica* Gmel. are used for distilling alcohol and seeds yield mohwa butter (oil) used for cooking, luminating, for adulterating ghee and to cure skin diseases and rheumatism.

Achras sapota L. (sapodilla plum) is cultivated for edible fruits. The resinous latex derived from the bark is used in making chewing gum (chickle gum).

Mimusops elengi L. (maulsiri) is an evergreen tree with dense foliage and fragrant flowers. It produces a very strong wood called bullet wood. The fruits are edible. Its bark yields tannin used for dyeing sails and fishing tackles.

Palaquium gutta (Hk. f.) Baill. yield milky latex, a chief source of *gutta parcha* of commerce, used for insulation.

Most trees of the family produce valuable timber. A few are grown for ornamental purposes (*e.g. Mimusops elengi* L.).

Systematics and Phylogeny

Sapotaceae are closely allied to *Ebenaceae* and related to *Strychaceae, Symplocaceae* and *Lissocarpaceae.* Hallier has placed the family *Sapotaceae* in a monotypic order **Sapotales.** Bentham and Hooker included the family in the order **Ebenales** which has been accepted by Hutchinson (1959), Takhtajan (1980, 91) and Cronquist (1981). Thorne (1991, 2001) kept it in the order **Styracales.** APG classification places the family *Sapotaceae* in APG Group 4 : **Eudicots-Asterids** in the order **Ericales.**

❑❑❑

Chapter **20**

Rosidae

(Rosaceae, Mimosaceae, Caesalpiniaceae, Fabaceae, Lythraceae, Myrtaceae, Combretaceae, Loranthaceae, Euphorbiaceae, Vitaceae, Sapindaceae, Anacardiaceae, Meliaceae, Rutaceae, Oxalidaceae, Geraniaceae, Apiaceae)

ROSACEAE (Rose Family)

Bentham & Hooker	Hutchinson	Cronquist	APG
Dicotyledons	Dicotyledones	Magnoliophyta	3 APG Group :
Polypetalae	Lignosae	Magnoliopsida	Eudicots-Rosids
Calyciflorae	Rosales	Rosidae	Eurosids
Rosales		Rosales	Rosales

The family comprises **110 genera** and **3100 species** distributed worldwide. In India, **26 genera** and about **257 species** are found growing mainly in Himalayas, of which 179 species are endemic. Common forms met in the country are *Rosa, Pyrus, Prunus, Eriobotrya, Spiraea, Potentilla, Cotoneaster* etc. *Rubus* (750 species), *Potentilla* (500 species) and *Prunus* (430 species) are some large genera of the family.

Vegetative Characters

Plants may be herbs, shrubs, trees, runner, sucker or climbers, *e.g. Fragaria* (creeping herb), *Potentilla* (erect herb), *Spiraea* (erect shrub), *Rosa* (erect or climbing shrub) and *Pyrus, Prunus* and *Eriobotrya* (trees). **Leaves** are alternate, stipulate, simple (*Pyrus* and *Prunus*), pinnate (*Rosa*) and palmate or digitate (*Fragaria* and *Rubus*). Stipules are adnate in *Rosa* and several other genera, but small and caducous in *Prunus* and *Pyrus* or absent, as in *Spiraea*.

Inflorescence and Flowers

Inflorescence shows variation. Flowers may be solitary (*Rosa*), grouped together in racemose (*Pygeum*), cymose (*Fragaria*), corymb (*Rubus*) or thyrsoid panicle (*Eriobotrya*).

Flowers are bracteate or rarely ebracteate (*Potentilla*), actinomorphic or rarely zygomorphic (*Parinarium* and *Parastemon*), bisexual, pentamerous, hypogynous (*Prunus*) or perigynous (*Rosa*) or rarely epigynous (*Pyrus*). Receptacle is large and cup-shaped in *Rosa, Pyrus* and *Almichilla* or flat or convex as in *Fragaria*. Flowers are unisexual in *Pygeum* and polygamous in *Cotoneaster*.

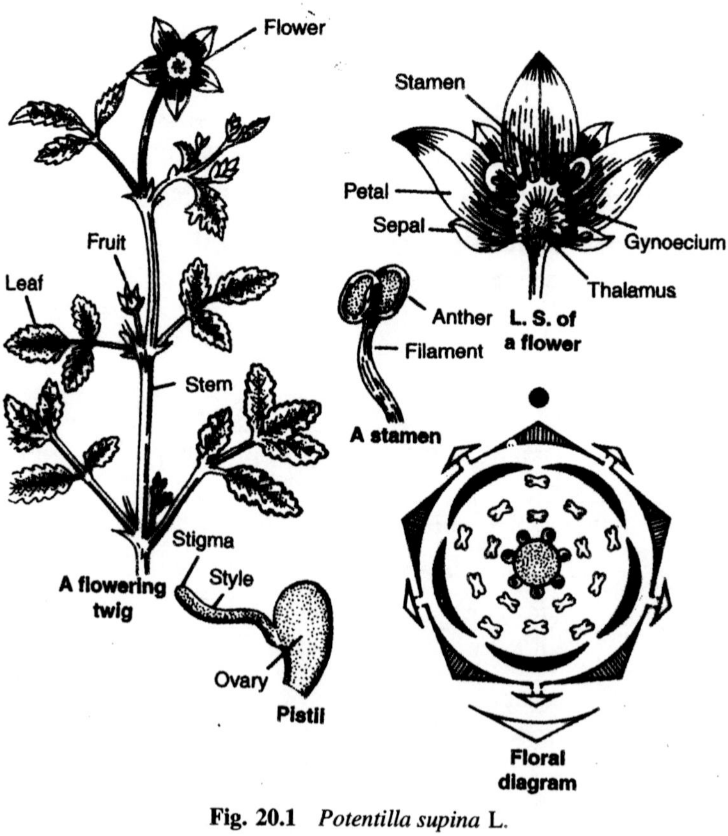

Fig. 20.1 *Potentilla supina* L.

Calyx consists of 5, sometimes 4, connate sepals (gamosepalous). Calyx tube may remain free or become adnate to the ovary wall or receptacle. Epicalyx, alternating with sepals are found in *Potentilla* and *Fragaria*. Aestivation of calyx may be imbricate or valvate. In *Rosa*, some sepals become large and foliaceous. **Corolla** consists of mostly 5, rarely 4, free petals (polypetalous). Cultivated species of *Rosa* have numerous petals which is due to outer one or more whorls of stamens getting modified into petals or petal-like structures. Petals are inserted on the receptacular cup or hypanthium. Petals are absent in *Alchemilla*.

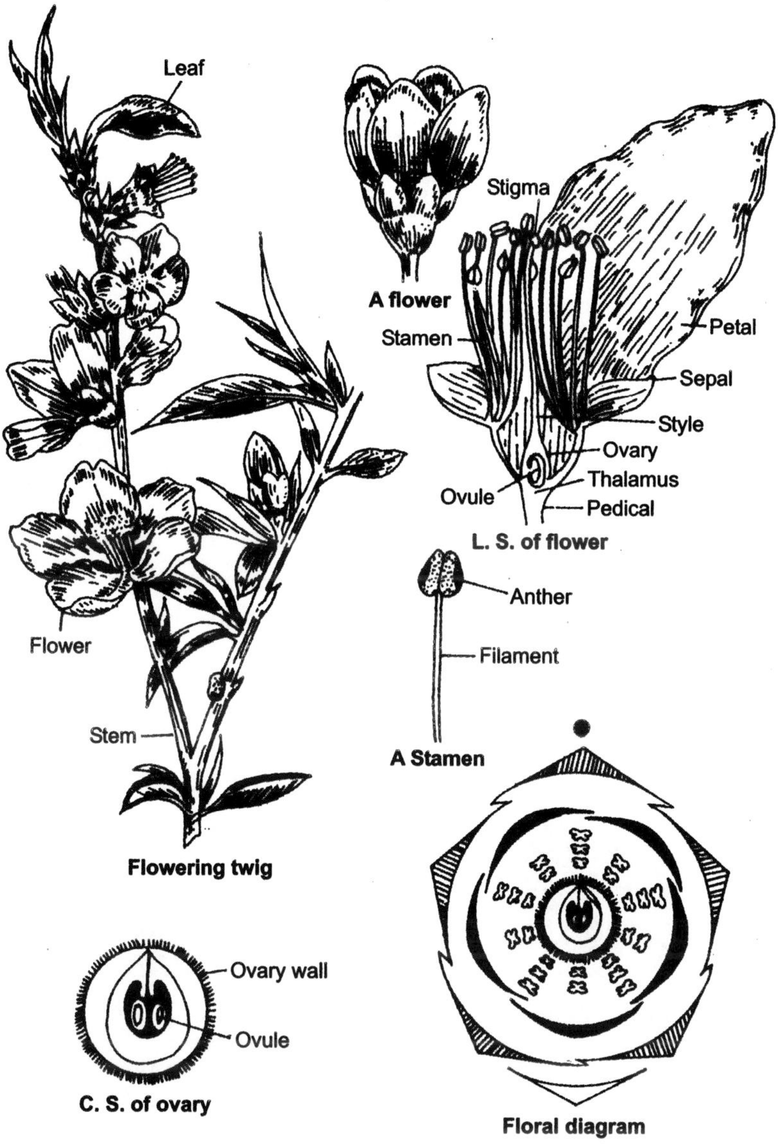

Fig. 20.2 *Prunus persica* Stokes.

Androecium shows variation in number of stamens. These may be 1-4 (*Alchemilla*) or 2-3 or more times to the number of petals. Stamens may be arranged in one (*Chrysobalanus*) or more whorls. Anthers are dithecous, introrse and dehiscing laterally. A cushion-shaped or ring-like nectar secreting disc is found between stamens and carpels.

Gynoecium shows a considerable variation and the number of carpels may vary from one (*Prunus*), 2-5 (*Pyrus*) to indefinite (*Rosa* and *Rubus*). It may be apocarpous (free carpels) or syncarpous. Ovary is mostly superior (*Prunus* and *Rubus*) but sometimes inferior (*Pyrus*) or semi-inferior (*Rosa*). Ovary is

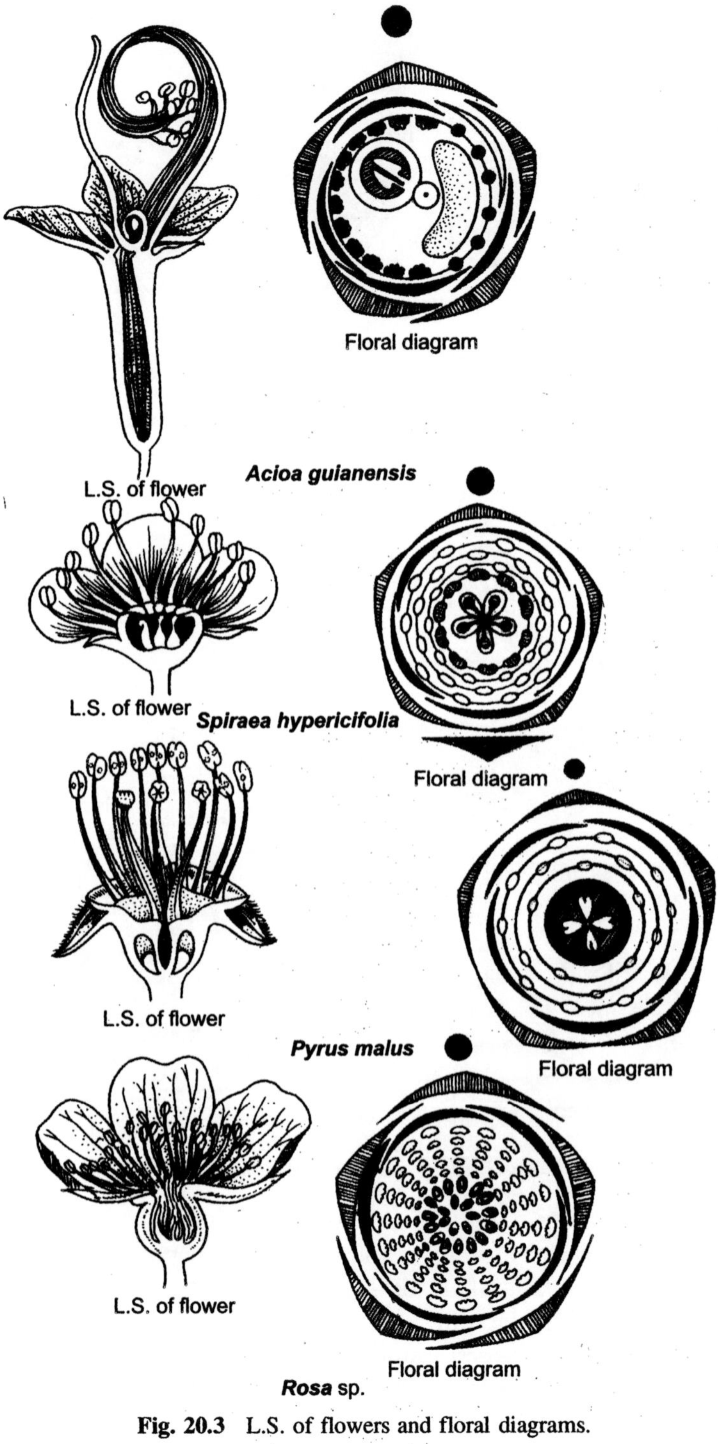

Fig. 20.3 L.S. of flowers and floral diagrams.

mostly unilocular but 2-5 locular in syncarpy. Placentation may be axile but often pendulous (*Rosa* and *Potentilla*). Style may be free or connate and is usually terminal, but lateral in *Fragaria* and basal in *Acioa* and *Alchemilla*. Stigma is capitate and number is equal to the number of carpels.

Fruits and Seeds

Several types of **fruits** are met in the family, *e.g.* single seeded drupe (*Prunus*), 2-5 locular pome (*Pyrus*), etaerio of achenes (*Potentilla*), etaerio of follicles (*Spiraea*), etaerio of barries (*Fragaria*) or pseudobaccate (*Rosa*).

Seeds are exalbuminous and with straight embryo.

Pollination and Dispersal

Flowers are insect pollinated. Seeds are dispersed by birds and animals.

Floral Formulae

Prunus	Br $\oplus$ ⚥ $K_{(5)}\ C_5\ A_\alpha\ \underline{G}_1$
Eriobotrya	Br $\oplus$ ⚥ $K_{(5)}\ C_5\ A_\alpha\ \underline{G}_{(2-5)}$
Potentilla	Ebr $\oplus$ ⚥ $K_{(5)}\ C_5\ A_\alpha\ \underline{G}_\alpha$

Economic Importance

Family has a great economic value.

Edibles. Famous fruit trees are : *Chrysobalanus icaco* L. (cocoplum), *Cydonia oblonga* Mill. (quince), *Eriobotrya japonica* Lindl. (loquat), *Fragaria vesca* L. (strawberry), *F. chiloensis* Duchesne (garden strawberry), *Prunus amygdalus* Baill. (almond), *P. armeniaca* L. (apricot), *P. persica* Stokes (peach, aaru), *P. domestica* L. subsp. *institia* (L.) Schneider (plum, alucha), *P. avium* L. (sweet cherry), *Pyrus communis* L. (pear), *P. malus* L. syn. *Malus sylvestris* Hort. non Mill. (apple, seb), *Rubus caesius* L. (raspberry) *R. lanatus* Wall. (hisalu), *R. niveus* Thunb. etc.

Medicinal plants. Root stocks of *Geum urbanum* L. are used in chronic dysentery, diarrhoea and intermittent fever. *Potentilla anserina* L. is used to cure kidney stones and arthritis. Female flowers of *Hagenia abyssinica* Gmel. produces anthelmintic drug **cusso**.

Timber. *Pyrus malus* L., *P. communis* L., *Rosa brunonii* Lindl., *Crataegus oxyacantha* L., *Cotoneaster bacillaris* Wall. etc. produce valuable wood used for making sticks and tool handles.

Ornamentals. Several species of *Rosa* (*R. multiflora* Thunb., *R. damascena* Mill., *R. indica* L., *R. banksiae* R. Br. and *R. centifolia* L.) and *Spiraea* (*S. cantoniensis* Lour., *S. canescens* Don. and *S. japonica* L.f.) are grown for their attractive flowers and inflorescence.

Other uses. Bark of *Quillaja saponaria* Malina (soap bark) is used as detergent and foaming agent and seeds of *Prinsepia utilis* Royle. yield oil used in cooking and as an illuminant. Petals of *Rosa damascena* Mill. are used to obtain perfume (otto of commerce) and of *Rosa centifolia* L. to make gulkand tonic.

Division of the Family

The family has been sub-divided into 6 sub-families (by Focke, *Pflanzenfamilien*) on the basis of floral characters.

(1) **Spiraeoideae.** Thalamus somewhat flat, stamens 10 to many, carpels 5, free, located in the centre. Examples *Spiraea* and *Quillaja*.

(2) **Pomoideae.** Thalamus deeply concave, stamens numerous, carpels 5, somewhat united and adnate to the thalamus, ovules 2 in each carpel, fruit—a pseudocarp (pome). Examples—*Pyrus* and *Cotoneaster*.

(3) **Rosoideae.** Thalamus concave, carpels many and free, 2 ovules in each carpel. Examples—*Rosa, Rubus, Potentilla* and *Fragaria*.

(4) **Neuradoideae.** Thalamus cup-shaped, carpels 5-10 and united. Examples—*Grielum* and *Neurada*.

(5) **Prunoideae.** Thalamus cup-shaped, stamens 10, 20 or more, carpel 1 containing two ovules, fruit a drupe. Example—*Prunus*.

(6) **Chrysobalanoideae.** Thalamus deeply cup-shaped, flowers irregular, stamens sometimes monadelphous and carpel 1. Examples—*Chrysobalanus* and *Hirtella*.

Systematics and Phylogeny

The family shows several primitive characters both in vegetative and floral structures, *e.g.* solitary vessels, trachieds and fibre tracheids, flat or convex receptacle (in some genera), many free stamens and free carpels. Fossil history also reveals that *Rosaceae* are one of the ancient families among Dicotyledons. *Rosaceae* resemble *Ranunculaceae* in number of floral parts but differ due to cup-shaped thalamus with carpels standing up or embedded in the bottom of cup. In *Ranunculaceae*, thalamus takes a shape of cone. *Rosaceae* are allied to the *Leguminosae* due to presence of perigynous flowers and monocarpellary pistil. *Acioa* shows connection with *Leguminosae* by the pesence of zygomorphic flowers and monadelphous stamens with one carpel. The family shows close resemblance to *Leguminosae* and *Saxifragaceae* with regard to the presence of stipules, pentamerous flowers, presence of torus etc. Phenolic make up, especially in ellagic acid and leucoanthocyanins, brings it close to *Saxifragaceae*.

Engler included several families in order **Rosales**. Many sub-families or tribes of *Rosaceae* have been raised to separate families, especially *Chrysobalanaceae* and *Neuradaceae* which have been well recognised by several workers. But Hutchinson (1973) has expressed his view that if one or two tribes of the family are taken out, atleast 18 or 19 should follow suit, and *Rosaceae* would be reduced to the genus *Rosa* only.

Family *Rosaceae* is a well recognised group whose monophyletic set up has been supported by Morgan *et al.* (1994) while working on *rbc*L sequences. Cronquist (1981) has retained these two tribes together with *Rosaceae*, but

Dahlgren (1983, 89) has placed *Chrysobalanaceae* in the order **Theales**, and Takhtajan (1991) kept it in a distinct order **Chrysobalanales**. Thorne (1999), recognising both tribes as separate families, placed *Neuradaceae* under order **Malvales** and *Chrysobalanaceae* under order **Dilleniales**. However, family *Rosaceae* in the order **Rosales** has been recognised by Cronquist (1981), Takhtajan (1991), Thorne (1999) and several others. APG classification places family *Rosaceae* in the order **Rosales** of sub-group I. **Eurosids** of monophyletic familial group 3 : **Eudicots—Rosids**. The system recognises *Chrysobalanaceae* in the order **Malpighiales** of the same group but shifted family *Neuradaceae* in the order **Malvales** of sub-group 2 : **Eurosids II**. Only 4 subfamilies of *Rosaceae* are now recognised - *Maloideae, Amygdaloideae, Rosoideae* and *Spiraeoideae* (Potter *et al.*, 2002).

MIMOSACEAE (*Acacia* Family) or (Sub-family Mimosoidae)

Bentham & Hooker	Hutchinson	Cronquist	APG
Dicotyledons	Dicotyledones	Magnoliophyta	3 APG Group :
Polypetalae	Lignosae	Magnoliopsida	Eudicots-Rosids
Calyciflorae	Leguminales	Rosidae	Eurosids I
Rosales		Fabales	Fabales

The family comprises **40 genera** and **2500 species** widely distributed in the tropical and sub-tropical regions of the world. Only **15 genera** and about **72 species** are found in India. The common examples are—*Mimosa, Acacia, Albizzia, Pithecolobium, Calliandra, Entada, Prosopis, Neptunia* etc. *Acacia* (1000 species), *Mimosa* (500 species) and *Inga* (250 species) are some large genera.

Vegetative Characters

Plants are mostly trees (*Acacia* and *Albizzia*), sometimes shrubs (*Colliandra*), rarely herbs (*Mimosa pudica*), climbers (*Entada*–leaf tendrils) or aquatic (*Neptunia*). **Stems** are mostly erect, solid and woody. Tannin sacs and gum passages are found in several forms. **Leaves** are mostly alternate, bipinnately compound, stipulate, rarely simple pinnate (*Inga*). Stipules of *Acacia* and *Mimosa* are modified into spines. Leaves of *Mimosa* and *Neptunia* are sensitive to touch. In some Australian acacias, petioles are modified into phyllode (*A. auriculiformis*).

Inflorescence and Flowers

The minute flowers are condensed into dense and globose heads (*Mimosa* and *Acacia*), or sometimes spicate (*Prosopis*) or racemose type (*Adenanthera*). In *Dichrostachys*, spikes are dimorphic, the upper half with perfect flowers and lower half with long filiform staminodes.

Flowers are small bracteate, sessile, actinomorphic, bisexual (hermaphrodite), pentamerous and hypogynous or perigynous. In *Mimosa*,

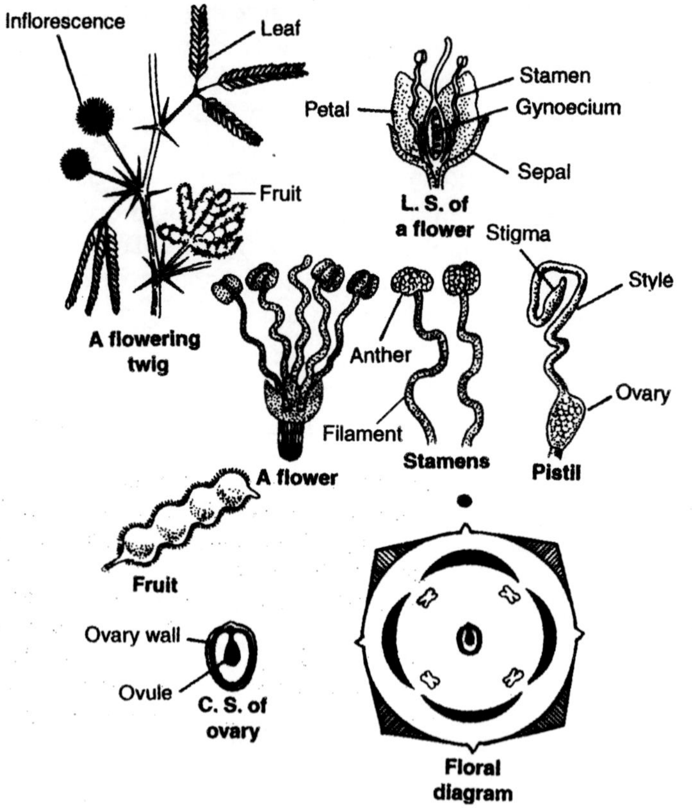

Fig. 20.4 *Mimosa pudica* L.

flowers are tetramerous. **Calyx** consists of 4 or 5 sepals, which are gamosepalous and tubular with distinct lobes or teeth. **Corolla** consists of 4 or 5 free petals (polypetalous), rarely connate at the base. Petals are usually minute, set in valvate aestivation.

Androecium consists of usually indefinite number of stamens with variation in number and cohesion. In *Acacia*, stamens are numerous and free; in *Albizzia*, indefinite stamens are monadelphous; in *Prosopis*, diplostemonous condition is found and in *Mimosa*, only four free stamens are found. Filaments are long, filiform and often exerted. Anthers are dithecous, introrse and dehiscing longitudinally. Filaments and anthers are often coloured and attractive (*Calliandra, Mimosa*).

Fig. 20.5 F.D. of *Acacia nilotica* (L.) Del.

Gynoecium consists of a single carpel (monocarpellary). Ovary is superior (or semi-inferior), unilocular with marginal placentation, and usually with several ovules along the suture, style one and stigma one.

Fruits and Seeds

Fruit is a legume or lomentum (*Acacia*). Lomentum of *Entada* is very long and woody. **Seeds** are ex-albuminous.

Pollination and Dispersal

Flowers are insect pollinated as insects are attracted by exerted gland-dotted, coloured stamens. Seeds are dispersed by wind, birds and animals.

Floral Formula Br ⊕ ⚥ $K_{(4) \text{ or } (5)}\ C_{4 \text{ or } 5}\ A_{\alpha}\ \underline{G}_1$

Economic Importance

Family has great economic value.

Edibles. Pods of *Neptunia oleracea* Lour. are used as vegetable and the aril (pod part) of *Pithecolobium dulce* Benth. is sweet and edible. The wood of *Acacia catechu* Willd. is used to extract catechu (kathha) which is used as medicine and eaten with betel.

Timbers. A large number of trees produce valuable wood, *e.g. Acacia nilotica* (L.) Del., *A. catechu* Willd., *A. modesta* Wall., *Albizzia procera* Benth., *A. lebbek* Benth., *A. lucida* Benth., *Xylia xylocarpa* Taub. etc.

Tannins. The bark of *Acacia nilotica* (L.) Del. (kikar), A. *catechu* Willd. (khair) and *A. pinnata* Willd. etc. yield tannin.

Gums. *Gum arabic* is obtained from *Acacia senegal* Willd. used in medicine and confectionary. *A. nilotica* (L.) Del. and *A. catechu* Willd. also yield valuable gum.

Ornamentals. Several plants are grown in the gardens, avenues and roadsides, *e.g. Mimosa pudica* L. (sensitive plant), *Leucaena leucophloea* Willd. (white popinae), *Acacia nilotica* (L.) Der., *Albizzia lebbek* Benth., *A. procera* Benth. etc.

CAESALPINIACEAE (*Cassia* Family) or (Sub-family-Caesalpinioidae)

Bentham & Hooker	Hutchinson	Cronquist	APG
Dicotyledons	Dicotyledones	Magnoliophyta	3 APG Group :
Polypetalae	Lignosae	Magnoliopsida	Eudicots-Rosids
Calyciflorae	Leguminales	Rosidae	Eurosids I
Rosales		Fabales	Fabales

The family is represented by **150 genera** and **2700 species** found in tropical and temperate regions of the world. Some **23 genera** and **80 species** are found in India, of which *Cassia, Saraca, Caesalpinia, Delonix, Tamarindus, Haematoxylon, Bauhinia, Parkinsonia* etc. are common. *Chamaecrisia* (260 species), *Bauhinia* (250 species) and *Senna* (250 species) are some large genera of the family.

Vegetative Characters

Plants are mostly trees (*Delonix, Tamarindus*), sometimes shrubs (*Cassia occidentalis*) or rarely herbs (*Cassia obtusa*). Some species of *Bauhinia* and *Pterolobium* are woody climbers. Prickles and spines are found in *Parkinsonia* and *Pterolobium*. Stipules are free and of different types. In *Parkinsonia*, rachis becomes flattened to form phyllode.

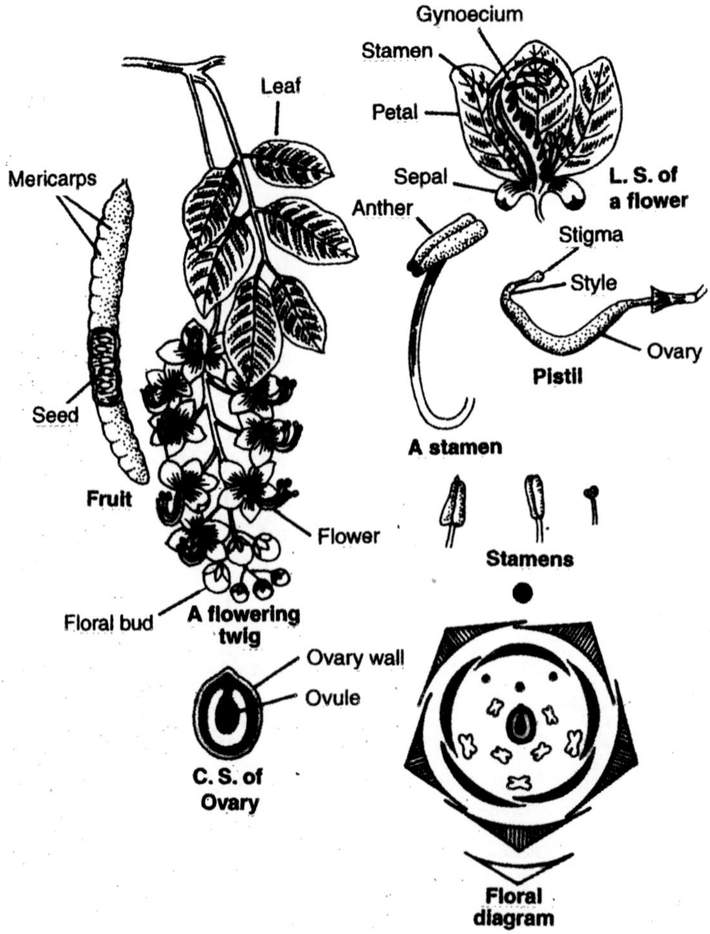

Fig. 20.6 *Cassia fistula* L.

Inflorescence and Flowers

Flowers are usually set in racemose type but may be panicle, or rarely single.

Flowers are mostly showy, bracteate, pedicillate, zygomorphic, bisexual, pentamerous, complete and hypogynous or perigynous. Zygomorphy is due to unequal insertion of petals on thalamus and not due to their form and size.

Calyx consists of 5 sepals (4 in *Amherstia*), free (*Cassia, Caesalpinia* and *Delonix*) or basally connate (*Bauhinia*) or may form a tube. The odd sepal is anterior. Aestivation is imbricate or valvate. **Corolla** consists of 5 free petals, alternate with sepals. Aestivation is **ascending imbricate**, the adaxial petal is always inside. Petals are 3-5 in *Humboldtia*, 3 in *Amherstia* and *Tamarindus* and absent in *Hardwickia*. Petals may be reduced to scales (as in *Tamarindus*).

Androecium consists mainly of 10 free stamens, or sometimes, variously connate. Number of stamens is often reduced, rarely these are numerous. In *Caesalpinia* and *Parkinsonia*, all 10 stamens are perfect, but in *Cassia* 5 to 7, in *Bauhinia* 3 to 5 and in *Tamarindus* only 3 stamens are perfect, the remaining are staminodes. Wherever, ten stamens are present, the **diplostemonous condition** is found, *i.e.*, two alternating whorls of 5 stamens each. Anthers are ditchecous and introrse, dehiscing longitudinally or by apical pores.

Gynoecium is monocarpellary. The ovary is superior or semi-inferior, unilocular with marginal placentation. The carpel is straight, curved or sickle-shaped, hairy or tomentose. Style is simple and stigma simple or capitate.

Fig. 20.7 F.D. of *Bauhinia variegata* L.

Fig. 20.8 F.D. of *Tamarindus indica* L.

Fruits and Seeds

Fruits are mostly legumes, sometimes indehiscent. **Seeds** are non-endospermic and embryo is large.

Pollination and Dispersal

Flowers are entomophilous and seeds are dispersed by animals or wind.

Floral Formula Br % ⚥ $K_{5 \text{ or } (5)}\ C_5\ A_{10}\ \underline{G}_1$

Economic Importance

The family has a large number of economically important plants.

Edibles. Fruit pulp of *Tamarindus indica* L. (tamarind) is used in the preparation of curries, chutneys and some beverages and floral buds of *Bauhinia variegata* L. are used as vegetable.

Medicinal plants. Fruits of *Cassia fistula* L. are used as purgative. Fruits and leaves of *C. angustifolia* Vahl. (senna) are used as laxatives, cathartic, vermifuge and purgative. Leaves of *Cassia alata* L. (ringworm cassia) are used to cure ringworm and skin diseases. Seeds of *Cassia absus* L. are used in treatment of ophthalmia and skin affections. Seeds of *Caesalpinia crista* L. are used in intermittent fever.

Dyes and Tannins. Bark of *Bauhinia malabarica* Roxb., *B. racemosa* Lamk., *B. purpurea* L. and *B. vahlii* W. & A. yield tannin of commercial value. Pods and bark of *Caesalpinia coriaria* Willd., *C. digyna* Rottl. and *C. sepiaria* Roxb. also yield tannin. Heartwood of *Haematoxylon campechianum* L. (logwood) yields blue dye and stain haematoxylin and of *Caesalpinia sappan* L. yields red dye used for colouring cotton and silk.

Gum. Gum of commercial value is obtained from *Bauhinia racemosa* Lamk. and *B. variegata* L. Seeds of *Ceratonia siliqua* L. yield tragasol (caroh gum) used in fabrics and printing.

Timber. Several tree species produce wood used for various purposes. *Hardwickia binata* Roxb. (anjun) produces hard and heavy wood used for piles, railway keys, carts and carriages, tool-handles etc.

Fibres. Barks of *H. binata* Roxb., *Bauhinia anguina* Roxb., *B. racemosa* Lamk. etc. are the source of fibre used for ropes and cordages.

Ornamental plants. Several species are grown for the showy, coloured and attractive flowers in the gardens, avenues and along roadsides, *e.g.* *Delonix regia* (Boj.) Raf. syn. *Poinciana regia* Boj. (royal poinciana), *Saraca indica* Roxb. (ashok, a sacred tree), *Bauhinia racemosa* Lamk., *B. variegata* L., *B. purpurea* L., *Cassia alata* L., *C. fistula* L., *C. glauca* Lamk., *Parkinsonia aculeata* L. (jerusalem thorn, vilayati kikar), *Caesalpinia pulcherrima* (L.) Swtz. syn. *Poinciana pulcherrima* L. (peacock flower, gul-mohur), *C. gilliesii* Wall. (bird-of-paradise) etc.

FABACEAE (PAPILIONACEAE) (Pea Family) or (Sub-family Papilionatae)

Bentham & Hooker	Hutchinson	Cronquist	APG
Dicotyledons	Dicotyledones	Magnoliophyta	3. APG Group :
Polypetalae	Lignosae	Magnoliopsida	Eudicots-Rosids
Calyciflorae	Leguminales	Rosidae	Eurosids I
Rosales		Fabales	Fabales

It is a fairly large family containing **440 genera** and some **12,800 species** growing world wide, primarily in warm temperate regions of the world. About **100 genera** and over **748 species** are found in India with common examples like *Pisum, Cicer, Vigna, Lathyrus, Phaseolus, Indigofera, Cajanus, Arachis, Glycine* etc. *Astragalus* (2000 species), *Indigofera* (700 species) and *Crotalaria* (600 species) are some large genera of the family.

Vegetative Characters

Plants are mostly herbs (*Medicago* and *Melilotus*), but a few are climbers (*Lathyrus*), shrubs (*Indigofera* and *Alhagi*) or trees (*Dalbergia* and *Erythrina*). Commonly, plants have nitrogen-fixing bacterial nodules in lateral roots. **Stem** is usually erect, branched and solid, but weak in climbers. **Leaves** are usually alternate, simple (*Crotalaria*), pinnately (*Sesbania* and *Dalbergia*) or

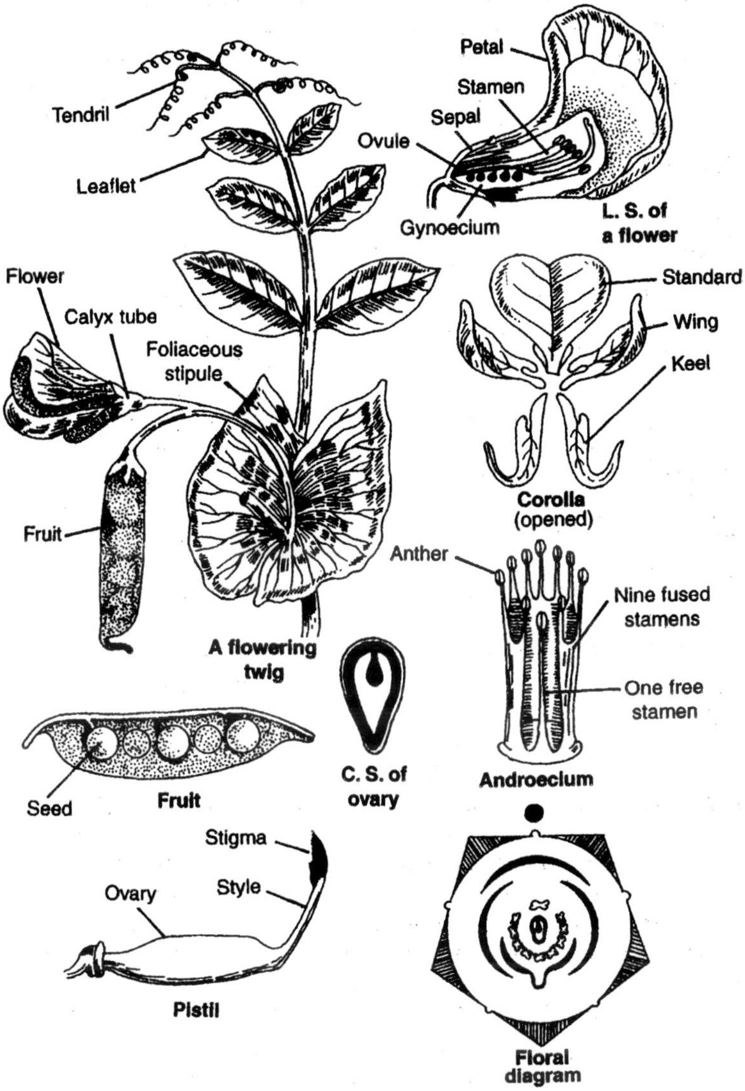

Fig. 20.9 *Pisum sativum* L.

digitately compound (*Trifolium* and *Melilotus*). In *Lathyrus*, *Vicia* and *Pisum*, upper leaflets are modified into tendrils. In *L. aphaca*, entire leaf is modified into tendrils. Stipules are foliaceous in *Lathyrus* and *Pisum*, adnate in *Trifolium* and free in several other genera. Leaflets are often stipulate. Leaves of several genera (*Phaseolus*, *Trifolium* and *Desmodium*) show nyctinasty. Leaflets of *Desmodium montarium* (telegraph plant) show continuous spontaneous movement.

Inflorescence and flowers

Inflorescence is usually racemose but variations are found, *e.g.* panicles (*Dalbergia*), solitary (*Lathyrus*), spikes (*Ononis*) or heads (*Trifolium*).

Flowers are zygomorphic, hermaphrodite, complete, pentamerous, bracteate and hypogynous or perigynous. **Calyx** consists of 5-sepals, more or less connate into a tube (gamosepalous), persistent with valvate aestivation. **Corolla** consists of 5 free petals (polypetalous), **papilionaceous** or *butterfly shaped* (one posterior petal large called **standard** or *vexillum*, two lateral ones called **wings** or *alae* and two anterior or inner most ones, more or less united to form a boat-shaped structure called **keel** or *carina*). Aestivation is **descending imbricate** or *vexillary*. Corolla is gamopetalous in *Trifolium*.

Androecium consists of ten stamens that are mostly **diadelphous** (9 fused and 1 posterior free) or 5 + 5. However, in *Crotalaria*, these are monadelphous and in *Sophora* and *Ormosia*, stamens are free. Anthers are dithecous, introrse and dehiscing usually longitudinally.

Fig. 20.10 F.D. of *Crotalaria juncea* L.

Gynoecium is monocarpellary. Ovary is superior, unilocular with marginal placentation.

Fruits and Seeds

Fruit is usually legume (*Pisum*) or indehiscent (*Dalbergia*) or sometimes breaking up into one seeded segments or lomentum (*Arachis hypogea*). The legume in *Astragalus* is incompletely bilocular due to the formation of a septum by incurved dorsal suture. *Dipteryx* and *Andira* have drupaceous fruits. Fruits are geocarpic in *Arachis hypogaea*, *Trifolium subterraneum* and *Vicia amphicarpa* due to elongation of ovary stalk, the carpophore, after fertilisation. **Seeds** are non-endospermic or with very little endosperm.

Pollination and Dispersal

Flowers are entomophilous, rarely self pollinated. Seeds are usually dispersed by birds, animals and human beings. The persistent calyx of *Anthyllis* becomes bladder-like and encloses the nut and falls along with it, thus helping in seed dispersal.

Floral Formula Br brl % ⚥ $K_{(5)}\ C_{1+2+(2)}\ A_{(9)+1}\ \underline{G}_1$

Economic Importance

The family is of great economic value.

Pulses. Important pulses are *Cicer arietinum* L. (gram), *Pisum sativum* L. (garden pea), *P. arvense* L. (field pea), *Cajanus cajan* (L.) Millsp. (red gram, arhar), *Vigna mungo* (L.) Hepper (urd, black gram), *V. radiata* (L.) Wilc. (mung, green gram), *Lens culinaris* Medic. (lentil, masur), *Lathyrus sativus* L. (grass pea) etc. Because of a large number of pulses, the family is commonly called **Pulse family**.

Vegetables. Fruits of *Canavalia ensiformis* (L.) DC. (jack bean), *Phaseolus vulgaris* L. (french bean), *P. lunatus* L. (lima bean), *Vicia faba* L. (broad bean), *Glycine max* Merr. (soybean), *Lablab purpureus* (L.) Sweet. syn. *Dolichos lablab* L. (sem, hyacinth bean), *Dolichos biflorus* L. (horse gram, kulat), *Vigna aconitifolia* (Jacq.) Marechel (moth), *V. unguiculata* (L.) Walp. (cowpea), *Cyamopsis tetragonoloba* L. Taubert (cluster bean, gwar) etc. are commonly used as vegetables.

Oils. Seeds of *Arachis hypogea* L. (ground nut) and *Glycine max* Merr. (soybean) are major sources of edible oils in the country.

Fibres. *Crotalaria juncea* L. (sunnhemp) yields commercial bast fibre.

Timber. *Dalbergia sissoo* Roxb. (shisham), *D. latifolia* Roxb. (east Indian rosewood), *Pterocarpus dalbergioides* Roxb., (Andman redwood), *Pterocarpus marsupium* (India kino tree), *P. santalinus* L.f. (red sandalwood), *P. indicus* Wild. (Malay padauk) etc. are important timber trees. These produce valuable wood, used variously.

Dyes. *Pterocarpus santalinus* L.f. (red resinous santalin dye from wood), *Crotalaria striata* DC (black dye), *Clitoria ternatea* L. (blue dye from flowers and seeds), *Butea monosperma* (Lamk.) Taub. (orange red dye from flowers), *Indigofera tinctoria* L. (indigo) (neel, indigo dye) etc. are sources of valuable dyes.

Gums. Various gum yielding plants of the family are : *Astragalus gummifer* Labill. (commercially important gum tragacanth), *Myroxylum balsamum* L. Taub. (fragrant gum resin), *Cyamopsis tetragonoloba* L. Taub. (commercially important gwar gum) etc.

Medicinal plants. Some valuable medicinal plants are : *Abrus precatorius* L. (ratti) (decoction of leaves and roots used against cold and cough), *Krameria triandra* Ruiz. & Pav. (dried roots yield drug **krameria** used to cure chronic diarrhoea), *Derris elliptica* Benth. (tuba or derris-roots seeds used as insecticides and fish poison), *Psoralea corylifolia* L. (seeds used in treatment of leucoderma and leprosy), *Moghania grahamiana* Kuntze. (resinous powder used as an anthelmintic), *M. strebilifera* (L.) St. Hil. (roots used to induce sleep and relieve pain), *Glycyrrhiza glabra* L. (liquorice) (roots and stem used in cough syrups), *Crotalaria albida* Heyne. (roots used as purgative), *Teramnus labialis* Spreng. (used in rheumatism and nervous disoders).

Ornamentals. *Lathyrus odoratus* L. (sweet pea), *Clitoria ternatea* L. (butterfly pea), *Wisteria sinensis* Sweet., *W. floribunda* DC., *Erythrina variegata* L. (coral tree), *Robinia pseudo-acacia* L. (black locust), *Pongamia pinnata* Merr., *Sesbania grandiflora* Pers. (sesban), *Sophora japonica* L. (Japanese pagoda tree), *Milletia ovalifolia* Kurz. etc. are various ornamental plants grown for their attractive flowers and foliage.

Systematics and Phylogeny of Leguminosae

Bentham and Hooker kept the family *Leguminosae* in the order **Rosales**, due to presence of perigynous flowers and absence of syncarpy. He recognised three subfamilies—*Papilionoideae, Caesalpinioideae* and *Mimosoideae* based on the differences between androecium and corolla, but kept them under one family due to basic similarities in the features of gynoecium. However, most taxonomists treated the three sub-families as independent families. Hutchinson (1959, 91) recognised them as families considering their vast diversity of corolla, androecium and fruits. Takhtajan (1980, 91) called *Papilionaceae* as *Faboideae* and kept it in the family *Fabaceae* (equivalent to *Leguminosae*) in the order **Fabales.** Cronquist (1981) recognised three separate families in the order **Fabales** and named *Papilionaceae* as *Fabaceae.* Dahlgren (1983) endorsed Cronquist's view point. Thus, Takhtajan (1980, 91), Thorne (1999, 2001), APG II and APweb have used the term *Fabaceae,* equal to the *Leguminosae,* and for three sub-families as *Faboideae, Mimosoideae* and *Caesalpinioideae.* The evolutionary sequence of these three varied in the list of different taxonomists, *e.g.* Bentham and Hooker proposed the evolutionary sequence as *Papilionoideae, Caesalpinioideae* and *Mimosoideae* whereas Hutchinson (1969) considered *Caesalpinaceae* as the most primitive family, as it is close to *Rosaceae* and seems to have been derived from the rosaseous stock. Cronquist (1981) considered *Fabaceae* as most advanced and *Mimosaceae* as most primitive in the order **Fabales.** APG system of classification recognises family *Fabaceae* (equivalent to *Leguminosae*) in the order **Fabales** of sub-group **1. Eurosids I** of monophyletic familial group—3 APG Group : **Eudicots-Rosids.** This system takes in account the recent studies on gene nucleotide sequences.

Table 20.1 Comparison between Sub-families

No.	Faboideae	Caesalpinioideae	Mimosoideae
1.	**Plants**-mostly herbs, a few trees.	Mostly shrubs and trees, rarely herbs.	Mostly shrubs or trees, rarely herbs.
2.	**Root nodules** present.	**Root** nodules absent.	**Root** nodules absent.
3.	**Stem** erect or climbing.	**Stem** erect.	**Stem** erect.
4.	**Leaves** usually imparipinnate or trifoliate.	**Leaves** paripinnate.	**Leaves** bipinnate.
5.	**Inflorescence** mostly raceme.	**Inflorescence** raceme or panicle.	**Inflorescence** cymose, heads or cylindric spikes.
6.	**Flowers** zygomorphic, papilionaceous, pentamerous, hypogynous or perigynous.	**Flowers** zygomorphic, hypogynous, pentamerous.	**Flowers** actinomorphic, penta - or tetramerous, peri - or hypogynous.
7.	**Corolla** consists of 5, free petals with descending imbricate (vexillary) aestivation. Petals unequal.	**Corolla** with 5 petals, set in ascending imbricate aestivation. Petals unequal.	**Corolla** with 4 or 5 petals, sub-connate at the base or form a short tube and set in valvate aestivation. Petals equal.
8.	**Stamens** 10, diadelphous, (9) + 1 or 5 + 5.	**Stamens** 10 in two whorls of 5 each, free, some may be reduced to staminodes.	**Stamens** 4, 10 or numerous free and brightly coloured.

9.	**Fruits** small legume (pod) and seeds ex-albuminous.	**Fruits** long, and legume and seeds in some cases endospermic.	**Fruits** lometum and seeds exalbuminous.
10.	**Floral Formula**	**Floral Formula**	**Floral Formula**
	% ⚥ $K_{(5)}\ C_{1+2+(2)}\ A_{(9)+1}\ \underline{G}_1$	% ⚥ $K_{5\ or\ (5)}\ C_5\ A_{10}\ \underline{G}_1$	⊕ ⚥ $K_{(4)\ or\ (5)}\ C_{4\ or\ 5}\ A_{\alpha}\ \underline{G}_1$

LYTHRACEAE (Lawsonia Family)

Bentham & Hooker	Hutchinson	Cronquist	APG
Dicotyledons	Dicotyledones	Magnoliophyta	3. APG Group :
Polypetalae	Herbaceae	Magnoliopsida	Eudicots-Rosids
Calyciflorae	Lythrales	Rosidae	Eurosids II
Myrtales		Myrtales	Myrtales

The family comprises about **26 genera** and **552 species** found in tropical and subtropical regions with majority in tropical America. In India, about **13 genera** and **50 species** are found with common examples like—*Lawsonia, Lagerstroemia, Woodfordia* etc.

Vegetative Characters

Plants are herbs (*Ammannia*), shrubs (*Lawsonia, Woodfordia*) or trees (*Lagerstroemia*). *Hydrolythrum* is a submerged herb and *Lawsonia* bears spines. **Leaves** are simple, opposite or whorled (*Rotala*), entire, exstipulate or with minute stipules. In stem, intraxylary phloem is invariably present.

Inflorescence and Flowers

Flowers are solitary and terminal or axillary, or arranged in racemes (*Axinandra*) or in panicles (*Lagerstroemia*) or cymose.

Flowers are bracteate, bisexual, actinomorphic or sometimes zygomorphic (*Cuphea*), perigynous, often large and showy or minute. Flowers are unisexual in *Crypteronia*. Flowers may be 3-, 4-, 5-8 merous. Receptacle is cup-shaped. **Sepals** are 4, 6 or 8, rarely more, on the margin of receptacle, free or united and valvate. The hypanthium is often subtended by an epicalyx of connate bracts. **Petals** are as many as sepals, free, usually crumpled, inserted on the margin of the receptacular cup, alternating with sepals, usually clawed and show imbricate or valvate aestivation. In some species of *Ammannia* and *Rotala*, petals are absent.

Stamens are twice as many as petals in two whorls and are inserted below the petals, sometimes the inner whorl is absent. In *Lagerstroemia*, an indefinite number of stamens is found but in some species of *Rotala*, only a single stamen is found. Filaments are unequal and bent in the bud. Anthers are dithecous, introrse and dehiscing longitudinally.

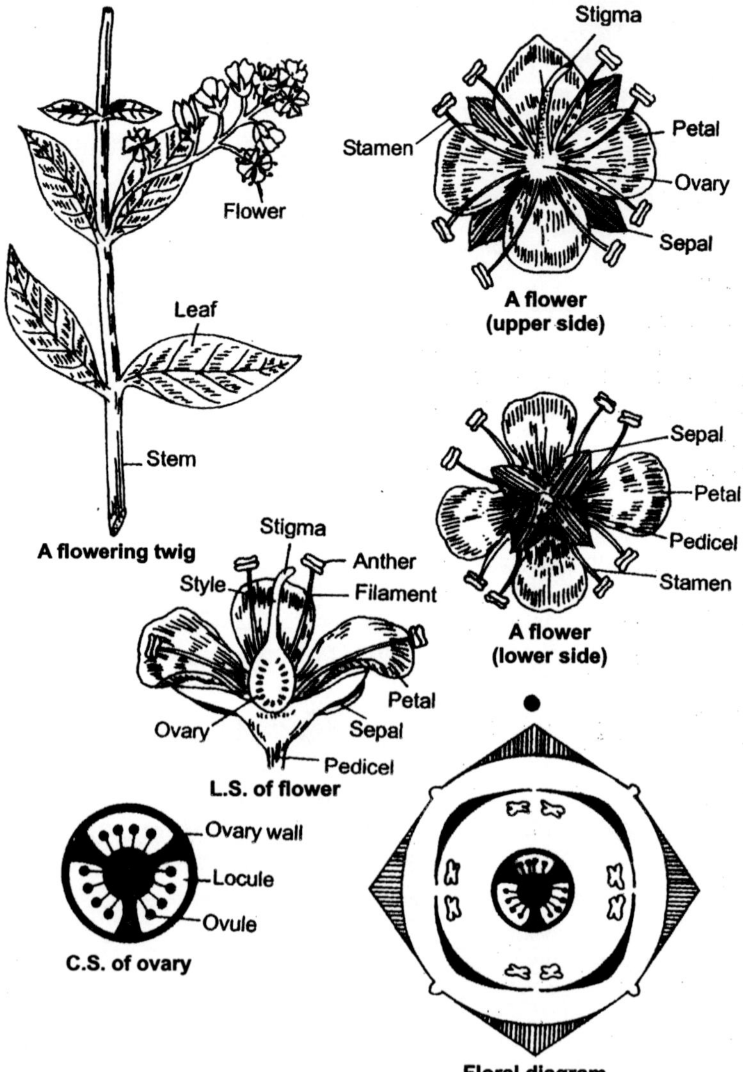

Fig. 20.11 *Lawsonia inermis* L.

Gynoecium consists of 2-6 carpels, united (syncarpous) to form a semi-inferior or superior, 2-6 loculed ovary. Ovules are usually many in each chamber, anatropous and ascending on axile placentae. Rarely ovary is unilocular and the placentation is parietal. Style is simple with capitate stigma.

Fruits and Seeds

Fruit is usually a capsule dehiscing transversely (*Pemphis*), or by valves (*Lagerstroemia*) or irregularly (*Lawsonia*). **Seeds** are non-endospermic and with a straight embryo.

Pollination and Dispersal

Flowers are insect pollinated but cleistogamous flowers of *Ammannia* are self pollinated. **Seeds** are dispersed by wind (winged seeds in *Lagerstroemia*), water *(Rotala)* or by birds and animals.

Floral Formula Br ⊕ ⚥ K_{4-8} C_{4-8} A_{8-16} $G_{(\bar{2}-6)}$

Economic Importance

Few species of the family have considerable economic value.

Timber. Some species of *Lagerstroemia* like *L. speciosa* (L.) Pers., *L. parviflora* Roxb., *L. indica* L., *L. lanceolata* Wall. etc. yield hard and durable timber of commercial value.

Dye and Perfume. A largely used famous dye, henna or mehendi, is obtained from leaves of *Lawsonia* (*L. inermis* L. syn. *L. alba* Lamk.). Flowers of *Lawsonia* yield an essential oil used in perfumery. Mehendi is used in dyeing hairs, beards and hands (palm) etc.

Ornamentals. *Lagerstroemia indica* L. (crape myrtle), *L. speciosa* (L.) Pers., *Lythrum salicaria* L., *Cuphea platycentra* Lem. are grown for their beautiful flowers of various colours.

Systematics and Phylogeny

Lythraceae are closely allied to *Myrtaceae, Onagraceae* and *Melastomaceae* and also to *Punicaceae* and *Sonneratiaceae*. Genus *Punica* was included in *Lythraceae* by Bentham and Hooker but in modern treatments, it has been included in a monogeneric family *Punicaceae* (Cronquist, 1981). Cronquist (1981), Takhtajan (1991), Thorne (1999, 2001) etc., have kept the family *Lythraceae* in the order **Myrtales.** APG classification recognises the family *Lythraceae* in the order **Myrtales** of sub-group 2. **Eurosids II** of the monophyletic familial group, 3. APG Group : **Eudicots-Rosids.**

MYRTACEAE (Myrtle Family)

Bentham & Hooker	Hutchinson	Cronquist	APG
Dicotyledons	Dicotyledones	Magnoliophyta	3. APG Group :
Polypetalae	Lignosae	Magnoliopsida	Eudicots-Rosids
Calyciflorae	Myrtales	Rosidae	Eurosids II
Myrtales		Myrtales	Myrtales

The family comprises **144 genera** and about **3000 species** widely distributed in tropical regions of the world. In India, **14 genera** and about **165 species** are found. Common examples met in the country are *Callistemon, Eugenia, Eucalyptus, Psidium, Syzygium* etc. *Eugenia* (600 species), *Eucalyptus* (500 species) and *Syzygium* (300 species) are some large genera of the family.

Vegetative Characters

Plants are mostly trees or shrubs with aromatic fragrance. **Leaves** are simple, entire, exstipulate, shortly petiolate, mostly opposite (*Psidium*), rarely alternate (*Eucalyptus*) and with oil glands.

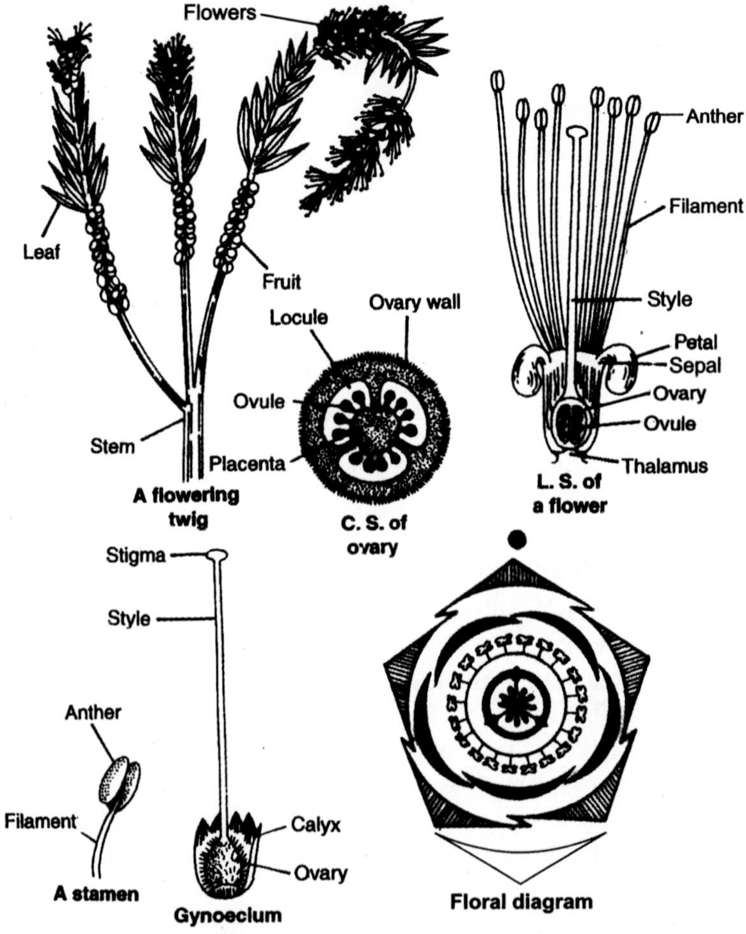

Fig. 20.12 *Callistemon lanceolatus* D.C.

Inflorescence and Flowers

Inflorescence may be raceme (*Barringtonia*), spike (*Callistemon*), cymose, panicle or rarely flowers may be solitary (*Myrtus* and *Psidium*).

Flowers are bracteate, complete, hermaphrodite, actinomorphic and usually epigynous (*Myrtus* and *Psidium*), rarely perigynous. **Calyx** consists of 4 or 5 sepals, connate at the base and united with receptacular cup to form hypanthium, lobes free and quincuncial, rarely valvate. In *Eucalyptus*, sepals are minute or absent. **Corolla** consists of 4 or 5 petals, usually free, imbricate often quincuncial or connate to form operculum (*Eucalyptus*). Petals in *Eucalyptus* fall off quickly to expose stamens.

Androecium consists of indefinite number of stamens, all free or slightly connate at the base, arranged in several whorls, sometimes in bundles opposite the petals. Anthers are usually versatile or dorsally fixed, dithecous, introrse and dehiscing by longitudinal slits or sometimes apically. In *Callistemon*, the filaments are scarlet red and form a brush-like structure. Stamens in *Melaleuca* are polyadelphous and in one whorl and bundles are equal to the number of petals.

Gynoecium consists of 2-5 or several carpels, united to form an inferior or semi-inferior, 2-5 locular ovary (number equal to the number of carpels) with 2 to many obliquely pendulous, campylo- or anatropous ovules in each chamber in axile, rarely parietal, placentation. Style is one and stigma is simple.

Fruits and Seeds

Fruit is loculicidal or septicidal dehiscent capsule but may be drupe, nut or berry (*Psidium*). **Seeds** are angular, cylindric or compressed and non-endospermic or with scanty endosperm.

Pollination and Dispersal

Flowers are insect pollinted. The insects are attracted by nectar-secreting disc around the ovary. Seeds are dispersed by wind, birds or animals. In *Callistemon*, flowers are pollinated by birds.

Floral Formula Br $\oplus$ ⚥ $K_{(4-5)}\ C_{4-5}\ A_{\alpha}\ \overline{G}_{(2-5 \text{ or } \alpha)}$

Economic Importance

The family includes many economically important plants.

Edibles. Many species produce edible fruits, *e.g. Psidium guajava* L. (guava), *P. cattleianum* Sabine (strawberry guava), *Eugenia uniflora* Berg. (Surinam cherry), *Syzygium cumini* (L.) Skeels syn. *E. jambolana* Lamk. (jambolan, jamun), *Syzygium jambos* Alston (rose apple), *S. aromaticum* (L.) Merr. & Perry (the clove, also used as spice) etc.

Timbers. Valuable timber is obtained from many species of *Eucalyptus* (*E. cerebra* F. Muell., *E. paniculata* Sm., *E. punctata* DC., *E. leucoxylon* F. Muell.) and *Syzygium cumini* (L.) Skeels.

Medicinal plants. Some plants are medicinally important, *e.g. Syzygium cumini* (L.) Skeels (seeds useful in diarrhoea, dysentry and diabetes), *Melaleuca leucadendron* L. (cajaput tree) (yields cajaput oil used in laryngitis and bronchitis), *Eucalyptus leucoxylon* F. Muell., *E. elaeophora* F. Muell., *E. globulus* Labill (medicinally important eucalyptus oil extracted from leaves) etc. Clove oil extracted from cloves of *Syzygium aromaticum* Merr. & Perry is used in toothaches and oil of baby rum.

Ornamentals. Several plants are grown for their ornamental value, *e.g. Callistemon* (bottle brush) (*C. lanceolatus* DC., *C. viminalis* G. Don ex Loud.,

C. linearis DC etc.), *Eucalyptus* (species grown for timber value), *Myrtus communis* L. (myrtle) etc.

Systematics and Phylogeny

Almost every system of taxonomic classification recognised the family *Myrtaceae* and placed it in the order **Myrtales.** Cronquist (1981), Takhtajan (1980, 91) and Thorne (1999, 2001) kept the order **Myrtales** in the sub-class **Rosidae** of class **Magnoliopsida**. The family is considered as closely related to *Rosaceae* and the order **Myrtales** being derived from **Rosales.** The molecular analysis based on nucleotide sequences of *rbc*L, *mat* K and *ndh* F (genes) reveals the monophyletic familial group – Sub-group 2. **Eursoids II** of 3. APG Group : **Eudicots-Rosids** (APG II, 2003).

COMBRETACEAE

Bentham & Hooker	Hutchinson	Cronquist	APG
Dicotyledons	Dicotyledones	Magnoliophyta	3. APG Group :
Polypetalae	Lignosae	Magnoliopsida	Eudicots-Rosids
Calyciflorae	Myrtales	Rosidae	Subgroup 2.
Myrtales		Myrtales	Eurosids II
			Myrtales

The family comprises about **20 genera** and **600 species** found in tropical and sub-tropical regions of the world. In India, **8 genera** with about **44 species** are found with common examples like *Terminalia, Quisqualis* and *Combretum.* A few are mangroves, *e.g. Lumnitzera, Conocarpus* and *Laguncalaria.*

Vegetative Characters

Plants are trees (*Terminalia*) or shrubs, often lianous (*Quisqualis*), armed or bare. **Leaves** are simple, alternate or opposite, entire and exstipulate. In *Illigera,* leaves are trifoliate.

Inflorescence and Flowers

Inflorescence may be spicate, racemose or paniculate. **Flowers** are bracteate, bisexual or rarely unisexual or polygamo-dioecious (*Combretum*), actinomorphic or zygomorphic, usually pentamerous (sometimes 3-8 merous) and epigynous. The tubular receptacle together with the adnate tubular base of calyx forms hypanthium. **Calyx** consists of 4-5 or more persistent lobes with valvate or imbricate aestivation. **Corolla** consists of 5, free lobes (petals) or lobes equal to the number of sepals, small, imbricate or valvate. Petals are often absent (*Terminalia* and *Anogeissus*).

Andrꝏcium comprises 4-5 stamens in one whorl or double the number in two whorls, outer whorl antisepalous and inserted on the calyx-tube, and inner whorl, sometimes, reduced to staminodes. Filaments are inflexed in the

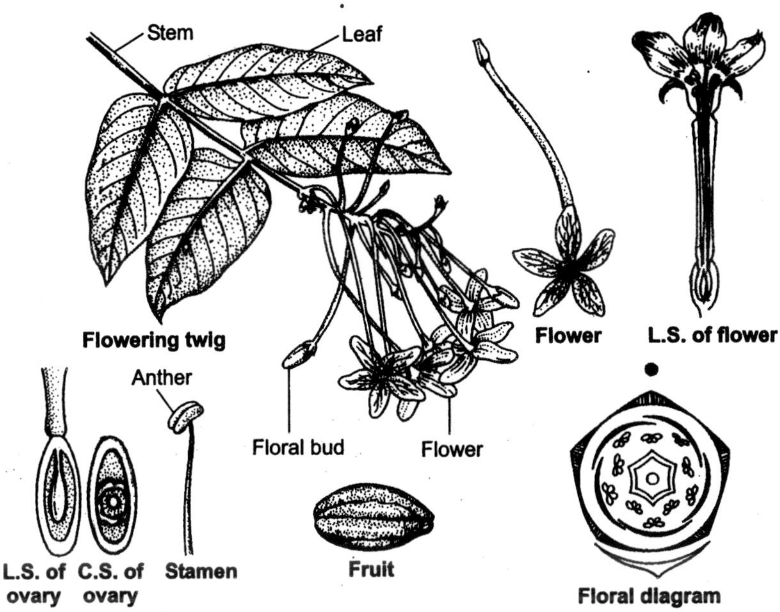

Fig. 20.13 *Quisqualis indica* L.

bud and anthers are dithecous and versatile, dehiscing by longitudinal slits (*Terminalia*) or by recurved lateral valves (*Illigera*).

Gynoecium has 4 or 5, or often less carpels and united (syncarpous). Ovary is inferior, unilocular, angular or ribbed–ribs equal to the number of calyx lobes, and with 2-6 anatropous ovules pendulous from the top of the locule. Style is slender with a capitate or pointed stigma. A nectariferous epigynous disc is present.

Fruits and Seeds

Fruit is a single seeded drupe with angled or winged pericarp. **Seeds** are ex-albuminous and embryo has convolute (*Terminalia*) or folded, rarely flat cotyledons.

Pollination and Dispersal

Flowers are insect pollinated. Fruits are dispersed by water (*Terminalia arjuna*), wind or by animals.

Floral Formula Br ⊕ ⚥ $K_{(4-5)}\ C_{4-5}\ A_{4-5 \text{ or } 8 \text{ or } 10}\ \overline{G}_{(4-5)}$

Economic Importance

Family has few genera with great economic value.

Timber. An useful timber is obtained from *Anogeissus latifolia* (Roxb.) Wall, *A. acuminata* (Roxb.) Wall, *Terminalia tomentosa* W. & A., *T. paniculata* Roth., *T. bialata* Stend., *T. catappa* L., *T. myriocarpa* Heurck & Small, *T. ivorensis* A. Chev., (indigo timber), *T. chebula* Retz., *T. citrina* Roxb. ex Flem. (harira),

T. manii King (black chuglam), *T. myriocarpa* Heurck & Muell. (hollock), *T. procera* Roxb. etc. Wood is largely used in construction works.

Medicinal plants. The bark of *Terminalia arjuna* (Roxb.) W. & A. and *T. bialata* Kurz. is highly valued as cardiac tonic. Fruits of *Terminalia chebula* Retz. (black myrobalan, hararh), *T. bellerica* Roxb. (belleric myrobalans, bahera) and *T. arjuna* (Roxb.) W. & A. are laxative and are used for stomachic and alterative properties. Leaves of *Calycopteris floribunda* Lamk. are used in colic, dysentery and malarial fever, and of *Combretum pilosum* Roxb. as anthelmintic.

Ornamentals. *Quisqualis indica* L. (Rangoon creeper) and *Combretum roxburghii* Spreng. (punk) are popular ornamental plants.

Systematics and Phylogeny

Family has been kept in the order **Myrtales** almost in every system of classification. It is closely related to *Myrtaceae* and *Rhizophoraceae*. It has been regarded as an advanced family of the order due to absence of petals (*Terminalia*), aggregation of flowers into heads and adaptation for wind dispersal. APG classification has placed the family in monophyletic familial group - 3. APG—**Eudicots : Rosids.**

LORANTHACEAE (Mistletoe Family)

Bentham & Hooker	Hutchinson	Cronquist	APG
Dicotyledons	Dicotyledones	Magnoliophyta	2. APG Group :
Monochlamydeae	Lignosae	Magnoliopsida	Core Eudicots
Achlamydosporeae	Santalales	Rosidae	Santalales
		Santalales	

The family consists of **30 genera** and **1100 species** found in tropical and temperate regions of the world. In India, **11 genera** and about **50 species** are found with common representatives like *Dendrophthoe, Viscum, Helixanthera, Scurrula* etc.

Vegetative Characters

Plants are hemiparasitic herbs or shrubs with sympodial, often dichotomous stem. They are attached to the branches of trees by means of haustoria (sucking roots). Leafless *Arceuthobium minutissimum* parasitising on *Pinus excelsa* is the **smallest dicotyledon** (found in West Himalayas). **Leaves** are simple, entire, exstipulate, coriaceous, opposite or rarely alternate, persistent, sometimes reduced to connate scales.

Inflorescence and Flowers

Inflorescence is axillary, usually a biparous cyme or of mixed type or flowers are solitary.

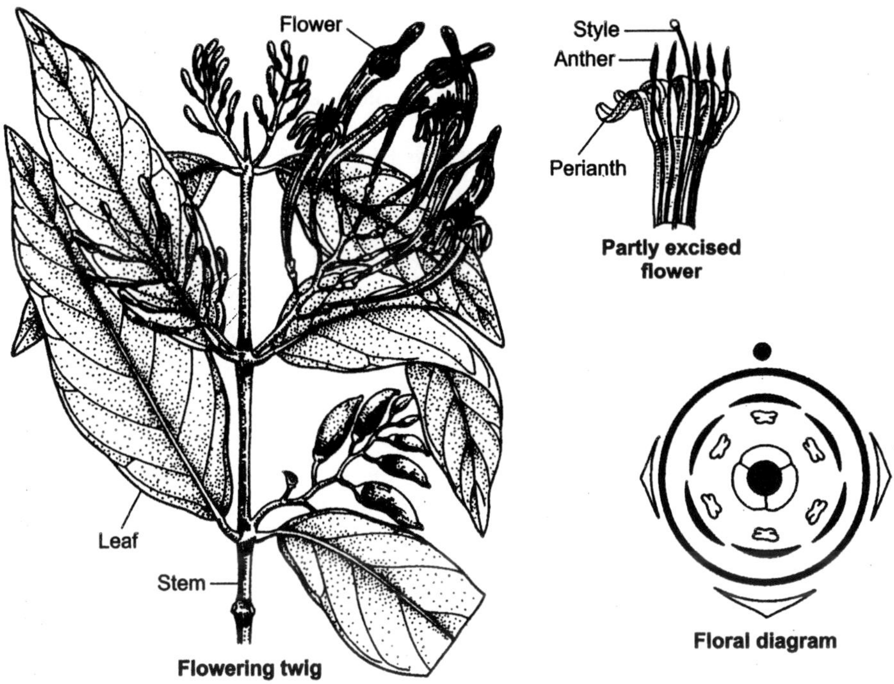

Fig. 20.14 *Dendrophthoe falcata* (L.f.) Etting.

Flowers are regular (actinomorphic), bisexual or unisexual (*Viscum*), brightly coloured, di-or trimerous, and epigynous with cup-shaped receptacle, bracts present.

Perianth is found in two whorls with 2-3 segments (tepals) in each whorl, both similar, *i.e.*, petaloid or sepaloid or in one whorl with six segments. Segments may be free or connate. The inner segments are valvate, and the outer ones are valvate or imbricate.

Androecium consists of stamens equal the number and opposite and adnate to the perianth lobes (epiphyllous). Anthers are dithecous (sometimes monothecous by the confluence), or transversely polythecous, dehiscing longitudinally or by transverse slits or by apical pores (*Viscum*).

Gynoecium comprises 3-4 carpels, united (syncarpous) to form a unilocular and inferior ovary. Ovules are several, indistinct on a large central placenta and only 1-3 attaining maturity. Style is columnar or filiform and stigma is obtuse or capitate. Ring like outgrowth of receptacle called *calyculus*, is found in flowers with uniseriate perianth.

Fruits and Seeds

Fruit is a pseudocarp or drupaceous and 1-3 seeded. **Seeds** are mucilagenous, without testa and with a large straight embryo and fleshy endosperm. Seeds in *Nuytsia* have three cotyledons.

Pollination and Dispersal

Flowers are brightly coloured and insect pollinated. Fruits/seeds are dispersed by birds. *Arceuthobium* has explosive fruits and the seeds are dispersed mechanically.

Floral Formula $Br \oplus ⚥ \widehat{P_{2+2 \text{ or } 3+3}\ A_{4 \text{ or } 6}}\ \overline{G}_{(3-4)}$

Economic Importance

The family is of little economic value. *Viscum album* L. (mistletoe) is commonly used for decoration during christmas in European countries. Seeds are laxative and cardio-tonic.

Systematics and Phylogeny

Bentham and Hooker considered the family a place in the series **Achlamydosporeae** while most taxonomists have kept it in the order **Santalales** (Hutchinson, 1959, 81 and Cronquist, 1981). *Loranthaceae* are closely related to *Santalaceae* but can be distinguished by the aerial-parasitic habit, cup-shaped receptacle, inferior ovary and absence of distinct ovules. APG classification has kept it in monophyletic familial group, 2. APG Group : **Core Eudicots**, and in the order **Santalales**.

EUPHORBIACEAE (Spurge Family)

Bentham & Hooker	Hutchinson	Cronquist	APG
Dicotyledons	Dicotyledones	Magnoliophyta	Group
Monochlamydeae	Lignosae	Magnoliopsida	3. APG Group
Unisexuales	Euphorbiales	Rosidae	Eudicots-Rosids
		Euphorbiales	Subgroup 1.
			Eurosids I
			Malpighiales

The family is quite large comprising **321 genera** and **7770 species**, widely distributed in tropical and sub-tropical regions of the world. In India, it is represented by **70 genera** and some **450 species** with common forms being *Acalypha, Bischofia, Croton, Euphorbia, Jatropha, Phyllanthus, Putranjiva, Manihot, Emblica, Ricinus* etc. *Euphorbia* is a large genus with 2100 species whereas *Croton* (720 species), *Phyllanthus* (500 species) and *Acalypha* (350 species) are other major genera.

Vegetative Characters

Plants are annual or perennial herbs (*Euphorbia prostata, E. hirta, Phyllanthus niruri, Acalypha indica*), unarmed or armed (*Euphorbia splendens*), shrubs (*Ricinus, Jatropha, Buxus*), trees (*Putranjiva, Bischofia, Hevea, Emblica*) or woody twinners (*Bridelia, Dalechampia*) or cactus-like (*Euphorbia royleana*).

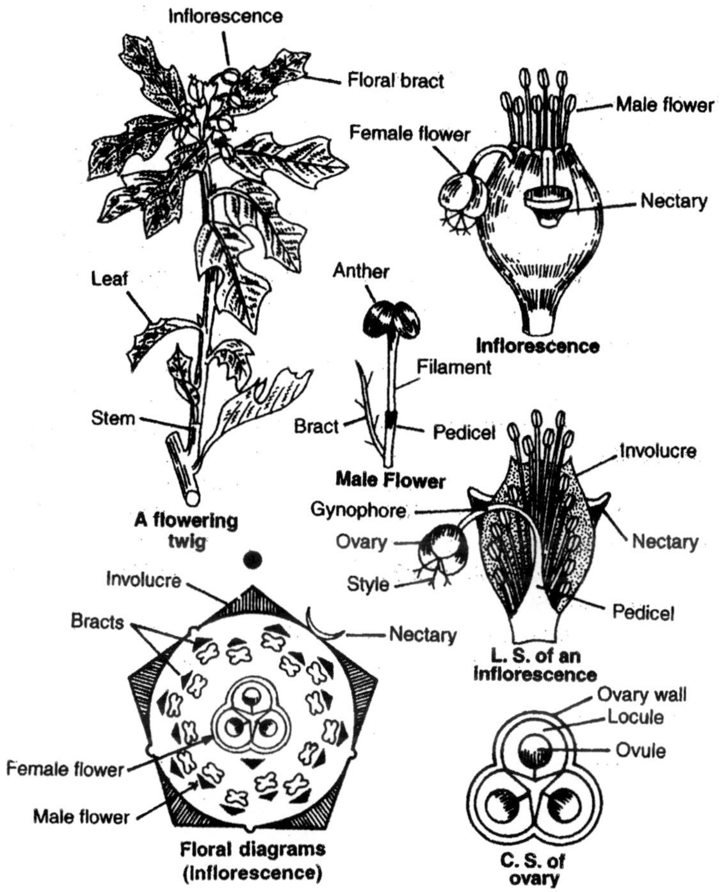

Fig. 20.15 *Euphorbia pulcherrima* Willd.

Stem is erect or prostrate (*Euphorbia prostrata*), thick, fleshy, unarmed (*Ricinus, Phyllanthus*) or armed with spines (*Euphorbia splendens*), hairy (*E. hispida*) or glabrous, pubescent (*Putranjiva*), cylindrical, angular, hollow (*Ricinus*) or solid, branched and usually containing milky latex. Stem of xerophytic species of *Euphorbia* (*E. royleana*) is modified (phylloclade), succulent and becomes cactus like.

Leaves are generally alternate, sometimes opposite (*Euphorbia hirta*), mostly simple, stipulate or exstipulate, and entire but in *Ricinus* and *Jatropha*, these are deeply palmately lobed. In several species of *Euphorbia*, leaves are reduced or scale-like, caducous with spiny stipules. Leaves are trifoliate and compound in *Hevea*.

Inflorescence and Flowers

Inflorescence is variable, as spike or catkin (*Acalypha*), raceme (*Ricinus* and *Croton*), compound (*Manihot*) or terminal cyme (*Jatropha*). In *Euphorbia*, the inflorescence apparently appears like a single flower and is called *cyathium*. Each cyathium is surrounded by an involucre of 4 or 5 connate

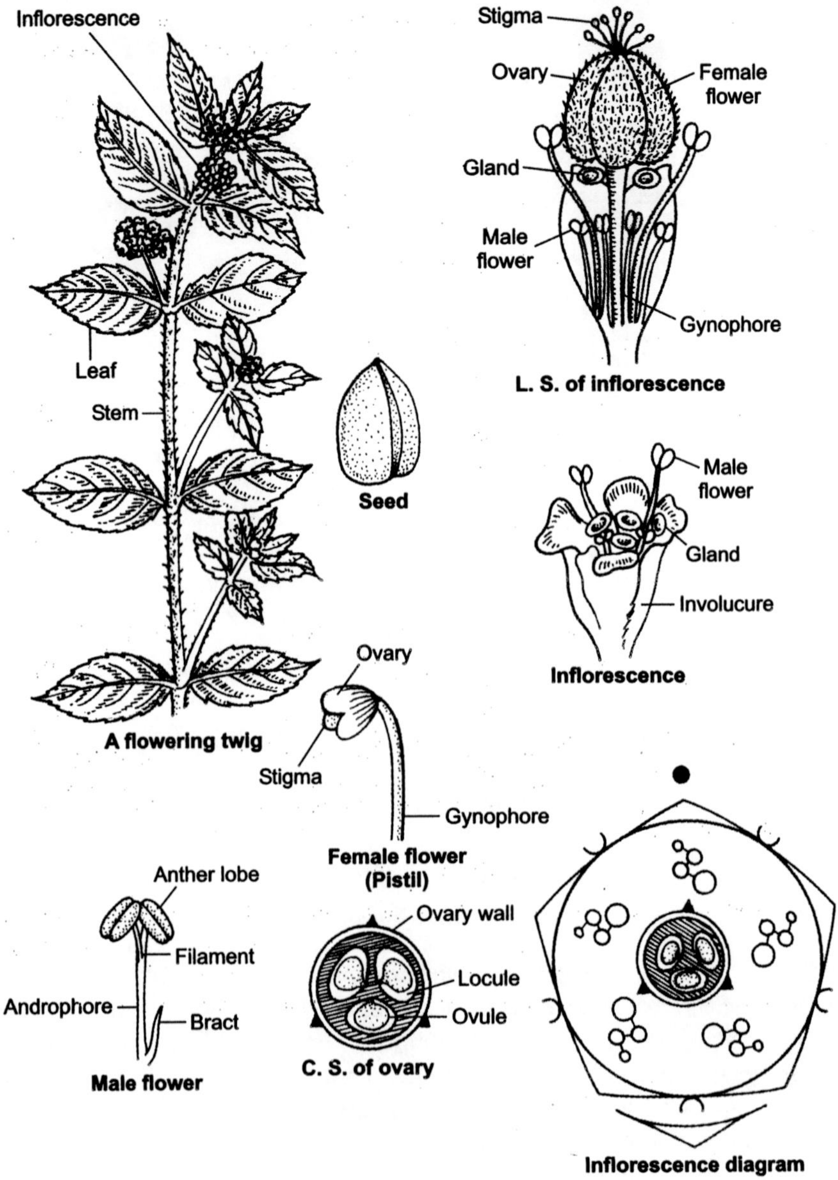

Fig. 20.16 *Euphorbia hirta* L.

bracts and between these, large and often coloured glands are present. Cyathium comprises a single central female flower borne on gynophore surrounded by a large number of male flowers, each represented by a single stalked stamen.

Flowers are minute, bracteate, actinomorphic, regular, unisexual and hypogynous. Plants are normally *monoecious,* rarely *dioecious* (species of *Excoecaria, Sureganda, Trewia*).

Perianth is mostly in one whorl, green or rarely petaloid, rarely two whorls (*Jatropha*), sometimes absent (*Euphorbia*).

F.D. of male flower F.D. of female flower

Jatropha gossypifolia L.

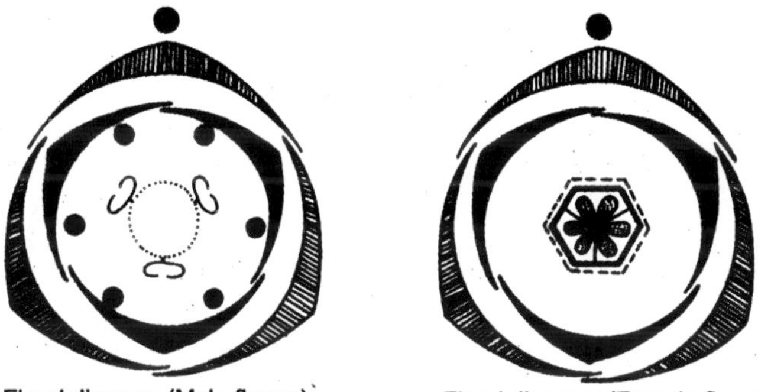

Floral diagram (Male flower) Floral diagram (Female flower)

Phyllanthus niruri Hook. f.

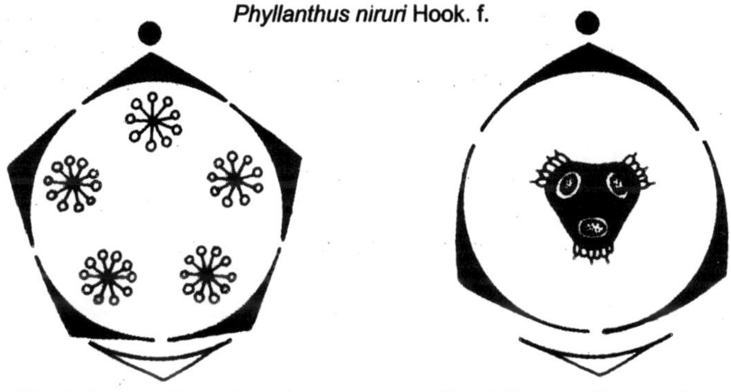

Floral diagram (Male flower) Floral diagram (Female flower)

Ricinus communis L.

Fig. 20.17 Floral diagrams of various flowers.

In *Jatropha*, flowers are dichlamydeous, and calyx consists of 5 sepals, usually valvate or imbricate and corolla of 5 petals, valvate or contorted. In *Phyllanthus* and *Ricinus*, flowers are apetalous and only sepaloid perianth is present while in *Euphorbia*, perianth is altogether absent and flowers are achlamydeous (naked), rarely may be seen as minute scale-like structures.

Androecium comprises variable number of stamens. Stamen is one in *Euphorbia* and *Pedilanthus*, 3-5 in *Phyllanthus* and *Breynia*, or 5, profusely branched in *Ricinus* and *Homonoia*, 8 in *Acalypha*, 10 in *Jatropha* and 80-100 in

Croton. Anthers are monothecous (*Ricinus* and *Phyllanthus*) or dithecous (*Euphorbia*), dehiscing by longitudinal or transverse slit or by pores. In *Manihot, Jatropha, Chrozophora* etc. stamens are in more than one whorl while in *Breynia* and *Phyllanthus*, there are 3 stamens, monaldelphous, forming a central column.

Gynoecium, found in female flowers, consists mostly of three united carpels (syncarpous). The ovary is superior, trilocular with axile placentation having one or two pendulous and anatropous ovules in each locule beneath a prominent obturator (placental growth). Styles are three, often bipartite, free or more or less united. Stigmas are 3 or 6, linear or broadened, often papillate or dissected into filiform stigma.

Mercurialis has usually a bicarpellate bilocular ovary while in *Antidesma* and *Drypetes*, the ovary is unilocular due to abortion. In *Hura*, a multicarpellate multilocular ovary is found. Androphore and gynophore may sometimes be found in *Bridelia*.

Fruits and Seeds

Fruit is mostly a schizocarpic capsule or regma (*Ricinus*) with spines, splitting into three one-seeded cocci. Rarely, it is a drupe (*Phyllanthus emblica*) or berry (*Bischofia*).

Seeds are endospermic and often with a conspicuous caruncle.

Pollination and Dispersal

Flowers are entomophilous, rarely anemophilous. Seeds in *Ricinus* are dispersed by explosive mechanism (autochory), but in many other cases, these are dispersed by birds and animals (zoochory).

Floral Formulae

Male flower $Br \oplus ♂ P_{3-5 \text{ or } 0} A_{1-\alpha} G_0$

Female flower $Br \oplus ♀ P_{3-5 \text{ or } 0} A_0 \underline{G}_{(3)}$ or more

Euphorbia

Male flower $Br \oplus ♂ K_0 C_0 A_1 G_0$

Female flower $Br \oplus ♀ K_0 C_0 A_0 \underline{G}_{(3)}$

Economic Importance

The family has several economically important plants, useful in various ways.

Edibles. The tuberous roots of *Manihot esculenta* Crantz. syn. *M. utilissima* Pohl. (cassava) are rich source of starch and are eaten-like potato. These produce tapioca. Fruits of *Emblica officinalis* Gaertn. syn. *Phyllanthus emblica* L. (emblic myrobalan, amla) are rich source of vitamin C and used in pickles, chutneys, murabba, scurvy etc. It is an important ingredient of famous ayurvedic drug *Chyawanprash*.

Rubber. Latex from the stem of Brazilean para-rubber tree *Hevea brasiliensis* (H.B.K.) Muell-Arg. is the major source of commercial rubber. In India, its trees are grown extensively in Kerala and Tamil Nadu. The latex of *Manihot glaziovii* Muell-Arg. (manicoba rubber tree) is the source of ceara-rubber.

Medicinal plants. The seeds of *Ricinus communis* L. (castor) yield castor oil which is used as a cathartic and mild laxative. Roots of *Jatropha gossypifolia* L. are used to cure leprosy and as an antidote in snake bite. Seeds of *Jatropha curcas* L. (purging nut) are purgative and anthelmintic and oil from seeds is used to cure rheumatism. Oil obtained from the seeds of *J. curcas* L. and *Croton tiglium* L. (purging croton, jaipal) is a strong purgative. The bark of *Croton oblongifolius* Roxb. is used externally in sprains.

Timbers. Some important timber yielding trees are *Bischofia javanica* Blume (malkot), *Trewia nudiflora* L. (false white teak), *Hura crepitans* (sandbox tree), *Putranjiva roxburghii* Wall. (Putijia) syn. *Drypetes roxburghii* (Wall.) Hurus, *Aporosa dioica* Muell. etc. The wood is used in construction, making appliances (agricultural) and packing boxes.

Oils. Seed oil of *Ricinus communis* L. (castor) is used in paints and plastic industries and also as lubricants. The famous essential oil *tung oil*, obtained from the seeds of *Aleurites fordii* Hemsl. and *A. moluccana* Willd., is used in the preparation of water proof paper, fabrics and linoleum. Oil of medicinal value is also procured from several other plants.

Tannins and Dyes. The bark of *Emblica officinalis* Gaertn. is a source of tannin used in leather industry and that of *Jatropha curcas* L. yields deep blue dye. *Crozophora tinctoria* A. Juss. yields purple dye. The capsule of *Mallotus philippinensis* Muell. (kamela tree) yields kamela dye used in colouring silk and wool.

Ornamentals. Some important plants grown as ornamentals are : *Jatropha podagrica* Hook., *J. gossypifolia* L., *Acalypha hispida* Burm. f., *Euphorbia pulcherrima* Willd. syn. *Poinsettia pulcherrima* R. Grah (poinsettia), *E. tirucalli* L. (milkbush), *E. milii* Ch. des-Mollins syn. *E. splendens* Boj. (crown of thorn), *Codiaeum variegatum* Blume. (croton), *Pedilanthus tithymaloides* Poit. (red bird cactus) etc. These are grown for attractive foliage and inflorescence.

Systematics and Phylogeny

The family *Euphorbiaceae* has been placed in **Unisexuales** by Bentham and Hooker, in **Geraniales** by Engler and Prantl and in **Euphorbiales** by Hutchinson, Cronquist, Takhtajan etc. It was included in **Unisexuales** due to unisexual flowers and in **Geraniales** due to ovule character which has a ventral raphe. However, the family appears more allied to *Sterculiaceae* due to monadelphous or polyadelphous stamens and presence of gynophore and androphore in some cases. Presence of schizocarpic capsule and carunculate seeds in some members of *Sterculiaceae* recall close relationship between the two families. Recognising the affinities between these two, Rendle has placed it in a small order **Tricoccae** in between **Malvales** and **Geraniales**.

Hutchinson kept it in the oder **Euphorbiales** with a single family. He believes that the members of *Euphorbiaceae* seem to be highly specialised due to the presence of highly reduced flowers in many genera, flower-like inflorescence (cyathium) in *Euphorbia* and milky latex in the plant body. These features led him to place **Euphorbiales** nearer to **Malvales** under **Malpighiales** in **Lignosae**. Takhtajan also accepted closeness of *Euphorbiaceae* with *Sterculiaceae* and places **Euphorbiales** just after **Malvales**. Cronquist holds a different view. He considers that **Euphorbiales** is closer to **Sapindales** as well as to **Geraniales, Linales** and **Polygalales**, all being derived from rosaceous stock while **Malvales** from **Dilleniales** through **Theales** and being not close to **Euphorbiales.**

Kare Bremer and coworkers (1998) and APG II classification (2003) place the family *Euphorbiaceae* in the monophyletic familial group - 3. APG Group : **Eudicots** - **Rosids** in the order **Malpighiales.** They further proposed that cladistic studies reveal to separate the genus *Purtranjiva* and to place it in the family *Putranjivaceae* (a new family). Webster (1994) and Thorne (1999, 2003) recognised five sub-families in the *Euphorbiaceae*—Phyllanthoideae, Oldfieldioideae, Acalyphoideae, Crotonoideae and Euphorbioideae. The first two sub-families on the basis of evidences from *rbc*L sequences have been separated into a distinct family *Phyllanthaceae*, as they do not form a clade with other members of *Euphorbiaceae*. Thus, Thorne (2003) and APweb hold the opinion to develop two families - *Phyllanthaceae* and *Putranjivaceae*, separated from *Euphorbiaceae*. It appears that groupings in the family need more elaborate studies based on recent findings and cladistic analyses.

VITACEAE (Grape Family)

Hutchinson	Cronquist	APG
Dicotyledones	Magnoliophyta	2. APG Group :
Lignosae	Magnoliopsida	Core Eudicots
Rhamnales	Rosidae	
	Rhamnales	Order not assigned

The family is small, comprising **11 genera** and about **750 species** found in tropics and warm temperate regions of the world. About **8 genera** and **95 species** are found in India, wherein the common examples are—*Vitis vinifera, Ampelocissus latifolia, Cissus repens, C. quadrangularis, Cayratia pedata* etc.

Vegetative Characters

Plants are extensive climbers with leaf-opposed tendrils, rarely shrubs or trees (*Carameriana*). Genus *Leea* is represented by trees, shrubs or erect herbs. Tendrils have been interpreted as metamorphosed apical parts of sympodial stem (Eichler). **Leaves** are alternate, the lower often being opposite (*Leea*), simple or pinnately or palmately compound, often pellucid punctate with petiolar stipules or exstipulate, usually distichous.

Inflorescence and Flowers

Inflorescence is leaf-opposed cymose panicle often very large with minute flowers, rarely raceme or spike or corymbose and terminal. Occasionally, the peduncles are flattened (ribbon-like) and expanded (*Pterisanthes*). Sometimes, inflorescence replaces the tendril.

Flowers are bracteate, actinomorphic, bisexual or unisexual in polygamo-monoecious or polygamo-dioecious plants, hypogynous, minute and tetra- or pentamerous. **Calyx** is gamosepalous, cupular, entire or shortly 5-4 toothed, sometimes reduced to an inconspicuous ring. **Corolla** consists of 5 or 4 minute petals, free or connate, caducous, flat or with hooded incurved apex and valvate. In *Vitis*, petals are free near the base and united by the apices and falling off early as a cap.

Androecium consists of stamens as many as petals and opposite to them. Filaments are short, free or connate at the base, inserted below the intrastaminal disc. Anthers are free or connate, dithecous, introrse and dehiscing longitudinally. Filaments in *Leea* are united to form a 5-lobed tube.

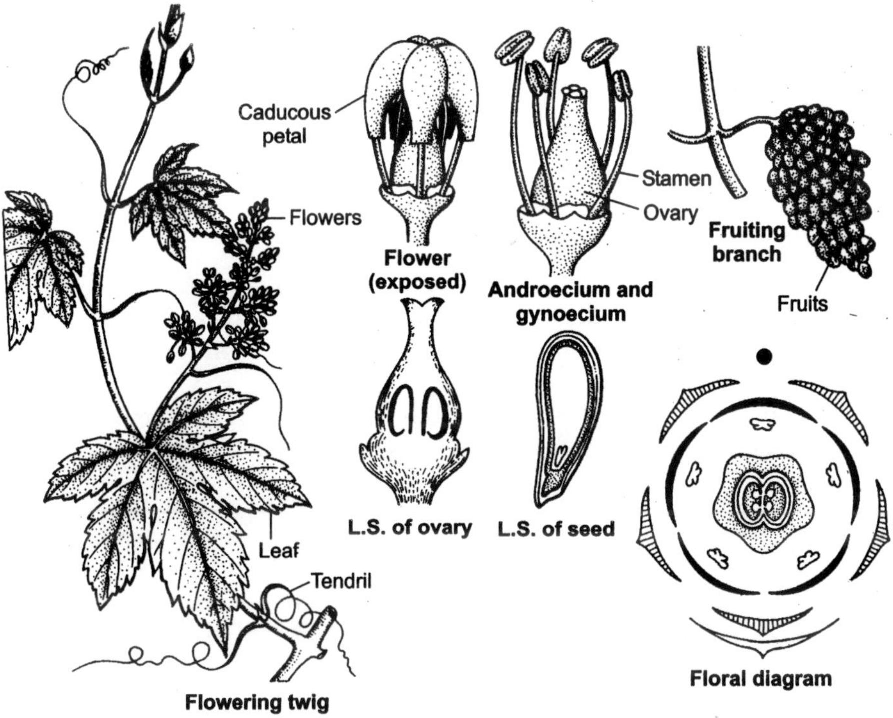

Fig. 20.18 *Vitis vinifera* L.

Gynoecium has 2-8 united carpels (syncarpous). Ovary is superior with 2-8 locules, each locule has 1-2 anatropous ovules set in axile placentation. Style is short and stigma is discoid or capitate.

Fruits and Seeds

Fruit is 2-8 chambered juicy berry with 1 or 2 seeds in each chamber. **Seed** has hard testa, copious ruminate endosperm and small straight embryo.

Pollination and Dispersal

Flowers are entomophilous, insects are attracted by nectaries present in intrastaminal discs. Seeds are dispersed by birds and other animals.

Floral Formula Br ⊕ ⚥ K_{4-5} $C_{(4-5)}$ A_{4-5} $\underline{G}_{(2-8)}$

Economic Importance

The family is popular mainly due to economic value of grape-vine, *Vitis vinifera* L. which is grown throughout the world for its use in the preparation of wines and brandies, and as raisins (dried fruits) and currants (dried tiny fruits of corinthian variety). Fruits of *Leea cripsa* L. and *L. macrophylla* Roxb. ex Hornem are often eaten by local people.

Medicinal plants. A few plants have medicinal value, *e.g.* the tubers of *Cissus adnata* Roxb. are blood purifier and its root powder is applied to cuts and fractures. The root tubers of *Leea macrophylla* Roxb. ex Hornem are applied to wounds, sores and ringworms, and that of *L. indica* (Burm.) Merr. are used in diarrhoea, colic pains and dysentery. A fomentation with hot decoction of roots and leaves of *Vitis trifolia* L. (amal-bel) is used in high fever to cause perspiration.

Ornamentals. Some species of *Ampelopsis, Cissus* and *Parthenocissus* are grown as ornamental climbing vines.

Systematics and Phylogeny

The family is closely allied to family *Rhamnaceae*, as such both are included in the order **Rhamnales** by most taxonomists. It differs from *Rhamnaceae* in climbing habit, berried fruits and seeds with copious ruminate endosperm. Some recent workers have separated *Leea* and placed it in the family *Leeaceae. Vitaceae* differ from *Leeaceae* in presence of climbing tendrils, leaf-opposed cymes, flowers polygamous, filaments free, anthers free or connate and bilocular ovary with two ovules in each locule. The recent cladistic APG system places it in 2-APG Group : **Core Eudicots** without assigning it to any order.

SAPINDACEAE (Soapberry Family)

Bentham & Hooker	Hutchinson	Cronquist	APG
Dicotyledons	Dicotyledones	Magnoliophyta	3. APG Group :
Polypetalae	Lignosae	Magnoliopsida	Eudicots-Rosids
Disciflorae	Sapindales	Rosidae	Sub-Group 2
Sapindales		Sapindales	Eurosids II
			Sapindales

The family with its **92 genera** and **900 species** is widely distributed in the tropical regions of the world. In India, it is fairly represented by **24 genera** and **72 species** with common examples like *Litchi chinensis, Aesculus indica, Sapindus mukorossi* and species of *Acer,* mainly found in Western Himalayas. *Litchi* and *Sapindus* are grown in most part of the country.

Vegetative Characters

Plants are trees or shrubs, rarely woody vines (*Cardiospermum* climbs by tendrils or modified inflorescence axes). The stem contains resin-passages with plenty of gum and acrid juice. **Stem** shows often extra-stelar vascular bundles (anomalous condition). **Leaves** are alternate or rarely opposite (*Acer*), usually pinnately compound, rarely ternate and very rarely simple (*Cardiospermum*), palmately lobed (*Acer*) or digitate (*Aescules*), exstipulate or in climbing species often stipulate, stipules are small and deciduous. Latex or resin cells are present in leaves. *Cardiospermum* is a herbaceous climber but *Serjania* and *Paullinia* are lianas.

Inflorescence and Flowers

Inflorescence is usually a cymose panicle or a thyrsus, rarely racemose. **Flowers** are small, regular or obliquely zygomorphic, bisexual or unisexual or polygamous, pentamerous or rarely tetramerous and hypogynous. **Calyx** comprises usually 5 free sepals or in regular flowers with sepals 4 due to union of 3rd and 5th sepals and imbricate in bud. **Corolla** is made up of 5 free petals, but 4 petals in actinomorphic flowers as a result of suppression.

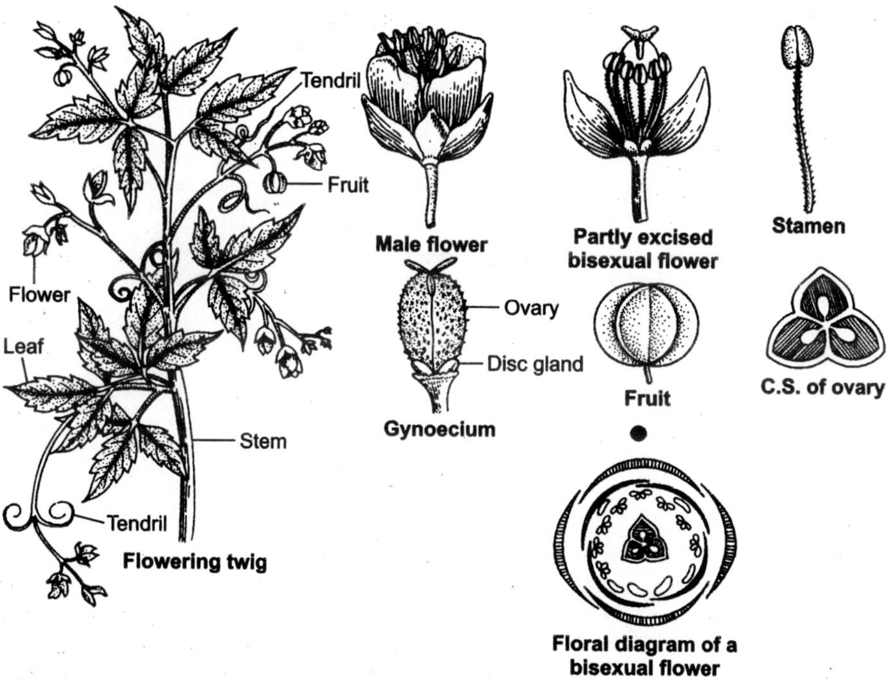

Fig. 20.19 *Cardiospermum halicacabum* L.

Petals are imbricate with hairy or scale-like appendages. In *Sapindus* and *Litchi*, flowers are actinomorphic but in *Erioglossum* and *Cardiospermum*, zygomorphic flowers with eccentric androecium are found. In many cases, flowers although apparently bisexual, are functionally unisexual (pollens do not germinate). Petals are absent in *Schleichera* and *Dodonaea*.

Androecium consists of 8 or 10 stamens in two whorls but *Turpinia* has 5 and *Deinbollia* has numerous stamens. Stamens are free and often eccentric (*Erioglossum*), but in *Cardiospermum*, filaments are basally connate. An extra-staminal disc is usually present, often represented by glands.

Gynoecium is usually tricarpellary and syncarpous. Ovary is trilocular and superior, lobed or not, with 1 or 2 ovules in each locule inserted on the central axis (axile placentation). The ovary is unilocular with a single ovule in *Nephelium* and *Litchi*. Style is simple and terminal.

Fruits and Seeds

Fruit is dry or fleshy; dry fruit may be a capsule or nut and fleshy fruit a drupe or berry. It is double samara in *Acer* and nut-like in *Litchi*. **Seeds** are usually arillate (*Nephelium litchi*) with curved embryo and without endosperm.

Pollination and Dispersal

Flowers are often protandrous and entomophilous. Fruits may show anemochory (*Dodonaea* and *Acer*) or zoochory (*Litchi*).

Floral Formula $\oplus$ or % ⚥ $K_5\ C_{5\ or\ 0}\ A_{8-10}\ \underline{G}_{(3)}$

Economic Importance

Some plants are economically quite important.

Edibles. Delicious edible fruits are produced by several species, *e.g.* *Litchi chinensis* Sonner. syn. *Nephelium litchi* Camb. (the litchi of Eastern Asia), *Nephelium lappaceum* L. (the Rambutan of Malaya), *Melicocca bijuga* L. (of West Indies), *Blighia sapinda* Koen. (the Akee of Guinea), *Dimocarpus longan* Lam. (the longan, Anshphal), *Schleichera oleosa* Merr. etc. *Acer saccharinum* L. is the source of maple sugar prepared from stem juice.

Timber. A good quality of timber is obtained from *Sapindus laurifolius* Vahl. (the soapnut tree of S. India), *Acer saccharinum* L. (sugar maple), *Aesculus indica* Colebr., *Schleichera oleosa* Merr., *Dodonaea viscosa* Jacq. etc.

Other uses. *Sapindus mukorossi* Gaertn. (the soapnut tree, ritha) produces *saponin*-rich fruits, largely used as substitute for soap. Fruits are used for washing clothes (woolen) and hairs. The seed oil of *Acer indica* Colelor. is used in rheumatic complaints. Several trees are grown for ornamental purposes, *e.g.* species of *Schleichera, Dodonaea, Acer, Sapindus* etc.

The seeds of *Nephelium lappaceum* L. (rambutan) yield a solid fat used for making soaps and candles.

Systematics and Phylogeny

The family shows resemblace with *Anacardiaceae, Aceraceae* and *Hippocastanaceae.* The earlier taxonomists included *Aceraceae* and *Hippocastanaceae* in *Sapindaceae* but these two have been maintained as separate families by modern workers (Cronquist etc.) *Aceraceae* differ from *Sapindaceae* in presence of opposite and palmate leaves, actinomorphic flowers and samaroid fruits, and *Hippocastanaceae* differ from *Sapindaceae* in presence of palmately compound leaves, a thyrsoid panicle inflorescence and leathery capsule with single seed. *Sapindaceae* differ from *Anacardiaceae* in presence of juice not irritant, compound leaves, irregular flowers, extra-staminal disc, sepals free, carpels 3, trilocular ovary with one ascending ovule in each locule, arillate seeds and drupe, berry, nut or capsule fruits. To some extent, *Sapindaceae* resemble *Euphorbiaceae* in having a single ovule in each locule, ventral raphe, tricarpellary ovary, often unisexual flowers and arillate seeds.

The family has been kept in the order **Sapindales** by most taxonomists but Thorne (1999, 2001) has kept it in the order **Rutales** close to *Rutaceae.* APG classification (2003) places the family in monocladistic group — 3. APG Group : **Eudicots-Rosids** and sub-group 2. **Eurosids II** and in the order **Sapindales**. The order **Sapindales** includes 10 families. *Rutaceae, Meliaceae, Anacardiaceae, Simaroubaceae* have been kept in the same order. Cronquist has recognised 15 families in the order **Sapindales.**

ANACARDIACEAE (Cashew Family)

Bentham & Hooker	Hutchinson	Cronquist	APG
Dicotyledons	Dicotyledones	Magnoliophyta	3. APG Group :
Polypetalae	Lignosae	Magnoliopsida	Eudicots-Rosids
Disciflorae	Sapindales	Rosidae	Eurosids II
Sapindales		Sapindales	Sapindales

The family has **77 genera** and about **600 species** distributed in tropical and sub-tropical regions of the world. In India, **22 genera** and about **110 species** are found, and the common forms are—*Mangifera, Anacardium, Pistacia, Rhus, Buchanania* etc. *Mangifera indica* (mango), is grown in most part of the country and *Anacardium occidentale* (cashew nut) in wet Western coastal regions of India.

Vegetative Characters

Plants are trees or shurbs, rarely woody vines, containing resin passages with plenty of gum and acrid juice. **Leaves** are simple or pinnate, alternate or rarely opposite (*Bouea*) and exstipulate.

Inflorescence and Flowers

Inflorescence is an axillary or terminal cymose panicle.

Flowers are bracteate, small and numerous, actinomorphic, bisexual or polygamous and hypogynous, or rarely perigynous (*Holigarna*), with a glandular or cup-shaped disc. The thalamus may be convex bearing nectar secreting outgrowths (cupular or cushion-like discs) between the stamens (*Rhus*) or outside the stamens (*Mangifera*) or elongated to form a gynophore as in *Melanorrhoea*. **Calyx** consists of 3-5 sepals, free or more or less connate, and imbricate. Calyx may be deciduous (*Semecarpus*) or persistent (*Rhus* and *Lannea*). **Corolla** is made up of 3-5 petals, free or rarely connate and imbricate, rarely absent (*Pistacia*). Petals are persistent and enlarged in fruit of *Melanorrhoea*.

Androecium shows variation, both in number and sterile and fertile stamens. In *Mangifera*, 4 staminodes and one fertile stamen; in *Anacardium*, 8-10 stamens in two whorls of which some are fertile and some are sterile; in *Spondias* and *Buchanania*, 10 fertile stamens in two alternate whorls and in *Rhus* species, 5 stamens in a single whorl, are found.

Filaments are usually subulate, free or basally connate as in *Anacardium*. Anthers are basi- or dorsifixed, dithecous, introrse, opening lengthwise.

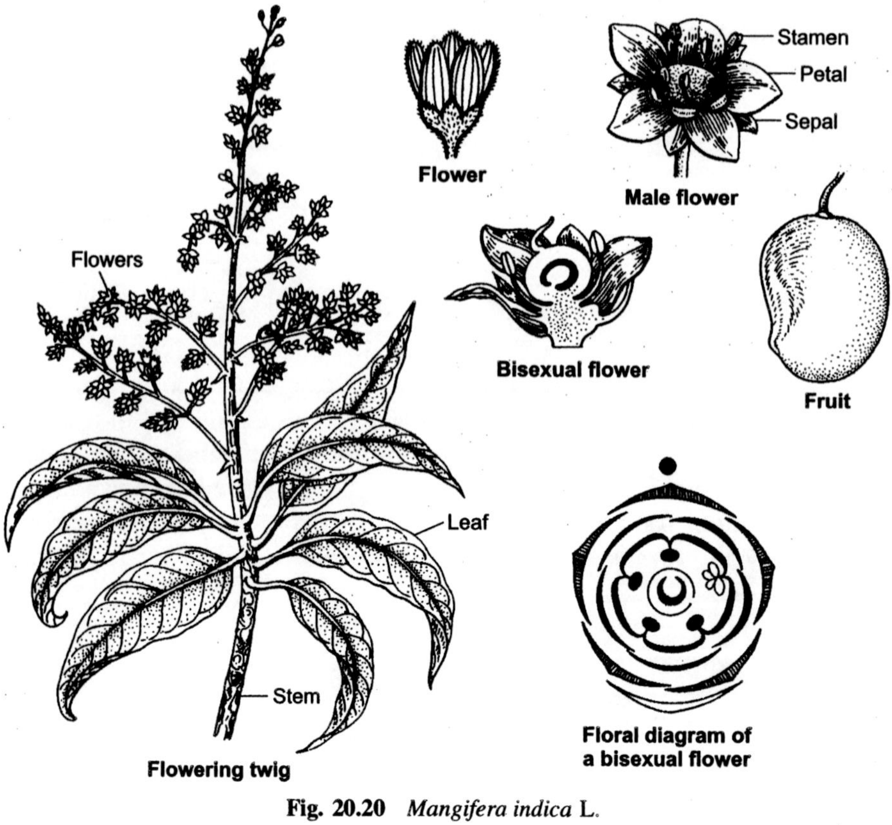

Fig. 20.20 *Mangifera indica* L.

Gynoecium consists of 1-5 carpels, one carpel and one locule in *Mangifera* and *Anacardium*. Ovary is superior and syncarpous with 1 to many locules (*Spondias*), each with a single pendulous or ascending ovule arising from the side of wall. Rarely carpels are free *(Buchanania)*. *Drimycarpus* has inferior ovary. Styles are united or free and usually terminal but in *Mangifera* and *Anacardium*, they are lateral. In *Lannea*, four free and persistent styles are found. Stigma is simple or capitate. Ovule is anatropous, pendulous or ascending with dorsal raphe. *Holigarna* has 3-5 fused carpels with unilocular inferior ovary.

Fruits and Seeds

In most genera, **fruit** is one seeded drupe but in *Spondias*, 4-5 seeds are found. The fruit is nut in *Anacardium* situated on pyriform fleshy structure derived from the accrescent disc and top of pedicel (thalamus). In *Mangifera*, mesocarp is fleshy and edible. **Seeds** are non-endospermic with a straight or curved embryo with plano-convex cotyledons.

Pollination and Dispersal

Flowers are entomophilous. Seeds are dispersed by animals or human agencies but fruits of *Parishia* and *Swintonia* are winged and dispersed by wind.

Floral Formula Br ⊕ ⚥ $K_{5 \text{ or } (5)}$ C_5 A_{5-10} $\underline{G}_{(1-5)}$

Economic Importance

Economically some species are quite valuable.

Edibles. *Mangifera indica* L. produces delicious mango fruits for which it is cultivated throughout India and tropical countries. *Anacardium occidentale* L., the cashewnut tree, is a native of Brazil and now cultivated in India and several other tropical countries. The kernels are eaten raw or roasted and also used in preparation of wine (locally called feni). *Pistacia vera* L. (pistachio nut) of West Asia (Afghanistan) is cultivated for the seeds which are the pistaceo nuts of commerce. The nuts are edible and used in sweet meats, ice creams and confectionaries. The fruits of *Spondias pinnata* (L.f.) Kurz. are edible and used in the peparation of pickles. Seeds of *Buchanania lanzan* Spr. (the piyal tree) are used as substitute of pistaceo nut. Fruits of *Rhus parviflora* Roxb. are used as tamarinds.

Timber. A good quality of timber is produced by *Mangifera indica* L., *Anacardium occidentale* L., *Swintonia floribunda* Giff., *Spondias axillaris* Roxb. etc. The timber is used for construction work, furniture, agricultural implements, packing cases etc.

Resins, Tannins, Turpentine etc. *Pistacia lentiscus* L. yields a mastic resin (used to flavour alcoholic beverages) while *P. terebinthes* L. yields turpentine. Resinous juice of *Rhus styphina* L., *R. glabra* L. and *R. coriaria* L. is used for tanning. *R. vernicifera* DC. yields lacquer. Gum produced by

Anacardium occidentale L. possesses insecticidal properties. The bark of *Buchanania lanzan* Spr. (chironji) and *Rhus mysurensis* Heyne ex wt. & Arn. is used for tanning. The pericarp of *Semecarpus anacardium* L.f. is largely used for dyeing textiles. *Rhus toxicodendron* L. and *R. venenata* DC are very poisonous.

Ornamentals. Some species of *Rhus, Cotinus, Spondias* etc. are cultivated as ornamentals.

Systematics and Phylogeny

Almost all taxonomists accept the affinity of *Anacardiaceae* with *Sapindaceae* and place both in the same order **Sapindales** (as discussed under *Sapindaceae*). However, Takhtajan (1991) has kept it in the order **Burserales** and Thorne (1999, 2001) in the order **Rutales**. APG classification (2003) places the family in the order **Sapindales** in sub-group 2. **Eurosids II** of monophyletic clade APG Group 3 : **Eudicots-Rosids.**

The family has been divided into 5 tribes :

(i) ***Mangiferae :*** Leaves simple, 1-6 carpels (*Mangifera, Buchanania*).

(ii) ***Spondieae :*** Leaves pinnately compound, 4-5 carpels, ovules as many as carpels (*Spondias*).

(iii) ***Rhoideae :*** Leaves usually pinnate, 4-5 carpels, ovary free from floral axis, ovule only one (*Rhus, Pistacia*).

(iv) ***Semecarpeae :*** Leaves simple, ovary is adnate to and sunken in the thalamus (*Semecarpus*).

(v) ***Dobineae :*** Leaves pinnate, perianth absent, carpel only one (*Dobinea, Compylopetalum*).

Some modern taxonomists developed new family *Pistaciaceae* (Airy Shaw 1973) for *Pistacia* and a distinct family *Podoaceae* by raising the tribe *Dobineae.*

Table 20.2 Comparison between Anacardiaceae and Sapindaceae

Anacardiaceae	Sapindaceae
1. **Members** are commonly trees or shrubs.	**Members** are trees, shrubs or climbers.
2. Resinous irritant juice is present.	Juice present is not irritant.
3. **Leaves** are simple or compound.	**Leaves** are usually compound.
4. **Flowers** are regular, bisexual or polygamous.	**Flowers** are regular or irregular, bisexual, unisexual or polygamous.
5. **Disc** present may be extrastaminal or intrastaminal.	**Disc** present is extrastaminal.
6. **Calyx** consists of 3–5 united sepals and **Corolla** comprises 4–5 free petals.	**Calyx** consists of 4–5 free sepals and **Corolla** comprises 4–5 free petals.
7. **Stamens** are 10 or less and free.	**Stamens** are 5+5 or fewer and free.
8. **Gynoecium** consists of 1–5 carpels. Ovary is syncarpous, 1–5 locular with one ascending or pendulous ovule in each locule.	**Gynoecium** is tricarpellary. Ovary is syncarpous, trilocular with one ascending ovule in each locule.

9. **Ovule** is with dorsal raphe.	**Ovule** is with ventral raphe.
10. **Style** is often lateral.	**Style** is terminal.
11. **Fruits** are usually drupacious.	**Fruits** are drupe, berry, nut or capsule.
12. Aril is absent.	**Seeds** are arillate.
13. **Floral formula**	**Floral formula**
Br ⊕ ⚥ $K_{(5)}\ C_5\ A_{5-10}\ A_{5-10}\ \underline{G}_{(1-5)}$	⊕ or % ⚥ $K_5\ C_5\ A_{8-10}\ \underline{G}_{(3)}$

MELIACEAE (Mahogany Family)

Bentham & Hooker	Hutchinson	Cronquist	APG
Dicotyledons	Dicotyledones	Magnoliophyta	3. APG Group :
Polypetalae	Lignosae	Magnoliopsida	Eudicots-Rosids
Disciflorae	Meliales	Rosidae	Sub-group 2. Eurosids II
Geraniales		Sapindales	Sapindales

The family consists of **50 genera** and **1400 species** distributed exclusively in tropical parts of the world. In India, some **19 genera** and **72 species** are found and common examples are *Melia, Azadirachta, Swietenia, Cedrela, Toona, Chloroxylon* etc. *Melia* and *Toona* are deciduous trees.

Vegetative Characters

Plants are mostly trees, sometimes shrubs emiting a very characteristic smell. **Leaves** are alternate, exstipulate, compound, pinnate (twice or thrice pinnate in *Melia*) and oblique at the base. In *Turraea*, leaves are simple.

Inflorescence and Flowers

Inflorescence is usually axillary, cymose panicle. **Flowers** are actinomorphic, bisexual or rarely unisexual and hypogynous. In *Amoora* and *Aphanamixis*, these are polygamo-dioecious. **Calyx** consists of 3–6 sepals which are united (gamosepalous) or rarely free and usually imbricate, rarely valvate. **Corolla** comprises 4 or 5 petals, free or rarely connate at the base. Sometimes, petals are adnate to the lower half of the staminal tube. These show imbricate, rarely valvate aestivation.

Androecium has 4-12 stamens inserted outside the base of a hypogynous disc. The filaments cohere to form a columnar tube (monadelphous) at the mouth of which anthers are inserted. Stamens are free in *Cedrela, Toona* and *Chloroxylon*. Anthers are dithecous, introrse and dehiscing longitudinally.

Gynoecium consists of 2-5 united carpels (syncarpous). Ovary is superior, 2-5 locular with 1 or 2 ovules in each locule in axile placentation. In *Soymida, Swietenia* and *Toona*, each locule has many ovules. Style is short or absent and the stigma is discoid or capitate.

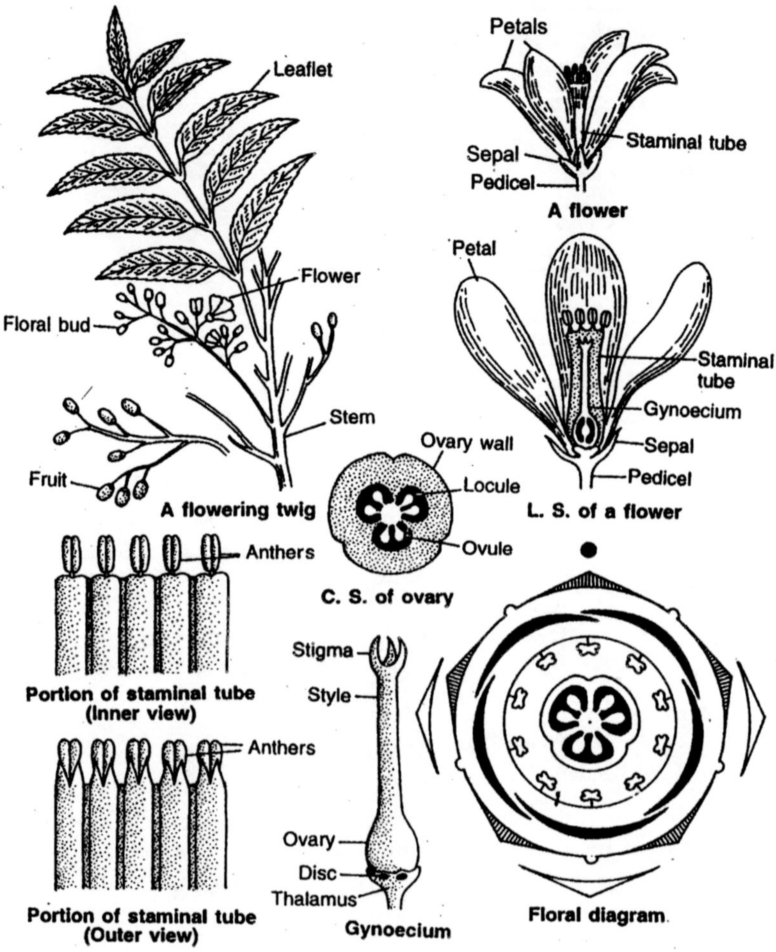

Fig. 20.21 *Azadirachta indica* Juss.

Fruits and Seeds

Fruit is berry or drupe or a capsule (drupaceous in *Melia*). **Seeds** are with or without endosperm and are winged in *Toona, Soymida, Swietenia, Khaya* etc. and wingless in *Azadirachta, Melia, Carapa* etc. In *Aphanamixis,* seeds are arillate.

Pollination and Dispersal

Flowers are usually insect pollinated. Winged seeds are dispersed by wind but in other cases, these are dispersed by birds, bats or other animals.

Floral Formula Br $\oplus$ ⚥ K_5 C_5 $A_{(10)}$ $\underline{G}_{(5)}$

Economic Importance

The family has considerable economic value.

Timber. The family comprises a good number of excellent timber trees, *e.g. Azadirachta indica* Juss. syn. *Melia azadirachta* L. (neem, margosa), *Melia azedarch* L. (Bokain, Pride of India), *Swietenia mahogani* Jacq. (West Indies Mahogany), *Khaya senegalensis* A. Juss. (African mahogany), *Toona ciliata* Roem. syn. *Cedrela toona* Roxb. (toon wood), *Chloroxylon swietenia* DC. (satin wood), *Dysoxylum malabaricum* Bedd. (white cedar), *Chickrassia tabularis* A. Juss. (Chittagong wood), *Cedrela odorata* L., *Soymida febrifuga* A. Juss., *Amoora wallichi* King. etc.

Medicinal plants. *Azadirachta indica* Juss. (neem or margosa) is an important plant of which almost every part has medicinal value. Leaves are used as insect repellant, twigs are widely used like tooth brush (to prevent pyrrhoea), seeds yield 'margosa oil' of commerce used in skin diseases and as anthelmintic and bark in the treatment of malarial fever. A drying oil obtained from the seeds of *Melia azedarch* L., is used in making soaps, hair oil and in the treatment of rheumatism.

Ornamentals. The important ornamental plants are *Algaia odorata* Lour. (white scented flowers), *Melia composita* Benth., *M. azedarch* L., *Azadirachta indica* Juss., *Swietenia mahogani* Jacq. etc.

Cedar oil obtained from *Cedrela odorata* L. is used in microscopy.

Systematics and Phylogeny

The family stands close to *Rutaceae, Sapindaceae, Burseraceae* and *Simaroubaceae,* but is singled out due to the presence of staminal tube and absence of glands on the leaves. Bentham and Hooker and Engler placed the family in the order **Geraniales** for the character of ovule with ventral raphe. Bessey considerd its origin from **Malvales** due to the presence of monadelphous stamens. Hutchinson placed *Meliaceae* in an order of its own, *i.e.,* **Meliales.** Cronquist has kept the family in **Sapindales.** Takhtajan (1991) and Thorne (1999, 2001) have kept it in the order **Rutales.** APG classification (2003) places the family in the monophyletic clade APG Group 3 : **Eudicots-Rosids** and it has been kept in the subgroup 2 : **Eurosids II** in the order **Sapindales,** which affirms Cronquist's view.

RUTACEAE (Citrus Family)

Bentham & Hooker	Hutchinson	Cronquist	APG
Dicotyledons Polypetalae Disciflorae Geraniales	Dicotyledones Lignosae Rutales	Magnoliophyta Magnoliopsida Rosidae Sapindales	3. APG Group : Eudicots - Rosids Sub-group 2 Eurosids II Sapindales

The family contains **162 genera** and **1650 species** widely distributed in the tropical and sub-tropical regions of the world. About **23 genera** with **70 species** are found in India with common examples as *Citrus, Aegle,*

Murraya, Feronia, Atalantia etc. *Zanthoxylum* (200 species), *Agathosma* (180 species) and *Citrus* (65 species) are some large genera of the family.

Vegetative Characters

Plants may be herbs, shrubs (*Murraya*) or trees (*Aegle, Citrus, Feronia*) and are often armed.The presence of pellucid glands filled with essential oils is a characteristic feature of the plants. **Leaves** are opposite or alternate, exstipulate and simple (*Calodendron, Diosnia*) or compound, trifoliate (*Aegle* and *Triphasia*) or pinnately compound (*Murraya*). In *Citrus*, leaves are unifoliate compound with winged petioles (some taxonomists consider them simple).

Inflorescence and Flowers

Inflorescence is racemose (*Atalantia*) or cymose or the flower may be axillary or terminal solitary (*Triphasia*). **Flowers** are bracteate, hermaphrodite,

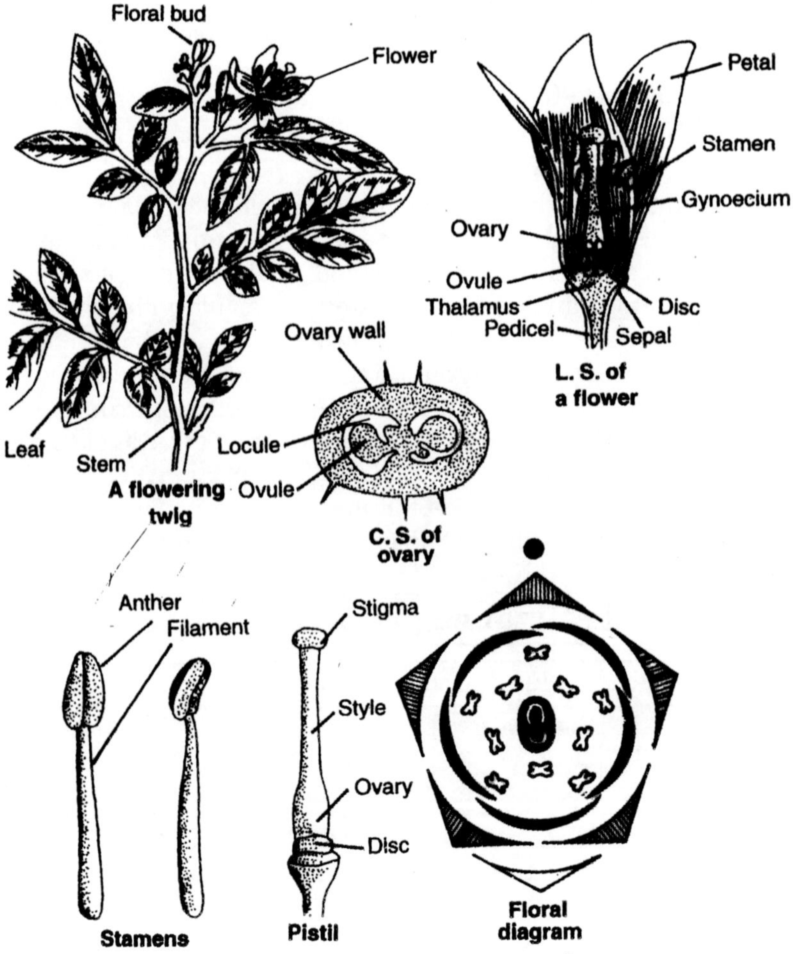

Fig. 20.22 *Murraya paniculata* (L.) Jacq.

actinomorphic, pentamerous (trimerous in *Triphasia* and tetramerous in *Achronychia*) and hypogynous. Flowers are unisexual in *Zanthoxylum* and *Toddalia* and zygomorphic in *Dictamnus, Galipea* and *Alnicida.* **Calyx** consists of 4 or 5 sepals, free or variously connate. Aestivation is imbricate, rarely valvate. **Corolla** has 4 or 5 free or rarely connate petals with aestivation valvate or imbricate.

Androecium consists of 5, 8 or 10 stamens (double the number of petals) or indefinite. When stamens are 10, these show **obdiplostemonous** condition and when 5, these are antisepalous. In *Citrus,* stamens are numerous and polyadelphous, *i.e.,* arranged in several bundles. An annular disc or a gynophore is present above the androecium. Anthers are dithecous, extrorse and dehiscing longitudinally. In *Atalantia,* monadelphy is found.

Gynoecium consists of 2, 4, 5 or indefinite number of carpels. Ovary is more or less deeply grooved, superior, syncarpous, multilocular, each locule with 1 or 2 ovules with ventral raphe and set in axile placentation. Sometimes, carpels are free at the base and united above the style. Ovary in *Feronia* is unilocular with several parietal placentae. Carpels are free in *Zanthoxylum, Correa, Evodia* etc. Style is short, often somewhat gynobasic, stigma is capitate.

Fruits and Seeds

Fruit is capsule (*Ruta*) or berry called hesperidium (characteristic to the family). In *Spathalia,* fruit is a winged drupe. In *Feronia* and *Aegle,* the fruit is amphisarca with woody rind. **Seeds** may or may not have endosperm. These are winged in *Chloroxylon.* In *Citrus,* polyembryony is common.

Pollination and dispersal

Flowers are protandrous and insect pollinated. Seeds are dispersed by human agency and animals.

Floral Formulae

Citrus Br $\oplus$ ⚥ $K_{(5) \text{ or } (4)}\ C_{5 \text{ or } 4}\ A_{\alpha \text{ (polyad.)}}\ \underline{G}_{(\alpha)}$

Murraya Br $\oplus$ ⚥ $K_{(5)}\ C_{5}\ A_{5+5}\ \underline{G}_{(2)}$

Economic Importance

Several species of the family are economically important.

Edibles. The family is important for its vitamin C rich fruits of *Citrus* and a few others like *Citrus aurantifolia* Swingle (lime, nimbu), *C. aurantifolia* var. *limetta* W. & A. (sweet lime), *C. aurantium* var. *aurantium* L. (bitter orange), *C. decumana* L. (shadock, chakotra), *C. limon* (L.) Burm. f. syn. *C. medica* L. var. *limon* L. (lemon, bara nimbu), *C. medica* L. syn. *C. medica* var. *medica proper* Hook. f. (citron, bijaura), *C. paradisi* Macf. syn. *C. racemosa* Marc. (grape fruit), *C. reticulata* Blanco. syn. *C. nobilis* Andrews non Lour. (mandarin, santara), *C. sinensis* (L.) Osbeck. syn. *C. aurantium* L. var.

sinensis L. (sweet orange, musambi), *Aegle marmelos* Correa (bael), *Feronia elephantum* Correa syn. *F. limoni* (L.) Swingle (wood apple) etc. Most fruits are eaten raw or after making their preparation in confectionaries, pickles etc.

Medicinal plants. Aromatic fruit pulp of *Aegle marmelos* Correa is a good remedy for dysentery and other stomach disorders. It is also used as drinks. Its root and bark are given in the intermittent fever. The fruit pulp of *Feronia elephantum* Correa is used as a substitute of bael in treating diarrhoea and dysentery. The root bark of *Toddalia asiatica* (L.) Lamk. is used in the treatment of malaria, cough and influenza, and that of *Dictamnus albus* L. in intermittent fever and nervous diseases. *Glycosmis pentaphylla* Corr. is used to cure cough, jaundice, rheumatism etc.

Spices. The bark of *Zanthoxylum armatum* DC. syn. *Z. alatum* Roxb. (tejpat) is carminative and anthelmintic and used as condiment. Fruits of *Z. nitidum* DC are used as condiments. Leaves of *Murraya koenigii* Spreng. (meetha neem) are aromatic and are used for flavouring curries, especially in South India.

Ornamentals. Several species are grown in gardens for their foliage and fragrant flowers, *e.g. Murraya exotica* L. (orange jasmine) syn. *M. paniculata* (L.) Jacq., *Ruta graveolens* L. (common rue), *Skimmia arborescens* T. Anders. ex Gamble, *Dictamnus albus* L. etc.

Systematics and Phylogeny

The family is allied to *Simaroubaceae, Burseraceae* and *Meliaceae,* mainly due to presence of woody habit, oil glands and a disc between the androecium and gynoecium. Bentham and Hooker and Engler placed the family in the order **Geraniales.** Hallier (1905) included the family *Rutaceae* in the order **Terebinthales.** Hutchinson (1973), Takhtajan (1980, 91), Dahlgren (1980, 83) and Thorne (1999, 2001) have accepted its position in its own order **Rutales.** Cronquist (1988) and APG system assigned the order **Sapindales** to the *Rutaceae.* APG classification placed it in the monophyletic clade 3. APG Group : **Eudicots-Rosids** and in the subgroup **Eurosids II.**

Table 20.3 Differences between Malvaceae and Rutaceae

Malvaceae	Rutaceae
1. **Leaves** are simple.	**Leaves** may be simple or compound.
2. **Inflorescence** is usually solitary axillary or terminal.	**Inflorescence** is usually axillary or terminal cyme.
3. **Calyx** consists of five free or united sepals with valvate aestivation.	**Calyx** consists of 4 or 5 free sepals with imbricate aestivation.
4. **Corolla** consists of five free petals but basally connate with staminal tube.	**Corolla** consists of four or five petals.
5. **Androecium** comprises numerous monadelphous stamens. Filaments unite to form staminal tube around the style.	**Androecium** consists of ten or indefinite number of stamens. Filaments are free. More often obdiplostemonous condition is found.

6. **Stamens** are epipetalous.	**Stamens** are free.
7. **Anthers** are monothecous and reniform.	**Anthers** are dithecous.
8. **Gynoecium** consists of five to several carpels.	**Gynoecium** consists of 2–5 carpels.
9. **Fruit** type shows variation and may be loculicidal capsule, schizocarpic, carcerulus or succulent type.	**Fruit** is simple or capsule. It is called **hesperidium**, which is characteristic to the family.
10. **Floral formula**	**Floral formula**
Br ⊕ ⚥ $E_{pi3\text{-}8}$ $K_{(5)}$ $\overset{\frown}{C_5\ \ A_{(a)}}$ $\underline{G}_{(5-\infty)}$	Br brl ⊕ ⚥ O $K_{(5)}$ C_5 A_{5+5} $\underline{G}_{(2)}$

OXALIDACEAE

Hutchinson	Cronquist	APG
Dictoyledones	Magnoliophyta	3. APG Group :
Herbaceae	Magnoliopsida	Eudicots-Rosids
Geraniales	Rosidae	Sub-Group 1. Eurosids I
	Geraniales	Oxalidales

The family consists of about **8 genera** and **900 species** distributed in tropical and sub-tropical regions of the world mainly in South America. In India, **3 genera** with about **12 species** are found, of which *Oxalis corniculata*, and *Biophytum sensitivum* are welknown. *Oxalis* is a large genus with about 800 species.

Vegetative Characters

Plants are perennial herbs with underground rhizomes, bulbs or corms, rarely trees (*Connaropsis*). **Leaves** are radical (*Oxalis*) or cauline, whorled or alternate, pinnately (*Biophytum*) or digitately (*Oxalis*) compound and exstipulate, rarely simple and stipulate. Leaves of most species show sleep movement and in *Biophytum*, these are sensitive to touch. In *Oxalis bupleurifolia*, phyllodes are found.

Inflorescence and Flowers

Inflorescence is cyme or cymose umbel or flowers are solitary axillary or terminal. **Flowers** are bracteate, complete, actinomorphic, hermaphrodite, pentamerous and hypogynous. **Calyx** consists of 5, free or united, persistent sepals, imbricate or rarely twisted. **Corolla** has 5, shortly clawed, free or basally connate and contorted petals.

Androecium consists of 10 stamens, set in two alternate whorls of five each and obdiplostemonous, *i.e.*, outer antipetalous and inner antisepalous,

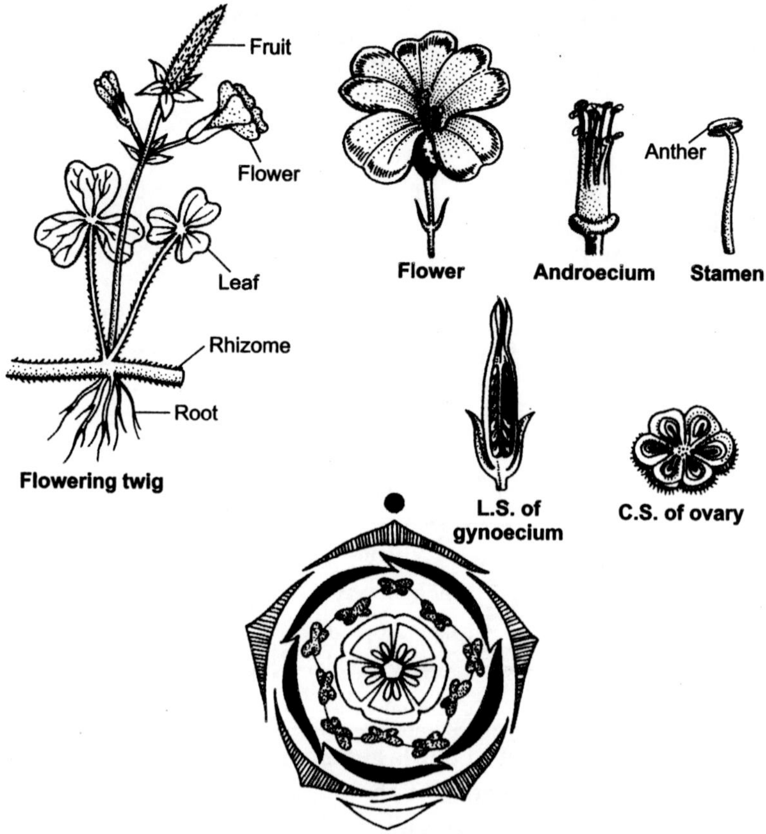

Fig. 20.23 *Oxalis corniculata* L.

often stamens are 15 in 3 whorls. Filaments are connate at the base. Anthers are dithecous, introrse and dehiscing lengthwise.

Gynoecium is pentacarpellary, syncarpous, pentalocular and superior. Placentation is axile with 1 or 2 ovules in each locule. Styles are 5, free and persistent and stigmas are capitate. Rarely trilocular ovary with one style is found (*Lepidobotrys* and *Hypseocharis*).

Fruits and Seeds

Fruit is a loculicidal capsule dehiscing elastically or a berry. **Seed** has straight embryo and fleshy endosperm. An explosive aril is often present on the seeds.

Pollination and Dispersal

Flowers are protandrous and entomophilous. Some species of *Oxalis* produce cleistogamous flowers. Seeds are dispersed mechanically (autochory). The aril separates from testa and shot off the seeds explosively, on slight disturbances.

Floral Formula $Br \oplus ⚥ K_5 C_5 A_{10 \text{ or } 5+5} \underline{G}_{(5)}$

Economic Importance

Family has little economic value. Some species of *Oxalis, e.g. O. pes-caprae* L., *O. corymbosa* DC., *O. latifolia* H.B. & K. etc. are grown for their small beautiful flowers. *Oxalis tetraphylla* Cav. produces edible tuberous roots. Plants of *Oxalis corniculata* L. (Indian sorrel) possess vermifuge and antiseptic properties. *Biophytum sensitivum* DC. is used in the treatment of chest complaints and stomach-ache.

Systematics and Phylogeny

Bentham and Hooker recognised it as a tribe of family *Geraniaceae*. Hutchinson has kept the family *Oxalidaceae* in the order **Geraniales** of **Herbaceae**. Cronquist (1991) and Thorne (1999, 2001) have also placed the family in the order **Geraniales.** Takhtajan (1980, 91) and APG classification (1997, 2003) have placed the family in its own order **Oxalidales.** APG classification finds its place in 3 APG Group : **Eudicots** - **Rosids** and in subgroup 1. **Eurosids I.** Hutchinson separated genera *Connaropsis* and *Averrhoa* (due to tree habit) and placed them in a separate family *Averrhoaceae.*

Families *Oxalidaceae* and *Geraniaceae* show resemblance in the structure of flowers and fruits, thus both have been kept in the order **Geraniales.**

GERANIACEAE

Bentham & Hooker	Hutchinson	Cronquist	APG
Dicotyledons	Dicotyledones	Magnoliophyta	3. APG Group :
Polypetalae	Herbaceae	Magnoliopsida	Eudicots
Disciflorae	Geraniales	Rosidae	Rosids
Geraniales		Geraniales	Geraniales

A small family *Geraniaceae* is represented by **11 genera** and **750 species** distributed in temperate and tropical regions of the world. A few forms are alpine. In India, **3 genera** and about **25 species** are found, mostly in Himalayas, specially species of *Geranium* and *Pelargonium.*

Vegetative Characters

Plants are herbs and undershrubs, rarely shrubs (*Dirachma*). **Leaves** are simple, palmately lobed or incised or compound, stipulate, alternate or rarely opposite. Leaves are glandular hairy, glands contain fragrant volatile oil. Large stipules are found in some species (*Geranium wallichianum* and *Pelargonium zonale*).

Inflorescence and Flowers

Inflorescence is commonly cymose (scorpoid cyme or umbellate cyme), rarely flowers may be solitary. In *Pelargonium,* it is like an umbel.

Flowers are bracteate, usually bracteolate, bisexual, complete, actinomorphic, rarely zygomorphic (*Pelargonium*), pentamerous and hypogynous. **Calyx** consists of 5 sepals, rarely 4 (*Vivania*) or 8, free or rarely

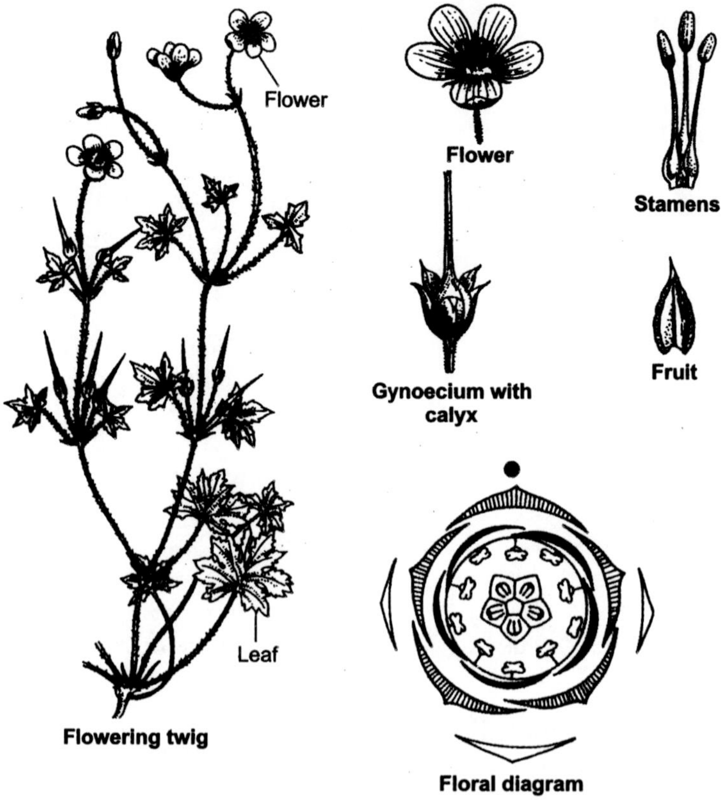

Fig. 20.24 *Geranium ocellatum* Jacq.

connate, persistent, imbricate or rarely valvate. In *Pelargonium*, posterior sepal is spurred. **Corolla** consists of petals as many as sepals, free, imbricate or rarely twisted. Nectar glands are found alternating with petals. Petals are alike except in zygomorphic flowers.

Androecium comprises 10 stamens set in two alternate whorls of 5 each, *i.e.*, obdiplostemonous. Sometimes, outer stamens are smaller (*Geranium*) or reduced to scale-like staminodes (*Erodium*). Rarely, only 2-4 fertile stamens are found (*Pelargonium*). Filaments are united at the base, rarely free. Anthers are pendulous, dithecous, introrse and dehiscing longitudinally.

Gynoecium consists of 5 united carpels (rarely 2 or 8). Ovary is superior, syncarpous and pentalocular, each locule has 1 or 2 (rarely more) pendulous, superposed, campylotropous ovules set in axile placentation. Style is one with number of stigmatic branches corresponding to locules or carpels.

Fruits and Seeds

Fruit is a capsule, dehiscing into as many parts (mericarps) as the number of carpels and rolling up or becoming spirally twisted along the beak-like style. **Seeds** are non-endospermic or with scanty endosperm. Cotyledons are folded. In *Biebersteinia*, fruits are 5 small nuts.

Pollination and Dispersal

Flowers are insect pollinated, mostly by bees, due to presence of nectar glands. Seeds are dispersed mechanically (autochory).

Floral Formula Br brl ⊕ ⚥ K_5 C_5 A_{5+5} $\underline{G}_{(5)}$

Economic Importance

Several species of *Pelargonium* (garden geraniums), (*P. zonale* L., *P. hortorum* Bailey), *Geranium* (*G. wallichianum* D. Don) and *Erodium* (*E. cicutarium* L. Her) are grown for their showy flowers and aromatic foliage. Geranium-oil (an essential oil), obtained from *Pelargonium graveolens* Ait., *P. rodule* L. Her etc., is a substitute for otto of roses.

Some plants have medicinal importance, *e.g. Geranium robertianum* L. is used in the treatment of diarrhoea and ulcers; *G. wallichianum* D. Don - root stock is used in eye troubles; *G. nepalensis* Sweet. is used in renal infections.

Systematics and Phylogeny

Geraniaceae are closely related to *Oxalidaceae* and *Tropaeolaceae. Oxalidaceae* are distinguished by compound leaves, regular flowers, calyx not spurred, style usually free and fleshy endosperm while *Tropaeolaceae* are distinguished from *Geraniaceae* by presence of exstipulate leaves, zygomorphic flowrs, calyx spurred, capsule dehiscing into 3-fleshy schizocarps and no endosperm.

De Candolle and Bentham and Hooker have not distinguished *Oxalidaceae* and *Tropaeolaceae* from *Geraniaceae* but Engler splitted the large family into 4 families (*Geraniaceae, Oxalidaceae, Tropaeolaceae* and *Balsaminaceae*), of which first three were placed in the order **Geraniales** and the last one in **Sapindales.** Order **Geraniales** has been recognised by most workers. In APG classification, the family has been kept in the order **Geraniales** of the monophyletic familial group **Eudicots-Rosids.** Cronquist believed that the order has been derived from **Sapindales**, and not from **Ranales** as proposed by earlier workers.

APIACEAE (UMBELLIFERAE) (Carrot Family)

Bentham & Hooker	Hutchinson	Cronquist	APG
Dicotyledons	Dicotyledones	Magnoliophyta	4. APG Group :
Polypetalae	Herbaceae	Magnoliopsida	Eudicots-Asterids
Calyciflorae	Umbellales	Rosidae	Euasterids II
Umbellales		Apiales	Apiales

The family consists of **421 genera** and **3220 species** distributed throughout the north temperate regions of the world. In India, some **30 genera** and about **180 species** are found, mainly in temperate Himalayas. Common forms met in the country are *Anethum, Centella, Coriandrum, Cumium, Daucus, Ferula, Foeniculum, Hydrocotyle, Peucedanum* etc. *Eryngium* (230 species), *Ferula* (180 species) and *Pimpinella* (150 species) are some large genera of the family.

Vegetative Characters

Plants are mostly herbs with hollow internodes. These may be annual, biennial or perennial herbs, shrubs (*Eryngium giganteum*), climbers (*Pseudocarum*), creepers (*Hydrocotyle*) or small trees (*Eryngium sarcophyllum* and *E. bupleuroides*). Plants have aromatic smell due to resin or oil canals present in the cortex, pericycle and sometimes in the phloem. **Leaves** are alternate and pinnately compound or decompound, rarely simple (*Hydrocotyle, Centella* and *Bupleurum*). *Eryngium* has spiny leaf margin, and looks-like a spiny thistle with compact floral heads. Leaves are ex-stipulate and with sheathing bases. However, scarious stipules are found in *Hydrocotyle. Bupleurum* resembles monocots due to parallel venation in leaves.

Inflorescence and Flowers

Inflorescence is characteristic to the family. In most cases, it is **compound umbel**, rarely simple umbel (*Hydrocotyle*) surrounded by an

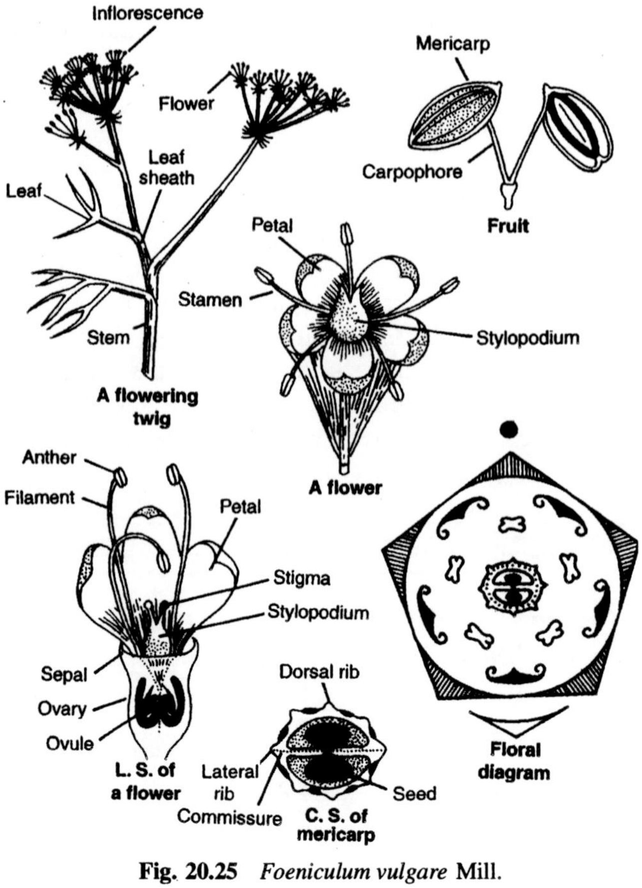

Fig. 20.25 *Foeniculum vulgare* Mill.

involucre of bracts. In some species of *Hydrocotyle*, umbel is reduced to a single flower and in *Petangia*, inflorescence is a dichasial cyme.

Flowers are minute, actinomorphic, bisexual, pentamerous and epigynous, white or yellow, rarely pink or violet. Species of *Heracleum* have zygomorphic flowers. Unisexual flowers are found in *Echinophora, Arctopus* and *Mathiasella*. In *Astrania*, flowers are polygamo-monoecious. In *Foeniculum*, peripheral flowers of the umbel have unequal petals (zygomorphy). In *Mathiasella bupleurioides*, female flowers are apetalous. **Calyx** consists of 5 inconspicuous, teeth- like sepals which remain adnate to the ovary wall. Calyx teeth are absent in *Foeniculum vulgare*. **Corolla** consists of 5 distinct petals (polypetalous) alternating with sepals. Petals are usually apically inflexed and these fall off soon. Aestivation is imbricate or valvate.

Androecium consists of 5 free stamens, alternating with petals and arising from an epigynous disc. Anthers are introrse, dithecous, basi- or dorsifixed, rarely versatile, dehiscing longitudinally.

Gynoecium has two united carpels (syncarpous). Ovary is inferior, bilocular with axile placentation and one pendulous, anatropous ovule in each locule. An epigynous, two-lobed nectar disc (called **stylopodium**) is found at the top of the ovary and is prolonged into two styles which are thickened at the base. Stigmas are capitate and distinct.

Fruits and Seeds

Fruit is characteristic to the family. It is schizocarpic which splits into two dry one-seeded mericarps. Such fruit is called **cremocarp.** The two mericarps usually hang for sometimes on a very thin, simple or forked stalk called **carpophore.** Mericarps are variously marked with ribs, grooves and bristles. **Seed** comprises a hard oily endosperm and a small embryo.

Pollination and Dispersal

Flowers are insect pollinated. Usually these are protandrous but in *Sanicula* and *Hydrocotyle*, flowers are protogynous. Mericarps, marked with ridges and bristles, are dispersed by birds and animals.

Floral Formula Br ⊕ ⚥ $K_{(5)}\ C_5\ A_5\ \overline{G}_{(2)}$

Economic Importance

The family has considerable economic value.

Edibles. Several plants are source of various spices and a few are used as vegetables. Important **spice** yielding plants are :

Apium graveolens L. (celery seed, ajmud)
Anethum graveolens L. (dill, sowa)
Carum copticum Benth. & Hook. (azwain)
C. bulbocastanum Koch. (black caraway)
Coriandrum sativum L. (coriander, dhania)

Cuminum cyminum L.	(cumin, jeera)
Foeniculum vulgare Mill.	(fennel, saunf)
Ferula asafoetida L., *F. foetida* Regel and	
F. narthox Boiss.	(oleo-resin-hing of commerce extracted as root exudate) etc.

Important **vegetable** yielding plants are :

Daucos carota L. (carrot)	—	fleshy tap root used as vegetable and eaten raw.
Apium graveolens L. (celery)	—	turnip-like roots used as vegetable.
Anetham graveolens L. (Sowa)	—	leaves used as vegetables.

Medicinal Plants. *Hydrocotyle asiatica* L. syn. *Centella asiatica* (L.) Urban. is used as diuretic, alterative and tonic. The root exudate, the hing (asafoetida), is used to stimulate respiratory tract and nervous system.

Ornamentals. Some species of *Pimpinella, Heracleum* and *Angelica* are grown as ornamental plants.

Systematics and Phylogeny

Apiaceae are closely related to *Araliaceae.* According to earlier workers, both were included in the same order. Hutchinson developed two orders, one placed in **Lignosae** and other in **Herbaceae**. This separation was arbitrary as most modern taxonomists placed two orders closer together under **Araliales** (Dahlgren, 1989; Takhtajan, 1991; Thorne, 2001) or **Apiales** (Cronquist, APG II, AP web). Hutchinson considered **Araliales** as more primitive and **Umbellales** as more advanced. *Umbelliferae* resemble *Compositae* (*Asteraceae*) to some extent in the reduction of calyx, bicarpellate, syncarpous ovary with two distinct styles. Hutchinson considered that *Umbelliferae* are derived from **Saxifragales** while Cronquist believes its derivation from **Sapindales**. Monophyly of the family is supported by morphology, nature of secondary metabolites and nucleotide sequences of *rbc*L and *mat*K genes. In APG system, the family is placed in **4. APG Group, Eudicots–Asterids, Euasterids II** and in order **Apiales**. In recent classification, *Hydrocotyle* is transferred to *Araliaceae* and *Centella* and *Trachymene* to *Mackinlayaceae.*

Table 20.4 Differences between Umbelliferae and Compositae

Umbelliferae (Apiaceae)	Compositae (Asteraceae)
1. **Leaves** are pinnately compound or decompound.	**Leaves** are usually simple.
2. **Inflorescence** is **umbel**, a characteristic to the family.	**Inflorescence** is **Capitulum**, a characteristic to the family.

3. **Flowers** are actinomorphic, bisexual and epigynous.	**Flowers** are actinomorphic (disc florets) or zygomorphic (ray florets), neuter or unisexual (ray florets) or bisexual (disc florets) and epigynous.
4. **Calyx** consists of five sepals.	**Calyx** is modified into pappus.
5. **Corolla** shows imbricate or valvate aestivation.	**Corolla** has valvate aestivation.
6. **Androecium** consists of five free and epipetalous stamens.	**Androecium** consists of 5, syngenesious, epipetalous stamens.
7. **Ovary** is bilocular and each locule has a single ovule in **axile placentation**.	**Ovary** is unilocular with a single ovule in **basal placentation**.
8. **Stylopodium** is present.	It is not found.
9. **Fruit** is either **cremocarp** or berry.	**Fruit** is **cypsella**.
10. **Floral formula**	**Floral formula**
Br ⊕ ⚥ $K_{(5)}\ C_5\ A_5\ \bar{G}_{(2)}$	**Ray floret :** Br % ♀ or ○ $K_{(pappus)}\ C_{(2/3)}\ A_0\ \bar{G}_{(2)}$ **Disc floret :** Br ⊕ ⚥ $K_{(pappus)}\ \widehat{C_{(5)}\ A_5}\ \bar{G}_{(2)}$

❑❑❑

Chapter **21**

Asteridae
(Apocynaceae, Asclepiadaceae, Solanaceae, Convolvulaceae, Boraginaceae, Verbenaceae, Lamiaceae, Scrophulariaceae, Acanthaceae, Pedaliaceae, Rubiaceae, Asteraceae)

APOCYNACEAE (Nerium Family)

Bentham & Hooker	Hutchinson	Cronquist	APG
Dicotyledons	Dicotyledones	Magnoliophyta	4. APG Group :
Gamopetalae	Lignosae	Magnoliopsida	Eudicots-Asterids
Bicarpellatae	Apocynales	Asteridae	Euasterids I
Gentianales		Gentianales	Gentianales

The family contains about **300 genera** and **1400 species** found mostly in tropical and sub-tropical regions of the world. Some **29 genera** and **85 species** are found in India and common examples are species of *Alstonia, Tabernaemontana, Holarrhena, Plumeria, Carissa, Catharanthus, Nerium, Thevetia, Rauvolfia* etc. *Tabernaemontana* (220 species) and *Rauvolfia* (105 species) are large genera of the family.

Vegetative Characters

Plants may be herbs (*Catharanthus*), shrubs (*Nerium, Thevetia, Carissa*) or trees (*Alstonia*). *Adenium* is a succulent xerophyte and *Landolphia* is a climber (tendrillar peduncles). Presence of milky latex in the leaves and stem is characteristic to the family. **Leaves** are usually exstipulate, opposite decussate or alternate (*Thevetia*) or whorled (3 in *Nerium* and *Rauvolfia* and 3-7 in *Alstonia*). In *Tabernaemontana*, interpetiolar stipules are found.

Inflorescence and Flowers

Inflorescence is usually axillary or terminal cyme, sometimes panicled raceme or flower solitary. **Flowers** are usually bracteate, sometimes ebracteate, bisexual, actinomorphic, pentamerous (rarely tetramerous) and hypogynous. **Calyx** consists of 4 or 5 united sepals (gamosepalous) with valvate or quincuncial aestivation, often with basal glands. Odd sepal is posterior. **Corolla** consists of 5, sometimes 4, united and lobed petals (gamopetalous). Corolla may be salverform, campanulate, urceolate or funnel-shaped. Aestivation is twisted, sometimes valvate. Some hairy or scaly corona-like outgrowths are found at the throat of corolla or within corolla tube, but not in *Plumeria*.

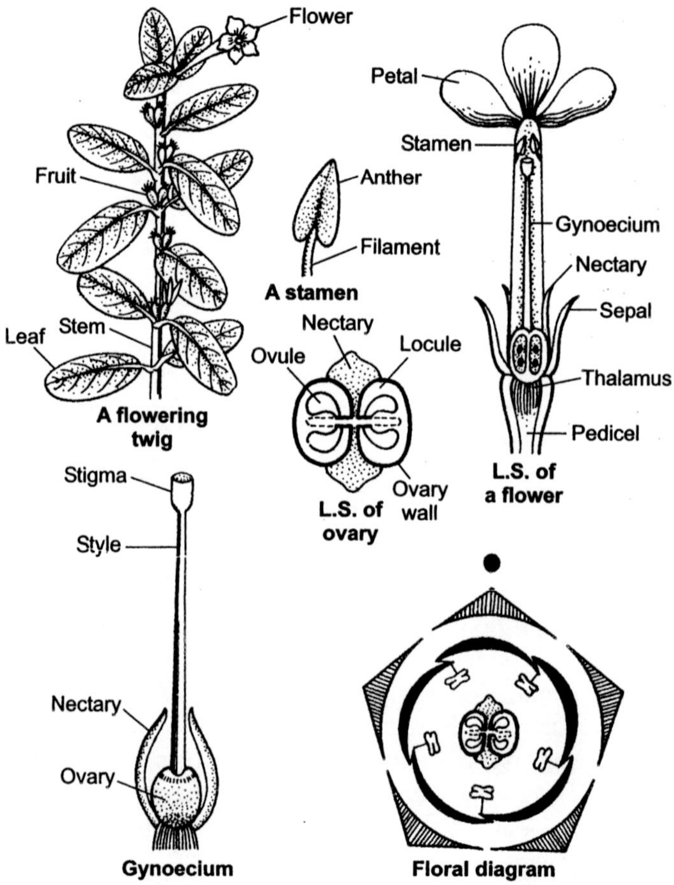

Fig. 21.1 *Catharanthus roseus* (L.) G. Don.

Androecium has stamens equal to the number of petals and alternate with corolla lobes. Stamens are epipetalous and attached at the mouth or deeper in corolla tube. Filaments are short or almost absent and anthers remain included in the corolla tube. Anthers are sagittate, dithecous, introrse, dehiscing longitudinally and free or connivent in the cone over the stigma and often adnate to it. In *Nerium*, anthers are tipped with long hairy appendages.

Gynoecium consists of two united carpels (syncarpous) with superior or semi-inferior (*Plumeria*) ovaries. Ovaries of the two carpels are frequently free or connate only at the base with united style and stigma. Each ovary is unilocular with ventral placenta. In case of fused carpels with unilocular ovary, the placentation is either parietal or axile. Stigma is often bilobed or capitate or dumb-bell shaped. Ovules are many, anatropous and pendulous.

Fruits and Seeds

Fruit consists of two follicles dehiscing by ventral suture, sometimes berry or capsule or two indehiscent mericarps. **Seeds** are flat, winged or comose, often provided with a terminal tuft of long silky hairs. Endosperm is either copious or absent.

Pollination and Dispersal

Flowers are insect pollinated. Winged seeds and presence of crown of hairs facilitate wind dispersal.

Floral Formulae

Nerium indicum $\text{Br brl} \oplus \text{⚥}\ K_5\ \widehat{C_{(5)}\ A_5}\ \underline{G}_{(2)}$

Thevetia peruviana $\text{Br brl} \oplus \text{⚥}\ K_{(5)}\ \widehat{C_{(5)}\ A_5}\ \underline{G}_{(2)}$

Economic Importance

Family has several economically important plants.

Medicinal plants. Dried roots of *Rauvolfia serpentina* (L). Benth. & Kurz. (sarpagandha) yield drug **reserpine**, used to cure hypertension. Bark and roots of *Holarrhena antidysentrica* Wall (indrajau) are used for the treatment of dysentry. Flowers of *Catharanthus roseus* (L.) G. Don. syn. *Vinca rosea* Reichb. (periwinkle) yield anticancer drugs **vincoblastin** and **vincristin** used in the treatment of leukaemia. Latex from *Plumeria alba* L. is used to relieve toothache. Latex of *Cerbera manghas* L. is the source of drug **cerberin** which is emetic and purgative.

Ornamental plants. Several plants are grown for their ornamental values especially for their showy flowers and foliage, *e.g. Allamanda neriifolia* Hook. and *A. cathartica* L., *Alstonia scholaris* R. Br. (satni), *Plumeria alba* L. (white champa), *Villaris solanacea* (Roth.) Ktze, *Thevetia peruviana* Schum. (yellow oleander), *Nerium indicum* Mill. (Indian oleander), *Tabernaemontana divaricata* (L.) R. Br., *Catharanthus roseus* (L.) G. Don. (periwinkle) etc.

Other uses. Unripe fruits of *Carissa carandas* L. (karaunda) are eaten as salad and used in pickles. Latex of *Rhynchodia verrucosa* (Blume) Woodson is used for preparing rubber.

Systematics and Phylogeny

Apocynaceae are closely related to the family *Asclepiadaceae* from which it differs : (i) by the presence of single style, (ii) absence of pollinia, and

(iii) presence of free stamens from the stigma. Bentham and Hooker, Engler and Prantl, Hutchinson, Cronquist, Dahlgren etc. considered it as a separate family. distinct from *Asclepiadaceae* but Thorne (1983), Takhtajan (1987, 97), Judd *et al.* (2002), APG II and APweb merged *Asclepiadaceae* as one sub-family of *Apocynaceae*. Thorne kept the family in the order **Rubiales**, Takhtajan and Hutchinson in **Apocynales** while Cronquist, Dahlgren, Bentham and Hooker and APG system kept it in the order **Gentianales**. In APG system, the family has been placed in **4. APG Group : Eudicots-Asterids**, Subgroup **Euasterids I** and order **Gentianales.**

Table 21.1 Differences between Apocynaceae and Asclepiadaceae.

Apocynaceae	Asclepiadaceae
1. **Flowers** are usually pentamerous and rarely tetramerous.	**Flowers** are pentamerous.
2. Number of **stamens** are equal to the number of petals. Stamens are epipetalous.	**Androecium** consists of five epipetalous stamens.
3. **Gynostegium** is not found.	**Anthers** are fused with stigmatic disc and form **gynostegium.**
4. **Pollinia** are not found.	**Pollinia** are found.
5. **Anthers** are sagittate, dithecous, and dehiscing transversely.	**Anthers** form pollinia and do not dehisce.
6. **Stigma** is drum-shaped, peltate or forked.	**Stigma** is peltate and pentangular.
7. **Fruit** is usually follicle, but berry or capsule may also be found.	**Fruit** is a follicle.
8. **Seeds** are usually winged, sometimes tuft of hairs is found.	**Seeds** are flat and bear tuft of hairs terminally.
9. **Floral formula** $\mathrm{Br} \oplus ⚥ K_{(5)} \overparen{C_{(5)}\ A_5}\ \underline{G}_{(2)}$	**Floral formula** $\mathrm{Br\ brl} \oplus ⚥ K_5 \overparen{C_{(5)}\ A_5}\ \underline{G}_{(2)}$

ASCLEPIADACEAE (Milkweed Family)

Bentham & Hooker	Hutchinson	Cronquist	APG
Dicotyledons	Dicotyledones	Magnoliophyta	Family merged
Gamopetalae	Lignosae	Magnoliopsida	as subfamily of
Bicarpellatae	Apocynales	Asteridae	Apocynaceae
Gentianales		Gentianales	

Family comprises about **250 genera** and **1800 species** growing in tropical and sub-tropical regions of the world. Some **53 genera** and **300 species** are found in India. Common forms met in the country are species of *Calotropis, Pergularia, Dischidia* (in Assam), *Hemidesmus, Cryptostegia, Gymnema, Leptadenia* etc. *Asclepias* is a large genus (220 species) of the family.

Vegetative Characters

Plants are usually perennial herbs (*Asclepias*), shrubs (*Leptadenia*) and sometimes woody climbers (*Cryptostegia, Tylophora*) or epiphytic (*Dischidia*) or succulent (*Stapelia*) or small tree (*Ulteria, Calotropis*). These contain milky latex and show xerophytic characters. **Stem** is erect, branched and woody in lower region. Bicollateral vascular bundles are characteristically found. **Leaves** are usually simple, opposite, decussate (*Calotropis*), sometimes alternate (*Ulteria*), entire and exstipulate. A thick waxy covering on the leaf surface is found in *Calotropis*. In *Stapelia*, leaves are reduced to spines or scales and plants appear cactus-like. In *Dischidia*, one of a pair of leaves is often transformed into a pitcher which has a few internally produced adventitious roots to absorb nutrients from debries fallen inside the pitcher with rain water.

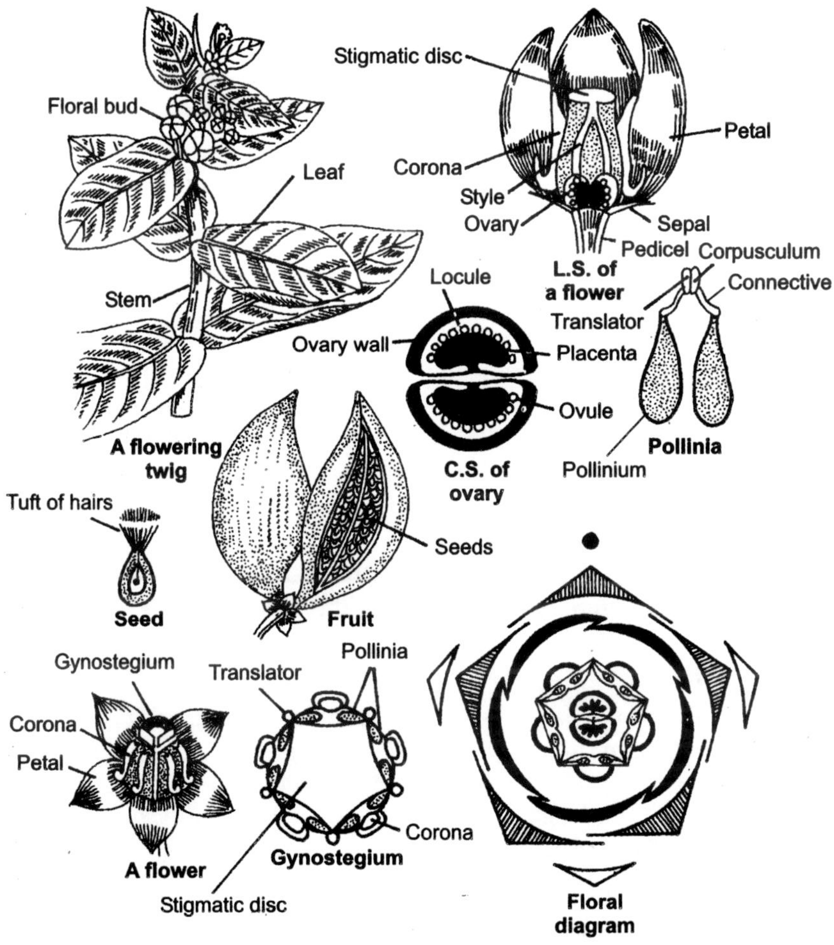

Fig. 21.2 *Calotropis procera* (Willd.) Dryand ex. W. Ait.

Inflorescence and Flowers

Inflorescence is usually dichasial cyme. In *Calotropis*, it is umbellate polychasial cyme.

Flowers are bracteate, usually bracteolate, complete, actinomorphic, bisexual, pentamerous and hypogynous. *Stapelia* produces largest flowers in the family. **Calyx** consists of 5, free sepals (polysepalous) or slightly united at the base with valvate, imbricate or quincuncial aestivation. **Corolla** has 5, united petals (gamopetalous) which may be rotate (*Calotropis*), campanulate (*Gymnema*), or funnel-form (*Cryptostegia*) and often provided with appendages forming corona. Aestivation is twisted or valvate.

Androecium consists of 5, epipetalous stamens. The relationship of stamens with two carpels (gynoecium) is complex and characteristic to the family. Stamens and carpels arise separately on the floral axis. Filaments of stamens are quite short or entirely absent. Anthers are united usually to the margins of the pentangular stigmatic head. The structure thus formed by the union of anthers and the stigmatic head is called **gynostegium**. The pollen grains of each half of the anther are agglutinated (adhering together) into a sac-like structure known as **pollinium**. Thus, in each anther, there are two **pollinia**, and the two pollinia are united together in pairs by means of a structure called **translator**. The translator consists of two parts, the **corpusculum**, a gland-like body which gets attached to the margin of the stigmatic surface between the anthers, and a pair of arms, called **retinaculi** or **connectives** by means of which the pollinia of the adjacent anther-halves are attached to the carpusculum.

Back parts of the anthers usually possess horn-like projections called **cuculi** which form corona of the stamens. Cuculi secrete and store nectar.

Gynoecium consists of two carpels, the ovary of which remains separate from each other (apocarpous) and terminates above in two distinct styles, which, however, unite to form a common style. Stigma is dilated to form a pentangular head with which are united the anthers of five epipetalous stamens. Ovary is superior with marginal placentation bearing several anatropous ovules, pendulous from the ventral suture.

Fruits and Seeds

Fruit consists of a pair of follicles. **Seeds** are usually flat, compressed and covered with large number of terminal long silky hairs which help in dispersal.

Pollination and Dispersal

Flowers are insect pollinated (**translator mechanism**) and seeds are dispersed by wind (**anemochory**). Seeds have terminal tuft of hairs to facilitate anemochory.

Floral Formula Br brl $\oplus$ ⚥ $K_5 \; \overset{\frown}{C_{(5)} \; A_5} \; \underline{G}_{(2)}$

Economic Importance

Only few plants of the family are economically important.

Medicinal plants. Roots of *Tylophora indica* (Burm. f.) Merr. (antamul) yield drug for the treatment of asthma, whooping cough and bronchitis. Dried stems of *Sarcostemma acidum* (Roxb.) Voigt. have emetic properties and its roots serve as antidote to snake-bite. Plants of *Cynanchum arnottianum* Wight. are used as an insecticide. Roots of *Hemidesmus indicus* R.Br. (anantmul, kauri, Indian sarsaparilla) are blood purifier and are also used in chronic rheumatism and skin diseases. Roots of *Gymnema sylvestre* (Retz.) Schult. (merasingi) are used as a cure for snake-bite and leaves as antidiabetic agent and in glycosuria. Plants are laxative and are used in stomachic.

Ornamental plants. Several plants are grown for foliage and showy flowers, *e.g. Stapelia gigantea* N.E.Br., *Cryptostegia grandiflora* (Roxb.) R.Br., *Asclepias curassavica* L., *Ceropegia woodii* Schlecht, *Calotropis gigantea* R.Br. etc.

Other Uses. Latex of *Calotropis gigantea* R. Br. and *C. procera* R.Br. is used in tanneries for deodorising and removing hairs of hides. Latex of *Cryptostegia grandiflora* (Roxb.) R.Br. is the commercial source of rubber. Bast fibres obtained from *Asclepias curassavica* L., *Tylophora tenuis* Blume, *Leptadenia pyrotechnica* (Forsk) Decne, *Calotropis gigantea* R.Br. etc. are used for cordages and fishing nets. The seed hairs of different species are used as insulating material in textile business.

Systematics and Phylogeny

Asclepiadaceae are advanced and close to *Apocynaceae*.

(For more details see family Apocynaceae.)

Bentham and Hooker divided the family into two sub-families.

(i) **Periploceae** : Filaments usually free and pollen masses granular, *e.g. Cryptostegia, Hemidesmus* etc.

(ii) **Euasclepiadaceae** : Filaments connate and pollen masses waxy and free, *e.g. Calotropis, Asclepias, Gymnema* etc.

SOLANACEAE (Datura Family)

Bentham & Hooker	Hutchinson	Cronquist	APG
Dicotyledons	Dicotyledones	Magnoliophyta	4. APG Group :
Gamopetalae	Herbaceae	Magnoliopsida	Eudicots-Asterids
Bicarpellatae	Solanales	Asteridae	Euasterids I
Polemoniales		Solanales	Solanales

The family comprises **83 genera** and **2925 species** found in tropical and temperate regions of the world. In India, some **15 genera** and **88 species** are found, of which about 60 species are wild. Common Indian forms are *Solanum, Cestrum, Nicotiana, Physalis, Capsicum, Hyoscyamus, Petunia, Datura,*

Lycopersicum etc. *Solanum* (1350 species), *Cestrum* (160 species) and *Nicotiana* (110 species) are some large genera of the family.

Vegetative Characters

Plants are usually annual (*Petunia*) or perennial (*Lycopersicum*) herbs, a few are shrubs (*Cestrum*) or rarely small trees (*Dunalia*, *Cyphonandra*) or climbers (*Solanum jasminoides*). **Stem** is often solid, branched and smooth. Stem tubers are found in *Solanum tuberosum* (potato). Stem anatomy shows presence of intra-xylary phloem. **Leaves** are mostly alternate, exstipulate, simple or rarely pinnately divided (*S. tuberosum* and *Lycopersicum*). Leaf arrangement in some species is slightly peculiar, and often in a pair, one is slightly larger or elevated. In *S. xanthocarpum*, leaves are spinous.

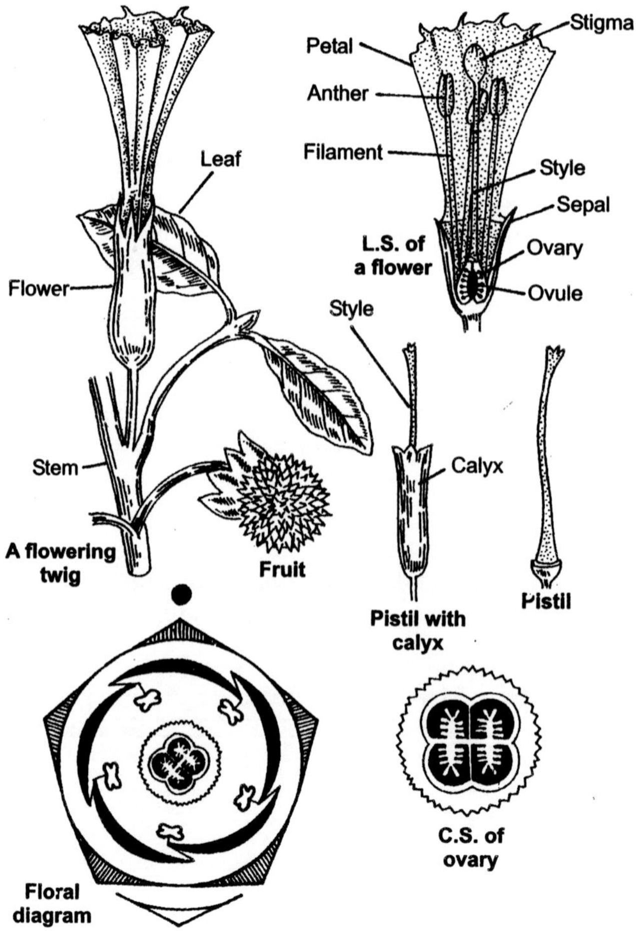

Fig. 21.3 *Datura stramonium* L.

Inflorescence and Flowers

Inflorescence is usually cymose or cymose panicle, often helicoid cyme or may be axillary or extra-axillary in nature (*S. nigrum*). Sometimes flowers are borne singly (solitary) in the axil of leaf (*Datura*) or terminal.

Flowers are ebracteate, hermaphrodite, actinomorphic, complete, pentamerous and hypogynous with a flat thalamus. However, in *Salpiglossis* and *Schizanthus*, flowers are bilabiate and zygomorphic and in *Withania coagulans*, these are unisexual. **Calyx** consists of 5, united persistent sepals (gamosepalous), enlarging with fruits. Aestivation is valvate. In *Physalis peruviana*, persistent calyx forms a baloon-like structure enveloping the whole fruit. In *Datura*, calyx falls off leaving a disc-like basal portion that enlarges in fruit. **Coralla** consists of 5, united petals (gamopetalous) which may be campanulate (*Physalis*), rotate (*Solanum*), trumpet-shaped (*Datura*), infundibulliform (*Petunia*) or bilabiate (*Schizanthus*). Aestivation is contorted, imbricate or valvate, the lobes often being folded. In *Brunfelsia latifolia*, corolla changes colour from violet (beginning) and then to white to yellow–that is why species is called *Yesterday-Today & Tomorrow* plant.

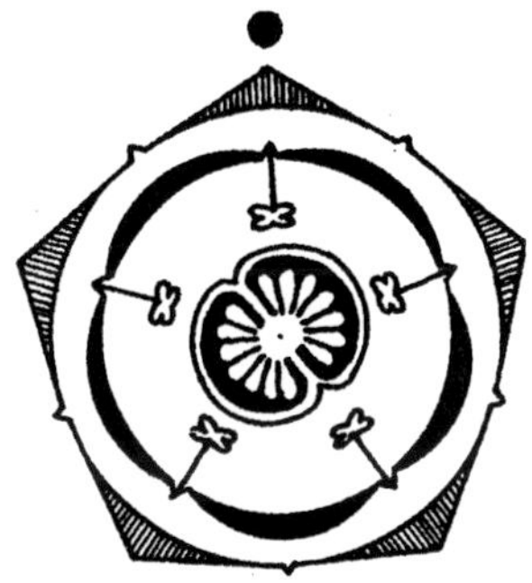

Fig. 21.4 F.D. of *Solanum nigrum* L.

Androecium has 5, epipetalous stamens, unequal in length and alternating with petals. In *Salpiglossis*, only 4 stamens–two sterile and two fertile, are found. Anthers are conniving, dithecous, introrse and dehiscing by longtitudinal slits or by apical pores. A hypogynous disc is often present. In *Solanum*, anthers set close together in a cone over the style and dehisce by apical pores.

Gynoecium is typically bicarpellary and syncarpous. Ovary is superior, obliquely placed, bilocular with swollen placenta and numerous shining anatropous ovules set in axile placentation. However, ovary in *Datura* becomes tetralocular due to the formation of false partition. Style is simple and stigma is bilobed or capitate. In *Henoonia*, ovary is unilocular while in *Capsicum* it is bilocular below and unilocular in upper part. Nectar disc is usually present at the base of ovary.

Fruits and Seeds

Fruit is either a berry or a capsule. In *Physalis* and *Withania*, berry remains enclosed in a persistent, swollen baloon-like calyx. **Seeds** are compressed, discoid or reniform with fleshy translucent endosperm (endospermic seeds).

Pollination and Dispersal

Flowers are insect pollinated (entomophilous) but in *Salpiglossis*, these are cleistogamous and self-pollinated. Seeds are dispersed by birds and animals (zoochory).

Floral Formula $\oplus$ ⚥ $K_{(5)}\ \overset{\frown}{C_{(5)}\ A_5}\ \underline{G}_{(2)}$

Economic Importance

Several plants of the family are of great economic value.

Edibles. Stem tubers of *Solanum tuberosum* L. (potato, also called King of vegetables) and fruits of *S. melongena* L. (egg plant, baigun), *Lycopersicum esculentum* Karst. (tomato), *Capsicum annum* (red pepper, chilly, mirch), *C. frutescens* L. are commonly used as vegetables. Fruits of *Physalis peruviana* L. (cape-goose-berry, rasbari) are edible.

Medicinal Plants. Several plants are alkaloid rich, hence quite valuable for various health problems. Roots of *Atropa belladonna* L. (belladonna) are source of alkaloids **belladonine** and **atropine** which are sedative and pain reliever and dilate the pupils of eyes, respectively. *A. acuminata* Royle (Indian belladona) is the commercial source of Indian belladona. Dried leaves and floral tops of *Hyoscyamus niger* L. (henbane) are the source of drug **henbane** of commerce. The drug is used in nervous disorders, asthma and whooping cough. Roots of *Withania somnifera* Dun. (asgandh) are the source of drug **asagandh** used in rheumatism, asthma, hiccup etc. Fruits of *W. coagulans* (Stock) Dun. are used in the treatment of asthma and as blood purifier. Leaves and fruits of *Datura stramonium* L. (dhatura, thorn-apple) and *Solanum xanthocarpum* Schrad. & Wendl. syn. *S. surattense* Burm. f. are expectorant–used in cough, asthma and pain in chest. Fruits yield glyco-alkaloids which can be a source material for cartisone and sex hormone preparations. Seeds are deadly poisonous and used for suicidal and homicidal purposes. Leaves of *Nicotiana tabacum* L. (tobacco) are the source of virginia tobacco used as fumitories and masticatories. These contain alkaloids, such as **nicotine** and **anabasine**, used in medicine as sedative, antispasmodic and vermifuge. Roots of *Mandragora autumnalis* Spreng. are sedative and hypnotic.

Ornamentals. Some plants are grown for their showy flowers and their fragrance. Important examples are *Petunia hybrida* Vilm., *P. violacea* Lindl., *Cestrum nocturnum* L. (night jessamine, raat-ki-rani), *C. diurnum* L. (day jessamine, din-ka-raja) *Salpiglossis sinuata* Ruiz. & Pav. (pointed tongue), *Schizanthus pinnatus* Ruiz & Pav. (butterfly flower), *Brunfelsia latifolia* Benth. (yesterday-today-tomorrow plant) etc.

Systematics and Phylogeny

Family bears a close relationship with *Boraginaceae* in the presence of alternate leaves, regular flowers and pentamerous stamens. It resembles *Convolvulaceae* in the presence of intra-xylary phloem, persistent calyx, twisted corolla and false septum in the ovary. It is also akin to *Scrophulariaceae* in the presence of exstipulate leaves, persistent and united 5 sepals, united petals, superior ovary etc. *Solanaceae* differ from *Scrophulariaceae* in the presence of generally actinomorphic flowers, typically pentamerous stamens and oblique position of carpels. The tribe *Salpiglossidae* with its zygomorphic flowers and two or four stamens form a connecting link between *Solanaceae* and *Scrophulariaceae*. *Solanaceae* differ from *Convolvulaceae* in the presence of ebracteate flowers, oblique position of carpels, numerous ovules on swollen placentae, and fruit a berry. Most taxonomists (Hutchinson, Cronquist, Thorne, Takhtajan) and APG groupings have placed it in its own order **Solanales** except Bentham & Hooker who placed it under **Polemoniales.** APG and APweb placed the family in the order **Solanales** of sub-group **Euasterids I of 4. APG Group : Eudicots-Asterids.**

Table 21.2 Differences between Solanaceae and Convolvulaceae

	Solanaceae	Convolvulaceae
1.	**Plants** are usually herbaceous.	**Plants** are climbers or herbaceous.
2.	**Flowers** are ebracteate.	**Flowers** are bracteate.
3.	**Gynoecium** is obliquely placed (obliquely septate). The ovary bears numerous shinning ovules on swollen placenta.	Such character is not found.
4.	Each **locule** of the pistil has a large number of ovules.	Each **locule** of the pistil has two ovules.
5.	**Fruit** is usually a berry.	**Fruit** is usually 2-4 valved capsule.
6.	**Floral formula** Ebr $\oplus$ ⚥ $K_{(5)}\ \overset{\frown}{C_{(5)}\ A_5}\ \underline{G}_{(2)}$	**Floral formula** Br brl $\oplus$ ⚥ $K_{(5)}\ \overset{\frown}{C_{(5)}\ A_5}\ \underline{G}_{(2)}$

CONVOLVULACEAE (Ipomoea Family)

Bentham & Hooker	Hutchinson	Cronquist	APG
Dicotyledons	Dicotyledones	Magnoliophyta	4. APG Group :
Gamopetalae	Herbaceae	Magnoliopsida	Eudicots-Asterids
Bicarpellatae	Solanales	Asteridae	Euasterids I
Polemoniales		Solanales	Solanales

The family contains **50 genera** and **1200 species** found in tropical and subtropical regions of the world. Some **20 genera** and **177 species** are found in India. Common forms met in the country are *Convolvulus, Evolvulus, Ipomoea, Argyreia, Porana* etc.

Vegetative Characters

Plants are predominantly twining or trailing (*Ipomoea*), herbs or shrubs (*Erycibe*) or trees (*Ipomoea arborescens* and *I. wolcottiana*). **Stem** is usually herbaceous, erect or trailing, branched and hairy. Internally, intra-xylary phloem is found in the stem. **Leaves** are alternate, exstipulate, simple or sometimes pinnately or palmately divided (*I. cairica*).

Inflorescence and Flowers

Inflorescence is axillary, biparous cyme or cymose or racemose panicles, or solitary axillary (*Convolvulus* and *Evolvulus* species).

Flowers are bracteate, complete, bisexual, actinomorphic, pentamerous and hypogynous, usually showy. In *Mina lobata,* flowers are slightly zygomorphic and in *Hildenbrandtia,* these are unisexual. **Calyx** consists of 5, persistent, united or free sepals set in valvate or imbricate aestivation. Outermost sepal is sometimes larger. **Corolla** consists of 5, united petals (gamopetalous) which may be campanulate (*Convolvulus*) or funnel-shaped (*Ipomoea*) or subglobose with induplicate valvate (*Ipomoea* sp.) or imbricate aestivation. Lobes are often with median folds.

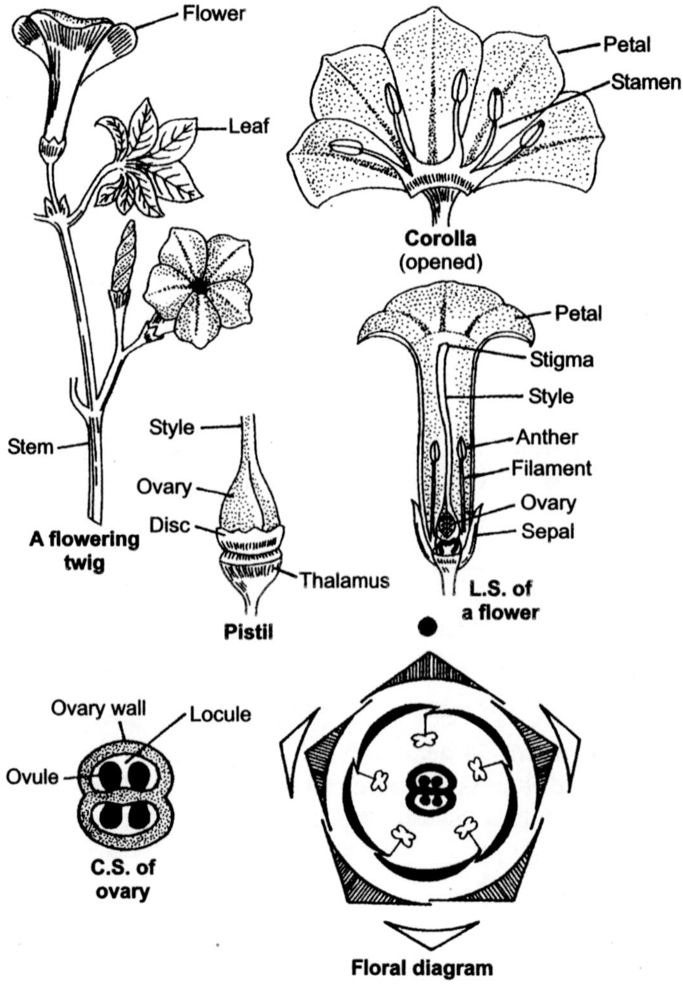

Fig. 21.5 *Ipomoea cairica* (L.) Sweet.

Androecium consists of 5 stamens, distinct, epipetalous and alternating with the corolla lobes (petals). Anthers are dithecous, oblong, introrse and dehiscing longitudinally. Pollen grains are ellipsoid or globose, smooth or spinulose. Annular or cupular intrastaminal disc is often present.

Gynoecium consists of two united carpels (syncarpous), with superior bilocular ovary. Placentation is axile with typically two anatropous ovules in each locule. Sometimes, the ovary becomes tetralocular due to the formation of false septum, with one ovule in each. Style is simple, sometimes ovary with two biclefted (*Evolvulus mummularius*) style. Stigma is bilobed and capitate.

Fruits and Seeds

Fruit is mostly a loculicidal capsule, sometimes berry or a nut. **Seeds** are endospermic (cartilaginous) or non-endospermic, may be smooth or hairy.

Pollination and Dispersal

Flowers are insect pollinated. In *Dichondra repens*, flowers are cleistogamous and self pollinated. Seeds are dispersed by wind or animals.

Floral Formula $Br \oplus ⚥ K_{5 \text{ or } (5)} \overset{\frown}{C_{(5)} A_5} \underline{G}_{(2)}$

Economic Importance

Family has moderate economic value.

Edibles. The tuberous roots of *Ipomoea batatas* L. (sweet potato, shakarkandi) are rich in starch and sugar and are eaten raw, boiled or roasted. These roots are also a commercial source of starch, pectin and alcohol. Leaves and tender shoots of *I. aquatica* Forsk. syn. *I. reptans* (L.) Poir are used as vegetables.

Medicinal Plants. Leaves and stem of *I. aquatica* Forsk. are cooling and given in nervous and general debility and also used to cure piles. Shoots of *Convolvulus pluricaulis* Chois. (sankhapushpi) are used in ayurvedic medicine as a mental tonic. Plants of *Merremia tridentata* (L.) Hallier f. are used in rheumatism, piles and urinary disorders. Plants of *I. purpurea* (L.) Rath. are used as purgative and antisyphilitic. Roots of *Argyreia nervosa* (Burm. f.) Boj. are used as tonic, and also in rheumatism and nervous disorders.

Ornamental Plants. Several plants are grown in gardens for their foliage and showy flowers, sometimes with fragrance. Common examples are *Ipomoea purpurea* (L.) Roth. (morning glory), *I. cairica* (L.) Sweet. syn. *I. palmata* Forsk. (railway creeper), *I. quamoclit* L. syn. *Quamoclit pennata* (Desr.) Boj. (linear, pinnetisect leaves and small scarlet flowers), *I. violacea* L. (heavenly blue), *I. alba* L. syn. *Calonyction aculeatum* (L.) House etc., *Argyreia nervosa* (Burm. f.) Boj. (wooly morning glory), *Porana paniculata* Roxb. etc.

Systematics and Phylogemy

The family is close to *Solanaceae* in several common features, both vegetative and reproductive, but it differs from *Solanaceae* in features like presence of (i) twinning or trailing habits in most plants, (ii) bracteate flowers, (iii) two ovules in each locule, and (iv) 2-4 valved capsule. Bentham and Hooker placed the family in the order **Polemoniales** but most taxonomists like Hutchinson, Cronquist, Thorne etc. and APG II/AP web placed it in the order **Solanales**. Takhtajan has created a new order **Convolvulales** to include the family. APG system includes the family in monophyletic group **4. Eudicots-Asterids** and subgroup I. **Euasterids I** and **Solanales.**

BORAGINACEAE (Borage Family)

Bentham & Hooker	Hutchinson	Cronquist	APG
Dicotyledons	Dicotyledones	Magnoliophyta	4. APG Group :
Gamopetalae	Herbaceae	Magnoliopsida	Eudicots-Asterids
Bicarpellatae	Boraginales	Asteridae	Euasterids I
Polemoniales		Lamiales	Order not assigned

The family comprises **100 genera** and **2000 species** widely distributed in temperate and tropical regions of the world. Some **37 genera** and **145 species** are found in India, of which *Heliotropium, Cordia, Myosotis, Ehretia* and *Coldenia* are common.

Vegetative Characters

Plants are herbs (*Heliotropium*), shrubs or trees (*Cordia, Ehretia*). **Leaves** are simple, entire, alternate and exstipulate. In *Trichodesma*, leaves show opposite arrangement.

Inflorescence and Flowers

Inflorescence is usually scorpoid cyme. However, in *Cordia* and *Ehretia*, it is corymb or panicle and in *Trichodesma*, flowers are solitary axillary.

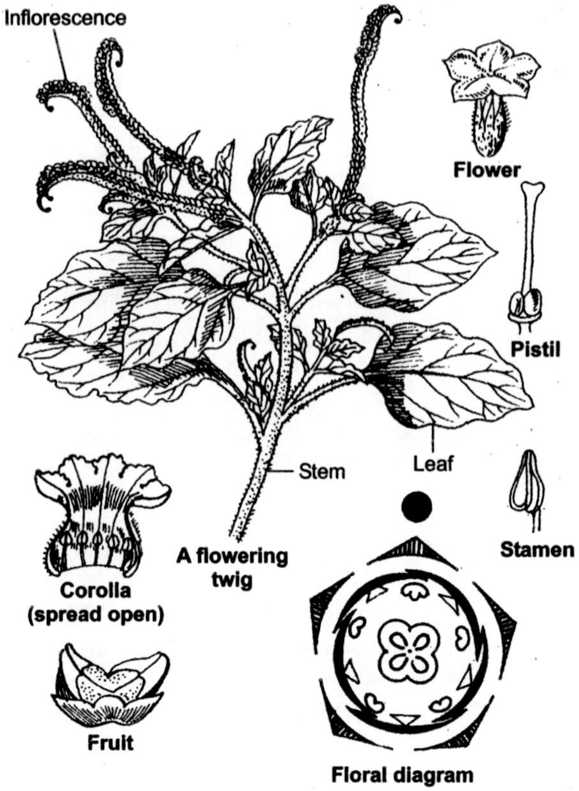

Fig. 21.6 *Heliotropium indicum* L.

Flowers are bracteate, bisexual, hypogynous, pentamerous and actinomorphic. A change in normal symmetry produces zygomorphy in *Echium* and *Lycopis.* **Calyx** contains 5 sepals with imbricate, rarely valvate aestivation. These may be almost free to the base or united to form a short (*Heliotropium*) or long (*Cordia*) tube. Sepals are acrescent in the fruit. **Corolla** consists of 5 petals, united below with spreading limb, arranged in imbricate aestivation. Corolla may be rotate (*Trichodesma*), salver-form (*Anchusa*), funnel-form (*Cynoglossum*) or tubular (*Heliotropium*). The throat of corolla is provided with scales (5 scales, alternate with stamens).

Androecium consists of stamens as many as lobes of corolla (5 stamens). Stamens are epipetalous, alternating with the petals, inserted or exerted, introrse and dehiscing longitudinally.

Gynoecium has two united carpels (syncarpous) seated on ring-like secreting disc. Ovary is superior, bilocular with axile placentation, sometimes becomes tetralocular (due to the formation of false septum) with a single ovule in each locule in the early stages. Owing to this, ovary appears four lobed and style becomes gynobasic. Style is simple but may be forked at the tip. Stigma is capitate.

Fruits and Seeds

Fruit is a drupe, often deeply four-parted into single seeded drupaceous segments (nutlets). **Seeds** are non-endospermic.

Pollination and Dispersal

Flowers are insect pollinated. Fruits, being covered with hooked or barbed bristles, are dispersed by animals. Fruits of *Myosotis* and *Anchusa* are dispersed by ants.

Floral Formula $Br \oplus ⚥ K_{(5)} \overset{\frown}{C_{(5)} A_5} \underline{G}_{(2)}$

Economic Importance

Some plants are economically quite valuable.

Edibles. Important plants grown for their edible fruits are : *Cordia dichotoma* var. *wallichi* (Cl.) Maheshwari (lassora), *C. gharaf* (Forsk) Ehrenb. & Asch. (gondi), *Ehretia acuminata* R.Br. (puna) etc.

Medicinal plants. Plants of *Heliotropium indicum* L. are used as a local application to boils and swellings and their decoction is used in fever. Leaves of *Coldenia procumbens* L. are applied to rheumatic swellings, and those of *Ehretia macrophylla* Wall. are used as a substitute for tea which is useful in cough and stomach trouble. Seeds of *Lithospermum officinale* L. (gromwell) are useful in treatment of gout and bladder diseases. Dried leaves and flowers of *Onosma bracteatum* Wall. are used as tonic, alterative, demulcent, diuratic and spasmolytic.

Ornamental plants. Several plants are grown in gardens for their foliage and showy flowers, *e.g.* species of *Heliotropium, Cynoglossum, Myosotis* (forget-me-nots), *Borago, Pulmonaria* and *Echium.*

Other uses. Wood of *Ehretia acuminata* R. Br. (puna), and *Cordia macleodii* Hook f. & Thoms. (dohipalas) is used for making carts and agricultural implements and that of *E. wallichiana* Hook. f. & Thoms. for building purposes.

Systematics and Phylogeny

Family bears close relationship with *Polemoniaceae* on one hand and *Hydrophyllaceae* on the other. It resembles *Solanaceae* in the presence of alternate leaves, regular flowers and pentamerous stamens. The family further shows link with *Lamiaceae* (Labiatae) in having a nectar secreting hypogynous disc, gynobasic style and nutlets, and with *Verbenaceae* in producing drupaceous fruits.

Bentham and Hooker placed the family in the order **Polemoniales**, Hutchinson and Takhtajan in **Boraginales** and Cronquist in **Lamiales.** Thorne has kept it in **Solanales.** APG classification does not assign any order to the family and placed it in 4. APG Group : **Eudicots-Asterids** (monophyletic affiliation) and subgroup **Euasterids I.**

Boraginaceae have been divided in 5 sub families.

I. Style terminal; fruit a drupe

1. **Cordioideae** : Style 4-fid and drupe 4 celled.
2. **Ehretioideae** : Style 2 or bifid, drupe 2 seeded or 4 one seeded pyrenes.
3. **Heliotropioideae** : Style simple or bifid, a ring of hairs below stigma, fruit drupe or nutlets.

II. Style gynobasic; fruit of 4 nutlets

4. **Boraginoideae** : Corolla often bent, zygomorphic, nutlets 4 or less.
5. **Wellstedioideae** : Flowers tetramerous, ovary 2–celled with one ovule in each locule.

VERBENACEAE (Verbain Family)

Bentham & Hooker	Hutchinson	Cronquist	APG
Dicotyledons	Dicotyledones	Magnoliophyta	4. APG Group :
Gamopetalae	Lignosae	Magnoliopsida	Eudicots-Asterids
Bicarpellatae	Verbenales	Asteridae	Euasterids I
Lamiales		Lamiales	Lamiales

The family contains **36 genera** and over **1035 species** found commonly in tropical and subtropical parts of the world. Some **21 genera** and **165 species** are found in India, mostly in southern and western parts and subtropical Himalayas, of which common examples are *Verbena, Lippia, Tectona, Duranta, Gmelina, Phyla, Clerodendrum* etc. *Verbena* (200 species), *Lippia* (180 species) and *Lantana* (140 species) are some large genera of the family.

Vegetative Characters

Plants are herbs (*Phyla, Lippia*), shrubs (*Lantana, Duranta*) or trees (*Tectona, Gmelina, Avicennia*) or rarely woody climbers (*Petrea*). **Leaves** are exstipulate, opposite or alternate, simple or rarely pinnately (*Peronema*) or palmately (*Vitex*) compound. In *Verbena*, leaves are highly dissected.

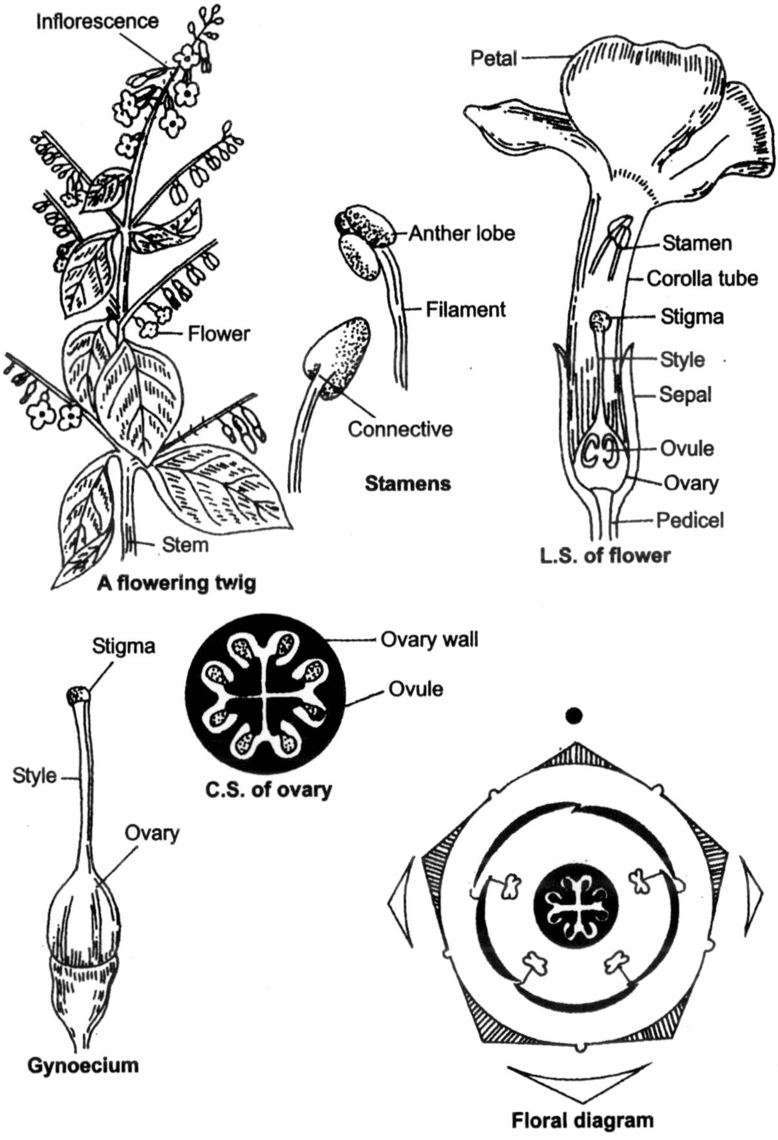

Fig. 21.7 *Duranta plumeri* Jacq.

Inflorescence and Flowers

Inflorescence may be cymose (*Clerodendrum*), racemose, spicate (*Lippia*), or panicled (*Tectona*).

Flowers are bisexual, more or less zygomorphic, usually pentamerous and hypogynous. In *Aegiphila,* flowers are unisexual. **Calyx** consists of 5, united sepals (gamosepalous), persistent, campanulate or tubular. **Corolla** has 5, united petals (gamopetalous), tubular and bilipped (*Duranta* and *Verbena*) or unequally 5 lobed, with imbricate aestivation.

Androecium consists of four, didynamous or 2, epipetalous stamens. The fifth posterior stamen is either represented by a staminode or is completely suppressed. In *Tectona,* all 5 stamens are perfect and in *Stachytarpheta,* only two anterio-lateral stamens are fertile and remaining three are reduced to staminodes. Anthers are dithecous, introrse and dehiscing lengthwise.

Gynoecium is bicarpellary and syncarpous. Ovary is four-lobed or entire, superior, 2 to 4 locular with one or two ovules in each locule set in axile placentation. Carpels are 3-5 in *Geunsia* and 4 in *Duranta.* In *Avicennia,* ovary is unilocular with free central placentation. Style is terminal and stigma are as many lobed as the number of carpels.

Fruits and Seeds

Fruit is a berry or drupe. In *Avicennia,* fruit is a capsule, dehiscing by two thick valves. **Seeds** are non-endospermic. *Avicennia* shows vivipary.

Pollination and Dispersal

Flowers are pollinated by bees and butterflies (entomophilous). Seeds are dispersed by birds and animals.

Floral Formulae

Duranta Br % ⚥ $K_{(5)}\ \overset{\frown}{C_{(2/3)}\ A_{2+2}}\ \underline{G}_{(4)}$

Lantana Br % ⚥ $K_{(5)}\ \overset{\frown}{C_{(4/1)}\ A_{2+2}}\ \underline{G}_{(2)}$

Economic Importance

Some plants of the family are economically important.

Ornamental plants. Important ornamental plants are *Verbena hybrida* Voss., *V. bipinnatifida* Sch., *Lantana camara* L., *Duranta plumeri* Jacq. (golden dew drop), *Petrea volubilis* L. (queens wreath), *Clerodendrum inerme* (L.) Gaertn., *C. fragrans* (Vent.) Willd., *C. speciosum* D'Ombrain, *Caryopteris incana* Miq. (blue beard) etc.

Timber Yielding plants. Important timber producing plants are *Gmelina arborea* Roxb. (gumhar)–the wood is used for making drums, sitars and other musical instruments; *Tectona grandis* L. f. (teak, sagwan)–the most valuable hard wood used for ship building, bridge and heavy constructions, railway carriages, furniture, cabinets etc.; *Premna bengalensis* Clarke–the wood is used for making bridges and posts; *Vitex altissima* L.f. and *V. pinnata* L.–the wood is used for construction works etc.

Other uses. Bark of *Avicennia* is used for tanning. *Lippia alba* (Mill.) N.E. Br. ex Britton and Wilson produces a valuable essential oil. *L. citridora*

H.B. & K. yields verbena oil. Leaves of *Vitex negundo* L. (nirgundi) serve as vermifuge and are used for preserving stored grains against insect attacks, and those of *V. trifolia* L. are reported to possess anticancer property and are also used against rheumatism and tuberculosis.

Systematics and Phylogeny

Family is related with *Lamiaceae* by the presence of persistent calyx and bilabiate corolla. It resembles *Boraginaceae* by the nature of calyx, inflorescence and fruits. It also shows relationship with *Acanthaceae* from which it differs in the presence of tetralocular ovary with one ovule in each locule or bilocular ovary with two ovules in each locule.

Hutchinson has placed the family in its own order, the **Verbenales** but most taxonomists (Bentham and Hooker, Cronquist, Thorne, Takhtajan) and APG classification recognised it under the order **Lamiales**. APG has placed the family under the monophyletic familial group **4. APG : Eudicots-Asterids** and subgroup **Euasterids I**. APG II, AP web, Judd *et al.* (2002) placed some genera like *Clerodendrum, Callicarpa, Vitex* and *Tectona* in the family *Lamiaceae* but *Avicennia* is separated. Thorne (2003) recognised *Avicenniaceae* as a separate family.

LAMIACEAE (LABIATAE) (Mint Family)

Bentham & Hooker	Hutchinson	Cronquist	APG
Dicotyledons	Dicotyledones	Magnoliophyta	4. APG Group :
Gamopetalae	Herbaceae	Magnoliopsida	Eudicots-Asterids
Bicarpellatae	Lamiales	Asteridae	Euasterids I
Lamiales		Lamiales	Lamiales

The family comprises **264 genera** and **6990 species** found worldwide. Some **64 genera** and **350 species** are found in India, mainly in dry areas and moderate altitudes. Common examples found in the country are *Salvia, Ocimum, Mentha, Coleus, Leucas, Nepeta* etc. *Salvia* (700 species), *Thymus* (340 species) and *Stachys* (300 species) are large genera of the family.

Vegetative Characters

Plants are usually annual or perennial herbs, sometimes, shrubs (*Ocimum gratissimum*) or rarely trees (*Canum* and *Leucosceptrum*). **Leaves** are simple, exstipulate, opposite, decussate or whorled (*Ocimum*) or pinnately compound (*Salvia hydrangea, S. plectanthoides*). Leaf margin may be entire, toothed, lobed or finely dissected. Glandular hairs found on all parts of body secrete volatide oils (essential oils).

Inflorescence and Flowers

Inflorescence is specialised and called **verticillaster**, which is a kind of condensed dichasial cyme situated in the axil of leaves. In *Scutellaria*, flowers are solitary axillary and in *Notochaete hamosa*, axillary inflorescence is a globose head.

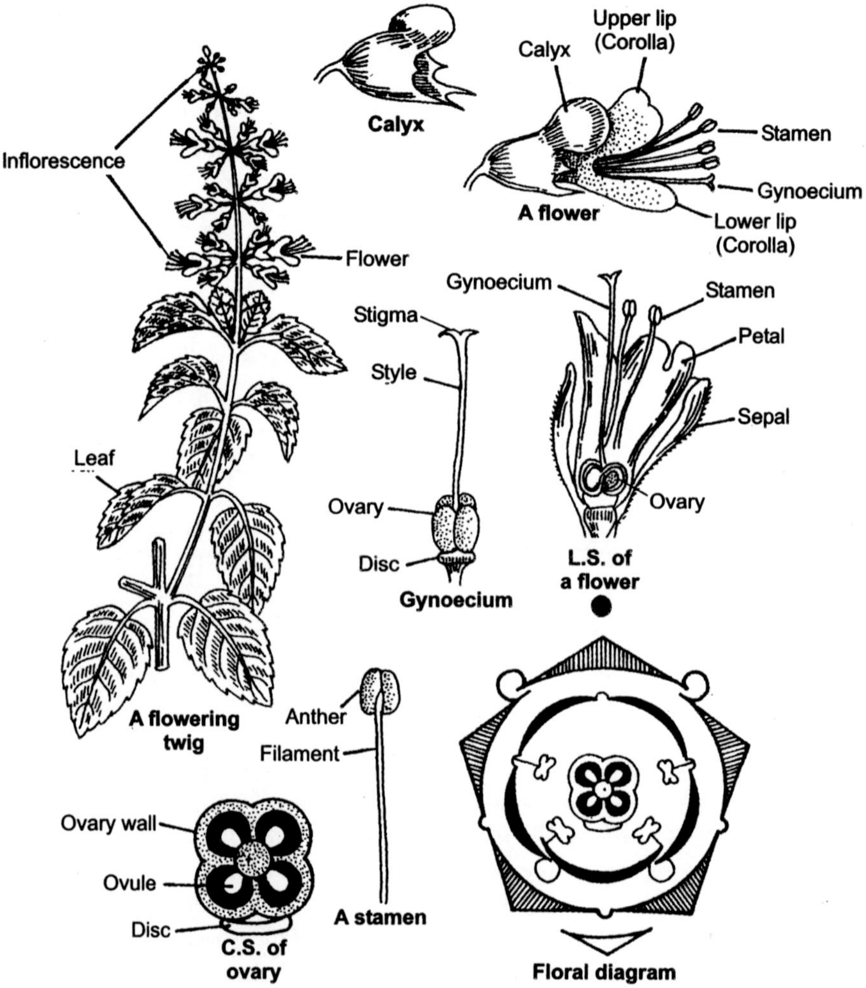

Fig. 21.8 *Ocimum sanctum* L.

Flowers are bracteate, complete, zygomorphic, bisexual, pentamerous and hypogynous. In *Mentha*, flowers are actinomorphic. **Calyx** consists of 5, persistent, united sepals (gamosepalous). Aestivation is valvate, rarely imbricate. In *Ocimum* and *Salvia*, calyx is bilipped with unequal teeth or tubular with teeth equal in others. **Corolla** consists of 5, united petals (gamopetalous) and usually bilabiate. In *Salvia*, the upper lip is composed of two posterior petals and lower lip of three petals, but in *Ocimum*, the upper lip is formed by four petals and the lower of one petal. Aestivation is imbricate or contorted.

Fig. 21.9 F.D. of *Salvia*.

Androecium consists of 4 stamens, the fifth one remains suppressed or reduced to staminode. These are usually didynamous. The anterior pair of stamens is usually longer. In *Salvia* and *Lycopus*, only two (anterior pair) stamens are fertile, whereas, other two are reduced to staminodes. In *Salvia*, two anther cells of a stamen are separated apart due to the enlargement of connective, and only posterior cell is fertile while anterior one is reduced to a sterile knob. In *Coleus*, stamens are monadelphous, the filaments connate below in a sheath separate from corolla. Normally anthers are dithecous, introrse, dehiscing longitudinally. In *Leucospectrum*, anther is unilocular and reniform.

Gynoecium is bicarpellary, syncarpous with superior and bilocular ovary. The ovary is sometimes tetralocular due to the formation of a false septum. Each locule has two ovules set in axile placentation. Ovule is erect, orthotropous, beaked with inferior micropyle. Style is often gynobasic arising between the lobes of the ovary, and stigma is bifid.

Fruits and Seeds

Fruit is usually a group of four equally developed achenes or nutlets, with basal or lateral aerola. **Seeds** are non-endospermic or with very scanty and fleshy endosperm.

Pollination and Dispersal

Flowers are usually dichogamous and insect pollinated. Seeds are dispersed by wind with the help of persistent swollen bladder–like calyx or in some cases by animals.

Floral Formulae

Ocimum Br % ⚥ $K_{(4/1)}\ \overparen{C_{(1/4)}\ A_{2+2}}\ \underline{G}_{(2)}$

Salvia Br % ⚥ $K_{(3+2)}\ \overparen{C_{(2/3)}\ A_{2}}\ \underline{G}_{(2)}$

Economic Importance

Family has several economically important plants.

Medicinal plants. Essential oil extracted from *Ocimum sanctum* L. (basil, holy basil) and *O. basilicum* L. (kali tulsi) is used in perfumery and medicine and their leaf decoctions are used in cough and fever. Juice of leaves of *O. sanctum* L. is used in catarrhal affections of children. Leaves of *Mentha piperita* L. (peppermint) yield peppermint oil (essential oil) used in pharmacy, confectionary and in medicine. Essential oil extracted from the leaves of *Majorana hortensis* Moench. is used externally for treating sprains, and stiff and paralytic limbs, and that from *Lavandula angustifolia* Mill., *L. hybrida* Raverchron and *L. latifolia* Vill. is carminative, stimulant and insect repellant.

Leaves and stem of *Melissa parviflora* Benth. are antipyretic and are used in brain, liver and heart diseases.

Ornamental Plants. Species of *Lamium, Salvia* and *Coleus* are grown for their showy leaves and attractive flowers.

Essential oils. Essential oils extracted from several plants are used in perfumery, cosmetics, pharmacy and confectionary. Common sources are *Lavandula angustifolia* Mill. and *L. latifolia* Vill. (yield lavander oil), *Mentha piperita* L. (peppermint oil), *Pogostemon plectranthoides* Desf. syn. *P. cablin* (Blanco) Kunth. (patchouli oil), *Rosmarinus officinalis* L. (rosemary oil) and *Thymus serpyllum* L. (thyme oil) etc. An essential oil extracted from dried tops of *Plectranthus mollis* Spreng. is used as cardiac depressant.

Some herbs are used to add flavour and aroma to foods., *e.g. Majorana hortensis* Moench. (Sweet majorum), *Marrubium vulgare* L. (pahari gandana), *Melissa axillaris* (Benth.) Bakh f., *Mentha arvensis* L., *Origanum vulgare* L. etc.

Systematics and Phylogeny

The family is closely related to *Acanthaceae*, but it differs from the latter in the presence of verticillaster inflorescence, gynobasic style and nature of ovary and fruits. It also shows relationship with *Boraginaceae*, but differs in characters of ovule. In *Labiatae*, the micropyle is inferior and raphe inward but in *Boraginaceae*, the micropyle is superior and raphe outward.

Among gamopetalous families, *Lamiaceae* are considered as highly advanced, especially owing to elaborate floral structure and adaptation to entomophily.

Lamiaceae have been placed in its own order **Lamiales** by all taxonomists including Bentham and Hooker, Hutchinson, Cronquist, Thorne, Takhtajan etc. and APG, AP web etc. APG/AP web placed the family in the monophyletic familial group : **4. APG Group, Eudicots-Asterids** and subgroup **Euasterids I.**

SCROPHULARIACEAE (Snapdragon Family)

Bentham & Hooker	Hutchinson	Cronquist	APG
Dicotyledons	Dicotyledones	Magnoliophyta	4. APG Group :
Gamopetalae	Herbaceae	Magnoliopsida	Eudicots-Asterids
Bicarpellatae	Personales	Asteridae	Euasterids I
Personales		Scrophulariales	Lamiales

Family is fairly large and is represented by **220 genera** and **3000 species** preferably growing in temperate regions of the world. In India, some **57 genera** and **275 species** are found and common examples are species of *Antirrhinum, Majus, Linaria, Lindenbergia, Limnophila, Russelia, Scoparia, Scrophularia, Veronica* etc. An introduced plant *Digitalis purpurea* is commonly grown in hill stations for the large purple flowers.

Vegetative Characters

Plants are annual or perennial herbs (*Lindenbergia, Majus, Linaria*), shrubs (*Russelia, Scrophularia, Verbascum*) or rarely trees (*Wrightia gigantea*). *Limnophila* is an aquatic herb; *Maurandia* is a petiole climber; *Pedicularis, Striga* and *Sopubia* are root parasites; *Melamporum* and *Castilleja* are saprophytes. **Stem** is usually erect, branched, solid and hairy, sometimes, creeping or prostrate. **Leaves** may be alternate, opposite or whorled and are simple, exstipulate, entire or with incised margin. Some species of *Limnophila* and *Aubulia* are heterophyllous. *Verbascum* has decurrent leaves which give stem a winged appearance.

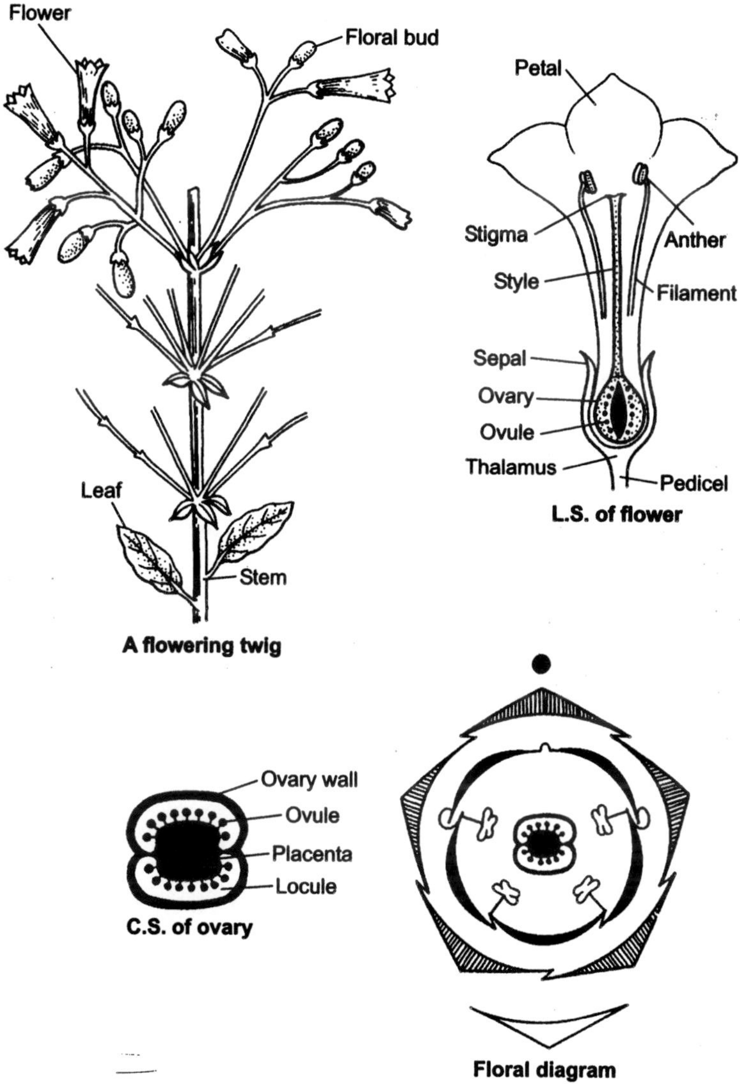

Fig. 21.10 *Russelia equisetiformis* Schl. & Cham..

Inflorescence and Flowers

Inflorescence is usually a racemose (*Linaria*) but cyme in *Scrophularia* and solitary in *Mimulus, Bacopa* and *Kickxia.*

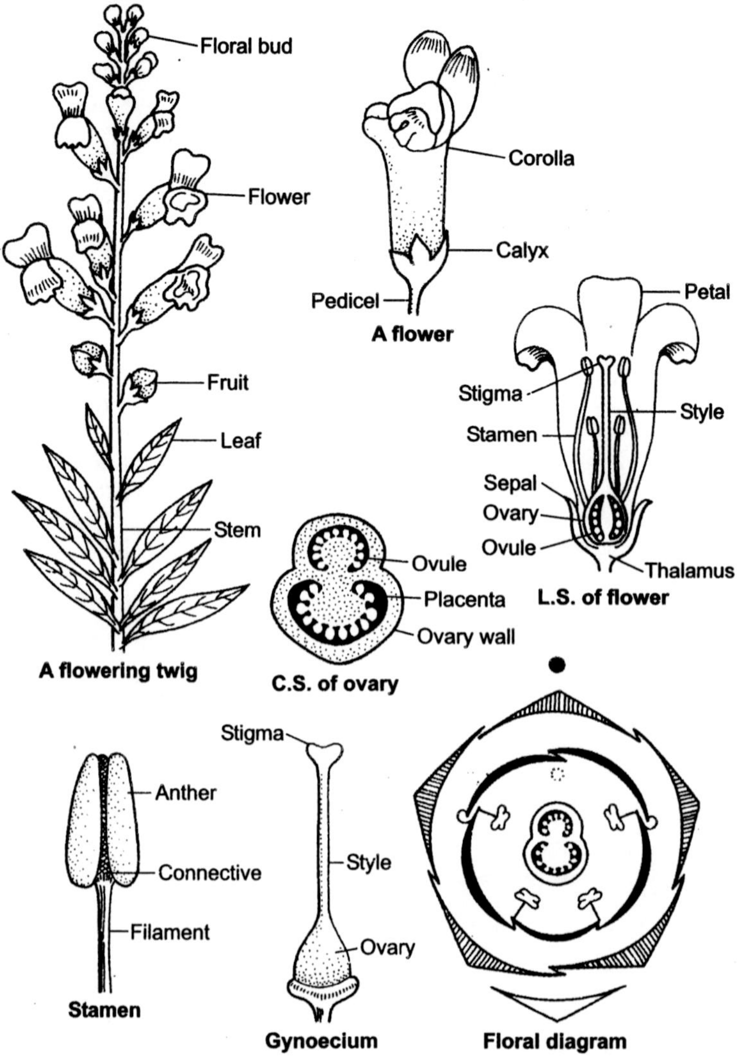

Fig. 21.11 *Antirrhinum majus* L.

Flowers are bracteate, zygomorphic, bisexual, pentamerous and hypogynous. *Scoparia, Bacopa* and *Verbascum* have nearly actinomorphic flowers. **Calyx** consists of 5, united sepals (gamosepalous). Sepals are persistent and show imbricate (*Antirrhinum*) or valvate (*Majus*) aestivation. **Corolla** consists of 5, united petals (gamopetalous), medianely zygomorphic and often bilabiate. Corolla is usually bilipped with upper bilobed lip and lower trilobed lip. Anterior petal is sometimes spurred (*Linaria*) or gibbous (*Antirrhinum*). Aestivation is imbricate.

Androecium consists of usually four stamens which are didynamous and epipetalous. The fifth stamen is often reduced, suppressed or aborted. *Verbascum* and *Capraria* have five fertile stamens. In *Vernonica,* only 2 and in *Stemotria,* only 3 stamens are found. Anthers are dithecous (often one larger than other), basifixed and introrse dehiscing by longitudinal slits. A hypogynous disc is often present. Pollen grains are loose and powdery.

Gynoecium is bicarpellary and syncarpous. Carpels are medianely placed. Ovary is superior and bilocular with several anatropous ovules on thick axile placenta. In *Antirrhinum,* anterior carpel is larger. Style is usually simple and terminal with bilobed stigma.

Fruits and Seeds

Fruit is generally a capsule, rarely a berry. **Seeds** are small and endospermic.

Pollination and Dispersal

Flowers are insect pollinated (entomophilous). Seeds are dispersed by water, birds or other animals.

Floral Formula Br % ⚥ $K_{(5)}\ \overline{C_{(5)\text{ or }(2/3)}\ A_{2+2}}\ \underline{G}_{(2)}$

Economic Importance

The family has limited economic importance but for some medicinal and ornamental plants.

Medicinal plants. Mature leaves of *Digitalis purpurea* L. are source of a valuable drug **digitoxin** used as cardiac stimulant, and those of *Scoparia dulcis* L. (sweet broom weed) for the treatment of fever, cough and respiratory disorders. Drug **picrorhiza,** obtained from dried rhizomes and roots of *Picrorhiza kurroa* Benth., is laxative and cathartic and is used in the treatment of dropsy. Plant extract of *Bacopa monnieri* (L.) Pennell. is used as nerve tonic to cure epilepsy and insanity, and of *Torenia polygonoides* Benth. smeared as poultice on sores and ulcers.

Ornamental plants. Important ornamental plants are *Antirrhinum majus* L. (snapdragon), *Russelia equisetiformis* Schl. & Cham. (coral blow), *Digitalis purpurea* L. (common foxglove), *Mimulus luteus* L. (monkey flower), *Nemesia versicolor* E. Mey. (nemesia), *Linaria vulgaris* Mill. (foadflax) etc.

Systematics and Phylogeny

Scrophulariaceae resemble *Solanaceae* in general floral plan but differ in possession of collateral vascular strands, non-plicate zygomorphic corolla, reduction of posterior stamen (staminode) and medianely placed carpels. While the family shows affinity with *Pedaliaceae* and *Bignoniaceae,* it differs basically in presence of endospermic seeds. *Scrophulariaceae* are related with *Acanthaceae* and *Lamiaceae* but differ from them in the presence of alternate leaves, and characters of corolla, ovary and fruit.

Bentham and Hooker and Hutchinson placed the family in the order **Personales** while Cronquist and Takhtajan placed it in its own order **Scrophulariales.** Thorne and APG/AP web have kept the family within the order **Lamiales.** According to APG, **Lamiales** is placed in the 4th monophyletic group **Eudicots–Asterids** under the subgroup **Euasterids I.**

ACANTHACEAE (Acanthus Family)

Bentham & Hooker	Hutchinson	Cronquist	APG
Dicotyledons	Dicotyledones	Magnoliophyta	4. APG Group :
Gamopetalae	Herbaceae	Magnoliopsida	Eudicots-Asterids
Bicarpellatae	Personales	Asteridae	Euasterids I
Personales		Scrophulariales	Lamiales

The family comprises **221 genera** and **3650 species** growing in tropics and warm temperate regions of the world. In India, some **68 genera** and **500 species** are found with common forms like *Acanthus, Adhatoda, Andrographis, Asystasia, Barleria, Beloperone, Justicia, Peristrophe, Ruellia, Rungia, Strobilanthus, Thunbergia* etc. *Justicia* (400 species), *Beloperone* (300 species) and *Barleria* (240 species) are the large genera of the family.

Vegetative Characters

Plants are usually annual (*Rungia*) or perennial (*Peristrophe*) herbs, shrubs (*Barleria, Adhatoda*) or rarly small trees (*Strobilanthus*) or lianas (*Thunbergia*). Some plants show xerophytic habitats. **Stem** is erect, herbaceous or woody, branched, angular and smooth surfaced. **Leaves** are mostly simple, opposite, usually decussate, entire and exstipulate, rarely alternate (*Nelsonia*). Axillary spines are found in *Barleria* and *Hygrophila*. In leaves, cystoliths are usually present.

Inflorescence and Flowers

Inflorescence is mostly of cymose type either dichasial or cincinus, sometimes racemose. *Thunbergia* bears solitary axillary flowers.

Flowers are large with prominent bracts, bracteolate, strongly zygomorphic, bisexual, showy, variously coloured, and hypogynous. **Calyx** consists of 5, united sepals (gamosepalous) with valvate or imbricate aestivation. In *Thunbergia*, calyx is redued to a narrow wing and in *Blepharis, Acanthus* and *Barleria*, calyx is four-lobed. **Corolla** consists of 5, united petals (gamopetalous). It is usually bilabiate (sometimes oblique) with the upper lip usually erect and bifid, and the lower lip almost horizontal and three lobed. In *Ruellia, Hemigraphis* and *Thunbergia*, corolla is almost five-lobed and in *Acanthus* and *Blepharis*, the upper lip is absent. Aestivation is imbricate or twisted.

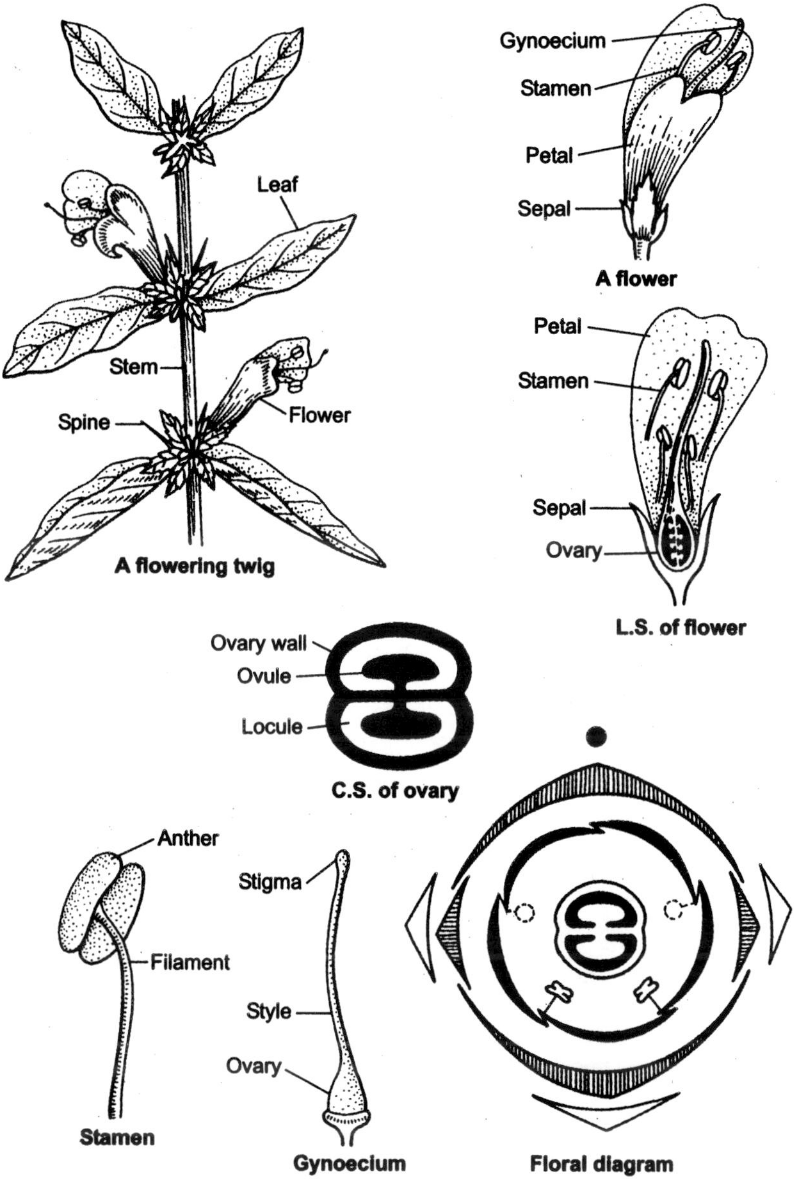

Fig. 21.12 *Barleria prionitis* L.

Androecium comprises 2-5 stamens. When number of stamens is four, these are **didynamous** and fifth posterior stamen is either completely suppressed or is represented by a staminode. In *Pentstemonacanthus,* five fertile stamens are found. Stamens are epipetalous with extrorse anthers. Anthers may be dithecous or monothecous, dehiscing by longitudinal slits or by apical pores. Pollen grains are variously sculptured.

Gynoecium is bicarpellary and syncarpous with bilocular, superior ovary having one or more anatropous ovules in each locule set in axile placentation. Style is long and stigma is bilobed, sometimes spathulate.

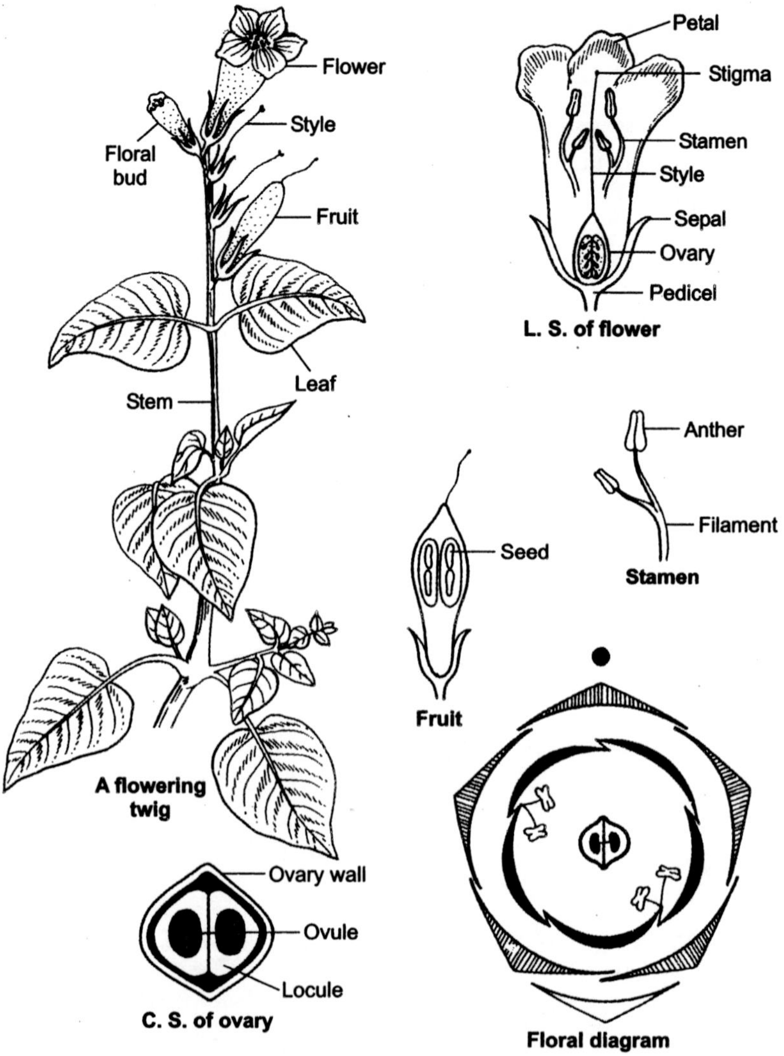

Fig. 21.13 *Asystasia gangetica* (L.) T. Anders.

Fruits and Seeds

Fruit is a loculicidal capsule, hard or leathery, usually supported on hook-like outgrowth of the funicle, called **retinacula** or **jaculator**. **Seeds** are non-endospermic.

Pollination and Dispersal

Flowers are protandrous and insect pollinated. Seeds are usually dispersed by explosive mechanism **(jaculator mechanism)**.

Floral Formula $Br \ \% \ ⚥ \ K_{(5)} \ \overparen{C_{(2/3)} \ A_{4 \text{ or } 2}} \ \underline{G}_{(2)}$

Economic Importance

Some plants have medicinal and ornamental values.

Medicinal plants. Leaves and roots of *Adhatoda vasica* Nees (vasaka) are used to cure bronchitis, asthma and cough, and those of *Hygrophila spinosa* T. Anders to cure jaundice and rheumatism. Plants of *Barleria buxifolia* L., *B. prionitis* L. (kala bansa) and *B. longifolia* L. are used against cough and inflammations.

Ornamental plants. Important ornamental plants valued for foliage and showy flowers are *Barleria montana* Nees and *B. gibsoni* Dalz., *Beloperone guttata* Brandegee (shrimp plant), *Eranthemum bicolor* Shrank and *E. reticulatum* Hort. ex Gard., *Justicia gendarussa* Burm., *Ruellia tuberosa* L., *R. brittoniana* Leonard., *Thunbergia grandiflora* Roxb. (clock vine), *T. alata* Boj. etc.

Systematics and Phylogeny

Acanthaceae are related to *Bignoniaceae* and *Scrophulariaceae* but differ in the presence of prominent bracts and bracteoles, imbricate calyx of unequal sepals, beaked ovary and exalbuminous seeds. The family is closely related to *Lamiaceae* (*Labiatae*) from which it shows differences in the bracteate inflorescence, bilocular unlobed ovary, terminal style and many seeded capsule with jaculators.

Bentham & Hooker and Hutchinson placed the family in the order **Personales** but Cronquist and Takhtajan have assigned it to **Scrophulariales.** Thorne and APG II/AP web recognised its position in the order **Lamiales.** APG grouping placed the family in monophyletic familial group : **4. APG Group : Eudicots–Asterids** in the subgroup **Euasterids I** and order **Lamiales.** It is considered as one of the advanced families of dicotyledons.

Table 21.3 Differences between Labiatae and Acanthaceae

Labiatae (Lamiaceae)	Acanthaceae
1. **Leaves** contain essential oil glands.	**Cystoliths** are found in leaves of some plants.
2. **Stem** is quadrangular.	**Stem** is more or less cylindrical. In some plants, cells have cystoliths.
3. **Inflorescence** is **verticillaster** type, which is characteristic to the family.	**Inflorescence** is usually **dichasial cyme.**
4. **Flowers** have **gynobasic** style, *i.e.*, style develops from the base of the ovary.	**Style** is normal type, *i.e.*, arising from top of the ovary.
5. **Ovary** possesses anteriorly placed nectary.	A hypogynous nectariferous disc is found at the base of ovary.
6. The **fruit** is of four one-seeded nutlets, enclosed by the persistent calyx.	The **fruit** is usually a loculicidal capsule.
7. **Floral formula** $Br\ \%\ ⚥\ K_{(1/4)}\ \overparen{C_{(4/1)}\ A_{2+2}}\ \underline{G}_{(2)}$	**Floral formula** $Br\ brl\ \%\ ⚥\ K_{(5)}\ \overparen{C_{(2/3)}\ A_{4\ or\ 2}}\ \underline{G}_{(2)}$

PEDALIACEAE (Sesamum Family)

Bentham & Hooker	Hutchinson	Cronquist	APG
Dicotyledons	Dicotyledones	Magnoliophyta	4. APG Group :
Gamopetalae	Lignosae	Magnoliopsida	Eudicots-Asterids
Bicarpellatae	Bignoniales	Asteridae	Euasterids I
Personales		Scrophulariales	Lamiales
Pedalineae			

This small family includes **17 genera** and **60 species** found mostly in South Africa, Indo-Malaya and Australia. In India, **3 genera** and about **7 species** are found, chiefly growing in South India. *Martynia annua* grows wild in the plains and *Sesamum indicum* is widely cultivated in warmer parts of the country.

Vegetative Characters

Plants are mostly either annual (*Martynia*) or perennial herbs, rarely shrubs, usually viscid. Some species of *Trapella* are aquatic herbs and have dimorphic leaves. **Leaves** are opposite, upper leaves may be alternate, simple, entire, lobed or divided (trifoliate), exstipulate. Hydathodes are usually present.

Inflorescence and Flowers

Flowers are either solitary axillary or in simple three flowered cymes with glands at the base of stalk. Glands are regarded as metamorphosed flowers. In *Martynia*, inflorescence is terminal raceme.

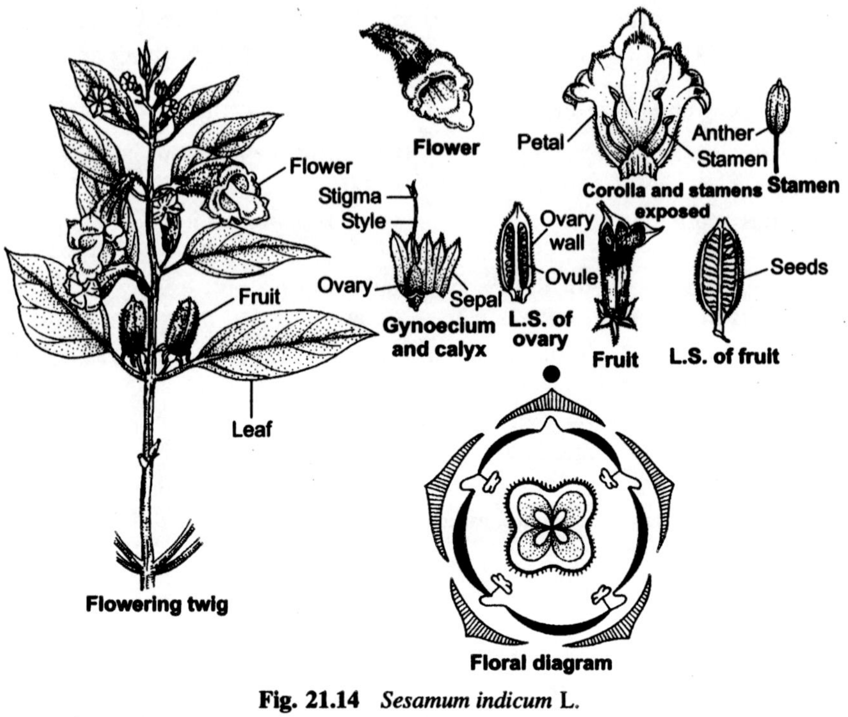

Fig. 21.14 *Sesamum indicum* L.

Flowers are shortly pedicellate, hermaphrodite, complete, zygomorphic and hypogynous, but epigynous flowers are found in *Trapella*. Bracts are minute or absent. **Calyx** consists of 5 (rarely 4), united sepals (gamosepalous) and these are tubular, or campanulate or spathe-like, persistent, hairy and imbricate. Distinct sepals are found in *Martynia*. **Corolla** comprises 5 (rarely 4) united petals (gamopetalous), indistinctly bilabiate and tubular ventricose, with imbricate aestivation.

Androecium consists of 4, epipetalous, didynamous stamens. A short staminode is often present. *Martynia* has 2 fertile stamens and 3 staminodes. Anthers are dithecous, introrse and dehiscing longitudinally.

Gynoecium consists of 2-united carpels (syncarpous), forming a superior uni- or bi- or quadrilocular ovary due to presence of false septa. Placentation is axile with one or more anatropous ovules in each locule. In *Martynia*, ovary is unilocular with two parietal placentae. Style is long, filiform and persistent. Stigmas are 2, leafy and elliptical. A nectar secreting disc is present at the base of the ovary.

Fruits and Seeds

Fruit is either a capsule or a hard and indehiscent nut, often winged or provided with spines or nooks. **Seeds** have straight embryo and scanty endosperm.

Pollination and Dispersal

Flowers are protandrous and insect pollinated. Fruits are dispersed by large mammals. Fruit of *Martynia* characteristically develops two long curved horns, helpful in dispersal by animals.

Floral Formula $\% \; ⚥ \; K_{(5)} \; \overparen{C_{(5)} \; A_{2+2}} \; \underline{G}_{(2)}$

Economic Importance

The family includes some important oil yielding plants, *e.g. Sesamum indicum* L., *S. angustifolium* Engl. and *Pedalium murex* L. *Sesamum indicum* L. is largely cultivated in the warmer parts of the country. Valuable sesame or til oil, extracted from its seeds, is used in confectionary, margarine preparation and in the manufacture of soaps, cosmetics and insecticides. Oil cake is used as cattle feed.

Systematics and Phylogeny

Family is closely allied to *Acanthaceae* and *Scrophulariaceae* but it is distinguished by the features like non-endospermic seeds and the structure of fruit and winged seeds in many cases. Hallier included the family in **Tubiflorae** as derived from *Scrophulariaceae*. The family has been included in the order **Personales** (Bentham & Hooker), **Scrophulariales** (Cronquist,

Takhtajan) or in **Lamiales** (Thorne, APG II/AP web). APG grouping placed the family in monophyletic familial group, *i.e.*, **4 APG Group : Eudicots-Asterids** and in subgroup-**Euasterids I** and order **Lamiales**. Some taxonomists separated *Martynia* and placed it in its own family *Martyniaceae* mainly due to parietal placentation.

RUBIACEAE (Ixora Family)

Bentham & Hooker	Hutchinson	Cronquist	APG
Dicotyledons	Dicotyledones	Magnoliophyta	4. APG Group :
Gamopetalae	Lignosae	Magnoliopsida	Eudicots-Asterids
Inferae	Rubiales	Asteridae	Euasterids I
Rubiales		Rubiales	Gentianales

It is a large family comprising **630 genera** and **10,400 species** growing mostly in tropical parts of the world. In India, some **76 genera** and **550 species** are found, of these, *Rubia, Galium* and *Luculia* grow in Himalayas and *Anthocephalus, Adina, Coffea, Gardenia, Ixora, Pavetta, Hamelia, Oldenlandia, Mussaenda* etc. are found in plains. *Psychotria* (1450 species), *Galium* (410 species) and *Ixora* (370 species) are the large genera of the family.

Vegetative Characters

Plants are mostly trees (*Adina, Anthocephalus*) or shrubs (*Ixora, Gardenia*). Some species of *Galium* are herbs and of *Rubia* and *Uncaria* are hook climbers. Species of *Myrmecodia* are myrmecophilous (ant-loving). **Leaves** are simple, usually opposite, decussate, rarely whorled, and stipulate. Stipules may be inter-or intrapetiolar. In *Gardenia*, 4, intrapetiolar, two each of opposite leaves, stipules are united into a conical cap, which are thrown off as the bud opens, and in *Galium* and *Rubia*, stipules are like ordinary leaves.

Inflorescence and Flowers

Inflorescence is mostly dichasial cyme, rarely panicled cyme. In *Mussaenda glabra*, the flowers are arranged in dense terminal cymes, in *Anthocephalus* and *Adina*, dichasial cymes are condensed together into heads and in *Hamelia*, 2-3 chotomous subscorpoid cyme is found. *Cinchona* has paniculate inflorescence.

Flowers are mostly bisexual, complete, bracteate, actinomorphic, pentamerous or tetramerous and epigynous, sometimes semiepigynous. Flowers are unisexual in *Anthospermum*. **Calyx** consists of 5, sometimes 4, persistent, united sepals (gamosepalous). In some species of *Mussaenda*, one sepal in few flowers becomes enlarged (leaf like) and showy being variously coloured. Aestivation is valvate. **Corolla** consists of 5, or sometimes 4, united

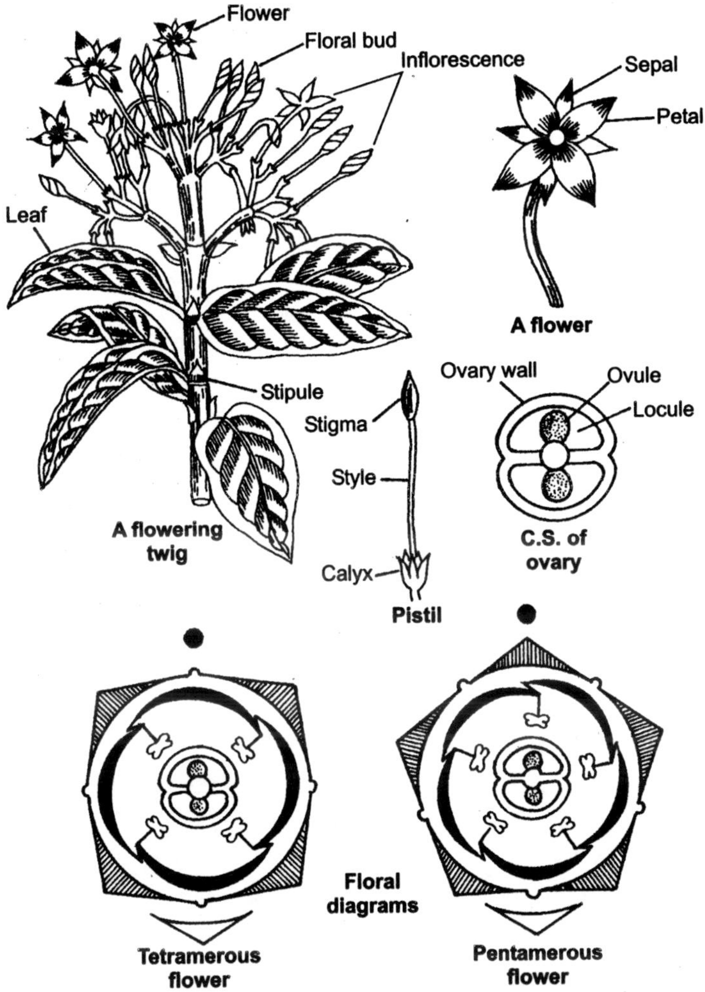

Fig. 21.15 *Ixora coccinea* L.

petals (gamopetalous) and is salver-form, rotate or funnel–form with valvate or contorted aestivation. In zygomorphic flowers, it is imbricate or contorted.

Androecium consists of 5, sometimes 4, stamens, alternate to petals, the number corresponds to number of petals. Stamens are epipetalous and are inserted together at or near the throat of corolla tube. Anthers are dithecous, introrse and opening lengthwise.

Gynoecium consists usually of 2 united carpels (syncarpous). Ovary is inferior and bilocular with one to several anatropous, unitegmic ovules in each locule set in axile placentation. Style is simple and filiform and stigma is capitate or bilobed. Ovary is half inferior in *Synaptantha*, superior in *Pagamea*, unilocular with parietal placentation in *Gardenia* and pentalocular in *Hamelia* and *Hamiltonia*.

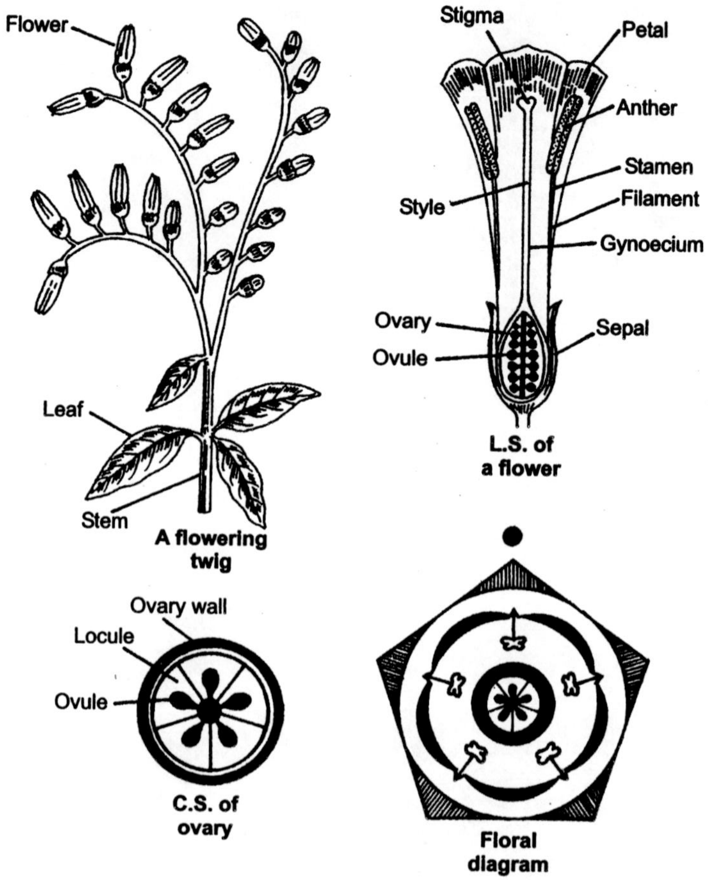

Fig. 21.16 *Hamelia patens* Jacq.

Fruits and Seeds

Fruit is usually a septicidal or loculicidal capsule, sometimes berry (*Mussaenda*) or drupe (*Coffea*). **Seeds** are mostly endospermic and often winged (*Cinchona*).

Pollination and Dispersal

Flowers are usually insect pollinated. Seeds are sticky and dispersed by birds. Winged seeds are wind dispersed.

Floral Formula Br ⊕ ⚥ $K_{(5 \text{ or } 4)} \overbrace{C_{(5 \text{ or } 4)}\ A_{(5 \text{ or } 4)}} \overline{G}_{(2)}$

Economic Importance

Some plants have great economic value.

Beverages. *Coffea arabica* L. is grown commercially in Brazil, Jamaica, Java etc. and in Nilgiris, Karnataka and Travancore in India. Seeds (roasted, ground, brewed) are source of world famous arabic coffee. Similarly, seeds of *C. liberica* Bull. ex Hievn. and *C. canephora* Pierre ex Forehner are source of liberian and robusta coffee (instant coffee), respectively.

Medicinal plants. Several plants are source of important drugs. The bark of *Cinchona* (*C. calisaya* Wedd., *C. ledgeriana* Moens ex Trimen, *C. officinalis* L.) is a source of valuable drug **quinine** used to cure malaria. *Cinchona* plants are grown in India (Tamil Nadu), Sri Lanka, Java etc. Roots of *Cephaelis ipecacuanha* (Stokes) Baill. are the source of drug **ipecacuanha** used for the treatment of amoebic dysentry. Fruit pulp of *Randia tinctoria* L. is emetic and anthelmintic.

Ornamental plants. Important ornamental plants grown for foliage and showy flowers are *Anthocephalus cadamba* Miq., *Morinda tinctoria* Roxb., *Ixora coccinea* L., *Gardenia jasminoides* Ellis. (gandha raj), *Mussaenda frondosa* L., *Hamelia patens* Jacq., *Hamiltonia suaveolens* Roxb. etc.

Timber yielding trees. Some plants yield valuable timber, *e.g. Adina cordifolia* Hook. f. (haldu), *Randia spinosa* Poir. (mainphal), *Anthocephalus cadamba* Miq. (kadam), *Mitragyna parviflora* (Roxb.) Korth. (kaddam), *M. rotundifolia* (Roxb.) Kuntze etc.

Dyes. Roots of *Rubia tinctoria* L. (madder) are source of dye alizarin. *Uncaria gambier* Roxb. yields gambercatechu, used as dye. A yellow dye is obtained from *Morinda tinctoria* Roxb.

Systematics and Phylogeny

Family is closely related to *Cornaceae* in features like cymose umbel-like inflorescence, epigyny and suppressed calyx. It is allied to *Adoxaceae* and *Caprifoliaceae* in the presence of inferior ovary. It establishes a link with *Asteraceae*, since capitulum inflorescence can be derived from the head-like inflorescence met in some members of *Rubiaceae*. The family seems to be related with *Umbelliferae* in having epigynous flowers, epigynous disc, dense cymose inflorescence, bicarpellate ovary and unitegmic ovule.

Rubiaceae have been kept in its own order **Rubiales** by most taxonomists (Bentham & Hooker, Hutchinson, Cronquist, Thorne, Takhtajan) but APG classification places the family in **4. APG Group : Eudicots-Asterids** with which it makes monophyletic familial group. It is placed in the order **Gentianales** within the subgroup **Euasterids I.**

ASTERACEAE (COMPOSITAE) (Sunflower Family)

Bentham & Hooker	Hutchinson	Cronquist	APG
Dicotyledons	Dicotyledones	Magnoliophyta	4. APG Group :
Gamopetalae	Herbaceae	Magnoliopsida	Eudicots-Asterids
Inferae	Asterales	Asteridae	Euasterids II
Asterales		Asterales	Asterales

It is a fairly large family comprising **1528 genera** and **23,840 species** growing worldwide. In India, some **138 genera** and **708 species** are found, of which common wild forms are *Ageratum, Eclipta, Eupatorium, Parthenium, Triodex, Vernonia, Xanthium* etc. and cultivated ones are *Aster, Calendula,*

Centaurea, Chrysanthemum, Dahlia, Helianthus, Lactuca, Tagetus, Zinnia etc. *Senecio* (1470 species), *Vernonia* (1050 species) and *Eupatorium* (590 species) are large genera of the family.

Vegetative Characters

Plants in majority are annual or perennial herbs, a few are shrubs (*Helianthus annuus*) or trees (*Vernonia arborea*). **Stem** is generally branched, solid and hairy. Stem tubers of *Helianthus tuberosus* and fasciculated roots of *Dahlia* are enriched with inulin. **Leaves** are normally alternate, exstipulate, simple or pinnately or palmately lobed. In *Helianthus*, leaves are opposite and simple and in *Haplophyllum*, these are modified into spines.

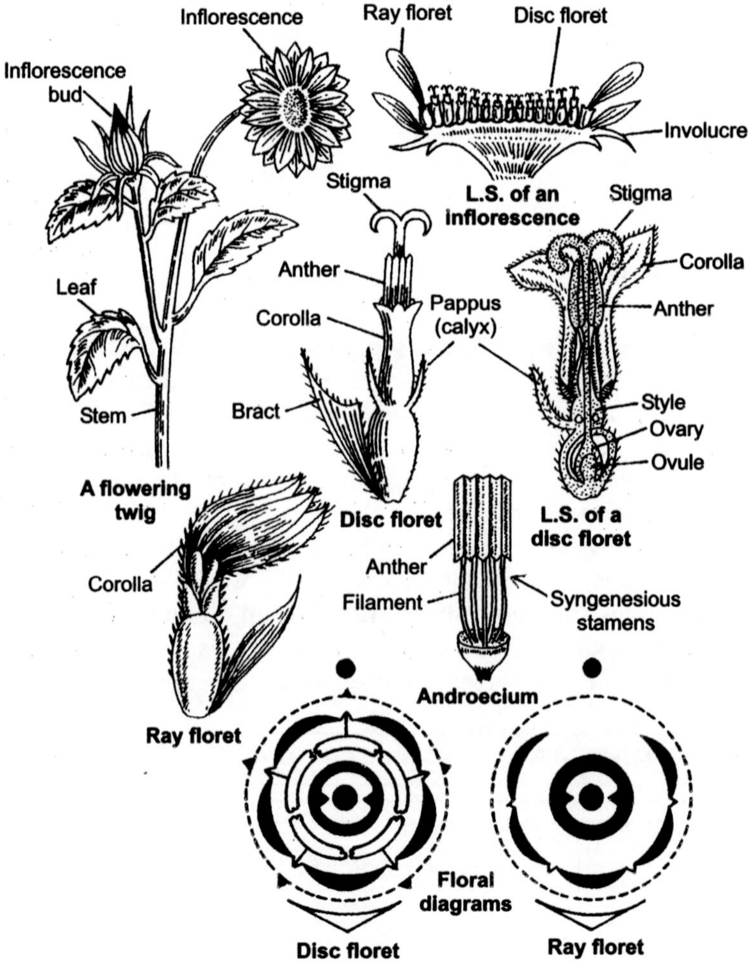

Fig. 21.17 *Helianthus annuus* L.

Inflorescence and Flowers

Inflorescence, characteristic to the family, is a special racemose head, commonly called **capitulum** or **head**. Each head is subtended by an involucre

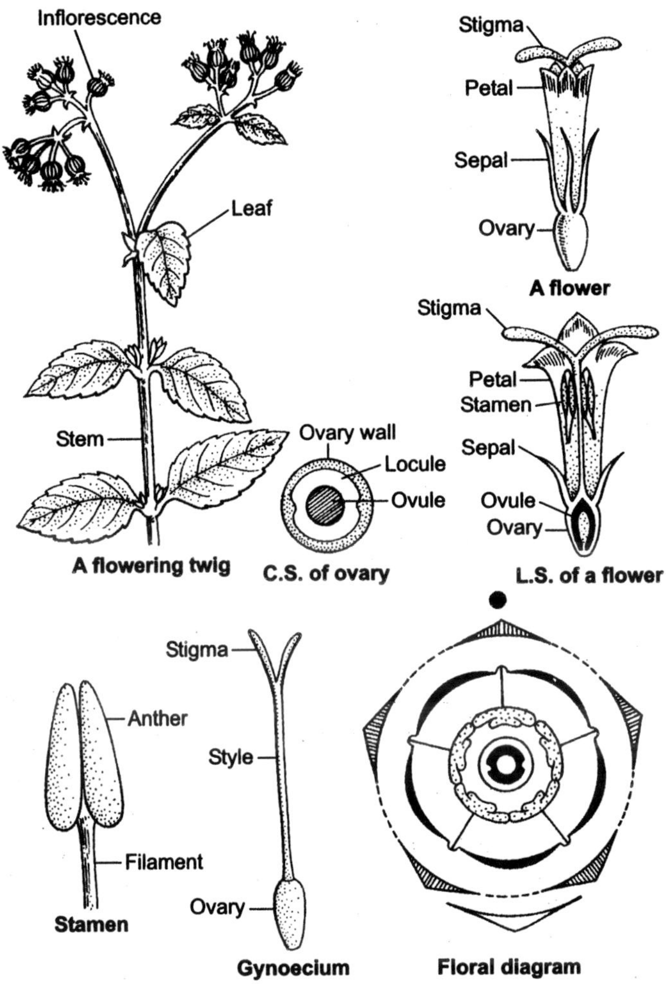

Fig. 21.18 *Ageratum conyzoides* L.

of green bracts, which may be uni-, bi- or multiseriate. These are spine tipped in *Carthamus.* Receptacle may be flat, concave or convex. In *Echinops* and *Lagascea,* one flowered capitulum is found. Capitulum of *Xanthium* is unisexual, the female flower consists of 2-florets in an urn-shaped 2-beaked involucre clothed with hooked spines.

Flowers are bracteate, sessile, pentamerous and epigynous. In most cases, two types of flowers are found, the tubular or DISC FLORETS and the ligulate or RAY FLORETS. On the basis of presence of type of flowers, three types of capitula are recognised.

(1) Heterogamous : Capitula bearing two types of florets, (i) centrally set actinomorphic, tubular, bisexual **disc florets**, and (ii) peripheral ligulate, zygomorphic, pistillate (female) or neutral **ray florets**, as in *Helianthus.*

(2) Homogamous tubular : Capitula bearing only actinomorphic tubular and bisexual florets, as in *Ageratum.*

(3) Homogamous ligulate : Capitula bearing only zygomorphic, ligulate and unisexual florets, as in *Sonchus.*

All florets are ligulate in *Lactuca* and *Cichorium.*

Calyx is usually represented by scales or hairs (**pappus**) which are persistent. **Corolla** consists of 5, united petals (gamopetalous), **tubular** with five small equal lobes or **ligulate** (strap-shaped) with three to five teeth at the end and unequally bilabiate, the upper lip comprising three and lower of two lobes. In *Sonchus,* all petals are united to form lower lip. Aestivation is valvate.

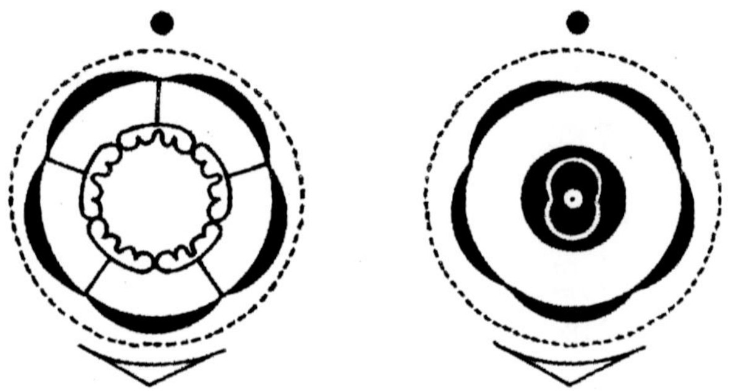

A B

Fig. 21.19 A-F.D. of male flower and B-F.D. of female flower of *Aster amellus* L.

Androecium consists of 5, epipetalous stamens. Anthers are **syngenesious**, dithecous, introrse and dehiscing longitudinally.

Gynoecium is bicarpellary and syncarpous. Ovary is inferior and unilocular with a single (solitary), unitegmic ovule set in a basal placentation. Style is simple and stigma is bilobed (recurved or spreading). A ring-like tubular nectary is found at the base of style.

Fruits and Seeds

Fruit is **cypsella**, a characteristic to the family, often compressed and crowned with hairy **pappus**. **Seeds** are non-endospermic.

Pollination and Dispersal

Flowers are insect pollinated. A single insect can pollinate all flowers of an inflorescence in one visit. In absence of cross pollination, self pollination occurs. Fruits show anemochory (parachute mechanism) or zoochory.

Floral Formulae

Ray floret Br % ♀ or Ω $K_{(pappus)}$ $C_{(5)}$ A_0 $\overline{G}_{(2)}$

Disc floret Br ⊕ ⚥ $K_{(pappus)}$ $\widehat{C_{(5)}\ A_{(5)}}$ $\overline{G}_{(2)}$

Economic Importance

Several plants of the family are economically important.

Edibles. Tubers of *Helianthus tuberosus* L. (Jerusalem artichoke) are eaten like potato. Leaves of *Lactuca sativa* L. (garden lettuce) and *Cichorium endivia* L. are eaten as salad. Roots of *Tragopogon porrifolius* L. (vegetable oyster) are edible and those of *Cichorium intybus* L. (chickory) are roasted, powdered and mixed with coffee for flavour.

Oils. Seeds of *Helianthus annuus* L. are source of valuable edible oil (sunflower oil). Seed oil of *Carthamus tinctorius* L. (safflower, kusum) is useful for patients suffering from cardiac problems. Whole plant body of *Eclipta prostrata* (L.) L. (bhangra) yield valuable oil used in the preparation of medicinal oil.

Medicinal plants. Some plants are medicinally important. *Artemisia cina* Berg. yields drug **santonin** used to repel worms from human intestine. *Cynara scolymus* L. (globe artichoke) is a source of **insulin** used to cure diabetes. Roots and rhizomes of *Taraxacum officinale* Wigg. (common dandelion) are diuretic and purgative and are used to cure liver disorders. Dried ligulate florets of *Calendula officinalis* L. (pot marigold) are used for the treatment of sprains and bruises. Juice of *Enhydra fluctuans* Lour. kills worms and induces sleep.

Dyes. Florets of *Carthamus tinctorius* L. (safflower) yield scarlet red dye used for colouring various edibles and clothes.

Ornamentals. Several plants are grown in the garden for their showy and attractive inflorescences. Important examples are *Dahlia pinnata* Cav., *Chrysanthemum coronarium* L., *C. cinerariaefolium* Vis., *C. morifolium* Ramat., *Aster grandiflorus* L., *A. amellus* L., *Cosmos bipinnatus* Cav., *Zinnia elegans* Jacq., *Z. linearis* Benth., *Centaurea moschata* L. (sweet sultan), *Dimorphotheca sinuata* DC., *Helianthus annuus* L., *Tagetus erecta* L. (marigold), *Calendula officinalis* L. etc.

Insecticides. Natural insecticides, **pyrethrum** and **cinerarin** are extracted from the inflorescence of *Chrysanthemum cinerariaefolium* Vis.

Rubber. *Parthenium argentatum* A. Gray yields rubber.

Asteraceae (Compositae), the Most Evolved Among Dicot Families

Among dicotyledons, Asteraceae are considered as the most or highest evolved family due to following advanced characters.

1. It is the largest family comprising 1528 genera and 23,840 species, amounting about ten per cent of all flowering plants. Cosmopolitan distribution of its members shows its success and suitability to the present time, recalling *survival* of *fittest.*
2. Majority of the plants are herbs and annuals. Shrubs and trees are extremely rare.
3. Capitulum is considered as a most evolved type of inflorescence. It is so well adapted that hundreds of flowers are set in a small space which can be pollinated by a single visit of an insect.
4. Flowers are protandrous, hence self pollination is avoided. However, if cross pollination fails, self pollination can occur successfully, as stigma curls back and comes in contact with pollen of the same flower.

5. Flowers are epigynous (ovary inferior) in conjunction with syngenesious anthers.
6. Calyx is reduced and modified into pappus (hairy calyx). Thus, flower shows oligomery by the reduction of the floral parts.
7. Corolla is gamopetalous and inconspicuous in disc florets.
8. Androecium comprises epipetalous stamens.
9. Ovary is inferior and uniovulate, ovule is set in basal placentation, though gynoecium is bicarpellary.
10. Fruits are wind dispersed (anemochory). Pappus forms a parachute for fruit dispersal. Thus the plants have an efficient mechanism of fruit (seed) dispersal.

Systematics and Phylogeny

Position of the family *Asteraceae* has always been at or very close the peak of dicot evolution. It offers similarities with *Dipsacaceae* and *Valerianaceae* and is akin to *Rubiaceae* in the aggregation of flowers into heads or head-like structure, in the reduction of calyx lobes and presence of inferior ovary. While accepting polyphyletic origin of *Asteraceae*, the origin of **Rubiales** and **Asterales** is traced through **Umbellales** where small flowers tend to aggregate into crowded umbels. *Asteraceae* are related to *Companulaceae* and *Goodeniaceae* in the presence of pentamerous epigynous flowers.

Most taxonomists, *e.g.* Bentham & Hooker, Hutchinson, Cronquist, Thorne, Takhtajan etc. have kept the family in its own order **Asterales**. APG II/AP web also place *Asteraceae* in the order **Asterales** in the monophyletic familial group–**4. APG Group : Eudicots-Asterids** and sub-group **Euasterids II.** Lundberg and Bremer (2002) proposed that *Asteraceae, Calyceraceae, Goodeniaceae* and *Manyanthaceae* together form a monophyletic group, but this relationship is not very clear in various genic (*rbc*L, ndhF, atpB) similarities (APG II).

Table 21.4 Differences between Umbelliferae and Compositae

Umbelliferae (Apiaceae)	Compositae (Asteraceae)
1. **Leaves** are pinnately compound or decompound.	**Leaves** are usually simple.
2. **Inflorescence** is **umbel**, a characteristic to the family.	**Inflorescence** is **Capitulum**, a characteristic to the family.
3. **Flowers** are actinomorphic, bisexual and epigynous.	**Flowers** are actinomorphic (disc florets) or zygomorphic (ray florets), neuter or unisexual (ray florets) or bisexual (disc florets), and epigynous.
4. **Calyx** consists of five sepals.	**Calyx** is modified into pappus.
5. **Corolla** shows imbricate or valvate aestivation.	**Corolla** has valvate aestivation.
6. **Androecium** consists of five free and epipetalous stamens.	**Androecium** consists of 5, syngenesious, epipetalous stamens.

7.	**Ovary** is bilocular and each locule has a single ovule set in **axile placentation.**	**Ovary** is unilocular with a single ovule set in **basal placentation.**
8.	**Stylopodium** is present.	It is not found.
9.	**Fruit** is either **cremocarp** or berry.	**Fruit** is **cypsella.**
10.	**Floral formula**	**Floral formula**
	Br $\oplus$ ⚥ $K_{(5)} \overset{\frown}{C_{(5)} A_5} \bar{G}_{(2)}$	**Ray floret :** Br % ♀ or Ω $K_{(pappus)} C_{(2/3)} A_0 \bar{G}_{(2)}$ **Disc floret :** Br $\oplus$ ⚥ $K_{(pappus)} \overset{\frown}{C_{(5)} \quad A_{(5)}} \bar{G}_{(2)}$

Table 21.5 Differences between Lamiaceae and Asteraceae

	Labiatae (Lamiaceae)	Compositae (Asteraceae)
1.	**Stem** is quadrangular.	**Stem** is cylindrical.
2.	**Leaves** are opposite.	**Leaves** are alternate or whorled.
3.	**Inflorescence** is special–**verticillaster** type which is a condensed dischasial cyme.	**Inflorescence** is specific–**capitulum** which is a racemose type.
4.	**Flowers** are zygomorphic, bisexual and hypogynous.	Two types of flowers are found. **Ray florets** are zygomorphic, neuter or female whereas **Disc florets** are actinomorphic, bisexual and epigynous.
5.	**Calyx** consists of five, united and bilipped sepals.	**Calyx** is modified into **pappus** (hairy).
6.	**Corolla** consists of five united bilipped petals.	**Corolla** consists of five united petals which may be ligulate (ray floret) or tubular (disc floret).
7.	**Aestivation** in corolla is contorted or imbricate.	**Aestivation** in corolla is valvate type.
8.	**Androecium** consists of four, **didynamous** stamens.	**Androecium** consists of five **syngenesious** stamens, *i.e.*, anthers are fused and filaments are free.
9.	**Ovary** is superior and bilocular, but may become tetralocular due to false septum.	**Ovary** is inferior and unilocular.
10.	**Placentation** is axile.	**Placentation** is basal.
11.	**Gynobasic style** is found.	Such condition is not found.
12.	**Fruit** is achene or nutlets.	**Fruit** is a cypsella.
13.	**Floral formula**	**Floral formulae**
	Br % ⚥ $K_{(1/4)} \overset{\frown}{C_{(4/1)} \quad A_{2+2}} \underline{G}_{(2)}$	**Ray floret :** Br % ♀ or Ω $K_{(pappus)} C_{(2/3)} A_0 \bar{G}_{(2)}$ **Disc floret :** Br $\oplus$ ⚥ $K_{(pappus)} \overset{\frown}{C_{(5)} \quad A_{(5)}} \bar{G}_{(2)}$

□□□

Chapter **22**

Liliopsida (Monocots)
(Alismataceae, Arecaceae, Araceae, Commelinaceae, Cyperaceae, Poaceae, Musaceae, Zingiberaceae, Liliaceae, Orchidaceae)

ALISMATACEAE (Water Plantain Family)

Bentham & Hooker	Hutchinson	Cronquist	APG
Monocotyledons	Monocotyledones	Magnoliophyta	III. APG Group :
Apocarpae	Calyciferae	Liliopsida	Monocots
	Alismatales	Alismatidae	1. Non-Commelinoid
		Alismatales	Monocots
			Alismatales

It is a small family comprising **12 genera** and **80 species** growing worldwide with plants of fresh water marshes, swamps, lakes, rivers and streams. In India, **7 genera** and some **10 species** are found, of which *Sagittaria sinensis, Limnophyton obtusifolium,* and several species of *Alisma* are common. *Echinodorus* (35 species), *Sagittaria* (25 species) and *Alisma* (9 species) are some large genera of the family.

Vegetative Characters

Plants are either perennial or annual marshy or aquatic herbs with perennial rhizomes, often stoloniferous. **Leaves** are generally radical with sheathing leafbases and elongated petiole. Shape of leaf blades varies from linear to ovate or sagittate (arrow-shaped) or hastate (side lobes of sagittate leaves directed outwards). Venation is reticulate with a few primary parallel veins converging apically and large number of closed and parallel transverse veins. Some scales are present in the axils of leaves. Latex is often present.

Inflorescence and Flowers

Inflorescence is usually much branched, the primary branching being racemose and secondary cymose, which may be racemose or paniculate. In *Butomus*, flowers are set in apparent umbels, subtended by an involucre. **Flowers** are often bisexual, pedicellate, trimerous, regular, bracteate and hypogynous. In *Sagittaria*, flowers are unisexual. **Perianth** consists of 6, free segments (tepals) set in two series, the *outer* three imbricate, herbaceous, persistent, green or sepaloid and *inner* three larger, petaloid, imbricate and deciduous, rarely absent. In *Butomus*, segments of both series are petaloid.

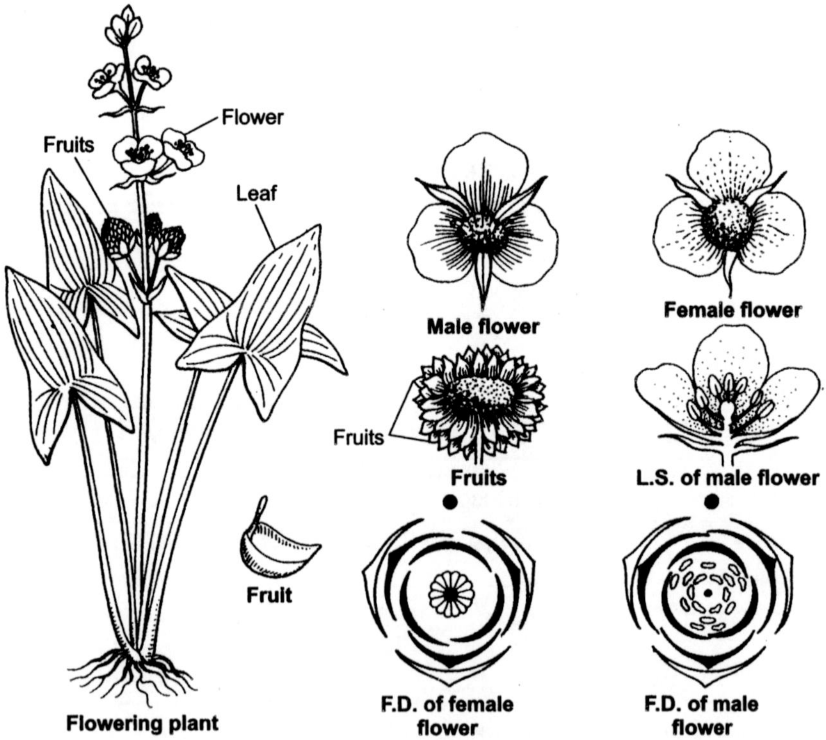

Fig. 22.1 *Sagittaria sinensis* L.

Androecium consists of 6 or more, rarely only 3 stamens (*Wisneria*). Six stamens in a single whorl is found in *Alisma* but in *Butomus* and *Echinodorus*, a second whorl of three stamens is also present. In *Sagittaria*, an outer whorl of six stamens is followed by numerous stamens and in *Hydrocleis* and *Limnocharis*, numerous stamens are surrounded by one or more whorls of staminodes. Filaments are free, often flattened, basifixed, dithecous, extrorse, dehiscing by lateral slits. Nectar is often secreted at the base of stamens.

Gynoecium comprises numerous ($3\text{-}\alpha$) free carpels, either spirally arranged or in a single whorl (*Alisma* and *Sagittaria*). In *Wisneria*, 3 carpels and in *Butomus*, 6 carpels in two whorls of three each are found. Carpels are free or united at the base in *Butomus*. Ovary is superior, unilocular with either solitary (rarely 2 or more), nearly basal ovule (*Alisma*). In *Butomus*, several ovules set in superficial placentation are found. Style is mostly acicular and persistent, ending in a indistinct single stigma.

Fruits and Seeds

Fruit is a bunch or whorl of achenes, rarely follicular dehiscing adaxially. **Seeds** are non-endospermic and embryo is curved or horse-shoe shaped.

Pollination and Dispersal

Flowers are insect pollinated, sometimes self pollinated. Seeds are dispersed by water or by wind.

Floral Formula $Br \oplus P_{3+3} A_{6-\alpha \text{ or } 3} \underline{G}_{3-\alpha}$

Economic Importance

Plants of the family have little economic value. Tuberous rhizomes of *Sagittaria sagittifolia* L. and *Alisma plantago-aquatica* L. are edible. Some species of *Sagittaria* and *Alisma* are grown in garden ponds or in aquaria.

Systematics and Phylogeny

Alismataceae are a primitive family as evident from the floral morphology, *e.g.* flowers are bisexual, hypogynous, with two whorls of free perianth segments, several free stamens and carpels. Hutchinson treated the family of one of two most primitive orders of **Calyciferae** and considered *Alismataceae* to be fairly closely related to the *Ranunculaceae*, and as a link to the **Ranales.** Origin of *Alismataceae* from *Ranunculaceae* is strongly supported by the genus *Ranalisma* which has flowers exactly like *Ranunculus* and since it belongs to *Alismataceae*, it was named as *Ranalisma* (Ran + Alisma). This plant was discovered in Malaya, about 60 years back. Genus *Ranalisma* has a great significance as it supports the theory of origin of *Alismataceae* from *Ranunculaceae*, or atleast from the order **Ranales** and origin of monocotyledons from the primitive dicots, probably **Ranales.**

Butomaceae, Hydrocharitaceae and *Alismataceae* form one aquatic clade of **Alismatales** (Judd *et al.*, 2002) owing to floral characters like, (i) perianth differentiated into sepals and petals, (ii) stamens more than six, (iii) carpels more than three, and (iv) ovules scattered over the inner surface of locules. Anatomical studies have, however, indicated that these numerous stamens are due to secondary increase from the ancestral six stamens.

Most taxonomists placed the family *Alismataceae* in the order **Alismales** or **Alismatales** (Bentham & Hooker, Hutchinson, Cronquist, Takhtajan, Thorne etc.). APG grouping also places the family in the order **Alismatales** in **Non-Commelinoid Monocots** of **III APG Group : Monocots** (a monophyletic familial group). APG proposes that monocots and Eudicots are not formally ranked and named because it is not yet clear at which level they should be recognised.

ARECACEAE (PALMAE) (Palm Family)

Bentham & Hooker	Hutchinson	Cronquist	APG
Monocotyledons Calycinae	Monocotyledones Corolliferae Palmales	Magnoliophyta Liliopsida Arecidae Arecales	III. APG Group : Monocots Commelinoid Arecales

The family is represented by **189 genera** and **2350 species**, widely distributed in tropical and subtropical regions of the world. In India, **28 genera** and about **93 species** are found, mainly growing in Eastern Himalayas, Assam and coastal regions of which important examples are coconut palm (*Cocos nucifera*), date palm (*Phoenix dactylifera*), betel nut palm (*Areca catechu*), cane palm (*Calamus rotang*), *Borassus* etc. *Chamaerops humilis, Trachycarpus martiana* and *Plectocomia himalayana* are found growing at an altitude of 2000 m in Himalayas. The large genera of the family are *Calamus* (350 species), *Bactris* (180 species) and *Pinanga* (120 species).

Vegetative Characters

Plants are shrubs or trees, solitary or gregarious with an umbranched, erect, climbing or trailing stem. *Calamus* has slender stem attaining a great

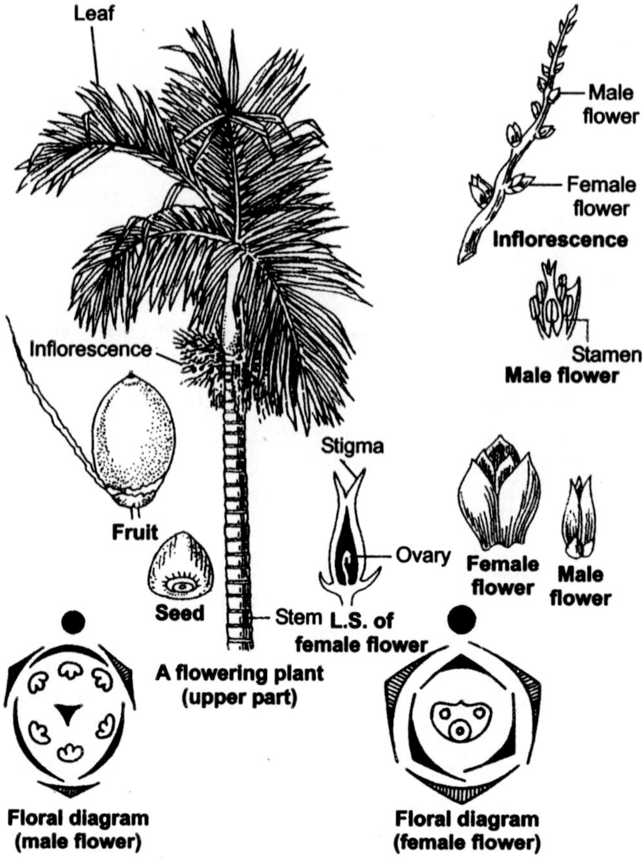

Fig. 22.2 *Areca catechu* L.

length with distinct nodes and internodes. Branching is rare. **Leaves** are usually in terminal clusters but in weak stemmed species (*Calamus* species), these are scattered alternately. In palms, these are often very large with a well developed petiole and persistent sheathing base (clasping the base of the stem). The petiole is often prickly at the margins and the blade is with pinnate (feather palms) or palmate (fan palms) segments. Stem has well marked leaf scars or in some cases petiole whither away leaving leaf bases intact on the stem (*Phoenix*).

Inflorescence and Flowers

Inflorescence is usually large, simple or compound spadix enclosed in one or more spathes, which splits longitudinally and panicles or spicates grow out exposing the flowers. It is simple in *Borassus* or richly branched, racemose in *Phoenix*. Inflorescence may appear either amongst, *i.e.*, interfoliar (*Borassus*) or below the leaves, *i.e.*, intrafoliar (*Areca*), rarely terminal. Terminal inflorescences are found in **monocarpic palms**, *e.g. Plectocomia, Corypha* and *Metroxylon*.

Flowers are unisexual, trimerous, actinomorphic, sessile, hypogynous and arranged in close spirals. Bracteoles are present and often connate below the flowers. Plants are monoecious (*Areca*) or dioecious (*Phoenix, Borassus, Lodoicea*). **Perianth** is represented by six small tepals arranged in two alternate and similar series of three each. These are somewhat leathery or fleshy, persistent, free or connate. Outer whorl shows imbricate or valvate aestivation but inner whorl is valvate in male and imbricate in female flowers.

Androecium consists of six stamens arranged in two series of three each, opposite the tepals. In *Nypa*, only 3 and in *Caryota*, several stamens are found. Anthers are dithecous, introrse, versatile and dehiscing by longitudinal slits.

Gynoecium is tricarpellary and syncarpous, rarely partly (*Nypa*) or completely (*Phoenix*) free. Ovary is superior, trilocular with one or two ovules in each locule set in axile placentation. In free carpel condition, each carpel is unilocular with a single ovule. Stigmas are usually three and sessile. Styles are absent or very short.

Fruits and Seeds

Fruit is berry or drupe with 1-3 seeds. The drupe develops from one of the three carpels, other two carpels degenerate. **Seeds** are endospermic. Endosperm is often large, sometimes ruminate (*Areca*) or hard (*Phoenix*).

Pollination and Dispersal

Flowers are mostly wind pollinated but a few are pollinated by insects. Fruits are commonly dispersed by water or animals.

Floral Formulae

Male flower $Br \oplus ♂ P_{3+3} A_{3+3} G_0$

Female flower $Br \oplus ♀ P_{3+3} A_0 \underline{G}_{(3)}$

Economic Importance

Some plants are of immense economic value.

Cocos nucifera L. (coconut palm, nariyal) is a man's most useful tree. Since long it has been popularly called **Tree of heaven** or **Kalpa Vriksha**. Every part of this plant is of great use. Coconut water (the **liquid endosperm**) is a nutritious refreshing drink. It is recommended for oral rehydration. It is also used in the tissue culture medium. The **solid endosperm** (*meat*) is eaten raw or cooked. When dried, it forms *copra* which on extraction yields edible oil (**coconut oil**). The oil is also used for applying on hair, in shampoos and for making soaps and margarine. The fibrous husk (*coir*, the **mesocarp**) is used for making a strong rope. *Coir* is also used for making mats, rugs, filters and for stuffing. India is world's largest producer of coir. The **young inflorescences** on tapping yield a sugary sap which may be boiled to prepare **jaggery** (a coarse sugar) or fermented to prepare *alcoholic beverages*. **Stem** is an ideal material for construction and furniture making. **Leaves** are commonly used for thatching and for making baskets and hats. **Coconut shells** serve as ladles or receptacles for collecting rubber or for preparing activated charcoal. Powdered shell is used as filters in plastics. The shell serves as a resonant baking for musical instruments and also as hubble-bubble (hookah).

Borassus flabellifer L. (palmyra palm, tar) is also an important plant. It's inflorescence is tapped for the sap which is consumed as a beverage or fermented into a palm wine or made into molasses. Young seedlings and ripen fruits are edible. The timber of the plant is valuable for making bows, joints, pillars, posts, rafters etc.

Nypa fruticans Wurmb. (nipa palm, golphal) grows in swampy mud in brackish water of Ganges Delta and in Andamans. The palm is much valued for the sweet sap obtained from the inflorescence and used for making jaggery, sugar, alcohol and venegar. Leaves are used as thatch and also for making coarse mats, baskets and bags. Young leaflets are used for wrapping cigarettes. Tender stem, young peduncles and unripe fruits are eaten raw or cooked. Burnt root ash is used in toothache and a lotion made from leaves is used in the treatment of ulcers.

Areca catechu L. (betel-nut palm, supari) is grown commercially in South India, West Bengal, Assam etc. Its nuts (seeds) are used chiefly as a masticatory and chewed along with paan (leaves of *Piper betle*) or used in paan masala. Powdered nut is useful in checking diarrhoea and urinary disorders. Husks are used for making hard-board and brown wrapping paper. An alkaloid, **arecoline** is used in ophthalmic treatment. Areca nut has prominent place in religious ceremonies. Nuts are also converted into fancy shaped beads and worn as necklaces.

Calamus rotang L. (ratan, bet), *C. andamanicus* Kurz., *C. tenuis* Roxb., *C. latifolius* Roxb. (East India cane) and other species are highly valued for their stem for preparing walking-sticks, baskets, cane furniture etc. Bamboos are often called **poor man's timber** for its utility in commercial purposes for earnings and for their dwellings.

Elaeis guineensis Jacq. (oil palm) produces two distinct oils, *palm oil* and *palm kernel-oil*. Source of palm oil is fleshy mesocarp while palm kernel-oil is obtained from the endosperm (kernel). Palm oil is largely used as edible oil. The palm kernel-oil is also edible, but is used chiefly in the manufacture of detergents and toilet soaps. Male inflorescences on tapping yield sap which is utilised for making alcoholic drinks. Leaves are used for thatching.

Caryota urens L. (toddy palm, wine palm) yields leaf fibre called **kittul fibres** used for making ropes, baskets, soft brushes and brooms. Inflorescence sap is used for preparing toddy or sugar. Stem pith yields sago or kittul flour which is used in making cakes.

Metroxylon rumphii Mart. (sago palm, sabudana) yields **sago** from stem pith which is chiefly eaten during convalescence or in fasting. Sago is also used for sizing and finishing cotton yarns.

Several palms are grown for ornamental purposes, *e.g. Caryota urens* L., *Livistona chinensis* R. Br. (fountain palm), *Oreodoxa regia* H.B. and K. syn. *Roystonea regia* O.F. Cook (royal palm, bottle palm), *Chamaerops humilis* L. (european fan palm) etc.

Systematics and Phylogeny

Engler and Prantl and Hutchinson kept the family in a monophyletic order **Palmales**, whereas most other taxonomists (Cronquist, Thorne, Takhtajan, APG II) placed the family in the monotypic order **Arecales**. However, Bentham and Hooker placed the family *Palmae (Arecaceae)* with *Flagellariaceae* and *Juncaceae* in the series **Calycinae**. Rendle included them in series **Spadiciflorae** together with *Aroideae* and *Lemnaceae*. It appears that Palms have been studied inadequately.

Arecaceae are closely related to *Araceae* in features like unisexual inconspicuous flowers borne in spadix, but Hutchinson, Cronquist, Thorne, APG II placed **Arales** far apart owing to their widely distinct morphological features. The presence of 2-trimerous whorls of perianth, six stamens and superior trilocular ovary bring *Arecaceae* close to *Liliaceae* and more so to *Agavaceae* (of Hutchinson) which have woody plants like *Yucca* and *Dracaena*. APG groupings placed it in the monophyletic familial group **Commelinoid Monocots** (III APG Group).

ARACEAE (Arum Family)

Bentham & Hooker	Hutchinson	Cronquist	APG
Monocotyledons	Monocotyledones	Magnoliophyta	III. APG Group :
Nudiflorae	Corolliferae	Liliopsida	Monocots
	Arales	Arecidae	Non-Commelinoid
		Arales	Alismatales

Family comprises **104 genera** and **3040 species** found globally but mainly in tropical and sub-tropical regions of the world, very common in wetlands and tropical forests. In India, some **24 genera** and **136 species** are found with familiar examples like *Colocasia, Amorphophallus, Monstera, Pistia, Arisaema, Pothos* etc. The large genera of the family are *Anthurium* (900 species), *Philodendron* (500 species) and *Arisaema* (150 species).

Vegetative Characters

Plants are mostly perennial and tuberous herbs (*Arisaema, Arum*), some are climbing shrubs rooting on trees (*Pothos*) or true epiphytes (*Anthurium*) or aquatic (*Pistia*) or marshy herbs (*Acorus* and *Lasia*). **Roots** are adventitious, may be climbing (*Pothos*) and absorbents. The **aerial roots** of some aroids develop velamen like tissue. **Stem** is underground and develops to form root stocks, tubers or corms (*Arisaema, Arum, Colocasia*) or aerial, monopodially (*Pothos*) or sympodially branched. It becomes woody in some forms. **Leaves** are basal (*Arisaema*) or cauline and alternate (*Pothos*) and simple but in *Arisaema*, these are compound. Leaves are very large and simple in *Alocasia* and *Xanthosoma*. Leaves are petiolate, often large, generally net-veined, with membraneous sheathing bases. Decompound leaves are found in *Amorphophallus* and *Drocontium*. Plants produce watery or acrid or milky latex.

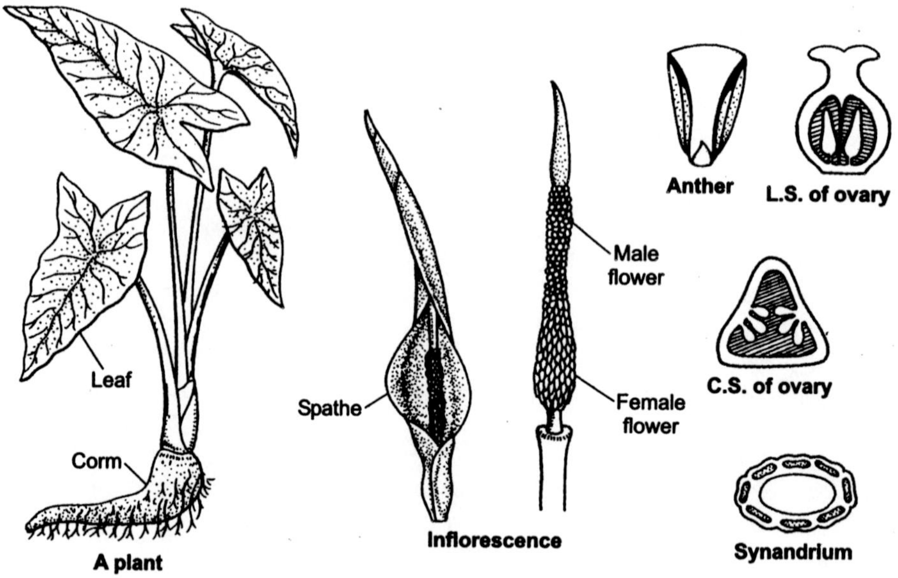

Fig. 22.3 *Alocasia formicata* Schott.

Inflorescence and Flowers

Inflorescence is a spadix, the subtending spathes usually enclose the spike, sometimes in the primary stage only. Spadix has male flowers at the top and female towards the base. Inflorescences of *Amorphophallus bulbifer* (devil's spitoon) coming out of the ground look like hoods of snakes, atleast from a distance. In *Arisaema* (snake or cobra plant), the spathe is greenish purple in colour and it expands over the spadix like the hood of the cobra. These are the good examples of **mimicking**.

Flowers are inconspicuous (small), ebracteate, regular, hypogynous or epigynous, bisexual or unisexual and monoecious. The axis of the spike may be wholly covered by flowers or the apical part is naked and is called appendage. Perianth is absent in unisexual flowers but present in most bisexual flowers in 2 trimerous or dimerous whorls. Tepals are free or united. In *Anadendron*, tepals unite to form cup.

Androecium consists of usually 2, 3 (in one whorl) or 6 stamens (in two whorls). Filaments are basally free or connate (*Arisaema*) forming a synandrium. Anthers are usually dithecous, introrse and opening by apical pores. Pollen grains are often in tetrad. Staminodes may be present in female flowers.

Gynoecium consists of one (*Arum, Pistia*) or more united carpels (syncarpous). Ovary is superior, one to many-locular with few to many ovules in each locule. The placentation may be axile (*Pothos*), parietal (*Arum*), basal (*Arisaema*) or apical (*Lasia*). Style is absent and stigma may be one or more.

Fruits and Seeds

Fruit is a berry and **seeds** are with or without endosperm.

Pollination and Dispersal

Flowers are protogynous and entomophilous. Insects are attracted by coloured spathe and the strong offensive odour emitted by the flowers (inflorescence). The spathe is often constricted above the female flowers from where it closes in the night to arrest the insects. The insects pollinate the female flowers while moving inside the spathial chamber. Some species of *Alocasia* and *Amorphophallus* are pollinated by snails. Fruits are dispersed by birds.

Floral Formula $\oplus$ ⚥ $P_{3+3}\, A_{3+3}\, \underline{G}_{1 \text{ or } (3)}$

Economic Importance

Some plants are economically useful.

Edibles. Tuberous rhizomes and corms of some plants are starch rich and are edible after boiling, *e.g. Alocasia indica* (Roxb.) Schott. (man-kanda or man kachoo), *Colocasia esculenta* (L.) Schott (arvi, ghuiyan), *Amorphophallus companulatus* Blume (elephant-foot, yam, zimikand) etc. Leaves and petioles of *Xanthosoma maximiliani* Schott., *Colocasia esculenta* (L.) Schott. syn. *C. antiquorum* Schott. and *Remusatia vivipara* Schott. are used as vegetables.

Medicinal Plants. Several plants of the family are medicinally valuable. Stem juice of *Alocasia macrorrhiza* Schott. is used to relieve pain from scorpion-sting and nettle-sting. Corm of *Amorphophallus campanulatus* Blume is used for the treatment of piles and dysentery and that of *Acorus calamus* L. (Sweet flag rhizome, bach) in curing dyspepsia and chronic diarrhoea. Dried fruits of *Scindapsus officinalis* Schott. are stimulant, carminative and anthelmintic.

Ornamental plants. A large number of plants are grown ornamentally for their attractive foliage and variously coloured spathes, *e.g. Monstera deliciosa* Liebm. (for perforated large leaves, climbers), *Caladium bicolor* Vent., *C. picturatum* Koch., *Dieffenbachia picta* Schott. (variegated large leaves), *Arisaema atrorubens* Blume (Indian turnip–curious spathes), *Scindapsus officinalis* Schott. (ivy arum) (maculate foliage), *Pothos aureus* Hort. (money plant–foliage) etc.

Systematics and Phylogeny

The *Araceae* are highly evolved family due to the presence of unisexual naked flowers borne in spadix. Bentham and Hooker included the family in the series **Nudiflorae** and Engler and Prantl under **Spathiflorae**. Cronquist, Takhtajan, Dahlgren, Thorne etc. have placed the family in its own order **Arales.** However, APG II/AP web in monophyletic affiliated grouping placed the family in the order **Alismatales** in **Non-Commelinoid Monocots** of **III. APG Group.**

Araceae show affinities with *Palmae* in the features of inflorescence and inconspicuous hypogynous regular flowers. Hutchinson believed that these two families are allied to each other both having been originated from **Liliaceous stock** in different lines. Closeness between the two families has also been supported by Cronquist and Takhtajan. APG II and APweb placed the family *Araceae* along with several other families under the order **Alismatales,** and merged *Lemnaceae* with *Araceae.*

COMMELINACEAE (Spiderwort Family)

Bentham & Hooker	Hutchinson	Cronquist	APG
Monocotyledons	Monocotyledones	Magnoliophyta	III. APG Group :
Coronarieae	Calyciferae	Liliopsida	Monocots
	Commelinales	Commelinidae	Commelinoid
		Commelinales	Commelinales

The family comprises **40 genera** and **650 species** distributed in tropical and subtropical regions of the world. Some **10 genera** and **76 species** are found in India with common forms like *Commelina, Tradescantia, Murdannia, Aneilema, Cyanotis, Rhoeo, Zebrina* etc. The large genera of the family are *Commelina* (170 species), *Tradescantia* (70 species) and *Aneilema* (65 species).

Vegetative Characters

Plants are annual or perennial herbs with more or less succulent and jointed stem, occasionally twinning (*Zebrina*). **Leaves** are alternate, entire, ovate-lanceolate or linear, usually with a closed basal membranous sheath.

Inflorescence and Flowers

Inflorescence is usually a helicoid cyme (monochasial cyme) or paniculate or cymose cluster, arising in the axil of leaf (*Tradescantia*) or of a spathe-like boat-shaped bract (*Commelina*) or terminal. In *Commelina benghalensis*, the subterranean flowers are *cleistogamous* while aerial flowers are *chasmogamous*.

Flowers are usually bisexual, regular or often medianely zygomorphic, trimerous and hypogynous. **Perianth** consists of 6 tepals, set in two whorls of three each. Outer whorl is distinctly sepaloid and the inner petaloid, white, blue or pinkish, equal or unequal and free or rarely united. Rarely one tepal (or petal) is reduced to scale-like structure. Aestivation is imbricate.

Androecium consists basically of 6 stamens, arranged in two alternate and trimerous whorls opposite the perianth segments. In *Tradescantia*, all six stamens are fertile, in *Floscopa* 5, in *Pollia* and *Commelina* 3 and in *Aneilema* only two stamens are fertile, and others are either represented by staminodes or absent. Filaments are often bearded with moniliform brightly coloured hairs. Anthers are basifixed, bilocular, loculi parallel and usually contiguous, opening by longitudinal slits or by apical pores.

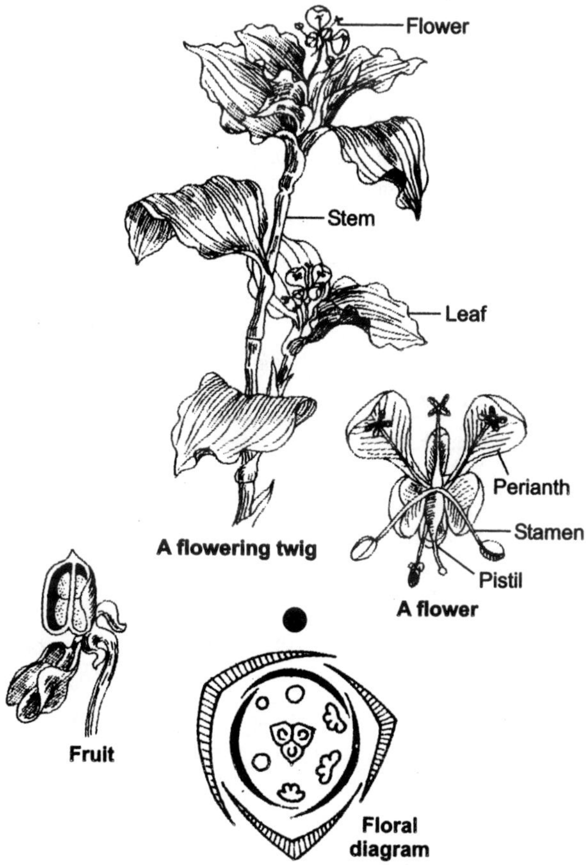

Fig. 22.4 *Commelina benghalensis* L.

Gynoecium consists of three united carpels (syncarpous). Ovary is superior, tri- or bilocular with few or one orthotropous ovule in each locule set in axile placentation. Style is one and stigma is capitate or trilobed.

Fruits and Seeds

Fruit is usually a loculicidal capsule or rarely a nut. **Seeds** are mostly muricate, ridged or reticulate and with copious mealy endosperm, rarely arillate. Embryo is small and located beneath a disc-like callosity (**embryostega**) on the seed coat.

Pollination and Dispersal

Flowers are entomophilous (insect pollinated) but for subterranean cleistogamous self pollinated flowers. Fruits are dispersed by birds and animals.

Floral Formula $\oplus$ or % ⚥ $P_{3+3}\ A_{3+3 \text{ or } (3)}\ \underline{G}_{(3)}$

Economic Importance

While family is of less economic value, a few plants are important.

Medicinal plants. *Aneilema scapiforum* is used as a substitute of *Curculigo orchioides* Gaert., the Kalimusli or tallura of Indian Kavirajes. Plants of *Commelina benghalensis* L. and several species of *Aneilema* are used in leprosy and leucoderma. Bruised plants of *Commelina nudiflora* L. relieve pain in burns and boils. Roots of *Cyanotis axillaris* Roem. & Schultz f. (soltraj) are used as remedy in dropsy and tympanites, and also to expel worms in cattles. Leaves of *C. benghalensis* L. are laxative and their juice neutralises the poisoning effect due to the touch of hairs of caterpillars.

Ornamental plants. Several plants are grown in gardens for their showy foliage and inflorescence, *e.g. Rhoeo discolor* Hance, *Zebrina pendula* Schnizl. (wandering jew), *Tradescantia virginiana* L. (spiderwort), *T. bracteata* Small. etc.

Edibles. Leaves and rhizomes of *C. benghalensis* L. and *C. diffusa* Burm. are sometimes used as vegetables. Fruits of *Curculigo latifolia* Dry. are edible.

Systematics and Phylogeny

Bentham and Hooker placed the family in **Coronarieae** but Hutchinson developed a new order **Commelinales** to include the family *Commelinaceae*. Order **Commelinales** has been supported by Cronquist, Takhtajan, Dahlgren, Thorne etc. and also by APG II/AP web groupings. APG grouped it in **Commelinoid : Monocots.** *Commelinaceae* are allied to *Pontederiaceae* in presence of mealy endosperm and ridged seeds, and also to *Flagellariaceae* and *Mayacaceae*.

CYPERACEAE (Sedge Family)

Bentham & Hooker	Hutchinson	Cronquist	APG
Monocotyledons	Monocotyledones	Magnoliophyta	III. APG Group :
Glumaceae	Glumiflorae	Liliopsida	Monocots
	Cyperales	Arecidae	Commelinoids
		Cyperales	Poales

Family is represented by **104 genera** and **4950 species** widely distributed, preferably in marshes and damp places throughout the world. In India, about **22 genera** and **450 species** are found in most parts of the country with common forms like species of *Cyperus, Carex, Scirpus, Fimbristylis, Eriophorum, Eleocharis* etc. Large genera of the family are *Carex* (1800 species), *Cyperus* (580 species) and *Fimbristylis* (290 species).

Vegetative Characters

Plants are perennial, rarely annual, grass-like herbs with sympodial rhizome and a solid triquetrous, usually unbranched stem (culm). These contain silica bodies. **Leaves** are usually in a basal tuft or crowded on the lower part of the culm. These are linear, long and tristichous (three ranked) and eligulate. Leaf sheath is closed.

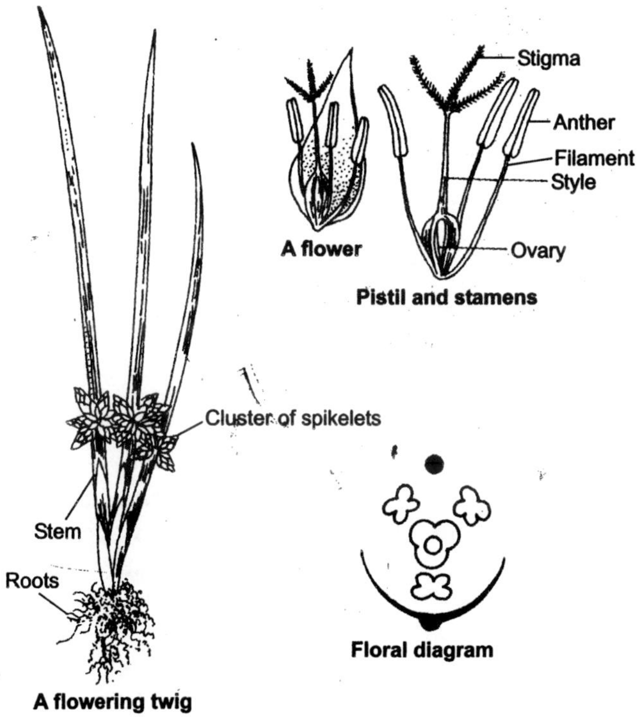

Fig. 22.5 *Scirpus articulatus* L.

Inflorescence and Flowers

Inflorescence is a spike or head of spikelets or panicles, often subtended by. one or more leafy involucral bracts. Spikelets have several flowers but female spikelets of *Carex* and *Diplacrum* have single flower.

Flowers are borne singly in the axils of subtending imbricating bracts. These are minute, usually bisexual or rarely unisexual (*Carex, Elyna*), mostly trimerous and hypogynous. In *Elyna,* the secondary axis bears male flowers above the female flowers (a monoecious plant), and in *Carex* and *Kobresia,* the bract completely enclose the naked ovary and is called *utricle.* **Perianth** is often absent (*Carex* and *Cyperus*) or represented by 3 or 6 scales or bristles (*Scirpus* and *Eleocharis*), or in two trimerous whorls (*Oreobolus*).

Androecium consists of mostly three stamens in one whorl but in *Rhynchospora,* these are six in two whorls. In *Hemicarpha* and *Bisboeckelera,* it is represented by a single stamen. Filaments are free and slender. Anthers are basifixed, dithecous and introrse, dehiscing by vertical slits.

Gynoecium consists of 2 or 3 united carpels (syncarpous). Ovary is superior, unilocular with one basal anatropous ovule. Style is one with three or two plumose or teeth-like stigma.

Fruits and Seeds

Fruit is achene, nut or utricle. The achene is trigonous or biconvex having a beak-like crown of the persistent base of the style. **Seeds** are albuminous with a small embryo surrounded by mealy endosperm.

Pollination and Dispersal

Flowers are usually wind pollinated. Plants commonly multiply by vegetative means.

Floral Formulae

Male flower Br % ♂ $P_0\ A_3\ G_0$

Female flower Br % ♀ $P_0\ A_0\ \underline{G}_{(3)}$

Bisexual flower Br % ⚥ $P_0\ A_{3\ or\ 3+3}\ \underline{G}_{(2-3)}$

Economic Importance

Family is of little economic value.

The tubers (rhizome) of *Cyperus esculentus* L. (kasaru) and *Eleocharis tuberosa* Schult. are edible; of *C. stoloniferous* L. are cardiac stimulant; of *C. articulatus* L., *C. longus* L. and *C. iria* L. are carminative, stimulant and used as tonic; and those of *C. rotundus* L. (motha) are used to treat insomnia. The rhizomes of *C. scariosus* R. Br. (nagar motha) and *C. rotundus* L. yield an oil used by perfumers as fixative.

The pith of papyrus plant (*Cyperus papyrus* L.) was in use by Egyptians for making the papers. Split stems of *C. articulatus* L., *C. iria* L., *C. malaccensis* Lamk., *C. corymbosa* Rottb., *Juncellus inundatus* Clarke etc. are woven into mats. The perianth bristles of *Eriophorum* (cotton sedge) are used for stuffing cushions and pillows.

Some plants are grown for their attractive structure, *e.g.* *Cyperus alternifolius* L. and *C. flabelliformis* Rott. (umbrella plant).

Systematics and Phylogeny

Cyperaceae closely resemble *Gramineae* (*Poaceae*), and hence several taxonomists included both families under **Glumiferae**. Both resemble in habit and general appearance but differ remarkably in features of flowers, fruits and seeds and in mode of germination. Spikelets of *Cyperaceae* are not at all homologous with those of *Gramineae* as these are widely arranged in *Cyperaceae* and fairly constant in *Gramineae*. Florets of *Cyperaceae* are axillary in position while these are terminal in *Gramineae*. A definite perianth can somewhat be demonstrated in *Cyperaceae* but the fruits with two or three carpels are less specialised than caryopsis of *Gramineae*. Basal placentation in *Cyperaceae* is believed to have originated from the free central type whereas in *Gramineae*, it has originated from the parietal condition. Ovules are mainly anatropous in *Cyperaceae* but orthotropous in *Gramineae*. In *Cyperaceae*, the terminal portion of cotyledon comes out of the seed whereas other portion remains in the seed to function as a sucker but in *Gramineae*, neither the cotyledon nor any other part, comes out of the seed.

These differences led taxonomists to place the two families in the separate orders. Hutchinson did not recognise any phyletic relationship between the two families and placed them in two distinct monotypic orders, the **Cyperales** and the **Graminales** in the division **Glumiflorae** and traced their origin from liliaceous ancestors via *Juncaceae* complex, but from distinct stocks on parallel lines. Wettstein and Takhtajan have supported Hutchinson.

Bremer (2002) strongly supported that *Thurniaceae*, *Juncaceae* and *Cyperaceae* are sister families of cyperid clade. APG Grouping/APweb placed both *Cyperaceae* and *Poaceae* (*Gramineae*) in the third order of **Commelinoid monocots**, the **Poales** alongwith 14 other families including *Centrolepidaceae*, *Juncaceae*, *Flagellariaceae*, *Typhaceae* etc.

POACEAE (GRAMINEAE) (Grass Family)

Bentham & Hooker	Hutchinson	Cronquist	APG
Monocotyledons	Monocotyledones	Magnoliophyta	III. APG Group :
Glumaceae	Glumiflorae	Liliopsida	Monocots
		Arecidae	Commelinoids
		Cyperales	Poales

It is a fairly large family comprising **656 genera** and **9975 species** with worldwide distribution, found in almost all conditions. In India, some **239 genera** and **850 species** are found, the common examples of wild forms are species of *Cynodon, Chrysopogon, Eragrostis, Setaria, Sporobolus, Vetiveria, Bambusa, Arundinaria, Dendrocalamus* etc. and of cultivated forms are cereals, millets, sugarcane etc. Large genera of the family are *Poa* (500 species), *Panicum* (450 species) and *Festuca* (430 species). Bamboos are arborescent grasses with about 30 genera and 550 species.

Vegetative Characters

Plants are annual, biennial or perennial herbs. Some bamboos are very tall attaining the size of tree and some species have life span upto 120 years. **Stem** is herbaceous, erect, prostrate or creeping. The aerial stems terminated by inflorescence, are known as culms (haulms). Nodes and internodes are distinct. Internodes are often hollow, rarely solid (*Zea, Saccharum*). Plants are often stoloniferous or may have suckers. Plants exhibit intercalary growth which is pronounced in bamboos. **Leaves** are distichous, simple, alternate and mostly with parallel venation. Each leaf usually consists of two parts, the *sheath* and the *blade* (lamina). The sheath often encircles the stem. At the junction of lamina and sheath, a membranous or hairy outgrowth, called **ligule**, is found. Leaves have silica cells. In *Echinochloa*, ligule is absent.

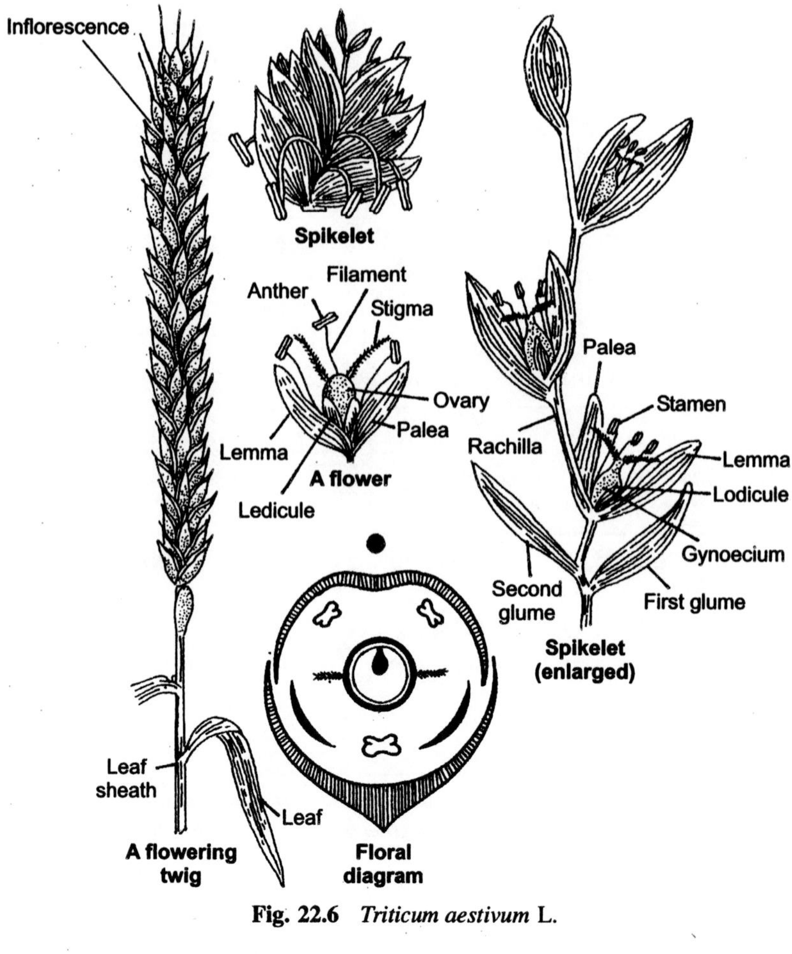

Fig. 22.6 *Triticum aestivum* L.

Inflorescence and Flowers

Inflorescence is of complex type. Its basic unit is a spikelet, and not a flower. Spikelets are arranged on the main axis in various ways forming a compound structure. Inflorescence may be *spike* of *spikelets* (*Triticum*) or *panicle of spikelets* (*Avena*). Each spikelet consists of a very short or minute axis (**rachilla**) which bears two opposite rows of bracts found close, one above the other. The two (sometimes 1-6) lowest bracts are sterile and are referred to as **sterile** or **empty glumes.** Sterile glumes are followed by a variable number (1-50) of **fertile glumes** or **lemma**. Lemma are often greenish, keeled or rounded and awned or awnless. A membranous structure present between the fertile glumes and rachilla is called **palea**, which represents bracteole below the flower. The lemma and palea bear a floret in their axil.

Florets (flowers) are small, inconspicuous, bisexual, rarely unisexual-monoecious (*Zea mays*) or dioecious, zygomorphic and hypogynous. **Perianth** is highly modified and reduced. It consists of two minute, fleshy, hyaline structures, called **lodicules.** Lodicules are found present anterio-laterally.

Androecium consists of three stamens in a whorl. However, in *Oryza* and *Bambusa*, there are two alternate trimerous whorls of stamens. In *Uniola,* a single stamen and in *Pariana,* 10 to 40 stamens are found. *Ochlandra* has 6 to 30 stamens which may be monadelphous or polyadelphous. Filaments are free and delicate and anthers are basifixed or versatile, dithecous and introrse dehiscing by longitudinal slits.

Gynoecium consists of a single carpel (monocarpellary). Ovary is superior and unilocular with a single ovule, adnate to the adaxial side (of the ovary). Placentation is basal. Styles are usually two. In bamboos, 3 styles are present while in *Nardus,* there is only one style with a simple stigma. In *Zea,* two simple styles unite to form a single silky thread-like persistent structure (the longest style in plant kingdom). Stigma are usually plumose.

Fruits and Seeds

Fruit is one-seeded **caryopsis** which is characteristic to the family. Rarely nut or berry may be found (some species of *Bambusa*). **Seeds** are endospermic. These have plenty of endosperm and the shield-shaped scutellum lying appressed. Radicle and plumule remain covered by sheaths, called *coleorrhiza* and *coleoptile,* respectively.

Pollination and Dispersal

Flowers are usually chasmogamous and wind pollinated, rarely cleistogamous and self pollinated. Seeds (grains) are dispersed by animals or wind.

Floral Formula $\quad Br \ \% \ ⚥ \ P_2 \ A_3 \ \underline{G}_1$

Economic Importance

Poaceae are economically most important family of the plant kingdom.

Edibles. Highly valued edible plants are commonly called cereals and millets etc. **Cereals** include *Triticum aestivum* L. (bread wheat), *Oryza sativa* L. (rice), *Hordeum vulgare* L. (barley), *Zea mays* L. (maize), *Avena sativa* L. (oat), *Secale cereale* L. (rye) etc. **Millets** are small grained cereals and include *Sorghum vulgare* Pers. (jowar), *Pennisetum typhoides* (Burm. f.) Stapf. & Hubbard (bajra, pearl-millet) and *Eleusine coracana* Gaertn. (ragi, finger millet). Rice is the highest cultivated crop of India and wheat is of the world.

Sugar and Alcohol. The first source of commercial sugar is the stem of sugarcane (*Saccharum officinarum* L.) hence the sugar is popularly called cane sugar or table sugar. Molasses, a by product formed during sugar manufacturing process, is used for the preparation of alcohol. Inferior quality of sugar is obtained from *Sorghum vulgare* Pers. var. *saccharatum* Koen. Some indigenous alcohol is also prepared from fermented rice (*Oryza sativa* L.)

Essential Oils. Some plants are source of valuable essential oil, *e.g.* leaves of *Cymbopogon citratus* Stapf. (lemon grass) are the source of lemon-grass oil, leaves of *C. nardus* Rendle are the source of citronella oil and roots of *Vetiveria zizanoides* Nash. are the source of vetiver oil or khas oil. Palmarosa oil is obtained from *Cymbopogon martinii* (Roxb.) Stapf var. *motia* and ginger grass oil from its *sofia* variety.

Paper and Timber. Culms of bamboos are used in the manufacture of paper and also as timber. Bamboos are commonly called **poor man's timber** as these are used in various constructions and in preparing commercial articles for earning livelihood, *e.g.* beds, tool handles, walking stickes, furnitures, mats, toys and novelty articles. Common bamboos found in India are *Dendrocalamus giganteus* Munro., *D. strictus* (Roxb.) Nees (for manufacture of papers), *Arundinaria racemosa* Munro., *Arundo donax* L., *Bambusa arundinacea* (Retz.) Willd., *B. polymorpha* Munro., *B. tulda* Roxb. etc.

Fibres. Several grasses are source of hard fibres (leaf fibres), *e.g.* *Erianthus munja* (Roxb.) Jesweit syn. *Saccharum munja* Roxb. (munj) and *S. spontaneum* L. (kans) largely used for making ropes or cordages.

Fodder. Important fodder grasses for domestic animals are *Cynodon dactylon* Pers., *Paspalum scrobiculatum* L., *Phalaris aquatica* Ait., *P. canariensis* L., *Cenchrus ciliaris* L., *Stipa orientalis* Trin., *Chrysopogon montanus* Duthie, *Sporobolus diander* Beauv., *Poa annua* L., *P. pratensis* L., *Dactylis glomerata* L. etc.

Systematics and Phylogeny

Poaceae resemble *Cyperaceae* in habit, minute flowers subtended by scaley bracts, spikelets with compound inflorescence. Engler placed both families in the same order, **Glumiflorae**, and proposed that lodicules of *Poaceae* are bracteoles and all other scales below the lodicules are bracts. Rendle supported his concept.

Hutchinson and Wettstein considered that *Poaceae* have been derived from *Restionaceae* of **Juncales** evolved from *Liliaceous stock*. They believed that flower of *Poaceae* is a reduction from a pentacyclic flower, where lodicules represent one of the two whorls of perianth, one whorl of stamens is suppressed and the ovary is tricarpellary but with abortive carpels. *Poaceae* differ considerably from *Cyperaceae* in features of flowers, fruits and seeds (see also *Cyperaceae*) led Hutchinson to put these two families in two different orders, **Graminales** and **Cyperales.** Takhtajan also placed them in two different orders.

Cronquist considered that *Poaceae* and *Cyperaceae* have originated from a *Commelinaceous stock* (not from *Liliaceous stock*), and two families are so closely related to be included in one order, *i.e.*, **Cyperales**. Only these two families have been placed in this order. APG grouping also placed both these families along with fourteen others including (*Centrolepidaceae, Typhaceae* etc.) in the order **Poales** of **Commelinoid Monocots**, a monophyletic familial group.

Table 22.1 Differences between Poaceae and Cyperaceae

	Poaceae (Gramineae)	Cyperaceae
1.	**Stem** is cylindrical, hollow with distinct nodes.	**Stem** is solid, triquetrous and without distinct nodes.
2.	**Leaves** are distichous, ligulate and sheath split on one side.	**Leaves** are tristichous, eligulate and sheath closed.
3.	**Bracts** (glumes, lemma, palea) are present but no involucral bracts.	Single floral **bract**, and leafy involucral bracts are present.
4.	**Perianth** is represented by 2 or 3 lodicules.	**Perianth** may be represented by hairs or bristles or may be absent.
5.	**Anthers** are versatile.	**Anthers** are basifixed.
6.	**Gynoecium** consists of 3 carpels of which 2 are abortive.	**Gynoecium** consists of three united carpels (syncarpous).
7.	**Fruit** is usually a caryopsis.	**Fruit** is achene, nut or utricle.
8.	**Embryo** lies outside the endosperm. Coleoptile and coleorrhiza are present.	**Embryo** lies embedded in the endosperm. Coleoptile and coleorrhiza are not found.
9.	During **germination**, no part of the cotyledon comes out.	During **germination**, terminal part of the cotyledon comes out.
10.	**Floral formula** Br % ⚥ $P_2\ A_3\ \underline{G}_1$	**Floral formula** Br % ⚥ $P_0\ A_{3\ or\ 3+3}\ \underline{G}_{(2-3)}$

MUSACEAE (Banana Family)

Bentham & Hooker	Hutchinson	Cronquist	APG
Monocotyledons	Monocotyledones	Magnoliophyta	III. APG Group :
Epigynae	Calyciferae	Liliopsida	Monocots
Scitamineae	Zingiberales	Zingiberidae	Commelinoids
(syn. Musaceae)		Zingiberales	Zingiberales

It is a small family comprising only **3 genera** and **40 species** found in tropical Africa, Asia and Australia. In India, **only genus** *Musa* with its **10 species** are found growing in Assam, Western Ghats, South India and Himalayas. *Musa* has 33 species, *Ensete,* 6 species and *Musella,* one species.

Vegetative Characters

Plants are perennial gigantic herbs, often tree-like in appearance and these persist by means of an underground rhizome and multiply by suckers. The aerial stem-like structure (pseudostem) is formed by inbricate, sheathing and stiff bases of the leaves. **Leaves** are large, alternate, entire or lacerated (due to mechanical injury by wind), distichous or spirally arranged with unicostate parallel venation.

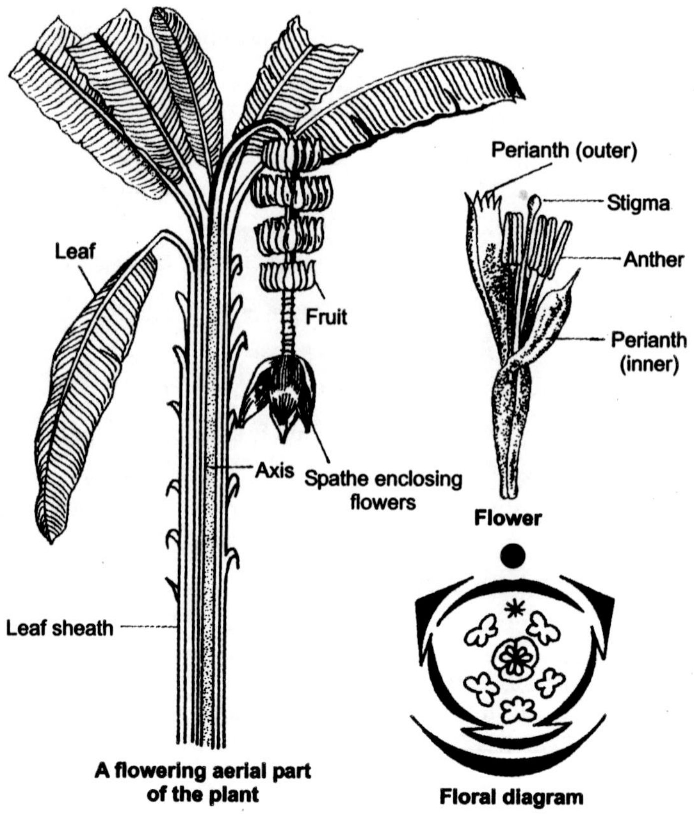

Fig. 22.7 *Musa sapientum* L.

Inflorescence and Flowers

Flowers are arranged in spike or panicle, subtended often by brilliantly coloured spathaceous bracts. The bracts fall off as the flowers open.

Flowers are bisexual (polygamous in *Musa*), slightly zygomorphic, trimerous and epigynous. Perianth consists of 6 segments (tepals) set in two trimerous whorls. These are petaloid and unequal in size and shape. The three outer tepals and two inner anterior tepals are united into a tube, the sixth tepal remaining free.

Androecium consists of 6 stamens in two whorls of three each, and opposite the tepals. Fertile stamens are 5, the posterior one is often represented by a staminode or absent (*Ensete*). Anthers are linear, dithecous and introrse, dehiscing by vertical slits.

Gynoecium is tricarpellary and syncarpous. Ovary is inferior and trilocular with numerous ovules set in axile placentation. Style is filiform and stigma is capitate or lobed.

Fruits and Seeds

Fruit is a fleshy berry. Most cultivated forms of *Musa* are seedless due to triploidy. **Seeds** are endospermic with thick and hard testa.

Pollination and Dispersal

Flowers are pollinated by insects. Seeds are dispersed by animals. Cultivated varieties of banana are propagated by rhizomes.

Floral Formula $\quad Br \ \% \ ⚥ \ P_{(3+2)+1} \ A_{3+2} \ \overline{G}_{(3)}$

Economic Importance

Some plants are valuable.

Musa paradisiaca L. (syn. *M. sapientum* L.) (banana, kela) is an important economic plant. It is grown in almost every part of India, but mostly in Assam, Maharashtra, Karnataka, Kerala and Tamil Nadu. Green bananas are eaten as vegetable and are useful in diarrhoea and dysentery. Ripe fruits are useful in diabetes, uraemia, nephritis, gout, hypertension and cardiac diseases. Juice of flowers, stems and roots is used in disorders of blood. Banana fruits are delicious and are eaten with taste. Banana powder is used as a baby food and in the manufacture of chocolates and biscuits. Leaves are commonly used as plates.

Fruits of *M. chinensis* Sweet syn. *M. acuminata* Colla (chinese or dwarf banana) are also sweet and edible.

Abaca fibres (hard fibres), obtained from the outer portion of leaf stalks of *M. textilis* Nees (Manila hemp, abaca), are used for making marine ropes, baggs and wrapping papers.

Systematics and Phylogeny

Bentham and Hooker recognised *Museae* as the tribe of the family *Scitamineae*, but Hutchinson and most others (Cronquist, Takhtajan, Dahlgren, Thorne, APG II etc.) recognised *Musaceae* as a distinct family. The family *Musaceae* is placed in the order **Zingiberales** by Cronquist, Dahlgren, and APG II/APweb. It was placed in **Musales** by Takhtajan and **Cannales** by Thorne who preferred this term over **Zingiberales.** *Musaceae, Cannaceae, Zingiberaceae* and *Marantaceae* are closely related families.

Table 22.2 Differences between Musaceae and Zingiberaceae

	Musaceae	Zingiberaceae
1.	**Leaves** are large with long petiole and prominent sheath. Ligule or pulvinus is absent.	**Leaves** are large or small and petiole and sheath are not very long. Ligule is present.
2.	**Spathe** is usually very conspicuous.	**Spathe** is not present, but flowers are bracteate.
3.	**Inflorescence** is a spadix or cincinus.	**Inflorescence** is raceme or spike.
4.	**Perianth** segments are 6 in 2 whorls, variously united.	**Perianth** segments are 6 in 2 whorls of 3 sepals and 3 united petals.
5.	**Fertile stamens** are 6 or 5. One staminode may be present or absent.	Only the median stamen of the inner whorl is fertile. Other 2 of the same whorl and 2 of outer whorl are modified into petaloid staminodes, and 3rd of outer whorl is suppressed.
6.	**Placentation** is axile or ovule solitary and basal.	**Placentation** axile or parietal.
7.	**Floral formula** Br % ⚥ $P_{(3+2)+1}\ A_{3+2}\ \bar{G}_{(3)}$	**Floral formula** Br brl % ⚥ $P_{(3)+(3)}\ A_1\ \bar{G}_{(3)}$

ZINGIBERACEAE (Ginger Family)

Bentham & Hooker	Hutchinson	Cronquist	APG
Monocotyledons	Monocotyledones	Magnoliophyta	III. APG Group :
Epigynae	Calyciferae	Liliopsida	Monocots
(Scitamineae)	Zingiberales	Zingiberidae	Commelinoids
		Zingiberales	Zingiberales

Family comprises **46 genera** and **1,275 species** distributed in tropical parts of Eastern Hemisphere, particularly in Indo-Malayan area. Some **17 genera** and **112 species** are found in India, mainly in Eastern Himalayas and Western Ghats. Popular cultivated examples are *Zingiber officinale* (ginger), *Curcuma longa* (turmeric), *Elettaria cardamomum* (Malabar cardamon), *Amomum subulatum* (Bengal cardamon) etc. Large genera of the family are *Alpinia* (165 species), *Amomum* (130 species) and *Zingiber* (95 species).

Vegetative Characters

Plants are perennial herbs with fleshy rhizomes. **Stems** are usually bracted, scapose or leafy, short or elongated. Scapose stems of *Curcuma* and *Zingiber* contain etherial oils, terpenes, phenyl propanoid complex etc. **Leaves** are radical or cauline, alternate, sessile or petiolate, distichous (two rows), and with a ligule at the junction of the blade and sheath. Blade is linear to elliptic.

Inflorescence and Flowers

Inflorescence is either terminal on the leafy shoot (as in *Globba*) or on a scale-leaf bearing scape (*Curcuma*) or produced directly from the root stock at the base of the stem (*Amomum*). Flowers are arranged in open raceme or panicle (*Globba*), paniculate (*Elettaria*), compact spike (*Curcuma*), dense heads (*Costus*) or solitary (*Gastrochilus, Monocostus*) subtended by prominent bracts. Inflorescence is surrounded by involucre (*Geanthus*) or without involucre (*Amomum*).

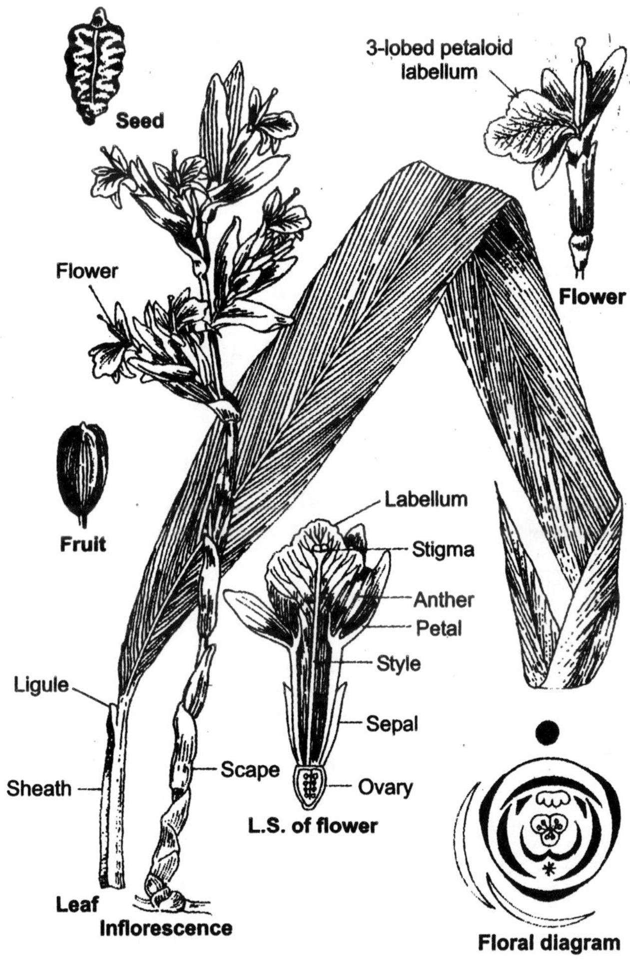

Fig. 22.8 *Elettaria cardamomum* Maton.

Flowers are bracteate, bracts are coloured and distichous or spirally arranged. These are bisexual, zygomorphic, trimerous and epigynous. Rarely flowers are regular or actinomorphic. **Perianth** consists of 6 segments arranged in two trimerous whorls, outer being sepaloid and inner petaloid, thus distinguishable into calyx and corolla. Sepals are united into a tube with valvate or imbricate aestivation. Petals are united below, similar or dissimilar–posterior segment is usually largest and covers the edges of the lateral segments. Petals are often showy and delicate.

Androecium, theoritically, consists of 6 stamens arranged in two whorls. In the outer whorl, the anterior stamen is absent and two lateral stamens are modified into staminodes. In the inner whorl, the posterior (or medium) stamen is fertile and other two are united to form a petaloid *labellum* which embraces the flower and is often most conspicuous part of the flower. Thus practically, androecium contains a single fertile stamen which has a slender, long and deeply grooved filament and a dithecous anther, dehiscing vertically.

Gynoecium is tricarpellary and syncarpous. The inferior ovary is unilocular with parietal or basal placentation (*Globba*) or trilocular with axile placentation. Style is simple and terminal and passes through the groove in between the anther lobes. Stigma is capitate or lobed. A pair of nectar-secreting epigynous glands are often present except in *Costus* where septal nectaries are found.

Fruits and Seeds

Fruit is generally a loculicidal capsule which opens by three valves but in *Amomum*, the fruit is fleshy and indehiscent. **Seeds** are arillate with a smaller endosperm and a larger perisperm. Embryo is straight.

Pollination and Dispersal

Flowers are protandrous and adapted for insect pollination. Fruits are dispersed by animals.

Floral Formula Br brl % ⚥ $P_{(3)+(3)}\ A_1\ \overline{G}_{(3)}$

Economic Importance

Some plants of the family have considerable economic value.

Spices. Important spice yielding plants are *Zingiber officinale* Rose. (ginger, adarak), *Curcuma longa* Koenig non L. syn. *C. domestica* Val. (turmeric, haldi), *C. angustifolia* Roxb. (East Indian arrowroot, tikhur), *C. zedoaria* (Berg.) Rose. (zedoary), *Amomum aromaticum* Roxb. (bari elaichi, Bengal cardamon), *A. subulatum* Roxb. (Indian or Nepal cardamom, bari elaichi), *Elettaria cardamomum* Maton. (chhoti elaichi, cardamon), *Alpinia galanga* (L.) Sw. (Siamese ginger, barakulanjan), *A. officinarum* Hance (kulanjan) etc.

Medicinal plants. Some plants are medicinally important. *Curcuma domestica* Val., largely grown in Tamil Nadu, A.P., Maharashtra, Bihar, Kerala and Orissa, is used in stomachic and as tonic, blood purifier and antiseptic. It is also applied to sprains and wounds. It contains orange red pigment **curcumin**. *Zingiber officinale* Rosc., largely grown in Kerala, U.P., West Bengal, Maharashtra, Andhra Pradesh etc., is used as carminative and digestive stimulant and in the treatment of cold and cough. Seeds of *Amomum aromaticum* Roxb. are used as stimulant, purgative and bitter. Decoction of seeds is used as gargle in affections of teeth and gums.

Ornamental plants. Commonly cultivated ornamental plants are *Alpinia speciosa* (Wendl.) Schum., *Globba racemosa* Sm., *Hedychium spicatum* Hamilt. ex Smith, *Roscoea purpurea* Sm., *Cautleya spicata* Baker etc.

Other uses. The root stock of *Hedychium spicatum* Hamilt. ex Smith (kafur-kachari) is used as a perfume and in preparing *abir.* The ginger oil obtained from *Zingiber officinale* Rosc. is used in perfumery.

Systematics and Phylogeny

Bentham and Hooker considered *Zingibereae* as a tribe of family *Scitamineae.* Lindley raised it to the rank of a family, which has also been followed by most taxonomists. The family *Zingiberaceae* has been included in the order **Zingiberales** by Hutchinson, Cronquist, Takhtajan, Dahlgren etc. and by APG II/AP web. Thorne has placed it in the order **Cannales**, a term prefered over **Zingiberales**. APG placed the family in **Commelinoids** and order **Zingiberales.** The family is considered to be related to *Musaceae* and liliaceous stock of monocots.

LILIACEAE (Lily Family)

Bentham & Hooker	Hutchinson	Cronquist	APG
Monocotyledons	Monocotyledones	Magnoliophyta	III. APG Group :
Coronarieae	Corolliferae	Liliopsida	Monocots
	Liliales	Liliidae	Non-Commelinoid
		Liliales	Liliales

Family comprises **250 genera** and **2500 species** (APG grouping system includes only **11 genera** and **545 species**, others are transferred to different families) found growing worldwide. In India, some **35 genera,** with about **189 species** are found, mainly in Himalayas. Common examples are *Asparagus, Dracaena, Smilax, Ruscus, Allium, Tulipa, Lilium, Gloriosa, Colchicum, Yucca, Sansevieria, Asphodelus* etc. Large genera are *Allium* (499 species), *Fritillaria* (90 species) and *Tulipa* (80 species).

Vegetative Characters

Plants are usually perennial herbs but a few are shrubs or undertrees (*Yucca, Cordyline, Dracaena*) or climbers (*Smilax, Gloriosa, Asparagus*). **Stem** may be herbaceous or woody, aerial or subterranean (rhizome, corm, bulb etc.) Rhizomes are found in *Polygonatum* and *Gloriosa*, corm in *Colchicum* and bulb in *Allium, Tulipa,* and *Lilium*. Stem is modified into cladode in *Asparagus* and Phylloclade in *Ruscus*. **Leaves** may be radical or cauline, either well developed (*Dracaena*) or extremely reduced to scales (*Rumex* and *Asparagus*) or fleshy (*Aloe*). Venation is often parallel but reticulate in *Smilax*. Stipules of *Smilax* are tendrillar. Leaves are usually alternate, rarely opposite (*Gloriosa*) or whorled (*Trillium*). In *Allium*, leaves are concentric.

Inflorescence and Flowers

Inflorescence is usually racemose type, the flowers being arranged in either simple or branched raceme. It is panicle in *Yucca*, monochasial

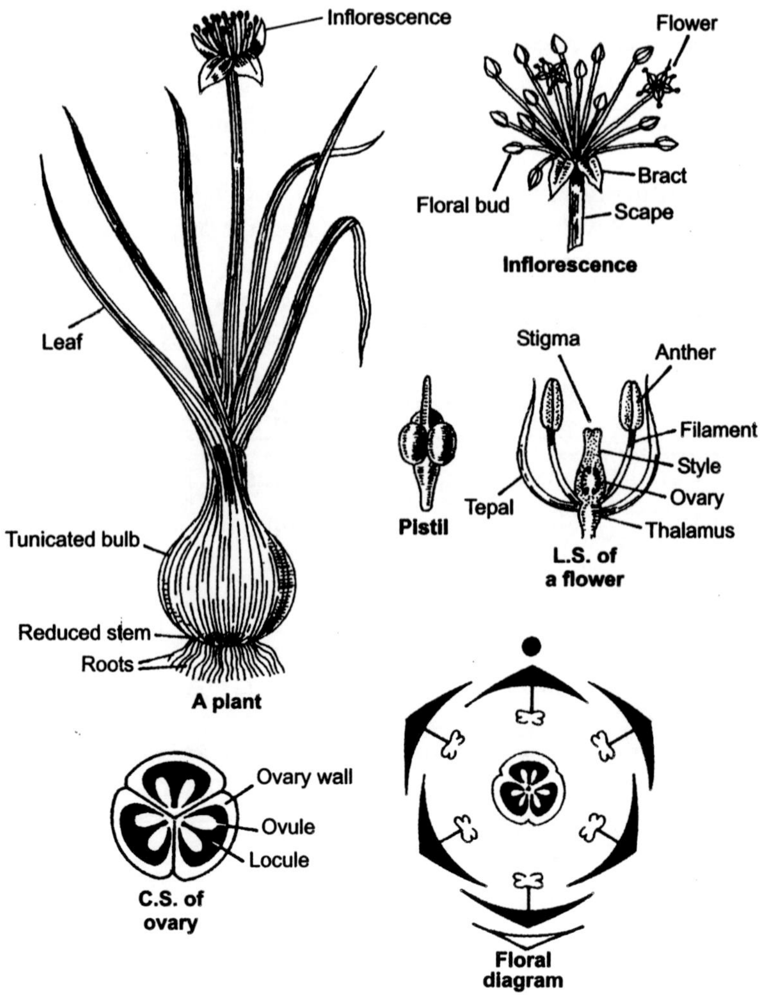

Fig. 22.9 *Allium cepa* L.

umbellate cyme in *Allium*, solitary terminal in *Tulipa*, *Fritillaria* and *Lilium* and solitary axillary in *Gloriosa*.

Flowers are mostly showy, bracteate, bisexual, actinomorphic, trimerous and hypogynous. Flowers are slightly zygomorphic in *Gellisea* and unisexual in *Ruscus* and *Smilax*. **Perianth** consists of usually 6 segments (tepals) arranged in two whorls of three each, which may be either free or united (*Polygonatum*). Tepals are often large and showy, set in valvate or imbricate aestivation.

Androecium usually consists of six stamens arranged in two whorls of three each, situated opposite the perianth leaves. Stamens may be free or epiphyllous, rarely united in a column (*Ruscus*, *Heterosmilax*, *Tricyritis*). Anthers are dithecous, basifixed, introrse or extrorse, dehiscing by longitudinal slits or terminal pores.

Gynoecium consists of three united carpels (syncarpous). Ovary is superior and trilocular with several ovules set in axile placentation. Style is normally simple and stigma is trilobed.

Fruits and Seeds

Fruit is usually a loculicidal or septicidal capsule, sometimes berry (*Smilax*). **Seeds** are endospermic, fleshy or hard.

Pollination and Dispersal

Flowers are insect pollinated (entomophilous). Pollination in *Yucca* is specialised intra-ovarian type and is carried out only by the female moth, *Pronuba yuccasella*. It furnishes a good example of parallel evolution of flowers and insects. In this case both are interdependent for multiplication. Fruits are dispersed by wind, human beings and animals.

Floral Formula $\text{Br} \oplus ⚥ \; P_{3+3 \text{ or } (3+3)} A_{3+3} \underline{G}_{(3)}$

Economic Importance

Family has considerable economic importance.

Edibles. Bulbs of *Allium cepa* L. (onion) are eaten raw or as vegetable and are stimulant, expectorant and diuretic. Bulblets (cloves) of *A. sativum* L. (garlic) are used as condiments and flavouring agents. Young shoots and roots of *Asparagus officinalis* L. (garden asparagus) are used as vegetables.

Medicinal plants. Bulblets of *Allium sativum* L. (garlic) are used in the treatment of rheumatism, chest congestion and to improve lungs. Juice is used as eardrops in earache, and in colic and flatulence. It has germicidal properties and is used in enteric fever and skin diseases. Oil extracted from the roots of *Asparagus racemosus* Willd. is used in nervous and rheumatic problems. Leaves of *Aloe vera* Tourn. ex. L. are laxative and used in cosmetics and several ayurvedic preparations. The tubers of *Gloriosa superba* L. are used for promoting labour pain and also as abortifacient. Dried rhizomes and roots of *Veratrum viride* Aiton. are used in hypertension associated with toximias of pregnancy. Corms and seeds of *Colchicum luteum* Baker are used in rheumatism and liver disorders. Bulbs of *Urginea indica* Kunth. (Indian squill) are used as cardiac stimulants and are used in dropsy, rheumatism and skin diseases. Roots of *Smilax ovalifolia* Roxb., *S. glabra* Roxb. and *S. china* L. are source of drug **sarsaparilla** used in skin and veneral diseases. Dried bulbs of *Fritillaria cirrhosa* D. Don. are used in asthma, bronchitis and T.B. Bulbs of *Scilla hyacinthiana* (Roth.) Mao are used as expectorant and cardiac stimulants.

Fibres. Leaves of certain plants such as of *Sansevieria roxburghiana* Schuld. f. (Indian bowstring hemp) and *Phormium tenax* Forst. yield commercial hard fibre used for making cordages, nets etc.

Ornamental plants Several plants are grown in the gardens for the ornamental purposes, *e.g. Gloriosa superba* L. (Malabar glory lily), *Lilium giganteum* Wall., *L. candidum* L., *Asparagus plumosa* Baker (asparagus fern), *Ruscus aculeatus* L. (butcher's broom), *Tulipa gesneriana* L., *Yucca gloriosa* L., *Y. aloifolia* L., *Y. filamentosa* L., *Dracaena* species etc.

Other uses. Corm and roots of *Colchicum luteum* Baker yield a famous alkaloid **colchicine** widely used to induce polyploidy in plants. Seeds of *Phormium tenax* Forst. yield a semi-drying oil used in paint and varnish industries.

Systematics and Phylogeny

Liliaceae are old family included in the series **Coronarieae** by Bentham and Hooker but in order **Liliales** by several workers (Hutchinson, Cronquist, Takhtajan etc.) The family shows so much diversity in the habit forms and the inflorescence, that most taxonomists believe that it can not be a natural one.. Hutchinson, therefore, tried to isolate this homogenous group into four sub-families, *e.g. Allioideae, Dracaenoideae, Smilacoideae* and *Luzuriagoideae.* Later he developed the order **Liliales** which also included *Liliaceae, Tecophilaeaceae, Trilliaceae, Smilacaceae, Ruscaceae* and *Pontederiaceae.* This step has been supported by many modern workers. APG Grouping supported such separation but further recognises *Asphodelaceae* and *Alliaceae.* APG II recognises order **Asparagales** in which some 29 families have been included, the important ones being *Agavaceae, Alliaceae, Asparagaceae, Asphodelaceae, Hyacinthaceae, Orchidaceae, Tecophilaeaceae* etc. **Liliales** of APG II contains 9 families including *Liliaceae, Luzuriagaceae, Smilaceae* etc. Takhtajan has recognised orders **Smilacales, Colchicales, Amaryllidales, Asparagales** and **Iridales** along with **Liliales.** By removing important genera like *Aloe, Asphodelus, Bulbine, Allium, Tulbhagia, Gilliesia, Crinum, Narcissus, Polianthes, Agave, Sansevieria, Dracaena, Coryline, Yucca, Smilax, Heterosmilax, Rhipogonum, Asparagus, Ruscus, Phormium* etc. from the family *Liliaceae,* it has now been tailored to comprise only **11 genera** and **545 species** (APG II/AP web).

Sub Family : AMARYLLIDOIDEAE

Bentham & Hooker	Hutchinson	Cronquist	APG
Monocotyledons	Monocotyledones	Magnoliophyta	III. APG Group :
Epigynae	Corolliferae	Liliopsida	Monocots
Amaryllidaceae	Amaryllidales	Liliidae	Non-Commelinoid
	Amaryllidaceae	Liliales	Asparagales
		Liliaceae	Amaryllidaceae

Family (or sub family) contains **66 genera** and **730 species** distributed in tropical and subtropical regions of the world. In India, some **9 genera** and **60 species** are found, mainly growing in Himalayas. Common examples are *Crinum, Amaryllis, Polianthes, Narcissus, Haemanthus, Zephyranthes* etc. Large genera of the family are *Crinum* (130 species), *Hippeastrum* (65 species) and *Zephyranthes* (55 species).

Vegetative Characters

Plants are perennial herbs with a bulbous or rhizomatous stem. **Leaves** are usually radical, linear or strap-shaped, thick, fleshy or stiff or grass-like and plicate.

Inflorescence and Flowers

Inflorescence is borne upon a leafless scape and is usually umbellate monochasial cyme subtended by a pair of spathe-like bracts, or more bracts.

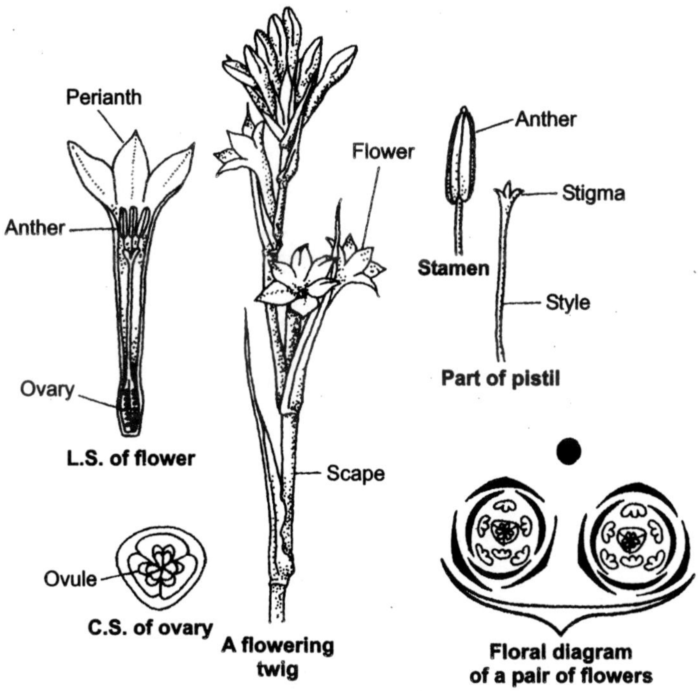

Fig. 22.10 *Polianthes tuberosa* L.

It is often racemose, spicate or paniculate. Flowers are solitary in *Galanthus* and *Zephyranthes.* **Flowers** are often large and showy, bisexual, actinomorphic sometimes zygomorphic (*Alstroemeria*), trimerous and epigynous. **Perianth** consists of 6 segments (tepals) in two trimerous whorls. Both whorls are similar and petaloid and united below into a tube with imbricate or valvate aestivation. A characteristic feature of the family is the presence of *corona* or *paracorolla* which is regarded as the basal outgrowth of filaments that are united to form the funnel-shaped structure or it is supposed to be the inner outgrowth of the perianth to constitute the funnel-shaped structure. Thus, corona are either staminal or perigonial in origin. Staminal nature is evident in *Pancratium* and perigonial in *Creptostephanus.*

Androecium consists of 6 stamens arranged in two whorls of three each. These are opposite the tepals and inserted on the perianth tube or towards the base of tepals (epiphyllous). Filaments are free or expanded at the base to form a tubular or funnel-shaped staminal cup (*Eucharis, Hymenocallis, Pancratium*). Anthers are basifixed or versatile, dithecous and introrse dehiscing by vertical slits.

Gynoecium is tricarpellary and syncarpous. Ovary is inferior, trilocular with many ovules arranged in two series set in axile placentation. Style is slender with a capitate or trilobed stigma. Septal nectaries are usually present. In *Astroemeria* and *Leontochir*, ovary is unilocular with three parietal placentae.

Fruits and Seeds

Fruit is a berry or loculicidal capsule. **Seeds** have fleshy endosperm (endospermic) and are often flattened or winged.

Pollination and Dispersal

Flowers are insect pollinated. Cross pollination is favoured by protandrous nature of flowers. Seeds are often dispersed by water.

Floral Formula $Br \oplus ⚥ P_{3+3} A_{3+3} \bar{G}_{(3)}$

Economic Importance

Some plants have considerable economic value.

Ornamental plants. Several plants bear beautiful showy flowers and are grown as ornamental plants, *e.g. Amaryllis belladona* L., *Narcissus* species (dafodils, nargis) (*N. tazetta* L., *N. poeticus* L., *N. jonquilla* L.), *Crinum* species (sudarshan, crinum lily) (*C. asiaticum* L., *C. latifolium* L.), *Zephyranthes candida* Herb., *Z. grandiflora* Lindl., *Z. tubispatha* Herb., *Polianthes tuberosa* L. (tuberose), (fragrant white flowers), *Eucharis grandiflora* Planch. (Amazon lily), *Hymenocallis americana* Roem. (spider lily), *Pancratium maritimum* L., *Haemanthus coccineus* L. (blood lily), *Curculigo latifolia* Dry. etc.

Other uses. An essential tuberose oil extracted from the flowers of *Polianthes tuberosa* L., is used in high grade perfumery. The tuberous roots of *Curculigo orchioides* Gaertn. are used as tonic and in the treatment of jaundice and asthma. Bulbs of *Hippeastrum* species are used for poulticing the swelling of the neck. Cooked bulbs of *Leucojum aestivum* L. are edible. *Pancratium zeylanicum* L. yields an alkaloid **lycocrine** which has emetic properties and causes paralysis.

Systematics and Phylogeny

Bentham and Hooker included the family in the series **Epigynae** along with other families which have inferior ovary. Hutchinson placed the family in its own order **Amaryllidales** and included other liliaceous genera that have scapose habit and umbellate inflorescence. Cronquist recognised it as a constituent of *Liliaceae*. Thorne kept the family in the order **Orchidales.** Takhajan and APG II groupings recognised the family *Amaryllidaceae* and kept it in the order **Asparagales.** APG considered it in monophyletic familial group **Non-Commelinoid Monocots** and in the order **Asparagales.**

Family is allied to *Iridaceae* by the nature of ovary but differs from the latter by the presence of six stamens. It bears relationship with *Liliaceae*, the genus *Agapanthus* forming a link between the two families. However, it has inferior ovary while *Liliaceae* has superior ovary. It is believed that both have a common origin.

ORCHIDACEAE (Orchid Family)

Bentham & Hooker	Hutchinson	Cronquist	APG
Monocotyledons	Monocotyledones	Magnoliophyta	III. APG Group :
Microspermae	Corolliferae	Liliopsida	Monocots
	Orchidales	Liliidae	Non-Commelinoid
		Orchidales	Asparagales

It is the second largest family (only after *Asteraceae*) of angiosperms comprising **788 genera** and over **18,500 species**, widely distributed but mostly in tropical humid forests of the world. Some **128 genera** and **700 species** are found in India distributed mainly in two zones : Eastern Himalayas and South-Western zone of Western Ghats and Nilgiri hills. Common forms found in the country are *Paphiopedilium, Dendrobium, Cymbidium, Coelogyne, Vanda* etc. and saprophytic forms like *Didymoplexis* and *Gastrodia*. Large genera of the family are *Pleorothallus* (1100 species), *Bulbophyllum* (970 species) and *Dendrobium* (900 species).

Vegetative Characters

Plants are perennial herbs, terrestrial or epiphytic or saprophytic. Main roots are always absent, instead three types of adventitious roots arise from the nodes : the aerial roots, the tuberous roots and the normal cylindrical earth roots. Some aerial roots have multiseriate epidermis of dead cells known as *velamen*. *Neottia, Gastrodia, Didymoplexis* and *Epigonium* are saprophytes. **Leaves** are thick, distichous or half alternate with a closed sheath. Shapes vary from linear to ovate. Venation is parallel. Stipules are absent.

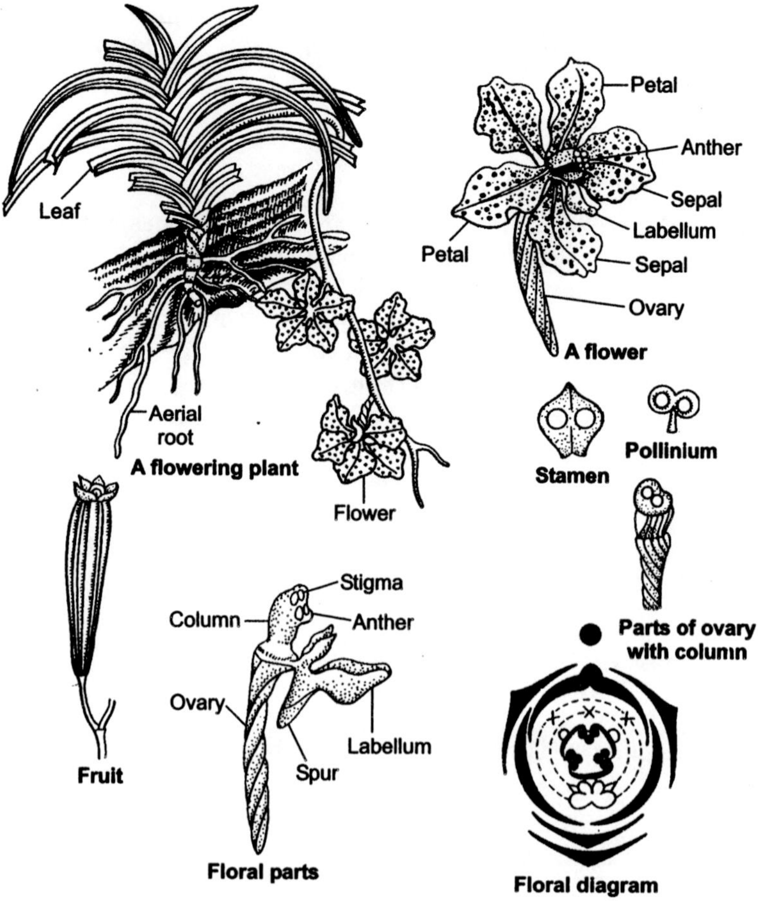

Fig. 22.11 *Vanda roxburghii* R. Br.

Inflorescence and Flowers

Flowers are either solitary (*Cypridium*) or borne in racemes, spikes or panicles. **Flowers** are showy, beautifully coloured, bracteate, bisexual or rarely unisexual, zygomorphic and epigynous. **Perianth** consists of 6, free segments arranged in two trimerous whorls, *outer* is often calyx-like (sometimes petaloid) nearly equal 3 sepals and *inner* of somewhat unequal 3 petals. These are imbricate or subovate in the bud. The two lateral tepals of the inner whorl are alike but the median one is very different in size, shape and colour and called **Labellum.** The labellum is posterior but rendered anterior by resupination, *i.e.*, torsion of the ovary through 180°.

Androecium consists of 1 or 2 stamens. Filament adheres to the style (**gynandrous**) and forms a column or **gynostemium** which springs from the top of the ovary and ends in a beak known as **rostellum.** Pollen grains remain united into masses called **pollinia.** Each pollinium bears a cord (*caudicle* or *translator*) which terminates in a sticky disc or gland called *retinaculum* or **viscidium.**

Gynoecium comprises three united carpels (syncarpous). Ovary is inferior, twisted, unilocular with many ovules on 3 double rows set in parietal placentation. Stigmas are 3, discoid, glutinous or situated below the rostellum and facing the labellum.

Fruits and Seeds

Fruit is a capsule opening laterally by 3 to 6 hygroscopically sensitive valves. **Seeds** are minute and non-endospermic with undifferentiated embryo.

Pollination and Dispersal

Flowers are specially adapted to insect pollination. The insects are attracted by beautiful form, colour and fragrance of flowers. Seeds are wind dispersed, sometimes winged.

Floral Formula $Br \% ⚥ P_{3+3} A_{1-2} \overline{G}_{(3)}$

Range of Floral Structure

Family *Orchidaceae* is known for a wide range of **floral variations.**

Sepals and petals may be similar or different in form. However, sepals are generally smaller and inconspicuous but becoming larger and distinct in *Masdevallia*. Sepals are free or somewhat united. In *Disa*, the odd sepal is spurred to form a hood-like structure.

Petals, *i.e.*, inner whorl of perianth, are very showy and attractive. These often show a great variation. Petals are usually filiform but smaller than labellum in *Epiranthes* and larger in *Oncidium*. These are ribbon-shaped in *Cypripedium*. In *Paphiopedilium*, petals may attain a length of 30 cm. The labellum is minute and narrowed or curved in *Cattleya* or large with spreading limbs in *Odontoglossum*, or slipper-like in *Cypripedium* or may form a cage in *Stanhopea* or bucket in *Coryanthes*. The labellum is simple in *Thelymitra* or trifid or tripartite or variously incised. In *Coryanthes*, the lateral petals are lost.

Androecium is represented usually by one or sometimes two sessile anthers. In monandrous forms, there is a single fertile stamen, terminal on the column. Outer whorl of stamens generally becomes fertile (as in *Arudina pentandra*) or may form a fleshy staminode (*Diuris*). Staminodes are leaf-like in *Orchis* or represented by small auricles in *Cypripedium*, or foliaceous in *Thelymitra*. The stamen is generally bent over the inner face of the column but may become erect and free on the top of the column as in *Ophrys* or it may hang down vertically. Anthers are 2-, 4- or 6 chambered, corresponding with the number of pollen masses. Pollen masses are free and granular in *Cephalanthera* or united into pockets by elastic nets but they are sessile in *Malaxis*.

Ovary is generally spindle-shaped or cylindrical but furnished with lines, ridges or wings. It is inferior, unilocular with parietal placentation but in *Apostasia*, it is trilocular with axile placentation.

Economic Importance

Family is known for the plants having most beautiful flowers, both in colour and shape and hence it has enormous ornamental value. Some important ornamental genera are *Cypripedium, Habenaria, Cattleya, Vanda, Cymbidium, Dendrobium, Epidendrum, Orchis, Epipactis, Coelogyne* etc.

The capsule of *Vanilla fragrans* Ames. (vanilla) and several other species are the source of vanilla extract of commerce, extensively used for flavouring food products, such as ice-cream.

Leaves of *Calanthe veratrifolia* contain a glycoside **indican** which yields indigo blue. Tubers of *Eulophia pratensis* Lindl. are used to remove scrophulous glands in neck. Root stocks of *Geodorum densiflorum* (Lamk.) Schl. are curshed and rubbed on cattle to kill flies. Dried leaves of *Jumellea fragrans* are the constituent of **Faham tea**. Roots of *Vanda roxburghii* R. Br. are used for the treatment of scorpion bites and rheumatism. Tubers of *Habenaria susannae* R. Br. are edible. Dried tuberous roots of *Orchis latifolia* L. are used as nervine tonic.

Systematics and Phylogeny

Bentham and Hooker placed the family under **Microspermae** in the beginning of **Monocotyledones.** Hutchinson recognised it as a separate order **Orchidales** at the end of **Corolliferae,** as he believed that the family has originated either from *Haemodoraceae* or from *Amaryllidaceae* as a result of complex development of floral parts and the reduction of stamens. Cronquist, Thorne and Takhtajan kept the family *Orchidaceae* in its own order **Orchidales.** APG II/AP web placed the family in the order **Asparagales** of **Non-Commelinoid Monocots**, which forms a monophyletic familial group.

Orchidaceae are related to the epigynous components of order **Liliales** and is believed to have originated through **Musaceae**, specially through genus like *Orchidantha.*

Most taxonomists believe that *Orchidaceae* represent the **Most Advanced Family in the Monocots** because of the following reasons :

(1) All orchids are herbaceous.
(2) Plants are found as ordinary rooted herbs, climbing vines, epiphytes or saprophytes.
(3) It is a large and most successful family comprising about 1/4th of the total monocotyledons.
(4) Flowers are epigynous and of diverse shape, size and colour.
(5) Flowers are zygomorphic accompanied by a wide variety of structural modifications in general and labellum formation with a spur containing nectar, in particular.
(6) A constant reduction in androecium is found.
(7) The stigmatic lobe is specialised, often developing into rostellum.
(8) All floral characters are related to the progressive specialisation for insect pollination and massive transfer of pollen grains.

Chapter **23**

Taxonomy and Phytogeography

Phytogeography or plant geography deals with the geographical distribution (ranges) of plants on or near the surface of earth and its waters, *i.e.*, biosphere. It is closely related with the plant taxonomy wherein, plants are arranged into taxa whereas in phytogeography, they are arranged in phytochoria. Although plant taxonomy is basic to phytogeography but phytogeographical findings can often be useful for better understanding of taxonomy. The overall significance of the interrelationships of plant taxa can be appreciated only after it is understood that the related plants have or have had through common ancestors and common or interrelated geographical distribution. Humboldt (1817) is generally credited to be the Father of Phytogeography.

Phytogeography or plant geography is a branch of botany that concerns with the distribution of plants in the present and the past (Good, 1947). Thus it becomes necessary for a plant taxonomist to be familiar with the spatial relationships of the plant groups concerned and the floras affecting distribution patterns. A taxonomist cannot remain satisfied by studying only the present day distribution of plants but is also expected to known their distribution in the past. Much has been discussed about great changes in the distribution of plants that have taken place during the course of geological time. It seems that primary cause of these changes can be attributed to the climate. The impact of climate is the most far reaching among all natural constraints controlling plant life. The climate change brings about change in ecological factors and consequently the physiological behaviour of the plants. The second, and probably equally responsible factor that has influenced plant distribution, is animal life on this planet. Man in particular, is apt to have the greatest influence on distribution of plants on the earth.

The climate has played a crucial role in the evolution of flora. Human beings have adapted to rapid climate changes in the past. The effect of persistent abrupt climate changes have been equally drastic on the flora that brought about shift from one type to another type of vegetation in the geological records. The most significant example of impact of man on the evolution of ecological niches, was domestication of plants. The domestification of plants was of the prime importance for agriculture. The early hominids were hunters and gatherers who relied on the naturally occurring vegetation, fruits, nuts etc. The climate ameliorated changes in vegetation brought about the changes in food habits and lifestyle of the people.

CLIMATIC REGIONS OF THE WORLD

Since climate refers to the atmospheric conditions of the earth over a long period of time, it is one of the most important aspects of the natural environment. It represents the living atmosphere with which lithosphere, hydrosphere and biosphere, constitute the natural environment. Climate affects and is also affected by the process and conditions of the other three components, *i.e.*, lithosphere, hydrosphere and biosphere. Climate sets limits to the distribution of different kinds of plant and animal lives. Thus, climate has a great significance for man.

Climatic factors that affect the life of plants and animals include mainly light, temperature, precipitation (rainfall), humidity of air and wind etc. Other factors such as radiation, cloudiness, storms etc., are also taken into account under special circumstances. These factors are often independent but their various combinations form the characteristic climate of various regions of the world. Broadly speaking, there are three main climatic regions of the world.

(i) Polar Regions. The climates of the polar regions and also of high altitudes, are mostly rigorous. The mean annual temperature is much below 10°C. The precipitation, that is mostly in the form of snow, is often less than 25 cm per annum. Due to low temperature and low evaporation rate, the relative humidity is fairly high. These regions exhibit wide seasonal and diurnal fluctuations. In extreme polar regions, there is continuous light during summer and continuous dark during winter.

(ii) Temperate Regions. The climates of temperate regions and adjacent lands are mostly fairly warm and moist, atleast during favourable periods. The mean annual temperature is slightly above 10°C and precipitation more than 76 cm. These regions show well marked seasonal and diurnal fluctuations from place to place.

(iii) Tropical Regions. The climate of the tropical and subtropical regions is warm and humid for most of the periods. The mean annual temperature is usually above 15°C and the precipitation is between 200-400 cm. The frost and the snow are usually unknown. The conditions are torrid and equable and thus in these regions often there is no marked seasonal or diurnal variation.

Other climatic regions include oceanic with even climate and continental regions with uneven climate. Irrespective of latitudinal temperature zones, these regions are found in all primary climatic zones.

World climatic regions fall under the following categories :

1. Humid Tropical Climate

The tropics having this kind of climate, occupy almost half of the earth's surface–over 20 per cent of the land and 40 per cent of the ocean surface. This climatic group is bound by the mean annual isotherm of 20°C. The hot deserts of North Africa and Arabia lying within this area, are excluded from this group because of their extreme aridity. Principal climatic types within this group are identified basically on the basis of their seasonal distribution of rainfall. These include :

(1) Tropical Rainforest Climate (Equational Climate). This type of climate is found in the belt extending from equator to 10° North and South latitudes and upto 20° North and South latitudes along the windward margin of the continents. The climate is a characteristic feature of Amazon basin, Congo basin and the South-West Asian islands. The temperature is high throughout the year, the mean annual being about 27°C. The daily temperature hovers between 10°C to 25°C. The rainfall is heavy and distributed throughout the year. The mean annual rainfall is about 250 cm. The high temperature and all the year round rainfall facilitates most luxuriant vegetation–the so called tropical rainforest or Selva.

(2) Savanna Climate. This type of climate is bound by tropical rainforest climate towards the equator and by dry climates towards the poles. Sudan and Veld Plateau of Africa and the Northern Australian tropical grasslands, the llanos of Guiana, highlands of South America and the campos of Brazil, are characterised by this type of climate. The climate represents a transition zone that gets convectional rainfall during the summer, whereas during rest of the year it remains dry under the dominating influence of the trade winds. The temperature remains high throughout the year. The annual variation of temperature is 5°C-6°C. The annual rainfall is about 160 cm. Wet summers alternate with dry winters and floods and drought are quite common. The duration of the rainy season and the annual rainfall and also the annual range of temperature increases with the increasing distance from the equator and thus tropical rainforests give way to Savanna, a tropical grassland with scattered deciduous trees.

(3) Monsoon Climate. The monsoon climate is a characteristic feature of India, South-East Africa, Northern Australia, the Guinea coast of West Africa and the Pacific coast of Columbia. The climate exhibits seasonal reversal of wind direction coupled with alternating periods of rainfall and drought. During summer, humid unstable air moves from the oceans towards the land, and causes the rain. In winter, dry wind, originating over the land, blows towards the seas. The monsoonal circulation system, having reversal of wind direction, develops in response to the difference in annual temperature variations between the continents and the oceans. During summer, the intense heating of Central Asia results an area of very low pressure, whereas in Southern Hemisphere there occurs the subtropical high pressure over the oceans. The south-east trade winds flow from the high pressure to low pressure and are deflected to the right and reach India and South-East Asia in the form of on–shore south and south-westernly winds laden with moisture and cause heavy rainfall in these parts of the Asian countries. During winter, the excessive cooling in Central Asia intensifies the high pressure belt and the off-shore north-east trade winds blow over South-East Asia. Due to the pressure of Himalayan range, these cold winds are prevented from descending into the Indian peninsula. During this season, the temperatures are low.

2. Dry Climatic Regions

The dry climatic regions are divided into the arid or desert type and the semi-arid or steppe type with many common features, the difference primarily being in the degree of dryness.

(1) Desert Climate. The deserts are characterised by scarcity of water mainly due to excessive evaporation, *i.e.,* rate of evaporation exceeds the rate of precipitation. The arid deserts lie in the low-latitude close to the Tropic of Cancer and Tropic of Capricon. The Sahara, the Arabia, the Thar, South-Western USA, South-Western Africa and Central Australia are characterised by this type of climate. The climate is dominated by subsidence of air masses and marked stability of sub-tropical anticyclones and hence the areas are nearly rainless. Precipitation is not only scanty but is also very erratic. The cloudless sky and low humidity, allow a great amount of solar radiation to reach the ground during the day and permit rapid terrestrial radiation during the night. As a result, low-latitude deserts in the interior of the continents have the greatest daily range of temperature even more than 15°C, on the earth. Mean annual temperature is 38°C. The tropical deserts, that are located on the west coasts of continents, reveal marked influence of cold ocean currents on their climates.

(2) Steppe. Unlike low-latitude deserts, the steppes are not controlled by the subsiding air masss of the sub-tropical anticyclones. Instead, steppes are dry lands principally because of their position in the deep interiors of large land masses away from the oceanic influences. Additionally, presence of mountain barriers across the path of prevailing winds, further check maritime influences. The middle latitude deserts having steppe climates are, therefore, most widespread in North America and Eurasia. Steppe climate type is characterised by meagre and unreliable precipitation. The annual mean temperatures (21°C) and annual range of temperatures (13°C), are comparatively lower. The annual rainfall is 30 cm. The steppes, that are located on the poleward side of the deserts, receive maximum rainfall during the cool season, while those located towards equator, receive it during the warm season.

3. Humid Mid-latitude Climatic Regions

Humid Mid-latitude climatic regions are sub-divided into the Mediterranean, China and West European climatic types. The temperatures, in coldest month, is below 18°C but it never goes below –3°C. The warmest month records temperature exceeding 10°C.

(1) Mediterranean Climate. This type of climate is found along the west coasts of continents between 30° and 45° latitudes. The climate is bound by the arid Steppe on the equatoward side and the Marine West European Climate on the poleward side. Such type of climate is experienced by Mediterranean Sea, Central California, Central Chile, Southern part of South Africa and South-Eastern and South-Western parts of Australia. Since these areas undergo seasonal shift of pressure belts with the annual movements of the sun, they come under the influence of sub-tropical high pressure conditions during the winter. And thus, the summers are warm and dry, the temperature being in between 20°C and 27°C. Winters are mild with temperature ranging between 4°C and 10°C. The annual range of temperature is about 10°C to 17°C. The rainfall is moderate and ranges between 40 cm to 60 cm annually, most of the rains occurring during winter season.

(2) Humid Sub-tropical Climate. This type of climate is found along the eastern coasts of continents between 25°C and 45° latitudes. The climate is experienced by South-Eastern United States, Uruguay, Argentina, Southern Brazil, Eastern China, Southern Japan and Eastern Coastal belt of Australia. In these areas, summers are hot and humid, winters are mild. The mean annual temperature is 20°C. Annual rainfall is usually more than 100 cm, well distributed throughout the year. These areas experience dreaded hurricanes and typhoons, mostly in the late summer or autumn.

(3) Marine West European Climate. This type of climate is found along western coasts of continents from about 40° and 65° north and south latitudes. The climatic region is dominated by the on–shore flow of oceanic air. Under the influence of maritime air masses, these areas experience mild winters, cool summers and ample amount of rainfall throughout the year. This climate is experienced by Europe, a narrow belt along the coasts of North and South America, New Zealand and South-Eastern tip of Australia. In these areas, weather remains highly variable and unpredictable throughout the year under the influence of low pressure system. The mean annual temperature is 10°C and the annual rainfall is about 140 cm. There occur more rains in winter than the summer. This climatic region, having low temperatures, have been sub-divided into the Taiga Climate, the Cool East Coast Climate and the Continental Climate.

(i) *Taiga Climate.* This climate type has been named after the coniferous forest cover occurring in these regions. Taiga climate region extends from Western Alaska to Newfoundland in North America, and from Norway to the Kamchatka peninsula in Eurasia. The climate is dominated by continental polar air masses and has short summers and long winters. The temperature during summer ranges between 10°C and 15°C and during winters it may be as low as –50°C. The rainfall is scanty and is concentrated in warmer months. Rainfall although low, is sufficient for tree growth because of low rate of evaporation. This climate supports soft-wood coniferous forests.

(ii) *Cool East Coast Climate.* This type of climate is located in the areas along North-Eastern Coast of United States, Northern China, Japan, Korea and lower Danube plains. The summers are long, hot and humid and the winters are cold. Mean temperature during summers, reaches around 25°C whereas average temperature during winter ranges between 4°C and 0°C.

Rainfall is variable as the summer precipitation is convectional in nature. Winter rainfall is generally less than in summer and falls in the form of snow.

(iii) *Continental Climate.* This climatic region occupies the position in the interior parts of the big continents between Taiga and the mid-latitude deserts. This type of climate is found in Eurasia-from Poland and Baltic States of the Central Russian Plain, North America-Northern States of the United States and Canada-Southern parts of the Central States. The temperature is usually higher than the Taiga type climate but is lower than Cool East Coast Climate. Summers are short and warm and the winters are moderate. Summer temperature ranges between 10°C and 21°C. Winters become long when temperature goes down freezing point. The precipitation is variable and results as rainfall during summer and as snow during winter.

4. Polar Climatic Regions

Polar climates are characterised by long cold condition and the absence of a warm period, the mean temperature of the warmest month being below 10°C. This climatic group, based on temperature range, has been sub-divided into two types-the Tundra Climate with the temperature of warmest month remaining above freezing point but less than 10°C and Ice-Cap Climate in which the temperature of warmest month is below 0°C.

(1) Tundra Climate. This type of climate is found only in Northern Hemisphere, occupying the coastal fringes of the Arctic Ocean besides many Arctic Islands and the ice-free shores of Iceland and Greenland. Winters are severe, precipitation is small and summers are cool. The temperature of the warmest month does rise above 0°C but never above 10°C and as such the ground remains free from the snow, albeit for a short period. The 10°C summer isotherm marks the equatoward limit of Tundra as well as the poleward limit of the tree growth. The conditions allow sparse growth of vegetation including grasses, mosses and lichens.

(2) Ice-cap Climate. Ice-cap climate does not permit increase of temperature above 0°C in any of the month of the year. Since it is a region of permanent ice and snow, the growth of the vegetation is not possible.

5. Highland Climatic Regions

In highlands or the mountainous regions, altitude and latitude play a significant role in controlling temperature and rainfall. On high mountains, the effect of attitudes is almost the same as that of high latitudes. Higher altitudes experience low temperature, low air pressure, large diurnal changes of temperature and relatively large amount of rainfall. Highland Climate is found in the Alps, the Himalayas, the Tibetan Plateau, the Rockies and the Andes. These areas exhibit altitudinal zonation of vegetation from the foot to the top of mountains.

PHYSICAL-GEOGRAPHIC REGIONS OF THE WORLD

While studying the physical geography of the world, it is essential to take into consideration the oceans, the islands and the continents and various landforms within the continents. Among various landforms, mountains and deserts require special attention.

1. Continents

There are six main land masses known as continents. These include Eurasia, Africa, Australia, North America, South America and Antarctica. Antarctica is completely devoid of plant life. Although in former ages it possessed considerable vegetation but at present it is covered with ice and snow. Majority of these land masses are mainly distributed in the Northern Hemisphere of the globe and are exactly opposite those of the Southern Hemisphere, *i.e.*, where there is land in the north, there is sea in the south and vice versa. The absolute level of edges of continents varies considerably and in most cases they stand clear of the water to their very boundaries but in others they pass gently below the sea level before their actual abrupt edges are reached. It is for this reason that the oceans or seas of the world are designated as 'deep sea'–that are proper bound by true edges of the continents and 'shallow sea' or marginal sea–that are really shallow flooding of the peripheral parts of the continents, appropriately called continental seas.

2. Islands

Apart from continents, there is a large number of lesser island landmasses. Islands take a large mass of land. They pose some interesting phytogeographical features as regards their size and distribution. Bigger islands include Aleutians, British Isles, the Canaries, Madagascar, Malaya, Formosa, Greenland, Howaiian Islands, Ireland, Japan, New Guinea, New Caledonia, New Zealand, Philippines, Seychelles, Tasmania and West Indies. These account for as much as 56 million sq km–greater than the area of the continent of Europe. The smaller islands count by the thousands.

For a plant taxonomist, the relative position of the continents and islands are of particular importance. The land masses, that are situated below the equator of South, *e.g.* South America and Africa and intermediate islands, are far more remote from one another than are similar land masses at an equivalent northern latitude. Islands are broadly divided into four types :

(1) Continental Islands. Islands, that rise from continental self, are called continental islands, *e.g.* British Isles or Newfoundland. They are an integral part of an adjoining continental plateform and have the same geographical structure as the continent to which they are related.

(2) Oceanic Islands. Oceanic islands are those that rise from the bottom of the oceans and are surrounded by deep seas. Their geological structure will have no relation to that of the nearest shores. They are very often the tops of submarine mountains, *e.g.* Ascension and Tristan-da-Cunha rise from the Central Atlantic ridge (mountain) or submarine volcanoes, such as St. Helena and Tenerifle islands that are formed by submarine volcanoes. Oceanic islands are more common in Pacific than in the Atlantic. Oceanic islands include Bonnins, the Mariannas, the Galapagos and the Hawaiian Islands. They are regarded older in age and more permanent than the less old and more fluctuating continental islands.

(3) Tectonic Islands. They are created by the movements in the earth's crust. The outermost layer of the earth, made of rigid plates, is in very slow but constant in motion. When one plate is pushed under another plate, the top plate may scrap off pieces of the bottom plate. Over millions of years, this material piles up to form an island. Barbados in the West Indies and Kodiak Island near Alaska, are good examples of tectonic islands.

(4) Coral Islands. Coral islands are the work of minute sea organisms called coral polyps which congregate in large colonies, die and leave their calcarious skeleton. Some of these rise above the water and give the islands. One type of coral thriving in warm tropical seas, excels in building reefs usually along the edges of islands–'fringing reefs.' These protect the islands from the ravages of the sea. Sometimes an island, with a coral fringe, begins to sink. Its shoreline goes down first, while coral building continues upwards. The sea invades the sinking shoreline and separates the coral reef from the rest of the island. Such a reef is called 'barrier reef'. The Great Barrier Reef, the biggest coral reef known that extends for more than 2000 km parallel with the coast of Queensland, Australia, is a good example of coral barrier reef. 'Atolls' are low circular coral islands each with a central lagoon of shallow water, *e.g.* Kwajelin in Marshall islands, Central Pacific and Christmas Island in Line Islands and Central Pacific.

3. Mountains

Most of the land masses are quite uneven rising at certain places into great heights so as to give rise to mountains. Mountains are uplifted portions of earth's surface. Mountains, in course of time, are worn into hills. Conventionally, on the basis of their origin and mode of formation, mountains are divided into four types : fold mountains, block mountains, residual mountains and volvanic mountains. Of these, fold, block and residual mountains are structural mountains whereas volcanic mountains are formed as a result of volcanic eruption.

All great mountain systems of earth are structural mountains. They lie near or parallel to continental coastlines and are thousands of kilometres long and hundreds of kilometres wide.

(1) Fold Mountains. Fold mountains arise because the rocks in them have been buckled and crumpled by pressure. Just as a table cloth when pushed along a table, wrinkles up into folds, the rocks of the earth's crust react to lateral pressure to form folds. In fact, it turns out that all our big mountain systems have been formed by colliding plates.

Major present day mountains, *viz.*, Alps in Europe, the Rockies in North America, the Andes of South America and the Himalayas of Asia, are structural fold mountains. These new fold mountains are still rising under the influence of earth's tectonic forces. They comprise a variety of rock structure, deep gorges and high pyramidal peaks. The Himalayan landform consists of long loops of three main ranges running from south to north of increasing heights and converging frequently to form knots. It has taken 7 million years for them to achieve present heights. The great rivers have been struggling to cut their way across high mountains. This has made deep canyons and narrow gorges separating the valleys from one another.

The Pennines of Europe, Appalachians of America and Aravallis of India, that have been weathered down to stumps long ago, were formed during early mountain forming period (dating back to Pangea), are old fold mountains. Aravalies, however, are considered to be one of the oldest mountains.

(2) Block Mountains. Block mountains arise as a result of vertical earth movements along cracks of faults. Such movements are also caused by the pressure generated by the plates. When such vertical earth movements leave a block of high elevation standing between two areas of low elevation, the high land area results a block mountain. Block mountains are usually steep-sided with fault scraps, flat topped and flat bottomed. The Vosges in France, Black Forest mountains in Germany and Salt Range in Pakistan, are mountains of this type.

(3) Residual Mountains. Residual mountains or dissected mountains are so deeply dissected and reduced by weathering and river action that they stand out alone in the surrounding area as skeletons. These mountains are the products of previously existing elevated regions. Erosional activities have changed them into hills, *e.g.* Nilgiris, the Parasnath, Girnar and Rajmahal hills in India. Catskill mountains of New York are typical of this category.

(4) Volcanic Mountains. Volcanic mountains came into being as a result of volcanic eruptions. When a volcano erupts, the ejected materials fall around a hole or crater and build up a mountain that is roughly conical in shape and has a crater at the top. Fujiyama in Japan, Vesuvius in Italy and the Chiborazo and Cotopaxi in the Andes (South America), are examples of such mountains.

Structural mountains have been modified considerably by the glaciers, rivers and streams. In India, some geographists are of the opinion that river Narmada flows through a faulted basin between Vindhyas and Satpuras that stand as block mountains to its north and south, respectively.

The mountains are of great significance to phytogeographer because their climate and consequently vegetation varies appreciably with altitude. There are many mountain ranges, that are situated in the tropical or subtropical latitudes, but show arctic climates on high altitudes.

4. Deserts

On the leeward side of mountain ranges, because of their great heights, the rain bearing clouds from the sea discharge before they pass further inland and create rain-shadow areas, the deserts. In general, deserts occur where the rainfall is less that 50 cm annually. The deserts, that occur in the tropics or at low elevations, are called hot deserts or warm deserts. Most of the deserts of the world are hot deserts. Hot deserts of the world include the Great Basin in North America, the North Mexican desert, Western Argentina, Peru and Chile in South America and the greatest desert of Africa– Sahara, and Asia, Arabian and Persian deserts. Sahara is one of the most peculiar among all hot deserts, because over much of its surface, there is sand bearing no plant life. Among other hot deserts, are deserts of West Pakistan and Kutch, southern Africa and Australia. They are the areas of earth's surface that are too dry to support plant life and are usually sparsely inhabited or uninhabited by man.

As regards the age of present deserts, there occurs a great deal of difference of opinions. According to one view, atleast some of the deserts have been a feature of the world throughout the geological time and have been here longer than the history of angiosperms. According to another view–the desert conditions may have occurred at intervals throughout the geological time but that all the present deserts are modern. According to some others desert conditions were unknown anywhere until the late geological time. Good (1947), however, is of the opinion that all present deserts are of different ages.

Cold deserts occur in the northern latitudes or at great elevations in the mountains of the tropics. Among cold deserts of the world are Turkestan, Tibet and Mongolia (Gobi) and cold deserts of Arctic and Antarctic. Cold deserts are extremely cold places because of very cold temperature (–20°C to 4°C) where every drop of water is frozen and water in any utilisable form is scarce.

5. Oceans

The oceans cover 70·8 per cent of the earth's total surface area and hold 1445 million cubic km of water. The ocean water contains about 3·5 per cent of dissolved salts. It is not potable and thus is of no use to man. Oceans, unlike continents, merge so naturally into one another that it is difficult to demarcate them. The geographers have divided the oceanic area into four oceans, *viz.*, the Pacific (being largest), the Altantic, the Indian and the Arctic. In 2000, International Hydrographic Organisation delimited Antarctic Ocean as the 5th world ocean. These oceans, by definition include seas, bays, gulfs and other ocean inlets attached to them.

PHYTOGEOGRAPHIC REGIONS OF THE WORLD (PHYTOCHORIA)

Phytogeographical approach of dividing the surface of earth by utilising biological systems to recognise floristic regions depending on the nature of the soil and climate and types of plants growing in such areas, has been made by several plant geographers. One of the earliest attempts was made by Schouw (1823), who divided the earth's surface into 25 phyto-kingdoms, each having several representives for most part of geographical areas, naming the phyto-kingdoms, wherever possible, after their most characteristic plants, such as kingdom of Umbelliferae and Cruciferae–North Eurasia (further subdivided into : Province of sedges–Arctic; Province of Primulaceae–Eurasian Alps; and Province of shrubby alpine composites–American alpine).

Hooker (1907) placed Indian Region under the Indo-Malayan subkingdom of the Palaeotropical kingdom. He divided Indian Region into nine provinces :

1. Eastern Himalayas extending from Sikkim to Mishmi mountain in Upper Assam.

2. Western Himalayas extending from Kumaon to Chitral.
3. Indus Plain including the Punjab, Rajasthan, West of Aravalli range, Kutch and Gujarat.
4. Gangetic Plain from Aravalli range to Bengal, plains of Assam and low country of Orissa.
5. Malabar including the humid belt of hilly or mountainous country along the western side with Konkan, Kanara, Malabar, Cochin, Travancore and the Lakhsdweep Islands.
6. Deccan which is dry elevated table land of India, east of Malabar and south of the Satpura range.
7. Assam and the hilly ranges of Khasi, Naga and Mizo hills, and the Andaman and Nicobar Islands.

Two more provinces Sri Lanka and Baluchistan are no more in the present Indian boundary.

Good (1947), recognised 6 kingdoms and 37 regions, atleast for theoretical purposes, and regarded them as equivalent in floristic values and importance. Turril (1959) preferred to use the term realm instead of kingdom. He recognised 4 realms and 19 regions.

Table 23.1 Classification of Earth by Turril (1959)

Realm	Regions
1. Northern Realm	1. Arctic and Subarctic Region 2. Boreal Region 3. Central European Region 4. Mediterranean Region 5. Central Asiatic Region 6. Regions of the Nearer East 7. Eastern Asiatic Region 8. North American Region
2. Palaeotropical Realm	1. Indian Region 2. Malaysian Region 3. Tropical African Region 4. Mascarene Region 5. Pacific Region
3. Neotropical Realm	1. Tropical and Subtropical American Region
4. Southern Realm	1. South African Region 2. New Zealand Region 3. Australian Region 4. Regions of Temperate South Africa 5. Antarctic Region

According to Turril (1959), Indian Region, which falls under Palaeotropical Realm, accounts for Indian subcontinent (excluding Pakistan) including Central and Lower Burma (Myanmar) and Sri Lanka. The common components of this region include conifers in the Himalayas; Rosaceae, Magnoliaceae, Primulaceae and rhododendrons in Assam; Shorea at the foot of hills and Dipterocarpaceae, *Ficus*, palms etc. in other parts; Rhizophoraceae and other halophytes in littoral forests; and ferns and orchids in Malabar.

In contrary to Tropical, Temperate and Polar Zones, commonly recognised, Hansen (1920), based on latitude, regarded seven zones on either side of the equator. These include :

1. Equatorial Zones	0° – 15°
2. Tropical Zones	15° – 23.5°
3. Subtropical Zones	23.5° – 34°
4. Warm Temperate Zones	34° – 45°
5. Cold Temperate Zones	45° – 58°
6. Sub-Arctic Zones	58° – 66.5°
7. Arctic Zones	66.5° – 72°
8. Polar Zones	72° – 90°

The vegetational zonation in terms of metre height above mean sea level can be summarised as :

0–600	Zone of palms and bananas
600–1200	Zone of tree ferns and figs
1200–1900	Zone of myrtles and laurels
1900–2600	Zone of evergreen trees
2600–3200	Zone of deciduous trees
3200–3800	Zone of coniferous trees
3800–4450	Zone of alpine shrubs
4450–5050	Zone of alpine herbs
5050 and above	Zone of permanent ice and snow

These altitudinal and latitudinal parameters of zonations do not harbour similar or same kinds of plants or plant species in different parts of the world because of small average range of species and various other factors, such as local soil conditions, day length, slopes of mountains etc., which play a significant role in determination of species types that can grow in a particular area. As such, this caused the need of a vegetational classification or floristic classification based on kind of plant species occurring in the area.

WORLD'S GREAT PLANT FORMATIONS (BIOMES)

Biomes represent the divisions of the major community types of the world, consisting of distinctive combinations of plant and animal species and are characterised by an approximately uniform life form of climatic climax vegetation such as grass, deciduous trees etc. The major biomes of the world are :

1. The Tundra

This type of vegetation covers large areas of the arctic region forming a circumpolar band between the Arctic Ocean and the polar ice caps to the north and coniferous forests to the south. Smaller but ecologically similar regions found above the tree line on high mountains, are 'alpine tundra'. The temperature falls below that required for plant growth for most of the year. Precipitation is low (usually less than 25 cm) and occurs mainly as snow. Below a certain depth, the ground remains permanently frozen forming permafrost. The low productivity and limited microbial activity result in thin soils. It has low productivity, but a high species richness. The long day length during summer, coupled with higher temperatures, allows primary productivity at this time of year to be an order of magnitude higher than winter. The vegetation and the soils are very slow to recover from disturbance. Since the discovery of oil in Tundra, vegetation has been lost and soils eroded. The control of the development and the protection of Tundra is an international problem.

Short growing season (about 60 days), with other adversities, allow only lichens (reindeer moss), grasses, sedges and dwarf woody plants to flourish in this region. Such type of vegetation is also common in high altitudes of mountains such as Himalayas, but these zones are free from permafrost.

2. Northern Coniferous Forests (Taiga)

This type of vegetation is stretched in a broad belt in the northern region of North America, Europe and Asia, extending southward in tropical mountains, particularly in Himalayas. It is characterised by cold climates and coniferous boreal forests. The boreal forests are dominated by coniferous tree species, mostly with needle-like leaves, such as species of *Pinus* (pine), *Picea* (spruce), *Juniperus, Abies* (fir), *Larix* (larch) etc. Intermixed with them, occur a smaller number of species of *Betula* (birch), *Populus* (polar) etc. The needle-like leaves of conifers minimise transpiration and reduce water loss. The conical growth form of many conifers is an adaptation against snow damage, preventing snow from settling on the crown of the tree. Their low evapo-transpiration rates and ability to produce a form of antifreeze in leaves allow them to remain in leaf throughout the winter. There is poor development of shrubs and herbs.

3. Moist Temperate Coniferous Forests

This type of vegetation is found along the west coast of North America and in the Eastern Himalayas. The fairly low temperature in summers, falling below freezing in winter and heavy rainfall, support distinctive type of forests consisting of some dominant coniferous species of *Tsuga, Pseudotsuga, Abies, Thuja, Cryptomeria* etc. There is a dense growth of shrubs in understory, and where light can not penetrate, there is a rich ground flora consisting of hebraceous plants, ferns, bryophytes and lichens.

4. Temperate Deciduous Forests

This type of vegetation is found in the areas with evenly distributed moderate rainfall and temperatures with distinct seasonal pattern. Such forests are found chiefly in Eastern North America, all of Europe, parts of Japan, Western Himalayas, Australia and tip of South America. These areas exhibit extreme seasonal variation, causing the trees and shrubs to shed their leaves during winter. In being scattered in distribution, such forests show a variety of species composition. Shrubs, herbs and ground vegetation is well developed. The forests are dominated by broad-leaved deciduous trees like oak, hickory, maple, ash, beech etc. Some coniferous species, such as pine and hemlock, are also present. The region is among one of the most biodiverse regions of the world.

5. Subtropical Evergreen Forests

This type of vegetation is found where there is relatively high humidity and less pronounced seasonal variation in temperature. Such forests are found in Central and Southern Japan, parts of Florida along the Gulf and South Atlantic coasts, and at moderate altitude in Eastern India. The dominant vegetation includes various species of *Quercus, Magnolia, Ficus,* certain palms etc.

6. Temperate Grasslands

Temperate grasslands cover vast tracts of temperate latitudes. Their climate is continental with hot summers and cold winters. The areas, occurring generally in the interior of continents, are moderately dry, having annual precipitation between 25-60 cm. In North America, grasslands are dominated chiefly by tall grasses and are called 'prairies' whereas in the Europe 'steppe' grasslands component are small. Dominant species vary from place to place, depending upon edaphic factors, moisture and seasonal variation in temperature. These regions are extremely important from agricultural point of view and are well suited for growing arable crops, such as corn and wheat. Associated with grasses occur broad-leaved perennial species.

7. Tropical Grasslands (Savannas)

These areas may receive upto 120 cm of annual rainfall in the wet season, but none during the prolonged dry season. This type of vegetation occurs throughout the tropics, covering large part of Africa and a sizable part in South America, South India and Australia. The vegetation exhibits scattered trees or lumps of trees. Species of *Acacia* are the characteristic feature of African Savannas, providing shade, food for browsing animals and support many invertebrates. Savanna trees have thick bark which insulates the living cambium against the heat of grassland fires which is a common annual feature in Africa and South America. The dominant genera of grasses include *Panicum, Pennisetum, Andropogon, Imperata* etc. The common plants are fire resistant species such as palms, euphorbias, legumes etc.

8. Tropical Rain Forests

This type of vegetation occurs in the low attitude zones near the equator. The vegetation experiences high temperature and well spread heavy rainfall (greater than 200 cm annually) and equable humid climate. This climate supports a great diversity of towering trees, the tallest being 60 m or so. The high temperature and rainfall provide optimal conditions for plant growth and the net productivity is highest of any terrestrial biome. Their high biodiversity is thought to be a result of their long history. These forests have never been glaciated, therefore, species have been able to accumulate, developing complex communities over time. Additionally, the complex physical environment provides many niches for specialist species.

Such forests occur mainly in the Amazon and Orinoco basins of South America and the Central American isthmus, the Congo, Nigeria and Zambezi basin of Central and Western Africa and Madagascar and the Indo-Malay-Borneo-New Guinea regions. Many of the hot spots of biodiversity occur in several megadiversity countries in these areas. The forests differ from each other in species composition but all are extremely rich in species diversity. The dense canopy of trees prevents light from reaching the forest floor, thus these forests are very dense. Rain forests are highly stratified. Trees generally form three layers, then are the shrubs and tall herbs and the ground vegetation. 'Montane rain forests' develop in mountaneous areas and are dominated by relatively dwarf tree species. 'Cloud forests' develop in still higher altitudes and possess a very high number of epiphytic species. 'Gallary forests' (riverine forests) are seen along banks of rivers.

9. Tropical Scrub and Deciduous Forests

The areas with moisture condition intermediate between Desert and Savanna on the one hand and the Rain Forests on the other, known as tropical scrub or thorn and tropical deciduous forests, may occur. This type of vegetation is scattered in many parts of the tropics including South America, India (Deccan) and Australia.

10. Deserts, Semideserts and Shrublands

Hot deserts are found around latitudes 30°N and 30°S, mainly in South Western Africa (the Sahara and Namib deserts), parts of Middle East and Asia (Gobi deserts), Australia, the Great-Basin and South Western United States and Northern Mexico. The vegetation is very sparse and may include widely spaced thorny shrubs which can shed their leaves and become dormant during dry periods. The desert annuals germinate immediately following rain and in being opportunistic, *i.e.*, 'ephemeral'–completing their life cycle within few weeks, grow and flower rapidly carpeting the desert floor for short period. 'Geophytes' survive underground as corms or bulbs whereas succulents, such as cacti, have developed water economy adaptations to survive long periods of drought.

Cool semi deserts occur in parts of North America, Central Asia and mountainous regions where climate is too dry. Temperature is too less (below 0°C) and not a single drop of water is available to living organisms due to freezing during winters. They have denser shrub vegetation remaining green all summer due to their extensive deep root system. Microflora-mosses, lichens and blue green algae, which remain dormant in the soil, are able to respond quickly to cool wet periods.

Semidesert ecosystems occur in less arid regions. Warm semidesert shrubs occur throughout the world. In dry warm and tropical climates, temperate shrub land is found around the shores of the Mediterranean sea where it is known as 'Maquis' and in Southern California it is called 'Chaparral'. Californian chaparrals contain species with small, thick and drought resistant leaves and are very flammable during dry periods.

PHYTOGEOGRAPHIC REGIONS OF INDIA
(Botanical Provinces of India)

India is the seventh largest country in the world as regards its geographical area. The country lies north of equator between 8°4′ and 37°6′ north latitude and 68°7′ and 97°25′ east longitude. It measures 3214 km from north to south and 2933 km from east to west with a total land area of 32,87,263 sq km. It is bounded on south-west by Arabian sea and in the south-east by the Bay of Bengal. On the north, north-east and north-west, lies the Himalayan ranges. Kanyakumari constitutes the southern tip of Indian peninsula where it gets narrower and loses itself into the Indian ocean.

India is endowed with rich biodiversity spread over 10 biogeographically distinct zones (Rodgers and Pawar, 1988) due to varying physical conditions and species groupings. Each zone has its own characteristic climate, soil and topography and contains distinct species of flora and fauna. Among these 10 biogeographical zones, Deccan peninsula has the most extensive coverage (about 42 per cent) of the land mass. The Western Ghats and North-East India account for 4 and 5·2 per cent of the geographical areas, respectively. Each biogeographical zone encompasses several habitats, biotic communities and ecosystems and depicts insight about distribution and environmental interactions of flora and fauna of the country.

The country occupies about 2·45 per cent of world's area and has 8·10 per cent of global biodiversity with a species count of about 0·131 million. A sizeable percentage of species are endemic, *i.e.*, species exclusive to India. About 33 per cent of flowering plant species (with 140 genera) recorded in India, are endemic. Endemics are mainly concentrated in North-East India, North-East and North-West Himalayas, Western Ghats and Andaman and Nicobar Islands. Western Ghat ecompasses a very high number of endemic species. The Gangetic plains are generally poor in endemism while Andaman and Nicobar Islands contribute atleast 220 species endemic to flora of India.

The Indian subcontinent has consistently been placed under Palaeotropic Realm (or kingdom). Recognition of botanical regions has been variable. It is customary to include the Indian subcontinent under the Indian Region and dividing it into a number of botanical provinces. Earlier, Clarke (1898) recognised 11, Hooker (1907) 7 and Chatterjee (1939) 10 botanical provinces in which they also included adjoining countries. Later Puri (1960) excluded Burma (Myanmar), Malay Peninsula, Pakistan, Nepal and Ceylon (Sri Lanka), thus including the present boundaries of India, and recognised 10 botanical provinces. Recently Wildlife Institute of India also recognised country's biological wealth as represented in 10 broad biogeographical zones. Following are the botanical provinces of India.

1. Western Himalayas. This province covers areas from Kashmir to Kumaon in Uttarakhand and Punjab, northern part of Haryana and Himachal Pradesh in between. The area receives low to moderate rainfall (usually less than 200 cm annually) and exhibits climatic zonation with altitude. It is further subdivided :

(i) *Submontane (Tropical and Subtropical) Zone.* The area extends from nearly 300 to 1500 m in Shiwalik ranges and adjoining areas. The forests are tropical and subtropical and are dominated by 'sal' (*Shorea robusta*), 'shisham' (*Dalbergia sissoo*) and 'toon' (*Cedrela toona*). Open areas are occupied by 'cutchh' or 'kathha' (*Acacia catechu*), 'dhaak' (*Butea monosperma*), *Zizyphus* spp. and *Euphorbia royleana* with isolated patches of grasses on exposed slopes. *Pinus roxburghii* (chir pine) occurs in the upper belt of this zone.

(ii) *Temperate (Submontane) Zone.* The area extends from 1500 to 3500 m. The common components of this zone are *Pinus wallichiana* with frequent pure strands of *Cedrus deodara* at places and at higher altitudes, *Aesculus indica, Abies pindrow* and *Taxus baccata.* At still higher altitudes occur *Betula utilis* (birch), species of *Salix* and *Rhododendron.*

(iii) *Alpine Zone.* This zone, which begins from the upper tree limit (tree line) at 3500 m and extends upto 4500 m, is dominated by dwarf bushy shrubs such as *Salix, Lonicera, Cassiope* and *Juniperus.* The alpine meadows are dominated by herbaceous species of *Potentilla, Iris, Saxifraga, Primula, Euphrasia, Pedicularis, Geranium* and *Polygonum.*

2. Eastern Himalayas. This is the most humid tract of the Himalayan ranges covering Darjeeling and Sikkim and extending in the east to Arunachal Pradesh. Similar climatic zone, as in Western Himalayas, occurs in this zone. The area has higher rainfall, less snow and higher temperature as compared to Western Himalayas and such zones extend about 300 to 500 m higher. This region has higher proportion of tropical elements (about 250 species of trees) with more of the species of *Rhododendron* and *Quercus* (oak). Ferns, bryophytes and orchids are also very common. It comprises three zones :

(i) *Submontane (Tropical and Subtropical) Zone.* The area extends from foothill plains to 1800 m and has tropical semi-evergreen and deciduous forests, dominated by *Shorea robusta.* Species of *Bauhinia, Anthocephalus, Lagestroemia* and *Bombax* are common. *Bambusa* species are the commonent element.

(ii) *Temperate (Montane) Zone.* The area extends between 1800 m to 3800 m with the forests which are dominated by *Quercus, Michelia, Eugenia* etc. at lower areas and *Cryptomeria, Juniperus, Rhododendron* etc. at higher areas. A common bamboo (*Arundinaria*) is also found at higher altitudes.

(iii) *Alpine Zone.* The area extends from 3800 m to 5000 m and thus both tree line and snow line are proportionately higher. Among common plants are dwarf shrubs of *Juniperus* and *Rhododendron.* Herbaceous plants are dominated by species of *Polygonum, Potentilla, Gentiana* and *Anemone.*

3. Indus Plains. The area includes Punjab, Rajasthan, part of Gujarat and other regions of west of Aravalli range including Delhi. The climate in general is dry, summer is very hot and rainfall is generally less than 70 cm per annum. The region is dominated by xerophytic shrubs and trees. The vegetation comprises tropical thorn forests in semi-arid regions and typical desert plants in arid regions. Most of the shrubs and small trees are totally leafless in hot season. Herbaceous flora is witnessed during the short growing season. Common shrubs and trees are *Acacia, Prosopis, Salvadora, Zizyphus, Capparis* etc. The common herbaceous plants are the species of *Saccharum, Cenchrus, Tephrosia, Euphorbia, Achyanthes* etc. The total species estimated are about 1500 belonging to 112 families.

4. Gangetic Plains. The area extends from eastern parts of Delhi, southern part of Uttar Pradesh, Bihar, West Bengal and the northern part of Orissa, the plain stretching towards the eastern side from river Yamuna formed by alluvium of the river Ganga. Rainfall ranges from less than 70 cm in the western boundary to more than 150 cm in West Bengal. Soils are fertile and the vegetation ranges from moist tropical forests to dry deciduous forests infested with tree species such as *Dalbergia sissoo, Butea monosperma, Madhuca indica, Terminalia arjuna, Diospyros melanoxylon* etc. and herbaceous plänts, such as species of *Xanthium, Amaranthus, Dichanthium, Peristrophe, Bothrichloa* etc. Gangetic delta has mangrove vegetation including species of *Rhizophora, Avicennia* etc. The region is roughly divisible into (i) moist dry western part, (ii) relatively humid Bihar and West Bengal and (iii) deltoid Sunderbans. The species number, although is more or less similar as in Indus plains, but species diversity differs in that, in relatively humid West Bengal. *Polyalthia longifolia, Pterospermum acerifolium, Artocarpus heterophyllus, Terminalia chebula* etc. are commonly found.

5. Central India. The region includes Madhya Pradesh, parts of Orissa and Gujarat. Annual rainfall ranges from 150 to 200 cm supporting mixed deciduous and thorny vegetation comprising species of *Tectona, Madhuca, Diospyros, Butea, Dalbergia* etc. as dominant trees and species of *Carissa, Zizyphus, Acacia* etc. as dominant thorny plants.

6. Malabar. The region is represented by a narrow strip of land west of the Western Ghats from southern Gujarat to Cape Comorin-covering parts of Maharashtra, Karnataka, Kerala and Tamil Nadu. The area receives an annual heavy rainfall, generally over 200 cm and is relatively more humid. It harbours one of the most luxuriant vegetations of the country. Evergreen as well as semievergreen tropical vegetation in this area has about 4000 species of flowering plants belonging to about 150 families. Among tropical and mixed deciduous forests also occur temperate evergreen forests. In Nilgiris, the forests are commonly called 'sholas.' The forests are dominated by *Eurya japonica, Michelia nilgirica* and species of *Dipterocarpus, Sterculia, Cedrela, Tectona* etc. Bamboos and orchids are also common.

7. Deccan. The area comprises largely a plateau zone extending from Satpura ranges in the north to the Nilgiri and Pulney hills in the south and eastward of Malabar. It includes Tamil Nadu, Andhra Pradesh and parts of Karnataka. Annual precipitation is low–about 100 cm, as the area falls under the rain shadow of Western Ghats. Hilly plateau supports tropical dry deciduous forests of *Boswellia, Tectona, Shorea, Hardwickia* etc. whereas the Coromandal coast has tropical dry evergreen forests of *Santalum, Cedrela* etc. supported by drought resistant species and thorny shrubs such as species of *Capparis, Euphorbia, Prosopis* etc.

8. Assam. This north–eastern land zone of the country has the richest vegetation due to very heavy annual rainfall upto 1000 cm in Cherapunji. The vegetation is dominated by evergreen tree species of *Dipterocarpus, Shorea, Ficus, Mesua* etc. Bamboos, ferns and orchids (world's highest concentration), are also very common in this region. The variety of plant life reaches its highest in these forests which harbour about 6000 species under 161 families. These forests have been considered as 'Cradle of Angiosperms.' chiefly because of rich representation of primitive taxa such as members of Magnoliaceae, Winteraceae, Annonaceae and many vesselless genera. In hilly areas, species of *Pinus, Alnus, Betula* and *Rhododendron* are commonly found. The area is characterised by a large percentage of endemic forms.

9. Andamans. The area is restricted to Andaman and Nicobar Islands. It is characterised by heavy rainfall and high temperature and as such this region has a mixed type of vegetation. Along the coasts, beach and mangrove forests are common whereas in the mainland evergreen forests are quite common. The area has a very high percentage of endemic species. Common plants include species of *Rhizophora, Mimusops, Calophylum, Dipterocarpus, Lagerstroemia* etc. Most of the area has been cleared for cultivation of paddy and sugarcane. In 2004 Tsunami, the area had a great loss to its endemic species of flora.

10. Laccadiv (Lakṣhdweep) and Mimicoy Group of Islands. This group of islands, occurring along the western coast of India, has largely scrub type of vegetation growing on rocky substrata coupled with dominant species of *Cocos, Piper, Acacia, Grewia, Ephedra,* ferns, and grasses such as species of *Aristida* and *Heteropogon.*

HOT SPOTS AND HOTTEST HOT SPOTS

The distribution of phytodiversity is not uniform across all phytogeographical regions of the world. An area usually exhibiting species richness coupled with richness of endemic species with its flora under constant threat of overexploitation, is called a **phytodiversity hot spot**. Certain areas of the world, where large number of species are found are called **megadiversity centres**, the concept introduced by Miltermeir and Werner (1990). The concept of hot spots of biodiversity was introduced by Norman Myers (1988) to designate priority areas for *in situ* conservation, identifying only 12 such areas. Later in 2000, he brought out an uptodate list of 25 terrestrial hot spots (fig. 23.1).

The hot spots are the richest and most threatened reservoirs of plant and animal life on the earth. They are highly targeted for conservation

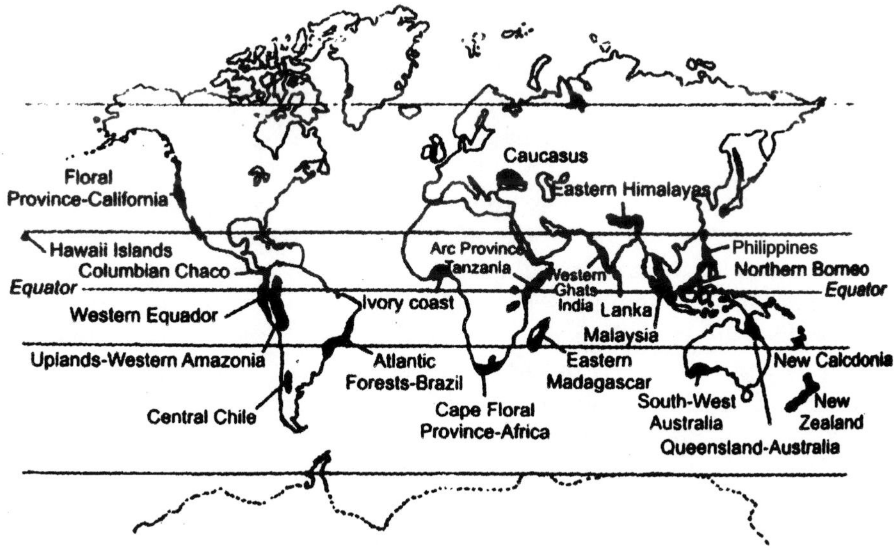

Fig. 23.1 Hot spots of biodiversity.

strategy world over. These hot spots together cover only 1·4 per cent of earth's land area yet harbour about 60 per cent of world's biodiversity. About 20 per cent of human population dwells in the hot spot regions. They together harbour world's more than 50 per cent of all known species including more than 50,000 plant species.

It is believed that hot spots might have arisen as a result of disturbances during the Ice Age of Pleistocene epoch during which climate fluctuated world wide, but the cooler conditions occurred in the tropics. This resulted in contraction in size of rainforests. It is thought that they got restricted to areas of milder climate mainly in the sheltered lowlands. The species that now form rainforests were, therefore able to survive in such areas even they got lost elsewhere. This geographical isolation also led to the evolution of new species in these areas. Thus today we not only find that these **refuse areas** not only have high species diversity but also show high level of endemism. These areas contain outstanding evolutionary process of speciation and extinction and are thus of great scientific interest.

The 25 hot spots of the world are : Caribbean; California Floristic Province; Mesoameria; Tropical Andes; Choco-Darien-Western Ecuador; Atlantic Forest; Brazilian Cerrado; Central Chile; Caucasus; Mediterranean Basin; Medagascar and Indian Ocean Islands; Eastern Arc Mountains and Coastal Forests; Guinean Forest of West Africa; Cape Floristic Region; Succulent Karoo in Africa; Mountains of Southwest China; Indo-Burma; Western Ghats and Sri Lanka; Philippines; Sundaland; Wallacea; Southwest Australia; New Zealand; New Caledonia; and Polynesia and Micronesia.

Hot Spots of India

Among 25 hot spots of the world, two regions of India : Western Ghats and Eastern Himalayas are included in the list. In 1981, Chilika was recognised as a biodiversity hot spot and a Ramsar Site. Andaman and Nicobar Islands also form a part of 25 recognised hot spots of the world.

Eastern Himalayas. Eastern Himalayan hot spot comes under Indo-Burma hot spot, which covers about 2 million sq km tropical Asia east of Indian continent. It had remained an active centre of evolution and is popularly known as **Cradle of Angiosperms**. Endowed with rich phytodiversity, it harbours numerous primitive angiosperm families such as Magnoliaceae, Winteraceae, Trochodendraceae, Tetracentraceae etc. and some primitive genera such as *Magnolia, Alnus, Exbucklandia, Butea, Tetracentron* etc. The region displays an ultra-varied topography, that fosters species richness, species diversity and endemism. There are many deep and semi-isolated valleys exceptionally rich in endemic plant species. Merely in an area of 7298 sq km of Sikkim, about 2550 (60 per cent of the total) species are endemic. Despite forest cover in this area has dwindled to about 1/3 of its original cover, the region is the home to some threatened species. During 2005, Botanical Survey of India (BSI) collected some plant species after 30 years or more including *Sapria himalayana*–a parasitic angiosperm sighted only twice in last 70 years. Some exceptionally phytodiversity rich places in this area are Namadapha, Tirap, Sirohi, Dzuko, Tura, Senchal and Tale and Buxa valleys. In Assam valley, *Dipterocarpus macrocarpus* and *Shorea assamica*, occur singly. Monsoon forests are mainly moist, sal-*Shorea robusta*, occurring widely in this region.

Western Ghats. Western Ghats hot spot comes under Western Ghats and Sri Lnka hot spot, which covers about 1,80,000 sq km area. Western Ghats extend as 1600 sq km long strip parallel to the western coast of India across Maharashtra, Karnataka, Tamil Nadu and Kerala. Fourty per cent of the total species in this area, are endemic. Forests at low elevations (500 m above mean sea level) are mostly evergreen and comprise about 1/5 of the entire forest expanse. Forests at an altitude between 500-1500 m are generally semi-evergreen. Out of 5 locations identified by International Union of Conservation of Nature and Natural Resources (IUCN) as threatened, three areas : Agasthyamalai Hills-Silent Valley; New Amarambalam Reserve Basin; and Periyar National Park, are found in this region. Only 6·8 per cent of the original forests are existing today while the rest have been deforested or degraded. In its south western part along the slopes of the mountains, there is heavy rainfall and thus there are montane forests. The eastern parts are drier with predominant vegetation with deciduous and tropical rain forests and grasslands and scrub forests in lower regions. Western Ghats harbour several tree species of great commercial value such as Indian red wood (*Dalbergia latifolia*), Malabar Kino (*Pterocarpus marsupium*), teak (*Tectona grandis*) etc. Southern 'Malabar' Western Ghats is the major genetic estate with an enormous phytodiversity of ancient lineage. In Nilgiris, the forests are commonly called 'sholas.'

Hottest Hot Spots

Hottest hot spots are highly rich biodiverse hot spots needing immediate attention for *in situ* conservation strategy. The areas are characterised by species diversity and species richness coupled with endemic and threatened species. There is an urgent need to bring these ecosystems to their original state. Most of them have undergone heavy infestation of grasses or weeds, depletion of soil seed bank and nutrients, local extinction of parent trees and chronic degradation. The Western Ghats in India, is one of the eight hottest hot spots of biodiversity. The area is suffering an annual deforestation rate of 1·16 per cent, despite 15 per cent of its land is being protected as wildlife sanctuaries and national parks.

Nature Conservation Foundation (NCF)–a Mysore based NGO, with cooperation from local tribals in an effort to restore tropical rainforests since 2001, has planted 10,000 saplings of 127 rainforest species and pioneer species of *Clerodendron*, *Macaranga* and *Mallotus* in Valparai area of Western Ghats.

ENDEMISM

Every plant species has its own particular distribution or range that is dependent on species history and its migration ability and adaptability. Generally, no two species have the same range of distribution. Their distributions continue to change all the time. The term area or range, in phytogeography is applied to the entire region of a particular taxon or vegetational unit. The local distribution within this range is termed as topography. Although climate plays the crucial role for limitation of distribution of a taxon, other considerations such as topography and edaphic and biotic factors also have their effects in delimitation of distribution range limits.

Migration of Plants

From their original abode, plants, especially angiosperms, covered the whole of the earth migrating to different countries along different routes. They moved northwards and covered the whole of Asia, and following a coastal route, reached the Bearing Straight and crossed over to America. Mountains of Malaya and Burma brought them to Himalayas by which route they had access to Europe. In North Vietnam, Burma, Assam, Yunan and Eastern Himalayas, a series of phytogenetic links do exist that connect the temperate forms of plants to the tropical and subtropical forms. The plants that reached Antarctica, travelled westwards and finally reached South America.

Migration of Plants depends chiefly on seeds. Although in some species, vegetative propagation is also responsible, but this must have played an insignificant role at the early stages of migration. The seeds are carried over to wider areas by different modes and agents. The young plants coming out from these widely dispersed seeds need to acclimatise themselves in the varied situations for survival. Their survival depends on the character of the seed (or fruit) that makes it easy or difficult for dispersal; natural barrier that stops further dispersal; and ecological conditions that make it easy or difficult for the seed to germinate and seedling to survive as a new comer to compete with local plants. Dispersal of seeds is a continuous process which enables a species to survive and spread. Thus, migration of plants, as a result of seed dispersal to wider areas, is a constant phenomenon.

The range of distribution of a species may be continuous or discontinuous. Discontinuous distribution refers to the presence of gaps in the range of distribution. Such gaps are largely due to ecological conditions, failure of survival in competition with other pre-established plants, changing of climate and composition of the soil in geological periods. Due to same reasons many plant species are restricted to smaller areas and are known as **endemic species**.

Endemism and Endemic Areas

Different taxa differ in the extent of their distribution. The taxa that occupy a fairly large geographical areas are known as cosmopolitan. There are taxa that are restricted to small areas. These taxa that have very restricted distribution or are confined to one particular area, region or a country, are known as endemic taxa or endemics, the phenomenon being endemism. The highest degree of endemism is found in remote oceanic islands, such as Hawaii (90 per cent). In Himalayas it is 29%.

The range of a taxon varies according to its size. And thus higher taxa such as families, normally occupy very large areas and are widely distributed. But when a particular family has distribution restricted to relatively a smaller area such as a single continent, it may be said to be endemic to that continent. Genera show smaller areas of their distribution. The species may show still smaller areas of their distribution. There are species that may be endemic to a very small area, such as the peak of a mountain. Thus smaller areas normally will have fewer endemic species whereas the larger ones will tend to have relatively more endemic species.

For example, Degeneriaceae, a monotypic family, is endemic in Fiji Islands, *Calacanthus*, a monotypic genus, is endemic in Western India and *Amherstia*, with a single species, is endemic in Burma (Myanmar). Many larger genera have some species distributed over wide areas but some of their species are endemic in smaller areas. *Primula, Impatiens, Gentiana, Rhododendron* etc. include many species distributed over wide areas, but some of their species are endemic. *Rhododendron nilgiricum* is endemic in Nilgiri hills of South India while quite a large number of species of *Rhododendron* occur in the Himalayas. An endemic area is the area of a taxon in its distribution, limited to some natural region or habitat.

The range of old endmics was once far more extensive than present day and that being remnant or survivers of former floras, may be called **relic endemics** or **epibiotics**. These include those taxa which were geologically well known with the extensive distribution in the past but have become restricted at present due to geological or geographical and climatic changes. Such taxa are particularly abundant on ancient islands and mountain massifs. The New Zealand flora has 72 per cent and St. Helena has 85 per cent of its species endemic. The percentage of endemic species is higher in islands and mountaneous regions than in wide landmasses because new species evolved by natural hybridisation between closely related species in a particular area, are not able to spread over wide areas as these areas are cut off by natural barriers.

According to **age of area theory** of Willis *very widely distributed species originated in the long past while the endemic species or a species with restricted distribution, is a recently evolved species, i.e., the span of existence of a species on the face of the earth is proportional to its range of distribution.* This theory, however, is not applicable in all cases as the genera like *Ginkgo* and *Metasequoia*, having restricted distribution, are not at all recent plants but are the representatives of very primitive plant groups. Thus two types of endemic taxa can be distinguished : those which are newly evolved (**Neoendemic**) and those primitive taxa, the distribution of which becomes restricted to small areas due to ecological constraints or due to competition with other taxa (**Paleoendemic**). *Sequoia* and *Metasequoia* of North America, *Ginkgo* of China, *Manglietia* and *Talauma* of South East Asia and *Degeneria* of Fiji islands, are the examples of paleoendemics. Neoendemics is relatively young, usually below the rank of a species. It results in a newer situation such as forest clearings etc. Under changed ecological conditions within the limits of some natural region, there is a tendency of new forms to evolve. The new forms naturally will show a little or more distributional change since they are physically unable to spread beyond their confines. Among other types of endemics, **Microendemics** refers to endemism at intraspecific level, **Pseudoendemics** to apparently mutants encountered in certain areas, **Ecological endemics** that have arisen in relation to particular habitat condition and **Local endemics** that are confined to very limited areas, such as small island or a mountain peak or other such place. Endemism may be qualitative when two floras, for instance may well have the same percentage of endemic taxa, but in one of these, the endemics may be closely related to each other and differ little from nonendemics or it may be quantitative when they may be very different from the other genera expressing high degree of distinctiveness and better quality of endemism.

There are two factors that are mainly responsible for endemism. Areas that are widely separated from other areas by natural barriers, such as islands in mid of the wide oceans, alpine regions of high mountains or oasis in vast deserts, form most suitable home for endemics and in these areas, when new species are evolved, they fail to spread over other areas due to natural barriers. The other factor, equally important, is the change of ecological conditions for a particular species making it too difficult for species to survive except in some niches and corners. Ridley has mentioned the examples of *Metasequoia* and *Ginkgo*, endemic in China, in favour of endemism of such type. These genera are considered to have had wide distribution in the past but due to change of conditions from favourable to unfavourable for their survival, they perished in most of the regions of the world. Among other factors, unsuccessful competition with other species might also have played an important role in their elimination from different areas. Conclusively, natural barriers in distribution and inability or poor adaptability of a species, make it to become endemic.

Thus, for analysis of age and history of a flora, it becomes necessary to understand and determine the proportion of various types of endemism. The relic endemics are of particular importance in indicating the antiquity, isolation and diversification of habitats, the factors that tend to produce additional endemics and add to their survival. Such endemics are lesser in biotypes and smaller amplitude of variation that leads to their restriction to narrow ecological conditions. They show retrogressive nature and have relatively small chromosome number. The secondarily derived endemics, the neoendemics, commonly have larger chromosome numbers and are relatively aggressive.

Endemic Plants of India

Global species composition reveals that India is tenth among the plant rich countries in the world. Out of 25 hot spots identified in the world, India has two : Eastern Himalayas and Western Ghats. Andaman and Nicobar Islands are no less than a hot spot and Chilika was recognised as a Ramsar Site in 1982. Both Eastern Himalayas and Western Ghats hotspots show a high degree of endemism and contain 5,332 endemic species of higher plants. Not only in these hot spots, on the whole, India has a rich element of endemism in its flora. Indian subcontinent has about 49,219 plant species, of which about 33 per cent are endemic. Andaman and Nicobar Islands contribute atleast 220 endemic plant species. Gangetic plains are generally poor in endemism. According to Botanical Survey of India report, about 7,000 species and 140 genera of the plants, mostly flowering angiosperms, are endemic to India. Of these, the Himalayas and the Khasi Hills account for about 3,000 and the Deccan peninsula for about 2,000 endemic species.

Sapria himalayana, Uvaria lurida, Alchimandra cathcartii, Magnolia gustavii, M. pealiana, Pachylarnax pleocarpa, Nepenthes khasiana, Dicentra roylei, several species of *Primula* and *Rhododendron* and the Lady's Slipper Orchid, *Paphiopodilum* species and *P. venustum,* are some of the noteworthy endemics of Himalayas and Khasi Hills. *Antiaris toxicaria, Campanula cytinoides, Pedicularis perrotteti* and some species of Podostemonaceae are endemic to Western Ghats–Nilgiri Hills in South India.

Many deep and semi-isolated valleys are exceptionally rich in endemic plant species such as in Sikkim with an area of only 7,298 sq km out of 4,250 plant species, 2,550 species are endemic.

The crops that first grew in India and then spread throughout the world, include rice, sugarcane, Asiatic vigna, jute, mango, citrus, banana, several species of millets, spices, medicinals, aromatics and ornamentals. According to Khoshoo (1990), India ranks sixth among the centres of diversity and origin as far as agro-biodiversity is concerned. The origin place of 167 cultivated plants and 320 species of wild relatives of cultivated crops is located in India. India is the centre of diversity of jackfruits, cucurbits, dioscoreas, colacasia, black pepper, ginger, turmeric, bamboos, brassicas, tree cotton and secondary centre of domestication of tobacco, potato and maize.

Whatever may be the cause of endemism, endemic plant species are also considered threatened. Use of endemic species is banned in different states of India. Any disturbance in their ecosystem would result in the total extinction of such plant species. Among 150 worldwide selected centres of rich plant diversity for conservation, five are in India. These include Agasthyamalai Hills, Silent Valley, New Amarambalam Reserve Basin and Periyar National Park (all in Western Ghats) and the Eastern and Western Himalayas.

INVASIONS AND INTRODUCTION

Invasion, *i.e.*, successful establishment of a species in a bare area, the species reaching the new site from any other area and introduction, *i.e.*, introduction of genotypes from the place where it is normally grown to an entirely new area, require knowledge of field based natural history.

Invasions

Successful establishment of a species reaching a new area from any other area, is completed in three successive stages, *i.e.*, migration (dispersal), ecesis (establishment) and aggregation (coming closer due to increase in number). Invasion of exotic plant species, *i.e.*, a plant growing where it is not desired, is facilitated by their reproductive capacity, easy adaptability, wide distribution and production of small-easily dispersed seeds. Such invasive species are commonly known as weeds. Though weeds come from diverse taxonomic groups, they are distinguished as land weeds and water weeds. Land weeds may be **ruderals**–occurring primarily on waste ground and along roadsides and **agrestal**–occurring in agro-ecosystems. It has been estimated that about one tenth out of total known plant species on the earth, are weeds but some 300 species of weeds cause severe damage to various ecosystems. Weeds are troublesome invasive, competitive, aggressive, undesirable and unsightly elements of the plant world. They thrive under even non-ideal conditions of growth and affect human welfare, adversely. It is perhaps due to this that the saying "*when weeds win all the people lose.*" has come into existence.

Invasive species cause disappearance of native species, perhaps the second greatest reason for habitat loss. There are many invasive weeds which are greatly responsible for degradation of habitats in many land and aquatic ecosystems. *Eichhornia crassipes* (Jalkumbhi), which was also labelled "million dollar weed", 'curse of Bengal', 'Blue-devil' etc., proved enemy number one for all the nuisance in clogging water ways, hampering navigation, interfering with fishing and agriculture. This water weed has degraded many aquatic ecosystems in tropical countries including India. *Lantana camara* invaded in many forest land areas of different parts of the country. Similarly *Opuntia* (Naagphani), *Argemone mexicana* (Satyanaasi), *Parthenium* (Gajar ghaas) etc. are other examples of invasive weeds that have competed with native species and established themselves successfully in many parts of the world and also in India. Because of their competitive ability, allowing them to exclude a broad spectrum of native plant species, the purple loosestrife (*Lythrum salacaria*)–a Euponean wetland species, invaded wetlands throughout temperate North America. Similarly a Japanese knotweed (*Fallopia japonica*) became a vigourous riverside weed in Europe.

The impact of invasive species is now recognised by experts around the world. Rapid escalation of the establishment of new global trade routes, triggered by falling trade barriers, is increasing the risk of new invasions. Intentional invasions, under country war strategies, are more serious. Controls have not kept pace with the spread of invasive species.

Introduction

Plant introduction refers to the introduction of genotypes from the place where it is normally grown to an entirely new area. The introductions may be from outside the country (**exotic collections**) or from other parts or states of the same country (**indigenous collections**). Early plant introductions in India, were not made in an organised manner. The invaders, early settlers and merchants, brought with them either the plant materials or their seeds and made the first introductions in this country. These included plant materials or seeds of cotton, mustard, apple, papaya, cashewnut, pineapple, litchi etc. Britishers introduced tea, litchi and loquat from China and cabbage, cauliflower and some other vegetables from Mediterranean. Botanical gardens in different parts of the world also played a significant role in plant introduction, since they collect and maintain all kinds of wild plants in living condition. At international level, role of Royal Botanic Gardens, Kew, England and at national level role of Indian Botanic Garden, Howrah; Llyod Botanic Garden, Darjeeling, and National Botanical Research Institute, Lucknow, are worth praising. In India, various agricultural and horticultural institutions and organisations have also played a significant role in plant introduction. With the establishment of Plant Introduction Division at Indian Agricultural Research Institute at New Delhi, in 1946, a concerted effort was made in this area. In 1976, in its effort made in germplasm collection, the Department was raised to National Bureau of Plant Genetic Resource (NBPGR).

Exotic species, often called **biological pollutants**, when introduced into a new areas where they had never been previously, cause alteration of habitats, resulting habitat degradation. These exert large impact in island ecosystems which harbour world's much of the threatened species. The great majority of introduced exotic species do not become established in new areas but successful species overdominate the native species, sometimes causing them to become threatened. Pathogenic species, if introduced to new virgin areas, may cause total elimination of native species. Introduced species often called **ecological cancer** have no predators, parasites or pathogens, and thus flourish well if established in new geographical region. A sizeable number of species have been introduced accidently or intentionally in coastal waters, wetlands and lakes. Deliberately introduced species to derive certain benefits, is a most serious cause of habitat degradation. In India, introduction of *Eucalyptus* spp., *Cryptomeria japonica, Casuarina equisetifolia, Acacia moniliformis, A. auriculiformis, Taxodium distichum, Barringtonia acutangula* etc. has altered ecological balance of many ecosystems.

PLANT EXPLORATIONS

Plant exploration is aimed at exploration of phytodiversity and plant resources of the country and to identify the species with economic virtues. It may be limited to a particular locality or region, district, state or may extend even to the entire country. The objectives of the plant exploration include identification, description and inventorisation of plants. Emphasis is given on identification of species with traditional economic uses and critically endangered taxa. Post-exploration activities involve preparation of inventory of live plants and their documentation, as respective floras. Monitoring of phytodiversity and preparation of protocols for conservation of species listed in Red Data Book of IUCN, also forms an integral part of plant exploration.

History of the Plant Explorations

History of plant exploration dates back to 372-287 BC when Theophrastus, the Greek philosopher naturalist, collected, named and described some 500 plants in his *De Historia plantarum*, the oldest work on plant exploration. Pliny (23-79 AD), travelled far and wide and collected plants of medicinal and other economic values. Dioscoides (62-127 AD) travelled a lot and gathered information about medicinal plants and compiled his famous book *Materia medica*, a monumental work in which he described about 600 medicinal plants. The progress in plant exploration was gradual as the explorers were largely interested in commercial plants. The earlier plant explorers, showed their interest in medicinal plants. Explorers of many countries, *viz.*, China, Egypt, Assyria etc., were particularly interested in cultivated plants. Explorations for taxonomic considerations were confined to a narrower range of plants.

A large number of enthusiastic students of Linnaeus set out to explore plant kingdom. Their easy method of plant identification and nomenclature, influenced many others who were engaged in collecting plants in different parts of the world. By the time, others had already developed interest in medicinal plants. Thus, there was an outbreak of plant explorations in the New as well as in Old World. History of plant explorations reveals that probably Hans Sloane (1660-1753) from England, was the first man to move out of his country to distant parts of the world for plant collection. His *Catalogue plantarum* contains 800 species new to plant world. Another English plant explorer Mark Catesby (1679-1749), went out to the parts of America and published the *Natural History of Carolina, Florida and the Bahama Islands* in two volumes which appeared in 1731 and 1743 and an appendix to it in 1748. It was around the same period that Rumphius (1628-1702) was engaged in the rich tropical flora of Indonesia. It was unfortunate that most of his collections and illustrations from island of Amboina during 1663-70, were destroyed in a fire and even the manuscript he had compiled was lost. Later on J Burmann published a reconstructed manuscript as *Herbarieum ambinense* (1741-55), containing descriptions of about 1,750 species, out of which 1,060 were illustrated.

The enthusiastic apostles of Linnaeus credited for discovering many new species, include Cristopher Ternstorm who selected West Indies, Peter Kalm–North America, Fedrik Hasselquist–Palestine, Petrus Loefling–Spain, Peter Forskal–Egypt and Arabia and Carl Peter Thunberg–Cape of Good Hope. In the year 1768, many other naturalists joined John Ellias and Peter Collinson on the ship *Endeavour* commanded by Lt. James Cook, accompained by Daniel Solander and Sir Joseph Banks. This team, after three years of exploration, made some marvellous scientific discoveries, apart from bringing a vast collection of plant specimens.

Among botanists who explored North America and made huge collections of plants, include John Torney (1796-1873) and Asa Gray (1810-88) who were also responsible for the publication of monumental works like *A Flora of North America* (1838-42), *Synoptical Flora of North America* (1878) and *Manual of Botany* (1848), the 8th edition of which appeared in 1950. The explorations in South America, which at that time were under control of Spain and Portugal, made by J de Jussieu, JC Mutis, Hipolito Ruiz, Jane Pavon, Alexander von Humboltt, Richard Spruce, Karl Friedrich, Philip von Martius and many others, resulted many significant publications, *viz.*, Ruiz and Paron's *Flora Peruvianae et Chilensis* (1778-1802), Humboldt and Kunth's *Nova Genera et species plantarum* (1788-1805) and Martiu's *Flora Basiliensis* (1829-1906), accounting for 22,767 species and 3,805 plates in 20,000 pages.

Michel Adanson (1727-1806) explored Tropical Africa and published *Historie naturelle de Senegal* (1757) and *Families des plantes* (1763). Source of the other plant explorers who selected Africa include Fedrich Martin Josel Welwitsch (1807-72), William Paterson (1755-1810) and William John Burchell (1782-1863).

Lt James Cook, Sir Joseph Banks, Daniel Solander, Sir Thomas Michel, Von Muller, George Bentham and others, made several explorations to Australia. Some of the well known plant explorers who contributed to the study of botany in South East Asia include Jaun de Loureiro (1715-69), Karl Ludwig Blume (1796-1862), Sir Stanford Raffles (1781-1826), Pierre Somnerat (1745-1814), E.D. Merril, Dr Joseph Arnold and others. China and many other parts were explored by several other botanists who contributed significantly to the knowledge of plants in 17th and 18th century. The European botanists and others, particularly those from England, came to India in various capacities and initiated the plant explorations.

Plant Explorations in India

Many of the Britishers, who came to India during pre-independence as rulers, doctors, engineers and in various other capacities, were posted in remote and isolated places of the country. Some of them, apart from their duties, started taking keen interest in plant world around them and became great plant explorers.

Nathaniel Wallich (1786-1854), an eminent taxonomist and the great plant explorer, was one of the Britishers who was appointed as a surgeon at

Kolkata with East India Company in India in 1806. This brought him close to then Royal Botanic Gardens, Sibpur, Howrah. He devoted his attention towards Indian plants and sacrificed his surgeon's job in 1815 for the post of Superintendent of Royal Botanic Garden and continued to be the head of this Garden for nearly 30 years. During this period, he utilised this opportunity to explore the vast flora of India and to add to the Garden. He travelled the vast tracts of Himalayas, Gangetic plains, Brahmputra Valley, Sylhet and the Khasi Hills and also Burma and Teanesserim. Wallich listed all his plant collections, popular today as *Wallich's Catalogue* which has been maintained at Wallachian Herbarium of Royal Botanic Gardens, Kew, England, JD Hooker, who later wrote *Flora of British India*, considered Wallich's collection as most significant contribution of its kind. Most of us may not know him but taxonomists frequently come across the plants which are named after Wallich, *e.g. Wallichiana densiflora, W. disticha* and so on and genera named by him, *e.g. Didymocarpus rottleriana* Wallich, *Centranthera humifusa* Wallich and so on.

Sir Joseph Dalton Hooker (1817-1911), a citadel of plant taxonomy and then Director of Royal Botanic Garden, Kew, England, 1948 made extensive plant explorations in the Himalayas and adjoining areas. He traced his way through Chhota Nagpur and via Patna and Purnea to the Sikkim Himalaya and established quarters at Darjeeling where he got himself acquinted with the plants of the area with the local people. Pooled plant collection by fellow British Collectors throughout the country could enable him to procure the material for *Flora of British India*, a masterpiece contribution contained in seven volumes, the publication of which was funded by East India Company. *Flora of British India*, the monumental compilation of this great plant explorer describes over 15,000 plants in span of just 25 years (1872-97). In the last volume of the Flora, Hooker expressed a wish that his *Flora of British India* would facilitate the preparation of local 'Floras'. Inspired with his pioneering work, emerged several local and regional 'Floras' namely *Flora of the Presidency of Madras, Flora of Bombay Presidency* etc., that serve as tools to know the names of plants useful to mankind.

Role of Botanical Survey of India in Plant Explorations

The Botanical Survey of India with its headquarter at Calcutta (now Kolkata), was established in 1890. It is aimed at exploration of plant resources of the country and to identify the plant species with economic virtues. Exploration, identification and inventorisation, *i.e.*, preparation of inventory of phytodiversity, survey, monitoring and documentation in the form of district, state and national floras, are its some activities. Since its initiation, particularly after 1954, Botanical Survey of India (BSI) has conducted several plant explorations and enriched its herbarium. The Regional Circles of BSI are at Jodhpur, Port Blair, Gangtok, New Itanagar, Dehradun, Allahabad, Pune, Coimbatore and Shillong. Deccan Circle has been opened in Hyderabad on 15th December 2005.

During the year 2005-06, fourty four field/exploration/collection tours were undertaken by different circle offices and units of BSI covering various parts of the country and some 4,500 specimens have been collected, of which 3,500 specimens belonging to 385 species were identified. Eighty four plant specimens, collected from Ranchi, were identified. Several new species/varieties, several new records for India and for States and some of the plants collected after 30 years or more, are some other significant contributions made during the year.

CENTRES OF ORIGIN

Centres of origin of plants have been a subject of considerable interest with regard to cultivated plants. When people acquired an in-depth knowledge of the environment they lived in, and started cultivation of food crops. The work of famous Russian scientist NI Vavilov is of special interest. He proposed the following eight regions or centres of origin of our crop plants : The China Centre; The Himalayan Centre; The Central Asia Centre; The Asia Minor Centre; The Mediterranean Centre; The Abyssinian Centre; The Central America Centre; and The South America Centre.

The centre of origin can be defined as the area of a taxon normally above the rank of a species in which the largest number of the constituent taxa of subordinate rank are concentrated. But this may not prove true in all cases unless extensive palaeobotanical evidences are avilable. Although a few species may have been originated separately at more than one place (**polytopy**), most plant species had their origin at a certain place (native place or place of origin), and then spread elsewhere. The exact location of origin can only be a matter of speculation based on indirect evidences. Turril (1939) gave a criteria for postulating the centres of origin. According to him :

1. The presence of a maximum variation, where the species is most polymorphic or the genus has a maximum number of species.
2. The polymorphism decreases outside the centre of origin, and in the course of migration, certain characters, such as genes and allelomorphs, are lost, at least in part through a change in action of natural selection. It follows that towards the margins of their migrations, species may often be less polymorphic and genera have fewer species than their centres of origin and development.
3. Species and genera occupy as a result of both greater genetic polymorphism and greater plasticity–a wider range of habitats near their centres (places) of origin than the margins of their ranges.
4. Towards the margins of their ranges, species may develop characters which are derived from more normal characters, such as apomixis, vegetative multiplication, sterility, polyploidy etc.
5. Known facts of geological history may suggest or confirm both the centres (places) of origin and the lines of migration.

Cain (1944), in his effort to postulate the centres of origin, highlighted the greatest abundance of individuals, location of closely related forms, location of maximum size of individuals, increase in number of dominant genes, location of least dependence upon a restricted habitat and location of greatest productivity and its relative stability in the centre (place) of origin as additional criteria. Camp (1947), based on distributional maps, concluded that most of the 103 plant families with restricted territories, as against to wide ranging families, are concentrated in Southern Hemisphere where they originated and spread north, and that is particularly true of primitive families. Centres of origin of the agriculture and some of the first crops is revealed in the table (23.2).

Table 23.2 Centres of Origin of Agriculture and Some of the First Crops

Centres of Origin	First Crops
1. India, Malaysia and Indonesia	Mango, Coconut, Banana, Mustard, Rice, Orange, Radish
2. Ethiopia	Beans, Okra
3. Mediterranean	Onion, Turnip, Lettuce, Parsnip, Celery, Rhuburb
4. China	Cucumber, Cherry, Apricot, Sugarcane, Tea, Egg plant
5. Central and South America	Bell Pepper, Papaya, Corn, Peanuts, Tomato, Sweet Potato
6. Middle East	Apple, Cotton, Grapes, Carrot, Spinach, Garlic Bulb, Wheat

It is currently accepted that agriculture reached Central and Western Europe by two routes : one from the Near East followed the Mediterranean reaching Italy, France and Switzerland and Germany and the other from the Near East, Greece to Hungary.

❑❑❑

BIBLIOGRAPHY

Adanson, M. (1763). *Familiesdes Plantes.* Vol **I** and **II**, Vincent, Paris.

Andrews, F.M. (1932). Preservation of dry plant material. *Ind. Acad. Sci. Proc.,* **41**: 80-81.

APG [= Angiosperm Phylogeny Group] II. 2003. An update of the Angiosperm Phylogeny Group classification for the orders and families of flowering plants: APG II. *Bot. J. Linnean Soc.* **141:** 399-436.

APG [= Angiosperm Phylogeny Group]. 1998. An ordinal classification for the families of flowering plants. *Ann.l Missouri Bot. Gard.* **85:** 531-553.

Arber, E. (1938). *Herbals : Their Origin and Evolution* (2nd ed.). Cambridge Univ. Press.

Arber, E. and J. Parkins. (1907). On the Origin of Angisperms. *Bot. J. Linn. Soc.* **38:** 29-80.

Ashlock, P.H. (1979). An evolutionary Systematits's view of classification, *Syst. Zool.* **28:** 441-450.

Asmussen, C.B., W.J. Baker and Dransfield. (2000). Phylogeny of the palm family (Arecaceae) based on *rps*16 intron and *trnL-trnF* plastid DNA sequences. pp. 525-535, in K.L. Wilson and D.A. Morrison (eds.), *Monocots : Systematics* and *Evolution.* CSIRO, Collingwood.

Axelrod, D.I. (1970). Mesozoic paleogeography and early angiosperm history. *Bot. Rev.* **36:** 277-319.

Bailey, I.W. (1944). The development of vessels in angiosperms and its significance in morphological research. *Am. J. Bot.* **31:** 421-428.

Bailey. L.H. (1949). *Manual of Cultiwated Plants* (rev. ed.). Macmillan, New York.

Bailey, I.W. and B.G.L. Swamy. (1951). The conduplicate carpel of dicotyledons and its initial trend of specialization. *Amer. J. Bot.* **38:** 373-379.

Benson, L. (1957). *Plant Classification.* Oxford and IBH Co., New Delhi.

Bentham, G. (1858). *Handbook of British Flora.* (7th ed., revised by A.B. Rendle in 1930). Ashford, Kent.

Bentham, G. (1863-1878). *Flora Australiensis.* London, 7 volumes.

Bentham, G. and J.D. Hooker (1862-83). *Genera Plantarum.* London, 3 vols.

Bessey, C.E. (1915). Phylogenetic taxonomy of flowering plants. *Ann. Mo. Bot. Gard.* **2:** 109-164.

Bhandari, M.M. (1978). *Flora of the Indian Desert.* Sc. Publ., Jodhpur.

Bhattacharya, B. and B.M. Johri. (1988) *Flowering Plants : Taxonomy and Phylogeny.* Narosa Publishing House, New Delhi.

Blake, S.F. (1961). *Geographical Guide to the Floras of the World.* Part II. Misc. Publ. 797. U.S. Dept. Agr., Washington. D.C.

Blake, S.F. and A.C. Atwood. (1941). *Geographical Guide to the Floras of the World.* Part I. Misc. Publ. 401, U.S. Dept. Agr., Washington, D.

Bremer, K. and H.E. Wanntorp. (1981). The cladistic approach to plant classification. In *Advances in Cladistics,* **1**, V.A. Funk and D.R. Brroks (eds.), 87-94, New York Botanical Garden, New York.

Bremer, K., A. Backlund, B. Sennblad, U. Sweson, K. Andreasen, M. Hjertson, J. Lundberg, M. Backlund and B. Bremer, (2001). A phylogentic analysis of 100+ genera and 50+ families of euasterids based on morphological and molecular data with notes on possible higher level morphological synapomorphies. *Plant Syst. Evol.* **229:** 137-169.

Brenner, G.H. and I. Bickoff. (1992), Palynology and age of the Lower Cretaceous basal Kurnub Group form the coastal plain to the northern Negev of Israel. *Palynology* **16:** 137-185.

Caesalpino, A. (1583). *De plantis libri.* Florentiae.

Cain, A.J. and G.A. Harrison. (1959). An analysis of the taxonomists's judgement of affinity. *Proc. Zool. Soc. Lond.* **131:** 85-98.

Cain, A.J. and G.A. Harrison. (1960). Phyletic weighting. *Proc. Zool. Soc. Lond.* **135:** 1-31.

Cain, S.A. (1944). *Foundations of Plant Geography.* New York & London.

Carlquist, S. (1987). Presence of vessels in *Sarcanda* (Chloranthaceae); comments on vessel origin in angiosperms. *Amer. J. Bot.* **64:** 1765-1771.

Carlquist, S. (1996). Wood anatomy of primitive Angiosperms : New perspective and syntheses. In: D.W. Taylor and L.J. Hickey (eds.) *Flowering Plant Origin, Evolution and Phylogeny.* Chapman and Hall Inc., New York, pp. 68-90.

Chase, M.W., D.E. Soltis, P.G. Rudalf, M.F. Fay, W.J. Hahn, S. Sullivan, J. Joseph, M. Malvray, P.J. Kores, T.J. Givnis, K.J. Systsma and J.C. Pires. (2000). Higher level systematics of monocotyledons : An assessment of current knowledge and a new classification. In K.L. Wilson and D.A. Morrison (eds.) *Systematics and evolution of monocots. Proceedings of the 2nd International monocot symposium.* Melbourne : CSIRO, 3-16.

Chase, M.W., S. Zmarzty, M.D. Lledo, K.J. Wurdack, S.M. Swansen and M.F. Fay. (2002). When in doubt, put it in Flacourtiaceae : A molecular phylogenetic analysis based on plastid *rbc*L DNA sequences. *Kew Bull.* **57:** 141-181.

Chatterjee, D. (1939). Studies on the endemic flora of India and Burma. Journ. Roy. *As. Soc. Beng. Sci.* **5:** 19-67.

Cheadle, V.I. (1953). Independent origin of vessels in the monocotyledons and dicotyledons. *Phytomorphology* **3:** 23-44.

Collett, H. (1921). *Flora Simlensis,* 2nd ed. Thacker, Spink, Calcutta.

Core, E.L. (1955). *Plant Taxonomy.* Prentice-Hall, Englewood Cliffs.

Crane. P.R., E.M. Friis and K.R. Pedersen. (1995). The origin and early diversification of angiosperms. *Nature* **374:** 27-33.

Cronquist, A. (1968). *Evolution and Classification of Flowering Plants.* Hoaghton Mifflin, New York.

Cronquist, A. (1977). On the taxonomic significance of secondary metabolites in Angiosperms. *Plant Syst. Evol.,* Suppl. **1:** 179-189.

Cronquist, A. (1981). *An Integrated system of Classification of Angiosperms.* Columbia Univ. Press, New York.

Cronquist, A. (1983). *Evolution and Classification of Flowering Plants.* (2nd ed.), New York Botanical Garden, Bronx, New York.

Dahlgren. R. (1975). A system of classification of angiosperms to be used to demonstrate the distribution of characters. *Bot. Notiser* **128:** 117-147.

Dahlgren, R. (1980). A revised system of classification of angiosperms. *Bot. J. Linn. Soc.* **80:** 91-124.

Dahlgren, R. (1983). General aspects of angiosperm evolution and macrosystemaics. *Nordic. J. Bot.* **3:** 119-149.

Dahlgren, G. (1989). An updated angiosperm classification. *Bot. J. Linn. Soc.* **100:** 197-203.

Dahlgren. G. (1989). The last Dahlgrenogram. System of classification of dicotyledons, pp. 249-260. In: K. Tan [ed.]. *The Davis and Hedge Festschrift.* Edinburgh Univ. Press. Edinburgh, pp. 249-260.

Dahlgren. G. (1991). Steps towards a natural system of the dicotyledons : embryological characters. *Aliso* **13** (1): 107-165.

Dahlgren, R. and F.N. Rasmussen. (1983). Monocotyledon evolution : characters and phylogenetic estimation. *Evol. Biol.* **16:** 255-395.

Davis, P.H. (1960). Materials for the Flora of Turkey, IV. Ranunculaceae, II. *Notes Roy. Bot. Gard. Edinburgh,* **23:** 103-161.

Davis, P.H. and V.H. Heywood. (1963). *Principles of Angiosperm Taxonomy.* Oliver and Boyd, London.

de Candolle, A.P. (1813). *Theorie elementaire de la botanique,* Paris.

de Candolle, A.P. (1824-73). *Prodromus systematis naturalis regni vegetabilis.* Paris, 17 vols.

Donoghue, M.J. and J.A. Doyle. (1989). Phylogenetic analysis of angiosperms and the relationships Hamamelidae. In: Pr. R. Crane and S. Blackmore (eds.). *Evolution, Systematics and Fossil History of the Hamamelidae.* Claredon Press, Oxford, pp. 17-45.

Doyle, J.A. (1969). Cretaceous angiosperm pollen of Atlantic Coastal Plain and its evolutionary significance. *J. Arnold Arboretum.* **50:** 1-35.

Doyle, J.A. (1978). Origin of angiosperms. *Ann. Rev. Ecol. Systematics.* **9:** 365-392.

Doyle, J.A. (2001). Significance of molecular phylogenetic analyses for paleobotanical investigations on the origin of angiosperms. *Palaeobotanist.* **50:** 167-188.

Doyle, J.A. and Donoghue. (1987). The origin of angiosperms : a cladistic approach. In: E.M. Friis, W.G. Chaloner and P.R. Crane (eds.). *The Origins of Angiosperms and Their Biological Consequences.* Cambridge University Press, U.K. pp. 17-49.

Doyle, J.A. and Donoghue. (1993). Phylogenies and angiosperm diversification. *Paleobiology* **19:** 141-167.

Eames, A.J. (1961). *Morphology of Angiosperms,* McGraw-Hill Book Co., New York.

Engler, A. (1892). *Syllabus der pflanzenfamilien.* Berlin.

Engler, A. (ed.), (1900-1953). *Das Pflanzenreich.* Regni vegetabilis conspectus Im Auftrage der Preus. Akademic der Wissenschaften. Leransgegeben von A. Engler. Berlin. (after Engler's death subsequent volumes, continuing upto 1953 were edited by other authors).

Engler, A. (H. Melchior and E. Werdermann. eds.). (1954). *Syllabus der pflanzen-familien.* 12th ed., vol. 1. Gebruder Borntraeger, Berlin.

Engler, A. (H. Melchior ed.). (1964). *Syllabus der pflanzenfamilien.* 12th ed., vol. 2. Gebruder Borntraeger, Berlin.

Engler, A. and Diels, L. (1936). *Syllabus der pflanzenfamilien,* 11th ed. Berlin.

Engler, A. and Prantl, K. (1887-1915). *Die naturlichen pflanzenfamilien.* Leipzig, 23 vols.

Fassett. N.C. (1957). *A Manual of Aquatic Plants.* Univ. Wisconin Press, Madison.

Frodin, D.G. (1984). *Guide to the Standard Floras of the World.* Cambridge Univ. Press.

Frost. F.H. (1930). Specialization in secondary xylem in dicotyledons. I. Irogin of vessels. *Botanical Gazette* **89:** 67-94.

Gershenzon, J. and T.J. Mabry. (1983). Secondary metabolites and the higher classification of angiosperms. *Nordic. J. Bot.* **3:** 5-34.

Gifford, E.M. and A.S. Foster. (1988). *Morphology and evolution of vascular plants.* 3rd ed. W.H. Freeman. New York.

Gleason, H.A. (1963). *The New Britton and Brown Illustrated Flora.* 3 Vols. Hafner, New York.

Good, R. (1931). 'A Theory of Plant Geography'. *New Phytologist* **30.**

Good, R. (1974). *The Geography of Flowering Plants.* 4th ed. (First in 1947). Longman, London.

Graham, S.W. and R.G. Olmstead. (2000a). Evolutionary significance of an unusual chloroplast DNA inversion found in two basal angiosperm lineages. *Curr. Genet.* **37:** 183-188.

Graham, S.W. and R.G. Olmstead. (2000b). Utility of 17 chloroplast genes for inferring the phylogeny of the basal angiosperms. *American J. Bot.* **87:** 1712-1730.

Grant. V. (1957). The plant species in theory and practice. In: E. Mayr (ed.). *The Species Problem.* Amer. Assoc. Adv. Sci. Washington. D.C., pp. 39-80.

Grant. V. (1981). *Plant Speciation* (2nd ed.). Columbia Univ. Press, New York.

Greuter, W., D.L. Hawksworth, J. Meneill. M.A. Mayo, A. Minelli, P.H.A. Sneath, B.J. Tindall, P. Trehane and P. Tubbs. (1998). Draft BioCode (1997); the prospective international rules for the scientific names of organisms. *Taxon* **47:** 127-150.

Greuter, W.l J. Mcneill, F.R. Barrie, H.M. Burdet, V. Demoulin, T.S. Filgueiras, D.H. Nicolson, P.C. Silva, J.E. Skog, P. Trehane N.J. Turland, D.L. Hawksworth (editors & compilers). (2000). International Code of Botanical Nomenclature (St. Louis Code) adopted by the Sixteenth International Botanical Congress St. Louis, Missouri, July-August 1999. *Regnum Veg.* Vol. **138.**

Henderson, D.M. (1983). *International Directory of Botanical Gardens IV.* Koeltz, Koenigstein.

Heywood, V.H. (ed.). (1985). *Flowering Plants of the World.* Prantice Hall, Englewood Cliffs, NJ.

Hickey, L.J. and J.A. Doyle, (1977). Early Cretaceous fossil evidence for angiosperm evolution. *Bot. Rev.* **43:** 1-104.

Hill, S.R. (1983). Microwave and the herbarium specimens : potential dangers. *Taxon* **32:** 614-615.

Hjelmquist, H. (1948). Studies on the floral morphology and phylogeny of the Amentiferae, *Bot. Notiser.* Suppl. **2:** 1-171.

Holmgren, P.K., N.H. Holmgren and L.C. Barnett. (1990). Index herbariorum. Part I: The Herbaria of the World (8th ed.). *Regnum Veg.* **120.**

Hooker, J.D. (1870). *Student's Flora of British Isles.* Macmillan and Co., London, (3rd ed. in 1884).

Hooker, J.D. (1872-97). *Flora of British India.* L. Reeve and Co., London, 7 vols.

Hooker, J.D. (1907). Botany Imperial Gazetter of India, **1.**

Hutchinson, J. and J.M. Dalzeil. (1927-1929). *Flora of West Tropical Africa.* London, 2 vols.

Hutchinson, J. (1946). *A Botanist in South Africa,* London.

Hutchinson, J. (1948). *British Flowering Plants.* London.

Hutchinson, J. (1964-67). *The Genera of Flowering Plants.* Claredon, Oxford, 2 vols.

Hutchinson, J. (1968). *Key to the Families of Flowering Plants of the World.* Clarendon, Oxford, 117 pp.

Hutchinson, J. (1969). *Evolution and Phylogeny of Flowering Plants.* Acad. Press. London.

Hutchinson, J. (1973). *The Families of Flowering Plants.* (3rd ed.). Oxford Univ. Press. (2nd ed. 1959; Ist ed. 1926, 1934).

Jain, S.K. and R.R. Rao (1977). *A Handbook of Field and Herbarium Methods.* Today and Tomorrow Publ. New Delhi.

Johri, B.M. (1967). Angiosperm embryology and taxonomy. *Bull. Natn. Inst. Sci.*, India. **34:** 263-268.

Judd, W.S., C.S. Cambell, E.A. Kellogg, P.F. Stevens and M.J. Donoghue. (2002). *Plant Systematics : A Phylogenetic Approach.* 2nd ed. Sinauer Associates. Inc., USA.

Judd, W.S., R.W. Sanders and M.J. Donoghue. (1994). Angiosperm family pairs : Preliminary cladistic analyses. *Harvard Pap. Bot.* No. **5:** 1-51.

Kapil, R.N. (1962). Some recent examples of the value of embryology in relation to taxonomy. *Bull. Bot. Surv. India.* **4:** 57-66.

Keating, R.C. (2003b). Vegetative anatomical data and its relationship to a revised classification of the genera of Araceae. *Missouri Bot. Gard. Monogr. Syst. Bot.* [in press; cited in APweb].

Keller, R.A., R.N. Boyd & Q.D. Wheeler. (2003). Illogical basis of Phylogenetic nomenclature. *Bot. Rev.* **69(1)**: 93-110.

Kellogg, E.A. (2000). The grasses : A case study in macroevolution. *Ann. Rev. Ecol. Syst.* **31**: 217-38.

Lawrence, G.H.M. (1951). *Taxonomy of Vascular Plants.* Macmillan, New York.

Lee, T.B. (1979). *Illustrated Flora of Korea.* Hyangmunsa, Seoul.

Linnaeus, C. (1730). *Hortus uplandicus.* Stockholm.

Linnaeus, C. (1735). *Systema naturae.* (2nd ed.). Lugduni Batavorum. Stockholm.

Linnaeus, C. (1737). *Critica botanica.* Leyden.

Linnaeus, C. (1737). *Flora Lapponica.* Amsterdam.

Linnaeus, C. (1737). *Genera plantarum.* Lugduni Batavorum.

Linnaeus, C. (1737). *Hortus Cliffortianus.* Amsterdam.

Linnaeus, C. (1751). *Philosophica botanica.* Stockholm, 362 pp.

Linnaeus, C. (1753). *Species plantarum.* Stockholm, 2 vols.

Linnaeus, C. (1762). *Fundamenta fructificationis.*

Löve, A., D. Löve and R.E. G Pichi-Sermoili. (1977). *Cytotaxonomic Atlas of Pteridophytes.* Cramer, Koenigstein.

Lundberg, J. (2001). A well resolved and supported phylogeny of Euasterids II based on a Bayesian inference, with special comphasis on Escalloniaceae and other incertae sedis. Chapter V, in Lundberg, *J. Phylogenetic studies in the Euasterids II with particular reference to Asterates and Escalloniaceae*. Acta Universitatis Upsaliensis. Uppsala.

Maheshwari, J.K. (1963). *Flora of Delhi.* CSIR, New Delhi.

Maheshwari, P. (1950). *An Introduction to Embryology of Angiosperms.* McGraw Hill Book Co. Inc. New York.

Maheshwari, P. (1964). Embryology in relation to taxonomy. In: W.B. Turril (ed.). *Vistas in Botany.* Pergamon Press, London, vol. **4**. pp. 55-07.

Martin, W., D. Lydiate, H. Brinkmann, G. Forkmann, H. Saedler and R. Cerff. (1993). Molecular phylogenies in angiosperm evolution. *Mol. Biol. Evol.* **10:** 140-162.

Mathew, K.M. (1983). *The Flora of the Tamil Nadu Carnatic.* The Rapinat Herbarium, St. Joseph's College, Tirucherapalli, India, 3 vols.

Mayr, E. (1957). *The Species Problem.* Amer. Assoc. Adv. Sci. Pub. No. 50.

Mayr, E. (1942). *Systematics and the Origin of Species.* Columbia Univ. Press, New York.

McMinn, H.E. and E. Maino. (1946). *Illustrated Manual of Pacific Coast Trees,* 2nd Ed. Univ. of Calif. Press, Berkeley, CA.

Meeuse, A.J.D. (1963). The multiple origins of the angiosperms. *Advancing Frontiers of Plant Sciences,* **1:** 105-127.

Meeuse, A.J.D. (1990). *All about Angiosperms.* Eburon, Delft.

Meglitsch, P.A. (1954). On the nature of species. *Zyst. Zool.* **3:** 49-65.

Melville, R. (1962). A new theory of the angiosperm flower, I. The gynoecium, *Kew Bull.* **16:** 1-50.

Melville, R. (1963). A new theory of the angiosperm flower, II. *Kew Bull.* **17:** 1-63.

Melville, R. (1983). Glossopteridae, Angiospermidae and the evidence of angiosperm origin. *Bot. J. Linn. Soc.* **86:** 279.

Metcalfe, C.R. and L. Chalk. (eds.) (1983). *Anatomy of Dicotyledons.* (2nd ed.). Claredon Press, Oxford. (Takhtajan's classification, vol., 2 pp. 258-300).

Michener, C.D. and R.R. Sokal. (1957). A quantitative approach to a problem in classification. *Evolution* **11:** 130-162.

Muhammad, A.F. and R. Sattler. (1982). Vessel structure of *Gnetum* and the origin of angiosperms. *Amer. J. Bot.* **69:** 1004-1021.

Naik, V.N. (1979). *The Flora of Osmanabad.* Venus Publishers, Aurangabad (India).

Nair, P.K.K. (1970). *Pollen Morphology of Angiosperms.* Vikas Publications, New Delhi.

Nandi, O.I., M.W. Chase and P.K. Endress (1998). A combined cladistic analysis of angiosperms using rbcL and non-molecular data sets. *Annals of the Missouri Botanical Garden.* **85:** 137-212.

Naik, V.N. (1984). *Taxonomy of Angiosperms.* Tata McGraw Hill. New Delhi.

Nixon, K.C. & J.M. Carpenter. (2000). On the other "Phylogenetic Systematics". *Cladistics* **16:** 298-318.

Nixon, K.C., J.M. Carpenter & D.W. Stevenson. (2003). The PhyloCode is fatally flawed and the "Linnaean" System can be easily fixed. *Bot. Rev.* **69(1):** 111-120.

Olmstead, R.G., K.J. Kim, R.K. Jansen and S.J. Wagstaff. (2000). The phylogeny of the Asteridae *sensu lato* based on chloroplast *ndh*F gene sequences. *Molecular Phylog. Evol.* **16:** 96-112.

Pankhurst, R.J. (1975). *Bioloogical Identification with Computers.* Academic Press, New York.

Porter, C.L. (1959). *Taxonomy of Flowering Plants.* W.H. Freeman, San Francisco.

Puri, G.S. (1960). *Indian Forest Ecology.* Oxford Bock & Stationary Co., New Delhi. 2 Vols.

Puri, V. (1967). The origin and evolution of angiosperms. *J. Indian bot. Soc.,* **46:** 1-14.

Qui, Y-L. J. Lee, F. Bernasconi-Quadroni, D.E. Soltis, P.S. Soltis, M. Zanis, E.A. Zimmer, Z. Chen, V. Savolainen and M.W. Chase. (1999). The earliest angiosperms : evidence from mitochondrial, plastid and nuclear genomes. *Nature* **402:** 404-407.

Qui, Y-L, J. Lee, F. Bernasconi-Quadroni, D.E. Soltis, P.S. Soltis, M. Zanis, E.A. Zimmer, Z. Chen, V. Savolainen and M.W. Chase. (2000). Phylogeny of basal angiosperms : Analyses of five genes from three genomes. *Int. J. Plant Sci.* **161** (6 : suppl.) : S3-S27.

Radford, A.E. (1986). *Fundamentals of Plant Systematics.* Harper and Row, New York.

Radford, A.E., W.C. Dickison., J.R. Massey and C.R. Bell. (1974). *Vascular Plant Systematics.* Harper and Row, New York.

Rammayya, N. and T. Rajgopal. (1968). Foliar epidermis as a taxonomc aid in the flora of Hyderabad. Part I. Portulacaceae and Aizoaceae. *J. Osmania Univ.*, **4:** 147-160.

Ray, J. (1686-1704). *Historia plantarum.* London, 3 vols.

Rendle, A.B. (1904). *Classification of flowering plants.* Cambridge, England, Vol. 2 1925; 2nd ed. Vol. 1 (1930).

Rogers, D.J. (1963). Taximetrics, New name, old concept, *Brittonia* **15:** 285-290.

Sahni, B. (1948). Pentoxyleae : A new group of Jurassic gymnosperms from the Rajmahal Hills of India. *Bot. Gaz.* **110:** 285-290.

Santapu, H. (1955). *Botanical Collector's Manual. Bot. Surv. India.*

Savolainen, V., M.W. Chase, S.B. Hoot, C.M. Morton, D.E. Soltis, C. Bayer, M.F. Fay, A.Y. de Bruijn, S. Sullivan and Y.L. Qiu. (2000). Phylogenetics of flowering plants based on combined analysis of plastid *atp*B and *rbc*L sequences. *Syst. Biol.* **49:** 306-362.

Shukla, P. and S.P. Misra. (1979). *An introduction to Taxonomy of Angiosperms.* Vikas Publishing House, New Delhi.

Singh, G. (1999). *Plant Systematics.* Science Publishers, New York.

Singh, G. (1999). *Plant Systematics : Theory and practice.* Oxford & IBH Publishing Co. Pvt. Ltd., New Delhi.

Singh, V., M. Sharma and D.K. Jain. (1974). Trichomes in *Salvia* (Labiatae) and their taxonomic significance. *Bull. Bot. Surv. India.* **16:** 27-34.

Sinnot, E.W. and I.W. Bailey. (1914). Investigations on the phylogeny of angiosperms. Part 3. *Amer. J. Bot.* **1:** 441-453.

Sneath, P.H. and R.R. Sokal. (1973). *Numerical Taxonomy.* W.H. Freeman and Company, San Francisco.

Sneath, P.H.A. (1957). The application of computers to taxonomy. *J. Gen. Microbiol.* **17:** 201-226.

Sokal, R.R. and P.H.A. Sneath. (1973). *Principles of Numerical Taxonomy.* W.H. Freeman and Company, San Francisco.

Solsbrig, O.T. (1970). *Principles and Methods of Plant Biosystematics.* Macmillan, London.

Soltis, D.E., P.S. Soltis, D.L. Nickrent, L.A. Johnson, W.J. Hahn, S.B. Hoot, J.A. Sweere, R.K. Kuzoff, K.A. Kron, M.W. Chase, S.M. Swensen, E.A. Zimmer, S.M. Chaw, L.J. Gillespie, W.J. Kress and K.J. Sytsma. (1997). Angiosperm phylogeny inferred from 18S ribosomal DNA sequences. *Ann. Missouri Bot. Garden* **84:** 1-49.

Soltis, D.E., P.S. Soltis, M.W. Chase, M.E. Mort, D.C. Albach, M. Zanis, V. Savolainen, W.H. Hahn, S.B. Hoot, M.F. Fay, M. Axtell, S.M. Swensen,

L.M. Prince, W.J. Kress, K.C. Nixon and J.A. Farris. (2000). Angiosperm phylogeny inferred from 18S *r*DNA, *rbc*L and *atp*B sequences. *Bot. Journ. Linn. Soc.* **133:** 381-461.

Soltis, P.S., D.E. Soltis and M.W. Chase. (1999). Angiosperm phylogeny inferred from multiple genes as a tool for comparative biology. *Nature* **90:** 461-470.

Soltis, P.S., D.E. Soltis, M.J. Zanis and S. Kim. (2000). Basal lineages of angiosperms. Relationships and implications for floral evolution. *International Journal of Plant Sciences* **161** (6, suppl.) : S97-S107.

Sporne, K.R. (1971). *The mysterious origin of flowering plants.* Oxford Biology Readers 3, F.F. Head and O.E. Lowenstein (eds.). Oxford Univ. Press, Oxford.

Sporne, K.R. (1974). *Morphology of Angiosperms.* Hutchinson Univ. Library, London.

Sporne, K.R. (1976). Character correlation among angiosperms and the importance of fossil evidence in assessing their significance. In: C.B. Beck (ed.). *Origin and Early Evolution of Angiosperms.* Columbia Univ. Press, New York.

Stebbins, G.L. (1950). *Variation and Evolution in Plants.* Columbia Univ. Press, NY.

Stebbins, G.L. (1974). *Flowering Plants : Evolution above the Species Level.* The Belknap Press, Harvard Univ. Press, Cambridge.

Steenis, C.G.G.J. van (ed.). (1948). *Flora Malesiana.* Series I : Spermatophyta. Groninger, Jakarta.

Stevens, P.F. (1998). What kind of classification should the practicing taxonomist use to be saved? *In Plant diversity in Malesia III.* J. Dransfield, M.J.E. Coode and D.A. Simpson (eds.), 295-319. Royal Botanical Gardens, Kew.

Stevens, P.F. (2003). Angiosperm Phylogeny Website. Version 4 May 2003 [and more or less continuously updated since]. http://www.mobot.org/MOBOT/research. Website developed and maintained by Hilary Davis (last updation incorporated 23/10/2003).

Stuessy, T.F. (1990). *Plant Taxonomy.* Columbia Univ. Press, New York.

Sun, G.G.L. Dilcher, S. Zheng and Z Zhou. (1998). In search of the first flower : A Jurassic angiosperm, *Archaefructus*, from northeast china. *Science.* **282:** 1692-1695.

Sun, G., Ji, Q., D. L. Dilcher, S. Zheng, K.C. Nixon and X. Wang. (2002). Archaefructaceae, a new basal angiosperm family. *Science* **296:** 899-904.

Takhtajan, A.L. (1958). *Origins of Angiospermous Plants.* Amer. Inst. Biol. Sci. (Translation Russian edition of 1954).

Takhtajan, A.L. (1959). *Die Evolution Der Angiospermen.* Gustav Fischer Verlag, Jena.

Takhtajan, A.L. (1966). *Systema et Phylogenia Magnoliophytorum.* Soviet Publishing Institution, Nauka.

Takhtajan, A.L. (1969). *Flowering Plants Origin and Dispersal.* (English translation by C. Jeffrey). Smithsonian Institution Press, Washington.

Takhtajan, A.L. (1980). Outline of classification of flowering plants (Magnoliophyta). *Bot. Rev.* **46:** 255-369.

Takhtajan, A.L. (1986). *Floristic Regions of the World.* Berkeley.

Takhtajan, A.L. (1987). *Systema Magnoliophytorum.* Nauka, Leningrad.

Takhtajan, A.L. (1991). *Evolutionary Trends in Flowering Plants*. Columbia Univ. Press, New York.

Takhtajan, A.L. (1997). *Diversity and Classification of Flowering Plants*. Columbia Univ. Press, New York, 642 pp.

Taylor, D.W. and L.J. Hickey. (1992). Phylogenetic evidence for herbaceous origin of angiosperms. *Plant Systematics and Evolution* **180:** 137-156.

Taylor, D.W. and L.J. Hickey. (1996). Evidence for and implications of an Herbaceous Origin of Angiosperms. In: D.W. Taylor and L.J. Hickey (eds.). *Flowering Plant Origin, Evolution and Phylogeny*. Chapman & Hall Inc., New York, pp. 232-266.

Theophrastus. *De causis plantarum* Translated by R.E. Dengler. Philadelphia (1927).

Theophrastus. *Enquiry into plants*. Translated by A. Hort. W. Heinemann, London (1916), 2 vols.

Thorne, R.F. (1968). Synopsis of a putative phylogenetic classification of flowering plants. *Aliso* **6(4):** 57-66.

Thorne, R.F. (1974). A phylogenetic classification of Annoniflorae. *Aliso* **8:** 147-209.

Thorne, R.F. (1976). A phylogenetic classification of angiosperms. *Evol. Biol.* **9:** 35-106.

Thorne, R.F. (1981). Phytochemistry and angiosperm phylogeny : A summary statement. In D.A. Young and D.S. Seigler (eds.). *Phytochemistry and Angiosperm Phylogeny*. Praeger, New York, pp. 233-295.

Thorne, R.F. (1983). Proposed new realignments in angiosperms. *Nordic J. Bot.* **3:** 85-117.

Thorne, R.F. (1992). An updated classification of the flowering plants. *Aliso* **13:** 365-389.

Thorne, R.F. (1992b). Classification and geography of the flowering plants. *Bot. Rev.* **58:** 225-348.

Thorne, R.F. (1996). The least specialized angiosperms. In: D.W. Taylor and L.J. Hickey (eds.). *Flowering Plant Origin, Evolution and Phylogeny*. Chapman & Hall Inc., New York, pp. 286-313.

Thorne, R.F. (1999). An Updated Classification of the Class Angiospermae. http://www.inform.umd.edu/PBIO/fam/thorneangiosp99.html. Website maintained (with nomenclatural additions) by Dr. J.L. Reveal of University of Maryland.

Thorne, R.F. (2000). The classification and geography of the monocotyledon subclasses Alismatidae, Lilidae and Comelinidae. pp. 75-124, in Nordenstam, B., El Ghazaly, G.

Thorne, R.F., (2000-2001). The classification and geography of flowering plants. Dicotyledons of the class Angiospermae (subclasses Magnoliidae, Ranunculidae, Caryophyllidae, Dilleniidae, Rosidae, Asteridae and Lamiidae). *Bot. Rev.* **66:** 441-647.

Thorne, R.F. (2003). *An Updated Classification of the Class Angiospermae*. www.rsabg.org/publications/angios.htm.

Tournefort, J.P. de. (1696). *Elements de botanique*. Parts, 3 volts.

Tournefort, J.P. de. (1700). *Institutiones ret herbariae*. Royale, Paris, 3 vols.

Trehane, P., *et al.* (1995). International Code of Nomenclature for Cultivated Plants. *Regnum Veg.* **133**.

Turrill, W.B. (1939). Principles of Plant Geography, *Kew Bull.* 208-237.

Turrill, W.B. (1959). On the Flora of St. Helena. *Kew Bull.* 358-362.

Turrill, W.B. (1959). Plant Geography. In W.B. Turrill (ed.). *Vistas in Botany* 171-229. Pergamon Press, London.

Turrill, W.B. (1964). Plant Taxonomy, Phytogeography and Ecology. In W.B. Turrill (ed.). *Vistas in Botany* Vol **4:** 187-224. Pergamon Press, London.

Tutin, T.G. *et al.* (ed.) (1964-1980). *Flora Europaea.* Cambridge University Press, Cambridge, 5 vols.

Vegter, I.H. (1988). Index herbariorum : A guide to the location and contents of the world's public herbaria. Part 2(7). Collectors T-Z. *Regnum Veg.* **117.** (Part 2(6). Collectors S. *Regum Veg.* **114** publ. 1986).

Wallace, R.S. and A.C. Gibson. (2002). Evolution and systematics, pp. 1-21 in Nobel, P.S. (ed.), *Cacti : Biology and Uses.* University of California Press, Berkeley.

Walters, D.R. and D.J. Keil. (1995). *Vascular Plant Taxonomy,* 4th ed. Kendall?Hunt, Dabuque, IA.

Watson L. and M.J. Dallwitz. (2000). *The families of flowering plants . Descriptions and illustration.* Website : *http://biodiversity.uno.edu/delta.*

Whitlock, B.A., J. Lee, O. Dombrovska and Y.L. Qiu. (2002). Effects of rate heterogeneity on estimates of the age of angiosperms. P. 158., in *Botany 2002 : Botany in the Curriculum.* Abstracts, [Madison, Wisconsin].

Willis, J.C. (1922). *Age and Area.* Cambridge.

Willis, J.C. (1973). *A Dictionary of Flowering Plants and Ferns.* 8th ed. (revised by H.K. Airy-Shaw). Cambridge Univ. Press, Cambridge, 1245 pp.

Wolfe, K.H.M. Gouy, Y.W. Yang, P.M. Sharp and W.H. Li. (1989). Date of monocot-dicot divergence estimated from Choloroplast DNA sequence data. *Proc. Nat. Acad. Sci.* USA **86:** 6201-6205.

Woodland, D.W. (1991). *Contemporary Plant Systematics.* Prenctice Hall. New Jersey.

Young, D.A. (1981). Are the angiosperms primitively vesselless? *Syst. Bot.* **6:** 313-330.

Young, D.J. and L. Watson. (1970). The classification of the dicotyledons : A study of the upper levels of hierarchy. *Aust. J. Bot.* **8:** 387-433.

Zanis, M.J., D.E. Soltis, P.S. Soltis, Y-L Qiu and E.A. Zimmer. (2003). Phylogenetic analyses and perianth evolution in basal angiosperms. *Ann. Missouri Bot. Gard.*

Zanis. M.J., D.E. Soltis, P.S. Soltis, Y-L. Qiu S. Mathews and M.J. Donoghue. (2002). The root of the angiosperms revised. *Proc. Nat. Acad. Sci.* **99:** 6848-6853.

Zomlefer, W.B. (1994). *Guide to flowering plant families.* University of Carolina Press, Chapel Hill.

❑❑❑

INDEX

A

B

C

I

J

K

S

T